Real Analysis Methods for Markov Processes

Kazuaki Taira

Real Analysis Methods for Markov Processes

Singular Integrals and Feller Semigroups

Springer

Kazuaki Taira
The College of Mathematics
University of Tsukuba
Tsuchiura, Ibaraki, Japan

ISBN 978-981-97-3658-4 ISBN 978-981-97-3659-1 (eBook)
https://doi.org/10.1007/978-981-97-3659-1

This Springer imprint is published by the registered company Springer Nature Singapore Pte Ltd.
The registered company address is: 152 Beach Road, #21-01/04 Gateway East, Singapore 189721, Singapore

Dedicated to Prof. Hiroki Tanabe
in appreciation of his constant
encouragement

Preface

If I have ever made any valuable discoveries,
it has been owing more to patient attention,
than to any other talent (Sir Isaac Newton).

This book is devoted to a careful and accessible exposition of the real analysis methods for the problem of construction of Markov processes with Ventcel' boundary conditions in probability. Analytically, a Markovian particle in a domain of Euclidean space is governed by an integro-differential operator, called the Waldenfels operator, in the interior of the domain, and it obeys a boundary condition, called the Ventcel' boundary condition, on the boundary of the domain. Probabilistically, a Markovian particle moves both by jumps and continuously in the state space and it obeys the Ventcel' boundary condition which consists of six terms corresponding to the diffusion along the boundary, the absorption phenomenon, the reflection phenomenon, the sticking (or viscosity) phenomenon and the jump phenomenon on the boundary and the inward jump phenomenon from the boundary.

In particular, second-order elliptic differential operators are called diffusion operators, which describe analytically strong Markov processes with continuous paths in the state space such as Brownian motion. We remark that second-order elliptic differential operators with *discontinuous* coefficients enter naturally in connection with the problem of construction of Markov processes in probability. Since second-order elliptic differential operators are pseudo-differential operators only if the coefficients are infinitely differentiable, we cannot make use of the theory of pseudo-differential operators as in the previous book: *Semigroups, Boundary Value Problems and Markov Processes* [154].

Our approach here is distinguished by the extensive use of the ideas and techniques characteristic of the recent developments in the theory of Calderón and Zygmund of singular integral operators with non-smooth (i.e., non-infinitely differentiable) kernels. It should be emphasized that singular integral operators with non-smooth kernels provide a powerful tool to deal with smoothness of solutions of partial differential equations, with minimal assumptions of regularity on the coefficients. The

Calderón–Zygmund theory of singular integrals continues to be one of the most influential works in modern history of analysis, and is a very refined mathematical tool whose full power is yet to be exploited. Several recent developments in the theory of singular integrals have made possible further progress in the study of elliptic boundary value problems with discontinuous coefficients and hence in the study of Markov processes. The presentation of these new results is the main purpose of this book.

In this book, we construct Feller semigroups associated with absorption, reflection, drift and sticking phenomena at the boundary. More precisely, we prove existence theorems for Feller semigroups with Dirichlet boundary condition, oblique derivative boundary condition and first-order Ventcel' boundary condition for second-order, uniformly elliptic differential operators with vanishing mean oscillation (VMO) coefficients. The contents of the book are divided into six principal parts.

To make the main results (Theorems 1.1, 1.3, and 1.5) of this book accessible to a broad spectrum of readers, I have added Chap. 9. In this introductory chapter, I have attempted to state our problems and results in such a fashion that the readers could understand.

(1) Part I (Chaps. 2–4) provides the elements of functional analysis, L^p spaces and real analysis. The material in these preparatory chapters is given for completeness, to minimize the necessity of consulting too many outside references. This makes the book fairly self-contained.
(2) Part II (Chaps. 5–8) is devoted to the L^p theory of Sobolev and Besov spaces, paying particular attention to Poisson integrals. Moreover, we study maximum principles in the framework of L^p Sobolev spaces, which are used throughout the book.
(3) Part III (Chaps. 9–11) is devoted to a concise and accessible exposition of the most elementary part of the Calderón–Zygmund theory of singular integral operators. This part is the heart of the subject. I have endeavored to present it in such a way as to make it accessible to undergraduates as well. Moreover, in order to make the book more up to date, additional references have been included in the bibliography.

Chapter 9 is devoted to a careful and accessible exposition of the most elementary part of the Calderón–Zygmund theory of singular integrals and the basic theory of the Hilbert transform, which is a special case of the singular integral of a single independent variable. The results discussed here are adapted from Calderón–Zygmund [23] and Tanabe [167].

The first main result in Chap. 10 (Theorem 10.1) asserts the existence of singular integral operators and the second main result (Theorem 10.2) concerns commutators of bounded mean oscillation (BMO) functions and singular integral operators. It should be emphasized again that singular integral operators with non-smooth kernels provide a powerful tool to deal with smoothness of solutions of partial differential equations, with minimal assumptions of regularity on the coefficients.

In Chap. 11 we consider singular integrals with kernels depending on a parameter, and prove theorems about singular integrals and commutators of L^∞ functions and singular integral operators (Theorems 11.2 and 11.3), generalizing Theorems 10.2 and 10.3. The main idea of proof is to reduce the variable kernel case to the constant kernel case. This is done by expanding the kernel into a series of spherical harmonics, each term defining a constant kernel operator treated in Chap. 10. Theorems about singular integrals and commutators are usually formulated in the whole space $\mathbf{R}^n$. However, our application to the theory of elliptic equations with discontinuous coefficients will require a local version of Theorems 11.2 and 11.3 (Theorems 11.4 and 11.5 and Corollary 11.7).

The Calderón–Zygmund theory of singular integral operators is a very refined mathematical tool whose full power is yet to be exploited.

(4) Our subject proper starts with the Part IV (Chaps. 12–15). Chapter 12 is devoted to the study of the *homogeneous* Dirichlet problem in the framework of L^p Sobolev spaces. We state interior and global a priori estimates for the Dirichlet problem (Theorems 12.1 and 12.2) which plays an essential role in the proof of the unique solvability theorem of the Dirichlet problem in Chap. 15.

Chapter 13 is devoted to the proof of Theorem 12.1 that is based on some interior estimates for the solutions of the Dirichlet problem (Lemma 13.2). The main idea of proof may be considered as an integral perturbation about the constant coefficient case, which goes back to Eugenio Elia Levi. The VMO assumption on the coefficients is of the greatest relevance in the study of an error term expressed by singular commutators.

In Chap. 14, we prove the global a priori estimate for the homogeneous Dirichlet problem stated in Theorem 12.2. The desired global estimate is consequences of the following two facts:

(i) The explicit representation formula (14.1) for the solutions of the Dirichlet problem, which is obtained from the half space Green function, involves the same integral operators as in the interior case.
(ii) An L^p boundedness of some singular integral operators and boundary commutators appearing in formula (14.1) (Theorems 14.2 and 14.6).

Chapter 15 is devoted to the study of the *homogeneous* Dirichlet problem for a second-order, uniformly elliptic differential operator with VMO coefficients in the framework of Sobolev spaces of L^p type. We prove an existence and uniqueness theorem for the Dirichlet problem (Theorem 15.1). Our proof is based on some interior and boundary a priori estimates for the solutions of problem (15.2) (Theorem 12.1 and Theorem 12.2). Both the interior and boundary a priori estimates are consequences of explicit representation formulas (12.1) and (13.1) for the solutions of problem (14.2) (Theorem 13.1 and Theorem 14.1) and also of the L^p-boundedness of Calderón–Zygmund singular integral operators and boundary commutators appearing in those representation formulas (Theorems 14.2 and 14.6). It should be emphasized that we make use of the method of continuity (Theorem 2.15) in order to prove the existence theorem for the homogeneous Dirichlet problem.

(5) Part V (Chaps. 16–20) is devoted to the study of the *regular* oblique derivative problem for a second-order, uniformly elliptic differential operator with discontinuous coefficients in the framework of L^p Sobolev spaces (Theorems 16.1 and 16.2). More precisely, we consider a second-order, uniformly elliptic differential operator with VMO coefficients and an oblique derivative boundary operator that is *nowhere* tangential to the boundary.

(6) Part VI (Chaps. 21–26) is devoted to the functional analytic approach to the problem of construction of Markov processes with first-order Ventcel' boundary condition for second-order, uniformly elliptic differential operators with *discontinuous* coefficients (Theorems 22.1, 23.1 and Theorem 24.9), which can be traced back to the pioneering work of Feller [47, 48] in the early 1950s. In particular, we prove a version of the Hille–Yosida theorem (Theorem 21.12) that forms a functional analytic background for the proof of Theorem 1.1 (Theorem 21.1). Our approach is distinguished by the extensive use of the ideas and techniques in the Calderón–Zygmund theory of singular integral operators with non-smooth kernels developed in Parts IV and V. Chapter 25 is devoted to the proof of Theorem 22.1 and Chapter 26 is devoted to the proof of Theorem 1. 1, respectively.

In Chap. 26, as concluding remarks, we give two overviews on generation theorems for Feller semigroups (Tables 26.1 and 26.2).

Our approach is distinguished by the extensive use of the ideas and techniques characteristic of the recent developments in the theory of *singular integral operators*. Moreover, we make use of an L^p variant of an estimate for the Green operator of the Dirichlet problem introduced in the study of Feller semigroups.

(7) In Appendix A, we set forth the basic concepts of Lebesgue integral theory on the Euclidean space $\mathbf{R}^n$. In particular, following faithfully Schwartz [130, Theorem] and Folland [51, Theorem 2.47] we prove two change-of-variables formulas under linear transformations and diffeomorphisms (Theorems A.1 and A.7). This appendix is included for the sake of completeness.

In Appendix B, following faithfully Gilbarg–Trudinger [59], we present a brief introduction to the classical *potential theoretic approach* to the Dirichlet problem for the Laplacian

$$\Delta = \sum_{j=1}^{n} \frac{\partial^2}{\partial x_j^2} = \frac{\partial^2}{\partial x_1^2} + \cdots + \frac{\partial^2}{\partial x_n^2}.$$

The approach here can be traced back to the pioneering work of Schauder, [127] and [128], on the Dirichlet problem for second-order elliptic partial differential operators. Ladyzhenskaya–Ural'tseva [84] and Miranda [106] are the classics for the theory of elliptic partial differential equations.

A bird's-eye view of the classical potential theoretic approach and the singular integral approach to the Dirichlet problem is given in Table B.1.

This book is addressed to advanced undergraduates or beginning-graduate students and also mathematicians with interest in real analysis, functional analysis, partial differential equations, and probability. For the former, it may serve as an effective introduction to these four interrelated fields of analysis. For the latter, it provides a method for the study of elliptic boundary value problems with discontinuous coefficients, a powerful method clearly capable of extensive further development. I have inserted several elementary formulas and examples in the text so that advanced undergraduates and graduate students can follow a step-by-step mathematical reasoning for the theory of singular integral operators founded by the Chicago school (USA). Moreover, this book is amply illustrated for easier understanding; 18 tables and 76 figures are provided with appropriate captions.

I hope that this book will lead to a better insight into the study of four interrelated subjects in analysis: Singular integrals, semigroups, elliptic boundary value problems, and Markov processes.

I am grateful to Profs. Hiroki Tanabe, Akihiko Miyachi, and Atsushi Yagi for fruitful conversations while working on this book. Especially, I am very happy to acknowledge the influence of Prof. Tanabe, from whose book [166] I first learned the theory of singular integral operators.

I would like to extend my warmest thanks to Prof. Jerome Arthur Goldstein who originally suggested that my work be published in book form.

I am sincerely grateful to the three anonymous referees and a copyeditor for their many valuable suggestions and constructive comments, which have substantially improved the exposition of this book, both mathematically and stylistically. I would like to extend my hearty thanks to the staff of Springer-Verlag (Tokyo), who have generously complied with all my wishes.

Last but not least, I owe a great debt of gratitude to my family who gave me moral support during the preparation of this book.

Tsuchiura, Japan Kazuaki Taira
April 2024

Contents

Chapter 1
Introduction and Main Results

This book provides a careful and accessible exposition of the functional analytic approach to the problem of construction of strong Markov processes with Ventcel' (Wentzell) boundary conditions in probability theory (see the last Chap. 26). More precisely, we prove existence theorems for Feller semigroups with Dirichlet boundary condition, oblique derivative boundary condition and first order Ventcel' boundary condition for second order, uniformly elliptic differential operators with *discontinuous* coefficients. Our approach here is distinguished by the extensive use of the ideas and techniques characteristic of the recent developments in the Calderón–Zygmund theory of singular integral operators with non-smooth kernels (see [33–35, 140, 166, 167]). It should be emphasized that there is an approximate functional calculus for such operators resembling the Fourier transform apparatus (see [22–26, 28, 137]).

Singular integral operators with non-smooth kernel may be used to represent differential operators [26]. This representation was used in proving uniqueness in the Cauchy problem for partial differential equations, with minimal assumptions of regularity on the coefficients [19, 107]. Moreover, singular integral operators provide a powerful tool to deal with smoothness of solutions of elliptic boundary value problems [20, 131–133].

The main results of generation theorems for Feller semigroups (Theorems 1.1, 1.3 and 1.5) may be summarized as in Table 1.1.

1.1 Formulation of the Problem

Now, let Ω be a bounded domain in Euclidean space $\mathbf{R}^N$, $N \geq 3$, with boundary $\partial\Omega$ of class $C^{1,1}$. We consider a second order, *uniformly elliptic differential operator A* with real *discontinuous* coefficients of the form

K. Taira, *Real Analysis Methods for Markov Processes*,
https://doi.org/10.1007/978-981-97-3659-1_1

Table 1.1 A bird's-eye view of strong Markov processes, Feller semigroups and boundary value problems

Boundary conditions (various phenomena)	Notation	Infinitesimal generators	Main theorems
Dirichlet case (absorption)	γ_0	$\mathfrak{A}_D$	Theorem 1.5
Oblique derivative case (absorption, reflection, drift)	L_0	$\mathfrak{A}_N$	Theorem 1.3
General case (absorption, reflection, drift, viscosity)	$L = L_0 - \delta A$	$\mathfrak{A}_L$	Theorem 1.1

$$Au(x) := \sum_{i,j=1}^{N} a^{ij}(x)\frac{\partial^2 u}{\partial x_i \partial x_j}(x) + \sum_{i=1}^{N} b^i(x)\frac{\partial u}{\partial x_i}(x) + c(x)u(x). \tag{1.1}$$

In the case of continuous coefficients $a^{ij}(x)$, an L^p Schauder theory has been elaborated for second order, uniformly elliptic differential operators (see [59]). However, the situation becomes rather difficult if we try to allow discontinuity on the $a^{ij}(x)$. In fact, the L^p Schauder theory breaks down under arbitrary discontinuity of the $a^{ij}(x)$ except for the two-dimensional case ($N = 2$), as is shown in Meyers [103], Talenti [165, Chap. 3] and Gilbarg–Trudinger [59, Chap. 9]. For elliptic boundary value problems for linear operators with discontinuous coefficients in the plane ($N = 2$), the reader is referred to Maugeri–Palagachev–Softova [99, Sect. 1.5].

In order to handle the multidimensional case ($N \geq 3$), additional conditions on the $a^{ij}(x)$ should be required. Here we shall see that the relevant condition is that the coefficients $a^{ij}(x)$ belong to the Sarason class VMO of functions with vanishing mean oscillation. We remark that the class VMO consists of the John–Nirenberg class BMO of functions with bounded mean oscillation whose integral oscillation over balls shrinking to a point converge uniformly to zero (see Chap. 4 for the precise definitions and references).

Throughout this book, we assume that the coefficients $a^{ij}(x)$, $b^i(x)$ and $c(x)$ of the differential operator A satisfy the following three conditions:

(1) $a^{ij} \in \mathrm{VMO} \cap L^\infty(\Omega)$, $a^{ij}(x) = a^{ji}(x)$ for almost all $x \in \Omega$ and there exist a constant $\lambda > 0$ such that (the uniform ellipticity condition)

$$\frac{1}{\lambda}|\xi|^2 \leq \sum_{i,j=1}^{N} a^{ij}(x)\xi_i\xi_j \leq \lambda\,|\xi|^2 \quad \text{for almost all } x \in \Omega \text{ and all } \xi \in \mathbf{R}^N.$$

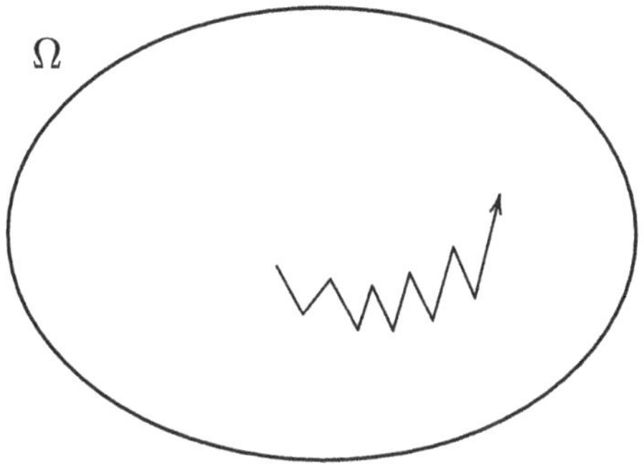

Fig. 1.1 A Markovian particle moves by continuous paths

It should be emphasized that a local $C^{1,1}$ diffeomorphism that straightens the boundary $\partial\Omega$ preserves the VMO character of the $a^{ij}(x)$. Hence, we can extend the coefficients $a^{ij}(x)$ to the whole $\mathbf{R}^N$ preserving their VMO character (see [1, Proposition 1.3]; [74, Theorem 1]; [169, Theorem 5.1]).

(2) $b^i \in L^\infty(\Omega)$ for $1 \leq i \leq N$.
(3) $c \in L^\infty(\Omega)$ and $c(x) \leq 0$ for almost all $x \in \Omega$.

The differential operator A is called a diffusion operator, which describes analytically a strong Markov process with continuous paths in the interior Ω such as Brownian motion (see Fig. 1.1).

Moreover, we consider a first order, boundary operator of the form

$$Lu(x') := \mu(x') \frac{\partial u}{\partial \mathbf{n}}(x') + \beta(x') \cdot u(x') + \gamma(x')u(x') - \delta(x') \big(Au(x')\big) \quad \text{on } \partial\Omega. \tag{1.2}$$

Throughout this book, we assume that the coefficients $\mu(x')$, $\beta(x')$, $\gamma(x')$ and $\delta(x')$ of the boundary operator L satisfy the following four conditions:

(4) $\mu(x')$ is a Lipschitz continuous function on $\partial\Omega$ and $\mu(x') \geq 0$ on $\partial\Omega$.
(5) $\beta(x')$ is a Lipschitz continuous vector field on $\partial\Omega$ (see Fig. 1.2).
(6) $\gamma(x')$ is a Lipschitz continuous function on $\partial\Omega$ and $\gamma(x') \leq 0$ on $\partial\Omega$.
(7) $\delta(x')$ is a Lipschitz continuous function on $\partial\Omega$ and $\delta(x') \geq 0$ on $\partial\Omega$.
(8) $\mathbf{n} = (n_1, n_2, \ldots, n_N)$ is the unit inward normal to the boundary $\partial\Omega$ (see Fig. 1.2).

The boundary condition L is called a first order *Ventcel' (Wentzell) boundary condition* (cf. [179]). The four terms of L

$$\gamma(x')u, \quad \mu(x')\frac{\partial u}{\partial \mathbf{n}}, \quad \beta(x') \cdot u, \quad \delta(x')(Au|_{\partial\Omega})$$

are supposed to correspond to the absorption phenomenon, the reflection phenomenon, the drift phenomenon along the boundary and the sticking (or viscosity) phenomenon, respectively (see Figs. 1.3 and 1.4).

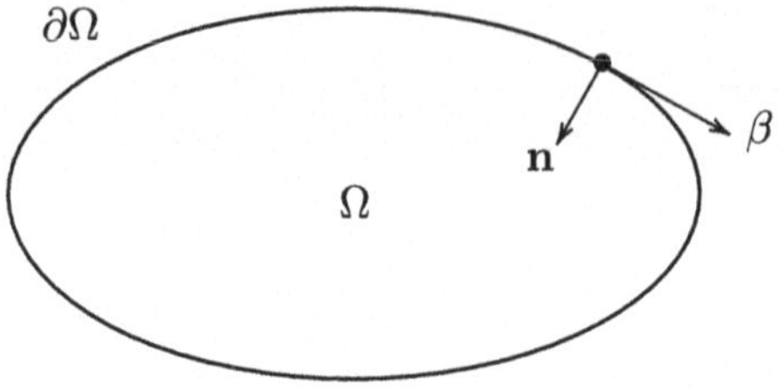

Fig. 1.2 The unit inward normal $\mathbf{n}$ to $\partial\Omega$ and the vector field β on $\partial\Omega$

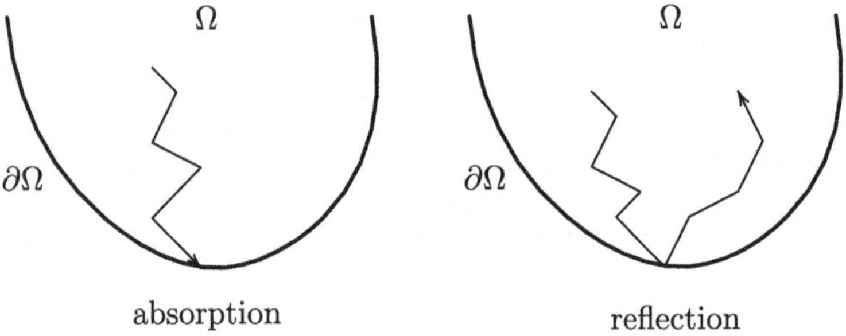

Fig. 1.3 The absorption phenomenon and the reflection phenomenon

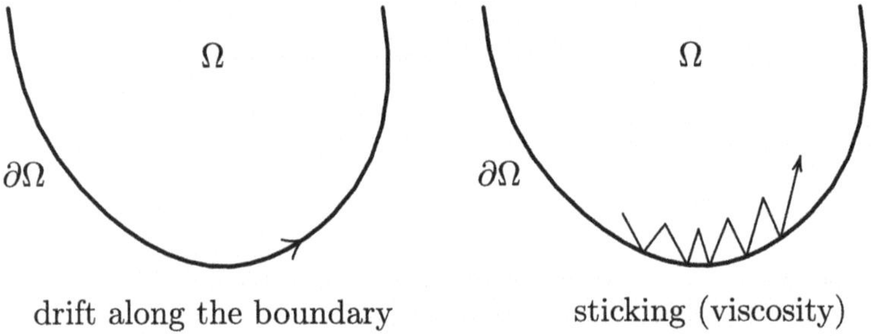

Fig. 1.4 The drift phenomenon along $\partial\Omega$ and the sticking (or viscosity) phenomenon

Let $C(\overline{\Omega})$ be the Banach space of real-valued, continuous functions on the closure $\overline{\Omega} = \Omega \cup \partial\Omega$, equipped with the maximum norm

$$\|f\|_{C(\overline{\Omega})} = \max_{x\in\overline{\Omega}} |f(x)| \quad \text{for } f \in C(\overline{\Omega}).$$

A strongly continuous semigroup $\{T_t\}_{t\geq 0}$ on the space $C(\overline{\Omega})$ is called a *Feller semigroup* if it is non-negative and contractive on $C(\overline{\Omega})$, that is,

$$f \in C(\overline{\Omega}),\ 0 \leq f(x) \leq 1 \ \text{ on } \overline{\Omega} \implies 0 \leq T_t f(x) \leq 1 \ \text{ on } \overline{\Omega}.$$

It is known (see [154, Chap. 6]) that if T_t is a Feller semigroup on $C(\overline{\Omega})$, then there exists a unique *Markov transition function* $p_t(x, \cdot)$ on $\overline{\Omega}$ such that

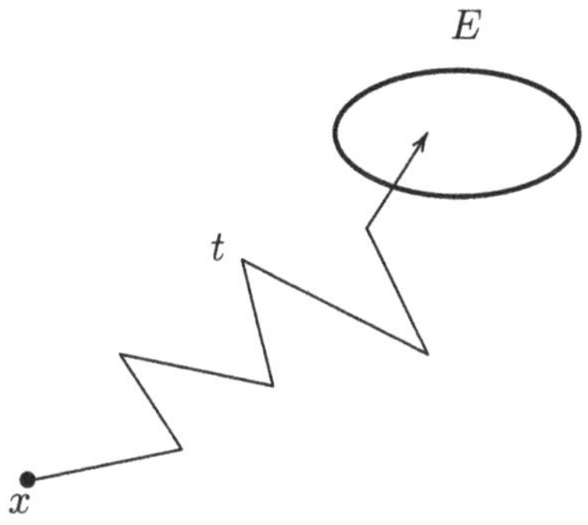

Fig. 1.5 A Markov transition function $p_t(x, E)$

$$T_t f(x) = \int_{\overline{\Omega}} p_t(x, dy) f(y) \quad \text{for all } f \in C(\overline{\Omega}).$$

Furthermore, it can be shown (see [42, Sect. 6.3]) that the function $p_t(x, \cdot)$ is the transition function of some *strong Markov process* whose paths are right-continuous and have no discontinuities other than jumps; hence the value $p_t(x, E)$ expresses the *transition probability* that a Markovian particle starting at position x will be found in the set E at time t (see Fig. 1.5).

The third purpose of this book is devoted to the functional analytic approach to the problem of existence of strong Markov processes in probability theory. More precisely, we consider the following *inverse problem*:

Problem Conversely, for any given analytic data (A, L), can we construct a Feller semigroup $\{T_t\}_{t \geq 0}$ whose infinitesimal generator $\mathfrak{A}$ is characterized by (A, L)?

1.2 Statement of Main Results

The next generation theorem for Feller semigroups asserts that there exists a Feller semigroup corresponding to such a diffusion phenomenon that a Markovian particle moves by continuous paths in the state space, with absorption, reflection, drift and sticking phenomena at the boundary (see Fig. 1.6):

Theorem 1.1 (the general case) *If $N < p < \infty$, we define a linear operator $\mathfrak{A}_L$ from $C(\overline{\Omega})$ into itself as follows:*

(a) The domain $D(\mathfrak{A}_L)$ is the set

$$D(\mathfrak{A}_L) = \left\{ u \in W^{2,p}(\Omega) : Au \in C(\overline{\Omega}), \ Lu = 0 \text{ on } \partial\Omega \right\}. \tag{1.3}$$

(b) $\mathfrak{A}_L u = Au$ for every $u \in D(\mathfrak{A}_L)$.

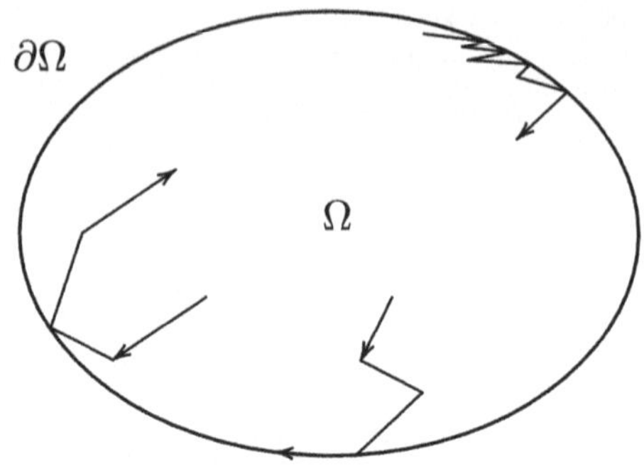

Fig. 1.6 The intuitive meaning of Theorem 1.1

Here Au and Lu are taken in the sense of distributions.

Assume that the function $\mu(x')$ satisfies the transversal condition

$$\boxed{\mu(x') > 0 \quad on\ \partial\Omega.} \tag{H.1}$$

Then the operator $\mathfrak{A}_L$ is the infinitesimal generator *of a* Feller semigroup *on $C(\overline{\Omega})$.*

Remark 1.2 The domain $D(\mathfrak{A}_L)$ does not depend on p, for $N < p < \infty$.

The situation of Theorem 1.1 can be visualized as follows:

$$\boxed{\begin{array}{ccc} C(\overline{\Omega}) & & \\ \uparrow & & \\ W^{2,p}(\Omega) & \xrightarrow{A} & L^p(\Omega) \\ \uparrow & & \uparrow \\ D(\mathfrak{A}_L) & \xrightarrow[\mathfrak{A}_L]{} & C(\overline{\Omega}) \end{array}}$$

The crucial point in the proof of Theorem 1.1 is that we consider the term $\delta(x')\left(Au(x')\right)$ of a sticking phenomenon in the boundary condition

$$Lu(x') = \mu(x')\frac{\partial u}{\partial \mathbf{n}}(x') + \beta(x') \cdot u(x') + \gamma(x')u(x') - \delta(x')\left(Au(x')\right) \quad \text{on } \partial\Omega$$

as a term of *perturbation* of the *oblique derivative boundary condition* ($\delta(x') \equiv 0$)

$$L_0u(x') := \mu(x')\frac{\partial u}{\partial \mathbf{n}}(x') + \beta(x') \cdot u(x') + \gamma(x')u(x') \quad \text{on } \partial\Omega.$$

To do this, we prove the following *generation theorem for Feller semigroups* with oblique derivative boundary condition (cf. [160, Theorem 1.2]):

Theorem 1.3 (the oblique derivative case) *If $N < p < \infty$, we define a linear operator $\mathfrak{A}_N$ from $C(\overline{\Omega})$ into itself as follows:*

(a) The domain $D(\mathfrak{A}_N)$ is the set

$$D(\mathfrak{A}_N) = \left\{u \in W^{2,p}(\Omega) : Au \in C(\overline{\Omega}),\ L_0 u = 0 \text{ on } \partial\Omega\right\}. \tag{1.4}$$

(b) $\mathfrak{A}_N u = Au$ for every $u \in D(\mathfrak{A}_N)$.

Here Au and $L_0 u$ are taken in the sense of distributions.

If the transversal condition (H.1) is satisfied, then the operator $\mathfrak{A}_N$ is the infinitesimal generator *of a* Feller semigroup *on $C(\overline{\Omega})$.*

Remark 1.4 The domain $D(\mathfrak{A}_N)$ does not depend on p, for $N < p < \infty$.

The situation of Theorem 1.3 can be visualized as follows:

$$\begin{array}{ccc}
C(\overline{\Omega}) & & \\
\uparrow & & \\
W^{2,p}(\Omega) & \xrightarrow{A} & L^p(\Omega) \\
\uparrow & & \uparrow \\
D(\mathfrak{A}_N) & \xrightarrow[\mathfrak{A}_N]{} & C(\overline{\Omega})
\end{array}$$

Rephrased, Theorem 1.3 asserts that there exists a Feller semigroup corresponding to such a diffusion phenomenon that a Markovian particle moves by continuous paths in the state space, with absorption, reflection and drift phenomena at the boundary (see Fig. 1.7).

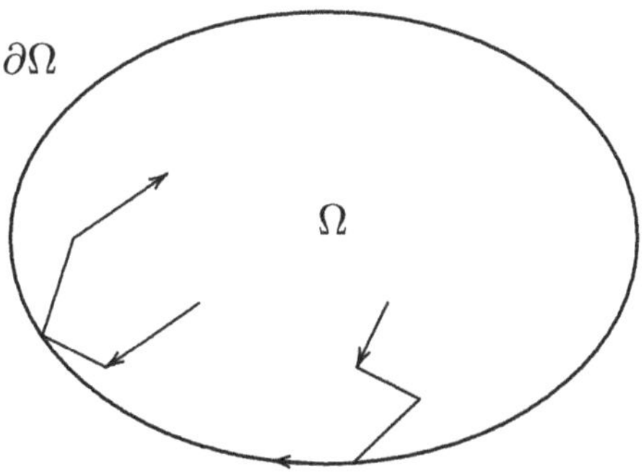

Fig. 1.7 The intuitive meaning of Theorem 1.3

Moreover, we construct a *semigroup* associated with absorption phenomenon at the boundary, which we shall formulate precisely.

We introduce a subspace of the space of continuous functions, which is associated with Dirichlet boundary condition ($\mu(x') \equiv 0$, $\beta(x') \equiv 0$, $\gamma(x') \equiv -1$, $\delta(x') \equiv 0$), by the formula

$$C_0(\overline{\Omega}) = \{u \in C(\overline{\Omega}) : u = 0 \text{ on } \partial\Omega\}.$$

A strongly continuous semigroup T_t on the space $C_0(\overline{\Omega})$ is called a *Feller semigroup* if it is non-negative and contractive on $C_0(\overline{\Omega})$, that is,

$$f \in C_0(\overline{\Omega}),\ 0 \leq f(x) \leq 1 \quad \text{on } \overline{\Omega} \implies 0 \leq T_t f(x) \leq 1 \quad \text{on } \overline{\Omega}.$$

It is known that if T_t is a Feller semigroup on $C_0(\overline{\Omega})$, then there exists a unique Markov transition function $p_t(x, \cdot)$ on Ω such that

$$T_t f(x) = \int_\Omega p_t(x, dy) f(y) \quad \text{for all } f \in C_0(\overline{\Omega}).$$

Furthermore, it can be shown (see [42, Sect. 6.3]) that the function $p_t(x, \cdot)$ is the transition function of some *strong Markov process* whose paths are right-continuous and have no discontinuities other than jumps; hence the value $p_t(x, E)$ expresses the transition probability that a Markovian particle starting at position x will be found in the set E at time t.

The next *generation theorem for Feller semigroups* asserts that there exists a Feller semigroup associated with *absorption phenomenon* at the boundary (see Fig. 1.8):

Theorem 1.5 (the Dirichlet case) *If $N < p < \infty$, we define a linear operator $\mathfrak{A}_D$ from $C_0(\overline{\Omega})$ into itself as follows:*

(a) The domain $D(\mathfrak{A}_D)$ is the set

$$D(\mathfrak{A}_D) = \{u \in W^{2,p}(\Omega) \cap C_0(\overline{\Omega}) : Au \in C_0(\overline{\Omega})\}. \tag{1.5}$$

(b) $\mathfrak{A}_D u = Au$ for every $u \in D(\mathfrak{A}_D)$.

Here Au is taken in the sense of distributions.

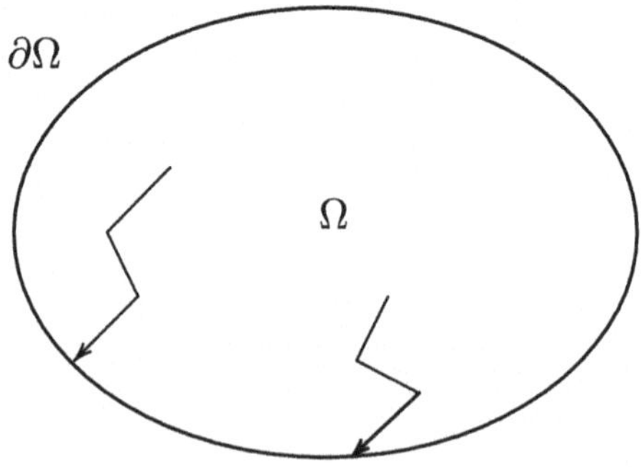

Fig. 1.8 The intuitive meaning of Theorem 1.5

Assume that

$$c(x) \leq 0 \quad \textit{for almost all } x \in \Omega.$$

Then the operator $\mathfrak{A}_D$ *is the infinitesimal generator of a Feller semigroup on* $C_0(\overline{\Omega})$.

Remark 1.6 The domain $D(\mathfrak{A}_D)$ does not depend on p, for $N < p < \infty$.

The situation of Theorem 1.5 can be visualized as follows:

$$\begin{array}{ccc}
C_0\left(\overline{\Omega}\right) & & \\
\uparrow & & \\
W^{2,p}(\Omega) \cap C_0\left(\overline{\Omega}\right) & \xrightarrow{A} & L^p(\Omega) \\
\uparrow & & \uparrow \\
D\left(\mathfrak{A}_D\right) & \xrightarrow[\mathfrak{A}_D]{} & C_0\left(\overline{\Omega}\right)
\end{array}$$

The semigroup approach to strong Markov processes can be traced back to the pioneering works of Kakutani [75], Feller [47] and [48] in the early 1950s. Our presentation here follows the book of Dynkin [43] and also part of Lamperti's [87].

Table 1.2 gives a bird's eye view of strong Markov processes, Feller semigroups and elliptic boundary value problems and how these relate to each other.

For more leisurely treatments of strong Markov processes and Feller semigroups, the reader is referred to Blumenthal–Getoor [12], Dynkin [42, 43], Itô–McKean, Jr. [72], Lamperti [87], Revuz–Yor [119] and also Taira [154].

1.3 Summary of the Contents

This introductory chapter is intended as a brief introduction to our problem and results presented in a way that a broad spectrum of readers should understand. The contents of the book are divided into seven principal parts.

1.3.1 Part I: A Short Course in Functional Analysis and Real Analysis

The first part (Chaps. 2–4) provides the elements of measure theory, functional analysis and real analysis. The material in these preparatory chapters is given for completeness, to minimize the necessity of consulting too many outside references. This makes the book fairly self-contained.

Table 1.2 A bird's eye view of strong Markov processes, Feller semigroups and elliptic boundary value problems

Probability (Microscopic approach)	**Functional Analysis** (Macroscopic approach)	**Elliptic Boundary Value Problems** (Mesoscopic approach)
Strong Markov process $\boxed{\mathcal{X} = (x_t, \mathcal{F}, \mathcal{F}_t, P_x)}$	**Feller semigroup** $T_t = e^{t\mathfrak{A}}$ $\boxed{T_t f(x) = \int_{\overline{\Omega}} p_t(x, dy)\, f(y)}$ (Kakutani)	**Infinitesimal generator** $\boxed{\mathfrak{A}}$ (Hille–Yosida theory)
Markov transition function $\boxed{P_x\{x_t \in dy\} = p_t(x, dy)}$ (Dynkin)	Radon measure $G_\alpha(x, dy) = \int_0^\infty e^{-\alpha t}\, p_t(x, dy)\, dt$ $\boxed{G_\alpha f(x) = \int_{\overline{\Omega}} G_\alpha(x, dy)\, f(y)}$ (Riesz–Markov)	Resolvent $\boxed{G_\alpha = (\alpha I - \mathfrak{A})^{-1}}$ (Hille–Yosida–Ray)
Chapman–Kolmogorov equation $p_{t+s}(x, dz)$ $= \int_{\overline{\Omega}} p_t(x, dy)\, p_s(y, dz)$	Semigroup property $\boxed{T_{t+s} = T_t \cdot T_s}$	Waldenfels operator $\boxed{W = A + S}$
Absorption, reflection, viscosity phenomena, two jump phenomena, diffusion along the boundary	Function spaces $C_0(\overline{\Omega})$ [Dirichlet case] $C(\overline{\Omega})$ [Other cases]	Ventcel' (Wentzell) boundary condition $\boxed{L}$

Chapter 2 is devoted to a review of standard topics from functional analysis. In Sect. 2.5 we formulate three pillars of functional analysis—Banach's open mapping theorem (Theorem 2.38), Banach's closed graph theorem (Theorem 2.39) and Banach's closed range theorem for closed operators (Theorem 2.42). Section 2.8 is devoted to the Riesz–Schauder theory for compact operators (Theorem 2.49). In Sect. 2.9 we state important properties of Fredholm operators (Theorems 2.50 through 2.55). In Sect. 2.10 we formulate the Riesz representation theorem for bounded linear functionals on a Hilbert space (Theorem 2.60). Finally, in Sect. 2.11 we formulate two fundamental theorems concerning spaces of continuous functions defined on a metric space—the Ascoli–Arzelà theorem (Theorem 2.69) and the Stone–Weierstrass theorem (Theorem 2.73). These topics form a necessary background for what follows.

In **Chap.** 3 we set forth the basic concepts of measure theory and develop the theory of integration on abstract measure spaces, paying particular attention to the Lebesgue integral on the Euclidean space $\mathbf{R}^n$. In particular, we prove Minkowski's inequality for integrals (Theorem 3.29) and Hardy's inequality (Theorem 3.31) in L^p spaces. In Sect. 3.9, we prove the Marcinkiewicz interpolation theorem (Theorem

3.30) that plays an important role in the proof of Theorem 9.6 in Sect. 9.3. In Sect. 3.10, as an application of Marcinkiewicz's interpolation theorem we study Riesz potentials in the classical potential theory (Theorem 3.48).

Chapter 4 is devoted to the precise definitions and statements, with some detailed proofs, of real analytic tools such as BMO and VMO functions, the Calderón–Zygmund decomposition (Theorem 4.9), the Hardy–Littlewood maximal function (Theorem 4.6), the John–Nirenberg inequality (Theorem 4.13), sharp functions (Theorem 4.19) and spherical harmonics (Theorem 4.41).

1.3.2 Part II: Elements of Function Spaces

In the second part (Chaps. 5–8) we study Sobolev spaces, Besov spaces and maximum principles in the framework of Sobolev spaces of L^p type that are used throughout the book.

The purpose of **Chap.** 5 is to study harmonic functions in the half-space $\mathbf{R}^{n+1}_+$ in terms of Poisson integrals of functions in L^p spaces, paying particular attention to the Poisson kernel $P(x, y)$ of the Dirichlet problem in the half-space (Lemma 5.11, Theorems 5.12, 5.15 and 5.16). In particular, we establish fundamental relationships between means of derivatives of Poisson integrals $u(x, y)$ taken with respect to the normal variable y and those taken with respect to the tangential variables x_i (Theorems 5.21 and 5.26).

In **Chap.** 6 we develop the theory of Besov spaces on the Euclidean space $\mathbf{R}^n$, paying particular attention to Poisson integrals. Besov spaces are function spaces defined in terms of the L^p modulus of continuity, and enter naturally in connection with boundary value problems in the framework of Sobolev spaces of L^p type. We prove a variety of equivalent norms for the Besov spaces on $\mathbf{R}^n$ via Poisson integrals (Theorems 6.4, 6.7 and 6.8). The results discussed here are adapted from Taibleson [146] and Stein [137].

Chapter 7 is devoted to the precise definitions and statements of function spaces with some detailed proofs. The function spaces of L^p type we treat are the generalized Sobolev spaces and Besov spaces on the boundary of a Lipschitz domain. Besov spaces enter in connection with boundary values of functions in L^p Sobolev spaces. Indeed, we need to make sense of the restriction $u|_{\partial\Omega}$ to the boundary $\partial\Omega$ as an element of a Besov space on $\partial\Omega$ when u belongs to a Sobolev space on the domain Ω. In particular, we formulate an important trace theorem (Theorem 7.7) that will be used in the study of the Dirichlet problem in Part IV. Finally, in Sect. 7.4 we prove that $W^{\theta,n/\theta}(\mathbf{R}^n) \subset \mathrm{VMO}$ for $0 < \theta \leq 1$ (Proposition 7.9).

In **Chap.** 8 we prove various maximum principles for second order, uniformly elliptic differential operators with *discontinuous* coefficients such as the weak and strong maximum principles (Theorems 8.1 and 8.9) and the Hopf boundary point lemma (Lemma 8.8) in the framework of Sobolev spaces of L^p type that plays an essential role in the proof of uniqueness theorems for the Dirichlet problem in Part IV.

1.3.3 Part III: Theory of Singular Integral Operators

The third part (Chaps. 9–11) is the heart of the subject. The Calderón–Zygmund theory of singular integral operators is a very refined mathematical tool whose full power is yet to be exploited (see Stein [140], Meyer [102]).

Chapter 9 is devoted to a concise and accessible exposition of the most elementary part of the Calderón–Zygmund theory of singular integral operators. We present a straightforward treatment of the Calderón–Zygmund theory necessary for the study of elliptic boundary value problems, assuming only basic knowledge of real analysis and functional analysis. In particular, we present the basic theory of the Hilbert transform H that is a special case of the singular integral of a single independent variable (Theorems 9.7 and 9.15):

$$Hf(x) := \frac{1}{\pi} \lim_{\varepsilon \downarrow 0} \int_{|x-t|>\varepsilon} \frac{f(t)}{x-t}\, dt = \frac{1}{\pi} \lim_{\varepsilon \downarrow 0} \int_{|s|>\varepsilon} f(x-s)\, \frac{ds}{s}.$$

The proofs of Theorems 9.7 and 9.15 are flowcharted.

Moreover, we study bounded kernels (Theorem 9.3), continuous kernels (Theorem 9.6), odd kernels (Theorem 9.16) and Riesz kernels

$$R_j(x) = -\frac{\Gamma((n+1)/2)}{\pi^{(n+1)/2}} \frac{x_j}{|x|^{n+1}} \quad \text{for } 1 \leq j \leq n$$

(Theorem 9.17) in great detail. The proof of the continuous kernel case is based on a version of the Calderón–Zygmund decomposition (Lemma 9.4). The study of *odd* kernels $K(x)$ and Riesz kernels $R_j(x)$ is reduced to that of the Hilbert transform H. More precisely, we have the following formula for the odd kernel $K(x)$ (see formula (9.86)):

$$\begin{aligned} K * f(x) &:= \lim_{\varepsilon \downarrow 0} \int_{|x-y|>\varepsilon} K(x-y)\, f(y)\, dy \\ &= \frac{1}{2} \int_{\Sigma_{n-1}} K(\sigma) \left(\lim_{\varepsilon \downarrow 0} \int_{|t|>\varepsilon} f(x-t\sigma) \frac{dt}{t} \right) d\sigma. \end{aligned}$$

On the other hand, singular integral operators with *even* kernel $K(x)$ can be expressed as a finite sum of products of singular integral operators with odd kernel (Theorem 9.26). We have the following decomposition formula:

$$K * f = -\sum_{j=1}^{n} R_j * (R_j * K) * f, \tag{1.6}$$

where the operators R_j and $R_j * K$ have respectively the odd kernels. This decomposition formula (1.6) may be rephrased as follows:

Table 1.3 A bird's eye view of Chap. 9

Integral kernels	Assumptions	Main theorems
The bounded case	Assumption 9.2	Theorem 9.3
The continuous case	Assumption 9.3	Theorem 9.6
Hilbert transform	$\dfrac{1}{\pi x}$	Theorems 9.7 and 9.15
The odd case	Assumption 9.4	Theorem 9.16
Riesz kernels	$R_j(x)$	Theorem 9.17
The even case	Assumption 9.5	Theorem 9.26
The general case	Conditions (9.168)	Theorem 9.27

$$\{\text{Even kernels}\} = \{\text{Riesz kernels}\} * \{\text{Odd kernels}\}.$$

In this way, we can prove the existence of the singular integral in the general case (Theorem 9.27). The proof of Theorem 9.27 is flowcharted.

The results discussed in Chap. 9 are adapted from Calderón–Zygmund [22–25, 28] and Tanabe [166] and [167]. Chapter 9 may be summarized as in Table 1.3.

The first main result in **Chap.** 10 (Theorem 10.2) asserts the existence of singular integral operators and the second main result (Theorem 10.3) concerns commutators of BMO functions and singular integral operators. It should be emphasized again that singular integral operators with non-smooth kernels provide a powerful tool to deal with smoothness of solutions of partial differential equations, with minimal assumptions of regularity on the coefficients.

In **Chap.** 11 we consider singular integrals with kernels depending on a parameter, and prove theorems about singular integrals and commutators of L^∞ functions and singular integral operators (Theorems 11.2 and 11.3), generalizing Theorems 10.2 and 10.3. The main idea of proof is to reduce the variable kernel case to the constant kernel case due to Calderón–Zygmund [28]. This is done by expanding the kernel into a series of spherical harmonics, each term defining a constant kernel operator treated in Chap. 10. Theorems about singular integrals and commutators are usually formulated in the whole space $\mathbf{R}^n$. However, our application to the theory of elliptic equations with discontinuous coefficients will require a local version of Theorems 11.2 and 11.3 (Theorems 11.4 and 11.5 and Corollary 11.7).

1.3.4 Part IV: Dirichlet Problems for Elliptic Differential Equations with Discontinuous Coefficients

Our proper subject starts with the fourth part (Chaps. 12–15).

Chapter 12 is devoted to the study of the *homogeneous* Dirichlet problem in the framework of Sobolev spaces of L^p type. We state interior and global a priori estimates for the Dirichlet problem (Theorems 12.1 and 12.2) that plays an essential role in the proof of the unique solvability theorem of the Dirichlet problem in Chap. 15. The results discussed here are adapted from Chiarenza–Frasca–Longo [34, 35] and Chiarenza [33]. Our approach can be traced back to the pioneering work of J. Schauder [127] and [128] on the Dirichlet problem for second order, uniformly elliptic differential operators.

The proofs of Theorems 12.1 and 12.2 are flowcharted.

Chapter 13 is devoted to the proof of Theorem 12.1 (Theorem 13.3) that is based on some local interior a priori estimates for the solutions of the Dirichlet problem (Lemma 13.2). The main idea of proof may be considered as an integral perturbation about the constant coefficient case, which goes back to Eugenio Elia Levi [89] (Theorem 13.1). The VMO assumption on the coefficients is of the greatest relevance in the study of an error term expressed by singular commutators (Corollary 11.7). The desired interior a priori estimate (12.2) follows in a standard way from Lemma 13.2 by a covering argument if we make use of Sobolev's imbedding theorem (Theorem 7.5), the contraction mapping principle (Theorem 2.1) and the interpolation inequality (Theorem 13.4).

In **Chap.** 14, we prove the global a priori estimate for the homogeneous Dirichlet problem stated in Theorem 12.2. This chapter is adapted from Chiarenza–Frasca–Longo [35] and Bramanti–Cerutti [17]. The desired global estimate is consequences of the following two facts:

(I) The explicit representation formula (14.2) for the solutions of the homogeneous Dirichlet problem, which is obtained from the half space Green function, involves the same integral operators as in the interior case.
(II) An L^p boundedness of some singular integral operators and boundary commutators appearing in formula (14.2) (Theorems 14.2 and 14.5).

The results of Chap. 14 are flowcharted.

Chapter 15 is devoted to the study of the *homogeneous* Dirichlet problem for a second order, uniformly elliptic differential operator with VMO coefficients in the framework of Sobolev spaces of L^p type. We prove an existence and uniqueness theorem for the homogeneous Dirichlet problem (Theorem 15.1) essentially due to Chiarenza–Frasca–Longo [35] and Chiarenza [33]. Our proof is based on some interior and boundary a priori estimates for the solutions of problem (15.2) (Theorems 12.1 and 12.2). Both the interior and boundary a priori estimates are consequences of explicit representation formulas (12.1) and (13.1) for the solutions of the homogeneous Dirichlet problem (Theorems 13.1 and 14.1) and also of the L^p-boundedness of Calderón–Zygmund singular integral operators and boundary commutators appearing in those representation formulas (Theorems 14.2 and 14.5). It should be emphasized that we make use of the *method of continuity* (Theorem 2.16) in order to prove the existence theorem for the homogeneous Dirichlet problem.

1.3.5 Part V: Oblique Derivative Problems for Elliptic Differential Equations with Discontinuous Coefficients

The fifth part (Chaps. 16–20) is devoted to the study of the *regular* oblique derivative problem for a second order, uniformly elliptic differential operator with discontinuous coefficients in the framework of Sobolev spaces of L^p type. More precisely, we consider a second order, uniformly elliptic differential operator with VMO coefficients and an oblique derivative boundary operator that is *nowhere* tangential to the boundary.

In **Chap.** 16 we state the global regularizing property of the oblique derivative problem in the framework of Sobolev spaces of L^p type (Theorem 16.1). Furthermore, we state an existence and uniqueness theorem for the oblique derivative problem in the framework of Sobolev spaces of L^p type (Theorem 16.2). The results discussed here are adapted from Di Fazio–Palagachev [40] and Maugeri–Palagachev–Softova [99].

In **Chap.** 17, for a given boundary function, we construct a special auxiliary function that satisfies an oblique derivative boundary condition (Lemma 17.1). This result will allow us to represent, locally near the boundary, the solution of the non-homogeneous oblique derivative problem in Chap. 19 (see formula (19.10)). In this way, we are reduced to the study of the *homogeneous* oblique derivative problem.

In **Chap.** 18 we prove boundary representation formulas for solutions of the homogeneous oblique derivative problem, by using the half-space Green function. The first step toward the boundary representation formulas is to derive a boundary representation formula for the solution of the homogeneous oblique derivative problem with constant coefficients operators and homogeneous boundary conditions (Lemma 18.1) just as in Gilbarg–Trudinger [59]. The second step is to derive integral representation formulas for the second derivatives of solutions of the homogeneous oblique derivative problem for variable coefficients differential operators and constant coefficients boundary operators (Theorem 18.3). The third step for the general couple $(\mathcal{L}, \mathcal{B})$ will be carried out in the next Chap. 19 (Lemma 19.1).

The purpose of **Chap.** 19 is to prove boundary Sobolev regularity of the solutions of the non-homogeneous oblique derivative problem toward the proof of Theorem 16.1 (Lemma 19.1). A combination of this regularity result with the interior regularity (Theorem 12.1) will prove the main result in Chap. 20.

Chapter 20 is devoted to the study of the non-homogeneous oblique derivative problem. We prove an existence and uniqueness theorem (Theorem 20.1) due to Di Fazio–Palagachev [40]. By Lemma 17.1, we are reduced to the study of the *homogeneous* oblique derivative problem. Our proof is based on some interior and boundary a priori estimates for the solutions of the homogeneous oblique derivative problem (Theorem 12.1 and Lemma 19.1). Both the interior and boundary a priori estimates are consequences of explicit representation formulas (19.10) and (19.11) for the solutions of homogeneous oblique derivative problem and also of the L^p-boundedness of Calderón–Zygmund singular integral operators and boundary commutators appearing in those representation formulas (Theorems 14.2 and 14.5). We make use of the *method of continuity* (Theorem 2.16) in order to prove the existence theorem for the homogeneous oblique derivative problem.

1.3.6 Part VI: Construction of Feller Semigroups with Discontinuous Coefficients

The sixth and final part (Chaps. 21–26) is devoted to the functional analytic approach to the problem of construction of strong Markov processes with first order Ventcel' (Wentzell) boundary condition for second order, uniformly elliptic differential operators with *discontinuous* coefficients. Our approach is distinguished by the extensive use of the ideas and techniques in the Calderón–Zygmund theory of singular integral operators with non-smooth kernels developed in Parts II and III.

Chapter 21 provides a brief description of the basic definitions and results about a class of semigroups (Feller semigroups) associated with strong Markov processes in probability theory, which forms a functional analytic background for the proof of Theorem 1.1. In particular, we formulate a version of the Hille–Yosida theorem adapted to the present context (Theorem 21.12). Moreover, we give two useful criteria in order that a linear operator is the infinitesimal generator of a Feller semigroup in terms of the *positive maximum principle* (Theorem 21.14 and Corollary 21.15).

Chapter 22 is devoted to the proof of Theorem 1.5. Namely, we construct a Feller semigroup associated with an absorption phenomenon at the boundary (see Fig. 1.8). To do so, we consider the Dirichlet problem for the diffusion operator with VMO coefficients in the framework of Sobolev spaces of L^p type, and prove an existence and uniqueness theorem for the Dirichlet problem (Theorem 22.1). The uniqueness result in Theorem 22.1 follows from a variant of the Bakel'man–Aleksandrov maximum principle in the framework of Sobolev spaces due to Bony [13] (Theorem 22.7). Section 22.4 is devoted to the proof of Theorem 1.5.

Chapter 23 is devoted to the proof of Theorem 1.3. Namely, we construct a Feller semigroup associated with absorption, reflection and drift phenomena at the boundary (see Fig. 1.7). To do so, we study the oblique derivative problem in the framework of Sobolev spaces of L^p type, and prove the existence and uniqueness theorem for the oblique derivative problem with VMO coefficients (Theorem 23.2). The uniqueness result in Theorem 23.2 follows from a variant of the Bakel'man–Aleksandrov maximum principle in the framework of Sobolev spaces due to Lieberman [92] (Theorem 23.5). Moreover, we construct a Feller semigroup associated with absorption, reflection and drift phenomena at the boundary. Section 23.4 is devoted to the proof of Theorem 1.3.

The purpose of **Chap.** 24 is to prove a general existence theorem for Feller semigroups with Ventcel' (Wentzell) boundary condition in terms of general boundary value problems (Theorem 24.9). Intuitively, Theorem 24.9 tells us that we can "piece together"a strong Markov process on the boundary with a diffusion in the interior to construct a strong Markov process on the closure of the domain (see Fig. 24.1).

It seems that our method of construction of Feller semigroups is, in spirit, not far away from the probabilistic method of construction of diffusion processes by means of Poisson point processes of Brownian excursions used by Watanabe [178].

Chapter 25 is devoted to the proof of Theorem 1.1. Namely, we construct a Feller semigroup corresponding to such a diffusion phenomenon that a Markovian particle moves by continuous paths in the state space, with absorption, reflection, drift and sticking phenomena at the boundary (see Fig. 1.6).

Section 25.1 is devoted to the proof of Theorem 1.1. The crucial point in the proof is that we consider the term $\delta(x')\left(Au(x')\right)$ of sticking phenomenon in the Ventcel' (Wentzell) boundary condition

$$Lu(x') = L_0 u(x') - \delta(x')\left(Au(x')\right) \quad \text{on } \partial\Omega$$

as a term of "perturbation" of the oblique derivative boundary condition

$$L_0 u(x') := \mu(x')\frac{\partial u}{\partial \mathbf{n}}(x') + \beta(x') \cdot u(x') + \gamma(x')u(x') \quad \text{on } \partial\Omega.$$

More precisely, we make use of a generation theorem for Feller semigroups with oblique derivative boundary condition L_0 to verify all the conditions of a version of the Hille–Yosida theorem (Theorem 21.14) adapted to the present context for the operator $\mathfrak{A}_L$ defined by formula (1.3).

Finally, in **Chap.** 26, as concluding remarks, we give two overviews on generation theorems for Feller semigroups (Tables 26.1 and 26.2).

Our real analysis approach to strong Markov processes can be visualized as in Figs. 1.9 and 1.10 (see [154, Chap. 10]).

1.3.7 Part VII: Appendices

In **Appendix** A we set forth the basic concepts of Lebesgue integral theory on the Euclidean space $\mathbf{R}^n$. In particular, following Schwartz [130] and Folland [51]

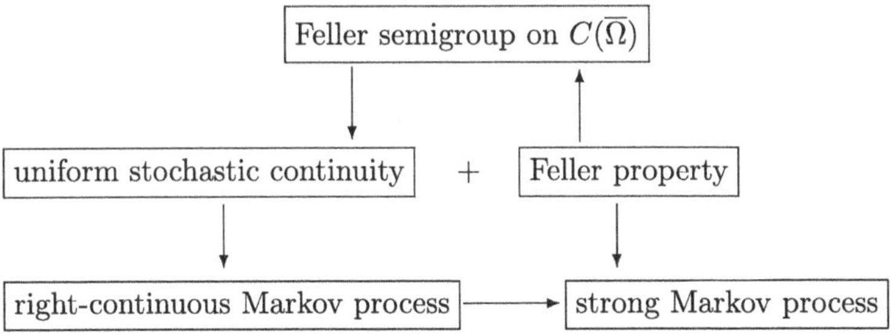

Fig. 1.9 A real analysis approach to strong Markov processes via Theorems 1.1 and 1.3

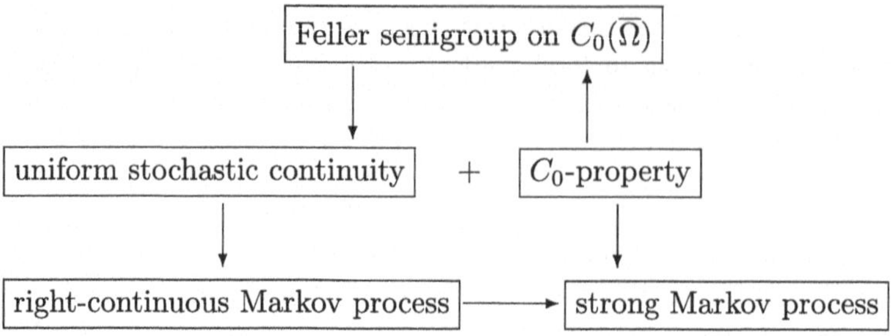

Fig. 1.10 A real analysis approach to strong Markov processes via Theorem 1.5

we prove two change-of-variables formulas under lin ear transformations and diffeomorphisms (Theorems A.1 and A.7). This appendix is adapted from Schwartz [130, Theorem] and Folland [51, Theorem 2.47].

In **Appendix** B, following faithfully Gilbarg–Trudinger [59], we present a brief introduction to the classical *potential theoretic approach* to the Dirichlet problem for the usual Laplacian Δ (Theorems B.9, B.11, B.18, B.20 and B.21). The approach here can be traced back to the pioneering work of Schauder, [127] and [128], on the Dirichlet problem for second order elliptic partial differential operators. Table B.1 gives a bird's eye view of the classical potential theoretic approach and the singular integral approach to the Dirichlet problem developed in this book.

Part I
A Short Course in Functional Analysis and Real Analysis

Chapter 2
Elements of Functional Analysis

This chapter is devoted to a review of standard topics from functional analysis such as quasinormed and normed linear spaces and closed and continuous (bounded) linear operators between Banach spaces. Most of the material will be quite familiar to the reader and may be omitted. This chapter, included for the sake of completeness, should serve to settle questions of notation and such.

In Sect. 2.5 we formulate three pillars of functional analysis for closed operators—Banach's open mapping theorem (Theorem 2.38), Banach's closed graph theorem (Theorem 2.39) and Banach's closed range theorem (Theorem 2.42). Section 2.8 is devoted to the Riesz–Schauder theory for compact operators (Theorem 2.49). In Sect. 2.9 we state important properties of Fredholm operators (Theorems 2.50 through 2.55). In Sect. 2.10 we formulate the Riesz representation theorem for bounded linear functionals on a Hilbert space (Theorem 2.60). Finally, in Sect. 2.11 we formulate two fundamental theorems concerning spaces of continuous functions defined on a metric space—the Ascoli–Arzelà theorem (Theorem 2.69) and the Stone–Weierstrass theorem (Theorem 2.73).

These topics form a necessary background for what follows. For more thorough treatments of this subject, the reader might refer to Brezis [18], Folland [51], Friedman [55], Kato [76], Kolmogorov–Fomin [80], Reed–Simon [118], Rudin [120], Schechter [129] and Yosida [182].

2.1 Metric Spaces and the Contraction Mapping Principle

A set X is called a *metric space* if there is defined a real-valued function ρ on the Cartesian product $X \times X$ such that:

(D1) $0 \leq \rho(x, y) < +\infty$;
(D2) $\rho(x, y) = 0$ if and only if $x = y$;

K. Taira, *Real Analysis Methods for Markov Processes*,
https://doi.org/10.1007/978-981-97-3659-1_2

(D3) $\rho(x, y) = \rho(y, x)$ (symmetry);
(D4) $\rho(x, y) \leq \rho(x, z) + \rho(y, z)$ (the triangle inequality).

The function ρ is called a *metric* or *distance function* on X.

If $x \in X$ and $\varepsilon > 0$, then $B(x; \varepsilon)$ will denote the open ball of radius ε about x, that is,

$$B(x; \varepsilon) = \{y \in X : \rho(x, y) < \varepsilon\}.$$

The countable family

$$\left\{B\left(x; \frac{1}{n}\right) : n \in \mathbf{N}\right\}$$

of open balls forms a fundamental neighborhood system of x; hence a metric space is a *topological space* which satisfies the first axiom of countability.

A topological space X is said to be *metrizable* if we can introduce a metric ρ on X in such a way that the induced topology on X by ρ is just the original topology on X.

Two metrics ρ_1 and ρ_2 on the same set X are said to be *equivalent* if, for each $\varepsilon > 0$ there exists $\delta > 0$ such that

$$\begin{cases} \rho_1(x, y) < \delta \Longrightarrow \rho_2(x, y) < \varepsilon, \\ \rho_2(x, y) < \delta \Longrightarrow \rho_1(x, y) < \varepsilon. \end{cases}$$

Equivalent metrics induce the same topology on X.

If x is a point of X and A is a subset of X, then we define the *distance* $\operatorname{dist}(x, A)$ from x to A by the formula

$$\operatorname{dist}(x, A) = \inf_{a \in A} \rho(x, a).$$

Let (X, ρ) be a metric space. A sequence $\{x_n\}$ in X is called a *Cauchy sequence* if it satisfies Cauchy's convergence condition

$$\lim_{n,m \to \infty} \rho(x_n, x_m) = 0.$$

A metric space X is said to be *complete* if every Cauchy sequence in X converges to a point in X.

Let (X, ρ) be a metric space. A map T from a subset X_0 of X into X is called a *contraction* on X_0 if there exists a number $0 < \theta < 1$ such that

$$\rho\,(T(x), T(y)) \leq \theta\, \rho(x, y) \quad \text{for all } x, y \in X_0. \tag{2.1}$$

The next theorem is the basis of many important existence theorems in analysis (cf. [55, Chap. 3, Theorem 3.8.2]).

Theorem 2.1 (the contraction mapping principle) *Let T be a map of a* complete *metric space (X, ρ) into itself. If T is a* contraction, *then there exists a unique point $z \in X$ such that $T(z) = z$.*

A point z for which $T(z) = z$ is called a *fixed point* of T. Hence Theorem 2.1 is also called a *fixed point theorem*.

2.2 Linear Operators and Functionals

Let X, Y be linear spaces over the same scalar field $\mathbf{K}$. A mapping T defined on a linear subspace $\mathcal{D}$ of X and taking values in Y is said to be *linear* if it preserves the operations of addition and scalar multiplication:

(L1) $T(x_1 + x_2) = Tx_1 + Tx_2$ for all $x_1, x_2 \in \mathcal{D}$.
(L2) $T(\alpha x) = \alpha Tx$ for all $x \in \mathcal{D}$ and $\alpha \in \mathbf{K}$.

We often write Tx, rather than $T(x)$, if T is linear. We let

$$\begin{aligned} D(T) &= \mathcal{D}, \\ R(T) &= \{Tx : x \in D(T)\}, \\ N(T) &= \{x \in D(T) : Tx = 0\}, \end{aligned}$$

and call them the *domain*, the *range* and the *null space* of T, respectively. The mapping T is called a *linear operator* from $D(T) \subset X$ into Y. We also say that T is a linear operator from X into Y with domain $D(T)$. In the particular case when $Y = \mathbf{K}$, the mapping T is called a *linear functional* on $D(T)$. In other words, a linear functional is a $\mathbf{K}$-valued function on $D(T)$ that satisfies conditions (L1) and (L2).

If a linear operator T is a one-to-one map of $D(T)$ onto $R(T)$, then it is easy to see that the inverse mapping T^{-1} is a linear operator on $R(T)$ onto $D(T)$. The mapping T^{-1} is called the *inverse operator* or simply the *inverse* of T. A linear operator T admits the inverse T^{-1} if and only if $Tx = 0$ implies that $x = 0$.

Let T_1 and T_2 be two linear operators from a linear space X into a linear space Y with domains $D(T_1)$ and $D(T_2)$, respectively. Then we say that $T_1 = T_2$ if and only if $D(T_1) = D(T_2)$ and $T_1x = T_2x$ for all $x \in D(T_1) = D(T_2)$. If $D(T_1) \subset D(T_2)$ and $T_1x = T_2x$ for all $x \in D(T_1)$, then we say that T_2 is an *extension* of T_1 and also that T_1 is a *restriction* of T_2, and we write $T_1 \subset T_2$.

2.3 Quasinormed Linear Spaces

Let X be a linear space over the real or complex number field $\mathbf{K}$. A real-valued function p defined on X is called a *seminorm* on X if it satisfies the following three conditions:

(S1) $0 \le p(x) < \infty$ for all $x \in X$.
(S2) $p(\alpha x) = |\alpha| p(x)$ for all $\alpha \in \mathbf{K}$ and $x \in X$.
(S3) $p(x+y) \le p(x) + p(y)$ for all $x, y \in X$.

Let $\{p_i\}$ be a countable family of seminorms on X such that

$$p_1(x) \le p_2(x) \le \cdots \le p_i(x) \le \cdots \quad \text{for each } x \in X, \tag{2.2}$$

and define

$$V_{ij} = \left\{ x \in X : p_i(x) < \frac{1}{j} \right\}, \quad i, j = 1, 2, \ldots.$$

Then it is easy to verify that a countable family of the sets

$$x + V_{ij} = \{ x + y : y \in V_{ij} \}$$

satisfies the axioms of a fundamental neighborhood system of x; hence X is a topological space which satisfies the first axiom of countability.

Furthermore, we have the following:

Theorem 2.2 *Let $\{p_i\}$ be a countable family of seminorms on a linear space X which satisfies condition* (2.2). *Assume that:*

$$\text{For every non-zero } x \in X \text{, there exists a seminorm } p_i \text{ such that } p_i(x) > 0. \tag{2.3}$$

Then the space X is metrizable *by the metric*

$$\rho(x, y) = \sum_{i=1}^{\infty} \frac{1}{2^i} \frac{p_i(x)}{1 + p_i(x)} \quad \text{for all } x, y \in X.$$

If we let

$$|x| := \rho(x, 0) = \sum_{i=1}^{\infty} \frac{1}{2^i} \frac{p_i(x)}{1 + p_i(x)} \quad \text{for } x \in X, \tag{2.4}$$

then the quantity $|x|$ enjoys the following four properties:

(Q1) $|x| \ge 0$; $|x| = 0$ *if and only if* $x = 0$.
(Q2) $|x + y| \le |x| + |y|$ *(the triangle inequality).*
(Q3) $\alpha_n \to 0$ *in* $\mathbf{K} \Longrightarrow |\alpha_n x| \to 0$ *for every* $x \in X$.
(Q4) $|x_n| \to 0 \Longrightarrow |\alpha x_n| \to 0$ *for every* $\alpha \in \mathbf{K}$.

This quantity $|x|$ is called a *quasinorm* of x, and the space X is called a *quasinormed linear space*.

Theorem 2.2 may be restated as follows:

Theorem 2.3 *A linear space X, topologized by a countable family $\{p_i\}$ of seminorms satisfying conditions* (2.2) *and* (2.3), *is a quasinormed linear space with respect to the quasinorm $|x|$ defined by formula* (2.4).

Let X be a quasinormed linear space. The convergence

$$\lim_{n\to\infty} |x_n - x| = 0$$

in X is denoted by $s - \lim_{n\to\infty} x_n = x$ or simply by $x_n \to x$, and we say that the sequence $\{x_n\}$ *converges strongly* to x. A sequence $\{x_n\}$ is called a *Cauchy sequence* if it satisfies Cauchy's condition

$$\lim_{m,n\to\infty} |x_m - x_n| = 0.$$

A quasinormed linear space X is called a *Fréchet space* if it is complete, that is, if every Cauchy sequence in X converges strongly to a point in X. If a quasinormed linear space X is topologized by a countable family $\{p_i\}$ of seminorms which satisfies conditions (2.2) and (2.3), then the above definitions may be reformulated in terms of seminorms as follows:

(i) A sequence $\{x_n\}$ in X converges strongly to a point x in X if and only if, for every seminorm p_i and every $\varepsilon > 0$, there exists a positive integer $N = N(i, \varepsilon)$ such that
$$n \geq N \Longrightarrow p_i(x_n - x) < \varepsilon.$$

(ii) A sequence $\{x_n\}$ in X is a Cauchy sequence if and only if, for every seminorm p_i and every $\varepsilon > 0$, there exists a positive integer $N = N(i, \varepsilon)$ such that
$$m, n \geq N \Longrightarrow p_i(x_m - x_n) < \varepsilon.$$

Let X be a quasinormed linear space. A linear subspace of X is called a *closed subspace* if it is a closed subset of X. For example, the closure $\overline{M}$ of a linear subspace M is a closed subspace. Indeed, the elements of $\overline{M}$ are limits of sequences in M; thus, if $x = \lim_n x_n$, $x_n \in M$ and $y = \lim_n y_n$, $y_n \in M$, then it follows that

$$x + y = \lim_n (x_n + y_n),$$
$$\alpha x = \lim_n \alpha x_n \quad \text{for } \alpha \in \mathbf{K}.$$

This proves that $x + y \in \overline{M}$ and $\alpha x \in \overline{M}$ for all $\alpha \in \mathbf{K}$.

2.3.1 Compact Sets

A collection $\{U_\lambda\}_{\lambda\in\Lambda}$ of open sets of a topological space X is called an *open covering* of X if $X = \cup_{\lambda\in\Lambda} U_\lambda$. A topological space X is said to be *compact* if every open covering $\{U_\lambda\}$ of X contains some finite subcollection of $\{U_\lambda\}$ which still covers X. A subset of a topological space X is itself a topological space with the topology induced from X. A subset of X is said to be *compact* if it is compact with respect to this induced topology.

A subset of a topological space X is said to be *relatively compact* (or *precompact*) if its closure is a compact subset of X. A topological space X is said to be *locally compact* if every point of X has a relatively compact neighborhood.

A subset of a topological space X is called a *σ-compact subset* if it is a countable union of compact sets.

Compactness is such a useful property that, given a non-compact space $(X, \mathcal{O})$, it is worthwhile constructing a compact space $(X', \mathcal{O}')$ with X being its dense subset. Such a space is called a *compactification* of $(X, \mathcal{O})$. The simplest way in which this can be achieved is by adjoining one extra point ∞ to the space X; a topology $\mathcal{O}'$ can be defined on $X' = X \cup \{\infty\}$ in such a way that $\left(X', \mathcal{O}'\right)$ is compact and that $\mathcal{O}$ is the relative topology induced on X by $\mathcal{O}'$. The topological space $\left(X', \mathcal{O}'\right)$ is called the *one-point compactification* of $(X, \mathcal{O})$, and the point ∞ is called the *point at infinity*.

Let X be a quasinormed linear space. A subset Y of X is called a *sequentially compact* if every sequence $\{y_n\}$ in Y contains a subsequence $\{y_{n'}\}$ which converges to a point y of Y:

$$\lim_{n'\to\infty} |y_{n'} - y| = 0.$$

Then we have the following criterion for compactness:

Theorem 2.4 *A subset of a quasinormed linear space X is compact if and only if it is sequentially compact.*

2.3.2 Bounded Sets

Let $(X, |\cdot|)$ be a quasinormed linear space. A set B in X is said to be *bounded* if it satisfies the condition

$$\sup_{x\in B} |x| < \infty.$$

We remark that every compact set is bounded.

A subset K of X is said to be *totally bounded* if, for any given $\varepsilon > 0$ there is a finite number of balls

$$B(x_i, \varepsilon) = \{x \in X : |x - x_i| < \varepsilon\}, \quad 1 \le i \le n,$$

of radius ε about $x_i \in X$ that cover K:

$$K \subset \bigcup_{i=1}^{n} B(x_i, \varepsilon).$$

Example 2.5 Let $(X, |\cdot|)$ be a quasinormed linear space. Assume that a subset A satisfies the following three conditions:

(a) For every $h > 0$, there exists a totally bounded subset A_h of X.
(b) For each point $x \in A$, there exists a point $y \in A_h$ such that $|x - y| \leq h$.
(c) For each point $z \in A_h$, there exists a point $w \in A$ such that $|z - w| \leq h$.

Then the subset A is totally bounded.

Proof For any given $\varepsilon > 0$, we assume that

$$A \subset \bigcup_{x \in A} B(x, \varepsilon).$$

Choose a number h_0 such that

$$0 < h_0 < \frac{\varepsilon}{3}.$$

Since we have the assertion

$$A_{h_0} \subset \bigcup_{y \in A_{h_0}} B(y, \varepsilon/3),$$

it follows from condition (a) that there is a finite number of points

$$\{y_1, y_2, \ldots, y_N\} \subset A_{h_0}$$

such that

$$A_{h_0} \subset \bigcup_{i=1}^{N} B(y_i, \varepsilon/3). \tag{2.5}$$

Moreover, by condition (c) we can find a finite number of points $\{x_1, \ldots, x_N\}$ of A such that

$$|x_i - y_i| \leq h_0, \quad 1 \leq i \leq N.$$

Then we have the assertion

$$A \subset \bigcup_{i=1}^{N} B(x_i, \varepsilon), \tag{2.6}$$

which proves that A is totally bounded.

Indeed, if x is an arbitrary point of A, by condition (b) we can find a point $y_0 \in A_{h_0}$ such that

$$|x - y_0| \leq h_0.$$

Moreover, by assertion (2.5), there exists a point $y_i \in A_{h_0}$ such that

$$|y_0 - y_i| \leq \frac{\varepsilon}{3}.$$

Hence we have the inequality

$$\begin{aligned}|x - x_i| &\leq |x - y_0| + |y_0 - y_i| + |y_i - x_i| \leq h_0 + \frac{\varepsilon}{3} + h_0 \\ &< \varepsilon.\end{aligned}$$

This proves the desired assertion (2.6).

The proof of Example 2.5 is complete. □

Finally, we have the following criterion for compactness:

Theorem 2.6 *Let X be a complete quasinormed linear space. A closed subset of X is compact if and only if it is totally bounded.*

Throughout the rest of this section, let X and Y be quasinormed linear spaces over the same scalar field $\mathbf{K}$, topologized respectively by countable families $\{p_i\}$ and $\{q_i\}$ of seminorms which satisfy conditions (2.2) and (2.3).

2.3.3 Continuity of Linear Operators

Let T be a linear operator from X into Y with domain $D(T)$. By virtue of the linearity of T, it follows that T is continuous everywhere on $D(T)$ if and only if it is continuous at one point of $D(T)$. Furthermore, we have the following:

Theorem 2.7 *A linear operator T from X into Y with domain $D(T)$ is continuous everywhere on $D(T)$ if and only if, for every seminorm q_j on Y, there exist a seminorm p_i on X and a constant $C > 0$ such that*

$$q_j(Tx) \leq C\, p_i(x) \quad \text{for all } x \in D(T).$$

2.3.4 Topologies of Linear Operators

We let

$$L(X, Y) = \text{the collection of continuous linear operators on } X \text{ into } Y.$$

We define in the set $L(X, Y)$ addition and scalar multiplication of operators in the usual way:

$$(T + S)x = Tx + Sx, \quad x \in X,$$
$$(\alpha T)x = \alpha(Tx), \quad \alpha \in \mathbf{K}, \ x \in X.$$

Then $L(X, Y)$ is a linear space.

We introduce three different topologies on the space $L(X, Y)$.

(1) *Simple convergence topology*: This is the topology of convergence at each point of X; a sequence $\{T_n\}$ in $L(X, Y)$ converges to an element T of $L(X, Y)$ in the simple convergence topology if and only if $T_n x \to Tx$ in Y for each $x \in X$.
(2) *Compact convergence topology*: This is the topology of uniform convergence on compact sets in X; $T_n \to T$ in the compact convergence topology if and only if $T_n x \to Tx$ in Y uniformly for x ranging over compact sets in X.
(3) *Bounded convergence topology*: This is the topology of uniform convergence on bounded sets in X; $T_n \to T$ in the bounded convergence topology if and only if $T_n x \to Tx$ in Y uniformly for x ranging over bounded sets in X.

The simple convergence topology is weaker than the compact convergence topology, and the compact convergence topology is weaker than the bounded convergence topology.

2.3.5 Product Spaces

Let X and Y be quasinormed linear spaces over the same scalar field $\mathbf{K}$. Then the Cartesian product $X \times Y$ becomes a linear space over $\mathbf{K}$ if we define the algebraic operations coordinatewise

$$\{x_1, y_1\} + \{x_2, y_2\} = \{x_1 + x_2, y_1 + y_2\},$$
$$\alpha \{x, y\} = \{\alpha x, \alpha y\}, \quad \alpha \in \mathbf{K}.$$

It is easy to verify that the quantity

$$|\{x, y\}| = \left(|x|_X^2 + |y|_Y^2\right)^{1/2} \tag{2.7}$$

satisfies axioms (Q1) through (Q4) of a quasinorm; hence the product space $X \times Y$ is a quasinormed linear space with respect to the quasinorm defined by formula (2.7). Furthermore, if X and Y are Fréchet spaces, then so is $X \times Y$. In other words, the completeness is inherited by the product space.

2.4 Normed Linear Spaces

A quasinormed linear space is called a *normed linear space* if it is topologized by just one seminorm that satisfies condition (2.3). We give the precise definition of a normed linear space.

Let X be a linear space over the real or complex number field $\mathbf{K}$. A real-valued function $\|\cdot\|$ defined on X is called a *norm* on X if it satisfies the following three conditions:

(N1) $\|x\| \geq 0$; $\|x\| = 0$ if and only if $x = 0$.
(N2) $\|\alpha x\| = |\alpha| \|x\|$, $\alpha \in \mathbf{K}$, $x \in X$.
(N3) $\|x + y\| \leq \|x\| + \|y\|$, $x, y \in X$ (the triangle inequality).

A linear space X equipped with a norm $\|\cdot\|$ is called a normed linear space. The topology on X is defined by the metric

$$\rho(x, y) = \|x - y\| .$$

The convergence

$$\lim_{n\to\infty} \|x_n - x\| = 0$$

in X is denoted by $s - \lim_{n\to\infty} x_n = x$ or simply $x_n \to x$, and we say that the sequence $\{x_n\}$ converges strongly to x. A sequence $\{x_n\}$ in X is called a *Cauchy sequence* if it satisfies the condition

$$\lim_{n,m\to\infty} \|x_n - x_m\| = 0.$$

A normed linear space X is called a *Banach space* if it is complete, that is, if every Cauchy sequence in X converges strongly to a point in X.

Two norms $\|\cdot\|_1$ and $\|\cdot\|_2$ defined on the same linear space X are said to be *equivalent* if there exist constants $c > 0$ and $C > 0$ such that

$$c\,\|x\|_1 \leq \|x\|_2 \leq C\,\|x\|_1 \quad \text{for all } x \in X.$$

Equivalent norms induce the same topology.

If X and Y are normed linear spaces over the same scalar field $\mathbf{K}$, then the product space $X \times Y$ is a normed linear space by the norm

$$\|\{x, y\}\| := \left(\|x\|_X^2 + \|y\|_Y^2\right)^{1/2}.$$

If X and Y are Banach spaces, then so is $X \times Y$.

Let X be a normed linear space. If Y is a closed linear subspace of X, then the factor space X/Y is a normed linear space by the norm

$$\|\widetilde{x}\| := \inf_{z \in \widetilde{x}} \|z\|. \tag{2.8}$$

If X is a Banach space, then so is X/Y. The space X/Y, normed by formula (2.8), is called a *normed factor space*.

2.4.1 Linear Operators on Normed Spaces

Throughout the rest of this section, the letters X, Y, Z denote normed linear spaces over the same scalar field $\mathbf{K}$.

The next theorem is a normed linear space version of Theorem 2.7.

Theorem 2.8 *Let T be a linear operator from X into Y with domain $D(T)$. Then T is continuous everywhere on $D(T)$ if and only if there exists a constant $C > 0$ such that*

$$\|Tx\| \leq C\|x\| \quad \textit{for all } x \in D(T). \tag{2.9}$$

Remark 2.9 In inequality (2.9), the quantity $\|x\|$ is the norm of x in X and the quantity $\|Tx\|$ is the norm of Tx in Y. Frequently, several norms appear together, but it is clear from the context which is which.

One of the consequences of Theorem 2.8 is the following extension theorem for a continuous linear operator.

Theorem 2.10 *If T is a continuous linear operator from X into Y with domain $D(T)$ and if Y is a Banach space, then T has a unique continuous extension $\widetilde{T}$ whose domain is the closure $\overline{D(T)}$ of $D(T)$.*

As another consequence of Theorem 2.8, we give a necessary and sufficient condition for the existence of the continuous inverse of a linear operator.

Theorem 2.11 *Let T be a linear operator from X into Y with domain $D(T)$. Then T admits a continuous inverse T^{-1} if and only if there exists a constant $c > 0$ such that*

$$\|Tx\| \geq c\|x\| \quad \textit{for all } x \in D(T).$$

A linear operator T from X into Y with domain $D(T)$ is called an *isometry* if it is norm-preserving, that is, if we have the formula

$$\|Tx\| = \|x\| \quad \text{for all } x \in D(T).$$

It is clear that if T is an isometry, then it is injective and both T and T^{-1} are continuous.

If T is a continuous, one-to-one linear mapping of X onto Y and if its inverse T^{-1} is also a continuous mapping, then it is called an *isomorphism* of X onto Y. Two normed linear spaces are said to be *isomorphic* if there is an isomorphism between them.

By combining Theorems 2.8 and 2.11, we obtain the following:

Theorem 2.12 *Let T be a linear operator on X onto Y. Then T is an isomorphism if and only if there exist constants $c > 0$ and $C > 0$ such that*

$$c\|x\| \leq \|Tx\| \leq C\|x\| \quad \text{for all } x \in X.$$

If T is a continuous linear operator from X into Y with domain $D(T)$, we let

$$\|T\| = \inf\{C : \|Tx\| \leq C\|x\|, \ x \in D(T)\}.$$

Then, in view of the linearity of T we have the formula

$$\|T\| := \sup_{\substack{x\in D(T)\\ x\neq 0}} \frac{\|Tx\|}{\|x\|} = \sup_{\substack{x\in D(T)\\ \|x\|=1}} \|Tx\| = \sup_{\substack{x\in D(T)\\ \|x\|\leq 1}} \|Tx\|. \tag{2.10}$$

This proves that $\|T\|$ is the smallest non-negative number such that

$$\|Tx\| \leq \|T\| \cdot \|x\| \quad \text{for all } x \in D(T). \tag{2.11}$$

Theorem 2.8 asserts that a linear operator T on X into Y is continuous if and only if it maps bounded sets in X into bounded sets in Y. Thus a continuous linear operator on X into Y is usually called a *bounded linear operator* on X into Y. We let

$$L(X, Y) = \text{the space of bounded (continuous) linear operators on } X \text{ into } Y.$$

In the case of normed linear spaces, the simple convergence topology on $L(X, Y)$ is usually called the *strong topology* of operators, and the bounded convergence topology on $L(X, Y)$ is called the *uniform topology* of operators. In view of formulas (2.10) and (2.11), it follows that the quantity $\|T\|$ satisfies axioms (N1), (N2) and (N3) of a norm; hence the space $L(X, Y)$ is a normed linear space by the norm $\|T\|$ given by formula (2.10). The topology on $L(X, Y)$ induced by the operator norm $\|T\|$ is just the uniform topology of operators.

We give a sufficient condition for the space $L(X, Y)$ to be complete:

Theorem 2.13 *If Y is a Banach space, then so is $L(X, Y)$.*

If T is a linear operator from X into Y with domain $D(T)$ and S is a linear operator from Y into Z with domain $D(S)$, then we define the product ST as follows:

(a) $D(ST) = \{x \in D(T) : Tx \in D(S)\}$,
(b) $(ST)(x) = S(Tx)$ for every $x \in D(ST)$.

As for the product of linear operators, we have the following:

Proposition 2.14 *If $T \in L(X, Y)$ and $S \in L(Y, Z)$, then it follows that $ST \in L(X, Z)$. Moreover, we have the inequality*

$$\|ST\| \leq \|S\| \cdot \|T\|.$$

We often make use of the following theorem in constructing the bounded inverse of a bounded linear operator:

Theorem 2.15 (C. Neumann) *If T is a bounded linear operator on a Banach space X into itself and satisfies $\|T\| < 1$, then the operator $I - T$ has a unique bounded linear inverse $(I - T)^{-1}$ which is given by* C. Neumann's series

$$(I - T)^{-1} = \sum_{n=0}^{\infty} T^n.$$

Here I is the identity operator: $Ix = x$ for every $x \in X$, and $T^0 = I$.

2.4.2 Method of Continuity

In Chap. 15 we make use of the following method of continuity (see [59, Chap. 5, Theorem 5.2]) in order to prove the *existence theorem* for the Dirichlet problem:

Theorem 2.16 (the method of continuity) *Let $\mathcal{B}$ be a Banach space and let $\mathcal{V}$ be a normed linear space. If $\mathcal{L}_0$ and $\mathcal{L}_1$ are two bounded linear operators from $\mathcal{B}$ into $\mathcal{V}$, we define a family of bounded linear operators*

$$\mathcal{L}_t = (1 - t)\mathcal{L}_0 + t\,\mathcal{L}_1 : \mathcal{B} \longrightarrow \mathcal{V}$$

for $0 \leq t \leq 1$. Assume that there exists a positive constant C, independent of x and t, such that

$$\|x\|_{\mathcal{B}} \leq C\,\|\mathcal{L}_t x\|_{\mathcal{V}} \quad \text{for all } x \in \mathcal{B}. \tag{2.12}$$

Then the operator $\mathcal{L}_1$ maps $\mathcal{B}$ onto $\mathcal{V}$ if and only if the operator $\mathcal{L}_0$ maps $\mathcal{B}$ onto $\mathcal{V}$.

Proof Assume that $\mathcal{L}_s$ is surjective for some $s \in [0, 1]$. By inequality (2.12), it follows that $\mathcal{L}_s$ is bijective, so that the inverse $\mathcal{L}_s^{-1} : \mathcal{V} \to \mathcal{B}$ exists. Here we remark that

$$\left\|\mathcal{L}_s^{-1}\right\| \leq C.$$

Now let t be an arbitrary point of the interval $[0, 1]$. For any given $y \in \mathcal{V}$, the equation

$$\mathcal{L}_t x = y$$

is equivalent to the equation

$$\mathcal{L}_s x = \mathcal{L}_t x + (\mathcal{L}_s - \mathcal{L}_t)x = y + (s-t)\,(\mathcal{L}_1 x - \mathcal{L}_0 x).$$

Hence we have the equivalent assertions

$$\begin{aligned}\mathcal{L}_t x = y \iff x &= \mathcal{L}_s^{-1}\,(y + (s-t)\,(\mathcal{L}_1 x - \mathcal{L}_0 x)) \\ \iff \left(I - (s-t)\mathcal{L}_s^{-1}\,(\mathcal{L}_1 - \mathcal{L}_0)\right) x &= \mathcal{L}_s^{-1} y.\end{aligned}$$

However, if $|t-s|$ is so small that

$$|s-t| < \delta := \frac{1}{C\,(\|\mathcal{L}_1\| + \|\mathcal{L}_0\|)},$$

then it follows that

$$\begin{aligned}\left\|(s-t)\,\mathcal{L}_s^{-1}\,(\mathcal{L}_1 - \mathcal{L}_0)\right\| &\le |s-t|\,\left\|\mathcal{L}_s^{-1}\right\|\,\|\mathcal{L}_1 - \mathcal{L}_0\| \\ &\le C\,|s-t|\,(\|\mathcal{L}_1\| + \|\mathcal{L}_0\|) = \frac{|s-t|}{\delta} \\ &< 1.\end{aligned}$$

This proves that the operator

$$\left(I - (s-t)\,\mathcal{L}_s^{-1}\,(\mathcal{L}_1 - \mathcal{L}_0)\right)$$

has, as a Neumann series (Theorem 2.15), the inverse

$$\left(I - (s-t)\,\mathcal{L}_s^{-1}\,(\mathcal{L}_1 - \mathcal{L}_0)\right)^{-1} = \sum_{n=0}^{\infty} (s-t)^n \left(\mathcal{L}_s^{-1}\right)^n (\mathcal{L}_1 - \mathcal{L}_0)^n.$$

Therefore, we obtain that, for all $t \in [0,1]$ satisfying $|t-s| < \delta$,

$$\mathcal{L}_t x = y \iff x = \left(I - (s-t)\mathcal{L}_s^{-1}\,(\mathcal{L}_1 - \mathcal{L}_0)\right)^{-1} \mathcal{L}_s^{-1} y.$$

By dividing the interval $[0,1]$ into subintervals of length less than δ, we find that the mapping $\mathcal{L}_t$ is surjective for all $t \in [0,1]$, provided that $\mathcal{L}_s$ is surjective for some $s \in [0,1]$. In particular, this proves that $\mathcal{L}_1$ maps $\mathcal{B}$ onto $\mathcal{V}$ if and only if $\mathcal{L}_0$ maps $\mathcal{B}$ onto $\mathcal{V}$.

The proof of Theorem 2.16 is complete. □

2.4.3 Finite-Dimensional Spaces

The next theorem asserts that there is no point in studying abstract finite-dimensional normed linear spaces:

Theorem 2.17 *All n-dimensional normed linear spaces over the same scalar field* $\mathbf{K}$ *are isomorphic to* $\mathbf{K}^n$ *with the maximum norm*

$$\|\alpha\| = \max_{1 \le i \le n} |\alpha_i|, \quad \alpha = (\alpha_1, \alpha_2, \ldots, \alpha_n) \in \mathbf{K}^n.$$

Topological properties of the space $\mathbf{K}^n$ applies to all finite-dimensional normed linear spaces.

Corollary 2.18 *All finite-dimensional normed linear spaces are complete.*

Corollary 2.19 *Every finite-dimensional linear subspace of a normed linear space is closed.*

Corollary 2.20 *A subset of a finite-dimensional normed linear space is compact if and only if it is closed and bounded.*

By Corollary 2.19, it follows that the closed unit ball in a finite-dimensional normed linear space is compact. Conversely, this property characterizes finite-dimensional spaces:

Theorem 2.21 *If the closed unit ball in a normed linear space* X *is compact, then* X *is finite-dimensional.*

2.4.4 The Hahn–Banach Extension Theorem

The Hahn–Banach extension theorem asserts the existence of linear functionals dominated by norms (see [182, Chap. IV, Sect. 5, Theorem 1 and Corollary]):

Theorem 2.22 (Hahn–Banach) *Let* X *be a normed linear space over the real or complex number field* $\mathbf{K}$ *and let* M *be a linear subspace of* X*. If* f *is a continuous linear functional defined on* M*, then it can be extended to a continuous linear functional* $\widetilde{f}$ *on* X *so that*

$$\|\widetilde{f}\| = \|f\|.$$

Let X be a real or complex, normed linear space. A closed subset M of X is said to be *balanced* if it satisfies the condition

$$x \in M, \ |\alpha| \le 1 \Longrightarrow \alpha x \in M.$$

The next two theorems assert the existence of non-trivial continuous linear functionals (see [182, Chap. IV, Sect. 6, Theorem 3]):

Theorem 2.23 (Mazur) *Let X be a real or complex, normed linear space and let M be a closed convex, balanced subset of X. Then, for any element $x_0 \notin M$ there exists a continuous linear functional f_0 on X such that*

$$f_0(x_0) > 1,$$
$$|f_0(x)| \leq 1 \quad \text{on } M.$$

Proof Since M is closed and $x_0 \notin M$, it follows that

$$\operatorname{dist}(x_0, M) > 0.$$

If $0 < d < \operatorname{dist}(x_0, M)$, we let

$$B\left(0, \frac{d}{2}\right) = \left\{x \in X : \|x\| \leq \frac{d}{2}\right\},$$
$$B\left(x_0, \frac{d}{2}\right) = x_0 + B\left(0, \frac{d}{2}\right) = \left\{x \in X : \|x - x_0\| \leq \frac{d}{2}\right\},$$
$$U = \left\{x \in X : \operatorname{dist}(x, M) \leq \frac{d}{2}\right\}.$$

Then we have the assertions

$$U \cap B\left(x_0, \frac{d}{2}\right) = \emptyset,$$
$$B\left(0, \frac{d}{2}\right) \subset U,$$

since $0 \in M$. Moreover, since M is convex and balanced, it is easy to verify the following three assertions:

(a) U is convex.
(b) U is balanced.
(c) U is absorbing, that is, for any $x \in X$, there exists a constant $\alpha > 0$ such that $\alpha^{-1}x \in U$.

Hence, we can define the *Minkowski functional* p_U of U by the formula

$$p_U(x) = \inf\left\{\alpha > 0 : \alpha^{-1}x \in U\right\} \quad \text{for every } x \in X.$$

Since U is closed, it is easy to verify the following assertions:

$$\begin{cases} p_U(x) > 1 & \text{if } x \notin U, \\ p_U(x) \leq 1 & \text{if } x \in U. \end{cases}$$

Therefore, by applying [182, Chap. IV, Sect. 6, Corollary 1 to Theorem 1] to our situation we can find a continuous linear functional f_0 on X such that

$$f_0(x_0) = p_U(x_0) > 1,$$
$$|f_0(x)| \leq p_U(x) \quad \text{on } X.$$

In particular, we have the assertion

$$|f_0(x)| = p_U(x) \leq 1 \quad \text{on } M,$$

since $M \subset U$.

The proof of Theorem 2.23 is complete. □

Theorem 2.24 *Let X be a normed linear space and let M be a closed linear subspace of X. Then, for any element $x_0 \notin M$ there exists a continuous linear functional f_0 on X such that*

$$\begin{cases} f_0(x_0) > 1, \\ f_0(x) = 0 \quad \text{on } M. \end{cases}$$

Proof Indeed, it suffices to note that

$$|f_0(x)| \leq 1 \quad \text{on } M \Longrightarrow f_0(x) = 0 \quad \text{on } M,$$

since M is a linear space.

The proof of Theorem 2.24 is complete. □

Finally, the next theorem asserts that, for each point $x_0 \neq 0$ there exists a continuous linear functional f_0 such that $f_0(x_0) \neq f_0(0) = 0$:

Theorem 2.25 *Let X be a normed linear space. For each non-zero element x_0 of X, there exists a continuous linear functional f_0 on X such that*

$$\begin{cases} f_0(x_0) = \|x_0\|, \\ \|f_0\| = 1. \end{cases}$$

2.4.5 Dual Spaces

Let X be a normed linear space over the real or complex number field $\mathbf{K}$. A continuous linear functional on X is usually called a *bounded linear functional* on X.

The space $L(X, \mathbf{K})$ of all bounded linear functionals on X is called the *dual space* of X, and is denoted by X'. We shall write

$$f(x) = \langle f, x \rangle$$

for the value of the functional $f \in X'$ and the vector $x \in X$. The bounded (resp. simple) convergence topology on X' is called the *strong* (resp. *weak**) *topology* on X' and the dual space X' equipped with this topology is called the *strong* (resp. *weak**) *dual space* of X.

It follows from an application of Theorem 2.13 with $Y := \mathbf{K}$ that the strong dual space X' is a Banach space with the norm

$$\|f\| = \sup_{x \in X \setminus \{0\}} \frac{|f(x)|}{\|x\|} = \sup_{\substack{x \in X \\ \|x\| \le 1}} |f(x)|.$$

We remark that

$$|\langle f, x \rangle| \le \|f\| \cdot \|x\| \quad \text{for all } x \in X.$$

Theorem 2.25 asserts that the dual space X' separates points of X, that is, for arbitrary two distinct points x_1, x_2 of X, there exists a functional $f \in X'$ such that $f(x_1) \neq f(x_2)$.

2.4.6 *Annihilators*

Let A be a subset of a normed linear space X. An element f of the dual space X' is called an annihilator of A if it satisfies the condition

$$f(x) = 0 \quad \text{for all } x \in A.$$

We let

$$A^0 = \left\{ f \in X' : f(x) = 0 \text{ for all } x \in A \right\}$$

be the set of all annihilators of A. This is not a one way proposition. If B is a subset of X', we let

$$^0B = \{x \in X : f(x) = 0 \text{ for all } f \in B\}$$

be the set of all annihilators of B.

Here are some basic properties of annihilators:

(i) The sets A^0 and ^0B are closed linear subspaces of X and X', respectively.
(ii) If M is a closed linear subspace of X, then $^0(M^0) = M$.
(iii) If A is a subset of X and M is the closure of the subspace spanned by A, then $M^0 = A^0$ and $M = {}^0(A^0)$.

2.4.7 Dual Spaces of Normed Factor Spaces

Let M be a closed linear subspace of a normed linear space X. Then each element f of M^0 defines a bounded linear functional $\widetilde{f}$ on the normed factor space X/M by the formula

$$\widetilde{f}(\widetilde{x}) = f(x) \quad \text{for all } \widetilde{x} \in X/M.$$

Indeed, the value $f(x)$ on the right-hand side does not depend on the choice of a representative x of the equivalence class $\widetilde{x}$, and we have the formula

$$\|\widetilde{f}\| = \|f\|.$$

Furthermore, it is easy to see that the mapping

$$\pi : f \longmapsto \widetilde{f}$$

of M^0 into $(X/M)'$ is linear and surjective; hence we have the following:

Theorem 2.26 *The strong dual space $(X/M)'$ of the factor space X/M can be identified with the space M^0 of all annihilators of M by the linear isometry π.*

2.4.8 Bidual Spaces

Each element x of a normed linear space X defines a bounded linear functional Jx on the strong dual space X' by the formula

$$Jx(f) = f(x) \quad \text{for all } f \in X'. \tag{2.13}$$

Then Theorem 2.25 asserts that

$$\|Jx\| = \sup_{\substack{f \in X' \\ \|f\| \leq 1}} |Jx(f)| = \|x\|,$$

so that the mapping J is a linear isometry of X into the strong dual space $(X')'$ of X'. The space $(X')'$ is called the *strong bidual* (or *second dual*) *space* of X.

Summing up, we have the following:

Theorem 2.27 *A normed linear space X can be embedded into its strong bidual space $(X')'$ by the linear isometry J defined by formula* (2.13).

If the mapping J is surjective, that is, if $X = (X')'$, then we say that X is *reflexive*. For example, we have the following:

Theorem 2.28 *The space $L^p(\Omega)$ is reflexive if and only if $1 < p < \infty$ (see [2, Theorem 2.46]).*

2.4.9 Weak Convergence

A sequence $\{x_n\}$ in a normed linear space X is said to be *weakly convergent* if a finite $\lim_{n\to\infty} f(x_n)$ exists for each f in the dual space X' of X. A sequence $\{x_n\}$ in X is said to *converge weakly* to an element x of X if $\lim_{n\to\infty} f(x_n) = f(x)$ for every $f \in X'$; we then write $w - \lim_{n\to\infty} x_n = x$ or simply $x_n \to x$ weakly. Since the space X' separates points of X, the limit x is uniquely determined. Theorem 2.27 asserts that X may be considered as a linear subspace of its bidual space $(X')'$; hence the *weak topology* on X is just the simple convergence topology on the bidual space $(X')' = L(X', \mathbf{K})$.

For weakly convergent sequences, we have the following:

Theorem 2.29 *(i) $s - \lim_{n\to\infty} x_n = x$ implies $w - \lim_{n\to\infty} x_n = x$.*

(ii) A weakly convergent sequence $\{x_n\}$ is bounded:

$$\sup_n \|x_n\| < +\infty.$$

Furthermore, if $w - \lim_{n\to\infty} x_n = x$, then the sequence $\{x_n\}$ is bounded and we have the inequality

$$\|x\| \leq \liminf_{n\to\infty} \|x_n\|.$$

Part (ii) of Theorem 2.29 has a converse:

Theorem 2.30 *A sequence $\{x_n\}$ in X converges weakly to an element x of X if the following two conditions are satisfied:*

(a) The sequence $\{x_n\}$ is bounded.
(b) $\lim_{n\to\infty} f(x_n) = f(x)$ for every f in some strongly dense subset of X'.

For the strong and weak closures of a linear subspace, we have the following (see [182, Chap. V, Sect. 1, Theorem 11]):

Theorem 2.31 (Mazur) *Let X be a normed linear space. If M is a closed linear subspace of X in the strong topology of X, then it is closed in the weak topology of X.*

Proof Our proof is based on a reduction to absurdity. Assume, to the contrary, that M is not weakly closed. Then there exists a point $x_0 \in X \setminus M$ such that x_0 is an accumulation point of the set M in the weak topology of X. Namely, there exists a sequence $\{x_n\}$ of M such that x_n converges weakly to x_0. However, by applying Mazur's theorem (Theorem 2.23) we can find a continuous linear functional f_0 on X such that

$$f_0(x_0) > 1,$$
$$|f_0(x)| \leq 1 \quad \text{on } M.$$

Hence we have the assertion

$$1 < |f_0(x_0)| = \lim_{n \to \infty} |f_0(x_n)| \leq 1.$$

This is a contradiction.

The proof of Theorem 2.31 is complete. □

Finally, the next Eberlein–Shmulyan theorem gives a necessary and sufficient condition for *reflexivity* of a Banach space in terms of *sequential weak compactness* (see [182, Appendix to Chap. V, Sect. 4, Theorem]):

Theorem 2.32 (Eberlein–Shmulyan) *A Banach space X is* reflexive *if and only if it is locally sequentially* weakly compact, *that is, X is* reflexive *if and only if every strongly bounded sequence of X contains a subsequence which converges* weakly *to an element of X.*

2.4.10 Weak* Convergence

A sequence $\{f_n\}$ in the dual space X' is said to be *weakly* convergent* if a finite $\lim_{n\to\infty} f_n(x)$ exists for every $x \in X$. A sequence $\{f_n\}$ in X' is said to *converge weakly** to an element f of X' if $\lim_{n\to\infty} f_n(x) = f(x)$ for every $x \in X$; we then write $w * -\lim_{n\to\infty} f_n = f$ or simply $f_n \to f$ weakly*. The weak* topology on X' is just the simple topology on the space $X' = L(X, \mathbf{K})$.

We have the following analogue of Theorem 2.29:

Theorem 2.33 *(i) $s - \lim_{n\to\infty} f_n = f$ implies $w * -\lim_{n\to\infty} f_n = f$.*

(ii) If X is a Banach space, then a weakly convergent sequence $\{f_n\}$ in X' converges weakly* to an element f of X' and we have the inequality*

$$\|f\| \leq \liminf_{n\to\infty} \|f_n\|.$$

One of the important consequences of Theorem 2.33 is the *sequential weak* compactness* of bounded sets:

Theorem 2.34 *Let X be a separable Banach space. Then every bounded sequence in the strong dual space X' has a subsequence which converges weakly* to an element of X'.*

2.4.11 Dual Operators

The notion of the *transposed matrix* may be extended to the notion of dual operators as follows: Let T be a linear operator from X into Y with domain $D(T)$ dense in X. Such operators are called *densely defined operators*. Each element g of the dual space Y' of Y defines a linear functional G on $D(T)$ by the formula

$$G(x) = g(Tx) \quad \text{for all } x \in D(T).$$

If this functional G is continuous everywhere on $D(T)$ in the strong topology on X', it follows from an application of Theorem 2.10 that G can be extended uniquely to a continuous linear functional g' on the closure

$$\overline{D(T)} = X,$$

that is, there exists a unique element g' of the dual space X' of X which is an extension of G. So we let

$$\begin{aligned} D(T') = {} & \text{the totality of those } g \in Y' \text{ such that the mapping} \\ & x \longmapsto g(Tx) \\ & \text{is continuous everywhere on } D(T) \text{ in the strong topology on } X', \end{aligned}$$

and define

$$T'g = g'.$$

In other words, the mapping T' is a linear operator from Y' into X' with domain $D(T')$ such that

$$g(Tx) = T'g(x) \quad \text{for all } x \in D(T) \text{ and } g \in D(T'). \tag{2.14}$$

The operator T' is called the *dual operator* or *transpose* of T.

Frequently, we write $\langle f, x \rangle$ or $\langle x, f \rangle$ for the value $f(x)$ of a functional f at a point x. For example, we write formula (2.14) as follows:

$$\langle Tx, g \rangle = \langle x, T'g \rangle \quad \text{for all } x \in D(T) \text{ and } g \in D(T'). \tag{2.14'}$$

The operators T and T' can be visualized as follows:

$$\boxed{\begin{array}{ccc} x \in D(T) & \xrightarrow{T} & Y \ni Tx \\ \updownarrow & & \updownarrow \\ T'g \in Y' & \xleftarrow{T'} & D(T') \ni g \end{array}}$$

The next theorem states that the continuity of operators is inherited by the transposes ([182, Chap. VII, Sect. 1, Theorem 2]):

Theorem 2.35 *Let X, Y be normed linear spaces and X', Y' their strong dual spaces, respectively. If T is a bounded linear operator on X into Y, then its transpose T' is a bounded linear operator on Y' into X', and we have the formula*

$$\|T'\| = \|T\|.$$

2.4.12 Adjoint Operators

We assume that a normed linear space X is equipped with a *conjugation*, that is, with a continuous, unitary operation

$$X \ni u \mapsto \overline{u} \in X$$

satisfying the following conditions:

$$\begin{aligned} \overline{u + v} &= \overline{u} + \overline{v}, \\ \overline{\alpha\, u} &= \overline{\alpha}\, \overline{u}, \\ \overline{(\overline{u})} &= u \end{aligned}$$

for all u, $v \in X$ and $\alpha \in \mathbf{K}$. For example, if X is a function space, then $\overline{u}$ is just the usual pointwise complex conjugate:

$$\overline{u}(x) = \overline{u(x)} \quad \text{for all } u \in X.$$

A conjugation on X induces a conjugation on the dual space X' by the formula

$$\langle \overline{f}, u \rangle = \overline{\langle f, \overline{u} \rangle} \quad \text{for all } f \in X' \text{ and } u \in X.$$

Hence we can define a bounded *sesquilinear* form $(\cdot, \cdot)$ on $X' \times X$ by the formula

$$(f, u) = \langle \overline{f}, u \rangle \quad \text{for all } f \in X' \text{ and } u \in X.$$

Now, following Kato [76, Chap. III, Sect. 5.5] and Yosida [182, Chap. VII] we consider the case where X and Y are each equipped with a conjugation. The notion of the *adjoint matrix* may be extended to the notion of adjoints as follows: Let $A \colon X \to Y$ be a linear operator with domain $D(A)$ dense in X. Each element v of Y' defines a linear functional V on $D(A)$ by the formula

$$V(u) = (Au, v) \quad \text{for every } u \in D(A).$$

If this functional V is continuous everywhere on $D(A)$ in the strong topology on X', by applying Theorem 2.10 we obtain that V can be extended uniquely to a continuous linear functional v^* on $\overline{D(A)} = X$. So we let

$$\begin{aligned} D(A^*) = &\text{the totality of those } v \in Y' \text{ such that the mapping} \\ &\quad u \longmapsto (Au, v) \\ &\text{is continuous everywhere on } D(A) \text{ in the strong topology on } X', \end{aligned}$$

and define

$$A^* v = v^*.$$

In other words, the mapping A^* is a linear operator from Y' into X' with domain $D(A^*)$ such that

$$\big(u, A^* v\big) = (Au, v) \quad \text{for all } u \in D(A) \text{ and } v \in D(A^*).$$

The operator A' is called the *adjoint operator* or *adjoint* of A. The operators A and A^* can be visualized as follows:

$$\boxed{\begin{array}{ccc} u \in D(A) & \xrightarrow{\ A\ } & Y \ni Au \\ \updownarrow & & \updownarrow \\ A^* v \in X' & \xleftarrow{\ A^*\ } & D(A^*) \ni v \end{array}}$$

2.5 Closed Operators

This section is adapted from Kato [76, Chap. III, Sects. 5.2 and 5.3] and Yosida [182, Chap. II, Sect. 6]. Let X and Y be normed linear spaces over the same scalar field $\mathbf{K}$. Let T be a linear operator from X into Y with domain $D(T)$. The *graph* $G(T)$ of T is the set

$$G(T) = \{\{x, Tx\} : x \in D(T)\}$$

in the product space $X \times Y$. Note that $G(T)$ is a linear subspace of $X \times Y$. We say that T is closed if its graph $G(T)$ is *closed* in $X \times Y$. This is equivalent to saying that

$$\begin{aligned} &\{x_n\} \subset D(T), \ x_n \longrightarrow x \text{ in } X, \ Tx_n \longrightarrow y \text{ in } Y \\ &\Longrightarrow x \in D(T), \ Tx = y. \end{aligned} \tag{2.15}$$

In particular, if T is continuous and its domain $D(T)$ is closed in X, then T is a closed linear operator.

We remark that if T is a closed linear operator which is also injective, then its inverse T^{-1} is a closed linear operator. Indeed, this follows from the fact that the mapping $\{x, y\} \longmapsto \{y, x\}$ is a homeomorphism of $X \times Y$ onto $Y \times X$.

A linear operator T is said to be *closable* if the closure $\overline{G(T)}$ in $X \times Y$ of $G(T)$ is the graph of a linear operator, say, $\overline{T}$, that is, $\overline{G(T)} = G(\overline{T})$.

A linear operator is called a *closed extension* of T if it is a closed linear operator which is also an extension of T. It is easy to see that if T is closable, then every closed extension of T is an extension of $\overline{T}$. Thus the operator $\overline{T}$ is called the *minimal closed extension* of T.

The next theorem gives a necessary and sufficient condition for a linear operator to be closable ([182, Chap. II, Sect. 6, Proposition 2]):

Theorem 2.36 *A linear operator T from X into Y with domain $D(T)$ is closable if and only if the following condition is satisfied:*

$$\{x_n\} \subset D(T), \; x_n \longrightarrow 0 \; in X, \; Tx_n \longrightarrow y \; in \, Y \implies y = 0. \tag{2.16}$$

The notion of adjoints introduced in Subsect. 2.4.12 gives a very simple criterion for closability. In fact, we can prove the following ([76, Chap. 3, Sect. 5, Theorem 5.29]):

Theorem 2.37 *Let X and Y be reflexive Banach spaces. If $T : X \to Y$ is a densely defined, closable linear operator, then the adjoint $T^* : Y' \to X'$ is closed and densely defined. Moreover, $T^{**} = (T^*)^*$ is the minimal closed extension of T. Namely, we have the formula*

$$G\left(T^{**}\right) = \overline{G(T)}.$$

Now we can formulate three pillars of functional analysis for closed operators:

(1) Banach's open mapping theorem ([182, Chap. II, Sect. 5, Theorem], [18, Theorem 2.6]).
(2) Banach's closed graph theorem ([76, Chap. III, Sect. 5.4, Theorem 5.20], [182, Chap. II, Sect. 6, Theorem 1], [18, Theorem 2.9]).
(3) Banach's closed range theorem ([76, Chap. IV, Sect. 5.2, Theorem 5.13], [182, Chap. VII, Sect. 5, Theorem], [18, Theorem 2.19]).

Theorem 2.38 (Banach's open mapping theorem) *Let X and Y be Banach spaces. Then every continuous linear operator on X onto Y is open, that is, it maps every open set in X onto an open set in Y.*

Theorem 2.39 (Banach's closed graph theorem) *Let X and Y be Banach spaces. Then every closed linear operator on X into Y is continuous.*

Corollary 2.40 *Let X and Y be Banach spaces. If T is a continuous, one-to-one linear operator on X onto Y, then its inverse T^{-1} is also continuous; hence T is an isomorphism.*

Indeed, the inverse T^{-1} is a closed linear operator, so that Theorem 2.39 applies.

We give useful characterizations of closed linear operators with closed range ([18, Exercise 2.14]):

Theorem 2.41 *Let X and Y be Banach spaces and T a closed linear operator from X into Y with domain $D(T)$. Then the range $R(T)$ of T is closed in Y if and only if there exists a constant $C > 0$ such that*

$$\operatorname{dist}(x, N(T)) \leq C \, \|Tx\| \quad \text{for all } x \in D(T).$$

Here

$$\operatorname{dist}(x, N(T)) = \inf_{z \in N(T)} \|x - z\|$$

is the distance from x to the null space $N(T)$ of T.

The key point of the next theorem is that if the range $R(T)$ is closed in Y, then a necessary and sufficient condition for the equation $Tx = y$ to be solvable is that the given right-hand side $y \in Y$ is annihilated by every solution $x' \in X'$ of the homogeneous transposed equation $T'x' = 0$ ([182, Chap. VII, Sect. 5, Theorem]):

Theorem 2.42 (Banach's closed range theorem) *Let X and Y be Banach spaces and T a densely defined, closed linear operator from X into Y. Then the following four conditions are equivalent:*

(i) The range $R(T)$ of T is closed in Y.
(ii) The range $R(T')$ of the transpose T' is closed in X'.
(iii) $R(T) = {}^0N(T') = \left\{x \in X : \langle x, x' \rangle = 0 \text{ for all } x' \in N(T')\right\}$.
(iv) $R(T') = {}^0N(T) = \left\{x' \in X' : \langle x', x \rangle = 0 \text{ for all } x \in N(T)\right\}$.

2.6 Complemented Subspaces

Let X be a linear space. Two linear subspaces M and N of X are said to be *algebraic complements* in X if X is the direct sum of M and N, that is, if $X = M \dotplus N$. Algebraic complements M and N in a normed linear space X are said to be *topological complements* in X if the addition mapping

$$\{y, z\} \longmapsto y + z$$

is an isomorphism of $M \times N$ onto X. We then write

$$X = M \oplus N.$$

As an application of Corollary 2.40, we obtain the following:

Theorem 2.43 *Let X be a Banach space. If M and N are closed algebraic complements in X, then they are topological complements.*

A closed linear subspace of a normed linear space X is said to be *complemented* in X if it has a topological complement. By Theorem 2.43, this is equivalent in Banach spaces to the existence of a closed algebraic complement.

The next theorem gives two criteria for a closed subspace to be complemented ([18, Sect. 2.4]):

Theorem 2.44 *Let X be a Banach space and M a closed subspace of X. If M has either finite dimension or finite codimension, then it is complemented in X.*

2.7 Compact Operators

Let X and Y be normed linear spaces over the same scalar field $\mathbf{K}$. A linear operator T on X into Y is said to be *compact* or *completely continuous* if it maps every bounded subset of X onto a relatively compact subset of Y, that is, if the closure of $T(B)$ is compact in Y for every bounded subset B of X. This is equivalent to saying that, for every bounded sequence $\{x_n\}$ in X, the sequence $\{Tx_n\}$ has a subsequence which converges in Y.

We list some facts which follow at once:

(i) Every compact operator is bounded.
Indeed, a compact operator maps the unit sphere onto a bounded set.
(ii) Every bounded linear operator with finite-dimensional range is compact.
This is an immediate consequence of Corollary 2.20.
(iii) No isomorphism between infinite-dimensional spaces is compact.
This follows from an application of Theorem 2.21.
(iv) A linear combination of compact operators is compact.
(v) The product of a compact operator with a bounded operator is compact.

The next theorem states that if Y is a Banach space, then the compact operators on X into Y form a closed subspace of $L(X, Y)$ (see [18, Theorem 6.1], [76, Chap. III, Sect. 4.2, Theorem 4.7], [182, Chap. X, Sect. 2, Theorem]):

Theorem 2.45 *Let X be a normed linear space and Y a Banach space. If $\{T_n\}$ is a sequence of compact linear operators which converges to an operator T in the space $L(X, Y)$ with the uniform topology, then T is compact.*

As for the transposes of compact operators, we have the following (see [18, Theorem 6.4], [76, Chap. III, Sect. 4.2, Theorem 4.10], [182, Chap. X, Sect. 4, Theorem]):

Theorem 2.46 (Schauder) *Let X and Y be normed linear spaces. If T is a compact linear operator on X into Y, then its transpose T' is a compact linear operator on Y' into X'.*

The operators T and T' can be visualized as follows:

$$\boxed{\begin{array}{ccc} X & \xrightarrow{T} & Y \\ \updownarrow & & \updownarrow \\ X' & \xleftarrow{T'} & Y' \end{array}}$$

2.8 The Riesz–Schauder Theory

Now we state the most interesting results on compact linear operators, which are essentially due to F. Riesz in the Hilbert space setting. The results are extended to Banach spaces by Schauder:

Theorem 2.47 *Let X be a Banach space and T a* compact *linear operator on X into itself. Set*

$$\boxed{S = I - T.}$$

Then we have the following three assertions:

(i) *The null space $N(S)$ of S is finite-dimensional and the range $R(S)$ of S is closed in X.*
(ii) *The null space $N(S')$ of the transpose S' is finite-dimensional and the range $R(S')$ of S' is closed in X'.*
(iii) $\dim N(S) = \dim N(S')$.

By combining Theorems 2.42 and 2.47, we can obtain an extension of the theory of linear mappings in finite-dimensional linear spaces ([18, Theorem 6.6]):

Corollary 2.48 *(the Fredholm alternative). Let T be a* compact *linear operator on a Banach space X into itself. If $S = I - T$ is either one-to-one or onto, then it is an isomorphism of X onto itself.*

Let T be a bounded linear operator on X into itself. The *resolvent set* of T, denoted by $\rho(T)$, is defined to be the set of scalars $\lambda \in \mathbf{K}$ such that $\lambda I - T$ is an isomorphism of X onto itself. In this case, the inverse $(\lambda I - T)^{-1}$ is called the *resolvent* of T. The complement of $\rho(T)$, that is, the set of scalars $\lambda \in \mathbf{K}$ such that $\lambda I - T$ is not an isomorphism of X onto itself is called the *spectrum* of T, and is denoted by $\sigma(T)$.

The set $\sigma_p(T)$ of scalars $\lambda \in \mathbf{K}$ such that $\lambda I - T$ is not one-to-one forms a subset of $\sigma(T)$, and is called the *point spectrum* of T. A scalar $\lambda \in \mathbf{K}$ belongs to $\sigma_p(T)$ if and only if there exists a non-zero element $x \in X$ such that $Tx = \lambda x$. In this case, λ is called an *eigenvalue* of T and x an *eigenvector* of T corresponding to λ. Also the null space $N(\lambda I - T)$ of $\lambda I - T$ is called the *eigenspace* of T corresponding to λ, and the dimension of $N(\lambda I - T)$ is called the *multiplicity* of λ.

By using C. Neumann's series (Theorem 2.15), we find that the resolvent set $\rho(T)$ is open in **K** and that

$$\{\lambda \in \mathbf{K} : |\lambda| > \|T\|\} \subset \rho(T).$$

Hence the spectrum $\sigma(T) = \mathbf{K} \setminus \rho(T)$ is closed and bounded in **K**.

If T is a compact operator and λ is a non-zero element of $\sigma(T)$, then, by applying Corollary 2.48 to the operator $\lambda^{-1}T$ we obtain that $\lambda I - T$ is not one-to-one, that is, $\lambda \in \sigma_p(T)$. Also note that if X is infinite-dimensional, then T is not an isomorphism of X onto itself; hence $0 \in \sigma_p(T)$. Therefore the scalar field **K** can be decomposed as follows (see [76, Chap. III, Sect. 4.7, Theorem 6.26], [182, Chap. X, Sect. 5, Theorem 2]):

$$\boxed{\mathbf{K} = \left(\sigma_p(T) \cup \{0\}\right) \cup \rho(T).}$$

We can say rather more about the spectrum $\sigma(T)$ in terms of transpose operators ([182, Chap. X, Sect. 5, Theorems 1, 2 and 3]):

Theorem 2.49 (Riesz–Schauder) *Let T be a compact linear operator on a Banach space X into itself. Then we have the following three assertions:*

(i) The spectrum $\sigma(T)$ of T is either a finite set or a countable set accumulating only at the zero 0; and every non-zero element of $\sigma(T)$ is an eigenvalue of T.
(ii) $\dim N(\lambda I - T) = \dim N\left(\overline{\lambda} - T'\right) < +\infty$ *for all* $\lambda \neq 0$.
(iii) Let $\lambda \neq 0$. The non-homogeneous equation

$$(\lambda I - T)x = y$$

has a solution if and only if y is orthogonal to the space $N(\overline{\lambda} - T')$. Similarly, the non-homogeneous transpose equation

$$\left(\overline{\lambda} I - T'\right)z = w$$

has a solution if and only if w is orthogonal to the space $N(\lambda I - T)$. Moreover, the operator $\lambda I - T$ is onto if and only if it is one-to-one.

2.9 Fredholm Operators

Throughout this section, the letters X, Y, Z denote Banach spaces over the same scalar field **K**.

A linear operator $T : X \to Y$ is called a *Fredholm operator* if the following five conditions are satisfied (see Kato [76, Chap. IV, Sect. 5.1]):

(i) The domain $D(T)$ of T is dense in X.
(ii) T is a closed operator.

(iii) The null space $N(T) = \{x \in D(T) : Tx = 0\}$ of T has finite dimension, i.e., $\dim N(T) < \infty$.
(iv) The range $R(T)$ of T is closed in Y.
(v) The range $R(T)$ has finite codimension, i.e., $\operatorname{codim} R(T) = \dim Y/R(T) < \infty$.

Then the *index* of T is defined by the formula

$$\operatorname{ind} T := \dim N(T) - \operatorname{codim} R(T). \tag{2.17}$$

By combining Theorems 2.42 and 2.47, we find from the definition formula (2.17) that if $X = Y$ and T is compact, then the operator $I - T$ is a Fredholm operator with

$$\operatorname{ind}(I - T) = 0.$$

We give a characterization of Fredholm operators. First, we have the following ([101, Theorem 2.24]):

Theorem 2.50 *If $T : X \to Y$ is a Fredholm operator with domain $D(T)$, then there exist a bounded linear operator $S : Y \to X$ and compact linear operators $P : X \to X$, $Q : Y \to Y$ such that*

(a) $ST = I - P$ on $D(T)$,
(b) $TS = I - Q$ on Y.

Furthermore, we have the formulas

$$R(P) = N(T),$$
$$\dim R(Q) = \operatorname{codim} R(T).$$

Theorem 2.50 has a converse:

Theorem 2.51 *Let T be a closed linear operator from X into Y with domain $D(T)$ dense in X. Assume that there exist bounded linear operators $S_1 : Y \to X$ and $S_2 : Y \to X$ and compact linear operators $K_1 : X \to X$, $K_2 : Y \to Y$ such that*

(a) $S_1 T = I - K_1$ on $D(T)$,
(b) $T S_2 = I - K_2$ on Y.

Then T is a Fredholm operator.

Now we state some important properties of Fredholm operators ([101, Theorem 2.21]):

Theorem 2.52 *If $T : X \to Y$ is a Fredholm operator and if $S : Y \to Z$ is a Fredholm operator, then the product $ST : X \to Z$ is a Fredholm operator, and we have the formula*

$$\operatorname{ind}(ST) = \operatorname{ind} S + \operatorname{ind} T.$$

The next stability theorem states that the index is *stable* under compact perturbations or small perturbations (see [76, Chap. IV, Sect. 5.5, Theorems 5.17 and 5.26], [101, Theorem 2.26]):

Theorem 2.53 (the stability theorem) *(i) If $T: X \to Y$ is a Fredholm operator and if $K: X \to Y$ is a* compact *linear operator, then the sum $T + K: X \to Y$ is a Fredholm operator, and we have the formula*

$$\operatorname{ind}(T + K) = \operatorname{ind} T.$$

(ii) The Fredholm operators form an open subset *of the space $L(X, Y)$ of bounded operators. More precisely, if $E: X \to Y$ is a bounded operator with* $\|E\|$ sufficiently small, *then the sum $T + E: X \to Y$ is a Fredholm operator, and we have the formula*

$$\operatorname{ind}(T + E) = \operatorname{ind} T.$$

As for the transposes of Fredholm operators, we have the following ([76, Chap. IV, Sect. 5.2, Corollary 5.14]):

Theorem 2.54 *If $T: X \to Y$ is a Fredholm operator and if Y is reflexive, then the transpose $T': Y' \to X'$ of T is a Fredholm operator, and we have the formula*

$$\operatorname{ind} T' = -\operatorname{ind} T.$$

Now we can state a generalization of the *Fredholm alternative* (Corollary 2.48) in terms of adjoint operators ([101, Theorem 2.27]):

Theorem 2.55 (the Fredholm alternative) *Let $A: X \to Y$ be a Fredholm operator with* $\operatorname{ind} A = 0$. *Then there are two, mutually exclusive possibilities:*

(i) The homogeneous equation $Au = 0$ has only the trivial solution $u = 0$. In this case, we have the following two assertions:

  (a) For each $f \in Y$, the non-homogeneous equation $Au = f$ has a unique solution $u \in X$.
  (b) For each $g \in X'$, the adjoint equation $A^* v = g$ has a unique solution $v \in Y'$.

(ii) The homogeneous equation $Au = 0$ has exactly p linearly independent solutions $u_1, u_2, \ldots, u_p$ for some $p \geq 1$. In this case, we have the following three assertions:

  (c) The homogeneous adjoint equation $A^* v = 0$ has exactly p linearly independent solutions $v_1, v_2, \ldots, v_p$.
  (d) The non-homogeneous equation $Au = f$ is solvable if and only if the right-hand side f satisfies the orthogonal conditions

$$(v_j, f) = 0 \quad \text{for all } 1 \leq j \leq p.$$

(e) *The non-homogeneous adjoint equation $A^* v = g$ is solvable if and only if the right-hand side g satisfies the orthogonal conditions*

$$\left(g, u_j\right) = 0 \quad \textit{for all } 1 \leq j \leq p.$$

2.10 Hilbert Spaces

A complex (or real) linear space X is called a *pre-Hilbert space* or *inner product space* if, to each ordered pair of elements x and y of X, there is associated a complex (or real) number (x, y) in such a way that:

(I1) $(y, x) = \overline{(x, y)}$.
(I2) $(\alpha x, y) = \alpha(x, y)$ for all $\alpha \in \mathbf{C}$ (or $\alpha \in \mathbf{R}$).
(I3) $(x + y, z) = (x, z) + (y, z)$ for all x, y and $z \in X$.
(I4) $(x, x) \geq 0$; $(x, x) = 0$ if and only if $x = 0$.

Here $\overline{(x, y)}$ denotes the complex conjugate of (x, y). In the real case condition (I1) becomes simply $(y, x) = (x, y)$. The number (x, y) is called the *inner product* or *scalar product* of x and y.

The following are immediate consequences of conditions (I1), (I2) and (I3):

(i) $(\alpha x + \beta y, z) = \alpha\,(x, z) + \beta\,(y, z)$ for all $\alpha, \beta \in \mathbf{C}$.
(ii) $(x, \alpha y + \beta z) = \overline{\alpha}\,(x, y) + \overline{\beta}\,(x, z)$ for all $\alpha, \beta \in \mathbf{C}$.

Properties (i) and (ii) are frequently called *sesquilinearity*. In the real case they are bilinearity.

We list some basic properties of the inner product:

(1) The *Schwarz inequality* holds true for all $x, y \in X$:

$$|\,(x, y)\,|^2 \leq (x, x)\,(y, y)\,.$$

Here the equality holds true if and only if x and y are linearly dependent.

(2) The quantity

$$\|x\| = \sqrt{(x, x)} \quad \text{(the non-negative square root)}$$

satisfies axioms (N1), (N2) and (N3) of a norm; hence a pre-Hilbert space is a normed linear space by the norm

$$\|x\| = \sqrt{(x, x)}.$$

(3) The inner product (x, y) is a continuous function of x and y:

$$\|x_n - x\| \longrightarrow 0, \ \|y_n - y\| \longrightarrow 0 \implies (x_n, y_n) \longrightarrow (x, y)\,.$$

(4) The *parallelogram law* holds true for all $x, y \in X$:

$$\|x + y\|^2 + \|x - y\|^2 = 2\left(\|x\|^2 + \|y\|^2\right). \tag{2.18}$$

Conversely, we assume that X is a normed linear space whose norm satisfies condition (2.18). We let

$$(x, y) = \frac{1}{4}\left(\|x + y\|^2 - \|x - y\|^2\right)$$

if X is a real normed linear space, and let

$$(x, y) = \frac{1}{4}\left(\|x + y\|^2 - \|x - y\|^2 + i\,\|x + iy\|^2 - i\,\|x - iy\|^2\right)$$

if X is a complex normed linear space. Then it is easy to verify that the number (x, y) satisfies axioms (I1) through (I4) of an inner product; hence X is a pre-Hilbert space.

A pre-Hilbert space is called a *Hilbert space* if it is complete with respect to the norm derived from the inner product.

If X and Y are pre-Hilbert spaces over the same scalar field $\mathbf{K}$, then the product space $X \times Y$ is a pre-Hilbert space by the inner product

$$(\{x_1, y_1\}, \{x_2, y_2\}) = (x_1, x_2) + (y_1, y_2) \quad \text{for all } x_1, x_2 \in X \text{ and } y_1, y_2 \in Y.$$

Furthermore, if X and Y are Hilbert spaces, then so is $X \times Y$.

2.10.1 Orthogonality

Let X be a pre-Hilbert space. Two elements x, y of X are said to be *orthogonal* if $(x, y) = 0$; we then write $x \perp y$. We remark that

$$x \perp y \iff y \perp x.$$
$$x \perp x \iff x = 0.$$

If A is a subset of X, we let

$$A^\perp = \{x \in X : (x, y) = 0 \text{ for all } y \in A\}.$$

In other words, $A^\perp$ is the set of all those elements of X which are orthogonal to every element of A.

We list some facts which follow at once:

(i) The set $A^\perp$ is a linear subspace of X.
(ii) $A \subset B \Longrightarrow B^\perp \subset A^\perp$.
(iii) $A \cap A^\perp = \{0\}$.
(iv) The set $A^\perp$ is closed.
(v) $A^\perp = \overline{A}^\perp = \overline{[A]}^\perp$, where $\overline{A}$ is the closure of A and $[A]$ is the space spanned by A, that is, the space of all finite linear combinations of elements of A.

Facts (iv) and (v) follow from the continuity of the inner product.

2.10.2 The Closest-Point Theorem and Applications

Theorem 2.56 (the closest-point theorem) *Let A be a closed convex subset of a Hilbert space X. If x is a point not in A, then there is a unique point a in A such that*

$$\|x - a\| = \mathrm{dist}(x, A) = \inf_{z \in A} \|x - z\|.$$

Theorem 2.56 can be proved by using the parallelogram law.

One of the consequences of Theorem 2.56 is that every closed linear subspace of a Hilbert space is complemented:

Theorem 2.57 *Let M be a closed linear subspace of a Hilbert space X. Then every element x of X can be decomposed uniquely in the form*

$$x = y + z, \quad y \in M, \ z \in M^\perp. \tag{2.19}$$

Moreover, the mapping $x \mapsto \{y, z\}$ is an isomorphism of X onto $M \times M^\perp$.

We shall write the decomposition (2.19) in the form

$$\boxed{X = M \oplus M^\perp,} \tag{2.19'}$$

emphasizing that the mapping $x \mapsto \{y, z\}$ is an isomorphism of X onto $M \times M^\perp$. The space $M^\perp$ is called the *orthogonal complement* of M.

Corollary 2.58 *If M is a closed linear subspace of a Hilbert space X, then it follows that $M^{\perp\perp} = (M^\perp)^\perp = M$. Furthermore, if A is a subset of X, then we have $A^{\perp\perp} = \overline{[A]}$, where $[A]$ is the space spanned by A.*

With the above notation (2.19), we define a mapping P_M of X into M by the formula

$$P_M x = y.$$

Since the decomposition (2.19) is unique, it follows that P_M is linear.

Furthermore, we easily obtain the following:

Theorem 2.59 *The operator P_M enjoys the following three properties:*

(i) $P_M^2 = P_M$ *(idempotent property).*
(ii) $\left(P_M x, x'\right) = \left(x, P_M x'\right)$ *(symmetric property).*
(iii) $\|P_M\| \leq 1$.

The operator P_M is called the *orthogonal projection* onto M.

Similarly, we define a mapping $P_{M^\perp}$ of X into $M^\perp$ by the formula

$$P_{M^\perp} x = z.$$

Then Corollary 2.58 asserts that $P_{M^\perp}$ is the orthogonal projection onto $M^\perp$. It is clear that

$$P_M + P_{M^\perp} = I.$$

The orthogonal projections P_M and $P_{M^\perp}$ can be visualized as in Fig. 2.1.

Now we give an important characterization of bounded linear functionals on a Hilbert space:

Theorem 2.60 (the Riesz representation theorem) *Every element y of a Hilbert space X defines a bounded linear functional $J_X y$ on X by the formula*

$$J_X y(x) = (x, y) \quad \textit{for all } x \in X, \tag{2.20}$$

and we have the formula

$$\|J_X y\| = \sup_{\substack{x \in X \\ \|x\| \leq 1}} |J_X y(x)| = \|y\|.$$

Conversely, for every bounded linear functional f on X, there exists a unique element y of X such that $f = J_X y$, that is,

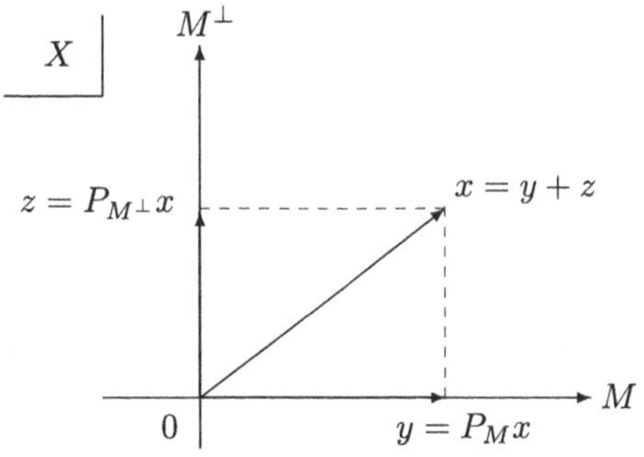

Fig. 2.1 The orthogonal projections P_M and $P_{M^\perp}$

$$f(x) = (x, y) \quad \textit{for all } x \in X,$$

and so

$$\|f\| = \|y\|.$$

In view of formula (2.20), it follows that the mapping J_X enjoys the following property:

$$J_X(\alpha y + \beta z) = \overline{\alpha} J_X y + \overline{\beta} J_X z \quad \text{for all } y, z \in X \text{ and } \alpha, \beta \in \mathbf{C}.$$

We express this by saying that J_X is *conjugate linear* or *antilinear*. In the real case, J_X is linear.

Let X' be the strong dual space of a Hilbert space X, that is, the space of bounded linear functionals on X with the norm

$$\|f\| = \sup_{\substack{x \in X \\ \|x\| \leq 1}} |f(x)|.$$

Then Theorem 2.60 may be restated as follows:

There is a conjugate linear, norm-preserving isomorphism J_X *of* X *onto* X'. (2.21)

In this case, we say that X' is *antidual* to X.

Recall that a sequence $\{x_n\}$ in a normed linear space X is said to converge weakly to an element x of X as $n \to \infty$ if $f(x_n) \to f(x)$ for every $f \in X'$ as $n \to \infty$. Assertion (2.21) tells us that a sequence $\{x_n\}$ in a Hilbert space X converges weakly to an element x of X as $n \to \infty$ if and only if $(x_n, y) \to (x, y)$ for every $y \in X$ as $n \to \infty$.

Another important consequence of Theorem 2.60 is the *reflexivity* of Hilbert spaces:

Corollary 2.61 *Every Hilbert space can be identified with its strong bidual space.*

2.10.3 Orthonormal Sets

Let X be a pre-Hilbert space. A subset S of X is said to be *orthogonal* if every pair of distinct elements of S is orthogonal. Furthermore, if each element of S has norm one, then S is said to be *orthonormal*. We remark that if S is an orthogonal set of non-zero elements, we can construct an orthonormal set from S by normalizing each element of S. If $\{x_1, x_2, \ldots, x_n\}$ is an orthonormal set and if $x = \sum_{i=1}^{n} \alpha_i x_i$, then we have the formula

$$\|x\|^2 = \sum_{i=1}^{n} |\alpha_i|^2, \quad \alpha_i = (x, x_i).$$

Therefore, every orthonormal set is linearly independent.

First, we state the Gram–Schmidt orthogonalization:

Theorem 2.62 (Gram–Schmidt) *Let $\{x_i\}_{i \in I}$ be a finite or countable infinite set of linearly independent vectors of X. Then we can construct an orthonormal set $\{u_i\}_{i \in I}$ such that, for each $i \in I$,*

(a) u_i is a linear combination of $\{x_1, x_2, \ldots, x_i\}$.
(b) x_i is a linear combination of $\{u_1, u_2, \ldots, u_i\}$.

Corollary 2.63 *Every n-dimensional pre-Hilbert space over the scalar field $\mathbf{K}$ is isomorphic to the space $\mathbf{K}^n$ with the usual inner product.*

Let $\{u_\lambda\}_{\lambda \in \Lambda}$ be an orthonormal set of a pre-Hilbert space X. For each $x \in X$, we let

$$\widehat{x}_\lambda := (x, u_\lambda) \quad \text{for } \lambda \in \Lambda.$$

The scalars $\widehat{x}_\lambda$ are called the *Fourier coefficients* of x with respect to $\{u_\lambda\}$.

Then we have the following:

Theorem 2.64 *For each $x \in X$, the set of those $\lambda \in \Lambda$ such that $\widehat{x}_\lambda \neq 0$ is at most countable. Furthermore, we have the*

Bessel inequality

$$\sum_{\lambda \in \Lambda} |\widehat{x}_\lambda|^2 \leq \|x\|^2.$$

An orthonormal set S of X is called a *complete orthonormal system* if it is not contained in a larger orthonormal set of X.

As for the existence of such systems, we have the following:

Theorem 2.65 *Let X be a Hilbert space having a non-zero element. Then, for every orthonormal set S in X, there exists a complete orthonormal system that contains S.*

The next theorem gives useful criteria for the completeness of orthonormal sets:

Theorem 2.66 *Let $S = \{u_\lambda\}_{\lambda \in \Lambda}$ be an orthonormal set in a Hilbert space X. Then the following five conditions are equivalent:*

(i) The set S is complete.
(ii) $S^\perp = \{0\}$.
(iii) The space $[S]$ spanned by S is dense in X: $\overline{[S]} = X$.
(iv) For every $x \in X$, we have the formula

$$\boxed{\|x\|^2 = \sum_{\lambda \in \Lambda} |\widehat{x}_\lambda|^2.} \tag{2.22}$$

(v) For every $x \in X$, we have the formula

$$\boxed{x = \sum_{\lambda \in \Lambda} \widehat{x_\lambda} u_\lambda \quad \text{in } X.} \tag{2.23}$$

Formula (2.22) is called the *Parseval identity* and formula (2.23) is called the *Fourier series expansion* of x with respect to $\{u_\lambda\}$.

2.10.4 Adjoint Operators

In this subsection, following Yosida [182, Chap. VII, Sect. 2] we consider adjoint operators in the framework of Hilbert spaces. Throughout this subsection, the letters X, Y, Z denote Hilbert spaces over the same scalar field $\mathbf{K}$.

Let T be a linear operator from X into Y with domain $D(T)$ dense in X. Each element y of Y defines a linear functional f on $D(T)$ by the formula

$$f(x) = (Tx, y) \quad \text{for every } x \in D(T).$$

If this functional f is continuous everywhere on $D(T)$, by applying Theorem 2.10 we obtain that f can be extended uniquely to a continuous linear functional $\widetilde{f}$ on $\overline{D(T)} = X$. Therefore, Riesz's theorem (Theorem 2.65) asserts that there exists a unique element y^* of X such that

$$\widetilde{f}(x) = \left(x, y^*\right) \quad \text{for all } x \in X.$$

In particular, we have the formula

$$(Tx, y) = f(x) = \left(x, y^*\right) \quad \text{for all } x \in D(T).$$

Hence we let

$$\begin{aligned} D\left(T^*\right) = &\text{ the totality of those } y \in Y \text{ such that the mapping} \\ &x \longmapsto (Tx, y) \\ &\text{is continuous everywhere on } D(T), \end{aligned}$$

and define

$$T^* y = y^*.$$

In other words, the mapping T^* is a linear operator from Y into X with domain $D\left(T^*\right)$ such that

$$(Tx, y) = \left(x, T^*y\right) \quad \text{for all } x \in D(T) \text{ and } y \in D\left(T^*\right).$$

The operator T^* is called the *adjoint operator* or simply the *adjoint* of T.

The operators T and T^* can be visualized as follows:

$$\boxed{\begin{array}{ccc} x \in D(T) & \xrightarrow{\;T\;} & Y \ni Tx \\ \updownarrow & & \updownarrow \\ T^*y \in Y & \xleftarrow{\;T^*\;} & D\left(T^*\right) \ni y \end{array}}$$

We list some basic properties of adjoints:

(i) The operator T^* is closed.
(ii) If $T \in L(X, Y)$, then $T^* \in L(Y, X)$ and $\|T^*\| = \|T\|$.
(iii) If $T, S \in L(X, Y)$, then $(\alpha T + \beta S)^* = \overline{\alpha}\, T^* + \overline{\beta}\, S^*$ for all $\alpha, \beta \in \mathbf{C}$.
(iv) If $T \in L(X, Y)$ and $S \in L(Y, Z)$, then $(ST)^* = T^* S^*$.

A densely defined linear operator T from X into itself is said to be *self-adjoint* if $T = T^*$. Note that every self-adjoint operator is closed.

As for the adjoints of closed operators, we have the following theorem (see Theorem 2.37):

Theorem 2.67 *If T is a densely defined, closed linear operator from X into Y, then the adjoint T^* is a densely defined, closed linear operator from Y into X and we have the formula*

$$T^{**} = \left(T^*\right)^* = T.$$

Corollary 2.68 *If T is a densely defined, closable linear operator, then the adjoint T^* is densely defined and the operator T^{**} coincides with the minimal closed extension $\overline{T}$ of T. Namely, we have the formula*

$$G\left(T^{**}\right) = \overline{G(T)}.$$

2.11 Continuous Functions on Metric Spaces

In this last section we formulate two fundamental theorems concerning spaces of continuous functions defined on a metric space.

2.11.1 The Ascoli–Arzelà Theorem

Let X be a subset of a metric space (S, ρ) and F a Banach space. We let

$$C(X, F) = \text{the space of continuous maps of } X \text{ into } F.$$

We say that a subset Φ of $C(X, F)$ is *equicontinuous* at a point x_0 of X if, for any given ε there exists a constant $\delta = \delta(x_0, \varepsilon) > 0$ such that we have, for all $f \in \Phi$,

$$x \in X, \ \rho(x, x_0) < \delta \implies |f(x) - f(x_0)| < \varepsilon.$$

We say that Φ is equicontinuous on X if it is equicontinuous at every point of X.

The next theorem provides a criterion for compactness of compact subsets of functions spaces:

Theorem 2.69 (Ascoli–Arzelà) *Let X be a compact subset of a metric space (S, ρ) and F a Banach space. Let Φ be a subset of the space $C(X, F)$ of continuous maps with supremum norm. Then Φ is* relatively compact *in $C(X, F)$ if and only if the following two conditions are satisfied:*

(I) Φ is equicontinuous.
(II) For each $x \in X$, the set $\Phi(x) = \{f(x) : f \in \Phi\}$ is relatively compact in F.

The next corollary is a version of the Ascoli–Arzelà theorem in the case where F is the real numbers $\mathbf{R}$ or complex numbers $\mathbf{C}$.

Corollary 2.70 *Let X be a compact subset of a metric space, and let Φ be a subset of the space $C(X)$ of continuous functions on X with supremum norm. Then Φ is relatively compact in $C(X)$ if and only if it is equicontinuous and bounded for the supremum norm.*

We remark that a subset Φ is *relatively compact* if and only if it has the property that any sequence in Φ has a convergent subsequence, converging in its closure. The Ascoli–Arzelà theorem is used mostly when X is a σ-compact space. In that case, we have the following version of the Ascoli–Arzelà theorem:

Corollary 2.71 *(Ascoli–Arzelà). Let X be a metric space that is a denumerable union of compact sets. Let $\{f_n\}$ be a sequence of continuous maps of X into a Banach space F. Assume that:*

(1) The sequence $\{f_n\}$ is equicontinuous as a family of maps.
(2) For each $x \in X$, the closure of the set $\{f_n(x) : n = 1, 2, \ldots\}$ is compact in F.

Then there exists a subsequence of $\{f_n\}$ that converges pointwise to a continuous map $f \in C(X, F)$, and this convergence is uniform on every compact subset of X.

2.11.2 *The Stone–Weierstrass Theorem*

Let (X, d) be a compact metric space. In the space $C(X)$ of continuous, real-valued functions on X, we define the *uniform metric*

$$\rho(f, g) = \max_{x \in X} |f(x) - g(x)| \quad \text{for } f, g \in C(X).$$

Moreover, we introduce the following operations on $C(X)$ for any $f, g \in C(X)$ and $\alpha \in \mathbf{R}$:

(S1) $(f + g)(x) = f(x) + g(x)$ for all $x \in X$.
(S2) $(fg)(x) = f(x) \cdot g(x)$ for all $x \in X$.
(S3) $(\alpha f)(x) = \alpha f(x)$ for all $x \in X$.

A subset $\mathcal{A}$ of $C(X)$ is called an *algebra* if it is a real linear subspace such that

$$f, g \in \mathcal{A} \implies fg \in \mathcal{A}.$$

Let $\mathcal{B}$ be a subset of $C(X)$. The intersection of all algebras containing $\mathcal{B}$ is an algbra. This algebra is called the *algebra generated by* $\mathcal{B}$.

Example 2.72 Let $X = [a, b]$ be the finite closed interval in $\mathbf{R}$. Let $C[a, b]$ be the space of continuous, real-valued functions on $[a, b]$. The algebra generated by functions 1 and x is the set P of all polynomials.

A subset $\mathcal{B}$ of $C(X)$ is said to *separate points* of X if, for any two distinct points $x, y \in X$ there exists a function $f \in \mathcal{B}$ such that

$$f(x) \neq f(y).$$

For example, the set P of all polynomials separates points of $X = [a, b]$.

The next theorem is a generalization of the classical *Weierstrass approximation theorem* (see [55, Chap. 3, Theorem 3.7.1], [51, Sect. 4.7, Theorem 4.45]):

Theorem 2.73 (Stone–Weierstrass) *Let X be a compact metric space and let $\mathcal{A}$ be an algebra in the space $C(X)$ of continuous, real-valued functions on X. If $\mathcal{A}$ contains the constant function* 1 *and separates points of X, then $\mathcal{A}$ is dense in $C(X)$, that is, $\overline{\mathcal{A}} = C(X)$.*

The classical Weierstrass theorem is a special case of the Stone–Weierstrass theorem (Theorem 2.73) if we take

$$X := C[a, b],$$
$$\mathcal{A} := P.$$

Chapter 3
Elements of Measure Theory and L^p Spaces

In this chapter we set forth the basic concepts of measure theory and develop the theory of integration on abstract measure spaces, paying particular attention to the Lebesgue integral on the Euclidean space $\mathbf{R}^n$. In particular, we prove Minkowski's inequality for integrals (Theorem 3.29) and Hardy's inequality (Theorem 3.31) in L^p spaces. In Sect. 3.9, we prove the Marcinkiewicz interpolation theorem (Theorem 3.47) which plays an important role in the proof of Theorem 9.3 in Sect. 9.2. In Sect. 3.10, as an application of Marcinkiewicz's interpolation theorem we study Riesz potentials in the classical potential theory (Theorem 3.48). This chapter is included for the sake of completeness, to minimize the necessity of consulting too many outside references. This makes the monograph fairly self-contained.

For more thorough treatments of this subject, the reader might refer to Duoandikoetxea [41], Federer [46], Folland [51], Friedman [55], Malý–Ziemer [97], Rudin [120], Stein–Shakarchi [141], Torchinsky [168] and Ziemer [183].

3.1 Measure Theory

This section is a summary of the basic definitions and results about measure theory which will be used throughout the book. Most of the material should be quite familiar to the reader and may be omitted. This section, included for the sake of completeness, should serve to settle questions of notation and such.

3.1.1 Measurable Spaces

Let X be a non-empty set. A non-empty collection $\mathcal{A}$ of subsets of X is said to be an *algebra* or a *finitely additive class* in X if it has the following three properties:

K. Taira, *Real Analysis Methods for Markov Processes*,
https://doi.org/10.1007/978-981-97-3659-1_3

(S1) The empty set $\emptyset$ belongs to $\mathcal{A}$.
(S2) If $A \in \mathcal{A}$, then its complement $A^c = X \setminus A$ belongs to $\mathcal{A}$.
(A3) If $\{A_j\}_{j=1}^N$ is a finite collection of members of $\mathcal{A}$, then the union $\bigcup_{n=1}^N A_j$ belongs to $\mathcal{A}$.

A collection $\mathcal{M}$ of subsets of X is said to be a *σ-algebra* in X if it has the following three properties:

(S1) The empty set $\emptyset$ belongs to $\mathcal{M}$.
(S2) If $A \in \mathcal{M}$, then its complement $A^c = X \setminus A$ belongs to $\mathcal{M}$.
(S3) If $\{A_n\}$ is an arbitrary countable collection of members of $\mathcal{M}$, then the union $\cup_{n=1}^\infty A_n$ belongs to $\mathcal{M}$.

The pair $(X, \mathcal{M})$ is called a *measurable space* and the members of $\mathcal{M}$ are called *measurable sets* in X.

For any collection $\mathcal{F}$ of subsets of X, there exists a smallest σ-algebra $\sigma(\mathcal{F})$ in X which contains $\mathcal{F}$. Namely, $\sigma(\mathcal{F})$ is the intersection of all σ-algebras containing $\mathcal{F}$. This $\sigma(\mathcal{F})$ is sometimes called the σ-algebra generated by $\mathcal{F}$.

3.1.2 Measurable Functions

We let

$$\overline{\mathbf{R}} = \{-\infty\} \cup \mathbf{R} \cup \{+\infty\}$$

with the obvious ordering. The topology on $\overline{\mathbf{R}}$ is defined by declaring that the open sets in $\overline{\mathbf{R}}$ are those which are unions of segments of the types

$$(a, b), \ [-\infty, a), \ (a, +\infty].$$

The elements of $\overline{\mathbf{R}}$ are called *extended real numbers*.

Let $(X, \mathcal{M})$ be a measurable space. An extended real-valued function f defined on a set $A \in \mathcal{M}$ is said to be *$\mathcal{M}$-measurable* or simply *measurable* if, for every $a \in \mathbf{R}$, the set

$$\{x \in A : f(x) > a\}$$

is in $\mathcal{M}$.

If A is a subset of X, we let

$$\chi_A(x) = \begin{cases} 1 & \text{if } x \in A, \\ 0 & \text{if } x \notin A. \end{cases}$$

The function χ_A is called the *characteristic function* of A.

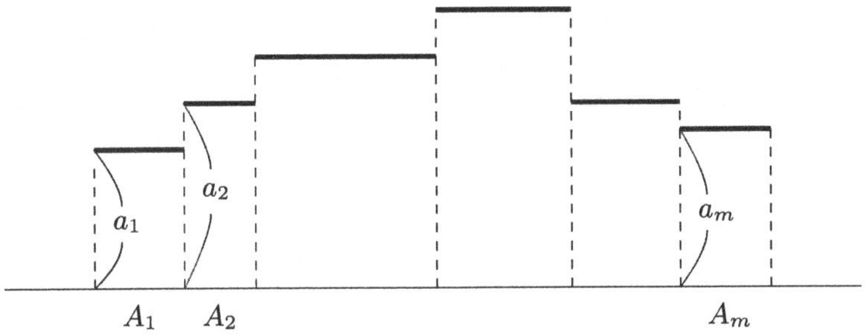

Fig. 3.1 A simple function $f(x) = \sum_{j=1}^{m} a_j \chi_{A_j}(x)$

A real-valued function f on X is called a *simple function* if it takes on only a finite number of values. Thus, if $a_1, a_2, \ldots, a_m$ are the distinct values of f, then $f(x)$ can be written in the form (see Fig. 3.1)

$$f(x) = \sum_{j=1}^{m} a_j \chi_{A_j}(x)$$

where $A_j = \{x \in X : f(x) = a_j\}$. We remark that the simple function f is measurable if and only if each A_j is measurable.

The next theorem characterizes measurable functions in terms of simple functions (see [51, Theorem 2.10]):

Theorem 3.1 *An extended real-valued function defined on a measurable set is measurable if and only if it is a pointwise limit of a sequence of measurable simple functions. Furthermore, every non-negative measurable function is a pointwise limit of an increasing sequence of non-negative measurable simple functions.*

3.1.3 Measures

Let $(X, \mathcal{M})$ be a measurable space. An extended real-valued function μ defined on $\mathcal{M}$ is called a *non-negative measure* or simply a *measure* if it has the following three properties:

(M1) $0 \leq \mu(A) \leq \infty$, $A \in \mathcal{M}$.
(M2) $\mu(\emptyset) = 0$.
(M3) The function μ is countably additive, that is,

$$\mu\left(\sum_{i=1}^{\infty} A_i\right) = \sum_{i=1}^{\infty} \mu(A_i)$$

for any disjoint countable collection $\{A_i\}$ of members of $\mathcal{M}$.

The triple $(X, \mathcal{M}, \mu)$ is called a *measure space*. In other words, a measure space is a measurable space which has a non-negative measure defined on the σ-algebra of its measurable sets. If $\mu(X) < \infty$, then the measure μ is called a *finite measure* and the space $(X, \mathcal{M}, \mu)$ is called a *finite measure space*. If X is a countable union of sets of finite measure, then the measure μ is said to be *σ-finite* on X. We also say that the measure space $(X, \mathcal{M}, \mu)$ is *σ-finite*.

Let $(X, \mathcal{M}, \mu)$ be a measure space. A set $E \in \mathcal{M}$ such that $\mu(E) = 0$ is called a *null set*. We remark that any countable union of null sets is a null set. A measure μ whose domain includes all subsets of null sets is said to be *complete*, that is, if it satisfies the condition

$$E \in \mathcal{M}, \ \mu(E) = 0 \ and \ F \subset E \implies F \in \mathcal{M}.$$

The next theorem asserts that, for a measure space $(X, \mathcal{M}, \mu)$ there exist a σ-algebra $\overline{\mathcal{M}}$ and a unique extension $\overline{\mu}$ of μ to a complete measure on $\overline{\mathcal{M}}$ [51, Theorem 1.9]:

Theorem 3.2 *Let $(X, \mathcal{M}, \mu)$ be a measure space. We let*

$$\begin{aligned} \mathcal{N} &:= \{N \in \mathcal{M} : \mu(N) = 0\}, \\ \overline{\mathcal{M}} &:= \{E \cup F : E \in \mathcal{M} \ and \ F \subset N \ for \ some \ N \in \mathcal{N}\}, \end{aligned}$$

and let

$$\overline{\mu}(E \cup F) = \mu(E) \quad for \ E \cup F \in \overline{\mathcal{M}}.$$

Then the family $\overline{\mathcal{M}}$ is a σ-algebra and the extension $\overline{\mu}$ of μ is a unique complete measure on $\overline{\mathcal{M}}$.

The measure $\overline{\mu}$ is called the *completion* of μ and the σ-algebra $\overline{\mathcal{M}}$ is called the *completion* of $\mathcal{M}$ with respect to μ. The situation of Theorem 3.2 can be visualized as in Table 3.1.

Table 3.1 A completion of the measure space $(X, \mathcal{M}, \mu)$

A non-empty set	Families of subsets	Measures
X	$\overline{\mathcal{M}}$ (complete σ-algebra)	$\overline{\mu}$ (complete measure)
X	$\mathcal{M}$ (σ-algebra)	μ (measure)

3.1.4 Signed Measures

Let $(X, \mathcal{M})$ be a measurable space. A real-valued function μ defined on $\mathcal{M}$ is called a *signed measure* or *real measure* if it is *countably additive*, that is,

$$\mu\left(\sum_{i=1}^{\infty} A_i\right) = \sum_{i=1}^{\infty} \mu(A_i)$$

for any disjoint countable collection $\{A_i\}$ of members of $\mathcal{M}$. It should be noticed that every rearrangement of the series $\sum_i \mu(A_i)$ also converges, since the disjoint union $\sum_i A_i$ is not changed if the subscripts are permuted. A signed measure takes its values in $(-\infty, +\infty)$, but a non-negative measure may take $+\infty$; hence the non-negative measures do not form a subclass of the signed measures.

If μ is a signed measure, we define a function $|\mu|$ on $\mathcal{M}$ by the formula

$$|\mu|(A) = \sup \sum_{i=1}^{n} |\mu(A_i)| \quad \text{for } A \in \mathcal{M}, \tag{3.1}$$

where the supremum is taken over all countable partitions $\{A_i\}$ of A into members of $\mathcal{M}$. Then the function $|\mu|$ is a finite non-negative measure on $\mathcal{M}$. The measure $|\mu|$ is called the *total variation measure* of μ, and the quantity $|\mu|(X)$ defined by formula (3.1) is called the *total variation* of μ. We remark that

$$|\mu(A)| \leq |\mu|(A) \leq |\mu|\,(X) \quad \text{for all } A \in \mathcal{M}.$$

3.1.5 Borel and Radon Measures

Let X be a locally compact Hausdorff space. There exists a smallest σ-algebra $\mathcal{B}$ in X that contains all open sets in X. The members of $\mathcal{B}$ are called *Borel sets* in X. A signed measure defined on $\mathcal{B}$ is called a *real Borel measure* on X. A non-negative Borel measure μ is said to be *regular* if we have, for every $B \in \mathcal{B}$,

$$\begin{aligned} \mu(B) &= \sup\{\mu(F) : F \subset B,\ F \text{ is compact}\} \\ &= \inf\{\mu(G) : B \subset G,\ G \text{ is open}\}. \end{aligned}$$

We give a useful criterion for the regularity of μ:

Theorem 3.3 *Let X be a locally compact Hausdorff space in which every open set is σ-compact. If μ is a non-negative Borel measure on X such that $\mu(K) < +\infty$ for every compact set $K \subset X$, then it is regular.*

A *Radon measure* on X is a Borel measure that is finite on all compact sets in X, and is *outer regular* on all Borel sets B in X and *inner regular* on all open sets G in X:

$$\mu(B) = \inf\{\mu(O) : B \subset O,\ O \text{ is open}\},$$
$$\mu(G) = \sup\{\mu(G) : F \subset G,\ F \text{ is compact}\}.$$

3.1.6 Lebesgue–Stieltjes Measures

In this subsection we develop how to construct measures on the real line **R** by using right-continuous functions.

(I) Sets of the form

$$(a, b] \text{ or } (a, \infty) \text{ or } \emptyset \quad \text{for } -\infty \leq a < b < \infty$$

are called *left-open and right-closed intervals* or simply *half-open intervals* in the real line **R**. The collection of such intervals in **R** is denoted by $\mathcal{I}_1$.

A set of finite disjoint unions of such intervals is called a *figure* in **R**. The collection of figures in **R** is denoted by $\mathcal{F}_1$.

It is easy to verify that the collections $\mathcal{I}_1$ and $\mathcal{F}_1$ are algebras. Moreover, we have the following claim [51, Proposition 1.2]:

Claim 3.4 *Both the σ-algebra $\sigma(\mathcal{I}_1)$ generated by $\mathcal{I}_1$ and the σ-algebra $\sigma(\mathcal{F}_1)$ generated by $\mathcal{F}_1$ are equal to the* Borel σ-algebra $\mathcal{B}(\mathbf{R})$ *on* **R**.

(I) Let $f(x)$ be a real-valued, *monotone increasing* function defined on the real line **R**. We assume that $f(x)$ is *right-continuous*:

$$\lim_{x \downarrow y} f(x) = f(y) \quad \text{for each } y \in \mathbf{R}.$$

First, if $I \in \mathcal{I}_1$, we can define a function

$$\mu_0 : \mathcal{I}_1 \longrightarrow [0, \infty]$$

as follows:

$$\mu_0(I) = \begin{cases} f(b) - f(a) & \text{if } I = (a, b], \\ f(\infty) - f(a) & \text{if } I = (a, \infty) \end{cases}$$

and

$$\mu_0(\emptyset) = 0.$$

Moreover, if $A \in \mathcal{F}_1$ is a union of disjoint intervals

$$A = I_1 + \cdots + I_N,$$

we let

$$\mu_0(A) = \mu_0\left(\bigcup_{j=1}^{N} I_j\right) := \sum_{j=1}^{N} \left(f(b_j) - f(a_j)\right),$$

where

$$I_j = (a_j, b_j] \quad \text{for } 1 \leq j \leq N.$$

Then it is easy to verify that μ_0 is *finitely additive* on the algebra $\mathcal{F}_1$. Namely, if A_1, ..., A_N are disjoint sets in $\mathcal{F}_1$, then it follows that

$$\mu_0\left(\bigcup_{j=1}^{N} A_j\right) = \sum_{j=1}^{N} \mu_0\left(A_j\right).$$

More precisely, we have the following proposition (see [51, Proposition 1.15]):

Proposition 3.5 *Assume that $f(x)$ be a real-valued, monotone increasing and* right-continuous *function on the real line* **R**. *Then the finitely additive measure μ_0 is a* premeasure *on the algebra $\mathcal{F}_1$. Namely, μ_0 enjoys the following property:*

If $\{A_j\}_{j=1}^{\infty}$ is a sequence of disjoint sets in $\mathcal{F}_1$ such that

$$\bigcup_{j=1}^{\infty} A_j \in \mathcal{F}_1,$$

then it follows that

$$\mu_0\left(\bigcup_{j=1}^{\infty} A_j\right) = \sum_{j=1}^{\infty} \mu_0\left(A_j\right).$$

This property is called the countable additivity *of μ_0 in the algebra $\mathcal{F}_1$.*

By virtue of Proposition 3.5, we can introduce an outer measure μ_f^* on the ral line **R** (see [51, Proposition 1.10]):

Proposition 3.6 *Assume that $f(x)$ be a real-valued, monotone increasing and* right-continuous *function on the real line* **R**. *For any subset $E \subset$* **R**, *we define*

$$\mu_f^*(E) := \inf\left\{\sum_{j=1}^{\infty} \mu_0(A_j) : A_j \in \mathcal{F}_1 \text{ and } E \subset \bigcup_{j=1}^{\infty} A_j\right\}.$$

Then μ_f^* *is an* outer measure. *Namely,* μ_f^* *enjoys the following three properties:*

- $\mu_f^*(\emptyset) = 0$.
- $\mu_f^*(E) \leq \mu_f^*(F)$ *if* $E \subset F$.
- $\mu_f^*\left(\bigcup_{j=1}^{\infty} A_j\right) \leq \sum_{j=1}^{\infty} \mu_f^*(A_j)$.

This outer measure μ_f^* determined by the pair $(\mu_0, \mathcal{F}_1)$ is called the *Lebesgue–Stieltjes outer measure* induced by f.

(II) A subset $A \subset \mathbf{R}$ is said to be μ_f^**-measurable* if we have the inequality

$$\mu_f^*(E) \geq \mu_f^*(E \cap A) + \mu_f^*\left(E \cap A^c\right)$$

for all $E \subset \mathbf{R}$ such that $\mu_f^*(E) < \infty$.

This notion is justified by the following theorem (see [51, Theorem 1.11]):

Theorem 3.7 (Carathéodory) *The collection* $\mathcal{M}$ *of* μ_f^**-measurable sets in* $\mathbf{R}$ *is a* σ*-algebra. Moreover, the restriction* μ_f *of* μ_f^* *to* $\mathcal{M}$ *is a complete measure. Namely,* μ_f *enjoys the following three properties:*

- $\mu_f(\emptyset) = 0$,
- $E \in \mathcal{M},\ \mu_f(E) = 0$ *and* $F \subset E \implies F \in \mathcal{M}$,
- $\mu_f\left(\bigcup_{j=1}^{\infty} E_j\right) = \sum_{j=1}^{\infty} \mu_f(E_j)$

if $\{E_j\}_{j=1}^{\infty}$ *is a sequence of disjoint sets in* $\mathcal{M}$.

The measure μ_f corresponding to the outer measure μ_f^* is called the *Lebesgue–Stieltjes measure* induced by f.

The situation of Theorem 3.7 can be visualized as in Table 3.2.

3.1.7 Lebesgue Measures

The Lebesgue–Stieltjes measure corresponding to the function $f(x) = x$ is just an ordinary Lebesgue measure on the real line $\mathbf{R}$.

More precisely, the next theorem is one of the fundamental theorems in measure theory (see [51, Theorem 1.21]):

Theorem 3.8 *There exist a* σ*-algebra* $\mathcal{L}$ *in* $\mathbf{R}$ *and a non-negative measure* m *on* $\mathcal{L}$ *having the following four properties:*

(i) Every open set in $\mathbf{R}$ *is in* $\mathcal{L}$.

Table 3.2 A bird's-eye view of Theorem 3.7

The real line	Families of subsets	Lebesgue–Stieltjes measures induced by $f(x)$
$\mathbf{R}$	$\mathcal{P}(\mathbf{R})$ (all subsets)	μ_f^* (outer measure)
$\mathbf{R}$	$\mathcal{M}$ (μ_f^*-measurable sets)	$\overline{\mu_f} = \mu_f^*\|_{\mathcal{M}}$ (complete measure)
$\mathbf{R}$	$\mathcal{B}(\mathbf{R}) = \sigma(\mathcal{F}_1)$ (Borel sets)	$\mu_f = \mu_f^*\|_{\mathcal{B}(\mathbf{R})}$ (Borel measure)
$\mathbf{R}$	$\mathcal{F}_1$ (figures)	μ_0 (premeasure)

Table 3.3 A bird's-eye view of Theorem 3.8

The real line	Families of subsets	Lebesgue–Stieltjes measures induced by $f(x) = x$
$\mathbf{R}$	$\mathcal{P}(\mathbf{R})$ (all subsets)	μ^* (outer measure)
$\mathbf{R}$	$\mathcal{L}$ (Lebesgue measurable sets)	$m = \overline{\mu} = \mu^*\|_{\mathcal{L}}$ (Lebesgue measure)
$\mathbf{R}$	$\mathcal{B}(\mathbf{R}) = \sigma(\mathcal{F}_1)$ (Borel sets)	$\mu = \mu^*\|_{\mathcal{B}(\mathbf{R})}$ (Borel measure)
$\mathbf{R}$	$\mathcal{F}_1$ (figures)	μ_0 (premeasure)

(ii) *If* $A \subset B$, $B \in \mathcal{L}$ *and* $\mu(B) = 0$, *then* $A \in \mathcal{L}$ *and* $m(A) = 0$.
(iii) *If* $A = \{x \in \mathbf{R} : a \leq x \leq b\}$, *then* $A \in \mathcal{L}$ *and* $m(A) = b - a$.
(iv) *The measure* m *is* translation invariant, *that is, if* $x \in \mathbf{R}$ *and* $A \in \mathcal{L}$, *then the set* $x + A = \{x + y : y \in A\}$ *is in* $\mathcal{L}$ *and* $m(x + A) = m(A)$.

The elements of $\mathcal{L}$ are called *Lebesgue measurable sets* in $\mathbf{R}$ and the measure m is called the *Lebesgue measure* on $\mathbf{R}$. The measure space $(\mathbf{R}, \mathcal{L}, m)$ is called the *Lebesgue measure space*.

The situation of Theorem 3.8 can be visualized as in Table 3.3.

3.1.8 Product Measures

Let $(X, \mathcal{M})$ and $(Y, \mathcal{N})$ be measurable spaces.

A set of the form

$$E = A \times B \quad \text{with } A \in \mathcal{M} \text{ and } B \in \mathcal{N}$$

is called a *rectangle*. It is easy to verify that the collection $\mathcal{A}$ of finite disjoint unions of rectangles is an algebra. We let

$$\begin{aligned}\mathcal{M} \otimes \mathcal{N} &:= \text{the smallest } \sigma\text{-algebra in } X \times Y \text{ that contains } \mathcal{A} \\ &= \text{the } \sigma\text{-algebra in } X \times Y \text{ generated by } \mathcal{A}.\end{aligned}$$

Then $(X \times Y, \mathcal{M} \otimes \mathcal{N})$ is a measurable space.

For the product of measure spaces, we have the following existence theorem (see [80, Sect. 35]):

Theorem 3.9 *Let $(X, \mathcal{M}, \mu)$ and $(Y, \mathcal{N}, \nu)$ be σ-finite measure spaces. Then there exists a unique σ-finite, non-negative measure λ on the product σ-algebra $\mathcal{M} \otimes \mathcal{N}$ such that*

$$\lambda(A \times B) = \mu(A)\nu(B) \quad \text{for } A \in \mathcal{M} \text{ and } B \in \mathcal{N}.$$

The measure λ is called the *product measure* of μ and ν, and is denoted by $\mu \times \nu$. The measure space $(X \times Y, \mathcal{M} \otimes \mathcal{N}, \mu \times \nu)$ is called the *product measure space* of $(X, \mathcal{M}, \mu)$ and $(Y, \mathcal{N}, \nu)$.

3.1.9 The n-Dimensional Lebesgue Measure

The Lebesgue measure space $(\mathbf{R}^n, \mathcal{L}^n, m^n)$ on the Euclidean space $\mathbf{R}^n$ is the completion of the n-fold product of the Lebesgue measure space on $\mathbf{R}$ with itself. Namely, the completion m^n of the product measure $m \times \cdots \times m$ of the product σ-algebra $\mathcal{L} \times \cdots \times \mathcal{L}$. The domain $\mathcal{L}^n$ is the class of Lebesgue measurable sets in $\mathbf{R}^n$. From now on, we omit the superscript n and write m for m^n.

More precisely, the next theorem is one of the fundamental theorems in measure theory (see [51, Theorem 2.42]):

Theorem 3.10 *There exist a σ-algebra $\mathcal{L}^n$ in $\mathbf{R}^n$ and a non-negative measure m on $\mathcal{L}^n$ having the following four properties:*

(i) *Every open set in $\mathbf{R}^n$ is in $\mathcal{L}^n$.*
(ii) *If $A \subset B$, $B \in \mathcal{L}^n$ and $m(B) = 0$, then $A \in \mathcal{L}^n$ and $m(A) = 0$.*
(iii) *If $A = \{x \in \mathbf{R}^n : a_j \le x_j \le b_j,\ 1 \le j \le n\}$, then $A \in \mathcal{L}^n$ and $m(A) = \prod_{j=1}^n (b_j - a_j)$.*
(iv) *The measure m is* translation invariant, *that is, if $x \in \mathbf{R}^n$ and $A \in \mathcal{L}^n$, then the set $x + A = \{x + y : y \in A\}$ is in $\mathcal{L}^n$ and $m(x + A) = m(A)$.*

The elements of $\mathcal{L}^n$ are called *Lebesgue measurable* sets in $\mathbf{R}^n$ and the measure m is called the *Lebesgue measure* on $\mathbf{R}^n$.

The situation of Theorem 3.10 can be visualized as in Table 3.4.

Table 3.4 Lebesgue measure m and Borel measure μ in Theorem 3.10

Euclidean space	Families of subsets	Measures induced by $f(x) = x$
$\mathbf{R}^n$	$\mathcal{P}(\mathbf{R}^n)$ (all subsets)	μ^* (outer measure)
$\mathbf{R}^n$	$\mathcal{L}^n$ (Lebesgue measurable sets)	$m = \overline{\mu} = \mu^*\vert_{\mathcal{L}^n}$ (Lebesgue measure)
$\mathbf{R}^n$	$\mathcal{B}(\mathbf{R}^n)$ (Borel sets)	$\mu = \mu^*\vert_{\mathcal{B}(\mathbf{R}^n)}$ (Borel measure)

3.1.10 *Integrals*

Let $(X, \mathcal{M}, \mu)$ be a measure space. If A is a measurable subset of X and if $f(x)$ is a non-negative measurable simple function on A of the form (see Fig. 3.1)

$$f(x) = \sum_{j=1}^{m} a_j \chi_{A_j}(x), \quad a_j \geq 0,$$

then we let

$$\int_A f(x)\, d\mu(x) = \sum_{j=1}^{m} a_j \mu\left(A_j\right). \tag{3.2}$$

The convention: $0 \cdot \infty = 0$ is used here; it may happen that $a_j = 0$ and $\mu\left(A_j\right) = \infty$. We remark that if $f(x)$ is a non-negative simple function, then the integral (3.2) does not depend on the expression of the simple function $f(x)$.

If $f(x)$ is a non-negative measurable function on A, we let

$$\int_A f(x)\, d\mu(x) := \sup \int_A s(x)\, d\mu(x), \tag{3.3}$$

where the supremum is taken over all measurable simple functions s on A such that $0 \leq s(x) \leq f(x), x \in A$. More precisely, by virtue of Theorem 3.1 we can define the integral (3.3) by approximating $f(x)$ from below by non-negative simple functions.

If $f(x)$ is a measurable function on A, we can write it in the form

$$f(x) = f^+(x) - f^-(x),$$

where

$$f^+(x) = \frac{1}{2}\left(|f(x)| + f(x)\right) = \max\{f(x), 0\},$$

$$f^-(x) = \frac{1}{2}(|f(x)| - f(x)) = \max\{-f(x), 0\}.$$

Both $f^+(x)$ and $f^-(x)$ are non-negative measurable functions on A. Then we define the integral of $f(x)$ by the formula

$$\int_A f(x)\,d\mu(x) = \int_A f^+(x)\,d\mu(x) - \int_A f^-(x)\,d\mu(x),$$

provided at least one of the integrals on the right-hand side is finite. If both integrals are finite, we say that $f(x)$ is *μ-integrable* or simply *integrable* on A.

For simplicity, we abbreviate

$$\int_A f\,d\mu = \int_A f(x)\,d\mu(x).$$

If μ is the Lebesgue measure on $\mathbf{R}^n$, we customarily write

$$\int_A f(x)\,dx,$$

instead of $\int_A f(x)\,d\mu(x)$.

A proposition concerning the points of a measurable set A is said to hold *μ-almost everywhere* (μ-a.e.) or simply *almost everywhere* (a.e.) in A if there exists a measurable set N of measure zero such that the proposition holds for all $x \in A \setminus N$. For example, if $f(x)$ and $g(x)$ are measurable functions on A and if $\mu(\{x \in A : f(x) \neq g(x)\}) = 0$, then we say that $f(x) = g(x)$ a.e. in A.

The next three theorems are concerned with the interchange of integration and limit process:

Theorem 3.11 (the monotone convergence theorem) *Let $\{f_n(x)\}$ be an increasing sequence of non-negative measurable functions defined on a measurable set A. Then we have the formula*

$$\lim_{n\to\infty} \int_A f_n(x)\,d\mu(x) = \int_A \left(\lim_{n\to\infty} f_n(x)\right) d\mu(x).$$

Lemma 3.12 (Fatou's lemma) *Let $\{f_n(x)\}$ be a sequence of non-negative measurable functions defined on a measurable set A. Then we have the inequality*

$$\int_A \left(\liminf_{n\to\infty} f_n(x)\right) d\mu(x) \leq \liminf_{n\to\infty} \int_A f_n(x)\,d\mu(x).$$

Theorem 3.13 (the Lebesgue dominated convergence theorem) *Let $\{f_n(x)\}$ be a sequence of measurable functions defined on a measurable set A which converges to some function $f(x)$ almost everywhere in A. Assume that there exists a non-negative integrable function $g(x)$ defined on A such that*

$$|f_n(x)| \leq g(x) \quad a.e. \text{ in } A, \quad n = 1, 2, \ldots.$$

Then the limit function $f(x)$ is integrable on A and we have the formula

$$\int_A f(x)\, d\mu(x) = \lim_{n\to\infty} \int_A f_n(x)\, d\mu(x).$$

Theorem 3.14 (Beppo Levi) *Let $\{f_n\}$ be an increasing sequence of nonnegative, $\mathcal{M}$-measurable functions defined on an $\mathcal{M}$-measurable set E. Assume that*

$$\sup_n \int_E f_n(x)\, d\mu(x) < \infty.$$

Then it follows that the limit function

$$f(x) = \lim_{n\to\infty} f_n(x)$$

is finite for μ-almost every $x \in E$, and is μ-integrable on E.

Moreover, we have the formula

$$\int_E f(x)\, d\mu(x) = \lim_{n\to\infty} \int_E f_n(x)\, d\mu(x) = \sup_{n\geq 1} \int_E f_n(x)\, d\mu(x).$$

Proof Assume that there exists a constant $K > 0$ such that

$$\sup_{n\geq 1} \int_E f_n(x)\, d\mu(x) \leq K.$$

Step (1): The case where $\mu(E) < \infty$: Theorem 3.14 will be proved in a series of claims. We begin by proving the following:

Claim 3.15 *For any $L > 0$, we have the inequality*

$$\mu(\{x \in E : f(x) > L\}) \leq \frac{K}{L}.$$

Proof Since we have, for all $n = 1, 2, \ldots$,

$$\begin{aligned} K \geq \int_E f_n(x)\, d\mu(x) &\geq \int_{\{x\in E: f_n(x)\geq L\}} f_n(x)\, d\mu(x) \\ &\geq L \cdot \mu(\{x \in E : f_n(x) \geq L\}), \end{aligned}$$

it follows that

$$\mu(\{x \in E : f_n(x) \geq L\}) \leq \frac{K}{L} \quad \text{for } n = 1, 2, \ldots.$$

On the other hand, it follows that

$$\{x \in E : f(x) > L\} \subset \bigcup_{n=1}^{\infty} \{x \in E : f_n(x) \geq L\} = \lim_{n \to \infty} \{x \in E : f_n(x) \geq L\}.$$

Therefore, we obtain that

$$\begin{aligned}\mu(\{x \in E : f(x) > L\}) &\leq \mu\left(\bigcup_{n=1}^{\infty} \{x \in E : f_n(x) \geq L\}\right) \\ &= \lim_{n \to \infty} \mu(\{x \in E : f_n(x) \geq L\}) \\ &\leq \frac{K}{L} \quad \text{for any } L > 0.\end{aligned}$$

The proof of Claim 3.15 is complete. □

Next we prove the following:

Claim 3.16 $\mu(\{x \in E : f(x) = +\infty\}) = 0.$

Proof Claim 3.16 is an immediate consequence of Claim 3.15. Indeed, it suffices to note that

$$\{x \in E : f(x) = +\infty\} = \bigcap_{L>0} \{x \in E : f(x) > L\},$$

so that

$$\mu(\{x \in E : f(x) = +\infty\}) = \lim_{L \to \infty} \mu(\{x \in E : f(x) > L\}) = 0.$$

The proof of Claim 3.16 is complete. □

Thirdly, we prove the following:

Claim 3.17 $f \in L^1(E)$. *More precisely, we have the inequality*

$$\int_E f(x)\, d\mu(x) \leq K.$$

Proof For each $L > 0$, the functions $\{f_n\}$ are uniformly bounded on the set

$$\{x \in E : f(x) \leq L\} = E \setminus \{x \in E : f(x) > L\}.$$

Thus it follows from an application of the Lebesgue dominated convergence theorem (Theorem 3.13) that

$$\int_{E\setminus\{x\in E:f(x)>L\}} f(x)\,d\mu(x) = \lim_{n\to\infty} \int_{E\setminus\{x\in E:f(x)>L\}} f_n(x)\,d\mu(x)$$
$$\leq K.$$

Therefore, by letting $L \to \infty$ we obtain that

$$\int_E f(x)\,d\mu(x) = \lim_{L\to\infty} \int_{E\setminus\{x\in E:f(x)>L\}} f(x)\,d\mu(x) \leq K.$$

The proof of Claim 3.17 is complete. □

Finally, we prove the following:

Claim 3.18 $\lim_{n\to\infty} \int_E f_n(x)\,d\mu(x) = \int_E f(x)\,d\mu(x)$.

Indeed, by virtue of Claim 3.17, Claim 3.18 follows by applying the Lebesgue dominated convergence theorem (Theorem 3.13).

Step (2): The case where $\mu(E) = \infty$: We may assume that the set E is a union of an increasing sequence of sets $\{E_j\}$ of finite measure:

$$E = \bigcup_{j=1}^{\infty} E_j, \quad \mu(E_j) < \infty.$$

Then, by Claim 3.17 with $E := E_j$ it follows that

$$\int_{E_j} f(x)\,d\mu(x) \leq K \quad \text{for } j = 1, 2, \ldots.$$

Hence we have the inequality

$$\int_E f(x)\,d\mu(x) = \lim_{j\to\infty} \int_{E_j} f(x)\,d\mu(x) \leq K.$$

This proves that $f \in L^1(E)$. Therefore, by applying Lebesgue's dominated convergence theorem (Theorem 3.13) we obtain that

$$\lim_{n\to\infty} \int_E f_n(x)\,d\mu(x) = \int_E f(x)\,d\mu(x).$$

Furthermore, by Claim 3.16 it follows that

$$\mu\left(\{x \in E : f(x) = +\infty\}\right) = \lim_{j\to\infty} \mu\left(\{x \in E_j : f(x) = +\infty\}\right) = 0.$$

Now the proof of Theorem 3.14 is complete. □

3.1.11 Fubini's Theorem

We consider integration on product spaces. Let $(X, \mathcal{M}, \mu)$ and $(Y, \mathcal{N}, \nu)$ be σ-finite measure spaces, and $\mu \times \nu$ the product measure of μ and ν. If $f(x, y)$ is an $\mathcal{M} \times \mathcal{N}$-measurable function on $X \times Y$ such that its integral exists, then we customarily write

$$\iint_{X \times Y} f(x, y)\, d(\mu \times \nu)(x, y).$$

This integral is called the *double integral* of f. If it happens that the function

$$g(x) = \int_Y f(x, y)\, d\nu(y) \quad \text{for } x \in X$$

is defined and also its integral exists, then we denote the integral

$$\int_X g(x)\, d\mu(x)$$

by any one of the following:

$$\int_X \left(\int_Y f(x, y)\, d\nu(y) \right) d\mu(x), \quad \int_X d\mu(x) \int_Y f(x, y)\, d\nu(y),$$
$$\iint_{X \times Y} f(x, y)\, d\nu(y)\, d\mu(x), \quad \iint_{X \times Y} f\, d\nu\, d\mu.$$

Similarly, we write

$$\int_Y \left(\int_X f(x, y) d\mu(x) \right) d\nu(y), \quad \int_Y d\nu(y) \int_X f(x, y)\, d\mu(x),$$
$$\iint_{X \times Y} f(x, y)\, d\mu(x)\, d\nu(y), \quad \iint_{X \times Y} f\, d\mu\, d\nu.$$

These integrals are called the *iterated integrals* of $f(x, y)$.

The next theorem describes the most important relationship between double integrals and iterated integrals:

Theorem 3.19 (Fubini) *Assume that $(X, \mathcal{M}, \mu)$ and $(Y, \mathcal{N}, \nu)$ are σ-finite measure spaces. If $(X \times Y, \mathcal{M} \otimes \mathcal{N}, \mu \times \nu)$ is the product measure space, then we have the following three assertions:*

(i) If $E \in \mathcal{M} \times \mathcal{N}$, then it follows that $E_x \in \mathcal{N}$ for all $x \in X$ and that $E^y \in \mathcal{M}$ for all $y \in Y$. Moreover, $\nu(E_x)$ is an $\mathcal{M}$-measurable function of x and $\mu(E^y)$ is an $\mathcal{N}$-measurable function of y, and we have the formula

$$\int_X \nu(E_x)\, d\mu(x) = \int_Y \mu(E^y)\, d\nu(y) = (\mu \times \nu)(E).$$

In particular, if $(\mu \times \nu)(E) < \infty$*, then it follows that* $\nu(E_x) < \infty$ *for* μ*-almost all* $x \in X$ *and that* $\mu(E^y) < \infty$ *for* ν*-almost all* $y \in Y$*.*

(ii) *If* $f(x, y)$ *is a non-negative,* $\mathcal{M} \otimes \mathcal{N}$*-measurable function on* $X \times Y$*, then it follows that* $f_x(y)$ *is an* $\mathcal{N}$*-measurable function of* y *for any* $x \in X$ *and that* $f^y(x)$ *is an* $\mathcal{M}$*-measurable function of* x *for all* $y \in Y$*. Moreover, the functions*

$$X \ni x \longmapsto g(x) = \int_Y f_x(y)\, d\nu(y) = \int_Y f(x, y)\, d\nu(y),$$

$$Y \ni y \longmapsto h(y) = \int_X f^y(x)\, d\mu(x) = \int_X f(x, y)\, d\mu(x)$$

are $\mathcal{M}$*-measurable and* $\mathcal{N}$*-measurable, respectively, and we have the formula*

$$\int_X g(x)\, d\mu(x) = \int_Y h(y)\, d\nu(y) = \iint_{X \times Y} f(x, y)\, d(\mu \times \nu).$$

(iii) *If* $f(x, y)$ *is a* $\mu \times \nu$*-integrable function on* $X \times Y$*, then it follows that the function* $f_x(y)$ *is* ν*-integrable for* μ*-almost all* $x \in X$ *and that the function* $f^y(x)$ *is* μ*-integrable for* ν*-almost all* $y \in Y$*. Furthermore, the function* $g(x)$ *is* μ*-integrable on* X *and the function* $h(y)$ *is* ν*-integrable on* Y*, and we have the formula*

$$\int_X g(x)\, d\mu(x) = \int_Y h(y)\, d\nu(y) = \iint_{X \times Y} f(x, y)\, d(\mu \times \nu).$$

The next theorem gives a useful version of Fubini's theorem:

Theorem 3.20 (Fubini) *Assume that* $(X, \mathcal{M}, \mu)$ *and* $(Y, \mathcal{N}, \nu)$ *are* σ*-finite measure spaces. If* $(X \times Y, \mathcal{M} \otimes \mathcal{N}, \mu \times \nu)$ *is the product measure space, then we have the following two assertions:*

(i) *If* $f(x, y)$ *is a* $\mu \times \nu$*-integrable function on* $X \times Y$*, then the function* $f_x(y)$ *on* Y *defined by* $f_x(y) = f(x, y)$ *for* $y \in Y$*, is* ν*-integrable for* μ*-almost all* $x \in X$*, and the function* $f^y(x)$ *on* X *defined by* $f^y(x) = f(x, y)$ *for* $x \in X$*, is* μ*-integrable for* ν*-almost all* $y \in Y$*. Furthermore, the function* $g(x)$*, defined by the formula*

$$g(x) = \int_Y f_x(y)\, d\nu(y) = \int_Y f(x, y)\, d\nu(y) \quad \text{for } \mu\text{-almost all } x \in X,$$

is μ*-integrable, and the function* $h(y)$*, defined by the formula*

$$h(y) = \int_X f^y(x)\,d\mu(x) = \int_X f(x,y)\,d\mu(x) \quad \textit{for } \nu\textit{-almost all } y \in Y,$$

is ν*-integrable; and we have the formula*

$$\iint_{X\times Y} f(x,y)d(\mu\times\nu) = \int_X g(x)\,d\mu = \int_Y h(y)\,d\nu.$$

(ii) *Conversely, if* $f(x,y)$ *is an* $\mathcal{M}\times\mathcal{N}$*-measurable function on* $X\times Y$*, then the functions*

$$\varphi(x) = \int_Y |f(x,y)|\,d\nu(y) \quad \textit{for } x\in X,$$
$$\psi(y) = \int_X |f(x,y)|\,d\mu(x) \quad \textit{for } y\in Y$$

are $\mathcal{M}$*-measurable and* $\mathcal{N}$*-measurable respectively; and we have the formula*

$$\iint_{X\times Y} |f(x,y)|d(\mu\times\nu) = \int_X \varphi(x)\,d\mu = \int_Y \psi(y)\,d\nu.$$

Furthermore, if either $\varphi(x)$ *or* $\psi(y)$ *is integrable, then* $f(x,y)$ *is integrable, and part (i) applies.*

Let $(X,\mathcal{M},\mu)$ and $(Y,\mathcal{N},\nu)$ be complete, σ-finite measure spaces, and let $(X\times Y,\mathcal{L},\lambda)$ be the completion of $(X\times Y,\mathcal{M}\otimes\mathcal{N},\mu\times\nu)$. The next theorem is a version of Fubini's theorem (Theorem 3.19) for *complete* measures:

Theorem 3.21 (Fubini) *Let* $(X\times Y,\mathcal{L},\lambda)$ *be the* completion *of the product measure space* $(X\times Y,\mathcal{M}\otimes\mathcal{N},\mu\times\nu)$*. Then we have the following three assertions:*

(i) *If* $E\in\mathcal{M}\times\mathcal{N}$*, then it follows that* $E_x\in\mathcal{N}$ *for all* μ*-almost* $x\in X$ *and that* $E^y\in\mathcal{M}$ *for all* ν*-almost* $y\in Y$*. Moreover,* $\nu(E_x)$ *is an* $\mathcal{M}$*-measurable function of* x *and* $\mu(E^y)$ *is an* $\mathcal{N}$*-measurable function of* y*, and we have the formula*

$$\int_X \nu(E_x)\,d\mu(x) = \int_Y \mu(E^y)\,d\nu(y) = \lambda(E).$$

In particular, if $\mu(E)<\infty$*, then it follows that* $\nu(E_x)<\infty$ *for* μ*-almost all* $x\in X$ *and that* $\mu(E^y)<\infty$ *for* ν*-almost all* $y\in Y$*.*

(ii) *If* $f(x,y)$ *is an* $\mathcal{L}$*-measurable function on* $X\times Y$ *such that* $f(x,y)\geq 0$ *for* λ*-almost all* $(x,y)\in X\times Y$*, then it follows that* $f_x(y)$ *is an* $\mathcal{N}$*-measurable function of* y *for* μ*-almost all* $x\in X$ *and that* $f^y(x)$ *is an* $\mathcal{M}$*-measurable function of* x *for* ν*-almost all* $y\in Y$*. Moreover, the functions*

$$X\ni x\longmapsto g(x) = \int_Y f_x(y)\,d\nu(y) = \int_Y f(x,y)\,d\nu(y),$$

$$Y \ni y \longmapsto h(y) = \int_X f^y(x)\, d\mu(x) = \int_X f(x, y)\, d\mu(x)$$

are $\mathcal{M}$-measurable and $\mathcal{N}$-measurable, respectively, and we have the formula

$$\int_X g(x)\, d\mu(x) = \int_Y h(y)\, d\nu(y) = \iint_{X\times Y} f(x, y)\, d\lambda.$$

(iii) *If $f(x, y)$ is an $\mathcal{L}$-measurable and λ-integrable function on $X \times Y$, then it follows that the function $f_x(y)$ is ν-integrable for μ-almost all $x \in X$ and that the function $f^y(x)$ is μ-integrable for ν-almost all $y \in Y$. Furthermore, the function $g(x)$ is μ-integrable on X and the function $h(y)$ is ν-integrable on Y, and we have the formula*

$$\int_X g(x)\, d\mu(x) = \int_Y h(y)\, d\nu(y) = \iint_{X\times Y} f(x, y)\, d\lambda.$$

The next corollary gives a useful version of Fubini's theorem for *complete* measures:

Corollary 3.22 *Let $(X \times Y, \mathcal{L}, \lambda)$ be the completion of the product measure space $(X \times Y, \mathcal{M} \otimes \mathcal{N}, \mu \times \nu)$. If $f(x, y)$ is an $\mathcal{L}$-measurable function on $X \times Y$, then the functions*

$$\varphi(x) = \int_Y |f(x, y)|\, d\nu(y) \quad \text{for } x \in X,$$
$$\psi(y) = \int_X |f(x, y)|\, d\mu(x) \quad \text{for } y \in Y$$

are $\mathcal{M}$-measurable and $\mathcal{N}$-measurable, respectively, and we have the formula

$$\iint_{X\times Y} |f(x, y)|\, d\lambda = \int_X \varphi(x)\, d\mu(x) = \int_Y \psi(y)\, d\nu(y).$$

Furthermore, if either $\varphi(x)$ or $\psi(y)$ is integrable, then $f(x, y)$ is integrable, and Theorem 3.21 applies.

3.2 L^p Spaces

Let $(X, \mathcal{M}, \mu)$ be a measure space. Two $\mathcal{M}$-measurable functions f and g are said to be equivalent if they are equal μ-almost everywhere in X with respect to the measure μ, that is, if $f(x) = g(x)$ for all x outside of a set of μ-measure zero:

$$f \sim g \Longleftrightarrow f(x) = g(x) \quad \text{for } \mu\text{-almost all } x \in X.$$

This is obviously an equivalence relation.

3.2.1 *Basic Properties of L^p-Spaces*

If $1 \leq p < \infty$, we let

$$\begin{aligned} L^p(X) = &\text{the space of equivalence classes of } \mathcal{M}\text{-measurable} \\ &\text{functions } f \text{ on } X \text{ such that } |f|^p \text{ is } \mu\text{-integrable on } X. \end{aligned}$$

We can define a norm by the formula

$$\|f\|_p := \left(\int_X |f(x)|^p \, d\mu(x) \right)^{1/p}.$$

An $\mathcal{M}$-measurable Lebesgue measurable function f on X is said to be *essentially bounded* if there exists a constant $C > 0$ such that

$$|f(x)| \leq C \quad \text{for } \mu\text{-almost all } x \in X.$$

For $p = \infty$, we let

$$\begin{aligned} L^\infty(X) = &\text{the space of equivalence classes of essentially bounded,} \\ &\mathcal{M}\text{-measurable functions } f \text{ on } X. \end{aligned}$$

We can define a norm by the formula

$$\begin{aligned} \|f\|_\infty &:= \operatorname{ess\,sup}_{x \in X} |f(x)| \\ &= \inf\{C : |f(x)| \leq C \text{ for } \mu\text{-almost all } x \in X\}. \end{aligned}$$

Then we have the following:

Theorem 3.23 *(i) The space $L^p(X)$, $1 \leq p < \infty$, is a Banach space with the norm $\|\cdot\|_p$.*

(ii) The space $L^\infty(X)$ is a Banach space with the norm $\|\cdot\|_\infty$.

The proof of Theorem 3.23 is given by the following series of theorems.

Step I: First, we show that the space $L^p(X)$ is a normed linear space. To do this, we need the following:

Theorem 3.24 (Young's inequality) *Let $0 < \theta < 1$. Then we have, for all $\varepsilon > 0$,*

$$a^\theta b^{1-\theta} \leq \theta \varepsilon^{1-\theta} a + (1-\theta)\, \varepsilon^{-\theta} b \quad \textit{for all } a \geq 0 \textit{ and } b \geq 0.$$

Moreover, the infimum of the right-hand side with respect to ε is equal to the left-hand side.

Proof We let

$$f(\varepsilon) = \theta\, \varepsilon^{1-\theta} a + (1-\theta)\, \varepsilon^{-\theta} b \quad \text{for } \varepsilon > 0.$$

Then we have the assertion

$$f'(\varepsilon) = \theta\,(1-\theta)\,\varepsilon^{-\theta}\left(a - \varepsilon^{-1} b\right),$$

so that the function $f(\varepsilon)$ attains its minimum at $\varepsilon = b/a$:

$$f\left(\frac{b}{a}\right) = a^{\theta}\, b^{1-\theta}.$$

The proof of Theorem 3.24 is complete. □

Theorem 3.25 (Hölder's inequality) *Let $1 \leq p, q \leq \infty$ and $1/p + 1/q = 1$. Then we have, for all $f \in L^p(X)$ and $g \in L^q(X)$,*

$$\left| \int_X f(x) g(x)\, d\mu(x) \right| \leq \left(\int_X |f(x)|^p\, d\mu(x) \right)^{1/p} \left(\int_X |g(x)|^q\, d\mu(x) \right)^{1/q}.$$

Proof It suffices to consider the case where

$$1 < p, q < \infty.$$

Then, by applying Young's inequality (Theorem 3.24) with $\theta := 1/p$ and $1 - \theta := 1/q$ we find that

$$|f(x)g(x)| \leq \frac{\varepsilon^{1/q}}{p} |f(x)|^p + \frac{\varepsilon^{-1/p}}{q} |g(x)|^q.$$

Hence, by integrating both sides over X we obtain that

$$\begin{aligned} \left| \int_X f(x) g(x)\, d\mu \right| &\leq \int_X |f(x) g(x)|\, d\mu \\ &\leq \frac{\varepsilon^{1/q}}{p} \int_X |f(x)|^p\, d\mu + \frac{\varepsilon^{-1/p}}{q} \int_X |g(x)|^q\, d\mu. \end{aligned}$$

However, we know that the infimum of the right-hand side with respect to ε is equal to the following:

$$\left(\|f\|_p^p\right)^{1/p} \left(\|g\|_q^q\right)^{1/q} = \|f\|_p \|g\|_q.$$

The proof of Theorem 3.25 is complete. □

Theorem 3.26 (Minkowski's inequality) *Let* $1 \leq p \leq \infty$. *Then we have, for all* f, $g \in L^p(X)$,

$$\left(\int_X |f(x)+g(x)|^p \, d\mu\right)^{1/p} \leq \left(\int_X |f(x)|^p \, d\mu\right)^{1/p} + \left(\int_X |g(x)|^p \, d\mu\right)^{1/p}. \tag{3.4}$$

Proof It suffices to consider the case where $1 < p < \infty$.

We remark that

$$f + g \in L^p(X).$$

Indeed, since we have the inequality

$$\left(\frac{|f(x)| + |g(x)|}{2}\right)^p \leq \frac{|f(x)|^p + |g(x)|^p}{2},$$

it follows that

$$\begin{aligned}\int_X |f(x)+g(x)|^p \, d\mu &\leq \int_X (|f(x)| + |g(x)|)^p \, d\mu \\ &\leq 2^{p-1}\left(\|f\|_p^p + \|g\|_p^p\right).\end{aligned}$$

Then, by Hölder's inequality (Theorem 3.25) it follows that

$$\begin{aligned}&\int_X |f(x)+g(x)|^p \, d\mu \\ &\leq \int_X |f(x)+g(x)|^{p-1}|f(x)| \, d\mu + \int_X |f(x)+g(x)|^{p-1}|g(x)| \, d\mu \\ &\leq \left(\int_X |f(x)+g(x)|^{(p-1)q} \, d\mu\right)^{1/q}\left(\int_X |f(x)|^p \, d\mu\right)^{1/p} \\ &+ \left(\int_X |f(x)+g(x)|^{(p-1)q} \, d\mu\right)^{1/q}\left(\int_X |g(x)|^p \, d\mu\right)^{1/p},\end{aligned}$$

so that

$$\|f+g\|_p^p \leq \|f+g\|_p^{p-1}\left(\|f\|_p + \|g\|_p\right). \tag{3.5}$$

If $\|f+g\|_p = 0$, then inequality (3.4) is valid.

If $\|f+g\|_p > 0$, we divide both sides of (3.5) by $\|f+g\|_p^{p-1}$ to obtain inequality (3.4).

The proof of Theorem 3.26 is complete. □

Step II: Secondly, we prove the completeness of $L^p(X)$. To do this, we need Beppo Levi's theorem (Theorem 3.14) and the following theorem:

Theorem 3.27 *Let $1 \le p \le \infty$, and assume that the sequence $\{f_n\}$ in $L^p(X)$ satisfies the condition*

$$\sum_{n=1}^{\infty} \|f_n\|_p < \infty.$$

Then we have the following three assertions:

(1) The sum

$$f(x) = \sum_{n=1}^{\infty} f_n(x) \tag{3.6}$$

converges absolutely for μ-almost every $x \in X$.

(2) The function f belongs to $L^p(X)$:

$$f = \sum_{n=1}^{\infty} f_n \in L^p(X). \tag{3.7}$$

(3) The finite sum $\sum_{n=1}^{N} f_n$ converges to f in $L^p(X)$:

$$\sum_{n=1}^{N} f_n \longrightarrow f \quad \text{in } L^p(X) \text{ as } N \to \infty. \tag{3.8}$$

Proof We only consider the case where $1 \le p < \infty$.

We let

$$g_N(x) = |f_1(x)| + |f_2(x)| + \ldots + |f_N(x)|,$$

and

$$g(x) = \lim_{N\to\infty} g_N(x) = \sum_{n=1}^{\infty} |f_n(x)|.$$

Then, by Minkowski's inequality (3.4) it follows that

$$\begin{aligned} \|g_N\|_p &\le \|f_1\|_p + \|f_2\|_p + \cdots + \|f_N\|_p \\ &\le \sum_{n=1}^{\infty} \|f_n\|_p < \infty. \end{aligned}$$

Hence we have, by Beppo Levi's theorem (Theorem 3.14),

$$\int_X g(x)^p \, d\mu(x) = \lim_{N\to\infty} \int_X g_N(x)^p \, d\mu(x) < \infty,$$

so that

$$g \in L^p(X).$$

This proves that

$$g(x) < \infty \quad \text{for } \mu\text{-almost every } x \in X,$$

that is, the sum

$$f(x) = \sum_{n=1}^{\infty} f_n(x)$$

converges absolutely for μ-almost every $x \in X$.

Furthermore, since we have the inequality

$$|f(x)| \leq g(x) \quad \text{for } x \in X,$$

it follows that

$$f \in L^p(X).$$

Finally, we remark that

$$\left| f(x) - \sum_{n=1}^{N} f_n(x) \right|^p \leq g(x)^p \quad \text{for } x \in X.$$

Hence we can apply the Lebesgue dominated convergence theorem (Theorem 3.13) to obtain that

$$\lim_{N \to \infty} \int_X \left| f(x) - \sum_{n=1}^{N} f_n(x) \right|^p \, d\mu(x) = 0,$$

so that

$$\sum_{n=1}^{N} f_n \longrightarrow f \quad \text{in } L^p(X) \text{ as } N \to \infty.$$

The proof of Theorem 3.27 is complete. □

Remark 3.28 Theorem 3.27 plays an important role in the proof of existence theorems of singular integrals in Sect. 11.1 (see Theorems 11.2 and 11.3).

3.2.2 End of Proof of the Completeness of $L^p(X)$

Let $\{f_n\}$ be an arbitrary Cauchy sequence in the normed space $L^p(X)$. Then we can find an increasing sequence $\{n(j)\}$ of positive integers such that

$$n > n(1) \Longrightarrow \left\| f_n - f_{n(1)} \right\|_p < \frac{1}{2},$$
$$n > n(2) \Longrightarrow \left\| f_n - f_{n(2)} \right\|_p < \frac{1}{2^2},$$
$$\vdots$$
$$n > n(k) \Longrightarrow \left\| f_n - f_{n(k)} \right\|_p < \frac{1}{2^k}.$$

In particular, we have the inequality

$$\left\| f_{n(k)} - f_{n(k+1)} \right\|_p < \frac{1}{2^k}.$$

This implies that

$$\sum_{k=1}^{\infty} \left\| f_{n(k+1)} - f_{n(k)} \right\|_p < 1.$$

Hence it follows from an application of Theorem 3.27 that there exists a function $f \in L^p(X)$ such that

$$f_{n(k)} = f_{n(1)} + \sum_{j=1}^{k-1} \left(f_{n(j+1)} - f_{n(j)} \right) \longrightarrow f \quad \text{in } L^p(X) \text{ as } k \to \infty.$$

However, since $\{f_n\}$ is a Cauchy sequence in $L^p(X)$, for each $\varepsilon > 0$ we can find a positive integer $N = N(\varepsilon)$ such that

$$n, m > N \Longrightarrow \|f_n - f_m\|_p < \varepsilon.$$

Therefore, summing up we obtain that

$$\begin{aligned} n, n(k) > N \Longrightarrow \|f_n - f\|_p &\leq \left\| f_n - f_{n(k)} \right\|_p + \left\| f_{n(k)} - f \right\|_p \\ &\leq \varepsilon + \left\| f - f_{n(k)} \right\|_p, \end{aligned}$$

so that

$$n > N \Longrightarrow \|f_n - f\|_p \leq \varepsilon.$$

This proves the completeness of the normed space $L^p(X)$.

Now the proof of Theorem 3.23 is complete. □

3.3 Minkowski's Inequality for Integrals

Minkowski's inequality (Theorem 3.25) can be generalized as follows:

Theorem 3.29 (Minkowski's inequality for integrals) *Let* $(X, \mathcal{M}, \mu)$ *and* $(Y, \mathcal{N}, \nu)$ *be* σ*-finite, complete measure spaces and let* $(X \times Y, \mathcal{L}, \lambda)$ *be the* completion *of the product measure space* $(X \times Y, \mathcal{M} \otimes \mathcal{N}, \mu \times \nu)$. *Let* $1 \leq p \leq \infty$ *and assume that* $f(x, y)$ *is an* $\mathcal{L}$*-measurable function on* $X \times Y$, *and satisfies the conditions*

$$f(\cdot, y) \in L^p(X) \quad \text{for } \nu\text{-almost all } y \in Y, \tag{3.9a}$$

$$\int_Y \|f(\cdot, y)\|_p \, d\nu(y) < \infty. \tag{3.9b}$$

Then we have the following three assertions (3.10a), (3.10b) *and* (3.10c)*:*

$$f(x, \cdot) \in L^1(Y) \quad \text{for } \mu\text{-almost all } x \in X. \tag{3.10a}$$

$$F(x) = \int_Y f(x, y) \, d\nu(y) \in L^p(X). \tag{3.10b}$$

$$\|F\|_p \leq \int_Y \|f(\cdot, y)\|_p \, d\nu(y). \tag{3.10c}$$

Inequality (3.10c) *is called* Minkowski's inequality for integrals.

Proof The proof is divided into two steps.

Step 1: First, we show that:

The function $y \mapsto \|f(\cdot, y)\|_p$ is $\mathcal{N}$-measurable.

Step 1-a: The case where $1 \leq p < \infty$. Since the function $|f(x, y)|^p$ is $\mathcal{L}$-measurable on $X \times Y$, it follows from an application of Fubini's theorem (Theorem 3.20) that the function

$$x \longmapsto |f(x, y)|^p$$

is $\mathcal{M}$-measurable for ν-almost all $y \in Y$, and that the function

$$y \longmapsto \int_X |f(x, y)|^p \, d\mu(x)$$

is $\mathcal{N}$-measurable. Hence we find that the function

$$y \longmapsto \|f(\cdot, y)\|_p = \left(\int_X |f(x, y)|^p \, d\mu(x) \right)^{1/p}$$

is $\mathcal{N}$-measurable.

Step 1-b: The case where $p = \infty$. We assume that

$$\mu(X) < \infty.$$

Then we have, by part (ii) of Theorem 3.33 in Sect. 3.5 below,

$$\|f(\cdot, y)\|_\infty = \lim_{q \uparrow \infty} \|f(\cdot, y)\|_q \quad \text{for } \nu\text{-almost all } y \in Y.$$

This proves the $\mathcal{N}$-measurability of the function $\|f(\cdot, y)\|_\infty$, since the function $y \mapsto \|f(\cdot, y)\|_q$ is $\mathcal{N}$-measurable.

Now we assume that X is σ-finite, that is, it is a countable union of sets X_n of finite measure:

$$X = \bigcup_{n=1}^{\infty} X_n, \quad \mu(X_n) < \infty.$$

Since we have the assertion

$$\|f(\cdot, y)\|_\infty = \sup_{n \geq 1} \|f(\cdot, y)\|_\infty \quad \text{for } \nu\text{-almost all } y \in Y,$$

it follows that the function $y \mapsto \|f(\cdot, y)\|_\infty$ is $\mathcal{N}$-measurable.

Step 2: Now we prove inequality (3.10c).

Step 2-a: The case where $p = \infty$. We remark that

$$|f(x, y)| \leq \|f(\cdot, y)\|_\infty$$
$$\text{for } \mu\text{-almost all } x \in X \text{ and for } \nu\text{-almost all } y \in Y.$$

Hence we have, by conditions (3.9a) and (3.9b),

$$\int_Y |f(x, y)| \, d\nu(y) \leq \int_Y \|f(\cdot, y)\|_\infty \, d\nu(y) < \infty \quad \text{for } \mu\text{-almost all } x \in X,$$

and so

$$f(x, \cdot) \in L^1(Y) \quad \text{for } \mu\text{-almost all } x \in X.$$

Furthermore, in view of Fubini's theorem (Theorem 3.20) it follows that the function

$$x \longmapsto F(x) = \int_Y f(x, y) \, d\nu(y)$$

is $\mathcal{M}$-measurable and that

$$|F(x)| \leq \int_Y \|f(\cdot, y)\|_\infty \, d\nu(y) \quad \text{for } \mu\text{-almost all } x \in X.$$

This proves the desired inequality (3.10c) for $p = \infty$:

$$\|F\|_\infty \leq \int_Y \|f(\cdot, y)\|_\infty \, d\nu(y).$$

Step 2-b: The case where $p = 1$. By Fubini's theorem (Theorem 3.20), we have the inequality

$$\begin{aligned}
\|F\|_1 &= \int_X \left| \int_Y f(x, y)\, d\nu(y) \right| d\mu(x) \\
&\le \int_X \int_Y |f(x, y)|\, d\nu(y)\, d\mu(x) = \int_Y \left(\int_X |f(x, y)|\, d\mu(x) \right) d\nu(y) \\
&= \int_Y \|f(\cdot, y)\|_1\, d\nu(y).
\end{aligned}$$

This proves the desired inequality (3.10c) for $p = 1$.

Step 2-c: The case where $1 < p < \infty$.

(i) First, we assume that

$$f(x, y) \ge 0 \quad \text{on } X \times Y. \tag{3.11}$$

Then, in view of Fubini's theorem (Theorem 3.20) it follows that the function

$$x \longmapsto F(x) = \int_Y f(x, y)\, d\nu(y)$$

is $\mathcal{M}$-measurable. Thus we have, by Hölder's inequality (Theorem 3.24),

$$\begin{aligned}
\int_X |F(x)|^p\, d\mu(x) &= \int_X |F(x)|^{p-1} \left(\int_Y f(x, y)\, d\nu(y) \right) d\mu(x) \\
&= \int_Y \left(\int_X |F(x)|^{p-1} f(x, y)\, d\mu(x) \right) d\nu(y) \\
&\le \int_Y \left(\int_X \|F\|_p^{p-1} \|f(\cdot, y)\|_p \right) d\nu(y) \\
&= \|F\|_p^{p-1} \int_Y \|f(\cdot, y)\|_p\, d\nu(y).
\end{aligned}$$

Therefore, if we have the inequality

$$\|F\|_p < \infty,$$

it follows that

$$\|F\|_p \le \int_Y \|f(\cdot, y)\|_p\, d\nu(y).$$

Now we assume that X and Y are σ-finite, that is, they are the countable unions of sets X_n and Y_n of finite measure, respectively:

$$X = \bigcup_{n=1}^{\infty} X_n, \quad \mu(X_n) < \infty,$$
$$Y = \bigcup_{n=1}^{\infty} Y_n, \quad \nu(Y_n) < \infty.$$

We let

$$f(x, y) = \begin{cases} \min\{f(x, y), n\} & \text{if } x \in X_n,\, y \in Y_n, \\ 0 & \text{otherwise.} \end{cases}$$

Then it follows that

$$f_n(x, y) \uparrow f(x, y) \quad \text{as } n \to \infty,$$

so that

$$F_n(x) = \int_Y f_n(x, y) d\nu(y) \uparrow F(x) = \int_Y f(x, y)\, d\nu(y) \quad \text{as } n \to \infty.$$

We remark that

$$\|F_n\|_p < \infty.$$

Hence it follows that

$$\|F_n\|_p \leq \int_Y \|f_n(\cdot, y)\|_p \, d\nu(y) \leq \int_Y \|f(\cdot, y)\|_p \, d\nu(y).$$

Therefore, by passing to the limit we obtain that

$$\|F\|_p \leq \int_Y \|f(\cdot, y)\|_p \, d\nu(y).$$

This proves the desired inequality (3.10c) for $1 < p < \infty$ under condition (3.11).

(ii) We consider the general case. By applying Step (i) to the function $|f(x, y)|$, we obtain that

$$G(x) = \int_Y |f(x, y)|\, d\nu(y) \in L^p(X),$$
$$\|G\|_p \leq \int_Y \|f(\cdot, y)\|_p \, d\nu(y).$$

Hence we have the assertion

$$G(x) < \infty \quad \text{for } \mu\text{-almost all } x \in X,$$

and so

$$f(x, \cdot) \in L^1(Y) \quad \text{for } \mu\text{-almost all } x \in X.$$

However, in view of Hölder's inequality (Theorem 3.24) it follows that

$$G(x) \in L^1(X_n), \quad n = 1, 2, \ldots.$$

By Fubini's theorem, this implies that

$$F(x) = \int_Y f(x, y)\, d\nu(y) \in L^1(X_n), \quad n = 1, 2, \ldots.$$

In particular, it follows that F is $\mathcal{M}$-measurable on each X_n, and so it is $\mathcal{M}$-measurable on $X = \bigcup_n X_n$.

Moreover, we have the inequality

$$\|F\|_p \leq \|G\|_p \leq \int_Y \|f(\cdot, y)\|_p\, d\nu(y).$$

This proves the desired inequality (3.10c) for $1 < p < \infty$.

The proof of Theorem 3.29 is now complete. □

3.4 Hardy's Inequality

First, we prove a general integral inequality on the interval $(0, \infty)$:

Theorem 3.30 *Let $K(x, y)$ be a Lebesgue measurable function defined on the interval $(0, \infty)$. Assume that $K(x, y)$ is positively homogeneous of degree -1, that is,*

$$K(\lambda x, \lambda y) = \lambda^{-1} K(x, y), \quad \lambda > 0,$$

and further that the integral

$$A_K := \int_0^\infty |K(1, y)|\, y^{-1/p}\, dy$$

is finite for some $1 \leq p \leq \infty$.

Then the operator Tf, defined by the formula

$$Tf(x) = \int_0^\infty K(x, y) f(y)\, dy,$$

is bounded from $L^p(0, \infty)$ into itself. More precisely, we have the inequality

$$\|Tf\|_p \leq A_K \|f\|_p, \quad f \in L^p(0, \infty).$$

Proof Since we have the formula

$$\begin{aligned} Tf(x) &= \int_0^\infty K(x, y) f(y)\, dy \\ &= \int_0^\infty K(x \cdot 1, x \cdot z) f(zx)\, x\, dz \\ &= \int_0^\infty x^{-1} K(1, z) f(zx)\, x\, dz \\ &= \int_0^\infty K(1, z) f(zx)\, dz, \end{aligned}$$

by applying Minkowski's inequality for integrals (Theorem 3.29) we obtain that

$$\|Tf\|_p \leq \int_0^\infty |K(1, z)|\, \|f(z\cdot)\|_p\, dz.$$

However, it is easy to see that

$$\begin{aligned} \|f(z\cdot)\|_p &= \left(\int_0^\infty |f(zx)|^p\, dx \right)^{1/p} \\ &= z^{-1/p} \left(\int_0^\infty |f(y)|^p\, dy \right)^{1/p} \\ &= z^{-1/p} \|f\|_p. \end{aligned}$$

Therefore, we have the inequality

$$\|Tf\|_p \leq \left(\int_0^\infty |K(1, z)| z^{-1/p}\, dz \right) \|f\|_p = A_K \|f\|_p.$$

The proof of Theorem 3.30 is complete. □

Now we can prove Hardy's inequality which will be used systematically:

Theorem 3.31 (Hardy's inequality) *Let $1 \leq p \leq \infty$ and $\gamma \neq 0$. If $f(x)$ is a non-negative, Lebesgue measurable function on the interval $(0, \infty)$, we define a function $F(x)$ by the formula*

$$F(x) = \begin{cases} \int_0^x f(y)\, dy & \text{if } \gamma < 0, \\ \int_x^\infty f(y)\, dy & \text{if } \gamma > 0. \end{cases}$$

Then we have the inequality

$$\left(\int_0^\infty (x^\gamma F(x))^p \frac{dx}{x}\right)^{1/p} \leq \frac{1}{|\gamma|}\left(\int_0^\infty (x^{\gamma+1} f(x))^p \frac{dx}{x}\right)^{1/p}. \tag{3.12}$$

Proof We only consider the case where $\gamma < 0$. The case where $\gamma > 0$ is proved similarly.

If we let

$$K(x,y) := \begin{cases} x^{\gamma-1/p} y^{-\gamma+1/p-1} & \text{if } 0 \leq y \leq x, \\ 0 & \text{if } x < y. \end{cases}$$

then it follows that $K(x, y)$ is positively homogeneous of degree -1 and satisfies the conditions

$$\int_0^\infty K(1,y) y^{-1/p}\, dy = \int_0^1 y^{-\gamma-1}\, dy = -\frac{1}{\gamma}.$$

If we introduce an integral operator

$$\begin{aligned} Tg(x) &:= \int_0^\infty K(x,y)\, g(y)\, dy \\ &= x^{\gamma-1/p} \int_0^x y^{-\gamma+1/p-1}\, g(y)\, dy, \quad g \in L^p(0,\infty), \end{aligned}$$

then, by applying Theorem 3.30 to our situation we obtain that

$$\left(\int_0^\infty \left(x^{\gamma-1/p} \int_0^x y^{-\gamma+1/p-1} g(y)\, dy\right)^p dx\right)^{1/p} \leq \frac{1}{|\gamma|}\left(\int_0^\infty g(y)^p\, dy\right)^{1/p}.$$

In particular, if we let

$$g(y) := y^{\gamma-1/p+1} f(y),$$

then it follows that

$$\begin{aligned} \left(\int_0^\infty (x^\gamma F(x))^p \frac{dx}{x}\right)^{1/p} &= \left(\int_0^\infty \left(x^{\gamma-1/p} \int_0^x f(x)\, dy\right)^p dx\right)^{1/p} \\ &\leq \frac{1}{|\gamma|}\left(\int_0^\infty \left(y^{\gamma-1/p+1} f(y)\right)^p dy\right)^{1/p} \\ &= \frac{1}{|\gamma|}\left(\int_0^\infty (x^{\gamma+1} f(x))^p \frac{dx}{x}\right)^{1/p}. \end{aligned}$$

The proof of Theorem 3.31 is complete. □

Example 3.1 If we let

$$\gamma =: \frac{1}{p} - 1, \quad 1 < p \leq \infty,$$

then we have, by inequality (3.14),

$$\left(\int_0^\infty \left(\frac{1}{x} \int_0^x f(y)\, dy \right)^p dx \right)^{1/p} \leq \frac{p}{p-1} \left(\int_0^\infty f(y)^p\, dy \right)^{1/p}.$$

3.5 The Generalized Hölder Inequality

Hölder's inequality (Theorem 3.24) can be generalized as follows:

Theorem 3.32 (the generalized Hölder inequality) *Let* $1 \leq p, q, r \leq \infty$ *such that*

$$\frac{1}{p} + \frac{1}{q} = \frac{1}{r}.$$

Then we have, for all $f \in L^p(X)$ *and* $g \in L^q(X)$,

$$\left(\int_X |f(x)g(x)|^r\, d\mu \right)^{1/r} \leq \left(\int_X |f(x)|^p\, d\mu \right)^{1/p} \left(\int_X |g(x)|^q\, d\mu \right)^{1/q}. \tag{3.13}$$

Proof We only consider the case where $1 \leq p, q < \infty$. Since we have the formula

$$\frac{1}{\frac{p}{r}} + \frac{1}{\frac{q}{r}} = 1,$$

the desired inequality (3.13) follows by applying Hölder's inequality (Theorem 3.24) to the functions $|f|^r \in L^{p/r}(X)$ and $|g|^r \in L^{q/r}(X)$.

The proof of Theorem 3.32 is complete. □

Some basic properties of the L^p spaces are summarized in the following theorem:

Theorem 3.33 *Assume that*

$$0 < \mu(X) < \infty.$$

Then we have the following two assertions:

(i) We have the inclusion

$$L^p(X) \subset L^q(X) \quad \textit{for } 1 \leq q < p \leq \infty,$$

with continuous injection.
More precisely, we have the inequality

$$\|f\|_q \leq \mu(X)^{1/q-1/p} \|f\|_p \quad \textit{for } f \in L^p(X). \tag{3.14}$$

(ii) The norm $\|\cdot\|_p$ is continuous for $p \geq 1$, that is,

$$\lim_{q \uparrow p} \|f\|_q = \|f\|_p \quad \text{for } f \in L^p(X).$$

Proof (i) By Hölder's inequality (Theorem 3.24), it follows that

$$\begin{aligned} \int_X |f(x)|^q \, d\mu &\leq \left(\int_X 1 \, d\mu \right)^{(p-q)/p} \left(\int_X |f(x)|^p \, d\mu \right)^{q/p} \\ &= \mu(X)^{1-p/q} \, \|f\|_p^q, \end{aligned}$$

so that

$$\|f\|_q \leq \mu(X)^{1/q-1/p} \, \|f\|_p.$$

This proves the desired inequality (3.14).

(ii) First, we remark that, by inequality (3.14),

$$\limsup_{q \uparrow p} \|f\|_q \leq \|f\|_p.$$

(ii-a) The case $p = \infty$: It suffices to show that

$$\liminf_{q \uparrow \infty} \|f\|_q \geq \|f\|_\infty. \tag{3.15}$$

We may assume that

$$\|f\|_\infty > 0.$$

For each $\varepsilon > 0$, we can find a set $X_1 \subset X$ such that

$$\begin{aligned} &X_1 \subset \{x \in X : |f(x)| \geq \|f\|_\infty - \varepsilon\}, \\ &\mu(X_1) > 0. \end{aligned}$$

Then we have the inequality

$$\begin{aligned} \|f\|_q &= \left(\int_X |f(x)|^q \, d\mu(x) \right)^{1/q} \\ &\geq \left(\int_{X_1} |f(x)|^q \, d\mu(x) \right)^{1/q} \\ &\geq \left(\|f\|_\infty - \varepsilon \right) \mu(X_1)^{1/q}, \end{aligned}$$

and so

$$\liminf_{q\uparrow\infty} \|f\|_q \geq \|f\|_\infty - \varepsilon.$$

This proves the desired inequality (3.17), since ε is arbitrary.

(ii-b) The case where $1 < p < \infty$: We let

$$X_1 = \{x \in X : |f(x)| > 1\},$$
$$X_2 = \{x \in X : |f(x)| \leq 1\}.$$

Then we have the formula

$$\|f\|_q^q = \int_{X_1} |f(x)|^q \, d\mu(x) + \int_{X_2} |f(x)|^q \, d\mu(x).$$

However, we have the inequalities

$$|f(x)|^q \leq |f(x)|^p, \quad x \in X_1,$$
$$|f(x)|^q \leq 1, \quad x \in X_2.$$

Hence, by applying the monotone convergence theorem (Theorem 3.11) to the first term and also the Lebesgue dominated convergence theorem (Theorem 3.13) to the second term we obtain that

$$\lim_{q\uparrow p} \|f\|_q^q = \int_{X_1} |f(x)|^p \, d\mu(x) + \int_{X_2} |f(x)|^p \, d\mu(x) = \|f\|_p^p.$$

Now the proof of Theorem 3.33 is complete. □

3.6 The Generalized Young Inequality

Now we prove a general theorem about integral operators on a measure space:

Theorem 3.34 (the generalized Young inequality) *Let $(X, \mathcal{M}, \mu)$ and $(Y, \mathcal{N}, \nu)$ be σ-finite, complete measure spaces and let $(X \times Y, \mathcal{L}, \lambda)$ be the* completion *of the product measure space $(X \times Y, \mathcal{M} \otimes \mathcal{N}, \mu \times \nu)$.*

Let $1 \leq p, q, r \leq \infty$ such that

$$\frac{1}{p} + \frac{1}{r} = 1 + \frac{1}{q}.$$

Assume that $K(x, y)$ is an $\mathcal{L}$-measurable function on the product space $X \times Y$ such that

$$\int_X |K(x, y)|^r \, d\mu(x) \leq M \quad \textit{for } \nu\textit{-almost all } y \in Y$$

and that

$$\int_Y |K(x, y)|^r \, d\nu(y) \le N \quad \textit{for } \mu\textit{-almost all } x \in X,$$

where M and N are positive constants.

If $f \in L^p(X)$, then the function $Tf(x)$, defined by the formula

$$Tf(x) = \int_Y K(x, y) f(y) \, d\nu(y),$$

is well-defined for μ-almost all $x \in X$, and is in the space $L^q(X)$.

Furthermore, we have the inequality

$$\|Tf\|_q \le M^{1/q} N^{1-(1/p)} \|f\|_p. \tag{3.16}$$

Proof The proof of Theorem 3.34 is divided into two steps.

Step 1: First, we assume that

$$\begin{aligned} &K(x, y) \ge 0 \quad \text{on } X \times Y, \\ &f(y) \ge 0 \quad \text{on } Y. \end{aligned}$$

Then we have, by Fubini's theorem (Theorem 3.20), the following two assertions:

(i) The function $y \mapsto K(x, y) f(y)$ is $\mathcal{N}$-measurable for μ-almost all $x \in X$.
(ii) The function $Tf(x) = \int_Y K(x, y) f(y) \, d\nu(y)$ is $\mathcal{M}$-measurable on X.

It follows from an application of the generalized Hölder inequality (Theorem 3.32) that

$$\begin{aligned} Tf(x) &= \int_Y K(x, y)^{1-r/q} K(x, y)^{r/q} f(y)^{p/q} f(y)^{1-p/q} \, d\nu(y) \\ &\le \left(\int_Y K(x, y)^r \, d\nu(y) \right)^{1-1/p} \left(\int_Y K(x, y)^r f(y)^p \, d\nu(y) \right)^{1/q} \\ &\quad \times \left(\int_Y f(y)^p \, d\nu(y) \right)^{1/p-1/q} \\ &\le N^{1-1/p} \|f\|_p^{1-p/q} \left(\int_X K(x, y)^r f(y)^p \, d\nu(y) \right)^{1/q}. \end{aligned} \tag{3.17}$$

Step I-a: The case where $q = \infty$: In view of inequality (3.17), it follows that

$$Tf(x) \le N^{1-1/p} \|f\|_p \quad \text{for } \mu\text{-almost all } x \in X,$$

so that

$$\|Tf\|_\infty \le N^{1-1/p} \|f\|_p.$$

This proves the desired inequality (3.16) for $q = \infty$.

Step I-b: The case where $1 \le q < \infty$: We have, by Fubini's theorem (Theorem 3.20),

$$\begin{aligned}\int_X Tf(x)^q \, d\mu(x) &\le \left(N^{1-1/p} \, \|f\|_p^{1-p/q}\right)^q \int_X \int_Y K(x, y)^r f(y)^p \, d\nu(y) \, d\mu(x) \\ &\le \left(N^{1-1/p} \, \|f\|_p^{1-p/q}\right)^q M \, \|f\|_p^p \\ &= \left(M^{1/q} N^{1-1/p} \, \|f\|_p\right)^q,\end{aligned}$$

and so

$$\|Tf\|_q \le M^{1/q} N^{1-1/p} \, \|f\|_p.$$

This proves the desired inequality (3.16) for $1 < q < \infty$.

Moreover, it follows that

$$Tf(x) < \infty \quad \text{for } \mu\text{-almost all } x \in X.$$

Step II: We consider the general case. To do this, we remark that

$$\begin{aligned}K(x, y) f(y) &= \left[(\mathrm{Re} K(x, y) f(y))^+ - (\mathrm{Re} K(x, y) f(y))^-\right] \\ &\quad + \sqrt{-1} \left[(\mathrm{Im} K(x, y) f(y))^+ - (\mathrm{Im} K(x, y) f(y))^-\right],\end{aligned}$$

and that

$$\begin{aligned}(\mathrm{Re} K(x, y) f(y))^\pm &\le |K(x, y) f(y)|, \\ (\mathrm{Im} K(x, y) f(y))^\pm &\le |K(x, y) f(y)|.\end{aligned}$$

Here

$$\begin{aligned}f^+(x) &= \max\{f(x), 0\} = \frac{1}{2}(|f(x)| + f(x)), \\ f^-(x) &= \max\{-f(x), 0\} = \frac{1}{2}(|f(x)| - f(x)).\end{aligned}$$

However, we have, by Step I,

$$\int_Y |K(x, y)||f(y)| \, d\nu(y) < \infty \quad \text{for } \mu\text{-almost all } x \in X.$$

Hence it follows that:

(1) The function $y \mapsto K(x, y) f(y)$ is $\mathcal{N}$-integrable for μ-almost all $x \in X$.
(2) The function $Tf(x) = \int_Y K(x, y) f(y) \, d\nu(y)$ is $\mathcal{M}$-measurable on X.

Furthermore, since we have the inequality

$$|Tf(x)| \leq \int_Y |K(x, y)||f(y)|\, d\nu(y) \quad \text{for } \mu\text{-almost all } x \in X,$$

we find that the desired inequality (3.16) remains valid for this case.
The proof of Theorem 3.34 is complete. □

The case where $r := 1$ and $p := q$ is useful in applications:

Corollary 3.35 (Schur's lemma) *Assume that $K(x, y)$ is an $(\mathcal{M} \times \mathcal{N})^*$-measurable function on the product space $X \times Y$ such that*

$$\int_X |K(x, y)|\, d\mu(x) \leq M \quad \text{for } \nu\text{-almost all } y \in Y$$

and that

$$\int_Y |K(x, y)|\, d\nu(y) \leq N \quad \text{for } \mu\text{-almost all } x \in X,$$

where M and N are positive constants.

If $f \in L^p(X)$, then the function $Tf(x)$, defined by the formula

$$Tf(x) = \int_Y K(x, y) f(y)\, d\nu(y),$$

is well-defined for μ-almost all $x \in X$, and is in $L^p(X)$.

Furthermore, we have the inequality

$$\|Tf\|_p \leq M^{1/p} N^{1-(1/p)}\, \|f\|_p.$$

3.7 Convolutions

In this section we prove a useful inequality for convolutions:

Theorem 3.36 (Young's inequality) *Let $1 \leq p, q, r \leq \infty$ such that*

$$\frac{1}{p} + \frac{1}{q} = \frac{1}{r} + 1.$$

If $f \in L^p(\mathbf{R}^n)$ and $g \in L^q(\mathbf{R}^n)$, then the function $(f * g)(x)$, defined by the formula

$$(f * g)(x) = \int_{\mathbf{R}^n} f(x - y)\, g(y)\, dy,$$

is well-defined for almost all $x \in \mathbf{R}^n$, *and is in* $L^r(\mathbf{R}^n)$. *Furthermore, we have the inequality*

$$\|f * g\|_r \leq \|f\|_p \|g\|_q . \tag{3.18}$$

The function $f * g$ *is called the* convolution *of* f *and* g.

Proof The proof of Theorem 3.36 is divided into two steps. **Step I**: First, we have to verify that

$$\text{The function } f(x \pm y) \text{ is Lebesgue measurable on } \mathbf{R}^n \times \mathbf{R}^n. \tag{3.19}$$

Step I-a: We prove assertion (3.19) under the condition that $f(x)$ is Borel measurable on $\mathbf{R}^n$, that is:

Claim 3.37 *If* $f(x)$ *is Borel measurable on* $\mathbf{R}^n$, *then* $f(x \pm y)$ *is Borel measurable on* $\mathbf{R}^n \times \mathbf{R}^n$.

The proof of Claim 3.37 is divided into five steps:

Step (1): If $f(x)$ is continuous on $\mathbf{R}^n$, then $f(x \pm y)$ is continuous on $\mathbf{R}^n \times \mathbf{R}^n$, and so it is Borel measurable on $\mathbf{R}^n \times \mathbf{R}^n$.

Step (2): Let G be an arbitrary open subset of $\mathbf{R}^n$, and let

$$f(x) := \text{the characteristic function } \chi_G(x) \text{ of } G.$$

If we let

$$f_N(x) := \min\{N \operatorname{dist}(x, G^c), 1\}, \quad N \in \mathbf{N},$$

then it follows that

$$f_N(x) \text{ is continuous on } \mathbf{R}^n,$$
$$f_N(x) \longrightarrow f(x), \quad x \in \mathbf{R}^n.$$

Hence we find from Step (1) that the function

$$f(x \pm y) = \lim_{N \to \infty} f_N(x \pm y)$$

is Borel measurable on $\mathbf{R}^n \times \mathbf{R}^n$, if $f = \chi_G$.

Step (3): We define

$$\mathfrak{A} = \{A \subset \mathbf{R}^n : \chi_A(x \pm y) \text{ is Borel measurable on } \mathbf{R}^n \times \mathbf{R}^n\}.$$

Then $\mathfrak{A}$ is a σ-algebra.

Indeed, we have the following three assertions:

(a) The empty set $\emptyset$ is in $\mathfrak{A}$, since $\chi_\emptyset = 0$.

(b) If $A \in \mathfrak{A}$, then we have $\chi_{A^c} = 1 - \chi_A$, so that the function

$$\chi_{A^c}(x \pm y) = 1 - \chi_A(x \pm y)$$

is Borel measurable on $\mathbf{R}^n \times \mathbf{R}^n$.

(c) If $A_j \in \mathfrak{A}$ and $A = \bigcup_{j=1}^{\infty} A_j$, then the function

$$\chi_A(x \pm y) = \sup_{j \geq 1} \chi_{A_j}(x \pm y)$$

is Borel measurable on $\mathbf{R}^n \times \mathbf{R}^n$.

Step (4): By Steps (2) and (3), it follows that $\mathfrak{A}$ is a σ-algebra which contains all open subsets of $\mathbf{R}^n$; hence $\mathfrak{A}$ contains all Borel subsets of $\mathbf{R}^n$. Namely, we find that the function $\chi_E(x \pm y)$ is Borel measurable on $\mathbf{R}^n \times \mathbf{R}^n$, for each Borel subset E of $\mathbf{R}^n$.

Step (5): Let $f(x)$ be an arbitrary Borel measurable function on $\mathbf{R}^n$. We may assume that

$$f(x) \geq 0 \quad \text{on } \mathbf{R}^n.$$

Then we can find an increasing sequence of non-negative, Borel measurable simple functions $f_N(x)$ on $\mathbf{R}^n$ such that

$$f_N(x) \longrightarrow f(x), \quad x \in \mathbf{R}^n.$$

Hence, by Step (4) it follows that the function

$$f(x \pm y) = \lim_{N \to \infty} f_N(x \pm y)$$

is Borel measurable on $\mathbf{R}^n \times \mathbf{R}^n$.

This completes the proof of Claim 3.37. □

Step I-b: Now we prove assertion (3.19).

Let $f(x)$ be an arbitrary Lebesgue measurable function on $\mathbf{R}^n$. We may assume that

$$f(x) \geq 0 \quad \text{on } \mathbf{R}^n.$$

Then we can find two Borel measurable functions $g(x)$ and $h(x)$ on $\mathbf{R}^n$ such that

$$0 \leq g(x) \leq f(x) \leq h(x) \quad \text{on } \mathbf{R}^n,$$
$$g(x) = f(x) = h(x) \quad \text{for almost all } x \in \mathbf{R}^n.$$

This follows from the fact that the Lebesgue measure is the completion of the Borel measure.

Hence we have the inequalities

$$0 \leq g(x \pm y) \leq f(x \pm y) \leq h(x \pm y) \quad \text{for } (x, y) \in \mathbf{R}^n \times \mathbf{R}^n.$$

However, we know from Step I-a that the functions $g(x \pm y)$ and $h(x \pm y)$ are Borel measurable on $\mathbf{R}^n \times \mathbf{R}^n$. Thus, by applying Fubini's theorem (Theorem 3.20) we obtain that

$$\begin{aligned}
&\int_{\mathbf{R}^n} \int_{\mathbf{R}^n} (h(x \pm y) - g(x \pm y))\, dx\, dy \\
&= \int_{\mathbf{R}^n} \left(\int_{\mathbf{R}^n} (h(x \pm y) - g(x \pm y))\, dx \right) dy \\
&= \int_{\mathbf{R}^n} \left(\int_{\mathbf{R}^n} (h(x) - g(x))\, dx \right) dy \\
&= 0.
\end{aligned}$$

This proves that

$$g(x \pm y) = h(x \pm y) \quad \text{for almost all } (x, y) \in \mathbf{R}^n \times \mathbf{R}^n,$$

so that

$$g\,(x \pm y) = f\,(x \pm y) = h\,(x \pm y) \quad \text{for almost all } (x, y) \in \mathbf{R}^n \times \mathbf{R}^n.$$

Therefore, we find that the function $f\,(x \pm y)$ is Lebesgue measurable on $\mathbf{R}^n \times \mathbf{R}^n$, since the functions $g\,(x \pm y)$ and $h\,(x \pm y)$ are Borel measurable on $\mathbf{R}^n \times \mathbf{R}^n$.

Step II: **End of Proof of Theorem** 3.36. The case where $1 \leq r < \infty$: It suffices to apply Theorem 3.34 with

$$\begin{aligned}
&(X, \mu) := \left(\mathbf{R}^n, dx\right), \quad (Y, \nu) := \left(\mathbf{R}^n, dy\right), \\
&p := q, \quad q := r, \quad r := p, \\
&K(x, y) := f(x - y), \quad f(y) := g(y).
\end{aligned}$$

Indeed, since we have the formula

$$\int_{\mathbf{R}^n} |f(x - y)|^p\, dx = \int_{\mathbf{R}^n} |f(x - y)|^p\, dy = \|f\|_p^p,$$

it follows from an application of Theorem 3.34 with $M = N := \|f\|_p^p$ that

$$\|f * g\|_r \leq \left(\|f\|_p^p\right)^{1/r} \left(\|f\|_p^p\right)^{1-1/q} \|g\|_p = \|f\|_p \|g\|_q.$$

This proves the desired inequality (3.18) for $1 \leq r < \infty$.

The case where $r = \infty$: If $f \in L^p(\mathbf{R}^n)$ and $g \in L^q(\mathbf{R}^n)$ with $1/p + 1/q = 1$, then, by applying Hölder's inequality (Theorem 3.24) we obtain that

$$\begin{aligned}|(f * g)(x)| &= \left|\int_{\mathbf{R}^n} f(x-y)g(y)\,dy\right| \\ &\leq \left(\int_{\mathbf{R}^n} |f(x-y)|^p\,dy\right)^{1/p} \left(\int_{\mathbf{R}^n} |g(y)|^q dy\right)^{1/q} \\ &= \|f\|_p \,\|g\|_q.\end{aligned}$$

Hence we have the desired inequality (3.18) for $r = \infty$:

$$\|f * g\|_\infty \leq \|f\|_p \,\|g\|_q.$$

The proof of Theorem 3.36 is now complete. □

3.7.1 *Approximations to the Identity*

Theorem 3.40 below underlies one of the most important uses of convolutions. Before coming to it, we need a technical lemma:

Lemma 3.38 *Assume that $f \in L^p(\mathbf{R}^n)$ with $1 \leq p < \infty$. If we let*

$$f_x(y) = f(x+y), \quad y \in \mathbf{R}^n,$$

then we have the assertion

$$f_x \longrightarrow f \quad \text{in } L^p(\mathbf{R}^n) \text{ as } x \to 0.$$

Proof (1) If g is a continuous function with compact support, then we have the assertion

$$g_x \longrightarrow g \quad \text{uniformly as } x \to 0,$$

and so

$$\|g_x - g\|_p \longrightarrow 0 \quad \text{as } x \to 0.$$

(2) Let $f \in L^p(\mathbf{R}^n)$. For each $\varepsilon > 0$, we can find a continuous function g with compact support such that

$$\|f - g\|_p < \frac{\varepsilon}{2}.$$

Then we have the inequality

$$\|f_x - g_x\|_p = \|f - g\|_p < \frac{\varepsilon}{2},$$

and so

$$\|f_x - f\|_p \le \|f_x - g_x\|_p + \|g_x - g\|_p + \|g - f\|_p < \varepsilon + \|g_x - g\|_p.$$

This implies that

$$\limsup_{x \to 0} \|f_x - f\|_p \le \varepsilon.$$

The proof of Lemma 3.38 is complete. □

Remark 3.39 Lemma 3.38 is not true for $p = \infty$. Indeed, the assertion

$$\|f_x - f\|_\infty = \sup_{y \in \mathbf{R}^n} |f(x+y) - f(y)| \longrightarrow 0 \quad \text{as } x \to 0$$

implies the uniform continuity of f on $\mathbf{R}^n$.

Theorem 3.40 *Let $\varphi(x)$ be a function in $L^1(\mathbf{R}^n)$ such that*

$$\int_{\mathbf{R}^n} \varphi(x)\, dx = 1.$$

For any $\varepsilon > 0$, we define

$$\varphi_\varepsilon(x) = \frac{1}{\varepsilon^n} \varphi\left(\frac{x}{\varepsilon}\right).$$

Then we have the following two assertions:

(i) If $f \in L^p(\mathbf{R}^n)$ with $1 \le p < \infty$, then it follows that

$$f * \varphi_\varepsilon \longrightarrow f \quad \text{in } L^p(\mathbf{R}^n) \text{ as } \varepsilon \downarrow 0.$$

(ii) If $f \in L^\infty(\mathbf{R}^n)$ is uniformly continuous, then it follows that

$$f * \varphi_\varepsilon \longrightarrow f \quad \text{in } L^\infty(\mathbf{R}^n) \text{ as } \varepsilon \downarrow 0.$$

Proof (i) Since we have the formula

$$\int_{\mathbf{R}^n} \varphi_\varepsilon(x)\, dx = \int_{\mathbf{R}^n} \varphi(x)\, dx = 1,$$

it follows that

$$\begin{aligned}(f * \varphi_\varepsilon)(x) - f(x) &= \int_{\mathbf{R}^n} (f(x-y) - f(x))\, \varphi_\varepsilon(y)\, dy \\ &= \int_{\mathbf{R}^n} (f(x - \varepsilon z) - f(x))\, \varphi(z)\, dz.\end{aligned}$$

Hence, by applying Minkowski's inequality for integrals (Theorem 3.29) we obtain that

$$\|f * \varphi_\varepsilon - f\|_p \leq \int_{\mathbf{R}^n} \|f_{-\varepsilon z} - f\|_p \, |\varphi(z)| \, dz.$$

However, we have the inequality

$$\|f_{-\varepsilon z} - f\|_p \leq \|f_{-\varepsilon z}\|_p + \|f\|_p = 2\,\|f\|_p\,,$$

and so

$$\|f_{-\varepsilon z} - f\|_p \, |\varphi(z)| \leq 2\,\|f\|_p \, |\varphi(z)|,$$

with

$$\varphi \in L^1(\mathbf{R}^n).$$

Furthermore, Lemma 3.38 tells us that

$$\lim_{\varepsilon \downarrow 0} \|f_{-\varepsilon z} - f\|_p = 0 \quad \text{for } z \in \mathbf{R}^n.$$

Therefore, it follows from an application of the Lebesgue dominated convergence theorem (Theorem 3.13) that

$$\limsup_{\varepsilon \downarrow 0} \|f * \varphi_\varepsilon - f\|_p \leq \lim_{\varepsilon \downarrow 0} \int_{\mathbf{R}^n} \|f_{-\varepsilon z} - f\|_p \, |\varphi(z)| \, dz = 0.$$

(ii) Assume that $f \in L^\infty(\mathbf{R}^n)$ is uniformly continuous. Since $\varphi(x)$ is integrable on $\mathbf{R}^n$, it follows that, for each $\delta > 0$, there exists a compact subset W of $\mathbf{R}^n$ such that

$$\int_{\mathbf{R}^n \setminus W} |\varphi(z)| \, dz < \delta.$$

Then we have the inequality

$$\begin{aligned}
&\sup_{x \in \mathbf{R}^n} |f * \varphi_\varepsilon(x) - f(x)| \\
&\leq \sup_{x \in \mathbf{R}^n} \left(\int_{\mathbf{R}^n} |f(x - \varepsilon y) - f(x)| \, |\varphi(y)| \, dy \right) \\
&= \sup_{x \in \mathbf{R}^n} \left(\int_W |f(x - \varepsilon y) - f(x)| \, |\varphi(y)| \, dy \right. \\
&\quad \left. + \int_{\mathbf{R}^n \setminus W} |f(x - \varepsilon y) - f(x)| \, |\varphi(y)| \, dy \right) \\
&\leq \sup_{x \in \mathbf{R}^n} (f(x - \varepsilon y) - f(x)|) \cdot \int_W |\varphi(y)| \, dy + 2\,\|f\|_\infty \cdot \int_{\mathbf{R}^n \setminus W} |\varphi(y)| \, dy
\end{aligned}$$

$$\leq \sup_{x \in \mathbf{R}^n} (|f(x - \varepsilon y) - f(x)|) \, \|\varphi\|_1 + 2\delta \, \|f\|_\infty .$$

However, we have the assertion

$$\lim_{\varepsilon \downarrow 0} \left[\sup_{\substack{x \in \mathbf{R}^n \\ x \in W}} |f(x - \varepsilon y) - f(x)| \right] = 0,$$

since f is uniformly continuous and since W is compact.

Summing up, we obtain that

$$\limsup_{\varepsilon \downarrow 0} \|f * \varphi_\varepsilon - f\|_\infty \leq 2\delta \, \|f\|_\infty .$$

This proves part (ii), since δ is arbitrary.

The proof of Theorem 3.40 is complete. □

The family of functions $\{\varphi_\varepsilon\}$ defined above is called an *approximation to the identity*. What makes these useful is that, by choosing φ appropriately we can obtain the functions $f * \varphi_\varepsilon$ that have nice properties.

Example 3.2 (*the heat kernel*) Let

$$K(x) := \frac{1}{(4\pi)^{n/2}} \, e^{-\frac{|x|^2}{4t}}, \quad t > 0,$$

and

$$K_t(x) = \frac{1}{(\sqrt{t}\,)^n} \, K\left(\frac{x}{\sqrt{t}}\right) = \frac{1}{(4\pi t)^{n/2}} \, e^{-\frac{|x|^2}{4t}} \quad \text{for } t > 0.$$

We define

$$\begin{aligned} u(x, t) &= f * K_t(x) \\ &= \frac{1}{(4\pi t)^{n/2}} \int_{\mathbf{R}^n} e^{-\frac{|x-y|^2}{4t}} \, f(y) \, dy \quad \text{for } t > 0. \end{aligned}$$

Then we have the following two assertions:

(i) If $f \in L^p(\mathbf{R}^n)$ with $1 \leq p < \infty$, then it follows that

$$u(\cdot, t) \longrightarrow f \quad \text{in } L^p(\mathbf{R}^n) \text{ as } t \downarrow 0.$$

(ii) If f is bounded and continuous on $\mathbf{R}^n$, then it follows that $u(x, t)$ is continuous on $\mathbf{R}^n \times [0, \infty)$ and further that

$$u(\cdot, t) \longrightarrow f \quad \text{in } L^\infty(\mathbf{R}^n) \text{ as } t \downarrow 0.$$

Proof (i) Note that

$$\int_{\mathbf{R}^n} K(x)\, dx = \frac{1}{(4\pi)^{n/2}} \int_{\mathbf{R}^n} e^{-\frac{|x|^2}{4t}}\, dx = 1.$$

Therefore, part (i) follows by applying Theorem 3.40 with

$$\varphi(x) := K(x),$$
$$\varepsilon := \sqrt{t}.$$

(ii) Similarly, part (ii) follows by applying Theorem 3.40 with

$$\varphi(x) := K(x),$$
$$\varepsilon := \sqrt{t}.$$

The proof of Example 3.2 is complete. □

Remark 3.41 It should be emphasized that the function $u(x,t) = f * K_t(x)$ is a solution of the following initial value problem for the heat equation:

$$\begin{cases} \frac{\partial u}{\partial t}(x,t) - \Delta u(x,t) = 0 & \text{for all } x \in \mathbf{R}^n \text{ and } t > 0, \\ u(x,0) = f(x) & \text{for all } x \in \mathbf{R}^n. \end{cases}$$

3.7.2 *Friedrichs' Mollifiers*

Let $\rho(x)$ be a non-negative, bell-shaped C^∞ function on $\mathbf{R}^n$ satisfying the following two conditions:

$$\operatorname{supp} \rho = \{x \in \mathbf{R}^n : |x| \leq 1\}, \tag{3.20a}$$

$$\int_{\mathbf{R}^n} \rho(x)\, dx = 1. \tag{3.20b}$$

For example, we may take

$$\rho(x) = \begin{cases} k \exp[-1/(1-|x|^2)] & \text{if } |x| < 1, \\ 0 & \text{if } |x| \geq 1, \end{cases}$$

where the constant factor k is so chosen that condition (3.20b) is satisfied.

For each $\varepsilon > 0$, we define

$$\rho_\varepsilon(x) := \frac{1}{\varepsilon^n} \rho\left(\frac{x}{\varepsilon}\right),$$

then $\rho_\varepsilon(x)$ is a non-negative, C^∞ function on $\mathbf{R}^n$, and satisfies the conditions

$$\operatorname{supp} \rho_\varepsilon = \{x \in \mathbf{R}^n : |x| \le \varepsilon\}, \tag{3.21a}$$

$$\int_{\mathbf{R}^n} \rho_\varepsilon(x)\, dx = 1. \tag{3.21b}$$

The functions $\{\rho_\varepsilon\}$ are called *Friedrichs' mollifiers* (see Fig. 3.2).

The next local version of Theorem 3.40 shows how mollifiers can be used to approximate a function by smooth functions in a domain of $\mathbf{R}^n$:

Theorem 3.42 *Let Ω be an open subset of $\mathbf{R}^n$. Then we have the following two assertions:*

(i) If $u \in L^p(\Omega)$ with $1 \le p < \infty$ and vanishes outside a compact subset K of Ω, then it follows that $\rho_\varepsilon * u \in C_0^\infty(\Omega)$ provided that $\varepsilon <$ dist $(K, \partial\Omega)$, and further that $\rho_\varepsilon * u \to u$ in $L^p(\Omega)$ as $\varepsilon \downarrow 0$.
(ii) If $u \in C_0^m(\Omega)$ with $0 \le m < \infty$, then it follows that $\rho_\varepsilon * u \in C_0^\infty(\Omega)$ provided that $\varepsilon <$ dist $(\operatorname{supp} u, \partial\Omega)$, and further that $\rho_\varepsilon * u \to u$ in $C_0^m(\Omega)$ as $\varepsilon \downarrow 0$.

Here

$$\operatorname{dist}(K, \partial\Omega) = \inf\{|x - y| : x \in K,\ y \in \partial\Omega\}.$$

The functions $\rho_\varepsilon * u$ are called *regularizations* of the function u.

Corollary 3.43 *The space $C_0^\infty(\Omega)$ is dense in $L^p(\Omega)$ for each $1 \le p < \infty$.*

Proof Corollary 3.43 is an immediate consequence of part (i) of Theorem 3.42, since L^p functions with compact support are dense in $L^p(\Omega)$. □

The next result gives another useful construction of smooth functions that vanish outside compact sets:

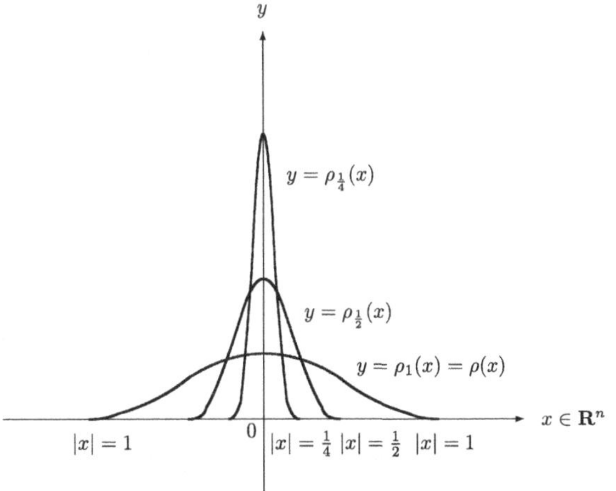

Fig. 3.2 Friedrichs' mollifiers $\{\rho_\varepsilon\}$

Corollary 3.44 *Let K be a compact subset of $\mathbf{R}^n$. If Ω is an open subset of $\mathbf{R}^n$ such that $K \subset \Omega$, then there exists a function $f \in C_0^\infty(\Omega)$ such that*

$$0 \le f(x) \le 1 \quad in\ \Omega,$$
$$f(x) = 1 \quad on\ K.$$

Proof Let

$$\delta = \text{dist}\,(K, \partial\Omega),$$

and define a relatively compact subset U of Ω, containing K, as follows:

$$U = \left\{x \in \Omega : |x - y| < \frac{\delta}{2} \text{ for some } y \in K\right\}.$$

Then it is easy to verify that the function

$$f(x) = \rho_\varepsilon * \chi_U(x) = \frac{1}{\varepsilon^n} \int_U \rho\left(\frac{x-y}{\varepsilon}\right) dy \quad \text{for } 0 < \varepsilon < \tfrac{\delta}{2}$$

satisfies all the conditions of Corollary 3.44. □

3.8 Distribution Functions

In this section we are interested in giving a concise expression for the L^p-norm of a function. Let $(\Omega, \mathcal{B}, \mu)$ be a σ-finite measure space. Let $f(x)$ be an $\mathcal{B}$-measurable function defined on Ω. If $\tau > 0$, we define the *distribution function* of f by the formula

$$\begin{aligned}\sigma(\tau) &:= \mu\left(\{x \in \Omega : |f(x)| > \tau\}\right)\\ &= \mu\left(f^{-1}((\tau, \infty))\right) + \mu\left(f^{-1}((-\infty, -\tau))\right) \quad \text{for } \tau \ge 0.\end{aligned}$$

We remark that the function $\sigma(\tau)$ is non-negative, non-increasing and right-continuous on the interval $[0, \infty)$. Indeed, it suffices to note that

$$\sigma(\tau + h) = \mu\left(f^{-1}((\tau + h, \infty))\right) + \mu\left(f^{-1}((-\infty, -\tau - h))\right) \quad \text{for } h > 0,$$

and further that

$$\mu\left(f^{-1}((\tau, \infty))\right) = \lim_{h \downarrow 0} \mu\left(f^{-1}((\tau + h, \infty))\right),$$
$$\mu\left(f^{-1}((-\infty, -\tau))\right) = \lim_{h \downarrow 0} \mu\left(f^{-1}((-\infty, -\tau - h))\right),$$

since we have, as $h \downarrow 0$,

$$f^{-1}((\tau + h, \infty)) \nearrow f^{-1}((\tau, \infty)),$$
$$f^{-1}((-\infty, -\tau - h)) \nearrow f^{-1}((-\infty, -\tau)).$$

The next theorem asserts that a measurable function is L^p-integrable over Ω with respect to μ if and only if its distribution function is integrable over the interval $[0, \infty)$ with respect to the *Lebesgue–Stieltjes measure* $d\tau^p = p\tau^{p-1}\,d\tau$:

Theorem 3.45 *Let $f(x)$ be a $\mathcal{B}$-measurable function defined on Ω, and let $1 \leq p < \infty$. Then we have the following assertions:*

(i) If $f(x)$ is a non-negative and $\mathcal{B}$-measurable function defined on Ω, then we have the formula

$$\int_\Omega f(x)^p\,d\mu(x) = \int_0^\infty \sigma(\tau)\,d\tau^p = p\int_0^\infty \tau^{p-1}\sigma(\tau)\,d\tau, \tag{3.22}$$

where

$$\sigma(\tau) = \mu(\{x \in \Omega : f(x) > \tau\}) \quad \textit{for } \tau \geq 0.$$

(ii) Let $f(x)$ be an $\mathcal{B}$-measurable function defined on Ω. Then $f \in L^p(\Omega)$ if and only if it satisfies the condition

$$\int_0^\infty \sigma(\tau)\,d\tau^p = p\int_0^\infty \tau^{p-1}\sigma(\tau)\,d\tau < \infty, \tag{3.23}$$

where

$$\sigma(\tau) = \mu(\{x \in \Omega : |f(x)| > \tau\}) \quad \textit{for } \tau \geq 0.$$

Moreover, in this case we have the formula

$$\int_\Omega |f(x)|^p\,d\mu = \int_0^\infty \sigma(\tau)\,d\tau^p = -\int_0^\infty \tau^p\,d\sigma(\tau). \tag{3.24}$$

Proof (i) If $f(x)$ is a non-negative and $\mathcal{B}$-measurable function defined on Ω, then we let

$$E_\tau = \{x \in \Omega : f(x) > \tau\} \quad \text{for } \tau \geq 0 \tag{3.25}$$

and

$$\sigma(\tau) = \mu(E_\tau) \quad \text{for } \tau \geq 0.$$

First, we consider a simple function of the form

$$f(x) = \sum_{j=1}^{N} c_j \chi_{A_j}(x),$$

where

$$A_j \in \mathcal{B} \quad \text{for } 1 \leq j \leq N,$$

and

$$0 < c_1 < c_2 < \ldots < c_{N-1} < c_N.$$

Then we have the formula

$$\int_\Omega f(x)^p \, d\mu(x) = \sum_{j=1}^{N} \int_{A_j} c_j^p \, d\mu(x) = \sum_{j=1}^{N} c_j^p \, \mu(A_j). \tag{3.26}$$

On the other hand, we have, by definition (3.25) of the set E_τ,

$$\begin{aligned} E_\tau &= \left\{ x \in \Omega : \sum_{j=1}^{N} c_j \chi_{A_j}(x) > \tau \right\} \\ &= \begin{cases} A_1 + \cdots + A_N & \text{for } 0 \leq \tau < c_1, \\ A_2 + \cdots + A_N & \text{for } c_1 \leq \tau < c_2, \\ \quad \vdots & \\ A_N & \text{for } c_{N-1} \leq \tau < c_N, \\ \emptyset & \text{for } \tau \geq c_N, \end{cases} \end{aligned}$$

and so

$$\sigma(\tau) = \mu(E_\tau) = \begin{cases} \mu(A_1) + \cdots + \mu(A_N) & \text{for } 0 \leq \tau < c_1, \\ \mu(A_2) + \cdots + \mu(A_N) & \text{for } c_1 \leq \tau < c_2, \\ \quad \vdots & \\ \mu(A_N) & \text{for } c_{N-1} \leq \tau < c_N, \\ 0 & \text{for } \tau \geq c_N. \end{cases} \tag{3.27}$$

Hence we obtain from formula (3.27)

$$\begin{aligned} &\int_0^\infty \sigma(\tau) \, d\tau^p \\ &= \int_0^{c_1} \sum_{j=1}^{N} \mu(A_j) \, dt^p + \int_{c_1}^{c_2} \sum_{j=2}^{N} \mu(A_j) \, dt^p + \cdots + \int_{c_{N-1}}^{c_N} \mu(A_N) \, dt^p \end{aligned} \tag{3.28}$$

$$= \sum_{j=1}^{N} \mu(A_j)\, c_1^p + \sum_{j=2}^{N} \mu(A_j)\left(c_2^p - c_1^p\right) + \cdots + \mu(A_N)\left(c_N^p - c_{N-1}^p\right)$$
$$= \sum_{j=1}^{N} c_j^p\, \mu(A_j).$$

Therefore, the desired formula (3.22) for simple functions is an immediate consequence of formulas (3.26) and (3.28):

$$\int_{\Omega} f(x)^p\, d\mu(x) = \sum_{j=1}^{N} c_j^p\, \mu(A_j) = \int_0^\infty \sigma(\tau)\, d\tau^p.$$

The general case of formula (3.22) follows by approximating non-negative, $\mathcal{B}$-measurable functions $f(x)$ from below by simple functions (see [51, Theorem 2.10]).

(ii) The proof of part (ii) is divided into two steps.

Step (I): If condition (3.23) holds true, then we have the following claim:

Claim 3.46 *$\tau^p\, \sigma(\tau) \to 0$ as $\tau \to 0$ and $\tau \to \infty$.*

Proof (a) First, we show that

$$\tau^p\, \sigma(\tau) \longrightarrow 0 \quad \text{as } \tau \to 0.$$

Our proof is based on a reduction to absurdity. Assume, to the contrary, that we have, as $\tau \to 0$,

$$\tau^p\, \sigma(\tau) \nrightarrow 0.$$

Then we can find a sequence $\{\tau_j\}$ and a positive constant δ such that

$$\tau_j^p\, \sigma(\tau_j) \geq \delta \quad \text{and} \quad 0 < \tau_{j+1} < \frac{1}{2}\tau_j.$$

Hence it follows that

$$p \int_{\tau_{j+1}}^{\tau_j} \tau^{p-1} \sigma(\tau)\, d\tau \geq p\, \sigma(\tau_j) \int_{\tau_{j+1}}^{\tau_j} \tau^{p-1}\, d\tau = \sigma(\tau_j)\left(\tau_j^p - \tau_{j+1}^p\right)$$
$$\geq \frac{\delta}{\tau_j^p}\left(\tau_j^p - \tau_{j+1}^p\right) = \delta\left(1 - \left(\frac{\tau_{j+1}}{\tau_j}\right)^p\right)$$
$$\geq \delta\left(1 - \frac{1}{2^p}\right) \quad \text{for } 1 \leq p < \infty.$$

This contradict condition (3.23), since we have the assertion

$$p \int_0^\infty \tau^{p-1} \sigma(\tau)\, d\tau = \infty \quad \text{for } 1 \leq p < \infty.$$

(b) Secondly, we show that

$$\tau^p \, \sigma(\tau) \longrightarrow 0 \quad \text{as } \tau \to \infty.$$

Our proof is based on a reduction to absurdity. Assume, to the contrary, that we have, as $\tau \to \infty$,

$$\tau^p \, \sigma(\tau) \nrightarrow 0.$$

Then we can find a sequence $\{\tau_j\}$ and a positive constant δ such that

$$\tau_{j+1}^p \, \sigma(\tau_{j+1}) \geq \delta \quad \text{and} \quad 0 < \tau_j < \frac{1}{2} \tau_{j+1}.$$

Hence it follows that

$$\begin{aligned} p \int_{\tau_j}^{\tau_{j+1}} \tau^{p-1} \sigma(\tau)\, d\tau &\geq p\, \sigma(\tau_{j+1}) \int_{\tau_j}^{\tau_{j+1}} \tau^{p-1}\, d\tau = \sigma(\tau_{j+1}) \left(\tau_{j+1}^p - \tau_j^p \right) \\ &\geq \frac{\delta}{\tau_{j+1}^p} \left(\tau_{j+1}^p - \tau_j^p \right) = \delta \left(1 - \left(\frac{\tau_j}{\tau_{j+1}} \right)^p \right) \\ &\geq \delta \left(1 - \frac{1}{2^p} \right) \quad \text{for } 1 \leq p < \infty. \end{aligned}$$

This contradict condition (3.23), since we have the assertion

$$p \int_0^\infty \tau^{p-1} \sigma(\tau)\, d\tau = \infty \quad \text{for } 1 \leq p < \infty.$$

The proof of Claim 3.46 is complete. □

In view of Claim 3.46, we can apply an integration by parts formula (see [51, Theorem 3.36]) with

$$\begin{aligned} f(\tau) &:= \tau^p, \\ g(\tau) &:= \sigma(\tau), \end{aligned}$$

we obtain from condition (3.23) that

$$\begin{aligned} \int_\Omega |f(x)|^p \, d\mu &= \int_0^\infty \sigma(\tau)\, d\tau^p = \lim_{\substack{c \to \infty \\ \varepsilon \downarrow 0}} \int_\varepsilon^c \sigma(\tau)\, d\tau^p \qquad (3.29) \\ &= \lim_{\substack{c \to \infty \\ \varepsilon \downarrow 0}} \left\{ \left[\tau^p \, \sigma(\tau) \right]_\varepsilon^c - \int_\varepsilon^c \tau^p \, d\sigma(\tau) \right\} \end{aligned}$$

$$= -\int_0^\infty \tau^p \, d\sigma(\tau).$$

This proves the desired formula (3.24).

Step (II): Conversely, we assume that $f \in L^p(\Omega)$. Then we have, for all $\tau > 0$,

$$\tau^p \, \sigma(\tau) \leq \int_{E(\tau)} |f(x)|^p \, d\mu, \tag{3.30}$$

where

$$E(\tau) := \{x \in \Omega : |f(x)| > \tau\}.$$

However, since $f \in L^p(\Omega)$ it follows that

$$\begin{aligned}\int_\Omega |f(x)|^p \, d\mu &= \lim_{\tau \to \infty} \int_{\Omega \setminus E(\tau)} |f(x)|^p \, d\mu \\ &= \int_\Omega |f(x)|^p \, d\mu - \lim_{\tau \to \infty} \int_{E(\tau)} |f(x)|^p \, d\mu,\end{aligned}$$

so that

$$\lim_{\tau \to \infty} \int_{E(\tau)} |f(x)|^p \, d\mu = 0.$$

Hence we have, by inequality (3.23),

$$\lim_{\tau \to \infty} \tau^p \, \sigma(\tau) \leq \lim_{\tau \to \infty} \int_{E(\tau)} |f(x)|^p \, d\mu = 0.$$

Therefore, by using an integration by parts formula (see [51, Theorem 3.36]) with

$$\begin{aligned} f(\tau) &:= \tau^p, \\ g(\tau) &:= \sigma(\tau), \end{aligned}$$

we obtain from formula (3.29) that

$$\begin{aligned}\int_\Omega |f(x)|^p \, d\mu &= -\int_0^\infty \tau^p \, d\sigma(\tau) \\ &= \lim_{\substack{c \to \infty \\ \varepsilon \downarrow 0}} \left\{ -\left[\tau^p \, \sigma(\tau)\right]_\varepsilon^c + \int_\varepsilon^c \sigma(\tau) \, d\tau^p \right\} \\ &= \lim_{\substack{c \to \infty \\ \varepsilon \downarrow 0}} \left\{ \varepsilon^p \, \sigma(\varepsilon) + \int_\varepsilon^c p \, \tau^{p-1} \, \sigma(\tau) \, d\tau \right\} \\ &\geq p \int_0^\infty \tau^{p-1} \, \sigma(\tau) \, d\tau.\end{aligned}$$

This proves the desired condition (3.23).

The proof of Theorem 3.45 is now complete. □

3.9 Marcinkiewicz's Interpolation Theorem

In this section we prove the Marcinkiewicz interpolation theorem (Theorem 3.47) for which we need some terminology. Let $(\Omega, \mathcal{B}, \phi)$ and $(\widehat{\Omega}, \widehat{\mathcal{B}}, \widehat{\phi})$ be two measure spaces, and the norms of $L^p(\Omega)$ and $L^p(\widehat{\Omega})$ will be denoted by the same notation $\|\cdot\|_p$, but it is clear from the context which is which.

(i) Let T be a not necessarily linear mapping of $L^p(\Omega)$ into $L^q(\widehat{\Omega})$ with $1 \leq p \leq \infty$ and $1 \leq q \leq \infty$. The mapping T is said to be *of type* (p, q) if there exists a positive constant K such that we have, for all $f \in L^p(\Omega)$,

$$\|Tf\|_q \leq K \, \|f\|_p.$$

(ii) Let T be a not necessarily linear mapping of $L^p(\Omega)$ into a set of measurable functions defined on $\widehat{\Omega}$ with $1 \leq p \leq \infty$. In the case where $q < \infty$, the mapping T is said to be *of weak type* (p, q) if there exists a positive constant K such that we have, for all $f \in L^p(\Omega)$ and all $s > 0$,

$$\widehat{\phi}\left(\left\{y \in \widehat{\Omega} : |(Tf)(y)| > s\right\}\right) \leq \left(\frac{K \, \|f\|_p}{s}\right)^q.$$

In the case where $q = \infty$ a mapping of type (p, ∞) is said to be *of weak type* (p, ∞).

It is easy to see that a mapping of type (p, q) is of weak type (p, q).

We denote the totality of functions which are expressed as a sum of a function in $L^p(\Omega)$ and a function in $L^q(\Omega)$ by $L^p(\Omega) + L^q(\Omega)$, that is:

$$L^p(\Omega) + L^q(\Omega) := \left\{f + g : f \in L^p(\Omega), \; g \in L^q(\Omega)\right\}.$$

Let $1 \leq p < q < r < \infty$. If $f \in L^q(\Omega)$, we let

$$f_1(x) := \begin{cases} 0 & \text{if } |f(x)| \leq 1, \\ f(x) & \text{if } |f(x)| > 1, \end{cases}$$

and

$$f_2(x) := \begin{cases} f(x) & \text{if } |f(x)| \leq 1, \\ 0 & \text{if } |f(x)| > 1. \end{cases}$$

Then it follows that $f(x)$ can be written in the form

$$f(x) = f_1(x) + f_2(x),$$

with

$$f_1 \in L^p(\Omega), \qquad f_2 \in L^r(\Omega),$$

Indeed, it suffices to note that

$$\begin{aligned} |f_1(x)|^p &\leq |f(x)|^q, \\ |f_2(x)|^r &\leq |f(x)|^q. \end{aligned}$$

Therefore, for $1 \leq p < q < r < \infty$, we have the inclusion

$$L^q(\Omega) \subset L^p(\Omega) + L^r(\Omega).$$

Marcinkiewicz's interpolation theorem reads as follows:

Theorem 3.47 (Marcinkiewicz) *Let $1 \leq p < q < r$ and let T be a linear mapping of the space $L^p(\Omega) + L^r(\Omega)$ into a set of measurable functions defined on $\widehat{\Omega}$. If T is of weak type (p, p) and also of weak type (r, r), then it is of type (q, q).*

Proof If $f(x)$ is a measurable function defined on Ω and if $\tau > 0$, we let

$$(f)_r := \phi(\{x \in \Omega : |f(x)| > \tau\}).$$

We shall use the same notation for a measurable function defined on $\widehat{\Omega}$ with ϕ replaced by $\widehat{\phi}$.

First, since T is of weak type (p, p) and of weak type (r, r), we can find two constants $K_1 > 0$ and $K_2 > 0$ such that we have, for any $\tau > 0$ and for any $f \in L^p(\Omega)$ or for any $f \in L^r(\Omega)$,

$$(Tf)_\tau \leq \left(\frac{K_1 \, \|f\|_p}{\tau}\right)^p \quad \text{or} \quad (Tf)_\tau \leq \left(\frac{K_2 \, \|f\|_r}{\tau}\right)^r, \tag{3.31}$$

respectively. If $f(x)$ is a function in $L^q(\Omega)$, we let

$$f_1(x) := \begin{cases} f(x) & \text{if } |f(x)| \leq \tau, \\ \tau \operatorname{sign} f(x) & \text{if } |f(x)| > \tau, \end{cases}$$

and

$$f_2(x) := f(x) - f_1(x).$$

Then, since $Tf(x) = Tf_1(x) + Tf_2(x)$, we have the inclusion

$$\{y : |(Tf)(y)| > \tau\} \subset \{y : |(Tf_1)(y)| > \tau/2\} \cup \{y : |(Tf_2)(y)| > \tau/2\}.$$

Hence, by condition (3.31) with $\tau := \tau/2$ it follows that

$$\begin{aligned}(Tf)_\tau &\le (Tf_1)_{\tau/2} + (Tf_2)_{\tau/2} \\ &\le \left(\frac{2K_1 \, \|f_1\|_r}{\tau}\right)^r + \left(\frac{2K_2 \, \|f_2\|_p}{\tau}\right)^p.\end{aligned} \tag{3.32}$$

However, we remark that

$$(f_1)_\sigma = \begin{cases} (f)_\sigma & \text{for } \sigma < \tau, \\ 0 & \text{for } \sigma \ge \tau, \end{cases}$$

and that

$$(f_2)_\sigma = (f)_{\sigma+\tau} \quad \text{for } \sigma \ge 0.$$

Thus, it follows from an application of Theorem 3.45 that

$$\|f_1\|_r^r = r \int_0^\infty \sigma^{r-1} (f_1)_\sigma \, d\sigma = r \int_0^\tau \sigma^{r-1} (f)_\sigma \, d\sigma, \tag{3.33}$$

and further that

$$\begin{aligned}\|f_2\|_p^p &= p \int_0^\infty \sigma^{p-1} (f_2)_\sigma \, d\sigma \\ &= p \int_0^\infty \sigma^{p-1} (f)_{\sigma+\tau} \, d\sigma = p \int_\tau^\infty (\sigma - \tau)^{p-1} (f)_\sigma \, d\sigma.\end{aligned} \tag{3.34}$$

By combining inequality (3.32) and two formulas (3.33) and (3.34), we obtain that

$$\begin{aligned}(Tf)_\tau \le \left(\frac{2K_1}{\tau}\right)^r r \int_0^\tau \sigma^{r-1} (f)_\sigma \, d\sigma \\ + \left(\frac{2K_2}{\tau}\right)^p p \int_\tau^\infty (\sigma - \tau)^{p-1} (f)_\sigma \, d\sigma.\end{aligned}$$

Therefore, we have, by this inequality and Theorem 3.45,

$$\begin{aligned}&\|Tf\|_q^q \\ &= q \int_0^\infty \tau^{q-1} (Tf)_\tau \, d\tau \\ &\le q \int_0^\infty \tau^{q-1} \left(\frac{2K_1}{\tau}\right)^r r \left(\int_0^\tau \sigma^{r-1} (f)_\sigma \, d\sigma\right) d\tau \\ &\quad + q \int_0^\infty \tau^{q-1} \left(\frac{2K_2}{\tau}\right)^p p \left(\int_\tau^\infty (\sigma - \tau)^{p-1} (f)_\sigma \, d\sigma\right) d\tau\end{aligned}$$

$$
\begin{aligned}
&= qr(2K_1)^r \int_0^\infty \tau^{q-r-1} \left(\int_0^\tau \sigma^{r-1} (f)_\sigma \, d\sigma \right) d\tau \\
&\quad + pq(2K_2)^p \int_0^\infty \tau^{q-p-1} \left(\int_\tau^\infty (\sigma - \tau)^{p-1} (f)_\sigma \, d\sigma \right) d\tau \\
&= qr(2K_1)^r \int_0^\infty \left(\int_\sigma^\infty \tau^{q-r-1} \, d\tau \right) \sigma^{r-1} (f)_\sigma \, d\sigma \\
&\quad + pq(2K_2)^p \int_0^\infty \left(\int_0^\sigma \tau^{q-p-1} (\sigma - \tau)^{p-1} \, d\tau \right) (f)_\sigma \, d\sigma \\
&= \frac{qr(2K_1)^r}{r-q} \int_0^\infty \sigma^{q-1} (f)_\sigma \, d\sigma + pq(2K_2)^p B(q-p, p) \int_0^\infty \sigma^{q-1} (f)_\sigma \, d\sigma \\
&= \left\{ \frac{r(2K_1)^r}{r-q} + p(2K_2)^p B(q-p, p) \right\} \| f \|_q^q .
\end{aligned}
$$

This proves that the operator T is of type (q, q), with

$$
K := \left(\frac{r(2K_1)^r}{r-q} + p(2K_2)^p B(q-p, p) \right)^{1/q} .
$$

The proof of Theorem 3.47 is complete. □

3.10 Riesz Potentials

In this section, as an application of Marcinkiewicz's interpolation theorem (Theorem 3.47) we study Riesz potentials in the classical potential theory.

Let Δ be the usual Laplacian

$$
\Delta = \sum_{j=i}^{n} \frac{\partial^2}{\partial x_j^2}.
$$

Then we have, for every $f \in \mathcal{S}(\mathbf{R}^n)$,

$$
\widehat{(-\Delta f)}(\xi) = |\xi|^2 \, \widehat{f}(\xi),
$$

and also

$$
\mathcal{F}((-\Delta)^{\alpha/2} f)(\xi) = |\xi|^\alpha \, \widehat{f}(\xi) \quad \text{for } \alpha > 0.
$$

If $0 < \alpha < n$, we define a *Riesz potential* I_α by the formula

$$
I_\alpha(f)(x) = (-\Delta)^{-\alpha/2}(f)(x) = \frac{1}{\gamma(\alpha)} \int_{\mathbf{R}^n} |x - y|^{\alpha - n} f(y) \, dy,
$$

where

$$\gamma(\alpha) = \frac{\pi^{n/2}\, 2^{\alpha}\, \Gamma(\alpha/2)}{\Gamma((n-\alpha)/2)}.$$

The function

$$\frac{1}{\gamma(\alpha)}\, |x|^{\alpha-n}$$

is called a *Riesz potential kernel.*

The purpose of this section is to prove the following theorem:

Theorem 3.48 *Let $0 < \alpha < n$, $1 < p < q$ and $1/q = 1/p - \alpha/n$. Then we have, for all $f \in L^p(\Omega)$,*

$$\|I_\alpha(f)\|_q \le A_{p,q}\, \|f\|_p, \tag{3.35}$$

with some constant $A_{p,q} > 0$.

Proof We have only to prove the theorem for the integral kernel

$$K(x) = |x|^{\alpha-n}.$$

We shall show that the mapping

$$f \longmapsto K * f$$

is of weak type (p, q). Namely, we have, for all $\lambda > 0$,

$$m\,\{x : |K * f(x)| > \lambda\} \le \left(A_{p,q}\, \frac{\|f\|_p}{\lambda}\right)^q, \tag{3.36}$$

where m is the Lebesgue measure on $\mathbf{R}^n$.

To do so, it suffices to prove that

$$m\,\{x : |K * g(x)| > \lambda\} \le \left(\frac{A_{p,q}}{\lambda}\right)^q \quad \text{for all } \|g\|_p = 1. \tag{3.37}$$

Indeed, by letting

$$g(x) = \frac{f(x)}{\|f\|_p},$$

we obtain from inequality (3.37) that

$$\begin{aligned} m\left\{x : |K * f(x)| > \lambda\, \|f\|_p\right\} &= m\,\{x : |K * g(x)| > \lambda\} \\ &\le \left(\frac{A_{p,q}}{\lambda}\right)^q. \end{aligned}$$

This proves the desired inequality (3.36) if we take

$$\lambda := \frac{\lambda}{\|f\|_p}.$$

In order to prove inequality (3.37), we let, for a positive constant μ,

$$K(x) := K_1(x) + K_\infty(x),$$

where

$$K_1(x) = \begin{cases} K(x) & \text{for } |x| \le \mu, \\ 0 & \text{for } |x| > \mu, \end{cases}$$

and

$$K_2(x) = \begin{cases} 0 & \text{for } |x| \le \mu, \\ K(x) & \text{for } |x| > \mu. \end{cases}$$

The constant μ will be chosen later on.

Since we have the formula

$$K * g = K_1 * g + K_\infty * g,$$

it follows that

$$\begin{aligned} & m\{x : |K * g(x)| > 2\lambda\} \\ & \le m\{x : |K_1 * g(x)| > \lambda\} + m\{x : |K_\infty * g(x)| > \lambda\}. \end{aligned} \tag{3.38}$$

(1) First, we have the inequality

$$\begin{aligned} & m\{x : |K_1 * g(x)| > \lambda\} \\ & \le \int_{\mathbf{R}^n} \left(\frac{|K_1 * g(x)|}{\lambda} \right)^p dx \\ & = \frac{\| |K_1 * g(x)| \|_p^p}{\lambda^p} \\ & \le \frac{1}{\lambda^p} \|K_1\|_1^p \|g\|_p^p = \frac{1}{\lambda^p} \|K_1\|_1^p. \end{aligned} \tag{3.39}$$

However, it follows that

$$\begin{aligned} \|K_1\|_1 &= \int_{|x| \le \mu} |x|^{\alpha - n} \, dx = \int_{\Sigma_{n-1}} \int_0^\mu r^{\alpha - n} r^{n-1} \, dr \, d\sigma \\ &= \frac{\omega_n}{\alpha} \mu^\alpha, \end{aligned} \tag{3.40}$$

where ω_n is the surface area of the unit sphere Σ_{n-1} in $\mathbf{R}^n$:

$$\omega_n := |\Sigma_{n-1}| = \frac{2\pi^{n/2}}{\Gamma(n/2)}.$$

Therefore, by combining inequality (3.39) and formula (3.40) we obtain that

$$m\{x : |K_1 * g(x)| > \lambda\} \leq C_1^p \frac{\mu^{\alpha p}}{\lambda^p} \tag{3.41}$$

with

$$C_1 := \frac{\omega_n}{\alpha}.$$

(2) Secondly, by using Young's inequality (Theorem 3.36) we have the inequality

$$\|K_\infty * g\|_\infty \leq \|K_\infty\|_{p'} \|g\|_p = \|K_\infty\|_{p'}, \quad p' = \frac{p}{p-1}. \tag{3.42}$$

However, it follows that

$$\begin{aligned}
\|K_\infty\|_{p'} &= \left(\int_{|x|\geq\mu} \left(|x|^{\alpha-n}\right)^{p'} dx\right)^{1/p'} && (3.43)\\
&= \left(\int_{\Sigma_{n-1}} \int_\mu^\infty r^{(\alpha-n)p'} r^{n-1}\, dr\, d\sigma\right)^{1/p'}\\
&= \left(\omega_n \left[\frac{1}{(\alpha-n)p'+n} r^{(\alpha-n)p'+n}\right]_\mu^\infty\right)^{1/p'}\\
&= \omega_n^{1/p'} \left(\frac{\mu^{n-(n-\alpha)p'}}{(n-\alpha)p'-n}\right)^{1/p'} = C_2\, \mu^{(\alpha-n)+n(1-1/p)}\\
&= C_2\, \mu^{-n/q}
\end{aligned}$$

with

$$C_2 := \left(\omega_n \frac{p-1}{p(n-\alpha)}\right)^{1/p'}.$$

If we choose the constant μ as

$$\mu := \left(\frac{C_2}{\lambda}\right)^{q/n},$$

or equivalently

$$\lambda = C_2\, \mu^{-n/q},$$

then it follows from inequalities (3.42) and (3.43) that

$$\|K_\infty * g\|_\infty \leq \|K_\infty\|_{p'} = \lambda.$$

This proves that

$$m\{x : |K_\infty * g(x)| > \lambda\} = 0. \tag{3.44}$$

(3) By combining inequalities (3.38), (3.41) and assertion (3.44), we have proved that

$$m\{x : |K * g(x)| > 2\lambda\} \leq C_1^p \left(\frac{1}{\lambda}\right)^q,$$

so that (by replacing 2λ by λ)

$$m\{x : |K * g(x)| > \lambda\} \leq \left(A_{p,q}\lambda\right)^q,$$

with

$$A_{p,q} := 2\, C_1^{p/q}.$$

Therefore, the desired inequality (3.37) (and hence (3.35)) follows from an application of Marcinkiewicz's Theorem 3.47. □

We can prove a more precise estimate for Riesz potentials with $\alpha := n\mu$ (see [97, Chap. 1, Lemma 1.34], [59, Chap. 7, Lemma 7.12]):

Theorem 3.49 *If $0 < \mu \leq 1$, we define the Riesz potential by the formula*

$$\left(I_{n\mu} f\right)(x) := \int_\Omega |x - y|^{n(\mu-1)}\, f(y)\, dy.$$

Then, for any p, q satisfying the conditions

$$1 \leq p \leq q \leq \infty,$$
$$0 \leq \delta = \delta(p, q) := \frac{1}{p} - \frac{1}{q} < \mu,$$

the operator $I_{n\mu}$ maps $L^p(\Omega)$ continuously into $L^q(\Omega)$. More precisely, we have, for all $f \in L^p(\Omega)$,

$$\left\|I_{n\mu} f\right\|_{L^q(\Omega)} \leq \left(\frac{1-\delta}{\mu-\delta}\right)^{1-\delta} V_n^{1-\mu}\, |\Omega|^{\mu-\delta}\, \|f\|_{L^p(\Omega)}.$$

Here V_n is the volume of the unit ball in $\mathbf{R}^n$

$$V_n = \frac{\omega_n}{n} = \frac{\pi^{n/2}}{\Gamma(n/2+1)}.$$

Chapter 4
Elements of Real Analysis

This chapter is devoted to the precise definitions and statements of real analytic tools such as bounded mean oscillation (BMO) and vanishing mean oscillation (VMO) functions, the Calderón–Zygmund decomposition (Theorem 4.9), the John–Nirenberg inequality (Theorem 4.13), the Hardy–Littlewood maximal function (Theorem 4.6), sharp functions (Theorem 4.19) and spherical harmonics (Theorem 4.41).

For more thorough treatments of this subject, the reader might refer to Duoandikoetxea [41], Garcia-Cuerva–Rubio de Francia [56], Garnett [57], Malý–Ziemer [97], Neri [108], Stein [137, 139] and Torchinsky [168].

4.1 BMO Functions

In this section we recall some basic definitions and results concerning BMO functions from real analysis.

(I) First, we let

$$L^1_{\mathrm{loc}}(\mathbf{R}^n) = \text{the space of equivalence classes of Lebesgue measurable functions on } \mathbf{R}^n \text{ which are integrable on every compact subset of } \mathbf{R}^n.$$

The elements of $L^1_{\mathrm{loc}}(\mathbf{R}^n)$ are called *locally integrable functions* on $\mathbf{R}^n$.

(II) A function $f \in L^1_{\mathrm{loc}}(\mathbf{R}^n)$ is said to be of *bounded mean oscillation*, $f \in \mathrm{BMO}$, if it satisfies the condition (see [73])

$$\|f\|_* := \sup_B \frac{1}{|B|} \int_B |f(x) - f_B| \, dx < \infty,$$

K. Taira, *Real Analysis Methods for Markov Processes*,
https://doi.org/10.1007/978-981-97-3659-1_4

where the supremum is taken over all balls B in $\mathbf{R}^n$, $|B|$ is the Lebesgue measure of B and f_B is the integral average of f over B

$$f_B := \frac{1}{|B|}\int_B f(y)\,dy.$$

The quantity $\|\cdot\|_*$ is called the BMO *norm*. This is not properly a norm, since any function which is constant almost everywhere has zero oscillation. However, it is easy to see that these are the only functions having zero oscillation. Therefore, we view the class BMO as the quotient space of the above space by the space of constant functions. In other words, two functions which differ by a constant coincide as functions in the class BMO. Then it should be emphasized (see [109]) that the class BMO is a Banach space equipped with the BMO norm $\|\cdot\|_*$. We remark that the BMO norm is scale invariant.

It is not important that we subtract exactly f_B in the definition of BMO. More precisely, we can obtain the following claim:

Claim 4.1 *Let $f \in L^1_{\mathrm{loc}}(\mathbf{R}^n)$. Assume that, for each ball B, there exists a constant α_B such that we have the inequality*

$$\frac{1}{|B|}\int_B |f(x) - \alpha_B|\,dx \le C,$$

with some positive constant C independent of B. Then it follows that $f \in$ BMO with $\|f\|_* \le 2C$.

Proof Indeed, we have, for all balls B,

$$|f_B - \alpha_B| = \frac{1}{|B|}\left|\int_B (f(y) - f_B)\,dy\right| \le \frac{1}{|B|}\int_B |f(x) - \alpha_B|\,dx \le C.$$

Hence it follows that

$$\begin{aligned}\frac{1}{|B|}\int_B |f(x) - f_B|\,dx &\le \frac{1}{|B|}\int_B |f(x) - \alpha_B|\,dx + \frac{1}{|B|}\int_B |\alpha_B - f_B|\,dx \\ &\le 2C.\end{aligned}$$

This proves that $f \in$ BMO with $\|f\|_* \le 2C$.

The proof of Claim 4.1 is complete. □

(III) Sometimes we will define the BMO norm with cubes in place of balls. If Q is a cube with sides parallel to the coordinate axes, then we can define an equivalent BMO norm $\|\cdot\|_{*'}$ by the formula

$$\|f\|_{*'} := \sup_Q \frac{1}{|Q|}\int_Q |f(x) - f_Q|\,dx,$$

where the supremum is taken over all cubes Q in $\mathbf{R}^n$, and f_Q is the integral average of f over Q

$$f_Q := \frac{1}{|Q|} \int_Q f(y)\, dy.$$

In fact, we can obtain the following lemma:

Lemma 4.2 *For any function $f \in L^1_{\mathrm{loc}}(\mathbf{R}^n)$, we have the inequalities*

$$\|f\|_{*'} \le \frac{n^{n/2} V_n}{2^{n-1}} \|f\|_*, \tag{4.1}$$

$$\|f\|_* \le \frac{2^{n+1}}{V_n} \|f\|_{*'}, \tag{4.2}$$

where

$$V_n := \frac{\omega_n}{n} = \frac{\pi^{n/2}}{\Gamma(n/2+1)}$$

is the volume of the unit ball in $\mathbf{R}^n$.

Proof (1) Let $Q = Q(x, r)$ be a cube with side length $2r$ and with center x, and let $B = B(x, \sqrt{n}\, r)$ be a ball of radius $\sqrt{n}\, r$ about x (see Fig. 4.1). Then it follows that

$$\begin{aligned} &Q(x, r) \subset B(x, \sqrt{n}\, r), \\ &|Q(x, r)| = (2r)^n, \quad |B(x, \sqrt{n}\, r)| = n^{n/2} r^n V_n. \end{aligned}$$

Hence, we have the inequality

$$\begin{aligned} &|f_Q - f_B| \\ &= \left| \frac{1}{|Q|} \int_Q (f(x) - f_B)\, dx \right| \le \frac{1}{|Q|} \int_Q |f(x) - f_B|\, dx \\ &\le \left(\frac{|B|}{|Q|} \right) \left(\frac{1}{|B|} \int_B |f(x) - f_B|\, dx \right) = \frac{n^{n/2} V_n}{2^n} \left(\frac{1}{|B|} \int_B |f(x) f_B|\, d \right) \\ &\le \frac{n^{n/2} V_n}{2^n} \|f\|_*. \end{aligned}$$

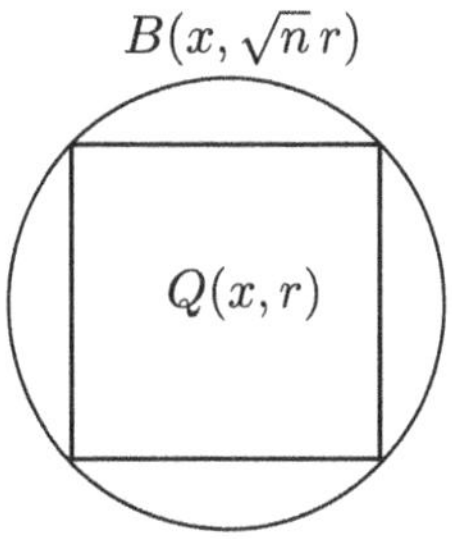

Fig. 4.1 The cube $Q(x, r)$ and the ball $B(x, \sqrt{n}\, r)$

Therefore, we obtain that

$$\begin{aligned}
&\frac{1}{|Q|}\int_Q |f(x)-f_Q|\,dx \\
&\le \frac{1}{|Q|}\int_Q |f(x)-f_B|\,dx + \frac{1}{|Q|}\int_Q |f_B-f_Q|\,dx \\
&\le \left(\frac{|B|}{|Q|}\right)\frac{1}{|B|}\int_B |f(x)-f_B|\,dx + |f_Q-f_B| \\
&\le \frac{n^{n/2}V_n}{2^n}\left(\frac{1}{|B|}\int_B |f(x)-f_B|\,dx\right) + \frac{n^{n/2}V_n}{2^n}\,\|f\|_* \\
&\le \frac{n^{n/2}V_n}{2^{n-1}}\,\|f\|_* .
\end{aligned}$$

This proves the desired inequality (4.1).

(2) Similarly, if $Q = Q(x,r)$ is a cube with side length $2r$ and with center x and if $B = B(x,r)$ is a ball of radius r about x (see Fig. 4.2), then it follows that

$$\begin{aligned}
&B(x,r) \subset Q(x,r), \\
&|B(x,r)| = r^n V_n, \quad |Q(x,r)| = (2r)^n.
\end{aligned}$$

Hence, we have the inequality

$$\begin{aligned}
&|f_B - f_Q| \\
&= \left|\frac{1}{|B|}\int_B \left(f(x)-f_Q\right)dx\right| \le \frac{1}{|B|}\int_B |f(x)-f_Q|\,dx \\
&\le \left(\frac{|Q|}{|B|}\right)\left(\frac{1}{|Q|}\int_Q |f(x)-f_Q|\,dx\right) = \frac{2^n}{V_n}\left(\frac{1}{|Q|}\int_Q |f(x)-f_Q|\,dx\right) \\
&\le \frac{2^n}{V_n}\,\|f\|_{*'} .
\end{aligned}$$

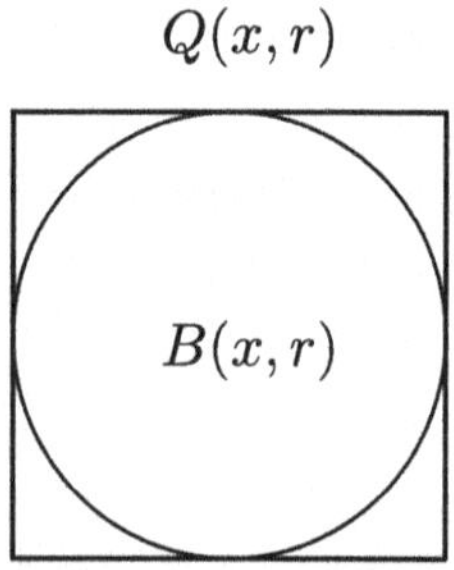

Fig. 4.2 The cube $Q(x,r)$ and the ball $B(x,r)$

Therefore, we obtain that

$$\begin{aligned}
&\frac{1}{|B|}\int_B |f(x)-f_B|\,dx \\
&\le \frac{1}{|B|}\int_B \big|f(x)-f_Q\big|\,dx + \frac{1}{|B|}\int_B \big|f_Q-f_B\big|\,dx \\
&\le \left(\frac{|Q|}{|B|}\right)\frac{1}{|Q|}\int_Q \big|f(x)-f_Q\big|\,dx + |f_B-f_Q| \\
&\le \frac{2^n}{V_n}\left(\frac{1}{|Q|}\int_Q \big|f(x)-f_Q\big|\,dx\right) + \frac{2^n}{V_n}\,\|f\|_{*'} \\
&\le \frac{2^{n+1}}{V_n}\,\|f\|_{*'}\,.
\end{aligned}$$

This proves the desired inequality (4.2).

The proof of Lemma 4.2 is complete. □

If Q is a cube with sides parallel to the coordinate axes, then we denote by δ_Q its side length and by x_Q its center, respectively. For each $\lambda > 0$, we denote by λQ the cube centered at x_Q with side length $\lambda\delta_Q$ (see Fig. 4.3).

Lemma 4.3 *Let $f \in$ BMO. For any positive integer j, there exists a constant $c(n) > 0$ such that*

$$|f_{2^j Q} - f_Q| \le c(n) j\, \|f\|_*\,. \tag{4.3}$$

For example, we may take

$$c(n) = 2n^{n/2} V_n = 2\frac{(n\pi)^{n/2}}{\Gamma(n/2+1)}.$$

Fig. 4.3 The cubes Q and λQ

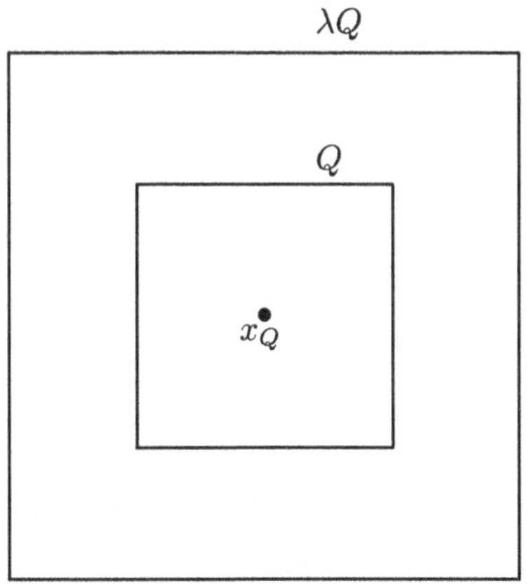

Proof First, it follows that

$$\begin{aligned}|f_{2Q} - f_Q| &= \left|\frac{1}{|Q|}\int_Q \big(f_{2Q} - f(x)\big)\,dx\right| \le \frac{1}{|Q|}\int_Q \big|f_{2Q} - f(x)\big|\,dx \\ &\le \frac{1}{|Q|}\int_{2Q} \big|f_{2Q} - f(x)\big|\,dx = 2^n\left(\frac{1}{|2Q|}\int_{2Q} \big|f_{2Q} - f(x)\big|\,dx\right) \\ &\le 2^n \,\|f\|_{*'}\,.\end{aligned}$$

Similarly, we have, for any positive integer j,

$$|f_{2^jQ} - f_{2^{j-1}Q}| \le 2^n \,\|f\|_{*'}\,.$$

Hence it follows that

$$\begin{aligned}|f_{2^jQ} - f_Q| &\le |f_{2^jQ} - f_{2^{j-1}Q}| + |f_{2^{j-1}Q} - f_{2^{j-2}Q}| + \ldots \\ &\quad + |f_{4Q} - f_{2Q}| + |f_{2Q} - f_Q| \\ &\le j2^n \,\|f\|_{*'}\,.\end{aligned} \tag{4.4}$$

However, we have, by Lemma 4.2,

$$\|f\|_{*'} \le \frac{n^{n/2}V_n}{2^{n-1}}\,\|f\|_*\,. \tag{4.5}$$

Therefore, the desired inequality (4.3) follows by combining inequalities (4.4) and (4.5):

$$|f_{2^jQ} - f_Q| \le j2^n \,\|f\|_{*'} \le j2^n \frac{n^{n/2}V_n}{2^{n-1}}\,\|f\|_* = 2jn^{n/2}V_n \,\|f\|_*\,.$$

The proof of Lemma 4.3 is complete. □

4.2 VMO Functions

(I) In this section, we introduce a subspace of BMO functions whose BMO norm over a ball vanishes as the radius of the ball tends to zero. More precisely, if $f \in \mathrm{BMO}$ and $r > 0$, then we let

$$\eta(r) := \sup_{\rho \le r} \frac{1}{|B|}\int_B |f(x) - f_B|\,dx,$$

where the supremum is taken over all balls B with radius $\rho \le r$. A function $f \in \mathrm{BMO}$ has *vanishing mean oscillation*, $f \in \mathrm{VMO}$, if it satisfies the condition (see Sarason [122])

$$\lim_{r \downarrow 0} \eta(r) = 0.$$

The function $\eta(r)$ will be referred to as the VMO modulus of f. The assumption that $f \in \text{VMO}$ means a kind of continuity in the average sense, not in the pointwise sense. First, we prove the following claim:

Claim 4.4 *(i) Uniformly continuous functions which belong to* BMO *are* VMO *functions.*

(ii) VMO *is a closed subspace of* BMO.

Proof (i) Assume that $f(x)$ is a uniformly continuous function which belongs to BMO. Then, for any $\varepsilon > 0$ there exists a constant $r = r(\varepsilon) > 0$ such that, for each ball B with radius r,

$$x, y \in B \implies |f(x) - f(y)| < \varepsilon,$$

or equivalently,

$$\sup_{x,y \in B} |f(x) - f(y)| \leq \varepsilon.$$

Hence it follows that we have, for all $x \in B$,

$$\begin{aligned}
|f(x) - f_B| &= \frac{1}{|B|} \left| \int_B (f(x) - f(y))\, dy \right| \leq \frac{1}{|B|} \int_B |f(x) - f(y)|\, dy \\
&\leq \sup_{x,y \in B} |f(x) - f(y)| \\
&\leq \varepsilon.
\end{aligned}$$

Therefore, we obtain that, for all balls B with radius $\rho \leq r$,

$$\frac{1}{|B|} \int_B |f(x) - f_B|\, dx \leq \varepsilon,$$

so that

$$\eta(r) \leq \varepsilon.$$

This proves that $\lim_{r \downarrow 0} \eta(r) = 0$, that is, $f \in \text{VMO}$.

(ii) Assume that f is the limit of a sequence $\{f_j\}$ of VMO functions in the BMO norm

$$\lim_{j \to \infty} \left\| f_j - f \right\|_* = 0.$$

Then it follows that, for all balls B with radius $\rho \leq r$,

$$\frac{1}{|B|} \int_B |f(x) - f_B|\, dx$$

$$\leq \frac{1}{|B|}\int_B |f_j(x) - f(x) - (f_j)_B + f_B|\,dx + \frac{1}{|B|}\int_B |f_j(x) - (f_j)_B|\,dx$$
$$\leq \|f_j - f\|_* + \eta_j(r),$$

where

$$\eta_j(r) = \sup_{\rho \leq r} \frac{1}{|B|}\int_B |f_j(x) - (f_j)_B|\,dx.$$

Hence we have the inequality

$$\eta(r) = \sup_{\rho \leq r} \frac{1}{|B|}\int_B |f(x) - f_B|\,dx \leq \|f_j - f\|_* + \eta_j(r). \tag{4.6}$$

However, for any given $\varepsilon > 0$ there exists a positive integer $N = N(\varepsilon)$ such that

$$\|f_N - f\|_* < \frac{\varepsilon}{2}. \tag{4.7}$$

Moreover, we can find a positive number $\delta = \delta(N, \varepsilon)$ such that

$$\eta_N(r) = \sup_{\rho \leq r} \frac{1}{|B|}\int_B |f_N(x) - (f_N)_B|\,dx < \frac{\varepsilon}{2} \quad \text{for all } 0 < r < \delta. \tag{4.8}$$

Therefore, by combining inequalities (4.6) with $j := N$, (4.7) and (4.8) we obtain that

$$\eta(r) = \sup_{\rho \leq r} \frac{1}{|B|}\int_B |f(x) - f_B|\,dx \leq \|f_N - f\|_* + \eta_N(r)$$
$$\leq \varepsilon \quad \text{for all } 0 < r < \delta.$$

This proves that $\lim_{r \downarrow 0} \eta(r) = 0$, that is, $f \in \text{VMO}$.

The proof of Claim 4.4 is complete. □

(II) The relationship between BMO and its subspace VMO is quite similar to the relationship between L^∞ and its subspace of bounded uniformly continuous functions (BUC). This can be visualized as in Fig. 4.4.

In fact, the next theorem collects some important results concerning VMO functions (see [57, Chap. VI, Theorem 5.1], [168, Chap. VIII]):

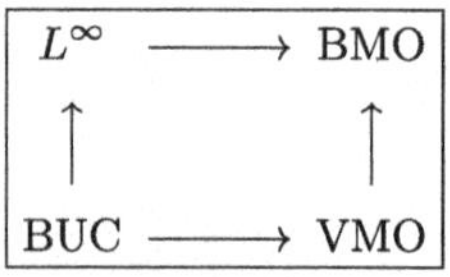

Fig. 4.4 The spaces L^∞, BMO, BUC and VMO

Theorem 4.5 (Sarason) *For a function $f \in \mathrm{BMO}$, the following three conditions (i), (ii) and (iii) are equivalent:*

(i) *f is in* VMO.

(ii) *f is in the* BMO *closure of uniformly continuous functions that belong to* BMO.

(iii) $\lim_{y \to 0} \|f(\cdot - y) - f(\cdot)\|_* = 0$, *where $f(x - y)$ is the translation of $f(x)$ by y-units. More precisely, there exists a positive constant $C = C(n)$ such that*

$$\|f(\cdot - y) - f(\cdot)\|_* \leq C\eta(r) \quad \textit{for } |y| < r. \tag{4.9}$$

Example 4.1 (i) $\ln |x| \in \mathrm{BMO}$, but $\ln |x| \notin \mathrm{VMO}$ in the case $n = 1$.

(ii) $\ln |\ln |x|| \in \mathrm{VMO}$ in the case $n = 1$ (see Fig. 4.5).

(iii) $L^\infty(\mathbf{R}^n) \subset \mathrm{BMO}$.

(iv) $W^{\theta, n/\theta}(\mathbf{R}^n) \subset \mathrm{VMO}$ for $0 < \theta \leq 1$ (see Proposition 7.9).

(v) The Sobolev space $W^{1,n}(\mathbf{R}^n)$ is a proper subspace of VMO. For example, $|\ln |x||^\alpha \in W^{1,n}(\mathbf{R}^n)$ if $0 < \alpha < 1 - 1/n$, while $|\ln |x||^\alpha \in \mathrm{VMO} \setminus W^{1,n}(\mathbf{R}^n)$ if $1 - 1/n \leq \alpha < 1$.

(III) Let Ω be a bounded domain with boundary $\partial\Omega$ in Euclidean space $\mathbf{R}^n$. By replacing the ball B above by the intersection $B \cap \Omega$, we obtain the definitions of the classes $\mathrm{BMO}(\Omega)$ and $\mathrm{VMO}(\Omega)$. Then we have the following localized version of Fig. 4.4:

$$\begin{array}{ccc} L^\infty(\Omega) & \longrightarrow & \mathrm{BMO}(\Omega) \\ \uparrow & & \uparrow \\ C\left(\overline{\Omega}\right) & \longrightarrow & \mathrm{VMO}(\Omega) \end{array}$$

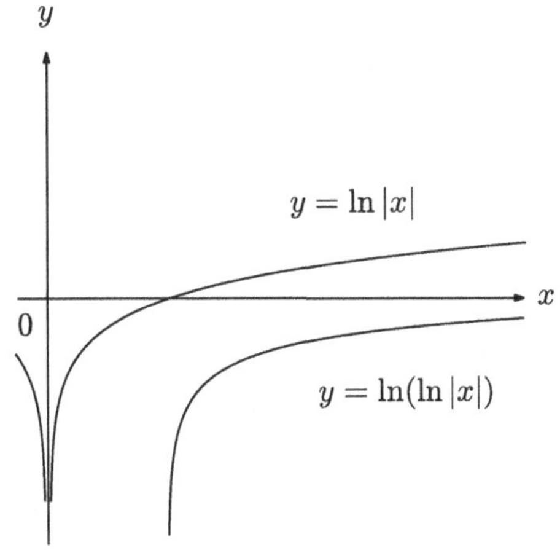

Fig. 4.5 The functions $\ln |x|$ and $\ln |\ln |x||$

Moreover, given a function on a $C^{1,1}$ domain Ω that belongs to BMO(Ω) (resp. VMO(Ω)), we can extend it to the whole $\mathbf{R}^n$ preserving its BMO (resp. VMO) character (see [1, Proposition 1.3]; [74, Theorem 1]; [169, Theorem 5.1]).

4.3 The Calderón–Zygmund Decomposition

Let $f \in L^1(\mathbf{R}^n)$. The purpose of this section is to describe the splitting of the space $\mathbf{R}^n$ into a subset Ω made up of non-overlapping cubes over each of which the average of $|f|$ is between t and $2^n t$ and the complement $\mathbf{R}^n \setminus \Omega$ where $|f(x)| \leq t$ (Theorem 4.9). This is known as the *Calderón–Zygmund decomposition*. The Calderón–Zygmund decomposition is a key step in real analysis. The idea behind this decomposition is that it is often useful to split an arbitrary integrable function into its "small" and "large" parts, and then we use different techniques to analyze each part.

In the Euclidean space $\mathbf{R}^n$ we define the unit cube, open on the right, to be the set $[0, 1)^n$, and we let

$$\begin{aligned}\mathcal{D}_0 = &\text{the collection of cubes in } \mathbf{R}^n \text{which are congruent}\\ &\text{to } [0,1)^n \text{ whose vertices lie on the lattice } \mathbf{Z}^n.\end{aligned}$$

If we dilate this family $\mathcal{Q}_0$ of cubes by a factor 2^{-k} for each integer $k \in \mathbf{Z}$, we obtain the family of cubes $\mathcal{D}_k$ as follows:

$$\begin{aligned}\mathcal{D}_k = &\text{the collection of cubes, open on the right, whose}\\ &\text{vertices are adjacent points of the lattice } (2^{-k}\,\mathbf{Z})^n.\end{aligned}$$

The cubes in $\mathcal{D} = \bigcup_{k \in \mathbf{Z}} \mathcal{D}_k$ are called *dyadic cubes*.

From this construction, we obtain the following three assertions:

(1) Given $x \in \mathbf{R}^n$, there exists a unique cube in each family $\mathcal{D}_k$ which contains x.
(2) Any two dyadic cubes are either disjoint or one is contained in the other.
(3) A dyadic cube in $\mathcal{D}_k$ is contained in a unique cube of each family $\mathcal{D}_j$, $j < k$, and contains 2^n dyadic cubes of $\mathcal{D}_{k+1}$.

If $f \in L^1_{\text{loc}}(\mathbf{R}^n)$, we define the Hardy–Littlewood *maximal function* Mf by the formula

$$Mf(x) := \sup_{x \in Q} \frac{1}{|Q|} \int_Q |f(y)|\, dy,$$

where the supremum is taken over all cubes Q containing x and $|Q|$ is the Lebesgue measure of Q. The Hardy–Littlewood maximal function will be studied in detail in Sect. 4.4 below.

In order to study the size of Mf, we shall look at its distribution function $|E_t|$. First, instead of looking at Mf, we try to obtain a family $\{Q_j\}$ of cubes such that the average of $|f|$ over each Q_j is comparable to t. This can be done most effectively by considering only dyadic cubes.

More precisely, we can prove the following theorem:

Theorem 4.6 *Let $f \in L^1(\mathbf{R}^n)$. Then, for every $t > 0$, there exists a family $\{Q_j\}$ of disjoint dyadic cubes which satisfies the following three conditions:*

$$E_t = \{x \in \mathbf{R}^n : Mf(x) > t\} \subset \bigcup_j 3Q_j. \tag{4.10}$$

$$\frac{t}{4^n} < \frac{1}{|Q_j|} \int_{Q_j} |f(y)|\, dy \leq \frac{t}{2^n} \quad \textit{for every } Q_j. \tag{4.11}$$

$$|E_t| \leq \frac{12^n}{t} \int_{\mathbf{R}^n} |f(y)|\, dy. \tag{4.12}$$

Here we recall that if Q is a cube with center x_Q and with side length δ_Q, then αQ, $\alpha > 0$, denotes the cube with the same center x_Q as Q and with side length $\alpha\delta_Q$.

Proof The proof is divided into two steps.

Step I: If $f \in L^1(\mathbf{R}^n)$ and $t > 0$, we let

$\mathcal{C}_t =$ the collection of cubes $Q \in \mathcal{D}$ which satisfy the condition

$$t < \frac{1}{|Q|} \int_Q |f(y)|\, dy \tag{4.13}$$

and are *maximal* among those which satisfy condition (4.13).

Then it is easy to verify the following three assertions:

(i) Every cube $Q \in \mathcal{D}$ satisfying condition (4.13) is contained in some cube $Q' \in \mathcal{C}_t$, since condition (4.13) imposes an upper bound on the size of Q, that is, since we have the inequality

$$|Q| \leq \frac{1}{t} \int_{\mathbf{R}^n} |f(y)|\, dy.$$

(ii) The cubes in $\mathcal{C}_t$ are non-overlapping.

(iii) For any cube $Q \in \mathcal{C}_t$, we have the inequalities

$$t < \frac{1}{|Q|} \int_Q |f(y)|\, dy \leq 2^n t.$$

Indeed, if $Q \in \mathcal{D}_k$ is in $\mathcal{C}_t$ and if Q' is the only cube in $\mathcal{D}_{k-1}$ containing Q, then it follows that

$$\frac{1}{|Q'|} \int_{Q'} |f(y)|\, dy \leq t.$$

However, since $|Q'| = 2^n |Q|$, we obtain that

$$\frac{1}{|Q|}\int_Q |f(y)|\,dy \le \frac{2^n}{|Q'|}\int_{Q'} |f(y)|\,dy \le 2^n t.$$

Therefore, we have constructed a family $\mathcal{C}_t = \{Q_j\}$ of disjoint dyadic cubes which satisfies the condition

$$t < \frac{1}{|Q_j|}\int_{Q_j} |f(y)|\,dy \le 2^n t. \tag{4.14}$$

Step II: (1) Assertion (4.10): If $\mathcal{C}_{t/4^n} = \{Q_j\}$, we show that

$$E_t = \{x \in \mathbf{R}^n : Mf(x) > t\} \subset \bigcup_j 3Q_j.$$

To do this, let x_0 be an arbitrary point of E_t. Then we can find a cube R containing x_0 in its interior and satisfying the condition

$$t < \frac{1}{|R|}\int_R |f(y)|\,dy.$$

We shall look for a dyadic cube of comparable size over which the average of $|f|$ is comparably large. If we take the only integer k such that

$$2^{-(k+1)n} < |R| \le 2^{-kn},$$

then there exists at most one point of the lattice $\Lambda_k = 2^{-k}\mathbf{Z}^n$ which is an interior point of R. Here it should be noticed that there exists a cube in $\mathcal{D}_k$, and at most 2^n cubes in $\mathcal{D}_k$, meeting the interior of R. Hence we can find a cube $Q \in \mathcal{D}_k$ which meets the interior of R and satisfies the condition

$$\int_{R\cap Q} |f(y)|\,dy > \frac{t|R|}{2^n}.$$

However, it follows that

$$|R| \le |Q| < 2^n|R|.$$

Hence we obtain that

$$\int_{R\cap Q} |f(y)|\,dy > \frac{t|R|}{2^n} > \frac{t}{2^n}\frac{|Q|}{2^n} = \frac{t|Q|}{4^n},$$

so that

$$\frac{1}{|Q|}\int_Q |f(y)|\,dy > \frac{t}{4^n}.$$

This implies that Q is contained in some cube $Q_j \in \mathcal{C}_{t/4^n}$.

Moreover, since R and Q meet and $|R| \le |Q|$, it follows that

$$x_0 \in R \subset 3Q \subset 3Q_j.$$

This proves the desired assertion (4.10).

(2) Assertion (4.11): This follows from inequalities (4.14) with $t := t/4^n$

$$\frac{t}{4^n} < \frac{1}{|Q_j|}\int_{Q_j} |f(y)|\, dy \le 2^n \cdot \frac{t}{4^n} = \frac{t}{2^n}.$$

(3) Assertion (4.12): By combining assertions (4.10) and (4.11), we obtain that

$$\begin{aligned}|E_t| &\le \sum_j |3Q_j| = 3^n \sum_j |Q_j| \\ &\le 3^n \sum_j \frac{4^n}{t}\int_{Q_j} |f(y)|\, dy = \frac{12^n}{t}\sum_j \int_{Q_j} |f(y)|\, dy \\ &\le \frac{12^n}{t}\int_{\mathbf{R}^n} |f(y)|\, dy.\end{aligned}$$

The proof of Theorem 4.6 is complete. □

As an application of inequality (4.12), we can obtain an extension of *Lebesgue's differentiation theorem*:

Theorem 4.7 (Lebesgue's differentiation theorem) *Let $f \in L^1_{\mathrm{loc}}(\mathbf{R}^n)$. If $x \in \mathbf{R}^n$ and $r > 0$, we let*

$$Q(x, r) := \prod_{k=1}^{n} [-r + x_k, r + x_k].$$

Then we have, for almost every point x of $\mathbf{R}^n$,

$$\lim_{r\downarrow 0} \frac{1}{|Q(x, r)|}\int_{Q(x,r)} |f(y) - f(x)|\, dy = 0. \tag{4.15}$$

Proof Without loss of generality, we may assume that $f \in L^1(\mathbf{R}^n)$. Since we have the formula

$$\begin{aligned}&\left\{x \in \mathbf{R}^n : \text{assertion (4.15) does not hold}\right\} \\ &= \bigcup_{j=1}^{\infty} \left\{x \in \mathbf{R}^n : \limsup_{r\downarrow 0} \frac{1}{|Q(x, r)|}\int_{Q(x,r)} |f(y) - f(x)|\, dy > \frac{1}{j}\right\},\end{aligned}$$

it suffices to show that, for every $t > 0$, the set

$$A_t := \left\{ x \in \mathbf{R}^n : \limsup_{r \downarrow 0} \frac{1}{|Q(x,r)|} \int_{Q(x,r)} |f(y) - f(x)|\, dy > t \right\}$$

has measure *zero*.

For any given $\varepsilon > 0$, we can find functions $g \in C_0(\mathbf{R}^n)$ and $h \in L^1(\mathbf{R}^n)$ such that

$$f(x) = g(x) + h(x),$$
$$\int_{\mathbf{R}^n} |h(x)|\, dx < \varepsilon.$$

Then we have the inequality

$$\begin{aligned}
&\frac{1}{|Q(x,r)|} \int_{Q(x,r)} |f(y) - f(x)|\, dy \\
&\le \frac{1}{|Q(x,r)|} \int_{Q(x,r)} |g(y) - g(x)|\, dy + |h(x)| + \frac{1}{|Q(x,r)|} \int_{Q(x,r)} |h(y)|\, dy \\
&\le \frac{1}{|Q(x,r)|} \int_{Q(x,r)} |g(y) - g(x)|\, dy + |h(x)| + Mh(x).
\end{aligned}$$

By the uniform continuity of g, it follows that

$$\lim_{r \downarrow 0} \frac{1}{|Q(x,r)|} \int_{Q(x,r)} |g(y) - g(x)|\, dy = 0.$$

Hence, by passing to the limit we obtain that

$$\limsup_{r \downarrow 0} \frac{1}{|Q(x,r)|} \int_{Q(x,r)} |f(y) - f(x)|\, dy \le |h(x)| + Mh(x).$$

This implies that the set A_t can be decomposed as follows:

$$A_t \subset \left\{ x \in \mathbf{R}^n : Mh(x) > \frac{t}{2} \right\} \bigcup \left\{ x \in \mathbf{R}^n : |h(x)| > \frac{t}{2} \right\}.$$

However, we have the inequality

$$\begin{aligned}
\left| \left\{ x \in \mathbf{R}^n : |h(x)| > \frac{t}{2} \right\} \right| &= \int_{\{x \in \mathbf{R}^n : |h(x)| > t/2\}} dy \\
&\le \int_{\{x \in \mathbf{R}^n : |h(x)| > t/2\}} \frac{2}{t} |h(y)|\, dy
\end{aligned}$$

$$\begin{aligned} &\le \int_{\mathbf{R}^n} \frac{2}{t}|h(y)|\,dy = \frac{2}{t}\int_{\mathbf{R}^n} |h(y)|\,dy \\ &< \frac{2}{t}\varepsilon, \end{aligned}$$

and also, by inequality (4.12) with $t := t/2$,

$$\begin{aligned} \left|\left\{x \in \mathbf{R}^n : Mh(x) > \frac{t}{2}\right\}\right| &\le \frac{12^n}{t/2}\int_{\mathbf{R}^n} |h(y)|\,dy = \frac{2\cdot 12^n}{t}\int_{\mathbf{R}^n} |h(y)|\,dy \\ &< \frac{2\cdot 12^n}{t}\varepsilon. \end{aligned}$$

Summing up, we have proved that, for any given $\varepsilon > 0$,

$$|A_t| \le \frac{2}{t}\varepsilon + \frac{2\cdot 12^n}{t}\varepsilon = \frac{2}{t}(1+12^n)\varepsilon.$$

This proves that $|A_t| = 0$ for every $t > 0$.

The proof of Theorem 4.7 is complete. □

Definition 4.1 Let $f \in L^1_{\mathrm{loc}}(\mathbf{R}^n)$. A point x of $\mathbf{R}^n$ is called a *Lebesgue point* for f if it satisfies the condition

$$\lim_{r\downarrow 0} \frac{1}{|Q(x,r)|}\int_{Q(x,r)} |f(y)-f(x)|\,dy = 0. \tag{4.15}$$

Rephrased, Theorem 4.7 asserts that almost every point of $\mathbf{R}^n$ is a Lebesgue point for $f \in L^1_{\mathrm{loc}}(\mathbf{R}^n)$.

Furthermore, we can prove the following corollary:

Corollary 4.8 *Let $f \in L^1_{\mathrm{loc}}(\mathbf{R}^n)$. Then, for every Lebesgue point x for f, and thus, for almost every point $x \in \mathbf{R}^n$, we have the following two assertions:*

$$f(x) = \lim_{r\downarrow 0} \frac{1}{|Q(x,r)|}\int_{Q(x,r)} f(y)\,dy. \tag{4.16}$$

$$|f(x)| \le Mf(x). \tag{4.17}$$

Proof (i) Let x be an arbitrary Lebesgue point for f. Then it follows from condition (4.15) that we have, as $r \downarrow 0$,

$$\begin{aligned} &\left|\frac{1}{|Q(x,r)|}\int_{Q(x,r)} f(y)\,dy - f(x)\right| \\ &= \left|\frac{1}{|Q(x,r)|}\int_{Q(x,r)} (f(y)-f(x))\,dy\right| \\ &\le \frac{1}{|Q(x,r)|}\int_{Q(x,r)} |f(y)-f(x)|\,dy \longrightarrow 0. \end{aligned}$$

This proves the desired assertion (4.16).

(ii) We remark that

$$\frac{1}{|Q(x,r)|}\left|\int_{Q(x,r)} f(y)\,dy\right| \le \frac{1}{|Q(x,r)|}\int_{Q(x,r)} |f(y)|\,dy \le Mf(x). \tag{4.18}$$

However, by assertion (4.16) it follows that we have, as $r \downarrow 0$,

$$\left|\frac{1}{|Q(x,r)|}\left|\int_{Q(x,r)} f(y)\,dy\right| - |f(x)|\right| \le \left|\frac{1}{|Q(x,r)|}\int_{Q(x,r)} f(y)\,dy - f(x)\right| \longrightarrow 0.$$

Therefore, by passing to the limit in inequality (4.18) we obtain that

$$|f(x)| = \lim_{r\downarrow 0}\frac{1}{|Q(x,r)|}\left|\int_{Q(x,r)} f(y)\,dy\right| \le Mf(x).$$

This proves the desired assertion (4.17).

The proof of Corollary 4.8 is complete. □

Summing up, we can obtain the following Calderón–Zygmund decomposition theorem:

Theorem 4.9 (the Calderón–Zygmund decomposition) *If $f(x)$ is an arbitrary function in the space $L^1(\mathbf{R}^n)$, then, for every $t > 0$ we can construct a family $\mathcal{C}_t = \{Q_j\}$ of disjoint maximal dyadic cubes over which the average of $|f|$ is greater than t. This family $\mathcal{C}_t$ satisfies the following two conditions* (4.14) *and* (4.19)*:*

$$t < \frac{1}{|Q_j|}\int_{Q_j} |f(y)|\,dy \le 2^n t \quad \textit{for every } Q_j. \tag{4.14}$$

$$|f(x)| \le t \quad \textit{for almost every } x \in \mathbf{R}^n \setminus \bigcup_j Q_j. \tag{4.19}$$

Moreover, we have, for every $t > 0$,

$$E_t = \{x \in \mathbf{R}^n : Mf(x) > t\} \subset \bigcup_j 3Q_j, \tag{4.10}$$

where the Q_j range over $\mathcal{C}_{t/4^n}$.

Proof By Theorem 4.6, we have only to prove assertion (4.19).

(i) Let x_0 be a Lebesgue point for f. If $\{Q_j\}$ is a sequence of cubes such that

$$Q_1 \supset Q_2 \supset \cdots \supset Q_j \supset \cdots,$$
$$\bigcap_j Q_j = \{x_0\},$$

then it follows that

$$f(x_0) = \lim_{j\to\infty} \frac{1}{|Q_j|} \int_{Q_j} f(y)\, dy. \tag{4.20}$$

Indeed, if the cube Q_j has side length r_j, then we have the assertion

$$Q_j \subset Q(x_0, r_j), \quad |Q_j| = r_j^n \downarrow 0,$$

Therefore, we obtain from condition (4.15) that, as $j \to \infty$,

$$\left| \frac{1}{|Q_j|} \int_{Q_j} f(y)\, dy - f(x_0) \right| = \left| \frac{1}{|Q_j|} \int_{Q_j} (f(y) - f(x_0))\, dy \right|$$
$$\leq \frac{1}{|Q_j|} \int_{Q_j} |f(y) - f(x)|\, dy$$
$$\leq \frac{2^n}{|Q(x_0, r_j)|} \int_{Q(x, r_j)} |f(y) - f(x_0)|\, dy \longrightarrow 0.$$

This proves the desired assertion (4.20).

(ii) Now, let $f \in L^1(\mathbf{R}^n)$, and let $\mathcal{C}_t = \{Q_j\}$ be a family of disjoint maximal dyadic cubes over which the average of $|f|$ is greater than t:

$$t < \frac{1}{|Q_j|} \int_{Q_j} |f(y)|\, dy \leq 2^n t \quad \text{for every } Q_j.$$

If x_0 is a point of $\mathbf{R}^n \setminus \bigcup_j Q_j$, then we have, for any dyadic cube Q containing x_0,

$$\frac{1}{|Q|} \int_Q |f(y)|\, dy \leq t.$$

Hence, by taking a sequence $\{R_k\}$ of dyadic cubes of decreasing size such that

$$\bigcap_k R_k = \{x_0\},$$

we obtain that

$$\frac{1}{|R_k|} \int_{R_k} |f(y)|\, dy \leq t. \tag{4.21}$$

If, in addition, x_0 is a Lebesgue point for f and so for $|f|$, then we have, by inequality (4.21) and assertion (4.20),

$$|f(x_0)| = \lim_k \frac{1}{|R_k|} \int_{R_k} |f(y)|\, dy \leq t.$$

Therefore, we have proved the desired assertion (4.19)

$$|f(x)| \leq t \quad \text{for almost every } x \in \mathbf{R}^n \setminus \bigcup_j Q_j.$$

The proof of Theorem 4.9 is complete. □

4.4 The Hardy–Littlewood Maximal Function

If $f \in L^1_{\mathrm{loc}}(\mathbf{R}^n)$, we define the Hardy–Littlewood *maximal function* Mf by the formula

$$Mf(x) := \sup_{x \in Q} \frac{1}{|Q|} \int_Q |f(y)|\, dy,$$

where the supremum is taken over all cubes Q containing x. It should be noticed that we can take only those cubes Q for which x is an interior point. This implies that the function $x \mapsto Mf(x)$ is lower semi-continuous, since the set

$$E_t := \left\{x \in \mathbf{R}^n : Mf(x) > t\right\}$$

is open for every $t > 0$. Indeed, it suffices to note that

$$E_t = \{x \in \mathbf{R}^n : Mf(x) > t\} = \bigcup \left\{ Q^o : \frac{1}{|Q^o|} \int_{Q^o} |f(y)|\, dy > t \right\},$$

where the Q^o range over all open cubes containing x.

If we define the *maximal function* $M'f$ by the formula

$$M'f(x) := \sup_{r>0} \frac{1}{|Q(x,r)|} \int_{Q(x,r)} |f(y)|\, dy,$$

where

$$Q(x,r) := \left\{ y \in \mathbf{R}^n : \max_{1 \leq j \leq n} |y_j - x_j| \leq r \right\} = \prod_{j=1}^n \left[x_j - r, x_j + r\right],$$

then we have the inequalities

$$M'f(x) \leq Mf(x) \leq 2^n M'f(x), \quad x \in \mathbf{R}^n.$$

Indeed, if Q is a cube with side length ℓ and containing x, then it follows that

$$Q \subset Q(x, \ell),$$
$$|Q(x, \ell)| = (2\ell)^n = 2^n |Q|.$$

Hence we obtain that

$$\frac{1}{|Q|} \int_Q |f(y)| \, dy \leq \frac{2^n}{|Q(x, \ell)|} \int_{Q(x,\ell)} |f(y)| \, dy \leq 2^n M'f(x), \quad x \in \mathbf{R}^n.$$

This proves that

$$Mf(x) \leq 2^n M'f(x).$$

Furthermore, if we define the *maximal function* $M''f$ by the formula

$$M''f(x) := \sup_{x \in B} \frac{1}{|B|} \int_B |f(y)| \, dy,$$

where the supremum is taken over all balls B containing x, then, by arguing as in the proof of Lemma 4.2 we can obtain the inequalities

$$\frac{V_n}{2^n} M''f(x) \leq Mf(x) \leq \frac{V_n}{n^{n/2}} M''f(x), \quad x \in \mathbf{R}^n,$$

where $V_n = \omega_n/n = \pi^{n/2}/\Gamma(n/2+1)$ is the volume of the unit ball in $\mathbf{R}^n$.

The next theorem gives a fundamental norm estimate for the maximal function $Mf(x)$ for $1 < p < \infty$ (cf. [137, Chap. I, Sect. 1.3, Theorem 1]):

Theorem 4.10 *If $f \in L^p(\mathbf{R}^n)$ with $1 < p < \infty$, then it follows that $Mf \in L^p(\mathbf{R}^n)$. More precisely, we have the inequality*

$$\|Mf\|_{L^p(\mathbf{R}^n)} \leq C_p \|f\|_{L^p(\mathbf{R}^n)}, \tag{4.22}$$

with a positive constant C_p. For example, we may take

$$C_p = 2 \left(\frac{12^n \cdot p}{p-1} \right)^{1/p}.$$

Proof Our proof is divided into two steps.

Step I: The essential step in the proof is the following theorem:

Theorem 4.11 *Let $f \in L^1_{\mathrm{loc}}(\mathbf{R}^n)$. If $t > 0$, we let*

$$E_t = \{x \in \mathbf{R}^n : Mf(x) > t\}.$$

Then we have the following two estimates for the Lebesgue measure $|E_t|$ of E_t:

$$|E_t| \leq \frac{2 \cdot 12^n}{t} \int_{\{x \in \mathbf{R}^n : |f(x)| > t/2\}} |f(y)| \, dy. \tag{4.23a}$$

$$|E_t| \geq \frac{1}{2^n t} \int_{\{x \in \mathbf{R}^n : |f(x)| > t\}} |f(y)| \, dy. \tag{4.23b}$$

Proof (i) First, we decompose the function $f(x)$ as follows:

$$f(x) = f_1(x) + f_2(x),$$

where

$$f_1(x) = \begin{cases} f(x) & \text{if } |f(x)| > \frac{t}{2}, \\ 0 & \text{otherwise,} \end{cases}$$

and

$$f_2(x) = \begin{cases} f(x) & \text{if } |f(x)| \leq \frac{t}{2}, \\ 0 & \text{otherwise.} \end{cases}$$

Then it follows that

$$\begin{aligned} Mf(x) &= \sup_{x \in Q} \frac{1}{|Q|} \int_Q |f(y)| \, dy \\ &\leq \sup_{x \in Q} \frac{1}{|Q|} \int_Q |f_1(y)| \, dy + \sup_{x \in Q} \frac{1}{|Q|} \int_Q |f_2(y)| \, dy \\ &= Mf_1(x) + Mf_2(x). \end{aligned}$$

However, since we have the inequality

$$|f_2(x)| \leq \frac{t}{2} \quad \text{in } \mathbf{R}^n,$$

it follows that

$$Mf_2(x) = \sup_{x \in Q} \frac{1}{|Q|} \int_Q |f_2(y)| \, dy \leq \frac{t}{2}.$$

Summing up, we have proved that

$$Mf(x) \leq Mf_1(x) + Mf_2(x) \leq \frac{t}{2}.$$

This implies that

$$Mf(x) > t \implies Mf_1(x) > \frac{t}{2}.$$

Therefore, by applying inequality (4.12) to the function f_1 we obtain that

$$\begin{aligned}
|E_t| &= \left|\{x \in \mathbf{R}^n : Mf(x) > t\}\right| \\
&\le \left|\left\{x \in \mathbf{R}^n : Mf_1(x) > \frac{t}{2}\right\}\right| \\
&\le \frac{12^n}{t/2} \int_{\mathbf{R}^n} |f_1(y)|\, dy = \frac{2 \cdot 12^n}{t} \int_{\mathbf{R}^n} |f_1(y)|\, dy \\
&= \frac{2 \cdot 12^n}{t} \int_{\{x \in \mathbf{R}^n : |f(x)| > t/2\}} |f(y)|\, dy.
\end{aligned}$$

This proves the desired estimate (4.23a).

(ii) Without loss of generality, we may assume that $f \in L^1(\mathbf{R}^n)$. Indeed, otherwise, we truncate and apply a limiting process.

Then, by using the Calderón–Zygmund decomposition for f (Theorem 4.9) we can construct a family $\{Q_j\}$ of disjoint dyadic cubes such that

- $t < \dfrac{1}{|Q_j|} \displaystyle\int_{Q_j} |f(y)|\, dy \le 2^n t$ for every Q_j.
- $|f(x)| \le t$ for almost every $x \in \mathbf{R}^n \setminus \displaystyle\bigcup_{j=1}^{\infty} Q_j$.

Then it follows that

$$x \in Q_j \implies Mf(x) \ge \frac{1}{|Q_j|} \int_{Q_j} |f(y)|\, dy > t \implies x \in E_t.$$

This proves that

$$\bigcup_{j=1}^{\infty} Q_j \subset E_t.$$

Hence, we obtain that

$$|E_t| \ge \sum_{j=1}^{\infty} |Q_j| \ge \frac{1}{2^n t} \sum_{j=1}^{\infty} \int_{Q_j} |f(y)|\, dy = \frac{1}{2^n t} \int_{\bigcup_{j=1}^{\infty} Q_j} |f(y)|\, dy.$$

However, we remark that

$$x \notin \bigcup_{j=1}^{\infty} Q_j \implies |f(x)| \le t,$$

or equivalently,

$$\{x \in \mathbf{R}^n : |f(x)| > t\} \subset \bigcup_{j=1}^{\infty} Q_j.$$

Therefore, we obtain that

$$\begin{aligned}|E_t| &\geq \sum_{j=1}^{\infty} |Q_j| \geq \frac{1}{2^n t} \int_{\bigcup_{j=1}^{\infty} Q_j} |f(y)|\, dy \\ &\geq \frac{1}{2^n t} \int_{\{x \in \mathbf{R}^n : |f(x)| > t\}} |f(y)|\, dy.\end{aligned}$$

This proves the desired estimate (4.23b).

The proof of Theorem 4.11 is complete. □

Step II: By applying Theorem 3.45 and Fubini's theorem (Theorem 3.20) to our situation, we obtain from estimate (4.23a) that

$$\begin{aligned}&\|Mf\|^p_{L^p(\mathbf{R}^n)} = \int_{\mathbf{R}^n} (Mf(x))^p\, dx \\ &= \int_0^{\infty} p t^{p-1} |E_t|\, dt \\ &\leq 2p \cdot 12^n \int_0^{\infty} t^{p-2} \left(\int_{\{y \in \mathbf{R}^n : |f(y)| > t/2\}} |f(x)|\, dx \right) dt \\ &= 2p \cdot 12^n \int_{\mathbf{R}^n} \left(\int_0^{2|f(x)|} t^{p-2}\, dt \right) |f(x)|\, dx \\ &= \frac{2p \cdot 12^n}{p-1} \int_{\mathbf{R}^n} (2|f(x)|)^{p-1} |f(x)|\, dx = \frac{2^p p \cdot 12^n}{p-1} \int_{\mathbf{R}^n} |f(x)|^p\, dx \\ &= \frac{2^p p \cdot 12^n}{p-1} \|f\|^p_{L^p(\mathbf{R}^n)}.\end{aligned}$$

This proves the desired estimate (4.22).

Now the proof of Theorem 4.10 is complete. □

Remark 4.12 Theorem 4.10 remains valid if the norm is taken on the half-space $\mathbf{R}^n_+$. In this case, the definition of BMO functions needs to be modified by taking only cubes Q contained in $\mathbf{R}^n_+$ (see Fig. 4.6).

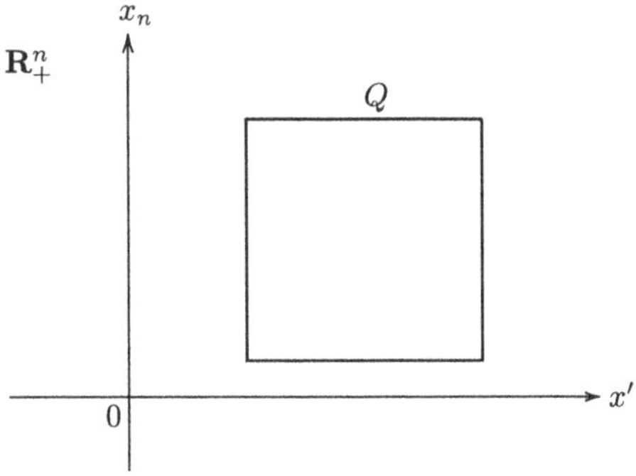

Fig. 4.6 The cube Q contained in the half-space $\mathbf{R}^n_+$

4.5 The John–Nirenberg Inequality

In this section we study the rate of growth of functions in BMO. The next theorem asserts that logarithmic growth is the maximum possible for BMO functions (see [73, Lemmas 1 and 1′]):

Theorem 4.13 (John–Nirenberg) *Let $f \in$ BMO. Then we can find constants $C_1 > 0$ and $C_2 > 0$, depending only on the dimension, such that we have, for any cube Q and any $t > 0$,*

$$\left|\left\{x \in Q : |f(x) - f_Q| > t\right\}\right| \leq C_1\, e^{-(C_2/\|f\|_*)t}\, |Q|\,. \tag{4.24}$$

For example, we may take

$$C_1 = \sqrt{e}\,, \quad C_2 = \frac{1}{2^{n+1}e}.$$

Proof First, we may assume that $\|f\|_* = 1$, since the John–Nirenberg inequality (4.24) is homogeneous. Indeed, if we let

$$g(x) := \frac{f(x)}{\|f\|_*}, \quad s := \frac{t}{\|f\|_*},$$

then we can rewrite inequality (4.24) in the form

$$\left|\left\{x \in Q : |g(x) - g_Q| > s\right\}\right| \leq C_1\, e^{-C_2 s}\, |Q|, \quad s > 0. \tag{4.24′}$$

The proof of inequality (4.24′) is divided into three steps.

Step I: We fix a cube Q and take the constant e. Then it follows that

$$\frac{1}{|Q|}\int_Q |f(x) - f_Q|\, dx \leq 1 < e.$$

We make use of the Calderón–Zygmund decomposition of Q for the function

$$f(x) - f_Q$$

relative to e (see Theorem 4.9). This is done just as in Theorem 4.9 except that we begin by bisecting the sides of Q to form 2^n equal cubes $\{Q_{1,j}\}$. Then we have the following assertions:

- $e < \dfrac{1}{|Q_{1,j}|} \displaystyle\int_{Q_{1,j}} \left| f(x) - f_Q \right| dx \leq 2^n e.$
- $\left| f(x) - f_Q \right| \leq e$ for almost every $x \in Q \setminus \bigcup_j Q_{1,j}$.
- $\left| f_{Q_{1,j}} - f_Q \right| \leq 2^n e.$
- $\displaystyle\sum_j |Q_{1,j}| \leq \frac{1}{e} |Q| .$

Indeed, it suffices to note that

$$\begin{aligned} |f_{Q_{1,j}} - f_Q| &= \frac{1}{|Q_{1,j}|} \left| \int_{Q_{1,j}} (f(y) - f_Q)\, dy \right| \\ &\leq \frac{1}{|Q_{1,j}|} \int_{Q_{1,j}} |f(y) - f_Q|\, dy \\ &\leq 2^n e, \end{aligned}$$

and further that

$$\begin{aligned} \sum_j |Q_{1,j}| &\leq \frac{1}{e} \sum_j \int_{Q_{1,j}} |f(y) - f_Q|\, dy \\ &\leq \frac{1}{e} \int_Q |f(y) - f_Q|\, dy \\ &\leq \frac{1}{e} |Q|. \end{aligned}$$

Step II: On each $Q_{1,j}$, we make use of the Calderón–Zygmund decomposition of $Q_{1,j}$ for the function

$$f(x) - f_{Q_{1,j}}$$

relative to e (see Theorem 4.9). This is done just as in Theorem 4.9 except that we begin by bisecting the sides of $Q_{1,j}$ to form 2^n equal cubes $\{Q_{2,k}\}$. Then we have the following assertions:

- $e < \dfrac{1}{|Q_{2,k}|} \displaystyle\int_{Q_{2,k}} \left| f(x) - f_{Q_{1,j}} \right| dx \leq 2^n e.$

- $\left|f(x) - f_{Q_{1,j}}\right| \le e$ for almost every $x \in Q_{1,j} \setminus \bigcup_k Q_{2,k}$.
- $\left|f_{Q_{2,k}} - f_{Q_{1,j}}\right| \le 2^n e$.
- $\sum_k |Q_{2,k}| \le \frac{1}{e}\left|Q_{1,j}\right|$.

Indeed, it suffices to note that

$$\begin{aligned}\sum_k |Q_{2,k}| &\le \frac{1}{e}\sum_k \int_{Q_{2,k}} |f(y) - f_{Q_{1,j}}|\,dy \\ &\le \frac{1}{e}\int_{Q_{1,j}} |f(y) - f_{Q_{1,j}}|\,dy \\ &\le \frac{1}{e}|Q_{1,j}|.\end{aligned}$$

Now we gather the cubes $Q_{2,k}$ corresponding to all the $Q_{1,j}$ and collectively rename them $Q_{2,k}$. Then we have the following assertions:

$$\sum_k |Q_{2,k}| \le \frac{1}{e}\sum_j |Q_{1,j}| \le \left(\frac{1}{e}\right)^2 |Q|.$$

$$\begin{aligned}\left|f(x) - f_Q\right| &\le \left|f(x) - f_{Q_{1,j}}\right| + \left|f_{Q_{1,j}} - f_Q\right| \le e + 2^n\, e \\ &\le 2 \cdot 2^n e \quad \text{for almost every } x \in Q \setminus \bigcup_k Q_{2,k}.\end{aligned}$$

Step III: By repeating this process for all cubes formed, we have, after N-steps, a family $\{Q_{N,j}\}$ of subcubes of Q which satisfies the following two conditions:

$$\sum_j |Q_{N,j}| \le \left(\frac{1}{e}\right)^N |Q|.$$

$$|f(x) - f_Q| \le N \cdot 2^n e \quad \text{for almost every } x \in Q \setminus \bigcup_j Q_{N,j}.$$

(a) If $N \cdot 2^n e \le t < (N+1) \cdot 2^n e$ with $N \in \mathbf{N}$, then it follows that

$$x \in Q,\ |f(x) - f_Q| > t \implies |f(x) - f_Q| > N \cdot 2^n e \implies x \in \bigcup_j Q_{N,j}.$$

This proves that

$$\left\{x \in Q : |f(x) - f_Q| > t\right\} \subset \bigcup_{j=1}^{\infty} Q_{N,j}.$$

Hence we have, for all $t \geq 2^n e$,

$$\begin{aligned}\left|\left\{x \in Q : |f(x) - f_Q| > t\right\}\right| &\leq \sum_{j=1}^{\infty} |Q_{N,j}| \leq e^{-N}\,|Q| \\ &\leq e^{-C_2 t}\,|Q|,\end{aligned}$$

with

$$C_2 := \frac{1}{2^{n+1}\,e}.$$

(b) On the other hand, if $0 < t < 2^n e$, then it follows that

$$C_2\,t = \frac{1}{2^{n+1} e} \cdot t < \frac{1}{2}.$$

Hence we have, for all $0 < t < 2^n e$,

$$\begin{aligned}\left|\left\{x \in Q : |f(x) - f_Q| > t\right\}\right| &\leq |Q| \\ &< e^{1/2 - C_2 t}\,|Q| = C_1\,e^{-C_2 t}\,|Q|,\end{aligned}$$

with

$$C_1 := \sqrt{e}, \quad C_2 := \frac{1}{2^{n+1} e}.$$

Summing up, we have proved that

$$\left|\left\{x \in Q : |f(x) - f_Q| > t\right\}\right| \leq \begin{cases} C_1\,e^{-C_2 t}\,|Q| & \text{for all } 0 < t < 2^n e, \\ e^{-C_2 t}\,|Q| & \text{for all } t \geq 2^n e. \end{cases}$$

This proves the desired inequality (4.24′).
The proof of Theorem 4.13 is complete. □

The next theorem asserts that if $f \in \mathrm{BMO}$, then it is locally in L^p for any $1 < p < \infty$ (see [139, Chap. IV, Sect. 1.3, Corollary]):

Theorem 4.14 *Let* $f \in \mathrm{BMO}$ *and* $1 < p < \infty$. *Then we have, for all cubes* Q,

$$\left(\frac{1}{|Q|}\int_Q |f(y) - f_Q|^p\,dy\right)^{1/p} \leq c_p\,\|f\|_*, \tag{4.25}$$

with a positive constant c_p. *For example, we may take*

$$c_p = \frac{(C_1\,p\,\Gamma(p))^{1/p}}{C_2}.$$

Proof We let

$$E_t := \{x \in Q : |f(x) - f_Q| > \lambda\}, \quad t > 0.$$

Then we have, by Theorem 3.45 and the John–Nirenberg inequality (4.24),

$$\begin{aligned}\int_Q |f(y) - f_Q|^p \, dy &= \int_0^\infty p\, t^{p-1} \, |E_t| \, dt \\ &\le C_1 \, |Q| \int_0^\infty p\, t^{p-1} \, e^{-C_2 t/\|f\|_*} \, dt.\end{aligned}$$

Therefore, by making the change of variables

$$s = \frac{C_2 t}{\|f\|_*},$$

we obtain that

$$\begin{aligned}\frac{1}{|Q|} \int_Q |f(y) - f_Q|^p \, dy &\le C_1 \, p \left(\frac{\|f\|_*}{C_2}\right)^p \int_0^\infty s^{p-1} \, e^{-s} \, ds \\ &= C_1 \, p \, \Gamma(p) \left(\frac{\|f\|_*}{C_2}\right)^p.\end{aligned}$$

This proves inequality (4.25) with

$$c_p := \frac{(C_1 \, p \, \Gamma(p))^{1/p}}{C_2}.$$

The proof of Theorem 4.14 is complete. □

Remark 4.15 Theorem 4.14 remains valid if the norm is taken on the half-space $\mathbf{R}^n_+$. In this case, the definition of BMO functions needs to be modified by taking only cubes Q contained in $\mathbf{R}^n_+$ (see Fig. 4.6).

Rephrased, Theorem 4.14 asserts that, for each $1 < p < \infty$, the quantity

$$\|f\|_{*,p} := \sup_Q \left(\frac{1}{|Q|} \int_Q |f(y) - f_Q|^p \, dy\right)^{1/p}$$

is a norm on the space BMO equivalent to $\|\cdot\|_*$.

Indeed, it suffices to note that we have, by Hölder's inequality (Theorem 3.24),

$$\frac{1}{|Q|} \int_Q |f(y) - f_Q| \, dy$$

$$\leq \frac{1}{|Q|}\left(\int_Q |f(y) - f_Q|^p \, dy\right)^{1/p} \left(\int_Q 1^q \, dy\right)^{1/q}$$
$$= \left(\frac{1}{|Q|}\int_Q |f(y) - f_Q|^p \, dy\right)^{1/p}$$
$$\leq \|f\|_{*,p} \, .$$

This proves that the reverse inequality

$$\|f\|_* = \sup_Q \frac{1}{|Q|}\int_Q \left|f(y) - f_Q\right| dy \leq \|f\|_{*,p}$$

holds true.

As a consequence of the proof of Theorem 4.14, we obtain the following two additional results:

Corollary 4.16 *Let $f \in \mathrm{BMO}$. Then we have, for all $0 < \lambda < C_2/\, \|f\|_*$,*

$$\frac{1}{|Q|}\int_Q e^{\lambda|f(y)-f_Q|} \, dy < \infty.$$

Proof If we expand the exponential function as a power series, we obtain from Theorem 3.45 and the John–Nirenberg inequality (4.24) that

$$\int_Q e^{\lambda|f(y)-f_Q|} \, dy = |Q| + \sum_{p=1}^{\infty} \frac{\lambda^p}{p!}\int_Q |f(y) - f_Q|^p \, dy$$
$$= |Q| + \sum_{p=1}^{\infty} \frac{\lambda^p}{p!}\int_0^{\infty} p\, t^{p-1} \, |E_t| \, dt$$
$$= |Q| + \sum_{p=1}^{\infty} \lambda \int_0^{\infty} \frac{(\lambda t)^{p-1}}{(p-1)!} |E_t| \, dt = |Q| + \int_0^{\infty} \lambda \, e^{\lambda t} \, |E_t| \, dt$$
$$\leq |Q|\left(1 + \int_0^{\infty} \lambda \, e^{\lambda t} \left(C_1 \, e^{-C_2 t/\|f\|_*}\right) dt\right)$$
$$= |Q|\left(1 + \frac{\lambda C_1}{C_2/\|f\|_* - \lambda}\right).$$

This proves that

$$\frac{1}{|Q|}\int_Q e^{\lambda|f(y)-f_Q|} \, dy \leq 1 + \frac{\lambda C_1}{C_2/\|f\|_* - \lambda}.$$

The proof of Corollary 4.16 is complete. □

The proof of Theorem 4.14 with $p := 1$ can be used to prove that the converse to the John–Nirenberg inequality (Theorem 4.13) holds true:

Corollary 4.17 *Let $f \in L^1_{\mathrm{loc}}(\mathbf{R}^n)$. Assume that there exist positive constants C_1, C_2 and K such that we have, for any cube Q and any $t > 0$,*

$$\left|\left\{x \in Q : \left|f(x) - f_Q\right| > t\right\}\right| \leq C_1\, e^{-C_2 t/K}\, |Q|. \tag{4.26}$$

Then it follows that $f \in \mathrm{BMO}$.

Proof Indeed, we have, by Theorem 3.45 and inequality (4.26),

$$\begin{aligned}\int_Q \left|f(y) - f_Q\right| dy &= \int_0^\infty |E_t|\, dt \\ &\leq C_1 \left(\int_0^\infty e^{-C_2 t/K}\, dt\right) |Q| \\ &= \frac{C_1\, K}{C_2}\, |Q|.\end{aligned}$$

Hence it follows that, for any cube Q,

$$\frac{1}{|Q|}\int_Q \left|f(y) - f_Q\right| dy \leq \frac{C_1\, K}{C_2}.$$

This proves that

$$\|f\|_* \leq \frac{C_1\, K}{C_2}.$$

The proof of Corollary 4.17 is complete. □

4.6 The Sharp Function and the Space BMO

If $f \in L^1_{\mathrm{loc}}(\mathbf{R}^n)$, we define the *sharp function* $f^\sharp(x)$ by the formula

$$f^\sharp(x) := \sup_{x \in Q} \frac{1}{|Q|}\int_Q \left|f(y) - f_Q\right| dy,$$

where the supremum is taken over all cubes Q containing x and $|Q|$ is the Lebesgue measure of Q. It should be noticed that we can take only those cubes Q for which x is an interior point.

If we define the sharp function $f^{\sharp\prime}(x)$ by the formula

$$f^{\sharp\prime}(x) := \sup_{x \in B} \frac{1}{|B|}\int_B |f(y) - f_B|\, dy,$$

where the supremum is taken over all balls B containing x, then, by arguing as in the proof of Lemma 4.2 we can obtain the inequalities

$$\frac{V_n}{2^n} f^{\sharp\prime}(x) \leq f^{\sharp}(x) \leq \frac{V_n}{2^{n/2}} f^{\sharp\prime}(x), \quad x \in \mathbf{R}^n,$$

where $V_n = \omega_n/n = \pi^{n/2}/\Gamma(n/2+1)$ is the volume of the unit ball in $\mathbf{R}^n$.

We begin by proving the following:

Claim 4.18 *For the maximal functions and sharp functions, we have the useful inequalities*

$$f^{\sharp}(x) \leq 2Mf(x), \quad x \in \mathbf{R}^n, \tag{4.27}$$

$$(|f|)^{\sharp}(x) \leq 2f^{\sharp}(x), \quad x \in \mathbf{R}^n. \tag{4.28}$$

Proof (i) First, we remark that

$$\left|f(y) - f_Q\right| \leq |f(y)| + \left|f_Q\right|,$$
$$\left|f_Q\right| \leq \frac{1}{|Q|}\int_Q |f(y)|\,dy \leq Mf(x).$$

Hence we have, for all cubes Q containing x,

$$\begin{aligned}\frac{1}{|Q|}\int_Q \left|f(y) - f_Q\right| dy &\leq \frac{1}{|Q|}\int_Q |f(y)|\,dy + \left|f_Q\right| \\ &\leq 2Mf(x).\end{aligned}$$

This proves the desired inequality (4.27).

(ii) Similarly, we remark that

$$\begin{aligned}\left||f(y)| - (|f|)_Q\right| &\leq \left||f(y)| - (|f_Q|)\right| + \left|(|f_Q|) - (|f|)_Q\right| \\ &\leq \left|f(y) - f_Q\right| + \left|(|f_Q|) - (|f|)_Q\right|,\end{aligned} \tag{4.29}$$

where

$$f_Q = \frac{1}{|Q|}\int_Q f(y)\,dy,$$
$$(|f|)_Q = \frac{1}{|Q|}\int_Q |f(y)|\,dy.$$

However, it follows that

$$\left|(|f|)_Q - |f_Q|\right| = \frac{1}{|Q|}\left|\int_Q \left(|f(y)| - |f_Q|\right) dy\right| \tag{4.30}$$

$$\leq \frac{1}{|Q|} \int_Q \big||f(y)| - |f_Q|\big| \, dy$$
$$\leq \frac{1}{|Q|} \int_Q \left|f(y) - f_Q\right| \, dy.$$

Therefore, by combining inequalities (4.29) and (4.30) we obtain that, for all cubes Q containing x,

$$\begin{aligned} \frac{1}{|Q|} \int_Q \big||f(y)| - (|f|)_Q\big| \, dy &\leq \frac{1}{|Q|} \int_Q \left|f(y) - f_Q\right| \, dy + \big||f_Q| - (|f|)_Q\big| \\ &\leq \frac{2}{|Q|} \int_Q \left|f(y) - f_Q\right| \, dy \\ &\leq 2 f^\sharp(x). \end{aligned}$$

This proves the desired inequality (4.28).

The proof of Claim 4.18 is complete. □

Secondly, it should be noticed that a function f is in BMO exactly when $f^\sharp$ is a bounded function in $\mathbf{R}^n$. More precisely, it is easy to verify that

$$\mathrm{BMO} = \left\{ f \in L^1_{\mathrm{loc}}(\mathbf{R}^n) : f^\sharp \in L^\infty(\mathbf{R}^n) \right\},$$

and further that

$$\|f\|_* = \left\|f^\sharp\right\|_{L^\infty(\mathbf{R}^n)} = \sup_Q \frac{1}{|Q|} \int_Q \left|f(y) - f_Q\right| \, dy.$$

In view of inequality (4.27), we obtain from Theorem 4.10 that, for $1 < p < \infty$,

$$\left\|f^\sharp\right\|_{L^p(\mathbf{R}^n)} \leq 2 \, \|Mf\|_{L^p(\mathbf{R}^n)} \leq 2C_p \, \|f\|_{L^p(\mathbf{R}^n)}.$$

On the other hand, by inequality (4.17) it follows that

$$|f(x)| \leq Mf(x) \quad \text{for almost every point } x \text{ of } \mathbf{R}^n,$$

so that

$$\|f\|_{L^p(\mathbf{R}^n)} \leq \|Mf\|_{L^p(\mathbf{R}^n)}.$$

Hence, we have the inequality

$$\left\|f^\sharp\right\|_{L^p(\mathbf{R}^n)} \leq 2C_p \, \|Mf\|_{L^p(\mathbf{R}^n)} \, 7!.$$

The next theorem asserts that the converse inequality holds true (see [137, Chap. IV, Sect. 2.2, Theorem 2], [183, Theorem 2.8.2], [97, Theorem 1.22]):

Theorem 4.19 *Let $1 < p < \infty$. If $f \in L^p(\mathbf{R}^n)$, then we have the inequality*

$$\|Mf\|_{L^p(\mathbf{R}^n)} \leq A_p \left\|f^\sharp\right\|_{L^p(\mathbf{R}^n)}, \tag{4.31}$$

with a positive constant A_p. For example, we may take

$$A_p := 2\left(2 \cdot 3^n\right)^{1/p} 4^{n+1} 2^{(n+1)p}.$$

Proof The proof is divided into four steps.

Step I: It suffices to show that

$$\|M(|f|)\|_{L^p(\mathbf{R}^n)} \leq c_p \left\|(|f|)^\sharp\right\|_{L^p(\mathbf{R}^n)}. \tag{4.31$'$}$$

Indeed, since we have, by Claim 4.18,

$$\begin{aligned} Mf(x) &= M(|f|)(x), \quad x \in \mathbf{R}^n, \\ (|f|)^\sharp(x) &\leq 2f^\sharp(x), \quad x \in \mathbf{R}^n, \end{aligned}$$

it follows from inequality (4.31′) that

$$\|Mf\|_{L^p(\mathbf{R}^n)} = \|M(|f|)\|_{L^p(\mathbf{R}^n)} \leq c_p \left\|(|f|)^\sharp\right\|_{L^p(\mathbf{R}^n)} \leq 2\,c_p \left\|f^\sharp\right\|_{L^p(\mathbf{R}^n)}.$$

This proves the desired inequality (4.31) with $A_p := 2c_p$.

Step II: Therefore, we may assume that

$$f(x) \geq 0 \quad \text{on } \mathbf{R}^n.$$

Our proof is based on the Calderón–Zygmund decomposition (Theorem 4.9). First, we see that this decomposition can be carried out for our function f. Let $t > 0$ and assume that Q is a cube such that

$$f_Q = \frac{1}{|Q|} \int_Q f(y)\,dy > t. \tag{4.32}$$

Then we have, for every $x \in Q$,

$$t < \frac{1}{|Q|} \int_Q f(y)\,dy \leq Mf(x),$$

and so

$$t^p \leq \frac{1}{|Q|} \int_Q (Mf(x))^p\,dx \leq \frac{1}{|Q|} \int_{\mathbf{R}^n} (Mf(x))^p\,dx.$$

This proves that

$$|Q| \leq \frac{1}{t^p} \int_{\mathbf{R}^n} (Mf(x))^p\,dx. \tag{4.33}$$

Hence, if $Q_1 \subset Q_2 \subset \cdots \subset Q_k \subset \cdots$ is an increasing family of dyadic cubes such that

$$f_{Q_k} = \frac{1}{|Q_k|} \int_{Q_k} f(y)\, dy > t,$$

then it follows from inequality (4.33) that the family is necessarily finite. Therefore, we find that each dyadic cube Q satisfying condition (4.32) is contained in a maximal one. If $\{Q_{t,j}\}$ is a family of these maximal dyadic cubes, it follows from an application of the Calderón–Zygmund decomposition (Theorem 4.9) that

- $t < f_{Q_{t,j}} = \dfrac{1}{|Q_{t,j}|} \displaystyle\int_{Q_{t,j}} f(y)\, dy \leq 2^n t.$
- $f(x) \leq t$ for almost every $x \in \mathbf{R}^n \setminus \bigcup_j Q_{t,j}$.

It should be noticed that if $t < s$, then each $Q_{s,j}$ is contained in some $Q_{t,k}$, since we have the inequality

$$f_{Q_{s,j}} = \frac{1}{|Q_{s,j}|} \int_{Q_{s,j}} f(y)\, dy > s > t.$$

For any given $t > 0$, we take

$$Q_0 := Q_{t/(2^{n+1}), j_0},$$
$$A := 2^{(n+1)p+2}.$$

Then there are two possibilities:

(A) $Q_0 \subset \{x \in \mathbf{R}^n : f^\sharp(x) > t/A\}$.
(B) $Q_0 \not\subset \{x \in \mathbf{R}^n : f^\sharp(x) > t/A\}$.

In the first case (A), we have the inequality

$$\sum_{\{j : Q_{t,j} \subset Q_0\}} |Q_{t,j}| \leq \left| \left\{ x \in \mathbf{R}^n : f^\sharp(x) > \frac{t}{A} \right\} \right|. \tag{4.34}$$

In the second case (B), we can find a point $x_0 \in Q_0$ such that

$$\frac{1}{|Q_0|} \int_{Q_0} |f(y) - f_{Q_0}|\, dy \leq f^\sharp(x_0) \leq \frac{t}{A}.$$

Here we remark the following inequalities:

$$\frac{t}{2^{n+1}} < f_{Q_0} = \frac{1}{|Q_0|}\int_{Q_0} f(y)\,dy \le 2^n \cdot \frac{t}{2^{n+1}} = \frac{t}{2}.$$

$$f_{Q_{t,j}} = \frac{1}{|Q_{t,j}|}\int_{Q_{t,j}} f(y)\,dy > t.$$

Then, since we have the inequality

$$\begin{aligned}\frac{t}{2}|Q_{t,j}| &= \left(t - \frac{t}{2}\right)|Q_{t,j}| \\ &\le \int_{Q_{t,j}} f(y)\,dy - f_{Q_0}|Q_{t,j}| = \int_{Q_{t,j}} (f(y) - f_{Q_0})\,dy \\ &\le \int_{Q_{t,j}} |f(y) - f_{Q_0}|\,dy,\end{aligned}$$

we obtain that

$$\begin{aligned}\frac{t}{2}\sum_{\{j:Q_{t,j}\subset Q_0\}} |Q_{t,j}| &\le \sum_{\{j:Q_{t,j}\subset Q_0\}} \int_{Q_{t,j}} |f(y) - f_{Q_0}|\,dy \\ &\le \int_{Q_0} |f(y) - f_{Q_0}|\,dy \\ &\le \frac{t}{A}|Q_0|.\end{aligned}$$

This proves that

$$\sum_{\{j:Q_{t,j}\subset Q_0\}} \left|Q_{t,j}\right| \le \frac{2}{A}\,|Q_0|\,. \tag{4.35}$$

Therefore, by combining inequalities (4.34) and (4.35) we have proved that

$$\sum_j |Q_{t,j}| \le \left|\left\{x \in \mathbf{R}^n : f^\sharp(x) > \frac{t}{A}\right\}\right| + \frac{2}{A}\sum_k \left|Q_{t/(2^{n+1}),k}\right|. \tag{4.36}$$

Step III: If we introduce a function

$$\alpha(t) := \sum_j |Q_{t,j}|,$$

then we can rewrite inequality (4.36) as follows:

$$\alpha(t) \le \left|\left\{x \in \mathbf{R}^n : f^\sharp(x) > \frac{t}{A}\right\}\right| + \frac{2}{A}\alpha\left(\frac{t}{2^{n+1}}\right). \tag{4.37}$$

If N is an arbitrary positive integer, we let

$$I_N := \int_0^N p\,t^{p-1}\,\alpha(t)\,dt.$$

In order to estimate the integral I_N, we need the following claim:

Claim 4.20 *If we introduce the* distribution function *of Mf by the formula*

$$\beta(t) := \left|\{x \in \mathbf{R}^n : Mf(x) > t\}\right|,$$

then we have the inequalities

$$\alpha(t) \leq \beta(t), \tag{4.38a}$$

$$\beta(t) \leq 3^n\,\alpha\left(\frac{t}{4^n}\right). \tag{4.38b}$$

Proof (i) Since we have, for any $x \in Q_{t,j}$,

$$t < \frac{1}{|Q_{t,j}|}\int_{Q_{t,j}} f(y)\,dy \leq Mf(x),$$

it follows that

$$\bigcup_j Q_{t,j} \subset \left\{x \in \mathbf{R}^n : Mf(x) > t\right\}.$$

This proves that

$$\alpha(t) = \sum_j \left|Q_{t,j}\right| \leq \left|\{x \in \mathbf{R}^n : Mf(x) > t\}\right| = \beta(t).$$

(ii) Just as in the proof of Theorem 4.9, we obtain that

$$\begin{aligned}\beta(t) = \left|\{x \in \mathbf{R}^n : Mf(x) > t\}\right| &\leq \sum_j \left|3Q_{t/4^n,j}\right| = 3^n \sum_j \left|Q_{t/4^n,j}\right| \\ &\leq 3^n\,\alpha\left(\frac{t}{4^n}\right).\end{aligned}$$

The proof of Claim 4.20 is complete. □

First, by inequality (4.38a) and Theorem 3.45 it follows that

$$\begin{aligned} I_N &= \int_0^N p\,t^{p-1}\,\alpha(t)\,dt \leq \int_0^N p\,t^{p-1}\,\beta(t)\,dt \\ &\leq \int_0^\infty p\,t^{p-1}\,\beta(t)\,dt = \int_{\mathbf{R}^n} Mf(x)^p\,dx \\ &= \|Mf\|_{L^p(\mathbf{R}^n)}^p\,. \end{aligned}$$

This proves that the finite limit

$$\lim_{N\to\infty} I_N = \int_0^\infty p\,t^{p-1}\,\alpha(t)\,dt \tag{4.39}$$

exists.

Secondly, we obtain from inequality (4.37) that

$$\begin{aligned} I_N &\leq \int_0^N p\,t^{p-1}\left|\left\{x \in \mathbf{R}^n : f^\sharp(x) > \frac{t}{A}\right\}\right| dt \\ &\quad + \frac{2}{A}\int_0^N p\,t^{p-1}\alpha\left(\frac{t}{2^{n+1}}\right) dt \\ &= \int_0^N p\,t^{p-1}\left|\left\{x \in \mathbf{R}^n : f^\sharp(x) > \frac{t}{A}\right\}\right| dt \\ &\quad + \frac{2}{A}\int_0^{N/(2^{n+1})} p\,(2^{n+1})^{p-1}\,s^{p-1}\alpha(s)\,2^{n+1}\,ds \\ &\leq \int_0^N p\,t^{p-1}\left|\left\{x \in \mathbf{R}^n : f^\sharp(x) > \frac{t}{A}\right\}\right| dt + \frac{1}{2}\int_0^N p\,s^{p-1}\alpha(s)\,ds \\ &= \int_0^N p\,t^{p-1}\left|\left\{x \in \mathbf{R}^n : f^\sharp(x) > \frac{t}{A}\right\}\right| dt + \frac{1}{2} I_N, \end{aligned} \tag{4.40}$$

since we have chosen

$$A = 2^{(n+1)p+2}.$$

Hence, by inequality (4.40) it follows that

$$I_N \leq 2\int_0^N p\,t^{p-1}\left|\left\{x \in \mathbf{R}^n : f^\sharp(x) > \frac{t}{A}\right\}\right| dt.$$

Therefore, in view of assertion (4.39) we can let $N \to \infty$ to obtain that

$$\int_0^\infty p\,t^{p-1}\,\alpha(t)\,dt \leq 2\int_0^\infty p\,t^{p-1}\left|\left\{x \in \mathbf{R}^n : f^\sharp(x) > \frac{t}{A}\right\}\right| dt. \tag{4.41}$$

Step IV: By inequalities (4.38b) and (4.41), it follows from an application of Theorem 3.45 that

$$\begin{aligned}
&\int_{\mathbf{R}^n} Mf(x)^p\,dx = \int_0^\infty p\,t^{p-1}\,\beta(t)\,dt \\
&\le 3^n \int_0^\infty p\,t^{p-1}\,\alpha\left(\frac{t}{4^n}\right)dt = 3^n \int_0^\infty p\,(4^n)^{p-1}\,s^{p-1}\,\alpha(s)\,4^n\,ds \\
&= 3^n \cdot 4^{np}\left(\int_0^\infty p\,s^{p-1}\,\alpha(s)\,ds\right) \\
&\le 3^n \cdot 4^{np}\left(2\int_0^\infty p\,t^{p-1}\left|\left\{x \in \mathbf{R}^n : f^\sharp(x) > \frac{t}{A}\right\}\right| dt\right) \\
&= 2 \cdot 3^n \cdot 4^{np} \int_0^\infty p\,A^{p-1}\,\sigma^{p-1}\left|\{x \in \mathbf{R}^n : f^\sharp(x) > \sigma\}\right| A\,d\sigma \\
&= 2 \cdot 3^n \cdot 4^{np} \cdot A^p \int_0^\infty p\,\sigma^{p-1}\left|\{x \in \mathbf{R}^n : f^\sharp(x) > \sigma\}\right| d\sigma \\
&= 2 \cdot 3^n \cdot 4^{np} \cdot A^p \int_{\mathbf{R}^n} f^\sharp(x)^p\,dx.
\end{aligned}$$

Summing up, we have proved the desired inequality

$$\|Mf\|_{L^p(\mathbf{R}^n)} \le c_p \left\|f^\sharp\right\|_{L^p(\mathbf{R}^n)}, \tag{4.31$'$}$$

with

$$c_p := \left(2 \cdot 3^n\right)^{1/p} 4^{n+1}\, 2^{(n+1)p}.$$

Now the proof of Theorem 4.19 is complete. □

Corollary 4.21 *Let $1 < p < \infty$. If $f \in L^p(\mathbf{R}^n)$, then we have the inequality*

$$\|f\|_{L^p(\mathbf{R}^n)} \le A_p \left\|f^\sharp\right\|_{L^p(\mathbf{R}^n)},$$

with a positive constant A_p.

Corollary 4.21 is an immediate consequence of inequality (4.17) and Theorem 4.19.

Remark 4.22 Theorem 4.19 and Corollary 4.21 remain valid if the norm is taken on the half-space $\mathbf{R}^n_+$. In this case, the definition of sharp functions need to be modified by taking only cubes Q contained in $\mathbf{R}^n_+$ (see Fig. 4.6).

4.7 Spherical Harmonics

In this section we introduce spherical harmonics in order to prove the fundamental results of singular integrals in Chaps. 10 and 11. The presentation here is a slightly expanded version of Neri [108] (see also [9, Chap. 5], [50, Chap. 2, Sect. H], [137, Chap. III, Sect. 3]).

If m is a non-negative integer, then we denote by Π_m the set of all polynomials in the variables

$$x = (x_1, x_2, \ldots, x_n), \quad n \geq 2,$$

which are homogeneous of degree m:

$$\Pi_m := \text{all homogeneous polynomials of degree } m.$$

A typical example of homogeneous polynomials of degree 2 is given by the formula

$$|x|^2 = x_1^2 + x_2^2 + \ldots + x_n^2.$$

For simplicity, we assume throughout this section that our polynomials have real coefficients, the complex case being of course identical except for the occasional presence of complex conjugates.

We begin by proving the following fundamental result:

Lemma 4.23 *The space Π_m is a finite-dimensional real vector space. Moreover, if we denote by $g(m)$ the dimension of the vector space Π_m, then we have the formula*

$$g(m) := \dim \Pi_m = \binom{m+n-1}{n-1} = \frac{(n+m-1)!}{m!(n-1)!}.$$

Proof (1) The first statement is obvious.

(2) Since the monomials x^α with $|\alpha| = m$ form a basis in the vector space Π_m, it follows that the dimension $g(m)$ is equal to the number of such monomials. However, it should be noticed that these monomials occur as coefficients in the formula

$$\prod_{j=1}^{n} \left(1 - x_j t\right)^{-1} = \prod_{j=1}^{n} \left(1 + x_j t + \ldots + x_j^k t^k + \ldots\right).$$

This implies that their number is equal to the coefficient of t^m in $(1-t)^{-n}$. Therefore, we obtain that

$$\begin{aligned} g(m) &= \frac{1}{m!} \frac{d^m}{dt^m}(1-t)^{-n}\bigg|_{t=0} \\ &= \frac{1}{m!} n(n+1)\cdots(n+m-1)(1-t)^{-(n+m)}\bigg|_{t=0} \end{aligned}$$

$$= \frac{(n+m-1)!}{m!(n-1)!} = \binom{n+m-1}{n-1}.$$

The proof of Lemma 4.23 is complete. □

We let

$$P(x) = \sum_{|\alpha|=m} a_\alpha x^\alpha \in \Pi_m,$$

and introduce the homogeneous differential polynomial

$$P\left(\frac{\partial}{\partial x}\right) = \sum_{|\alpha|=m} a_\alpha \left(\frac{\partial}{\partial x}\right)^\alpha,$$

which is obtained from $P(x)$ by replacing each monomial x^α by the corresponding differential monomial $(\partial/\partial x)^\alpha$. Now we can prove the following:

Lemma 4.24 *Every space Π_m is a real inner product space endowed with the inner product*

$$\langle P, Q\rangle := P\left(\frac{\partial}{\partial x}\right) Q(x), \quad P, Q \in \Pi_m,$$

Proof First, it should be noticed that we have, for $|\alpha| = |\beta|$,

$$\left(\frac{\partial}{\partial x}\right)^\alpha x^\beta = \begin{cases} \alpha! & \text{if } \alpha = \beta, \\ 0 & \text{otherwise.} \end{cases}$$

Hence it is easy to verify that the formula $\langle P, Q\rangle$, $P, Q \in \Pi_m$, defines a bilinear form on the product space $\Pi_m \times \Pi_m$. Moreover, if $P(x) = \sum_\alpha a_\alpha x^\alpha$ and $Q(x) = \sum_\beta b_\beta x^\beta$ with $|\alpha| = |\beta| = m$, then we obtain that

- $\langle P, Q\rangle = \sum_{|\alpha|=m} \alpha!\, a_\alpha b_\alpha = \sum_{|\alpha|=m} \alpha!\, b_\alpha a_\alpha = \langle Q, P\rangle,$
- $\langle P, P\rangle = \sum_{|\alpha|=m} \alpha!\, a_\alpha^2 \geq 0,$
- $\langle P, P\rangle = 0$ if and only if $P(x) \equiv 0$.

This proves that the quantity $\langle P, Q\rangle$ satisfies the axioms (I1) through (I4) of inner product in Sect. 2.10.

The proof of Lemma 4.23 is complete. □

Let $P \in \Pi_\ell$ and $Q \in \Pi_m$ with $\ell \leq m$. Then it is easy to see that the polynomial

$$\langle P, Q\rangle = P\left(\frac{\partial}{\partial x}\right) Q(x)$$

belongs to the space $\Pi_{m-\ell}$. Hence it follows that the associated differential polynomial $P(\partial/\partial x)$ defines a linear operator

$$P\left(\frac{\partial}{\partial x}\right) : \Pi_m \longrightarrow \Pi_{m-\ell}$$

by the formula $\langle P, Q\rangle$. Moreover, we have the following:

Lemma 4.25 *Let $P \in \Pi_\ell$ such that $P(x) \not\equiv 0$, and $m \geq \ell$. Then the linear map $P\left(\frac{\partial}{\partial x}\right) : \Pi_m \to \Pi_{m-\ell}$ is surjective.*

Proof Our proof is based on a reduction to absurdity. Assume, to the contrary, that the range $P(\partial/\partial x)\,\Pi_m$ is a proper subspace V of $\Pi_{m-\ell}$. Then, by using Lemma 4.24 we can find an element $R(x) \not\equiv 0$ in the inner product space $\Pi_{m-\ell}$ which is orthogonal to the space V. Therefore, we have, for all $Q \in \Pi_m$,

$$\left\langle R, P\left(\frac{\partial}{\partial x}\right) Q(x)\right\rangle = R\left(\frac{\partial}{\partial x}\right) P\left(\frac{\partial}{\partial x}\right) Q(x) \equiv 0.$$

In particular, by choosing $Q(x) := R(x)P(x)$ we obtain that $\langle Q, Q\rangle = 0$, so that $Q(x) = R(x)P(x) \equiv 0$. This implies that $R(x) \equiv 0$, since $P(x) \not\equiv 0$. This contradiction proves that $P(\partial/\partial x) : \Pi_m \to \Pi_{m-\ell}$ is surjective.

The proof of Lemma 4.25 is complete. □

Theorem 4.26 (the decomposition theorem) *Let $P \in \Pi_\ell$ such that $P(x) \not\equiv 0$, and $m \geq \ell$. Then, every element $T \in \Pi_m$ can be decomposed uniquely in the form*

$$T(x) = \sum_k P^k(x)\, R_k(x) \tag{4.42}$$

where

$$R_k \in \Pi_{m-k\ell},$$
$$P\left(\frac{\partial}{\partial x}\right) R_k(x) \equiv 0,$$

and the summation is taken over all non-negative integers k such that $k\ell \leq m$. Moreover, the $R_k(x) \not\equiv 0$ are not divisible by $P(x)$.

Proof (1) First, we show that every element $T \in \Pi_m$ can be decomposed uniquely in the form

$$T = PS_1 + R_0,$$

where $S_1 \in \Pi_{m-\ell}$ and $P(\partial/\partial x)\, R_0(x) \equiv 0$.

To do this, we introduce the subspace

$$M := \{PS : S \in \Pi_{m-\ell}\}$$

of all elements of Π_m which are divisible by $P(x)$, and the subspace

$$N := \left\{ R \in \Pi_m : P\left(\frac{\partial}{\partial x}\right) R(x) \equiv 0 \right\}$$

of all elements of Π_m which are annihilated by $P(\partial/\partial x)$. Then we have the assertion

$$M \cap N = \{0\}.$$

Indeed, if $PS \in M$ belongs to N, then it follows that

$$P\left(\frac{\partial}{\partial x}\right) P(x)S(x) \equiv 0,$$

so that

$$S\left(\frac{\partial}{\partial x}\right) P\left(\frac{\partial}{\partial x}\right) P(x)S(x) = \langle PS, PS \rangle = 0.$$

This implies that $PS = 0$. Hence, we have the formula

$$\dim(M \oplus N) = \dim(M) + \dim(N) = g(m - \ell) + \dim(N).$$

However, by Lemma 4.25 it follows that

$$g(m - \ell) + \dim(N) = \dim(\Pi_m),$$

so that,

$$\dim(M \oplus N) = \dim(\Pi_m).$$

Therefore, we have proved that

$$\Pi_m = M \oplus N,$$

and further that each element $T \in \Pi_m$ can be decomposed uniquely in the form

$$T = PS_1 + R_0,$$

with $S_1 \in \Pi_{m-\ell}$ and $R_0 \in N$. Moreover, it should be emphasized that $R_0(x)$ is not divisible by $P(x)$ unless $R_0(x) \equiv 0$, since we have $M \cap N = \{0\}$.

(2) Similarly, by repeating the same procedure for $S_1(x)$ (in place of $T(x)$) we obtain that

$$S_1 = PS_2 + R_1,$$

where $S_2 \in \Pi_{m-2\ell}$ and $R_1(x)$ is annihilated by $P(\partial/\partial x)$ and not divisible by $P(x)$ (unless $R_1(x) \equiv 0$). We remark that

$$T = PS_1 + R_0 = P\,(PS_2 + R_1) + R_0 = P^2 S_2 + PR_1 + R_0.$$

(3) This recursive process ends when we reach an element $S_k \in \Pi_{m-k\ell}$ of degree less than ℓ. In this case, it follows that $S_k(x)$ is not divisible by $P(x)$ and that $P\,(\partial/\partial x)\,S_k(x) \equiv 0$. Hence, we may take $S_k = R_k$. This proves the desired decomposition (4.42).

The proof of Theorem 4.26 is complete. □

It should be emphasized that the homogeneous polynomial

$$|x|^2 = x_1^2 + x_2^2 + \ldots + x_n^2$$

of degree 2 corresponds to the usual Laplacian

$$P\left(\frac{\partial}{\partial x}\right) = \frac{\partial^2}{\partial x_1^2} + \frac{\partial^2}{\partial x_2^2} + \ldots + \frac{\partial^2}{\partial x_n^2} = \Delta.$$

All polynomials $P \in \Pi_m$ are called *solid harmonics* of degree m if they satisfy the Laplace equation: $\Delta P = 0$ in $\mathbf{R}^n$. The restrictions of these solid harmonics to the unit sphere Σ_{n-1} are called *spherical harmonics* of degree m.

We shall use the following notation:

$$\mathcal{S}_m := \text{all solid harmonics of degree } m,$$
$$\mathcal{H}_m := \text{all spherical harmonics of degree } m.$$

Remark 4.27 (a) We find from the proof of Theorem 4.26 with $P(x) := |x|^2$ that (see [50, Proposition (2.49)])

$$\Pi_m = |x|^2\,\Pi_{m-2} \oplus \mathcal{S}_m.$$

(b) By Lemma 4.25, it follows that $\mathcal{S}_m$ is the nullspace of the Laplacian as a linear map of Π_m onto Π_{m-2}. If $m = 0$ and $m = 1$, then we have $\mathcal{S}_m = \Pi_m$. However, if $m \geq 2$, then we obtain from assertion (a) that $\mathcal{S}_m$ is a proper subspace Π_m of dimension

$$d(m) := \dim \mathcal{S}_m = g(m) - g(m-2).$$

This formula remains valid for all non-negative integers m if we let $g(-1) = g(-2) = 0$.

(c) If $S_m \in \mathcal{S}_m$, then, we have, for all $x \neq 0$ and $x' = x/|x|$,

$$S_m(x) = S_m\left(|x|\frac{x}{|x|}\right) = |x|^m\,S_m\left(\frac{x}{|x|}\right) = |x|^m\,Q_m(x'),$$

where $Q_m(x') = S_m(x')$ is the corresponding spherical harmonic in $\mathcal{H}_m$. Since this correspondence is an isomorphism, it follows from assertion (a) that $\mathcal{H}_m$ is a vector space of dimension $d(m) = g(m) - g(m-2)$.

(d) On the Euclidean space $\mathbf{R}^2$, we have, for any integer $m \geq 1$,

$$g(m) = \binom{m+1}{1} = m+1, \quad g(m-2) = m-1.$$

Hence it follows that $d(m) = g(m) = m+1$ for $m = 0$ and 1, and that $d(m) = g(m) - g(m-2) = 2$ for $m \geq 2$. In terms of the complex variable $z = x + iy = re^{i\theta}$, we obtain that $z^m = r^m \cos m\theta + ir^m \sin m\theta$. Therefore, we can conclude that, for every m, $\cos m\theta$ and $\sin m\theta$ are the only linearly independent spherical harmonics on the plane.

On the Euclidean space $\mathbf{R}^3$, it is easy to see that $d(m) = 2m+1$, so the dimension of $\mathcal{S}_m$, and hence that of $\mathcal{H}_m$, increases as the odd integers.

(e) On the general Euclidean space $\mathbf{R}^n$, we have, as $m \to \infty$,

$$\begin{aligned} g(m) &= \binom{m+n-1}{n-1} = \frac{(m+n-1)(m+n-2)\cdots(m+1)}{(n-1)!} \\ &\sim \frac{m^{n-1}}{(n-1)!}. \end{aligned}$$

Moreover, by using the mean value theorem we obtain that

$$\begin{aligned} d(m) &= g(m) - g(m-2) \\ &\sim \left[\frac{m^{n-1}}{(n-1)!} - \frac{(m-2)^{n-1}}{(n-1)!} \right] \sim 2\frac{m^{n-2}}{(n-2)!}. \end{aligned}$$

Summing up, we have proved that, for any dimension $n \geq 2$,

$$d(m) \sim c(n)\, m^{n-2}$$

as $m \to \infty$, where $c(n)$ is a positive constant depending only on the dimension n.

Corollary 4.28 *Any continuous function on the unit sphere Σ_{n-1} can be approximated uniformly by a finite linear combination of spherical harmonics.*

Proof Let $f(x')$ be an arbitrary continuous function on Σ_{n-1}. We may assume that $f(x')$ is real valued. By the Stone–Weierstrass theorem (Theorem 2.73), it follows that $f(x')$ can be approximated uniformly by the restriction to Σ_{n-1} of some polynomial $T(x)$. However, since $T(x)$ is a finite sum of homogeneous pieces, we have only to consider each piece $T_m \in \Pi_m$.

We remark that the assertion is obvious for $m = 0$ and $m = 1$.

If $m \geq 2$, by applying the decomposition theorem (Theorem 4.26) to $P(x) := |x|^2$ and $P(\partial/\partial x) := \Delta$ we obtain that

$$T_m(x) = \sum_{2k \le m} |x|^{2k} R_k(x),$$

where $R_k(x)$ are solid harmonics of degree $m - 2k$. Hence, we have, on the unit sphere Σ_{n-1},

$$T_m(x') = \sum_{2k \le m} R_k(x'),$$

where the $R_k(x')$ belong to $\mathcal{H}_{m-2k}$.

The proof of Corollary 4.28 is complete. □

Lemma 4.29 *Let $Q_j \in \mathcal{H}_j$ and $S_j(x) = |x|^j Q_j(x')$ its corresponding solid harmonics. Then we have, for $k \neq m$,*

$$\int_{\Sigma_{n-1}} Q_k(x')\, Q_m(x')\, d\sigma(x') = 0, \tag{4.43}$$

$$\int_{\{|x| \le 1\}} S_k(x)\, S_m(x)\, dx = 0. \tag{4.44}$$

Here $d\sigma$ is the surface measure on the unit sphere Σ_{n-1}.

Proof Since we have, in terms of polar coordinates,

$$\begin{aligned}
&\int_{\{|x| \le 1\}} S_k(x)\, S_m(x)\, dx \\
&= \left(\int_0^1 r^{k+m} r^{n-1}\, dr \right) \int_{\Sigma_{n-1}} Q_k(x')\, Q_m(x')\, d\sigma(x') \\
&= \frac{1}{k+m+n} \int_{\Sigma_{n-1}} Q_k(x')\, Q_m(x')\, d\sigma(x'),
\end{aligned}$$

it suffices to prove formula (4.43) for spherical harmonics.

By using Green's formula (see Theorem 5.4 in Chap. 5), we obtain that

$$\begin{aligned}
&\int_{\{|x| \le 1\}} (S_k(x) \cdot \Delta S_m(x) - S_m(x) \cdot \Delta S_k(x))\, dx \\
&= \int_{\Sigma_{n-1}} \left(S_k(x') \frac{\partial S_m}{\partial \boldsymbol{\nu}}(x') - S_m(x') \frac{\partial S_k}{\partial \boldsymbol{\nu}} \right) d\sigma(x'),
\end{aligned}$$

where $\boldsymbol{\nu}$ is the outward normal to the unit sphere Σ_{n-1}. However, since we have the formula

$$\frac{\partial}{\partial \boldsymbol{\nu}} = \frac{\partial}{\partial |x|},$$

it follows that

$$\frac{\partial}{\partial \boldsymbol{\nu}} \left(S_j(x) \right) = j |x|^{j-1} Q_j(x').$$

Moreover, by the definition of solid harmonics it follows that $\Delta S_j = 0$ in $\mathbf{R}^n$. Therefore, we obtain that

$$\begin{aligned} 0 &= \int_{\Sigma_{n-1}} \left(Q_k(x') \cdot m Q_m(x') - Q_m(x') \cdot k Q_k(x')\right) d\sigma(x') \\ &= (m-k) \int_{\Sigma_{n-1}} Q_k(x')\, Q_m(x')\, d\sigma(x'). \end{aligned}$$

This proves that we have, for $k \neq m$,

$$\int_{\Sigma_{n-1}} Q_k(x')\, Q_m(x')\, d\sigma(x') = 0.$$

The proof of Lemma 4.29 is complete. □

We consider the vector spaces $\mathcal{H}_m$ as linear subspaces of the real Hilbert space $L^2(\Sigma_{n-1})$ with inner product

$$(f, g) = \int_{\Sigma_{n-1}} f(x')\, g(x')\, d\sigma(x').$$

With respect to this inner product, we can construct an orthonormal basis $\{Y_{\ell m}\}$, $1 \leq \ell \leq d(m)$, in each space $\mathcal{H}_m$. Hence, we have, for each m,

$$\left(Y_{im}, Y_{jm}\right) = \int_{\Sigma_{n-1}} Y_{im}(x')\, Y_{jm}(x')\, d\sigma(x') = \delta_{ij} \quad \text{for } 1 \leq i, j \leq d(m).$$

Moreover, we have the following theorem (cf. Theorem 2.66):

Theorem 4.30 *(i) The Hilbert space $L^2(\Sigma_{n-1})$ can be decomposed into the infinite direct sum in the sense of Hilbert space theory:*

$$L^2(\Sigma_{n-1}) = \sum_{m=0}^{\infty} \mathcal{H}_m.$$

More precisely, any function $f \in L^2(\Sigma_{n-1})$ has the development

$$f\left(x'\right) = \sum_{m=0}^{\infty} Y_m(x'), \quad Y_m(x') \in \mathcal{H}_m,$$

where the convergence is in the $L^2(\Sigma_{n-1})$ norm, and we have the formula

$$\int_{\Sigma_{n-1}} |f(x')|^2\, d\sigma(x') = \sum_{m=0}^{\infty} \int_{\Sigma_{n-1}} |Y_m(x')|^2\, d\sigma(x').$$

(ii) Furthermore, the family $\{Y_{\ell m}\}_{\substack{m=0,1,\ldots,\\ \ell=1,2,\ldots,d(m)}}$ *forms a* complete orthonormal system *in the Hilbert space* $L^2(\Sigma_{n-1})$. *More precisely, any function* $f \in L^2(\Sigma_{n-1})$ *has the Fourier series expansion with respect to* $\{Y_{\ell m}\}$

$$f(x') = \sum_{m=0}^{\infty} \sum_{\ell=1}^{d(m)} a_{\ell m}\, Y_{\ell m}(x'),$$

where the Fourier coefficients $a_{\ell m}$ *are given by the formulas*

$$a_{\ell m} = \int_{\Sigma_{n-1}} f(y)\, Y_{\ell m}(y)\, d\sigma(y),$$

and satisfy the Parseval identity

$$\sum_{m=0}^{\infty} \sum_{\ell=1}^{d(m)} |a_{\ell m}|^2 = \int_{\Sigma_{n-1}} |f(x')|^2\, d\sigma(x').$$

Proof Since part (i) is an immediate consequence of part (ii), we have only to prove part (ii).

First, by Lemma 4.29 it follows that spherical harmonics of distinct degrees are orthogonal, so that the family $\{Y_{\ell m}\}$ is orthonormal. On the other hand, since the continuous functions are dense in $L^2(\Sigma_{n-1})$, it follows from an application of Corollary 4.28 that the set of all finite linear combinations of the $Y_{\ell m}$ is dense in $L^2(\Sigma_{n-1})$. This implies that the orthonormal system $\{Y_{\ell m}\}$ is complete in the Hilbert space $L^2(\Sigma_{n-1})$.

The proof of Theorem 4.30 is complete. □

Now we prove some important bounds on the spherical harmonics $\{Y_{\ell m}\}$ and their derivatives. These bounds can be deduced from the fact that we have, for each integer $m \geq 0$,

$$\sum_{\ell=1}^{d(m)} Y_{\ell m}^2(x') = \frac{d(m)}{\omega_n}, \quad x' \in \Sigma_{n-1}, \tag{4.45}$$

where

$$d(m) = g(m) - g(m-2),$$

and

$$\omega_n := |\Sigma_{n-1}| = \frac{2\pi^{n/2}}{\Gamma(n/2)}$$

is the surface area of the unit sphere Σ_{n-1}. In order to prove formula (4.45), we introduce the concept of zonal harmonics:

Lemma 4.31 *For each m and each point $x' \in \Sigma_{n-1}$, we can find a spherical harmonic $Z_{x'} \in \mathcal{H}_m$, called a* zonal harmonic *with pole x', such that we have, for all $Q \in \mathcal{H}_m$,*

$$Q\left(x'\right) = (Q, Z_{x'}). \tag{4.46}$$

Moreover, we have, for all $y \in \Sigma_{n-1}$,

$$Z_{x'}(y) = \sum_{\ell=1}^{d(m)} Y_{\ell m}(x')\, Y_{\ell m}(y). \tag{4.47}$$

Proof (1) First, we remark that the map $Q \to Q(x')$ is a linear functional over the finite-dimensional Hilbert space $\mathcal{H}_m$. Hence, the desired formula (4.46) follows from an application of the Riesz representation theorem (Theorem 2.60).

(2) Secondly, since the $\{Y_{\ell m}\}$, $1 \leq \ell \leq d(m)$, form an orthonormal basis in each space $\mathcal{H}_m$, we obtain that any element $Q \in \mathcal{H}_m$ can be written uniquely in the form

$$Q\left(x'\right) = \sum_{\ell=1}^{d(m)} a_\ell\, Y_{\ell m}(x') \quad \text{for } x' \in \Sigma_{n-1}, \tag{4.48}$$

where the Fourier coefficients a_ℓ are given by the formulas

$$a_\ell := a_{\ell m} = (Q, Y_{\ell m}) = \int_{\Sigma_{n-1}} Q(y)\, Y_{\ell m}(y)\, d\sigma(y).$$

Hence we have, by formula (4.46) and formula (4.48),

$$(Q, Z_{x'}) = \sum_{\ell=1}^{d(m)} (Q, Y_{\ell m}) Y_{\ell m}(x') = \left(Q, \sum_{\ell=1}^{d(m)} Y_{\ell m}(x')\, Y_{\ell m}\right). \tag{4.49}$$

This proves the desired formula (4.47), since formula (4.49) holds true for all Q in $\mathcal{H}_m$.

The proof of Lemma 4.31 is complete. □

Lemma 4.32 *If $u\colon \Sigma_{n-1} \to \Sigma_{n-1}$ is a rotation, then we have, for any zonal harmonic $Z_{x'}$,*

$$Z_{ux'}(uy) = Z_{x'}(y), \quad y \in \Sigma_{n-1}. \tag{4.50}$$

Proof If the function $Q(x')$ belongs to $\mathcal{H}_m$, then so does the function $Q(ux')$, since the Laplacian Δ is invariant under rotations. By applying formula (4.46) to the harmonic $Q_u(x') := Q(ux')$, we obtain that

$$\begin{aligned} Q\left(ux'\right) = Q_u(x') &= \int_{\Sigma_{n-1}} Q_u(y)\, Z_{x'}(y)\, d\sigma(y) \\ &= \int_{\Sigma_{n-1}} Q(uy)\, Z_{x'}(y)\, d\sigma(y). \end{aligned} \tag{4.51}$$

On the other hand, by applying again formula (4.46) to the harmonic $Q(x')$ and by changing variables we obtain that

$$Q\left(ux'\right) = \int_{\Sigma_{n-1}} Q(y)\, Z_{ux'}(y)\, d\sigma(y) = \int_{\Sigma_{n-1}} Q(uy)\, Z_{ux'}(uy)\, d\sigma(y), \qquad (4.52)$$

since the surface measure $d\sigma$ is also invariant under rotations.

Therefore, it follows from formula (4.51) and formula (4.52) that we have, for all $Q \in \mathcal{H}_m$,

$$\int_{\Sigma_{n-1}} Z_{x'}(y)\, Q(uy)\, d\sigma(y) = \int_{\Sigma_{n-1}} Z_{ux'}(uy)\, Q(uy)\, d\sigma(y).$$

This proves the desired formula (4.50).

The proof of Lemma 4.32 is complete. □

Remark 4.33 If the pole $x' \in \Sigma_{n-1}$ of a zonal harmonic $Z_{x'}$ lies on the axis of a rotation u, then we have $ux' = x'$ and so, by Lemma 4.32,

$$Z_{x'}(y) = Z_{x'}(uy), \quad y \in \Sigma_{n-1}.$$

In other words, a zonal harmonic $Z_{x'}$ is constant along each "parallel" relative to the pole x'.

The next lemma proves the desired formula (4.45):

Lemma 4.34 *For any integer $m \geq 0$, we have the formula*

$$\sum_{\ell=1}^{d(m)} Y_{\ell m}^2(x') = \frac{d(m)}{\omega_n}, \quad x' \in \Sigma_{n-1}. \qquad (4.45)$$

Proof By formula (4.46), we have, for any $x' \in \Sigma_{n-1}$,

$$Z_{x'}\left(x'\right) = \sum_{\ell=1}^{d(m)} Y_{\ell m}^2(x'),$$

and we have, for any rotation u,

$$Z_{ux'}\left(ux'\right) = \sum_{\ell=1}^{d(m)} Y_{\ell m}^2\left(ux'\right).$$

Hence, by applying Lemma 4.32 to our situation we obtain that

$$\sum_{\ell=1}^{d(m)} Y_{\ell m}^2 (x') = \sum_{\ell=1}^{d(m)} Y_{\ell m}^2(ux') \quad \text{for } x' \in \Sigma_{n-1}. \tag{4.53}$$

Since each point y of Σ_{n-1} is the image $y = ux'$ under a suitable rotation u, we find from formula (4.53) that

$$\sum_{\ell=1}^{d(m)} Y_{\ell m}^2(x') = A_m \quad \text{for } x' \in \Sigma_{n-1}, \tag{4.54}$$

where A_m is a constant depending only on the degree m and the dimension n.

Hence, by integrating formula (4.54) over Σ_{n-1} we obtain that

$$\int_{\Sigma_{n-1}} \sum_{\ell=1}^{d(m)} Y_{\ell m}^2(x')\, d\sigma(x') = A_m \int_{\Sigma_{n-1}} d\sigma(x') = A_m \omega_n. \tag{4.55}$$

On the other hand, since the $\{Y_{\ell m}\}$, $1 \leq \ell \leq d(m)$, form an orthonormal basis in the Hilbert space $L^2(\Sigma_{n-1})$, it follows that

$$\begin{aligned} \int_{\Sigma_{n-1}} \sum_{\ell=1}^{d(m)} Y_{\ell m}^2(x')\, d\sigma(x') &= \sum_{\ell=1}^{d(m)} \int_{\Sigma_{n-1}} Y_{\ell m}(x')\, Y_{\ell m}(x')\, d\sigma(x') \\ &= d(m). \end{aligned} \tag{4.56}$$

Therefore, by combining formula (4.55) and formula (4.56) we have proved that

$$A_m = \frac{d(m)}{\omega_n}.$$

The proof of Lemma 4.34 is complete. □

Now we can prove the desired bounds on the spherical harmonics $Y_{\ell m}$:

Theorem 4.35 *(i) For all $x' \in \Sigma_{n-1}$, we have the estimate*

$$\left| Y_{\ell m} (x') \right| \leq C_1\, m^{(n-2)/2}, \tag{4.57}$$

where C_1 is a positive constant depending only on the dimension n.

(ii) More generally, we have, for any multi-index α,

$$\left| \left(\frac{\partial}{\partial x} \right)^{\alpha} \left(|x|^m Y_{\ell m}(x') \right) \right| \leq C_2\, m^{(n-2)/2+|\alpha|} |x|^{m-|\alpha|}, \tag{4.58}$$

where C_2 is a positive constant depending only on $|\alpha|$ and n.

Proof (i) By Lemma 4.34 and assertion (e) of Remark 4.27, it follows that we have, for some positive constant C depending only on n,

$$Y_{\ell m}^2\left(x'\right) \leq \sum_{\ell=1}^{d(m)} Y_{\ell m}^2(x') = \frac{d(m)}{\omega_n} \leq C\, m^{n-2} \quad \text{for } x' \in \Sigma_{n-1}.$$

Hence, the desired estimate (4.57) follows by taking the square root of this inequality.

(ii) First, we remark that estimate (4.58) for $|\alpha| = 0$ is reduced to estimate (4.57).

Secondly, we consider the case where $|\alpha| = 1$. To do this, we let

$$P(x) := |x|^m\, Y_{\ell m}\left(x'\right) \quad \text{for } 1 \leq \ell \leq d(m),$$

and estimate the first partial derivatives $\partial P/\partial x_j$.

Since we have the assertion

$$P \in \mathcal{S}_m,$$

and so

$$\Delta\left(\frac{\partial P}{\partial x_j}\right) = \frac{\partial}{\partial x_j}(\Delta P) = 0 \quad \text{in } \mathbf{R}^n,$$

it follows that

$$\frac{\partial P}{\partial x_j} \in \mathcal{S}_{m-1}.$$

Hence, by applying Lemma 4.34 to the functions $\partial P/\partial x_j$ and by using Schwarz's inequality (Theorem 3.24 with $p = q := 2$) we obtain that

$$\begin{aligned}
\left|\frac{\partial P}{\partial x_j}(x')\right|^2 &= \left|\sum_{\ell=1}^{d(m-1)} b_\ell\, Y_{\ell m-1}(x')\right|^2 \\
&\leq \left(\sum_{\ell=1}^{d(m-1)} b_\ell^2\right)\left(\sum_{\ell=1}^{d(m-1)} Y_{\ell m-1}^2(x')\right) \\
&= A\,\frac{d(m-1)}{\omega_n},
\end{aligned}$$

where the Fourier coefficients b_ℓ are given by the formulas

$$\begin{aligned}
b_\ell &:= \left(\frac{\partial P}{\partial x_j}, Y_{\ell m-1}\right) \\
&= \int_{\Sigma_{n-1}} \frac{\partial P}{\partial x_j}(y)\, Y_{\ell m-1}(y)\, d\sigma(y) \quad \text{for } 1 \leq \ell \leq d(m-1),
\end{aligned}$$

and A is a positive constant given by the formula

$$A := \sum_{\ell=1}^{d(m-1)} b_\ell^2 = \int_{\Sigma_{n-1}} \left| \frac{\partial P}{\partial x_j}(x') \right|^2 d\sigma(x').$$

Therefore, we have the estimate

$$\begin{aligned} \left| \frac{\partial P}{\partial x_j}(x') \right|^2 &\leq \frac{d(m-1)}{\omega_n} \int_{\Sigma_{n-1}} \left| \frac{\partial P}{\partial x_j} \right|^2 d\sigma \\ &\leq \frac{d(m-1)}{\omega_n} \int_{\Sigma_{n-1}} |\text{grad } P|^2 \, d\sigma \quad \text{for } x' \in \Sigma_{n-1}. \end{aligned} \tag{4.59}$$

However, since grad P is homogeneous of degree $m-1$, by letting $r := |x|$ and by using polar coordinates we obtain that

$$\begin{aligned} \int_{\{|x|\leq 1\}} |\text{grad } P|^2 \, dx &= \left(\int_0^1 r^{2m+n-3} \, dr \right) \int_{\Sigma_{n-1}} |\text{grad } P|^2 \, d\sigma \\ &= \frac{1}{2m+n-2} \int_{\Sigma_{n-1}} |\text{grad } P|^2 \, d\sigma, \end{aligned}$$

so that

$$\int_{\Sigma_{n-1}} |\text{grad } P|^2 \, d\sigma = (2m+n-2) \int_{\{|x|\leq 1\}} |\text{grad } P|^2 \, dx. \tag{4.60}$$

Moreover, since $P(x)$ is harmonic in $\mathbf{R}^n$, it follows that

$$\text{div}\,(P \text{ grad } P) = |\text{grad } P|^2 + P \cdot (\Delta P) = |\text{grad } P|^2 .$$

Hence, we have, by the divergence theorem (see Theorem 5.3 in Chap. 5),

$$\begin{aligned} \int_{\{|x|\leq 1\}} |\text{grad } P|^2 \, dx &= \int_{\{|x|\leq 1\}} \text{div}\,(P \text{ grad } P) \, dx \\ &= \int_{\Sigma_{n-1}} P \, (\text{grad } P) \cdot \boldsymbol{\nu} \, d\sigma. \end{aligned} \tag{4.61}$$

However, since $P(x)$ is homogeneous of degree m, we have, by Euler's formula for homogeneous functions,

$$\begin{aligned} &(\text{grad } P) \cdot \boldsymbol{\nu} = m \, P\left(x'\right), \\ &P(x) = |x|^m \, Y_{\ell m}\left(x'\right), \end{aligned}$$

Hence it follows that

$$\int_{\Sigma_{n-1}} P\,(\operatorname{grad} P)\cdot \boldsymbol{\nu}\, d\sigma \tag{4.62}$$
$$= m \int_{\Sigma_{n-1}} P\left(x'\right)^2 d\sigma(x') = m \int_{\Sigma_{n-1}} Y_{\ell m}\left(x'\right)^2 d\sigma(x')$$
$$= m \quad \text{for } 1 \leq \ell \leq d(m-1).$$

By combining formulas (4.60), (4.61) and (4.62), we obtain that

$$\int_{\Sigma_{n-1}} |\operatorname{grad} P|^2\, d\sigma = \int_{\Sigma_{n-1}} \left|\operatorname{grad}\left(|x|^m\, Y_{\ell m}(x')\right)\right|^2 d\sigma$$
$$= m\,(2m+n-2) \quad \text{for } 1 \leq \ell \leq d(m-1).$$

Therefore, by carrying this formula into estimate (4.59) we obtain from assertion (e) with $m := m-1$ of Remark 4.27 that

$$\begin{aligned}\left|\frac{\partial P}{\partial x_j}(x')\right|^2 &\leq \frac{d(m-1)}{\omega_n}\, m\,(2m+n-2)\\ &\leq C(m-1)^{n-2}\, m\,(2m+n-2)\\ &\leq C'\, m^n \quad \text{for all } x' \in \Sigma_{n-1},\end{aligned}$$

where C and C' are positive constants depending on n. Moreover, since $\partial P/\partial x_j$ is homogeneous of degree $m-1$, we have, for some positive constant C'' depending on n,

$$\left|\frac{\partial P}{\partial x_j}(x)\right| \leq C''\, m^{n/2}\, |x|^{m-1} \quad \text{for all } x \in \mathbf{R}^n.$$

This proves the desired estimate (4.58) for $|\alpha| = 1$.

Repeating this argument, we obtain the desired estimate (4.58) for the general case.

The proof of Theorem 4.35 is complete. □

By part (ii) of Theorem 4.30, we know that any function $f \in L^2(\Sigma_{n-1})$ can be expanded by the Fourier series

$$f\left(x'\right) = \sum_{m=0}^{\infty} \sum_{\ell=1}^{d(m)} a_{\ell m}\, Y_{\ell m}\left(x'\right), \tag{4.63}$$

where the Fourier coefficients $a_{\ell m}$ are given by the formulas

$$a_{\ell m} = \int_{\Sigma_{n-1}} f(y)\, Y_{\ell m}(y)\, d\sigma(y). \tag{4.64}$$

and satisfy the Parseval identity

$$\sum_{m=0}^{\infty}\sum_{\ell=1}^{d(m)}|a_{\ell m}|^2 = \int_{\Sigma_{n-1}} \left|f\left(x'\right)\right|^2 \, d\sigma(x').$$

Remark 4.36 Any continuous function $f(x')$ on the unit sphere Σ_{n-1} has a continuous extension $\widetilde{f}(x) := f(x/|x|)$ to the space $\mathbf{R}^n \setminus \{0\}$ which is homogeneous of degree zero. Conversely, any function $g \in C(\mathbf{R}^n \setminus \{0\})$ which is homogeneous of degree zero is of the form $g(x) = \widetilde{f}(x)$, where $f(x)$ is its restriction to the unit sphere Σ_{n-1}. In the following, we shall not distinguish between $f(x')$ and $\widetilde{f}(x)$, and if we write $f(x') \in C^\infty(\Sigma_{n-1})$, then we mean that $\widetilde{f}(x) \in C^\infty(\mathbf{R}^n \setminus \{0\})$.

If $f(x') \in C^\infty(\Sigma_{n-1})$, then it is well known that the Fourier series (4.63) converges absolutely and uniformly to $f(x')$ in the unit sphere Σ_{n-1}. In fact, we shall prove (Theorem 4.40) that a necessary and sufficient condition for a sequence

$$\{a_{\ell m}\}_{\substack{m=0,1,\ldots, \\ \ell=1,2,\ldots,d(m)}}$$

to be the "harmonic Fourier coefficients" of some function $f(x') \in C^\infty(\Sigma_{n-1})$ is that the sequence $\{a_{\ell m}\}$ be rapidly decreasing. This result is a consequence of the formula (4.71) analogous to formula (4.64), which we shall prove in Theorem 4.39 below.

First, we introduce a second order differential operator $\mathcal{L}$, which preserves the degree of homogeneity of a function, given by the formula

$$\mathcal{L}f := |x|^2 \, \Delta f.$$

We remark that $\mathcal{L}f = \Delta f$ on the unit sphere Σ_{n-1}. Then we have the following lemma:

Lemma 4.37 *For any function $Y_{\ell m}(x')$ and any integer $r \geq 1$, we have the formula*

$$\mathcal{L}^r \, Y_{\ell m}(x') = (-m)^r \, (m+n-2)^r \, Y_{\ell m}(x') \quad \text{for } x' \in \Sigma_{n-1}. \tag{4.65}$$

Proof First, we prove formula (4.65) for $r = 1$.

If $P(x) := |x|^m Y_{\ell m}(x')$ is the corresponding solid harmonic, then it follows that

$$\mathcal{L}Y_{\ell m}(x') = |x|^2 \, \Delta Y_{\ell m} = |x|^2 \, \Delta \left(|x|^{-m} \, P(x)\right). \tag{4.66}$$

However, we remark the following elementary formulas:

- $\Delta(fg) = f \, \Delta g + 2(\text{grad } f) \cdot (\text{grad } g) + g \, \Delta f,$
- $\text{grad}\left(|x|^k\right) = k|x|^{k-2} \, x,$
- $\Delta\left(|x|^k\right) = k \, (k+n-2) \, |x|^{k-2},$
- $x \cdot (\text{grad } P) = m \, P(x).$

Hence, by the harmonicity of $P(x)$ it follows that

$$\begin{aligned}
&\Delta\left(|x|^{-m} P(x)\right) \\
&= 2\left(\text{grad}\,|x|^{-m}\right)\cdot(\text{grad}\,P) + (-m)\,(-m+n-2)\;|x|^{-m-2}\,P(x) \\
&= 2(-m)|x|^{-m-2}\,x\cdot(\text{grad}\,P) + (-m)\,(-m+n-2)\;|x|^{-m-2}\,P(x) \\
&= 2\left(-m^2\right)|x|^{-m-2}\,P(x) + (-m)\,(-m+n-2)\,|x|^{-m-2}\,P(x) \\
&= (-m)\,(m+n-2)\;|x|^{-m-2}\,P(x).
\end{aligned}$$

Therefore, by substituting this formula into formula (4.66) we obtain that

$$\begin{aligned}
\mathcal{L}Y_{\ell m}(x') &= (-m)\,(m+n-2)\;|x|^{-m}\,\Delta\left(|x|^{-m}\,P(x)\right) \\
&= (-m)\,(m+n-2)\,Y_{\ell m}(x').
\end{aligned}$$

This proves the desired formula (4.65) for $r = 1$.

The desired formula (4.65) for any integer $r \geq 2$ follows by iteration.

The proof of Lemma 4.37 is complete. □

Lemma 4.38 *If $f,\ g \in C^{2r}\left(\mathbf{R}^n \setminus \{0\}\right)$ are homogeneous of degree zero, then we have the formula*

$$\int_{\Sigma_{n-1}} f\cdot\mathcal{L}^r g\,d\sigma = \int_{\Sigma_{n-1}} g\cdot\mathcal{L}^r f\,d\sigma. \tag{4.67}$$

Proof Using polar coordinates, we obtain that, for some constant $C \neq 0$,

$$\begin{aligned}
&\int_{\{1\leq|x|\leq 2\}} \left(f(x)\cdot\Delta g(x) - g(x)\cdot\Delta f(x)\right)dx \qquad (4.68)\\
&= C\int_{\Sigma_{n-1}} \left(f\left(x'\right)\cdot\Delta g\left(x'\right) - g\left(x'\right)\cdot\Delta f\left(x'\right)\right)d\sigma(x').
\end{aligned}$$

However, we have, by Green's formula (see Theorem 5.4 in Chap. 5),

$$\begin{aligned}
&\int_{\{1\leq|x|\leq 2\}} \left(f(x)\cdot\Delta g(x) - g(x)\cdot\Delta f(x)\right)dx \qquad (4.69)\\
&= \int_{\{|x|=2\}} \left(f\left(x'\right)\frac{\partial g}{\partial \boldsymbol{\nu}}(x') - g\left(x'\right)\frac{\partial f}{\partial \boldsymbol{\nu}}(x')\right)d\sigma(x') \\
&\quad - \int_{\{|x|=1\}} \left(f\left(x'\right)\frac{\partial g}{\partial \boldsymbol{\nu}}(x') - g\left(x'\right)\frac{\partial f}{\partial \boldsymbol{\nu}}(x')\right)d\sigma(x') \\
&= 0.
\end{aligned}$$

Indeed, it suffices to note that

$$\frac{\partial}{\partial \boldsymbol{\nu}} = \frac{\partial}{\partial |x|}$$

is the radical derivative and that $f(x)$ and $g(x)$ are constant along radii.

Therefore, by combining formula (4.68) and formula (4.69) we obtain that

$$\int_{\Sigma_{n-1}} (f \cdot \Delta g - g \cdot \Delta f)\, d\sigma = 0. \tag{4.70}$$

Since we have $\mathcal{L} f = \Delta f$ and $\mathcal{L} g = \Delta g$ on Σ_{n-1}, it follows from formula (4.70) that

$$\int_{\Sigma_{n-1}} f \cdot \mathcal{L} g\, d\sigma = \int_{\Sigma_{n-1}} g \cdot \mathcal{L} f\, d\sigma.$$

Moreover, by replacing g by $\mathcal{L} g$ in this formula we obtain that

$$\int_{\Sigma_{n-1}} f \cdot \mathcal{L}^2 g\, d\sigma = \int_{\Sigma_{n-1}} \mathcal{L} g \cdot \mathcal{L} f\, d\sigma = \int_{\Sigma_{n-1}} g \cdot \mathcal{L}^2 f\, d\sigma.$$

Hence, the desired formula (4.67) follows by repeating this process.

The proof of Lemma 4.38 is complete. □

Theorem 4.39 *If $f \in C^\infty(\Sigma_{n-1})$ and if $\{a_{\ell m}\}$ are its Fourier coefficients with respect to the spherical harmonics $\{Y_{\ell m}\}$, then we have, for any integer $r \geq 0$,*

$$a_{\ell m} = (-m)^{-r} (m + n - 2)^{-r} \int_{\Sigma_{n-1}} Y_{\ell m} \cdot \mathcal{L}^r f\, d\sigma. \tag{4.71}$$

Proof By combining Lemma 4.37 and Lemma 4.38, we obtain that

$$\begin{aligned} a_{\ell m} &= \int_{\Sigma_{n-1}} f \cdot Y_{\ell m}\, d\sigma \\ &= (-m)^{-r} (m + n - 2)^{-r} \int_{\Sigma_{n-1}} f \cdot \mathcal{L}^r Y_{\ell m}\, d\sigma \\ &= (-m)^{-r} (m + n - 2)^{-r} \int_{\Sigma_{n-1}} Y_{\ell m} \cdot \mathcal{L}^r f\, d\sigma. \end{aligned}$$

This proves the desired formula (4.71). □

Theorem 4.40 *(i) If $f \in C^\infty(\Sigma_{n-1})$ and if $\{a_{\ell m}\}$ are its Fourier coefficients with respect to the spherical harmonics $\{Y_{\ell m}\}$, then we have, for any integer $r \geq 0$,*

$$\sum_{m=0}^{\infty} \sum_{\ell=1}^{d(m)} m^r \, |a_{\ell m}| < \infty. \tag{4.72}$$

(ii) Conversely, given any family $\{a_{\ell m}\}_{\substack{m=0,1,\ldots,\\ \ell=1,2,\ldots,d(m)}}$ of constants which satisfies condition (4.72) for any integer $r \geq 0$, we can construct a function $f \in C^\infty(\Sigma_{n-1})$ such that

$$f\left(x'\right) = \sum_{m=0}^{\infty} \sum_{\ell=1}^{d(m)} a_{\ell m}\, Y_{\ell m}(x').$$

Proof (i) We have only to prove estimate (4.72) for all r sufficiently large. By formula (4.71), it follows that

$$|a_{\ell m}| = m^{-r}\,(m+n-2)^{-r} \left| \int_{\Sigma_{n-1}} Y_{\ell m}(y)\, \mathcal{L}^r f(y)\, d\sigma(y) \right|.$$

Hence, by using Schwarz's inequality (Theorem 3.24 with $p = q := 2$) and the normality of $Y_{\ell m}$ we obtain that

$$\begin{aligned} &|a_{\ell m}| \qquad\qquad (4.73)\\ &\leq m^{-2r} \left(\int_{\Sigma_{n-1}} |Y_{\ell m}(y)|^2\, d\sigma(y) \right)^{1/2} \left(\int_{\Sigma_{n-1}} \left|\mathcal{L}^r f(y)\right|^2\, d\sigma(y) \right)^{1/2} \\ &= m^{-2r} \left(\int_{\Sigma_{n-1}} \left|\mathcal{L}^r f(y)\right|^2\, d\sigma(y) \right)^{1/2} \\ &= C(r)\, m^{-2r}, \end{aligned}$$

where $C(r)$ is a positive constant defined by the formula

$$C(r) := \left(\int_{\Sigma_{n-1}} \left|\mathcal{L}^r f(y)\right|^2\, d\sigma(y) \right)^{1/2}.$$

Since we have, by assertion (e) of Remark 4.27,

$$d(m) \leq C\, m^{n-2}$$

for some positive constant C, it follows from inequality (4.73) that we have, for all integers $r > n-1$,

$$\begin{aligned} \sum_{m=0}^{\infty} \sum_{\ell=1}^{d(m)} m^r\, |a_{\ell m}| &\leq C(r) \sum_{m=1}^{\infty} d(m)\, m^{-r} \\ &\leq C C(r) \sum_{m=1}^{\infty} m^{n-2-r} < \infty. \end{aligned}$$

This proves part (i).

(ii) Conversely, we assume that the constants $\{a_{\ell m}\}$ satisfy condition (4.72) for all integers $r \geq 0$. Then it follows from an application of Theorem 4.35 that we have the estimate

$$\left|Y_{\ell m}(x')\right| \leq C\, m^{(n-2)/2} \quad \text{for all } x' \in \Sigma_{n-1},$$

with a positive constant C independent of m. This proves that

$$\sum_{m=0}^{\infty} \sum_{\ell=1}^{d(m)} \left| a_{\ell m} Y_{\ell m}(x') \right| \leq C \sum_{m=0}^{\infty} \sum_{\ell=1}^{d(m)} |a_{\ell m}| \, m^{(n-2)/2} \quad \text{for all } x' \in \Sigma_{n-1}.$$

Therefore, we obtain from condition (4.72) that the series

$$\sum_{\ell,m} a_{\ell m} Y_{\ell m}(x')$$

converges absolutely and uniformly to some continuous function $f(x')$ in the unit sphere Σ_{n-1}.

Moreover, by repeating this argument and by using estimate (4.58) of Theorem 4.35 we obtain that the series $\sum_{\ell,m} a_{\ell m} (\partial/\partial x)^{\alpha} Y_{\ell m}(x')$ is absolutely and uniformly convergent to the continuous function $(\partial/\partial x)^{\alpha} f(x')$ in the unit sphere Σ_{n-1}. Summing up, we have proved that $f \in C^{\infty}(\Sigma_{n-1})$.

The proof of Theorem 4.40 is complete. □

By combining Theorems 4.30, 4.35 and 4.40, we have proved the following fundamental results for spherical harmonics:

Theorem 4.41 *(i) The space $\mathcal{H}_m$ of n-dimensional spherical harmonics of degree m has dimension*

$$d(m) = g(m) - g(m-2),$$

where

$$g(\ell) = \binom{\ell+n-1}{n-1} = \frac{(\ell+n-1)!}{(n-1)!\,\ell!}, \quad \ell \geq 0,$$
$$g(-1) = g(-2) = 0.$$

Moreover, we have the estimate

$$d(m) \leq c(n)\, m^{n-2}, \tag{4.74}$$

with a positive constant $c(n)$ depending on n.

(ii) The Hilbert space $L^2(\Sigma_{n-1})$ can be decomposed into the infinite direct sum in the sense of Hilbert space theory:

$$L^2(\Sigma_{n-1}) = \sum_{m=0}^{\infty} \mathcal{H}_m.$$

More precisely, any function $f \in L^2(\Sigma_{n-1})$ has the development

$$f\left(x'\right)=\sum_{m=0}^{\infty} Y_m(x') \quad with\ Y_m \in \mathcal{H}_m,$$

where the convergence is in the $L^2(\Sigma_{n-1})$ norm, and we have the formula

$$\int_{\Sigma_{n-1}} \left|f\left(x'\right)\right|^2 d\sigma\left(x'\right)=\sum_{m=0}^{\infty}\int_{\Sigma_{n-1}} \left|Y_m(x')\right|^2 d\sigma(x').$$

(iii) If $Y_m(x')$ is any n-dimensional spherical harmonic of degree m, then its partial derivatives $(\partial^\alpha Y_m/\partial x^\alpha)(x')$ satisfy the estimates

$$\sup_{x'\in\Sigma_{n-1}}\left|\frac{\partial^\alpha Y_m}{\partial x^\alpha}(x')\right| \leq c\, m^{|\alpha|+(n-2)/2}, \tag{4.75}$$

with a positive constant $c = c(n, |\alpha|)$ depending on the multi-indices α and n.

(iv) The spherical harmonics $\{Y_{km}\}_{\substack{m=0,1,\ldots,\\ k=1,2,\ldots,d(m)}}$ form a complete orthonormal system in the Hilbert space $L^2(\Sigma_{n-1})$. If $f \in C^\infty(\Sigma_{n-1})$ has the Fourier series expansion with respect to $\{Y_{km}\}$

$$f(x')=\sum_{m=0}^{\infty}\sum_{k=1}^{d(m)} a_{km}\, Y_{km}(x'),$$

$$a_{km}=\int_{\Sigma_{n-1}} f(y)\, Y_{km}(y)\, d\sigma(y),$$

then the Fourier coefficients a_{km} satisfy, for any integer $r > 1$, the estimate

$$|a_{km}| \leq c(r)\, m^{-2r}, \tag{4.76}$$

with a positive constant $c(r)$ depending on r.

Part II
Elements of Function Spaces

Chapter 5
Harmonic Functions and Poisson Integrals

The purpose of this chapter is to study harmonic functions in the half-space $\mathbf{R}^{n+1}_+$ in terms of Poisson integrals of functions in various L^p spaces, paying particular attention to the Poisson kernel $P(x, y)$ of the Dirichlet problem in the half-space (Lemma 5.11, Theorems 5.12, 5.15 and 5.16). We establish fundamental relationships between means of derivatives of Poisson integrals $u(x, y)$ taken with respect to the normal variable y and those taken with respect to the tangential variables x_i (Theorems 5.21 and 5.26). More precisely, Theorem 5.21 (resp. Theorem 5.26) establishes fundamental relationships between means of the first (resp. second) derivatives of $u(x, y)$ taken with respect to y and those taken with respect to x_i.

The results here are adapted from Taibleson [146] and Folland [50] (see also [9, Chap. 1]).

5.1 Lipschitz Domains and Green's Formulas

An open set Ω in $\mathbf{R}^n$ is called a *Lipschitz hypograph* if its boundary $\partial\Omega$ can be represented as the graph of a Lipschitz continuous function. Namely, there exists a Lipschitz continuous function $\zeta : \mathbf{R}^{n-1} \to \mathbf{R}$ such that (see Fig. 5.1)

$$\Omega = \left\{x = (x', x_n) \in \mathbf{R}^n : x_n < \zeta(x'), \quad x' \in \mathbf{R}^{n-1}\right\}. \tag{5.1}$$

Sometimes, a different smoothness condition will be needed, so we broaden the above terminology as follows (see [181, Sect. 2]): For any non-negative integer k and any $0 < \theta \leq 1$, we say that the domain Ω defined by formula (5.1) is a $C^{k,\theta}$ *hypograph* if the function ζ is of class $C^{k,\theta}$, that is, if ζ is of class C^k and its k-th order partial derivatives are Hölder continuous with exponent θ.

K. Taira, *Real Analysis Methods for Markov Processes*,
https://doi.org/10.1007/978-981-97-3659-1_5

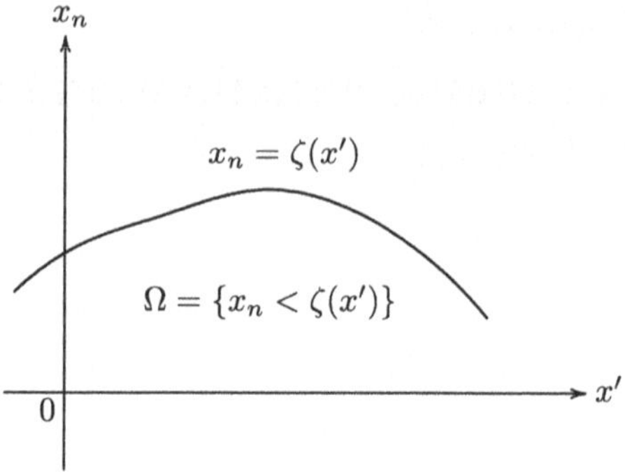

Fig. 5.1 The Lipschitz hypograph $\Omega = \{x_n < \zeta(x')\}$

The next definition requires that, roughly speaking, the boundary of Ω can be represented locally as the graph of a Lipschitz continuous function, by using different systems of Cartesian coordinates for different parts of the boundary:

Definition 5.1 Let Ω be a bounded domain in Euclidean space $\mathbf{R}^n$ with boundary $\partial\Omega$. We say that Ω is a *Lipschitz domain* if there exist finite families $\{U_j\}_{j=1}^J$ and $\{\Omega_j\}_{j=1}^J$ having the following three properties (see Fig. 5.2):

(i) The family $\{U_j\}_{j=1}^J$ is a finite open covering of $\partial\Omega$.
(ii) Each Ω_j can be transformed to a Lipschitz hypograph by a rigid motion, that is, by a rotation plus a translation.
(iii) The set Ω satisfies the conditions

$$U_j \cap \Omega = U_j \cap \Omega_j, \quad 1 \leq j \leq J.$$

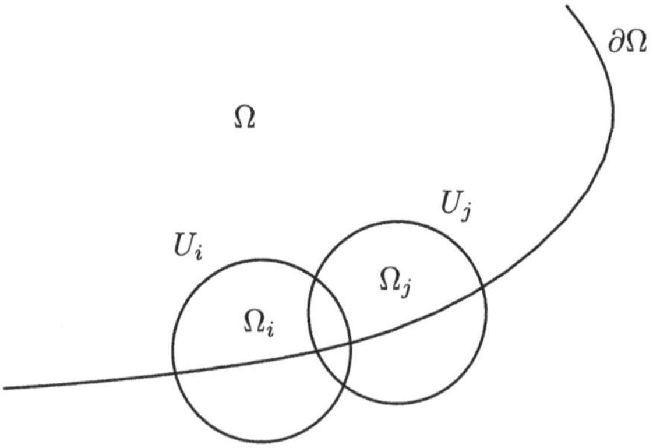

Fig. 5.2 The families $\{U_j\}$ and $\{\Omega_j\}$ of the Lipschitz domain Ω

In the obvious way, we define a $C^{k,\theta}$ *domain* by substituting "$C^{k,\theta}$" for "Lipschitz" throughout Definition 5.1. It should be emphasized that a Lipschitz domain is the same thing as a $C^{0,1}$ domain.

If Ω is a Lipschitz hypograph defined by formula (5.1), then we remark that its boundary

$$\partial\Omega = \left\{x = (x', \zeta(x')) : x' \in \mathbf{R}^{n-1}\right\}$$

is an $(n-1)$-dimensional, $C^{0,1}$ submanifold of $\mathbf{R}^n$. Hence we find that $\partial\Omega$ has a surface measure $d\sigma$ and an outward unit normal $\boldsymbol{\nu}$ which exists $d\sigma$-almost everywhere in $\mathbf{R}^{n-1}$, if we apply Rademacher's theorem [97, Chap. 1, Corollary 1.73], [138, Theorem] as follows:

Theorem 5.2 (Rademacher) *Any Lipschitz continuous function on* $\mathbf{R}^n$ *admits* L^∞ *first partial derivatives almost everywhere in* $\mathbf{R}^n$.

Indeed, it follows from an application of Rademacher's theorem that the function $\zeta(x')$ is Fréchet differentiable almost everywhere in $\mathbf{R}^{n-1}$ with

$$\|\nabla\zeta\|_{L^\infty(\mathbf{R}^{n-1})} \leq C, \tag{5.2}$$

where C is any Lipschitz constant for the function $\zeta(x')$. Then we have the following formulas for $d\sigma$ and $\boldsymbol{\nu}$ (see Fig. 5.3):

$$d\sigma = \sqrt{1 + |\nabla\zeta(x')|^2}\, dx',$$
$$\boldsymbol{\nu} = \frac{(-\nabla\zeta(x'), 1)}{\sqrt{1 + |\nabla\zeta(x')|^2}}.$$

Here it should be noticed that we have, by inequality (5.2),

$$1 \leq \sqrt{1 + |\nabla\zeta(x')|^2} \leq \sqrt{1 + C^2},$$

so that the surface measure $d\sigma$ is equivalent locally to the Lebesgue measure dx'.

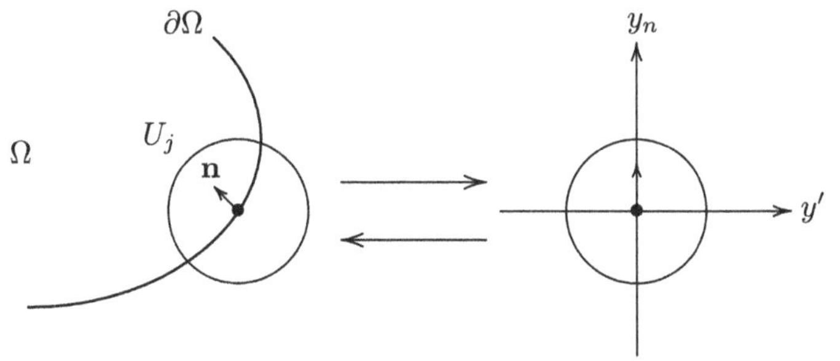

Fig. 5.3 The unit inward normal $\mathbf{n} = -\boldsymbol{\nu}$ to the boundary $\partial\Omega = \{x_n = \zeta(x')\}$

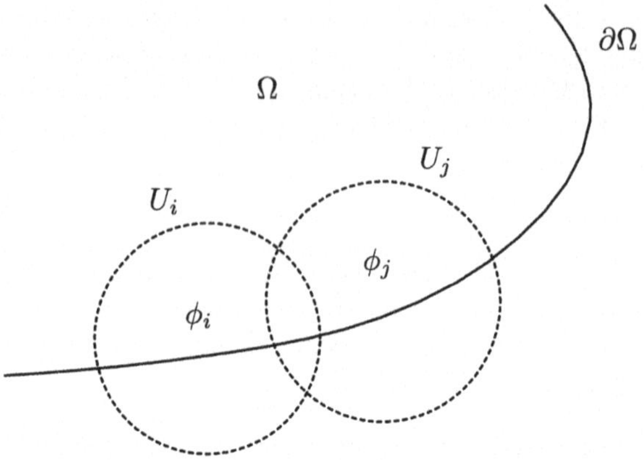

Fig. 5.4 The partition of unity $\{\phi_j\}$ subordinate to the open covering $\{U_j\}$ of $\partial\Omega$

We consider the case where Ω is a bounded Lipschitz domain. By using the notation of Definition 5.1, we choose a partition of unity $\{\phi_j\}_{j=1}^J$ subordinate to the open covering $\{U_j\}_{j=1}^J$ of $\partial\Omega$ (see Fig. 5.4), that is,

$$\phi_j \in C_0^\infty(U_j),$$
$$0 \le \phi_j(x) \le 1 \quad \text{in } U_j,$$
$$\sum_{j=1}^J \phi_j(x) = 1 \quad \text{on } \partial\Omega.$$

Then, upon using local coordinate systems flattening out the boundary $\partial\Omega$, together with a partition of unity, we can prove the following divergence theorem for Lipschitz domains:

Theorem 5.3 (the divergence theorem) *Let Ω be a bounded, Lipschitz domain of $\mathbf{R}^n$ with boundary $\partial\Omega$. If $F = (f_1, f_2, \ldots, f_n)$, is a C^1 vector field, then we have the formula*

$$\int_\Omega \sum_{i=1}^n \frac{\partial f_i}{\partial x_i}\, dx = \int_{\partial\Omega} \sum_{i=1}^n f_i \nu_i \, d\sigma, \quad f_i \in C^1(\overline{\Omega}), \tag{5.3}$$

or equivalently,

$$\int_\Omega \operatorname{div} F\, dx = \int_{\partial\Omega} (F, \boldsymbol{\nu})\, d\sigma,$$

where $\boldsymbol{\nu} = (\nu_1, \nu_2, \ldots, \nu_n)$ is the unit outward normal to $\partial\Omega$ and $d\sigma$ is the surface measure of $\partial\Omega$.

5.2 Harmonic Functions

We start with the following elementary result:

Let $I = (a, b)$ be an open interval of $\mathbf{R}$. If $u \in C^2(I)$ and $d^2u/dx^2 = 0$ in I, then $u(x)$ is a linear function. In particular, we have, for all sufficiently small $r > 0$,

$$u(x) = \frac{1}{2}[u(x+r) + u(x-r)] \quad \text{(the mean value property)},$$

$$u(x) = \frac{1}{2r}\int_{-r}^{r} u(x+z)\,dz = \frac{1}{2}\int_{-1}^{1} u(x+ry)\,dy.$$

The purpose of this section is to extend these results to the n-dimensional case by replacing the operator d^2/dx^2 by the usual Laplacian

$$\Delta = \frac{\partial^2}{\partial x_1^2} + \frac{\partial^2}{\partial x_2^2} + \ldots + \frac{\partial^2}{\partial x_n^2}.$$

To do this, we need the following Green formulas for Lipschitz domains:

Theorem 5.4 (Green's formula) *Let Ω be a bounded, Lipschitz domain of $\mathbf{R}^n$ with boundary $\partial\Omega$. For u, $v \in C^2(\overline{\Omega})$, we have the formulas*

$$\int_{\Omega} (v\Delta u + \nabla v \cdot \nabla u)\,dx = \int_{\partial\Omega} v\frac{\partial u}{\partial \boldsymbol{\nu}}\,d\sigma, \tag{5.4a}$$

$$\int_{\Omega} (v\Delta u - u\Delta v)\,dx = \int_{\partial\Omega} \left(v\frac{\partial u}{\partial \boldsymbol{\nu}} - \frac{\partial v}{\partial \boldsymbol{\nu}}u\right)\,d\sigma. \tag{5.4b}$$

Proof (i) Formula (5.4a) follows by applying the divergence theorem (Theorem 5.3) with

$$f = v\,\nabla u = \left(v\frac{\partial u}{\partial x_1}, v\frac{\partial u}{\partial x_2}, \ldots, v\frac{\partial u}{\partial x_n}\right).$$

(ii) Formula (5.4b) follows by interchanging u and v in formula (5.4a) and then by subtracting this formula from formula (5.4a).

The proof of Theorem 5.4 is complete. □

A function $u(x)$ defined in Ω is said to be *harmonic* if it is twice continuously differentiable in Ω and satisfies the Laplace equation

$$\Delta u = \frac{\partial^2 u}{\partial x_1^2} + \frac{\partial^2 u}{\partial x_2^2} + \ldots + \frac{\partial^2 u}{\partial x_n^2} = 0 \quad \text{in } \Omega.$$

Corollary 5.5 *If a function $u \in C^2(\Omega)$ is harmonic in Ω, then we have, for any Lipschitz subdomain Ω' of Ω,*

$$\int_{\partial\Omega'} \frac{\partial u}{\partial \boldsymbol{\nu}}\,d\sigma = 0.$$

Proof It suffices to take $v := 1$ in formula (5.4a). □

Theorem 5.6 (the mean value theorem) *Assume that a function $u \in C^2(\Omega)$ is harmonic in Ω. If $x \in \Omega$ and $r > 0$ is small enough so that $\overline{B(x,r)} \subset \Omega$, then we have the formulas*

$$u(x) = \frac{1}{r^{n-1}\omega_n} \int_{S(x,r)} u(y)\, d\sigma(y) \tag{5.5a}$$

$$= \frac{1}{\omega_n} \int_{S(0,1)} u(x + rz)\, d\sigma(z). \tag{5.5b}$$

Here

$$\omega_n := |\Sigma_{n-1}| = \frac{2\pi^{n/2}}{\Gamma(n/2)}$$

is the surface area of the unit sphere $\Sigma_{n-1} = S(0,1)$ in $\mathbf{R}^n$.

Proof For any $0 < \varepsilon < r$, we consider the annular domain (see Fig. 5.5)

$$\Gamma_{\varepsilon r} := \{y \in \Omega : \varepsilon < |y - x| < r\} = B(x,r) \setminus \overline{B(x,\varepsilon)}.$$

(i) The proof of the first formula (5.5a): We let

$$v(y) = \begin{cases} \frac{1}{|x-y|^{n-2}} & \text{if } n \geq 3, \\ -\log|x-y| & \text{if } n = 2. \end{cases}$$

Then, since we have the formula

$$\frac{\partial}{\partial \boldsymbol{\nu}} = \begin{cases} \frac{1}{r}\sum_{j=1}^{n}(y_j - x_j)\frac{\partial}{\partial y_j} & \text{on } S(x,r) = \{x \in \Omega : |y - x| = r\}, \\ \frac{1}{\varepsilon}\sum_{j=1}^{n}(x_j - y_j)\frac{\partial}{\partial y_j} & \text{on } S(x,\varepsilon) = \{x \in \Omega : |y - x| = \varepsilon\}, \end{cases}$$

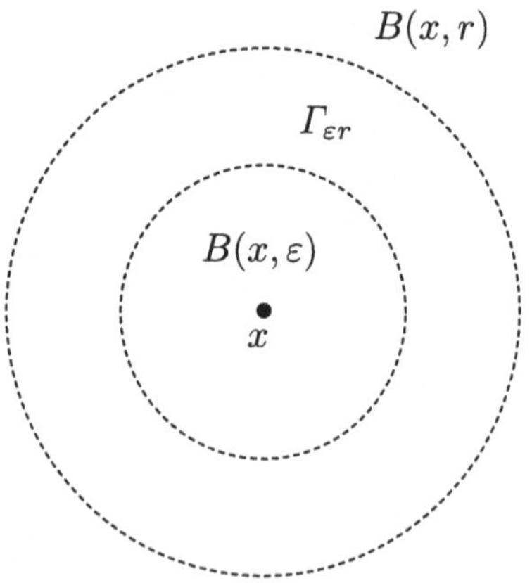

Fig. 5.5 The open balls $B(x,r)$, $B(x,\varepsilon)$ and the annular domain $\Gamma_{\varepsilon r}$

it follows that

$$\frac{\partial v}{\partial \boldsymbol{\nu}} = \begin{cases} (2-n)\frac{1}{r^{n-1}} & \text{on } S(x,r), \\ -(2-n)\frac{1}{\varepsilon^{n-1}} & \text{on } S(x,\varepsilon). \end{cases}$$

Furthermore, it is easy to see that

$$\Delta v(y) = 0 \quad \text{for all } y \in \Omega \setminus \{x\}.$$

Thus, by applying Green's formula (5.4b) with $\Omega := \Gamma_{\varepsilon r}$ we obtain that

$$\begin{aligned} 0 &= \int_{S(x,r)} \left(v\frac{\partial u}{\partial \boldsymbol{\nu}} - \frac{\partial v}{\partial \boldsymbol{\nu}} u \right) d\sigma + \int_{S(x,\varepsilon)} \left(v\frac{\partial u}{\partial \boldsymbol{\nu}} - \frac{\partial v}{\partial \boldsymbol{\nu}} u \right) d\sigma \\ &= \int_{S(x,r)} \frac{1}{r^{n-2}} \frac{\partial u}{\partial \boldsymbol{\nu}} \, d\sigma - \int_{S(x,r)} (2-n)\frac{1}{r^{n-1}} u \, d\sigma \\ &\quad + \int_{S(x,\varepsilon)} \frac{1}{\varepsilon^{n-2}} \frac{\partial u}{\partial \boldsymbol{\nu}} \, d\sigma + \int_{S(x,\varepsilon)} (2-n)\frac{1}{\varepsilon^{n-1}} u \, d\sigma. \end{aligned} \tag{5.6}$$

However, by Corollary 5.5 it follows that

$$\int_{S(x,r)} \frac{\partial u}{\partial \boldsymbol{\nu}} \, d\sigma = 0,$$
$$\int_{S(x,\varepsilon)} \frac{\partial u}{\partial \boldsymbol{\nu}} \, d\sigma = 0.$$

Hence we have, by formula (5.6),

$$\frac{1}{r^{n-1}} \int_{S(x,r)} u(y) \, d\sigma(y) = \frac{1}{\varepsilon^{n-1}} \int_{S(x,\varepsilon)} u(z) \, d\sigma(z). \tag{5.7}$$

Now we need the following:

Claim 5.7 *We have, as* $\varepsilon \downarrow 0$,

$$\frac{1}{\varepsilon^{n-1}} \int_{S(x,\varepsilon)} u(z) \, d\sigma(z) \longrightarrow \omega_n \, u(x).$$

Proof Indeed, by the continuity of $u(x)$ it follows that

$$\begin{aligned} &\frac{1}{\omega_n \varepsilon^{n-1}} \left| \int_{S(x,\varepsilon)} u(z) \, d\sigma(z) - u(x) \right| \\ &= \frac{1}{\omega_n \varepsilon^{n-1}} \left| \int_{S(x,\varepsilon)} (u(z) - u(x)) \, d\sigma(z) \right| \\ &\le \frac{1}{\omega_n \varepsilon^{n-1}} \int_{S(x,\varepsilon)} |u(z) - u(x)| \, d\sigma(z) \end{aligned}$$

$$\leq \sup_{z \in B(x,\varepsilon)} |u(z) - u(x)| \longrightarrow 0 \quad \text{as } \varepsilon \downarrow 0.$$

This proves Claim 5.7. □

Therefore, by letting $\varepsilon \downarrow 0$ in formula (5.7) we obtain from Claim 5.7 that

$$u(x) = \frac{1}{r^{n-1}\omega_n} \int_{S(x,r)} u(y)\, d\sigma(y).$$

(ii) The proof of the second formula (5.5b): We make the change of variables

$$y = x + rz, \quad z \in S(0, 1).$$

Then, since we have the formula

$$d\sigma(y) = r^{n-1}\, d\sigma(z),$$

it follows from formula (5.5a) that

$$\begin{aligned} u(x) &= \frac{1}{r^{n-1}\omega_n} \int_{S(x,r)} u(y)\, d\sigma(y) = \frac{1}{r^{n-1}\omega_n} \int_{S(0,1)} u(x + rz) r^{n-1}\, d\sigma(z) \\ &= \frac{1}{\omega_n} \int_{S(0,1)} u(x + rz)\, d\sigma(z). \end{aligned}$$

The proof of Theorem 5.6 is complete. □

Corollary 5.8 *Assume that $u(x)$ is harmonic in Ω. Then we have the formulas*

$$u(x) = \frac{n}{r^n \omega_n} \int_{B(x,r)} u(y)\, dy \tag{5.8a}$$

$$= \frac{n}{\omega_n} \int_{B(0,1)} u(x + rz)\, dz. \tag{5.8b}$$

Remark 5.9 The volume V_n of the unit ball $B(0, 1)$ is given by the formula

$$V_n = \int_0^1 \int_{S(0,1)} r^{n-1}\, d\sigma(y)\, dr = \omega_n \int_0^1 r^{n-1}\, dr = \frac{\omega_n}{n} = \frac{\pi^{n/2}}{\Gamma(n/2 + 1)}.$$

Hence we can rewrite formula (5.8b) as follows:

$$u(x) = \frac{1}{V_n} \int_{B(0,1)} u(x + rz)\, dz. \tag{5.9}$$

Proof (i) The proof of the first formula (5.8a): By formula (5.5b), it follows that, for $0 < \rho < 1$,

$$u(x) = \frac{1}{\omega_n} \int_{S(0,1)} u(x + \rho r y)\, d\sigma(y).$$

Therefore, integrating both sides from 0 to 1 with respect to $\rho^{n-1} d\rho$ we obtain that

$$\begin{aligned}
\frac{1}{n} u(x) &= \frac{1}{\omega_n} \int_0^1 \int_{S(0,1)} u(x + \rho r y) \rho^{n-1}\, d\rho\, d\sigma(y) \\
&= \frac{1}{\omega_n} \int_0^r \int_{S(0,1)} u(x + ty) \left(\frac{t}{r}\right)^{n-1} \frac{dt}{r}\, d\sigma(y) \\
&= \frac{1}{\omega_n} \frac{1}{r^n} \int_0^r \int_{S(0,1)} u(x + ty)\, t^{n-1}\, d\sigma(y)\, dt \\
&= \frac{1}{r^n \omega_n} \int_{B(x,r)} u(y)\, dy.
\end{aligned}$$

This proves formula (5.8a).

(ii) The proof of the second equality (5.8b): We make the change of variables

$$y = x + rz, \quad z \in B(0, 1).$$

Then, since we have the formula

$$dy = r^n\, dz,$$

it follows from formula (5.8a) that

$$\begin{aligned}
u(x) &= \frac{n}{r^n \omega_n} \int_{B(x,r)} u(y)\, dy = \frac{n}{r^n \omega_n} \int_{B(0,1)} u(x + rz) r^n\, dz \\
&= \frac{n}{\omega_n} \int_{B(0,1)} u(x + rz)\, dz.
\end{aligned}$$

This proves formula (5.8b).

The proof of Corollary 5.8 is complete. □

5.3 Poisson Kernels

To motivate the definition of Poisson kernels, we begin with some heuristic remarks on the Dirichlet boundary value problem

$$\begin{cases} \left(\frac{\partial^2}{\partial x_1^2} + \frac{\partial^2}{\partial x_2^2} + \cdots + \frac{\partial^2}{\partial x_n^2} + \frac{\partial^2}{\partial y^2}\right) u(x, y) = 0 & \text{in } \mathbf{R}_+^{n+1}, \\ u(x, 0) = f(x) & \text{on } \mathbf{R}^n \end{cases} \tag{5.10}$$

in the half-space $\mathbf{R}_+^{n+1} = \{(x, y) : x \in \mathbf{R}^n, y > 0\}$. If we consider the partial Fourier transform $\widetilde{u}(\xi, y)$ of a solution $u(x, y)$ of problem (5.10), then we have the following initial value problem:

$$\begin{cases} \left(\frac{d^2}{dy^2} - |\xi|^2\right) \widetilde{u}(\xi, y) = 0, & y > 0, \\ \widetilde{u}(\xi, 0) = \widehat{f}(\xi). \end{cases}$$

Hence it follows that the partial Fourier transform $\widetilde{u}(\xi, y)$ is given by the formula

$$\widetilde{u}(\xi, y) = \widehat{f}(\xi)\, e^{-|\xi| y}, \quad y > 0.$$

Therefore, by the partial Fourier inversion formula we obtain that the solution $u(x, y)$ of problem (5.10) may be "formally" expressed as follows:

$$\begin{aligned} u(x, y) &= \frac{1}{(2\pi)^n} \int_{\mathbf{R}^n} e^{ix\xi}\, e^{-|\xi| y}\, \widehat{f}(\xi)\, d\xi \\ &= \frac{1}{(2\pi)^n} \int_{\mathbf{R}^n} e^{-|\xi| y} \left(\int_{\mathbf{R}^n} e^{i(x-z)\xi}\, f(z)\, dz \right) d\xi \\ &= \int_{\mathbf{R}^n} \left(\frac{1}{(2\pi)^n} \int_{\mathbf{R}^n} e^{i(x-z)\xi}\, e^{-|\xi| y}\, d\xi \right) f(z)\, dz \\ &:= \int_{\mathbf{R}^n} P(x - z, y)\, f(y)\, dy, \end{aligned}$$

where

$$P(x, y) := \frac{1}{(2\pi)^n} \int_{\mathbf{R}^n} e^{ix\xi} e^{-|\xi| y}\, d\xi, \quad y > 0. \tag{5.11}$$

The function $P(x, y)$ is called the *Poisson kernel* for the half-space $\mathbf{R}_+^{n+1}$. However, we can calculate explicitly the Poisson kernel $P(x, y)$. Indeed, we have the following:

Claim 5.10 *The Poisson kernel $P(x, y)$ is given by the formula*

$$P(x, y) = \frac{1}{c_n} \frac{y}{(|x|^2 + y^2)^{(n+1)/2}}, \quad x \in \mathbf{R}^n, \; y > 0, \tag{5.12a}$$

where c_n is a positive constant given by the formula

$$c_n = \frac{\pi^{(n+1)/2}}{\Gamma((n+1)/2)} = \frac{\omega_{n+1}}{2}. \tag{5.12b}$$

Here

$$\omega_{n+1} := |\Sigma_n| = (n+1) V_{n+1} = \frac{2\pi^{(n+1)/2}}{\Gamma((n+1)/2)}$$

is the surface area of the unit sphere Σ_n in $\mathbf{R}^{n+1}$.

Proof First, we recall the following two well-known formulas (5.13) and (5.14):

$$e^{-\gamma} = \frac{1}{\sqrt{\pi}} \int_0^\infty \frac{e^{-u}}{\sqrt{u}} e^{-\frac{\gamma^2}{4u}} \, du, \quad \gamma > 0, \tag{5.13}$$

$$\frac{1}{(2\pi)^n} \int_{\mathbf{R}^n} e^{ix\xi} e^{-\alpha|\xi|^2} \, dx = \frac{1}{(4\alpha\pi)^{\frac{n}{2}}} e^{-\frac{|x|^2}{4\alpha}}, \quad \alpha > 0. \tag{5.14}$$

Therefore, by using Fubini's theorem (Theorem 3.21) we obtain from formula (5.13) with $\gamma := |\xi| y$ and formula (5.14) with $\alpha := y^2/(4u)$ that

$$\begin{aligned}
&\frac{1}{(2\pi)^n} \int_{\mathbf{R}^n} e^{ix\xi} e^{-|\xi| y} \, d\xi \\
&= \int_{\mathbf{R}^n} e^{ix\xi} \left(\frac{1}{\sqrt{\pi}} \int_0^\infty \frac{e^{-u}}{\sqrt{u}} e^{-|\xi|^2 y^2/(4u)} \, du \right) d\xi \\
&= \frac{1}{\sqrt{\pi}} \int_0^\infty \frac{e^{-u}}{\sqrt{u}} \left(\frac{1}{(2\pi)^n} \int_{\mathbf{R}^n} e^{ix\xi} e^{-|\xi|^2 y^2/(4u)} \, d\xi \right) du \\
&= \frac{1}{\sqrt{\pi}} \int_0^\infty \frac{e^{-u}}{\sqrt{u}} \left(\frac{u}{\pi y^2} \right)^{n/2} e^{-|x|^2 u/y^2} \, du \\
&= \frac{1}{\pi^{(n+1)/2}} \frac{1}{y^n} \int_0^\infty u^{(n-1)/2} e^{-(1+|x|^2/y^2)u} \, du \\
&= \frac{\Gamma((n+1)/2)}{\pi^{(n+1)/2}} \frac{y}{(|x|^2 + y^2)^{(n+1)/2}} \\
&= P(x, y).
\end{aligned}$$

The proof of Claim 5.10 is complete. □

Now we prove fundamental properties of the Poisson kernel $P(x, y)$:

Lemma 5.11 *The Poisson kernel $P(x, y)$ enjoys the following properties:*

$$\int_{\mathbf{R}^n} P(x, y) \, dx = 1 \quad \text{for all } x \in \mathbf{R}^n, \tag{5.15a}$$

$$P_y(x, y) = \frac{1}{c_n} \frac{|x|^2 - ny^2}{(|x|^2 + y^2)^{(n+3)/2}} \quad \text{for all } x \in \mathbf{R}^n, \tag{5.15b}$$

$$|P_y(x, y)| \leq \frac{n+1}{y} P(x, y) \quad \text{for all } x \in \mathbf{R}^n,$$

$$P_{x_i}(x, y) = \frac{1}{c_n} \frac{-(n+1)x_i y}{(|x|^2 + y^2)^{(n+3)/2}} \quad \text{for all } x \in \mathbf{R}^n, \tag{5.15c}$$

$$|P_{x_i}(x, y)| \leq \frac{n+1}{2y} P(x, y) \quad \text{for all } x \in \mathbf{R}^n \text{ and } 1 \leq i \leq n.$$

Here $P_y(x, y)$ and $P_{x_i}(x, y)$ indicate the partial derivative of $P(x, y)$ in the direction of the element subscripted, that is,

$$P_y(x,y) = \frac{\partial P}{\partial y}, \quad P_{x_i}(x,y) = \frac{\partial P}{\partial x_i}.$$

Proof It is easy to verify assertions (5.15b) and (5.15c). Moreover, assertions (5.15a) follow from a direct calculation. Indeed, we have, by formula (5.12a),

$$P(x,y) = \frac{1}{c_n} \frac{y}{(|x|^2 + y^2)^{(n+1)/2}} = \frac{1}{y^n} P\left(\frac{x}{y}, 1\right) \quad \text{for all } y > 0.$$

Hence, it follows from an application of Theorem A.14 or Corollary A.17 that

$$\begin{aligned}
&\int_{\mathbf{R}^n} P(z,1)\,dz \\
&= \frac{\Gamma((n+1)/2)}{\pi^{(n+1)/2}} \int_{\mathbf{R}^n} \left(1+|z|^2\right)^{-(n+1)/2} dz \\
&= \frac{\Gamma((n+1)/2)}{\pi^{(n+1)/2}} \int_0^\infty \int_{\Sigma_{n-1}} \left(1+r^2\right)^{-(n+1)/2} r^{n-1}\,dr\,d\sigma \\
&= \frac{\Gamma((n+1)/2)}{\pi^{(n+1)/2}} \left(\frac{2\pi^{n/2}}{\Gamma(n/2)}\right) \int_0^\infty \left(1+r^2\right)^{-(n+1)/2} r^{n-1}\,dr \\
&= \frac{2}{\sqrt{\pi}} \frac{\Gamma((n+1)/2)}{\Gamma(n/2)} \int_1^\infty s^{-(n+1)/2} (s-1)^{(n-1)/2} \cdot \frac{1}{2\sqrt{s-1}}\,ds \\
&= \frac{1}{\sqrt{\pi}} \frac{\Gamma((n+1)/2)}{\Gamma(n/2)} \int_1^\infty \left(1-\frac{1}{s}\right)^{(n-2)/2} s^{-3/2}\,ds \\
&= \frac{1}{\sqrt{\pi}} \frac{\Gamma((n+1)/2)}{\Gamma(n/2)} \int_0^1 (1-t)^{(n-2)/2} t^{-1/2}\,dt \\
&= \frac{1}{\sqrt{\pi}} \frac{\Gamma((n+1)/2)}{\Gamma(n/2)} B(1/2, n/2) \\
&= \frac{1}{\sqrt{\pi}} \frac{\Gamma((n+1)/2)}{\Gamma(n/2)} \frac{\Gamma(1/2)\Gamma(n/2)}{\Gamma((n+1)/2))} \\
&= 1.
\end{aligned}$$

Here we have made the following changes of variables:

(1) $z = r\sigma$ for $r > 0$ and $\sigma \in \Sigma_{n-1}$,
(2) $s = 1 + r^2$ for $r > 0$,
(3) $t = 1/s$ for $s \geq 1$,

and

$$\frac{2\pi^{n/2}}{\Gamma(n/2)} = \omega_n := |\Sigma_{n-1}|$$

is the surface area of the unit sphere Σ_{n-1} in $\mathbf{R}^n$.

Therefore, we obtain that

$$\int_{\mathbf{R}^n} P(x, y)\, dx = \int_{\mathbf{R}^n} \frac{1}{y^n} P\left(\frac{x}{y}, 1\right) dx = \int_{\mathbf{R}^n} P(z, 1)\, dz = 1.$$

The proof of Lemma 5.11 is complete. □

5.4 Poisson Integrals

If $f \in L^p(\mathbf{R}^n)$ with $1 \leq p \leq \infty$, then the *Poisson integral* of $f(x)$ is defined as the convolution of $f(x)$ with the Poisson kernel $P(x, y)$:

$$\begin{aligned} u(x, y) &= P(x, y) * f(x) \\ &= \frac{1}{c_n} \int_{\mathbf{R}^n} \frac{y}{(|x - z|^2 + y^2)^{(n+1)/2}} f(z)\, dz \quad \text{for all } y > 0. \end{aligned} \tag{5.16}$$

Then we have the following assertions for the Poisson integral $u(x, y)$:

Theorem 5.12 *(i) Let $f \in L^p(\mathbf{R}^n)$ with $1 \leq p \leq \infty$. Then the function*

$$u(x, y) = P(x, y) * f(x)$$

is well defined in the half-space $\mathbf{R}^{n+1}_+$, and satisfies the inequality

$$\|u(\cdot, y)\|_p \leq \|f\|_p \quad \textit{for all } y > 0.$$

Furthermore, the function $u(x, y)$ is harmonic in $\mathbf{R}^{n+1}_+$.

(ii) If $1 \leq p < \infty$, we have, as $y \downarrow 0$,

$$u(\cdot, y) \longrightarrow f \quad \textit{in } L^p(\mathbf{R}^n). \tag{5.17}$$

(iii) If, in addition, $f(x)$ is bounded and continuous on $\mathbf{R}^n$, then the function $u(x, y)$ is continuous on $\overline{\mathbf{R}^{n+1}_+} = \{(x, y) : x \in \mathbf{R}^n, y \geq 0\}$, and satisfies the Dirichlet condition

$$u(x, 0) = f(x) \quad \textit{for every } x \in \mathbf{R}^n. \tag{5.18}$$

Proof The proof is divided into three steps.

Step (i): Since we have, for all $y > 0$,

$$P(\cdot, y) \in \bigcap_{p=1}^{\infty} L^p(\mathbf{R}^n),$$

it follows from an application of Young's inequality (Theorem 3.36) with $p = r$ and $q = 1$ that the function $u(x, y) = P(x, y) * f(x)$ is well defined on $\mathbf{R}_+^{n+1}$.

Furthermore, we can differentiate formula (5.16) under the integral sign to obtain the following formulas:

$$\frac{\partial^2 u}{\partial y^2}(x, y) \tag{5.19}$$
$$= \frac{1}{c_n} \int_{\mathbf{R}^n} \left[\frac{n(n+1)y^3 - 3(n+1)y|x-z|^2}{(|x-z|^2 + y^2)^{(n+1)/2+1}} \right] f(z)\, dz,$$

$$\frac{\partial^2 u}{\partial x_i^2}(x, y) \tag{5.20}$$
$$= \frac{1}{c_n} \int_{\mathbf{R}^n} \left[\frac{(n+1)(n+3)y(x_i - z_i)^2 - (n+1)(t^2 + |x-z|^2)}{(|x-z|^2 + y^2)^{(n+1)/2+2}} \right] f(z)\, dz.$$

It should be noticed that the integrals converge absolutely on compact subsets of $\mathbf{R}_+^{n+1}$. Indeed, it suffices to note that the terms in the bracket $[\cdot]$ in formulas (5.19) and (5.20) are bounded by a function in $L^{p'}(\mathbf{R}^n)$ uniformly in (x, y) of a compact subset of $\mathbf{R}_+^{n+1}$, where $p' = p/(p-1)$ is the exponent conjugate to p.

Summing up, we find that

$$u(x, y) \in C^2(\mathbf{R}_+^{n+1}),$$

and that

$$\left(\sum_{j=1}^{n} \frac{\partial^2}{\partial x_j^2} + \frac{\partial^2}{\partial y^2} \right) u(x, y) = 0 \quad \text{in } \mathbf{R}_+^{n+1},$$

that is, the function $u(x, y)$ is harmonic in the half-space $\mathbf{R}_+^{n+1}$.

Step (ii): We recall that

$$\int_{\mathbf{R}^n} P(x, 1)\, dx = 1,$$

and further that

$$\frac{1}{y^n} P\left(\frac{x}{y}, 1 \right) = P(x, y) \quad \text{for all } y > 0.$$

Therefore, the desired assertion (5.17) follows from an application of Theorem 3.40 with

$$\varphi(x) := P(x, 1), \quad \varepsilon := y.$$

Step (iii): To prove assertion (5.18), we assume that $f(x)$ is bounded and continuous on $\mathbf{R}^n$. Then we have, for all $y > 0$,

$$\begin{aligned}P(x,y)*f(x)-f(x) &= \int_{\mathbf{R}^n} P(x-z,y)[f(z)-f(x)]\,dz \\ &= \int_{\mathbf{R}^n} P(z,y)[f(x-z)-f(x)]\,dz \\ &= \int_{\mathbf{R}^n} P(\xi,1)[f(x-`y\xi)-f(x)]\,d\xi.\end{aligned} \tag{5.21}$$

However, we remark that

$$|P(\xi,1)\,[f(x-y\xi)-f(x)]\,| \le 2\max|f|\cdot P(\xi,1),$$

and that

$$\int_{\mathbf{R}^n} P(\xi,1)\,d\xi = 1.$$

Therefore, by applying the Lebesgue dominated convergence theorem (Theorem 3.13) we obtain from formula (5.21) that

$$\lim_{y\downarrow 0} P(x,y)*f(x) = f(x) \quad \text{for every } x \in \mathbf{R}^n.$$

The proof of Theorem 5.12 is complete. □

Assertions (5.17) and (5.18) may be made precise. To do this, we recall some important results in the theory of differentiation of integrals of functions defined on $\mathbf{R}^n$ (Theorem 4.10 and Corollary 4.8).

First, Corollary 4.8 may be restated as follows:

Claim 5.13 *If $f(t)$ is a locally integrable function on $\mathbf{R}^n$, then we have, for almost every $x \in \mathbf{R}^n$,*

$$\lim_{r\downarrow 0} \frac{1}{r^n}\int_{|t|<r} (f(x-t)-f(x))\,dt = 0. \tag{5.22}$$

In particular, this is true for functions $f \in L^p(\mathbf{R}^n)$ with $1 \le p \le \infty$.

Indeed, it suffices to note that

$$\begin{aligned}&\left|\frac{1}{r^n}\int_{|t|<r} (f(x-t)-f(x))\,dt\right| \\ &= \left|\frac{1}{r^n}\int_{B(x,r)} (f(y)-f(x))\,dy\right| \le \frac{1}{r^n}\int_{B(x,r)} |f(y)-f(x)|\,dy \\ &= 2^n\,\frac{1}{(2r)^n}\int_{B(x,r)} |f(y)-f(x)|\,dy \\ &\le 2^n\,\frac{1}{|Q(x,r)|}\int_{Q(x,r)} |f(y)-f(x)|\,dy.\end{aligned}$$

Therefore, the desired assertion (5.22) follows from assertion (4.15).

We let

$$D(f) = \text{ the set of all points } x \in \mathbf{R}^n \text{for which condition 5.22 holds true.}$$

The set $D(f)$ is called the set of points where the integral of f is *differentiable*. There is a smaller but closely related set which is also naturally associated with each locally integrable function.

We let

$$L(f) = \text{the set of all Lebesgue points } x \in \mathbf{R}^n \text{ for } f, \text{ i.e.,}$$
$$\lim_{r \downarrow 0} \frac{1}{r^n} \int_{|t|<r} |f(x-t) - f(x)|\, dt = 0. \tag{5.23}$$

The set $L(f)$ is called the *Lebesgue set* of f (see Definition 4.1).

Secondly, Theorem 4.10 may be restated as follows:

Claim 5.14 *If $f(x)$ is a locally integrable function on $\mathbf{R}^n$, then the complement of the Lebesgue set $L(f)$ has measure* zero.

Now we can generalize assertions (5.17) and (5.18) as follows:

Theorem 5.15 *Let $f \in L^p(\mathbf{R}^n)$ with $1 \leq p \leq \infty$. Then we have the formula*

$$\lim_{\varepsilon \downarrow 0} P(x, \varepsilon) * f(x) = \lim_{\varepsilon \downarrow 0} \int_{\mathbf{R}^n} P(x-t, \varepsilon) f(t)\, dt = f(x), \tag{5.24}$$

whenever x belongs to the Lebesgue set $L(f)$ of f. In particular, the convergence (5.24) *holds true for almost all $x \in \mathbf{R}^n$.*

Proof If x is a point of the Lebesgue set $L(f)$, then it follows from condition (5.23) that, for each $\delta > 0$, there exists a constant $\eta = \eta(x, \delta) > 0$ such that

$$\frac{1}{r^n} \int_{|t|<r} |f(x-t) - f(x)|\, dt < \delta \quad \text{for all } 0 < r \leq \eta. \tag{5.25}$$

We let

$$\varphi(t) := P(t, 1) = \frac{1}{c_n} \frac{1}{(|t|^2 + 1)^{(n+1)/2}}, \quad t \in \mathbf{R}^n,$$
$$\varphi_\varepsilon(t) := P(t, \varepsilon) = \frac{1}{c_n} \frac{\varepsilon}{(|t|^2 + \varepsilon^2)^{(n+1)/2}}, \quad t \in \mathbf{R}^n, \quad \varepsilon > 0.$$

Then we have the inequality

$$|P(x, \varepsilon) * f(x) - f(x)| = \left| \int_{\mathbf{R}^n} P(t, \varepsilon) f(x-t)\, dt - f(x) \right| \tag{5.26}$$

$$
\begin{aligned}
&= \left| \int_{\mathbf{R}^n} (f(x-t) - f(x))\, \varphi_\varepsilon(t)\, dt \right| \\
&\le \int_{|t|\ge\eta} |f(x-t) - f(x)| \varphi_\varepsilon(t)\, dt \\
&\quad + \int_{|t|<\eta} |f(x-t) - f(x)| \varphi_\varepsilon(t)\, dt \\
&:= I_1^{(\varepsilon)}(x,\eta) + I_2^{(\varepsilon)}(x,\eta).
\end{aligned}
$$

We estimate the two integrals $I_1^{(\varepsilon)}(x,\eta)$ and $I_2^{(\varepsilon)}(x,\eta)$ on the right-hand of inequality (5.26).

(i) The estimate of $I_1^{(\varepsilon)}(x,\eta)$: We let

$$
\chi_\eta(x) = \text{ the characteristic function of the set } \{t \in \mathbf{R}^n : |t| \ge \eta\}.
$$

Then we have, by Hölder's inequality (Theorem 3.24),

$$
\begin{aligned}
I_1^{(\varepsilon)}(x,\eta) &= \int_{|t|\ge\eta} |f(x-t) - f(x)|\, \varphi_\varepsilon(t)\, dt \\
&= \int_{\mathbf{R}^n} |f(x-t) - f(x)|\, \chi_\eta(t)\varphi_\varepsilon(t)\, dt \\
&\le \int_{\mathbf{R}^n} |f(x-t)|\, \chi_\eta(t)\varphi_\varepsilon(t)\, dt + \int_{\mathbf{R}^n} |f(x)|\, \chi_\eta(t)\varphi_\varepsilon(t)\, dt \\
&\le \|f\|_p \left\|\chi_\eta\varphi_\varepsilon\right\|_{p'} + |f(x)| \left\|\chi_\eta\varphi_\varepsilon\right\|_1.
\end{aligned}
$$

However, it follows that, as $\varepsilon \downarrow 0$,

$$
\left\|\chi_\eta\varphi_\varepsilon\right\|_1 = \int_{|t|\ge\eta} \frac{1}{\varepsilon^n} \varphi\left(\frac{t}{\varepsilon}\right) dt = \int_{|z|\ge\frac{\eta}{\varepsilon}} \varphi(z)\, dz \longrightarrow 0,
$$

and further that

$$
\begin{aligned}
\left\|\chi_\eta\varphi_\varepsilon\right\|_{p'} &= \left(\int_{|t|\ge\eta} \varphi_\varepsilon(t)^{p'}\, dt \right)^{1/p'} \\
&= \left(\int_{\mathbf{R}^n} \left(\chi_\eta(t)\varphi_\varepsilon(t)\right) \left(\chi_\eta(t)\varphi_\varepsilon(t)\right)^{p/p'}\, dt \right)^{1/p'} \\
&\le \left\|\chi_\eta\varphi_\varepsilon\right\|_\infty^{1/p} \left\|\chi_\eta\varphi_\varepsilon\right\|_1^{1/p'} \longrightarrow 0,
\end{aligned}
$$

since we have the formula

$$
\left\|\chi_\eta\varphi_\varepsilon\right\|_\infty = \sup_{|t|\ge\eta} |\varphi_\varepsilon(t)| = \frac{1}{c_n} \frac{\varepsilon}{(\eta^2+\varepsilon^2)^{(n+1)/2}}.
$$

Hence we find that, as $\varepsilon \downarrow 0$,

$$I_1^{(\varepsilon)}(x, \eta) \leq \|f\|_p \left\|\chi_\eta \varphi_\varepsilon\right\|_{p'} + |f(x)| \left\|\chi_\eta \varphi_\varepsilon\right\|_1 \longrightarrow 0. \tag{5.27}$$

(ii) The estimate of $I_2^{(\varepsilon)}(x, \eta)$: We let

$$g(r) := \int_{\Sigma_{n-1}} |f(x - r\sigma) - f(x)| \, d\sigma,$$

where Σ_{n-1} is the unit sphere in $\mathbf{R}^n$. Then we find that condition (5.25) is equivalent to the following condition: For each $\delta > 0$, there exists a constant $\eta = \eta(x, \delta) > 0$ such that

$$\begin{aligned} G(r) := \int_0^r s^{n-1} g(s) \, ds = \int_{|t|<r} |f(x - t) - f(x)| \, dt \\ < \delta r^n \quad \text{for all } 0 < r \leq \eta. \end{aligned} \tag{5.25$'$}$$

Moreover, we introduce a function $\psi(r)$ (associated with the function $\varphi(x)$) by the formula

$$\psi(r) := \frac{1}{c_n} \frac{1}{(r^2 + 1)^{(n+1)/2}}, \quad r > 0.$$

By integration by parts, it follows that

$$\begin{aligned} I_2^{(\varepsilon)}(x, \eta) &\leq \int_{|t|<\eta} |f(x - t) - f(x)| \, \varphi_\varepsilon(t) \, dt \\ &= \int_0^\eta \int_{\Sigma_{n-1}} |f(x - r\sigma) - f(x)| \, \varepsilon^{-n} \, \psi\left(\frac{r}{\varepsilon}\right) r^{n-1} \, dr \, d\sigma \\ &= \int_0^\eta r^{n-1} g(r) \, \varepsilon^{-n} \, \psi\left(\frac{r}{\varepsilon}\right) dr \\ &= \left[G(r) \, \varepsilon^{-n} \, \psi\left(\frac{r}{\varepsilon}\right)\right]_0^\eta - \int_0^\eta G(r) \left(\varepsilon^{-n} \, \psi'\left(\frac{r}{\varepsilon}\right)\right) \varepsilon^{-1} \, dr \\ &= G(\eta) \, \varepsilon^{-n} \, \psi\left(\frac{\eta}{\varepsilon}\right) - \int_0^{\eta/\varepsilon} G(\varepsilon s) \, \varepsilon^{-n} \, \psi'(s) \, ds. \end{aligned}$$

Hence we have, by condition (5.25′),

$$I_2^{(\varepsilon)}(x, \eta) \leq \delta \, \eta^n \, \varepsilon^{-n} \, \psi\left(\frac{\eta}{\varepsilon}\right) - \delta \int_0^{\eta/\varepsilon} (\varepsilon s)^n \varepsilon^{-n} \, \psi'(s) \, ds \leq A\delta, \tag{5.28}$$

where A is a positive constant defined by the formula

$$A := \frac{1}{c_n}\left(\sup_{r>0}\left\{\frac{r^n}{(r^2+1)^{(n+1)/2}}\right\} + n\int_0^\infty \frac{s^{n-1}}{(s^2+1)^{(n+1)/2}}\,ds\right).$$

Therefore, by combining inequalities (5.26), (5.27) and (5.28) we obtain that

$$\limsup_{\varepsilon\downarrow 0} |P(x,\varepsilon) * f(x) - f(x)| \le \limsup_{\varepsilon\downarrow 0}\left\{I_1^{(\varepsilon)}(x,\eta) + I_2^{(\varepsilon)}(x,\eta)\right\} \le A\delta.$$

This proves the desired assertion (5.24), since δ is arbitrary.

The proof of Theorem 5.15 is complete. □

5.5 Manipulations of Harmonic Functions

Theorem 5.12 has a converse:

Theorem 5.16 *Assume that a function $u(x, y)$ is harmonic in the half-space $\mathbf{R}_+^{n+1}$ and that there exist constants $c > 0$ and $1 \le p < \infty$ such that*

$$\|u(\cdot, y)\|_p \le c \quad \textit{for all } y > 0. \tag{5.29}$$

Then we have the following two assertions (a) and (b):

(a) If $1 < p < \infty$, there exists a function $f \in L^p(\mathbf{R}^n)$ such that

$$\begin{aligned} u(x,y) &= P(x,y) * f(x) \\ &= \int_{\mathbf{R}^n} P(x-t,y)\, f(t)\, dt \quad \textit{for all } y > 0. \end{aligned} \tag{5.30}$$

(b) If $p = 1$, there exists a finite Borel measure μ on $\mathbf{R}^n$ such that

$$u(x,y) = \int_{\mathbf{R}^n} P(x-t,y)\, d\mu(t). \tag{5.31}$$

Furthermore, if $\{u(\cdot, y)\}$ is a Cauchy sequence in $L^1(\mathbf{R}^n)$ as $y \downarrow 0$, then there exists a function $f \in L^1(\mathbf{R}^n)$ such that $u(x, y) = P(x, y) * f(x)$.

The proof of Theorem 5.16 is divided into two steps.

Step 1: In order to prove Theorem 5.16, we need some lemmas:

Lemma 5.17 *Assume that a function $u(x, y)$ is harmonic in the half-space $\mathbf{R}_+^{n+1}$ and satisfies condition* (5.29). *Then we have the inequality*

$$\|u(\cdot, y)\|_\infty \le \left(\frac{2^{n+1}}{V_{n+1}}\right)^{1/p} c\, y^{-n/p} \quad \textit{for all } y > 0,$$

where V_{n+1} is the volume of the unit ball in $\mathbf{R}^{n+1}$

$$V_{n+1} := \frac{2\pi^{(n+1)/2}}{(n+1)\Gamma((n+1)/2)}.$$

In particular, the function $u(x, y)$ is bounded in each proper sub-half space of $\mathbf{R}^{n+1}_{+}$.

Proof By applying the mean value theorem for harmonic functions (Theorem 5.6) with

$$\Omega := \mathbf{R}^{n+1}, \quad r := \frac{y}{2},$$

we obtain that

$$\begin{aligned}
u(x, y) &= \frac{n+1}{(y/2)^{n+1}} \int_0^{y/2} u(x, y) r^n \, dr \\
&= \frac{(n+1)2^{n+1}}{y^{n+1}} \int_0^{y/2} \left(\frac{1}{\omega_{n+1}} \int_{\Sigma_n} u((x, y) + r\sigma) \, d\sigma \right) r^n \, dr \\
&= \frac{2^{n+1}}{V_{n+1}} \frac{1}{y^{n+1}} \int_{|t| \le y/2} u((x, y) + t) \, dt.
\end{aligned}$$

Here Σ_n is the unit sphere in $\mathbf{R}^{n+1}$ and

$$\omega_{n+1} := |\Sigma_n| = (n+1) V_{n+1} = \frac{2\pi^{(n+1)/2}}{\Gamma((n+1)/2)}$$

is its surface area. Therefore, we have, by Hölder's inequality (Theorem 3.24) and inequality (5.29),

$$\begin{aligned}
|u(x, y)| &\le \frac{2^{n+1}}{V_{n+1}} \frac{1}{y^{n+1}} \int_{|(x,y)-(\xi,\eta)| \le y/2} |u(\xi, \eta)| \, d\xi \, d\eta \\
&\le \frac{2^{n+1}}{V_{n+1}} \frac{1}{y^{n+1}} \left(\int_{|(x,y)-(\xi,\eta)| \le y/2} |u(\xi, \eta)|^p \, d\xi \, d\eta \right)^{1/p} \\
&\quad \times \left(V_{n+1} \left(\frac{y}{2} \right)^{n+1} \right)^{1-1/p} \\
&\le \left(\frac{2^{n+1}}{V_{n+1}} \right)^{1/p} y^{-(n+1)/p} \left(\int_{y/2}^{3y/2} \left(\int_{\mathbf{R}^n} |u(\xi, \eta)|^p \, d\xi \right) d\eta \right)^{1/p} \\
&\le \left(\frac{2^{n+1}}{V_{n+1}} \right)^{1/p} y^{-(n+1)/p} \left(\int_{y/2}^{3y/2} c^p \, d\eta \right)^{1/p} \\
&\le \left(\frac{2^{n+1}}{V_{n+1}} \right)^{1/p} c y^{-n/p}.
\end{aligned}$$

The proof of Lemma 5.17 is complete. □

The next lemma is a special case of the *Schwarz reflection principle* for hyperplanes (see [9, Chap. 4, Theorem 4.12]):

Lemma 5.18 *If $u(x, y)$ is harmonic in the half-space $\mathbf{R}_+^{n+1}$, everywhere continuous and bounded on the closure $\overline{\mathbf{R}_+^{n+1}} = \mathbf{R}_+^{n+1} \cup \mathbf{R}^n$ and is equal to zero on the boundary $\mathbf{R}^n$, then it follows that $u(x, y) \equiv 0$ on $\mathbf{R}_+^{n+1}$.*

Proof The proof is divided into two steps.

(1) We let

$$u^*(x, y) := \begin{cases} u(x, y) & \text{if } y \geq 0, \\ -u(x, -y) & \text{if } y < 0. \end{cases}$$

Then it follows that

$$\begin{cases} \Delta u^* = 0 & \text{in } \mathbf{R}_+^{n+1}, \\ \Delta u^* = 0 & \text{in } \mathbf{R}_-^{n+1}, \end{cases}$$

and further that

$$u^* = 0 \quad \text{on } \mathbf{R}^n.$$

Hence we have the assertion

$$u^* \in C\left(\mathbf{R}^{n+1}\right) \cap C^2\left(\mathbf{R}_+^{n+1}\right) \cap C^2\left(\mathbf{R}_-^{n+1}\right).$$

Now we show that u^* is harmonic near $\mathbf{R}^n$ in $\mathbf{R}^{n+1}$. To do so, let $(x_0, 0)$ be an arbitary point of $\mathbf{R}^n$ and let

$$B_r := B\left((x_0, 0), r\right) \subset \mathbf{R}^{n+1}$$

be an arbitrary small open ball of radius r about $(x_0, 0)$ (see Fig. 5.6).

Then we have the formula

$$\begin{aligned} &\int_{\partial B_r} u^*(x, y)\, d\sigma(x, y) \\ &= \int_{\partial B_r^+} u(x, y)\, d\sigma(x, y) - \int_{\partial B_r^-} u(x, -y)\, d\sigma(x, y) \\ &= \int_{\partial B_r^+} u(x, y)\, d\sigma(x, y) - \int_{\partial B_r^+} u(x, y)\, d\sigma(x, y) \\ &= 0. \end{aligned}$$

Hence there exists a constant $\varepsilon > 0$ such that

$$\int_{\partial B_r} u^*(x, y)\, d\sigma(x, y) = 0 \quad \text{for all } 0 < r < \varepsilon.$$

Therefore, by using Green's identity (5.4a) with $\Omega := B_r$ and $v := 1$ we obtain that

$$\begin{aligned}
0 &= \frac{d}{dr}\left(\frac{1}{r^n\,\omega_{n+1}}\int_{\partial B_r} u^*(x,y)\,d\sigma(x,y)\right)\\
&= \frac{d}{dr}\left(\frac{1}{\omega_{n+1}}\int_{S(0,1)} u^*\left((x_0,0)+rz\right)\,d\sigma(z)\right)\\
&= \frac{1}{\omega_{n+1}}\int_{S(0,1)} z\cdot\nabla u^*\left((x_0,0)+rz\right)\,d\sigma(z)\\
&= \frac{1}{\omega_{n+1}}\int_{S(0,r)} \frac{w}{r}\cdot\nabla u^*\left((x_0,0)+w\right)\,r^{-n}d\sigma(w)\\
&= \frac{1}{r^n\,\omega_{n+1}}\int_{S(0,r)} \frac{w}{r}\cdot\nabla u^*\left((x_0,0)+w\right)\,d\sigma(w)\\
&= \frac{1}{r^n\,\omega_{n+1}}\int_{S(0,r)} \frac{\partial u^*}{\partial \boldsymbol{\nu}}\left((x_0,0)+w\right)d\sigma(w)\\
&= \frac{1}{r^n\,\omega_{n+1}}\int_{B_r} \Delta u^*(x,y)\,dx\,dy \quad \text{for all } 0<r<\varepsilon,
\end{aligned}$$

where

$$S(0,r) = \left\{w\in\mathbf{R}^{n+1} : |w| = r\right\}.$$

Since the integral of Δu^* over any ball near the point $(x_0, 0)$ vanishes, it follows that

$$\Delta u^*(x,y) = 0 \quad \text{in a neighborhood of each point } (x_0,0) \text{ of } \mathbf{R}^n.$$

Summing up, we have proved that $u^*(x, y)$ is harmonic in the whole space $\mathbf{R}^{n+1}$.

(2) By Lemma 5.17, it follows that the function $u^*(x, y)$ is harmonic and bounded on the whole space $\mathbf{R}^{n+1}$. Hence, by applying Liouville's theorem for harmonic functions [9, Chap. 2, Theorem 2.1] we obtain that

$$u^*(x,y) \equiv 0 \quad \text{on } \mathbf{R}^{n+1},$$

so that

$$u(x,y) \equiv 0 \quad \text{on } \mathbf{R}^{n+1}_+.$$

The proof of Lemma 5.18 is complete. □

Lemma 5.19 *Assume that a function $u(x, y)$ is harmonic in the half-space $\mathbf{R}^{n+1}_+$ and bounded in a proper sub-half space $\{(x, y) \in \mathbf{R}^{n+1}_+ : y > y_0\}$ of $\mathbf{R}^{n+1}_+$, for each $y_0 > 0$. Then we have the formula*

$$u\left(x, y_1 + y_2\right) = u\left(x, y_1\right) * P\left(x, y_2\right) \quad \textit{for all } y_1, y_2 > 0.$$

Fig. 5.6 The open ball B_r of radius r about $(x_0, 0)$

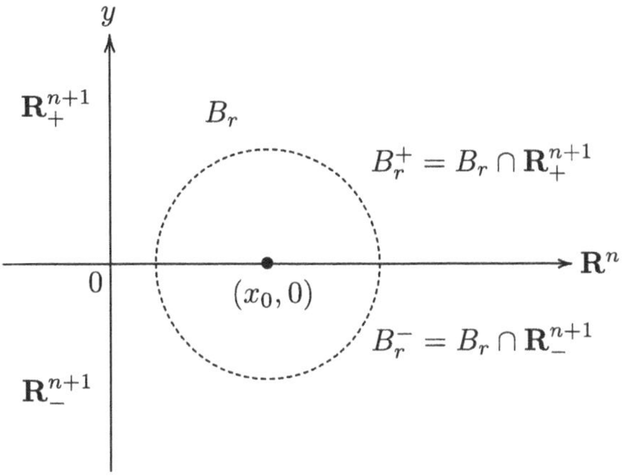

Proof For each $y_1 > 0$, we let

$$w_1(x, y) := u(x, y_1 + y) \quad \text{for all } y \geq 0,$$

and

$$w_2(x, y) := u(x, y_1) * P(x, y) = \int_{\mathbf{R}^n} P(x - t, y)\, u(t, y_1)\, dt \quad \text{for all } y > 0.$$

Then it suffices to show that

$$w_1(x, y) \equiv w_2(x, y) \quad \text{for all } (x, y) \in \mathbf{R}_+^{n+1}.$$

We find that the functions $w_1(x, y)$ and $w_2(x, y)$ are both harmonic in $\mathbf{R}_+^{n+1}$, everywhere continuous and bounded on $\mathbf{R}_+^{n+1}$, and further take the same boundary values $u(x, y_1)$ on $\mathbf{R}^n$. Hence, by applying Lemma 5.18 to the function

$$h(x, y) = w_1(x, y) - w_2(x, y),$$

we obtain that

$$h(x, y) \equiv 0 \quad \text{on } \mathbf{R}_+^{n+1},$$

that is,

$$w_1(x, y) \equiv w_2(x, y) \quad \text{for all } (x, y) \in \mathbf{R}_+^{n+1}.$$

The proof of Lemma 5.19 is complete. □

Step 2: Now, the proof of Theorem 5.16 may be carried out as follows.
Step 2-a: The case $1 < p < \infty$. We assume that

$$\|u(\cdot, y)\|_p \leq c \quad \text{for all } y > 0.$$

Recall (see Theorem 2.28) that $L^p(\mathbf{R}^n)$ is *reflexive* for $1 < p < \infty$. Hence, by the *weak compactness* of $L^p(\mathbf{R}^n)$ due to the Eberlein–Shmulyan theorem (Theorem 2.32) we can find a sequence $\{y_k\}$, $y_k \downarrow 0$, and a function $f \in L^p(\mathbf{R}^n)$ such that

$$u(\cdot, y_k) \longrightarrow f \quad \text{weakly in } L^p(\mathbf{R}^n) \text{ as } k \to \infty.$$

Namely, we have the assertion

$$\int_{\mathbf{R}^n} u(t, y_k) g(t)\, dt \longrightarrow \int_{\mathbf{R}^n} f(t) g(t)\, dt \quad \text{for all } g \in L^{p'}(\mathbf{R}^n) \text{as } k \to \infty.$$

However, we remark that

$$P(\cdot, y) = \frac{1}{c_n} \frac{y}{(|\cdot|^2 + y^2)^{(n+1)/2}} \in L^{p'}(\mathbf{R}^n) \quad \text{for each } y > 0.$$

Hence we have the assertion

$$\int_{\mathbf{R}^n} P(x - t, y) u(t, y_k)\, dt \longrightarrow \int_{\mathbf{R}^n} P(x - t, y) f(t)\, dt \quad \text{for all } y > 0. \tag{5.32}$$

Thus it remains to show the following:

Claim 5.20 *The Poisson integral formula* (5.30) *holds true for all* $y > 0$.

Proof By Lemma 5.19, it follows that

$$u(x, y + y_k) = \int_{\mathbf{R}^n} P(x - t, y)\, u(t, y_k)\, dt. \tag{5.33}$$

Hence, by letting $k \to \infty$ in formula (5.33) we obtain from assertion (5.32) that

$$\begin{aligned} u(x, y) &= \lim_{k \to \infty} u(x, y + y_k) \\ &= \lim_{k \to \infty} \int_{\mathbf{R}^n} P(x - t, y)\, u(t, y_k)\, dt \\ &= \int_{\mathbf{R}^n} P(x - t, y) f(t)\, dt \quad \text{for all } y > 0. \end{aligned}$$

The proof of Claim 5.20 is complete. □

Step 2-b: The case $p = 1$. Assume that

$$\|u(\cdot, y)\|_1 \leq c \quad \text{for all } y > 0.$$

Then, by the *compactness* of measures (Theorem 2.37) we can find a sequence $\{y_k\}$, $y_k \downarrow 0$, and a finite Borel measure μ on $\mathbf{R}^n$ such that

$$\int_{\mathbf{R}^n} u(t, y_k)\, g(t)\, dt \longrightarrow \int_{\mathbf{R}^n} g(t)\, d\mu(t) \quad \text{for all } g \in C_0(\mathbf{R}^n),$$

where

$$C_0(\mathbf{R}^n) = \text{the space of continuous functions on } \mathbf{R}^n \text{ vanishing at infinity.}$$

However, we find that

$$P(\cdot, y) = \frac{1}{c_n} \frac{y}{(|\cdot|^2 + y^2)^{(n+1)/2}} \in C_0(\mathbf{R}^n) \quad \text{for all } y > 0.$$

Hence we have the assertion

$$\int_{\mathbf{R}^n} P(x-t, y)\, u(t, y_k)\, dt \longrightarrow \int_{\mathbf{R}^n} P(x-t, y)\, d\mu(t) \quad \text{for all } y > 0. \tag{5.34}$$

By arguing just as in the proof of Claim 5.20, we obtain that

$$u(x, y) = \int_{\mathbf{R}^n} P(x-t, y)\, d\mu(t) \quad \text{for all } y > 0.$$

Furthermore, if $\{u(\cdot, y)\}$ is a Cauchy sequence in $L^1(\mathbf{R}^n)$ as $y \downarrow 0$, then there exists a function $f \in L^1(\mathbf{R}^n)$ such that

$$u(\cdot, y) \longrightarrow f \quad \text{in } L^1(\mathbf{R}^n) \text{ as } y \downarrow 0.$$

Hence it follows that

$$\int_{\mathbf{R}^n} u(t, y_k)\, g(t)\, dt \longrightarrow \int_{\mathbf{R}^n} f(t)\, g(t)\, dt \quad \text{for all } g \in L^\infty(\mathbf{R}^n).$$

However, we remark that

$$P(\cdot, y) = \frac{1}{c_n} \frac{y}{(|\cdot|^2 + y^2)^{(n+1)/2}} \in L^\infty(\mathbf{R}^n) \quad \text{for all } y > 0.$$

Thus we have the assertion

$$\int_{\mathbf{R}^n} P(x-t, y)\, u(t, y_k)\, dt \longrightarrow \int_{\mathbf{R}^n} P(x-t, y) f(t)\, dt \quad \text{for all } y > 0,$$

and so, by assertion (5.34),

$$u(x, y) = \int_{\mathbf{R}^n} P(x-t, y)\, d\mu(t) = \int_{\mathbf{R}^n} P(x-t, y)\, f(t)\, dt \quad \text{for all } y > 0.$$

The proof of Theorem 5.16 is now complete. □

The next theorem establishes fundamental relationships between means of the *first derivatives* of Poisson integrals $u(x, y)$ taken with respect to the variable y and those taken with respect to the variables x_i:

Theorem 5.21 *Assume that a function $u(x, y)$ is harmonic in the half-space $\mathbf{R}_+^{n+1}$ and bounded in a proper sub-half space $\{(x, y) \in \mathbf{R}_+^{n+1} : y > y_0\}$ of $\mathbf{R}_+^{n+1}$, for each $y_0 > 0$. Then we have, for $\alpha > 0$, $1 \leq p$ and $q \leq \infty$,*

$$\sup_{1 \leq i \leq n} \left\| y^\alpha u_{x_i}(\cdot, \cdot) \right\|_{pq} \leq M_\alpha \left\| y^\alpha u_y(\cdot, \cdot) | \right\|_{pq}, \tag{5.35a}$$

$$\left\| y^\alpha u_y(\cdot, \cdot) \right\|_{pq} \leq M_\alpha \sup_{1 \leq i \leq n} \left\| y^\alpha u_{x_i}(\cdot, \cdot) \right\|_{pq}. \tag{5.35b}$$

Here $u_y(x, y)$ and $u_{x_i}(x, y)$ indicate the partial derivative of $u(x, y)$ in the direction of the element subscripted, that is,

$$u_y(x, y) = \frac{\partial u}{\partial y}, \quad u_{x_i}(x, y) = \frac{\partial u}{\partial x_i},$$

and the mixed norm $\|\cdot\|_{pq}$ is defined by the formula

$$\|y^\alpha u(\cdot, \cdot)\|_{pq} := \begin{cases} \left(\int_0^\infty (y^\alpha \|u(\cdot, y)\|_p)^q \frac{dy}{y} \right)^{1/q} & \text{if } 1 \leq q < \infty, \\ \sup_{y>0} \left\{ y^\alpha \|u(\cdot, y)\|_p \right\} & \text{if } q = \infty. \end{cases}$$

The proof of Theorem 5.21 is divided into two steps.

Step 1: In order to prove Theorem 5.21, we need several lemmas.

Lemma 5.22 *Let $1 \leq p \leq \infty$. Then we have the inequality*

$$\|P(\cdot, y)\|_p \leq \left(\frac{1}{c_n} \right)^{1/p'} y^{-n/p'} \quad \text{for all } y > 0, \tag{5.36}$$

where p' is the exponent conjugate to p:

$$p' = \frac{p}{p-1}.$$

Proof We remark that, by formulas (5.12a) and (5.15a),

$$\|P(\cdot, y)\|_1 = 1,$$

$$\|P(\cdot, y)\|_\infty = \frac{1}{c_n} y^{-n} \quad \text{for all } y > 0.$$

Hence the desired inequality (5.36) follows from the logarithmic convexity of the L^p norm as a function of $1/p$, since we have the formula

$$\frac{1}{p} = 1 - \frac{1}{p'} = \left(1 - \frac{1}{p'}\right)\frac{1}{1} + \frac{1}{p'}\frac{1}{\infty}.$$

The proof of Lemma 5.22 is complete. □

Lemma 5.23 *If $f \in L^p(\mathbf{R}^n)$ with $1 \le p \le \infty$ and if $u(x, y) = P(x, y) * f(x)$ is its Poisson integral, then we have the inequality*

$$\|u(\cdot, y)\|_\infty \le \left(\frac{1}{c_n}\right)^{1/p} \|f\|_p \, y^{-n/p} \quad \textit{for all } y > 0.$$

Proof By Young's inequality (Theorem 3.36), it follows that

$$\begin{aligned}\|u(\cdot, y)\|_\infty &\le \|P(\cdot, y)\|_{p'} \, \|f\|_p \\ &\le \left(\frac{1}{c_n}\right)^{1/p} y^{-n/p} \, \|f\|_p \quad \text{for all } y > 0.\end{aligned}$$

The proof of Lemma 5.23 is complete. □

Lemma 5.24 *Assume that a function $u(x, y)$ is harmonic in the half-space $\mathbf{R}_+^{n+1}$ and that, for each $y_0 > 0$, there exists a constant $A = A(y_0) > 0$ such that*

$$\|u(\cdot, y)\|_p \le A \quad \textit{for all } y \ge y_0, \tag{5.37}$$

with $1 \le p \le \infty$. Then we have the following inequalities:

$$\left\|u_y(\cdot, y)\right\|_p \le 2(n+1)\, y^{-1} \, \|u(\cdot, y/2)\|_p \quad \textit{for all } y > 0, \tag{5.38a}$$

$$\left\|u_{x_i}(\cdot, y)\right\|_p \le (n+1)\, y^{-1} \, \|u(\cdot, y/2)\|_p \quad \textit{for all } y > 0. \tag{5.38b}$$

Proof By condition (5.37), we can apply Theorem 5.16 to obtain that there exists a function $h \in L^p(\mathbf{R}^n)$ such that

$$u(x, y + y_0) = P(x, y) * h(x) \quad \text{for all } y > 0.$$

Hence it follows from an application of Lemma 5.22 that

$$\|u(\cdot, y + y_0)\|_\infty \le \|P(\cdot, y)\|_{p'} \, \|h\|_p \le c_n^{-1/p} \, \|h\|_p \, y^{-n/p} \quad \text{for all } y > 0.$$

This proves that the function $u(x, y + y_0)$ is bounded in each proper sub-half space of $\mathbf{R}_+^{n+1}$. Hence we have, by Lemma 5.19,

$$u(x, y_1 + y_2 + y_0) = P(x, y_2) * u(x, y_1 + y_0) \quad \text{for all } y_1, y_2 > 0,$$

or equivalently,

$$u(x, y+z) = P(x, y) * u(x, z) \quad \text{for all } y, z > 0.$$

If we differentiate both sides with respect to y (resp. x_i), we find that

$$u_y(x, y+z) = P_y(x, y) * u(x, z) \quad \text{for all } y, z > 0,$$
$$u_{x_i}(x, y+z) = P_{x_i}(x, y) * u(x, z) \quad \text{for all } y, z > 0,$$

since the integrals defining the convolutions converge absolutely.

Therefore, inequality (5.38a) follows from part (b) of Lemma 5.11. Indeed, we have the inequality

$$\begin{aligned}\left\|u_y(\cdot, y)\right\|_p &= \left\|P_y(\cdot, y/2) * u(\cdot, y/2)\right\|_p \leq \left\|P_y(\cdot, y/2)\right\|_1 \|u(\cdot, y/2)\|_p \\ &\leq 2(n+1)y^{-1} \|u(\cdot, y/2)\|_p \quad \text{for all } y > 0.\end{aligned}$$

Similarly, inequality (5.38b) follows from part (c) of Lemma 5.11.

The proof of Lemma 5.24 is complete. □

Lemma 5.25 *Assume that a function $u(x, y)$ is harmonic in the half-space $\mathbf{R}_+^{n+1}$ and bounded in a proper sub-half space $\{(x, y) \in \mathbf{R}_+^{n+1} : y > y_0\}$ of $\mathbf{R}_+^{n+1}$, for each $y_0 > 0$.*

(i) Then we have the inequalities for $\alpha > 0$, $1 \leq p$ and $q \leq \infty$:

$$\left\|y^{\alpha+1} u_y(\cdot, \cdot)\right\|_{pq} \leq 2^{\alpha+1}(n+1) \|y^\alpha u(\cdot, \cdot)\|_{pq}, \tag{5.39a}$$
$$\left\|y^{\alpha+1} u_{x_i}(\cdot, \cdot)\right\|_{pq} \leq 2^{\alpha}(n+1) \|y^\alpha u(\cdot, \cdot)\|_{pq}, \quad 1 \leq i \leq n. \tag{5.39b}$$

(ii) If, in addition, $u(x, y) \to 0$ as $y \to \infty$, then we have the inequality

$$\|y^\alpha u(\cdot, \cdot)\|_{pq} \leq \frac{1}{\alpha} \left\|y^{\alpha+1} u_y(\cdot, \cdot)\right\|_{pq}. \tag{5.40}$$

Proof We recall that

$$u(x, y+z) = P(x, y) * u(x, z) \quad \text{for all } y, z > 0.$$

Hence, if we differentiate both sides with respect to y (resp. x_i), we obtain that

$$u_y(x, y+z) = P_y(x, y) * u(x, z) \quad \text{for all } y, z > 0,$$
$$u_{x_i}(x, y+z) = P_{x_i}(x, y) * u(x, z) \quad \text{for all } y, z > 0.$$

In particular, it follows that

$$u_y(x, y) = P_y(x, y/2) * u(x, y/2) \quad \text{for all } y > 0,$$
$$u_{x_i}(x, y) = P_{x_i}(x, y/2) * u(x, y/2) \quad \text{for all } y > 0.$$

Therefore, we have, by inequality (5.38b),

$$\begin{aligned}\left\|u_y(\cdot, y)\right\|_p &= \left\|P_y(\cdot, y/2) * u(\cdot, y/2)\right\|_p \\ &\le \left\|P_y(\cdot, y/2)\right\|_1 \|u(\cdot, y/2)\|_p \\ &\le 2(n+1)y^{-1} \|u(\cdot, y/2)\|_p \quad \text{for all } y > 0.\end{aligned} \tag{5.41}$$

Similarly, we have the inequality

$$\left\|u_{x_i}(\cdot, y)\right\|_p \le (n+1)y^{-1} \|u(\cdot, y/2)\|_p \quad \text{for all } y > 0. \tag{5.42}$$

(a-1) The case $1 \le q < \infty$: By inequality (5.41), it follows that

$$\begin{aligned}\left\|y^{\alpha+1} u_y(\cdot, \cdot)\right\|_{pq} &= \left(\int_0^\infty \left(y^{\alpha+1} \left\|u_y(\cdot, y)\right\|_p\right)^q \frac{dy}{y}\right)^{1/q} \\ &\le \left(\int_0^\infty \left(y^{\alpha+1} 2(n+1)\, y^{-1} \|u(\cdot, y/2)\|_p\right)^q \frac{dy}{y}\right)^{1/q} \\ &= 2^{\alpha+1}(n+1) \left(\int_0^\infty \left(z^\alpha \|u(\cdot, z)\|_p\right)^q \frac{dz}{z}\right)^{1/q} \\ &= 2^{\alpha+1}(n+1)\ \|y^\alpha u(\cdot, \cdot)\|_{pq}.\end{aligned}$$

(a-2) The case $q = \infty$: We have, by inequality (5.42),

$$\begin{aligned}\left\|y^{\alpha+1} u_y(\cdot, \cdot)\right\|_{p\infty} &= \sup_{y>0}\{y^{\alpha+1} \left\|u_y(\cdot, y)\right\|_p\} \\ &\le 2(n+1)\ \sup_{y>0}\{y^\alpha \|u(\cdot, y/2)\|_p\} \\ &= 2^{\alpha+1}(n+1)\ \sup_{z>0}\{z^\alpha \|u(\cdot, z)\|_p\} \\ &= 2^{\alpha+1}(n+1)\ \|y^\alpha u(\cdot, \cdot)\|_{p\infty}.\end{aligned}$$

This completes the proof of inequality (5.39a).

(b) Similarly, we have the inequality

$$\left\|y^{\alpha+1} u_{x_i}(\cdot, \cdot)\right\|_{pq} \le 2^\alpha (n+1)\, \|y^\alpha u(\cdot, \cdot)\|_{pq}, \quad 1 \le q \le \infty.$$

This proves inequality (5.39b).

(c) If $u(x, y) \to 0$ as $y \to \infty$, it follows that

$$u(x, y) = -\int_y^\infty u_s(x, s)\, ds \quad \text{for all } y > 0.$$

Hence, by applying Minkowski's inequality for integrals (Theorem 3.29) we obtain that

$$\|u(\cdot, y)\|_p \leq \int_y^\infty \|u_s(\cdot, s)\|_p \, ds \quad \text{for all } y > 0.$$

(c-1) The case $1 \leq q < \infty$: By Hardy's inequality (Theorem 3.31), it follows that

$$\begin{aligned}
\|y^\alpha u(\cdot, \cdot)\|_{pq} &= \left(\int_0^\infty (y^\alpha \|u(\cdot, y)\|_p)^{\alpha+1} \frac{dy}{y} \right)^{1/q} \\
&\leq \left(\int_0^\infty \left(y^\alpha \int_y^\infty \|u_s(\cdot, s)\|_p \, ds \right)^q \frac{dy}{y} \right)^{1/q} \\
&\leq \frac{1}{\alpha} \left(\int_0^\infty \left(y^{\alpha+1} \left\| u_y(\cdot, y) \right\|_p \right)^q \frac{dy}{y} \right)^{1/q} \\
&= \frac{1}{\alpha} \left\| y^{\alpha+1} u_y(\cdot, \cdot) \right\|_{pq}.
\end{aligned}$$

(c-2) The case $q = \infty$: We let

$$C_\alpha = \left\| y^{\alpha+1} u_y(\cdot, \cdot) \right\|_{p\infty}.$$

Then we have the inequality

$$\left\| u_y(\cdot, y) \right\|_p \leq C_\alpha y^{-\alpha-1} \quad \text{for all } y > 0,$$

so that

$$\|u(\cdot, y)\|_p \leq \int_y^\infty \|u_s(\cdot, s)\|_p \, ds \leq C_\alpha \alpha^{-1} y^{-\alpha} \quad \text{for all } y > 0.$$

This proves that

$$\|y^\alpha u(\cdot, \cdot)\|_{pq} = \sup_{y>0} \{ y^\alpha \|u(\cdot, y)\|_p \} \leq \frac{1}{\alpha} \left\| y^{\alpha+1} u_y(\cdot, \cdot) \right\|_{p\infty}.$$

The proof of inequality (5.40) is complete.

Now the proof of Lemma 5.25 is complete. □

Step 2: First, we remark that, for each $y_0 > 0$, there exists a constant $A_{y_0} > 0$ such that

$$\|u(\cdot, y)\|_\infty \leq A_{y_0} \quad \text{for all } y > y_0.$$

Hence, by virtue of Lemma 5.24 it follows that

$$\begin{aligned}\left\|u_y(\cdot, y)\right\|_\infty &\leq \frac{2(n+1)}{y} \left\|u(\cdot, y/2)\right\|_\infty \\ &\leq \frac{2(n+1)A_{y_0}}{y} \quad \text{for all } y \geq 2y_0,\end{aligned} \tag{5.43}$$

and that

$$\begin{aligned}\left\|u_{x_i}(\cdot, y)\right\|_\infty &\leq \frac{(n+1)}{y} \left\|u(\cdot, y/2)\right\|_\infty \\ &\leq \frac{(n+1)A_{y_0}}{y} \quad \text{for all } y \geq 2y_0.\end{aligned} \tag{5.44}$$

Step 2-a: Since the function $u_y(x, y)$ is harmonic in $\mathbf{R}_+^{n+1}$ and satisfies the inequality (5.43), it follows from an application of inequality (5.39b) that

$$\left\|y^{\alpha+1} u_{x_i y}(\cdot, \cdot)\right\|_{pq} \leq 2^\alpha (n+1) \left\|y^\alpha u_y(\cdot, \cdot)\right\|_{pq}. \tag{5.45}$$

Moreover, we remark by inequality (5.44) that $u_{x_i}(x, y) \to 0$ as $y \to \infty$. Hence, by applying inequality (5.40) to the function $u_{x_i}(x, y)$ we obtain that

$$\left\|y^\alpha u_{x_i}(\cdot, \cdot)\right\|_{pq} \leq \frac{1}{\alpha} \left\|y^{\alpha+1} u_{x_i y}(\cdot, \cdot)\right\|_{pq}. \tag{5.46}$$

Therefore, it follows from inequalities (5.45) and (5.46) that

$$\left\|y^\alpha u_{x_i}(\cdot, \cdot)\right\|_{pq} \leq \frac{2^\alpha}{\alpha}(n+1) \left\|y^\alpha u_y(\cdot, \cdot)\right\|_{pq}.$$

This proves the desired inequality (5.35a).

Step 2-b: Since the function $u_{x_i}(x, y)$ is harmonic in $\mathbf{R}_+^{n+1}$ and satisfies the inequality (5.44), it follows from an application of inequality (5.38b) that

$$\left\|y^{\alpha+1} u_{x_i x_j}(\cdot, \cdot)\right\|_{pq} \leq 2^\alpha (n+1) \left\|y^\alpha u_{x_i}(\cdot, \cdot)\right\|_{pq}. \tag{5.47}$$

However, we remark that

$$u_{yy}(x, y) = -\sum_{i=1}^n u_{x_i x_i}(x, y).$$

Hence we have, by inequality (5.47),

$$\left\|y^{\alpha+1} u_{yy}(\cdot, \cdot)\right\|_{pq} \leq \sum_{i=1}^n \left\|y^{\alpha+1} u_{x_i x_i}(\cdot, \cdot)\right\|_{pq} \tag{5.48}$$

$$\leq 2^\alpha (n+1) \sum_{i=1}^{n} \left\| y^\alpha u_{x_i}(\cdot,\cdot) \right\|_{pq}$$
$$\leq 2^\alpha n(n+1) \sup_{1\leq i\leq n} \left\| y^\alpha u_{x_i}(\cdot,\cdot) \right\|_{pq}.$$

Moreover, we remark by inequality (5.43) that $u_y(x,y) \to 0$ as $y \to \infty$. Hence, by applying inequality (5.40) to the function $u_y(x,y)$ we obtain that

$$\left\| y^\alpha u_y(\cdot,\cdot) \right\|_{pq} \leq \frac{1}{\alpha} \left\| y^{\alpha+1} u_{yy}(\cdot,\cdot) \right\|_{pq}. \tag{5.49}$$

Therefore, it follows from inequalities (5.48) and (5.49) that

$$\left\| y^\alpha u_y(\cdot,\cdot) \right\|_{pq} \leq \frac{2^\alpha}{\alpha} n(n+1) \sup_{1\leq i\leq n} \left\| y^\alpha u_{x_i}(\cdot,\cdot) \right\|_{pq}.$$

This proves the desired inequality (5.35b).

The proof of Theorem 5.21 is now complete. □

The next theorem establishes fundamental relationships between means of the *second derivatives* of Poisson integrals $u(x,y)$ taken with respect to the variable y and those taken with respect to the variables x_i:

Theorem 5.26 *Assume that a function $u(x,y)$ is harmonic in the half-space $\mathbf{R}_+^{n+1}$ and bounded in a proper sub-half space $\{(x,y) \in \mathbf{R}_+^{n+1} : y > y_0\}$ of $\mathbf{R}_+^{n+1}$, for each $y_0 > 0$. Then we have, for $\alpha > 0$, $1 \leq p$ and $q \leq \infty$,*

$$\sup_{1\leq i,j\leq n} \left\| y^\alpha u_{x_i x_j}(\cdot,\cdot) \right\|_{pq} \leq M_\alpha \left\| y^\alpha u_{yy}(\cdot,\cdot) \right\|_{pq}, \tag{5.50a}$$
$$\left\| y^\alpha u_{yy}(\cdot,\cdot) \right\|_{pq} \leq M_\alpha \sup_{1\leq i,j\leq n} \left\| y^\alpha u_{x_i x_j}(\cdot,\cdot) \right\|_{pq}. \tag{5.50b}$$

Proof Inequality (5.50b) follows from the equation

$$u_{yy}(x,y) = -\sum_{i=1}^{n} u_{x_i x_i}(x,y).$$

On the other hand, inequality (5.50a) follows by applying (5.35a) of Theorem 5.21 twice.

The proof of Theorem 5.26 is complete. □

Chapter 6
Besov Spaces via Poisson Integrals

In this chapter we develop the theory of Besov spaces on the Euclidean space $\mathbf{R}^n$, paying particular attention to Poisson integrals. Besov spaces are function spaces defined in terms of the L^p modulus of continuity, and enter naturally in connection with boundary value problems in the framework of Sobolev spaces of L^p type. We prove a variety of equivalent norms for the Besov spaces on $\mathbf{R}^n$ via Poisson integrals (Theorems 6.4, 6.7 and 6.8). The results discussed here are adapted from Taibleson [146] and Stein [137].

6.1 Definition of Besov Spaces

Let $\alpha > 0$ and $1 \leq p, q \leq \infty$. We let

$$\begin{aligned} B^{\alpha}_{p,q}(\mathbf{R}^n) &= \text{the space of functions } f \in L^p(\mathbf{R}^n) \text{ for which the quantity} \\ \|f\|_{\alpha;p,q} &= \|f\|_p + \left\| y^{\overline{\alpha}-\alpha} u_y^{(\overline{\alpha})}(\cdot,\cdot) \right\|_{pq} \qquad (6.1) \\ &= \|f\|_p + \left(\int_0^\infty \left(y^{\overline{\alpha}-\alpha} \left\| u_y^{(\overline{\alpha})}(\cdot, y) \right\|_p \right)^q \frac{dy}{y} \right)^{1/q} \\ &\text{is finite.} \end{aligned}$$

Here $\overline{\alpha}$ is the smallest integer greater than α and $u(x, y) = P(x, y) * f(x)$ is the Poisson integral of $f(x)$

$$\begin{aligned} u(x, y) &= P(x, y) * f(x) \\ &= \frac{1}{c_n} \int_{\mathbf{R}^n} \frac{y}{(|x-z|^2 + y^2)^{(n+1)/2}} f(z)\, dz \quad \text{for all } y > 0, \qquad (5.16) \end{aligned}$$

K. Taira, *Real Analysis Methods for Markov Processes*,
https://doi.org/10.1007/978-981-97-3659-1_6

where c_n is a positive constant given by the formula

$$c_n = \frac{\pi^{(n+1)/2}}{\Gamma((n+1)/2)}.$$

Moreover, we recall that the mixed norm $\| \cdot \|_{pq}$ is defined by the formula

$$\|y^\alpha u(\cdot, \cdot)\|_{pq} = \begin{cases} \left(\int_0^\infty (y^\alpha \|u(\cdot, y)\|_p)^q \frac{dy}{y} \right)^{1/q} & \text{if } 1 \leq q < \infty, \\ \sup_{y>0} \left\{ y^\alpha \|u(\cdot, y)\|_p \right\} & \text{if } q = \infty. \end{cases}$$

First, we prove the completeness of $B^\alpha_{p,q}(\mathbf{R}^n)$:

Theorem 6.1 *The space $B^\alpha_{p,q}(\mathbf{R}^n)$ is a Banach space.*

Proof We only consider the case:

$$0 < \alpha < 1, \quad \overline{\alpha} = 1.$$

Let $\{f^k\}$ be an arbitrary Cauchy sequence in $B^\alpha_{p,q}(\mathbf{R}^n)$, that is, we have, as k, $\ell \to \infty$,

$$\left\| f^k - f^\ell \right\|_{\alpha; p,q} = \left\| f^k - f^\ell \right\|_p + \left\| y^{1-\alpha} \left(u^k_y(\cdot, \cdot) - u^\ell_y(\cdot, \cdot) \right) \right\|_{pq} \longrightarrow 0, \quad (6.2)$$

where

$$u^k(x, y) = P(x, y) * f^k(x).$$

Then, since $L^p(\mathbf{R}^n)$ is complete, there exists a function $f \in L^p(\mathbf{R}^n)$ such that

$$\left\| f^k - f \right\|_p \longrightarrow 0 \quad \text{as } k \to \infty.$$

We let

$$u(x, y) = P(x, y) * f(x).$$

Then we obtain the following claim:

Claim 6.2 $\left\| y^{1-\alpha} \left(u^k_y(\cdot, \cdot) - u_y(\cdot, \cdot) \right) \right\|_{pq} \longrightarrow 0$ *as* $k \to \infty$.

Proof Since we have the formulas

$$u^\ell_y(x, y) = P_y(x, y) * f^\ell(x),$$
$$u_y(x, y) = P_y(x, y) * f(x),$$

it follows that we have, by Lemma 5.11,

$$\begin{aligned}\left\|u_y^\ell(\cdot, y) - u_y(\cdot, y)\right\|_p &\le \left\|P_y(\cdot, y)\right\|_1 \left\|f^\ell - f\right\|_p \\ &\le (n+1)y^{-1} \left\|f^\ell - f\right\|_p \quad \text{for all } y > 0.\end{aligned}$$

This implies that we have, for each $y > 0$,

$$\left\|y^{1-\alpha}\left(u_y^k(\cdot, y) - u_y^\ell(\cdot, y)\right)\right\|_p \longrightarrow \left\|y^{1-\alpha}\left(u_y^k(\cdot, y) - u_y(\cdot, y)\right)\right\|_p \quad \text{as } \ell \to \infty.$$

Hence, by applying Fatou's lemma (Lemma 3.12) we obtain from assertion (6.2) that

$$\begin{aligned}&\left\|y^{1-\alpha}\left(u_y^k(\cdot, \cdot) - u_y(\cdot, \cdot)\right)\right\|_{pq}^q \\ &= \int_0^\infty \left(y^{1-\alpha}\left\|u_y^k(\cdot, y) - u_y(\cdot, y)\right\|_p\right)^q \frac{dy}{y} \\ &= \int_0^\infty \lim_{\ell\to\infty}\left(y^{1-\alpha}\left\|u_y^k(\cdot, y) - u_y^\ell(\cdot, y)\right\|_p\right)^q \frac{dy}{y} \\ &\le \liminf_{\ell\to\infty}\int_0^\infty \left(y^{1-\alpha}\left\|u_y^k(\cdot, y) - u_y^\ell(\cdot, y)\right\|_p\right)^q \frac{dy}{y} \\ &\le \sup_{\ell\ge k}\left\|y^{1-\alpha}\left(u_y^k(\cdot, \cdot) - u_y^\ell(\cdot, \cdot)\right)\right\|_{pq}^q \\ &\le \sup_{\ell\ge k}\left\|f^k - f^\ell\right\|_{\alpha;p,q}^q \longrightarrow 0 \quad \text{as } k \to \infty.\end{aligned}$$

The proof of Claim 6.2 is complete. □

Therefore, we find from Claim 6.2 that

$$\begin{aligned}&f \in B^\alpha_{p,q}(\mathbf{R}^n), \\ &f^k \longrightarrow f \quad \text{in } B^\alpha_{p,q}(\mathbf{R}^n),\end{aligned}$$

since we have the assertion

$$\begin{aligned}&\left\|y^{1-\alpha}u_y(\cdot, \cdot)\right\|_{pq} \\ &\le \left\|y^{1-\alpha}\left(u_y(\cdot, \cdot) - u_y^k(\cdot, \cdot)\right)\right\|_{pq} + \left\|y^{1-\alpha}u_y^k(\cdot, \cdot)\right\|_{pq} < \infty,\end{aligned}$$

and also

$$\left\|f^k - f\right\|_{\alpha;p,q} = \left\|f^k - f\right\|_p + \left\|y^{1-\alpha}\left(u_y^k(\cdot, \cdot) - u_y(\cdot, \cdot)\right)\right\|_{pq} \longrightarrow 0 \quad \text{as } k \to \infty.$$

This proves the completeness of $B^\alpha_{p,q}(\mathbf{R}^n)$.

The proof of Theorem 6.1 is complete. □

Proposition 6.3 *Let $f \in L^p(\mathbf{R}^n)$ with $1 \le p \le \infty$, and $u(x, y) = P(x, y) * f(x)$ its Poisson integral. Then, for any integer $k > \alpha$, the norm*

$$\|f\|_p + \left\|y^{k-\alpha} u_y^{(k)}(\cdot, \cdot)\right\|_{pq}$$

is equivalent to the norm $\|f\|_{\alpha;p,q}$.

Proof It suffices to show that the ratio

$$\frac{\left\|y^{k-\alpha} u_y^{(k)}(\cdot, \cdot)\right\|_{pq}}{\left\|y^{\overline{\alpha}-\alpha} u_y^{(\overline{\alpha})}(\cdot, \cdot)\right\|_{pq}}$$

is bounded above and below by positive constants independent of f. However, this follows from a repeated application of part (a) and part (c) of Lemma 5.25.

The proof of Proposition 6.3 is complete. □

6.2 Various Norms of Besov Spaces

The next theorem gives a variety of equivalent norms for the Besov space $B^\alpha_{p,q}(\mathbf{R}^n)$:

Theorem 6.4 *Let f be a function in $L^p(\mathbf{R}^n)$ with $1 \le p, q \le \infty$, and let $u(x, y) = P(x, y) * f(x)$ be its Poisson integral. We define the eight norms A through H as follows:*

$$\begin{aligned}
A &:= \left\| \frac{f(\cdot + h) - f(\cdot)}{|h|^\alpha} \right\|_{pq} \\
&= \left(\int_{\mathbf{R}^n} \left(\frac{\|f(\cdot + h) - f(\cdot)\|_p}{|h|^\alpha} \right)^q \frac{dh}{|h|^n} \right)^{1/q}, \quad 0 < \alpha < 1. \\
B &:= \left\| \frac{f(\cdot + h) - 2f(\cdot) + 2f(\cdot - h)}{|h|^\alpha} \right\|_{pq} \\
&= \left(\int_{\mathbf{R}^n} \left(\frac{\|f(\cdot + h) - 2f(\cdot) + f(\cdot - h)\|_p}{|h|^\alpha} \right)^q \frac{dh}{|h|^n} \right)^{1/q}, \quad 0 < \alpha < 2. \\
C &:= \frac{1}{\omega_n} \left(\int_0^\infty \left(t^{-\alpha} \left\| \int_{\Sigma_{n-1}} (f(\cdot + t\sigma) - f(\cdot))\, d\sigma \right\|_p \right)^q \frac{dt}{t} \right)^{1/q}, \quad 0 < \alpha < 2. \\
D &:= \left(\int_0^\infty \left(y^{1-\alpha} \left\|u_y(\cdot, y)\right\|_p \right)^q \frac{dy}{y} \right)^{1/q} = \left\|y^{1-\alpha} u_y(\cdot, \cdot)\right\|_{pq}, \quad 0 < \alpha < 1. \\
E &:= \left(\int_0^\infty (y^{2-\alpha} \left\|u_{yy}(\cdot, y)\right\|_p)^q \frac{dy}{y} \right)^{1/q} = \left\|y^{2-\alpha} u_{yy}(\cdot, \cdot)\right\|_{pq}, \quad 0 < \alpha < 2.
\end{aligned}$$

$$F := \left(\int_0^\infty \left(t^{-\alpha} \cdot \sup_{0<|h|\le t} \|f(\cdot + h) - f(\cdot)\|_p \right)^q \frac{dt}{t} \right)^{1/q}, \quad 0 < \alpha < 1.$$

$$G := \left(\int_0^\infty \left(t^{-\alpha} \cdot \sup_{0<|h|\le t} \|f(\cdot + h) - 2f(\cdot) + f(\cdot - h)\|_p \right)^q \frac{dt}{t} \right)^{1/q},$$

$0 < \alpha < 2.$

$$H := \left(\int_0^\infty \left(y^{-\alpha} \|u(\cdot, y) - f(\cdot)\|_p \right)^q \frac{dy}{y} \right)^{1/q}, \quad 0 < \alpha < 1.$$

Here Σ_{n-1} is the unit sphere in $\mathbf{R}^n$, $d\sigma$ is its surface element and ω_n is its surface area

$$\omega_n := |\Sigma_{n-1}| = \int_{\Sigma_{n-1}} d\sigma = \frac{2\pi^{n/2}}{\Gamma(n/2)}.$$

Then the norms A, B, C, D, E, F, G and H are equivalent for $0 < \alpha < 1$, *while the norms B, C, E and G are equivalent for* $0 < \alpha < 2$.

Proof The proof is divided into ten steps.

Step 1: $E \le 2^{2-\alpha}(n+1)D$ and $D \le 1/(1-\alpha)E$ for $0 < \alpha < 1$.

Assume that

$$u(x, y) = P(x, y) * f(x) \quad \text{for } f \in L^p(\mathbf{R}^n).$$

Then we have, by part (i) of Theorem 5.12,

$$\|u(\cdot, y)\|_p = \|P(\cdot, y) * f\|_p \le \|f\|_p \quad \text{for all } y > 0.$$

Hence, by Lemma 5.17 it follows that

$$\|u(\cdot, y)\|_\infty \le \left(\frac{2^{n+1}}{V_{n+1}} \right)^{1/p} \|f\|_p \, y^{-n/p} \quad \text{for all } y > 0,$$

where V_{n+1} is the volume of the unit ball in $\mathbf{R}^{n+1}$

$$V_{n+1} = \frac{\omega_{n+1}}{n+1} = \frac{\pi^{(n+1)/2}}{\Gamma((n+1)/2 + 1)}.$$

By applying Lemma 5.24 with $p := \infty$, we obtain that, for each $y_0 > 0$,

$$\left\| u_y(\cdot, y) \right\|_\infty \le 2(n+1)\, A_{y_0}\, y^{-1} \quad \text{for all } y \ge 2y_0, \tag{6.3}$$

with

$$A_{y_0} := \left(\frac{2^{n+1}}{V_{n+1}} \right)^{1/p} \|f\|_p \, y_0^{-n/p}.$$

Therefore, it follows from an application of part (a) of Lemma 5.25 with $\alpha := 1 - \alpha$ that

$$
\begin{aligned}
E &= \left\| y^{2-\alpha} u_{yy}(\cdot,\cdot) \right\|_{pq} \leq 2^{2-\alpha}(n+1) \left\| y^{1-\alpha} u_y(\cdot,\cdot) \right\|_{pq} \\
&= 2^{2-\alpha}(n+1) D, \quad 0 < \alpha < 1.
\end{aligned}
$$

On the other hand, we find from inequality (6.3) that $u_y(x, y) \to 0$ as $y \to \infty$. Hence, by applying part (c) of Lemma 5.25 with $u(x, y) := u_y(x, y)$ and $\alpha := 1 - \alpha$ we obtain that

$$
D = \left\| y^{1-\alpha} u_y(\cdot,\cdot) \right\|_{pq} \leq \frac{1}{1-\alpha} \left\| y^{2-\alpha} u_{yy}(\cdot,\cdot) \right\|_{pq} = \frac{1}{1-\alpha} E, \quad 0 < \alpha < 1.
$$

Step 2: $C \leq (2\omega_n^{1/q})^{-1} B$.

Step 2-a: The case $q = \infty$. By Minkowski's inequality for integrals (Theorem 3.29), it follows that

$$
\begin{aligned}
C &= \frac{1}{\omega_n} \sup_{t>0} \left\{ t^{-\alpha} \left\| \int_{\Sigma_{n-1}} (f(\cdot + t\sigma) - f(\cdot))\, d\sigma \right\|_p \right\} \\
&= \frac{1}{2} \frac{1}{\omega_n} \sup_{t>0} \left\{ t^{-\alpha} \left\| \int_{\Sigma_{n-1}} (f(\cdot + t\sigma) - 2f(\cdot) + f(\cdot - t\sigma))\, d\sigma \right\|_p \right\} \\
&\leq \frac{1}{2} \frac{1}{\omega_n} \sup_{t>0} \left\{ \int_{\Sigma_{n-1}} \frac{\| f(\cdot + t\sigma) - 2f(\cdot) + f(\cdot - t\sigma) \|_p}{t^\alpha}\, d\sigma \right\} \\
&\leq \frac{1}{2} \frac{1}{\omega_n} \int_{\Sigma_{n-1}} \left(\sup_{h \in \mathbf{R}^n} \frac{\| f(\cdot + h) - 2f(\cdot) + f(\cdot - h) \|_p}{|h|^\alpha} \right) d\sigma \\
&= \frac{1}{2} B, \quad 0 < \alpha < 1.
\end{aligned}
$$

Step 2-b: The case $1 \leq q < \infty$. By applying Hölder's inequality (Theorem 3.24) with $q' = q/(q-1)$, we obtain that

$$
\begin{aligned}
& C \\
&= \frac{1}{\omega_n} \left(\int_0^\infty \left(t^{-\alpha} \left\| \int_{\Sigma_{n-1}} (f(\cdot + t\sigma) - f(\cdot))\, d\sigma \right\|_p \right)^q \frac{dt}{t} \right)^{1/q} \\
&= \frac{1}{2} \frac{1}{\omega_n} \left(\int_0^\infty \left\| \int_{\Sigma_{n-1}} (f(\cdot + t\sigma) - 2f(\cdot) + f(\cdot - t\sigma))\, d\sigma \right\|_p^q \frac{dt}{t^{1+\alpha q}} \right)^{1/q} \\
&\leq \frac{1}{2} \frac{1}{\omega_n} \left(\int_0^\infty \left(\int_{\Sigma_{n-1}} \| f(\cdot + t\sigma) - 2f(\cdot) + f(\cdot - t\sigma) \|_p\, d\sigma \right)^q \frac{dt}{t^{1+\alpha q}} \right)^{1/q}
\end{aligned}
$$

$$
\begin{aligned}
&\leq \frac{1}{2}\frac{1}{\omega_n}\left(\int_0^\infty \int_{\Sigma_{n-1}} \|f(\cdot+t\sigma)-2f(\cdot)+f(\cdot-t\sigma)\|_p^q \, d\sigma \cdot \omega_n^{q/q'} \frac{dt}{t^{1+\alpha q}}\right)^{1/q} \\
&\leq \frac{1}{2}\frac{1}{\omega_n^{1/q}}\left(\int_0^\infty \int_{\Sigma_{n-1}} \|f(\cdot+t\sigma)-2f(\cdot)+f(\cdot-t\sigma)\|_p^q \, d\sigma \frac{dt}{t^{1+\alpha q}}\right)^{1/q} \\
&= \frac{1}{2}\frac{1}{\omega_n^{1/q}}\left(\int_{\mathbf{R}^n} \left(\frac{\|f(\cdot+h)-2f(\cdot)+f(\cdot-h)\|_p}{|h|^\alpha}\right)^q \frac{dh}{|h|^n}\right)^{1/q} \\
&= \frac{1}{2\omega_n^{1/q}} B, \quad 0<\alpha<1.
\end{aligned}
$$

Step 3: $B \leq 2A$. We remark that

$$
\begin{aligned}
\|f(\cdot+h)-2f(\cdot)+f(\cdot-h)\|_p &\leq \|f(\cdot+h)-f(\cdot)\|_p + \|f(\cdot)-f(\cdot-h)\|_p \\
&\leq 2\,\|f(\cdot+h)-f(\cdot)\|_p .
\end{aligned}
$$

Hence we have the inequality

$$
\begin{aligned}
B &= \left(\int_{\mathbf{R}^n} \left(\frac{\|f(\cdot+h)-2f(\cdot)+f(\cdot-h)\|_p}{|h|^\alpha}\right)^q \frac{dh}{|h|^n}\right)^{1/q} \\
&\leq \left(\int_{\mathbf{R}^n} \left(\frac{2\,\|f(\cdot+h)-f(\cdot)\|_p}{|h|^\alpha}\right)^q \frac{dh}{|h|^n}\right)^{1/q} \\
&= 2A, \quad 0<\alpha<1.
\end{aligned}
$$

Step 4: $A \leq \omega_n^{1/q} F$. This is obvious. Indeed, we have the inequality

$$
\begin{aligned}
A &= \left(\int_{\mathbf{R}^n} \left(\frac{\|f(\cdot+h)-f(\cdot)\|_p}{|h|^\alpha}\right)^q \frac{dh}{|h|^n}\right)^{1/q} \\
&= \left(\int_0^\infty \int_{\Sigma_{n-1}} \|f(\cdot+t\sigma)-f(\cdot)\|_p^q \frac{dt}{t^{1+\alpha q}} \, d\sigma\right)^{1/q} \\
&\leq \omega_n^{1/q}\left(\int_0^\infty \left(t^{-\alpha} \cdot \sup_{0<|h|\leq t} \|f(\cdot+h)-f(\cdot)\|_p\right)^q \frac{dt}{t}\right)^{1/q} \\
&= \omega_n^{1/q} F, \quad 0<\alpha<1.
\end{aligned}
$$

Step 5: $F \leq M_\alpha D$. We recall (see Theorem 5.15) that

$$
\lim_{y \downarrow 0} u(x, y) = f(x),
$$

whenever x belongs to the Lebesgue set $L(f)$ of f. We let

$$\mathcal{L} = \{(x,h) \in \mathbf{R}^n \times \mathbf{R}^n : x + h \in L(f)\}.$$

By Fubini's theorem (Theorem 3.20), we find that the complement of $\mathcal{L}$ has measure zero.

Now we assume that

$$(x,0),\,(x,h) \in \mathcal{L} \quad \text{with } 0 < |h| \le t.$$

Then we have, for $h = s\sigma$, $0 < s \le t$ and $\sigma \in \Sigma_{n-1}$,

$$\begin{aligned}
& f(x+h) - f(x) \\
&= (f(x+h) - u(x+h,t)) + (u(x+h,t) - u(x,t)) + (u(x,t) - f(x)) \\
&= -\int_0^t u_y(x+s\sigma,y)\,dy + \int_0^s u_r(x+r\sigma,t)\,dr + \int_0^t u_y(x,y)\,dy.
\end{aligned}$$

Hence it follows from an application of Minkowski's inequality for integrals (Theorem 3.29) that

$$\begin{aligned}
\|f(\cdot+h) - f(\cdot)\|_p &\le \int_0^t \left\|u_y(\cdot + s\sigma, y)\right\|_p dy + \int_0^s \|u_r(\cdot + r\sigma, t)\|_p\, dr \\
&\quad + \int_0^t \left\|u_y(\cdot, y)\right\|_p dy \\
&\le 2\int_0^t \left\|u_y(\cdot, y)\right\|_p dy + t\,\|u_r(\cdot, t)\|_p \\
&\le 2\int_0^t \left\|u_y(\cdot, y)\right\|_p dy + t\sum_{i=1}^n \left\|u_{x_i}(\cdot, t)\right\|_p,
\end{aligned}$$

since we have the formula

$$u_r(x+r\sigma,t) = \sum_{i=1}^n \frac{\partial u}{\partial x_i}(x+r\sigma,t)\sigma_i.$$

Step 5-a: The case $q = \infty$. Since we have the inequality

$$\left\|u_y(\cdot, y)\right\|_p \le D y^{\alpha-1} \quad \text{for all } y > 0,$$

it follows from part (a) of Theorem 5.21 that

$$\sup_{1\le i\le n} \left\|u_{x_i}(\cdot, y)\right\|_p \le M'_\alpha \left\|u_y(\cdot, y)\right\|_p \le M'_\alpha D y^{\alpha-1} \quad \text{for all } y > 0.$$

Thus we obtain that

$$
\begin{aligned}
\|f(\cdot+h)-f(\cdot)\|_p &\leq 2\int_0^t \left\|u_y(\cdot,y)\right\|_p\,dy + t\sum_{i=1}^n \left\|u_{x_i}(\cdot,t)\right\|_p \qquad (6.4)\\
&\leq 2D\int_0^t y^{\alpha-1}\,dy + M'_\alpha D\,nt\,t^{\alpha-1}\\
&= \left(\frac{2}{\alpha}+nM'_\alpha\right) D\,t^\alpha \quad \text{for all } 0<|h|\leq t.
\end{aligned}
$$

This proves that

$$
t^{-\alpha}\cdot \sup_{0<|h|\leq t}\|f(\cdot+h)-f(\cdot)\|_p \leq \left(\frac{2}{\alpha}+nM'_\alpha\right) D,
$$

so that

$$
F=\sup_{t>0}\left(t^{-\alpha}\cdot \sup_{0<|h|\leq t}\|f(\cdot+h)-f(\cdot)\|_p\right)\leq M_\alpha D,
$$

with

$$
M_\alpha := \frac{2}{\alpha}+nM'_\alpha, \quad 0<\alpha<1.
$$

Step 5-b: The case $1\leq q<\infty$. In view of inequality (6.4), we have, for some positive constant M_q,

$$
\begin{aligned}
F &= \left(\int_0^\infty \left(t^{-\alpha}\cdot \sup_{0<|h|\leq t}\|f(\cdot+h)-f(\cdot)\|_p\right)^q \frac{dt}{t}\right)^{1/q} \qquad (6.5)\\
&\leq \left(\int_0^\infty t^{-\alpha q}\left(2\int_0^t \left\|u_y(\cdot,y)\right\|_p\,dy + t\sum_{i=1}^n \left\|u_{x_i}(\cdot,t)\right\|_p\right)^q \frac{dt}{t}\right)^{1/q}\\
&\leq M_q\Bigg[\left(\int_0^\infty \left(t^{-\alpha}\int_0^t \left\|u_y(\cdot,y)\right\|_p\,dy\right)^q \frac{dt}{t}\right)^{1/q}\\
&\quad + \sum_{i=1}^n \left(\int_0^\infty \left(t^{1-\alpha}\left\|u_{x_i}(\cdot,t)\right\|_p\right)^q \frac{dt}{t}\right)^{1/q}\Bigg].
\end{aligned}
$$

However, we have, by Hardy's inequality (Theorem 3.31),

$$\begin{aligned}
&\left(\int_0^\infty \left(t^{-\alpha} \int_0^t \left\|u_y(\cdot, y)\right\|_p \, dy\right)^q \frac{dt}{t}\right)^{1/q} \\
&\le \frac{1}{\alpha} \left(\int_0^\infty \left(y^{1-\alpha} \left\|u_y(\cdot, y)\right\|_p\right)^q \frac{dy}{y}\right)^{1/q} \\
&= \frac{1}{\alpha} D.
\end{aligned} \tag{6.6}$$

Furthermore, we have, by part (a) of Theorem 5.21,

$$\begin{aligned}
&\sum_{i=1}^n \left(\int_0^\infty \left(t^{1-\alpha} \left\|u_{x_i}(\cdot, t)\right\|_p\right)^q \frac{dt}{t}\right)^{1/q} \\
&\le M'_\alpha \left(\int_0^\infty \left(y^{1-\alpha} \left\|u_y(\cdot, y)\right\|_p\right)^q \frac{dy}{y}\right)^{1/q} \\
&= M'_\alpha D.
\end{aligned} \tag{6.7}$$

Therefore, by combining inequalities (6.5), (6.6) and (6.7) we find that

$$F \le M_\alpha D,$$

with

$$M_\alpha := M_q \left(\frac{1}{\alpha} + M'_\alpha\right), \quad 0 < \alpha < 1.$$

Step 6: $B \le \omega_n^{1/q} G$. This is obvious. Indeed, we have the inequality

$$\begin{aligned}
&B \\
&= \left(\int_{\mathbf{R}^n} \left(\frac{\|f(\cdot + h) - 2f(\cdot) + f(\cdot - h)\|_p}{|h|^\alpha}\right)^q \frac{dh}{|h|^n}\right)^{1/q} \\
&= \left(\int_0^\infty \int_{\Sigma_{n-1}} \|f(\cdot + t\sigma) - 2f(\cdot) + f(\cdot - t\sigma)\|_p^q \, d\sigma \frac{dt}{t^{1+\alpha q}}\right)^{1/q} \\
&\le \omega_n^{1/q} \left(\int_0^\infty \left(t^{-\alpha} \cdot \sup_{0<|h|\le t} \|f(\cdot + h) - 2f(\cdot) + f(\cdot - h)\|_p\right)^q \frac{dt}{t}\right)^{1/q} \\
&= \omega_n^{1/q} G, \quad 0 < \alpha < 2.
\end{aligned}$$

Step 7: $G \le M_\alpha E$. We have, for $h = s\sigma, 0 < s \le t, \sigma \in \Sigma_{n-1}$ and $\varepsilon > 0$,

$$\begin{aligned}
u(x \pm h, \varepsilon) &= \left\{\left[tu_y(x \pm h, t) - \varepsilon u_y(x \pm h, \varepsilon)\right] \right. \\
&\quad \left. - [u(x \pm h, t) - u(x \pm h, \varepsilon)]\right\} \\
&\quad - \left[tu_y(x \pm h, t) - \varepsilon u_y(x \pm h, \varepsilon)\right] + u(x \pm h, t) \\
&= \int_\varepsilon^t y u_{yy}(x \pm h, y)\, dy - tu_y(x \pm h, t) + u(x \pm h, t) \\
&\quad + \varepsilon u_y(x \pm h, \varepsilon),
\end{aligned}$$

and

$$u(x, \varepsilon) = \int_\varepsilon^t y u_{yy}(x, y)\, dy - tu_y(x, t) + u(x, t) + \varepsilon u_y(x, \varepsilon).$$

Hence it follows that

$$\begin{aligned}
&u(x + h, \varepsilon) - 2u(x, \varepsilon) + u(x - h, \varepsilon) \\
&= \int_\varepsilon^t y u_{yy}(x + h, y)\, dy + \int_\varepsilon^t y u_{yy}(x - h, y)\, dy - 2\int_\varepsilon^t y u_{yy}(x, y)\, dy \\
&\quad - t\left[u_y(x + h, t) - u_y(x, t)\right] - t\left[u_y(x - h, t) - u_y(x, t)\right] \\
&\quad + [u(x + h, t) - 2u(x, t) + u(x - h, t)] \\
&\quad + \varepsilon\left[u_y(x + h, \varepsilon) - 2u_y(x, \varepsilon) + u_y(x - h, \varepsilon)\right].
\end{aligned}$$

Thus, by applying Minkowski's inequality for integrals (Theorem 3.29) we obtain that

$$\begin{aligned}
&\|u(\cdot + h, \varepsilon) - 2u(\cdot, \varepsilon) + u(\cdot - h, \varepsilon)\|_p \qquad (6.8) \\
&\leq \int_\varepsilon^t y \left\|u_{yy}(\cdot + h, y)\right\|_p dy + \int_\varepsilon^t y \left\|u_{yy}(\cdot - h, y)\right\|_p dy \\
&\quad + 2\int_\varepsilon^t y \left\|u_{yy}(\cdot, y)\right\|_p dy \\
&\quad + t\left\|u_y(\cdot + h, t) - u_y(\cdot, t)\right\|_p + t\left\|u_y(\cdot - h, t) - u_y(\cdot, t)\right\|_p \\
&\quad + \|u(\cdot + h, t) - 2u(\cdot, t) + u(\cdot - h, t)\|_p \\
&\quad + \varepsilon\left\|u_y(\cdot + h, \varepsilon) - 2u_y(\cdot, \varepsilon) + u_y(\cdot - h, \varepsilon)\right\|_p \\
&\leq 4\int_\varepsilon^t y \left\|u_{yy}(\cdot, y)\right\|_p dy + 2t\left\|u_y(\cdot + h, t) - u_y(\cdot, t)\right\|_p \\
&\quad + \|u(\cdot + h, t) - 2u(\cdot, t) + u(\cdot - h, t)\|_p + 4\varepsilon\left\|u_y(\cdot, \varepsilon)\right\|_p \\
&\leq 4\int_\varepsilon^t y \left\|u_{yy}(\cdot, y)\right\|_p dy + 2t\sum_{i=1}^n \int_0^s \left\|u_{yx_i}(\cdot + r\sigma, t)\right\|_p dr
\end{aligned}$$

$$+\sum_{i,j=1}^{n}\int_0^s\int_0^r \left\|u_{x_ix_j}(\cdot+v\sigma,t)\right\|_p\,dv\,dr$$

$$+\sum_{i,j=1}^{n}\int_0^s\int_0^r \left\|u_{x_ix_j}(\cdot-v\sigma,t)\right\|_p\,dv\,dr+4\varepsilon\left\|u_y(\cdot,\varepsilon)\right\|_p$$

$$\leq 4\int_\varepsilon^t y\left\|u_{yy}(\cdot,y)\right\|_p\,dy+2t^2\sum_{i=1}^{n}\left\|u_{yx_i}(\cdot,t)\right\|_p$$

$$+t^2\sum_{i,j=1}^{n}\left\|u_{x_ix_j}(\cdot,t)\right\|_p+4\varepsilon\left\|u_y(\cdot,\varepsilon)\right\|_p.$$

Step 7-1: Here we need the following lemma:

Lemma 6.5 *Assume that $u(x,y)$ is harmonic in the half-space $\mathbf{R}_+^{n+1}$ and bounded in each proper subhalf space of $\mathbf{R}_+^{n+1}$, and that we are given $\alpha>0$, an integer $k>\alpha$, $D>0$ and $y_0>0$ such that*

$$\left\|y^{k-\alpha}u_y^{(k)}(\cdot,\cdot)\right\|_{pq}\leq D, \tag{6.9}$$

$$\|u(\cdot,y)\|_p\leq D \quad \textit{for all } y\geq y_0. \tag{6.10}$$

Then it follows that the function $u(x,y)$ is the Poisson integral of some function $f\in B_{p,q}^{\alpha}(\mathbf{R}^n)$. Moreover, we have the following two assertions:

$$\left\|u_y(\cdot,y)\right\|_p=o(y^{-1}) \quad \textit{as } y\downarrow 0. \tag{a}$$

$$\|f\|_{\alpha;p,q}\leq M_{\alpha,k,y_0}D. \tag{b}$$

Proof The proof of Lemma 6.5 is divided into two steps.

Step (I): First, we prove assertion (a).

(a-1) The case $0<\alpha<1$: We remark that

$$u(x,y)=O(1) \quad \text{as } y\to\infty,$$

uniformly in $x\in\mathbf{R}^n$. However, we have, by the mean value theorem,

$$u(x,y)=u(x,y_0)+u_y(x,z)(y-y_0) \quad \text{for some } z\in(y_0,y).$$

Hence it follows that

$$u_y(x,y)=o(1) \quad \text{as } y\to\infty,$$

uniformly in $x\in\mathbf{R}^n$.

Now, by a repeated application of part (c) of Lemma 5.25 we find from inequality (6.9) that

$$\begin{aligned}\left\| y^{1-\alpha} u_y(\cdot, y) \right\|_{pq} &\le M_\alpha \left\| y^{2-\alpha} u_{yy}(\cdot, y) \right\|_{pq} \\ &\le M_{\alpha,k} \left\| y^{k-\alpha} u^{(k)}(\cdot, y) \right\|_{pq} \\ &\le M_{\alpha,k}\, D \quad \text{for all } y > 0.\end{aligned} \tag{6.11}$$

However, it is easy to see that the function $\left\| u_y(\cdot, y) \right\|_p$ is non-increasing of y. Indeed, since $u_y(x, y)$ is harmonic in $\mathbf{R}_+^{n+1}$, it follows that

$$\begin{aligned}\left\| u_y(\cdot, y_1 + y_2) \right\|_p &= \left\| P(\cdot, y_1) * u_y(\cdot, y_2) \right\|_p \\ &\le \| P(\cdot, y_1) \|_1 \left\| u_y(\cdot, y_2) \right\|_p \\ &= \left\| u_y(\cdot, y_2) \right\|_p \quad \text{for all } y_1 \text{and} y_2 > 0.\end{aligned}$$

Thus we have, by inequality (6.11),

$$\begin{aligned}\frac{y^{1-\alpha}}{((1-\alpha)q)^{1/q}} \left\| u_y(\cdot, y) \right\|_p &= \left(\int_0^y \left(t^{(1-\alpha)} \left\| u_y(\cdot, y) \right\|_p \right)^q \frac{dt}{t} \right)^{1/q} \\ &\le \left(\int_0^y \left(t^{(1-\alpha)} \left\| u_y(\cdot, t) \right\|_p \right)^q \frac{dt}{t} \right)^{1/q} \\ &\le \left(\int_0^\infty \left(t^{(1-\alpha)} \left\| u_y(\cdot, t) \right\|_p \right)^q \frac{dt}{t} \right)^{1/q} \\ &= \left\| y^{1-\alpha} u_y(\cdot, \cdot) \right\|_{pq} \\ &\le M_{\alpha,k}\, D \quad \text{for all } y > 0.\end{aligned}$$

This proves that

$$y \left\| u_y(\cdot, y) \right\|_p \le M_{\alpha,k} \left((1-\alpha)q \right)^{1/q} D\, y^\alpha \quad \text{for all } y > 0. \tag{6.12}$$

(a-2) The case $\alpha \ge 1$: We remark that, by inequality (6.9),

$$\begin{aligned}&\left(\int_0^{2y_0} \left(y^{k-1/2} \left\| u_y^{(k)}(\cdot, y) \right\|_p \right)^q \frac{dy}{y} \right)^{1/q} \\ &= \left(\int_0^{2y_0} y^{q(\alpha-1/2)} \left(y^{k-\alpha} \left\| u_y^{(k)}(\cdot, y) \right\|_p \right)^q \frac{dy}{y} \right)^{1/q} \\ &\le (2y_0)^{(\alpha-1/2)} D.\end{aligned} \tag{6.13}$$

On the other hand, it follows from inequality (6.10) that, for all $y \ge 2y_0$,

$$\begin{aligned}\left\|u_y^{(k)}(\cdot, y)\right\|_p &= \left\|P_y^{(k)}(\cdot, y/2) * u(\cdot, y/2)\right\|_p \\ &\le \left\|P_y^{(k)}(\cdot, y/2)\right\|_1 \|u(\cdot, y/2)\|_p \le \frac{M_k}{y^k} \|u(\cdot, y/2)\|_p \\ &\le (M_k D) y^{-k}.\end{aligned}$$

Thus we have the inequality

$$\begin{aligned}&\left(\int_{2y_0}^{\infty} \left(y^{k-1/2} \left\|u_y^{(k)}(\cdot, y)\right\|_p\right)^q \frac{dy}{y}\right)^{1/q} \qquad (6.14)\\ &\le M_k D \left(\int_{2y_0}^{\infty} (y^{k-1/2} y^{-k})^q \frac{dy}{y}\right)^{1/q} \\ &= \frac{2M_k D}{q}(2y_0)^{-q/2}.\end{aligned}$$

Therefore, by combining inequalities (6.13) and (6.14) we obtain that

$$\begin{aligned}&\left(\int_0^{\infty} \left(y^{k-1/2} \left\|u_y^{(k)}(\cdot, y)\right\|_p\right)^q \frac{dy}{y}\right)^{1/q} \\ &= \left(\int_0^{2y_0} \left(y^{k-1/2} \left\|u_y^{(k)}(\cdot, y)\right\|_p\right)^q \frac{dy}{y} + \int_{2y_0}^{\infty} \left(y^{k-1/2} \left\|u_y^{(k)}(\cdot, y)\right\|_p\right)^q \frac{dy}{y}\right)^{1/q} \\ &\le M_{\alpha,k,y_0} D,\end{aligned}$$

where M_{α,k,y_0} is a positive constant given by the formula

$$M_{\alpha,k,y_0} := (2y_0)^{(\alpha-1/2)} + \frac{2M_k}{q}(2y_0)^{-q/2}.$$

Hence it follows from a repeated application of part (c) of Lemma 5.25 that

$$\begin{aligned}&\left(\int_0^{\infty} \left(y^{1/2} \left\|u_y(\cdot, y)\right\|_p\right)^q \frac{dy}{y}\right)^{1/q} \\ &\le M_k \left(\int_0^{\infty} \left(y^{k-1/2} \left\|u_y^{(k)}(\cdot, y)\right\|_p\right)^q \frac{dy}{y}\right)^{1/q} \\ &\le M_k M_{\alpha,k,y_0} D.\end{aligned}$$

This proves that

$$\begin{aligned}s^{1/2} \left\|u_y(\cdot, s)\right\|_p &= \left(\frac{q}{2}\right)^{1/q} \left(\int_0^{s} \left(y^{1/2} \left\|u_y(\cdot, s)\right\|_p\right)^q \frac{dy}{y}\right)^{1/q} \\ &\le M_k M_{\alpha,k,y_0} D \quad \text{for all } s > 0,\end{aligned}$$

or equivalently,

$$y \left\|u_y(\cdot, y)\right\|_p \leq M_k M_{\alpha,k,y_0} D y^{1/2} \quad \text{for all } y > 0. \tag{6.15}$$

Summing up, we obtain from inequalities (6.12) and (6.15) that we have, for some $\beta \in (0, 1)$,

$$\left\|u_y(\cdot, y)\right\|_p \leq (M'_{\alpha,k,y_0} D) y^{\beta-1} \quad \text{for all } y > 0, \tag{6.16}$$

In particular, it follows that

$$\left\|u_y(\cdot, y)\right\|_p = o(y^{-1}) \quad \text{as } y \downarrow 0.$$

This proves the desired assertion (a).

Step (II): Secondly, we prove assertion (b). To do this, we show that there exists a function $f \in B^{\alpha}_{p,q}(\mathbf{R}^n)$ such that

$$u(x, y) = P(x, y) * f(x).$$

We remark that

$$u(x, y) = u(x, y_0) + \int_y^{y_0} u_s(x, s)\, ds \quad \text{for all } 0 < y \leq y_0.$$

Hence we have, by inequality (6.16),

$$\begin{aligned} \|u(\cdot, y)\|_p &\leq \|u(\cdot, y_0)\|_p + \int_y^{y_0} \|u_s(\cdot, s)\|_p\, ds \\ &\leq D + M'_{\alpha,k,y_0} D \int_y^{y_0} s^{\beta-1}\, ds \leq D\left(1 + M'_{\alpha,k,y_0} \frac{y_0^{\beta}}{\beta}\right) \\ &:= M''_{\alpha,k,y_0} D \quad \text{for all } 0 < y \leq y_0, \end{aligned} \tag{6.17}$$

where M''_{α,k,y_0} is a positive constant given by the formula

$$M''_{\alpha,k,y_0} := 1 + M'_{\alpha,k,y_0} \frac{y_0^{\beta}}{\beta}, \quad 0 < \beta < 1.$$

By a similar argument, it follows that we have, for some positive constant M'''_{α,k,y_0},

$$\|u(\cdot, y_1) - u(\cdot, y_2)\|_p \leq \left|\int_{y_2}^{y_1} \|u_s(\cdot, s)\|_p\, ds\right| \leq M'''_{\alpha,k,y_0} D \left|y_2^{\beta} - y_1^{\beta}\right|.$$

This implies that $\{u(\cdot, y)\}$ is a Cauchy sequence in $L^p(\mathbf{R}^n)$ as $y \downarrow 0$. Thus, by arguing as in the proof of Theorem 5.16 we find that the function $u(x, y)$ is the Poisson integral of a function $f \in L^p(\mathbf{R}^n)$:

$$u(x, y) = P(x, y) * f(x).$$

Furthermore, by letting $y \downarrow 0$ in inequality (6.17) we obtain that

$$\|f\|_p = \lim_{y \downarrow 0} \|u(\cdot, y)\|_p \leq M''_{\alpha,k,y_0} D. \tag{6.18}$$

Therefore, in view of Proposition 6.3, it follows from inequalities (6.9) and (6.18) that

$$\|f\|_{\alpha;p,q} = \left\| y^{k-\alpha} u_y^{(k)}(\cdot, y) \right\|_{pq} + \|f\|_p \leq \left(1 + M''_{\alpha,k,y_0}\right) D.$$

This proves the desired assertion (b), with

$$M_{\alpha,k,y_0} := 1 + M''_{\alpha,k,y_0}.$$

The proof of Lemma 6.5 is complete. □

Step 7-2: Now we recall that

$$E = \left\| y^{2-\alpha} u_{yy}(\cdot, \cdot) \right\|_{pq},$$

and that

$$\|u(\cdot, y)\|_\infty \leq \|P(\cdot, y)\|_{p'} \|f\|_p \leq \left(\frac{1}{c_n}\right)^{1/p} y^{-n/p} \|f\|_p \quad \text{for all } y > 0.$$

Hence, by applying Lemma 6.5 to our situation we find that

$$\varepsilon \left\| u_y(\cdot, \varepsilon) \right\|_p \longrightarrow 0 \quad \text{as } \varepsilon \downarrow 0.$$

Furthermore, we remark that

$$\begin{aligned} &u(\cdot + h, \varepsilon) - 2u(\cdot, \varepsilon) + u(\cdot - h, \varepsilon) \\ &= P(\cdot, \varepsilon) * (f(\cdot + h) - 2f(\cdot) + f(\cdot - h)) \longrightarrow f(\cdot + h) - 2f(\cdot) + f(\cdot - h) \\ &\text{in } L^p(\mathbf{R}^n) \text{as} \varepsilon \downarrow 0. \end{aligned}$$

Therefore, by letting $\varepsilon \downarrow 0$ in inequality (6.8) we obtain that

$$\begin{aligned}&\|f(\cdot+h)-2f(\cdot)+f(\cdot-h)\|_p \\ &\leq 4\int_0^t y\left\|u_{yy}(\cdot,y)\right\|_p\,dy+2t^2\sum_{i=1}^n\left\|u_{yx_i}(\cdot,t)\right\|_p+t^2\sum_{i,j=1}^n\left\|u_{x_ix_j}(\cdot,t)\right\|_p.\end{aligned} \tag{6.19}$$

However, we have, by Theorems 5.21 and 5.26,

$$t^2\left\|u_{yx_i}(\cdot,t)\right\|_p \leq Mt^2\left\|u_{yy}(\cdot,t)\right\|_p \quad \text{for all } t>0, \tag{6.20}$$

$$t^2\left\|u_{x_ix_j}(\cdot,t)\right\|_p \leq Mt^2\left\|u_{yy}(\cdot,t)\right\|_p \quad \text{for all } t>0. \tag{6.21}$$

Therefore, by combining inequalities (6.19), (6.20) and (6.21) we obtain that

$$\begin{aligned}&\|f(\cdot+h)-2f(\cdot)+f(\cdot-h)\|_p \\ &\leq 4\int_0^t y\left\|u_{yy}(\cdot,y)\right\|_p\,dy+2Mt^2\left\|u_{yy}(\cdot,t)\right\|_p.\end{aligned} \tag{6.22}$$

Step 7-2a: The case $q=\infty$. Since we have the inequality

$$\left\|u_{yy}(\cdot,y)\right\|_p \leq y^{\alpha-2}E \quad \text{for all } y>0,$$

it follows from inequality (6.22) that

$$\begin{aligned}\|f(\cdot+h)-2f(\cdot)+f(\cdot-h)\|_p &\leq 4E\int_0^t y^{\alpha-1}dy+MEt^\alpha \\ &\leq M_\alpha Et^\alpha \quad \text{for all } 0<|h|=s\leq t.\end{aligned}$$

This proves that

$$\begin{aligned}G&=\sup_{t>0}\left(t^{-\alpha}\cdot\sup_{0<|h|\leq t}\|f(\cdot+h)-2f(\cdot)+f(\cdot-h)\|_p\right) \\ &\leq M_\alpha E, \quad 0<\alpha<2.\end{aligned}$$

Step 7-2b: The case $1\leq q<\infty$. By applying Hardy's inequality (Theorem 3.31), we obtain from inequality (6.22) that

$$\begin{aligned}G&=\left(\int_0^\infty\left(t^{-\alpha}\cdot\sup_{0<|h|\leq t}\|f(\cdot+h)-2f(\cdot)+f(\cdot-h)\|_p\right)^q\frac{dt}{t}\right)^{1/q} \\ &\leq\left(\int_0^\infty t^{-\alpha q}\left(4\int_0^t y\left\|u_{yy}(\cdot,y)\right\|_p\,dy+2Mt^2\left\|u_{yy}(\cdot,t)\right\|_p\right)^q\frac{dt}{t}\right)^{1/q}\end{aligned}$$

$$\leq \left[2^{3-3/q} \left(\int_0^\infty \left(t^{-\alpha} \int_0^t y \left\| u_{yy}(\cdot, y) \right\|_p dy \right)^q \frac{dt}{t} \right)^{1/q} \right.$$
$$\left. + 2^{1-1/q} M \left(\int_0^\infty \left(t^{2-\alpha} \left\| u_{yy}(\cdot, t) \right\|_p \right)^q \frac{dt}{t} \right)^{1/q} \right]$$
$$\leq \left(\frac{2^{3-3/q}}{\alpha} + 2^{1-1/q} M \right) \left(\int_0^\infty \left(t^{2-\alpha} \left\| u_{yy}(\cdot, t) \right\|_p \right)^q \frac{dt}{t} \right)^{1/q}$$
$$\leq \left(\frac{8}{\alpha} + 2M \right) E.$$

This proves that

$$G \leq M_\alpha E,$$

with

$$M_\alpha := \frac{8}{\alpha} + 2M, \quad 0 < \alpha < 2.$$

Step 8: $E \leq M_\alpha C$. We remark that

$$|P_{yy}(z, y)| \leq \frac{My}{\left(|z|^2 + y^2\right)^{(n+3)/2}} \quad \text{for all } y > 0,$$

and that

$$\int_{\mathbf{R}^n} P_{yy}(z, y)\, dz = \frac{d^2}{dy^2} \left(\int_{\mathbf{R}^n} P(z, y)\, dz \right) = 0.$$

Hence we have the formula

$$u_{yy}(x, y) = \int_{\mathbf{R}^n} [f(x - z) - f(x)]\, P_{yy}(z, y)\, dz,$$

and so

$$\begin{aligned} &\left\| u_{yy}(\cdot, y) \right\|_p \qquad (6.23) \\ &\leq M \int_{\mathbf{R}^n} \|f(\cdot - z) - f(\cdot)\|_p \frac{y}{\left(|z|^2 + y^2\right)^{(n+3)/2}}\, dz \\ &= M \int_0^\infty \int_{\Sigma_{n-1}} \|f(\cdot - t\sigma) - f(\cdot)\|_p \frac{y}{\left(t^2 + y^2\right)^{(n+3)/2}}\, t^{n-1}\, d\sigma\, dt \\ &= M\omega_n \int_0^\infty \|\phi_t(\cdot)\|_p \frac{y}{\left(t^2 + y^2\right)^{(n+3)/2}}\, t^{n-1}\, dt. \end{aligned}$$

Here

$$\phi_t(x) = \frac{1}{\omega_n} \int_{\Sigma_{n-1}} (f(x + t\sigma) - f(x))\, d\sigma.$$

Step 8-a: The case $q = \infty$. Since we have, by Minkowski's inequality for integrals (Theorem 3.29),

$$\|\phi_t(\cdot)\|_p \leq \frac{1}{\omega_n} \int_{\Sigma_{n-1}} \|f(\cdot + t\sigma) - f(\cdot)\|_p \leq C t^\alpha \quad \text{for all } t > 0,$$

it follows from inequality (6.23) that

$$\begin{aligned}
\left\|u_{yy}(\cdot, y)\right\|_p &\leq MC\,\omega_n \int_0^\infty \frac{y}{\left(t^2 + y^2\right)^{(n+3)/2}} t^{\alpha+n-1}\, dt \\
&\leq MC\,\omega_n \left(\int_0^y \frac{y \cdot y^{\alpha+n-1}}{y^{n+3}}\, dt + \int_y^\infty \frac{y}{t^{n+3}} t^{\alpha+n-1}\, dt \right) \\
&= MC\,\omega_n \left(\int_0^y y^{\alpha-3}\, dt + \int_y^\infty y t^{\alpha-4}\, dt \right) \\
&= \frac{4-\alpha}{3-\alpha} M\,\omega_n\, C\, y^{\alpha-2} \quad \text{for all } y > 0.
\end{aligned}$$

This proves that

$$E = \sup_{y>0} \left\{ y^{2-\alpha} \left\|u_{yy}(\cdot, y)\right\|_p \right\} \leq M_\alpha C,$$

with

$$M_\alpha := \frac{4-\alpha}{3-\alpha} M\,\omega_n, \quad 0 < \alpha < 2.$$

Step 8-b: The case $1 \leq q < \infty$. By applying Hardy's inequality (Theorem 3.31), we obtain that

$$\begin{aligned}
E &= \left(\int_0^\infty \left(y^{2-\alpha} \left\|u_{yy}(\cdot, y)\right\|_p \right)^q \frac{dy}{y} \right)^{1/q} \\
&\leq \left(\int_0^\infty \left(y^{\alpha-2} M\,\omega_n \int_0^\infty \|\phi_t(\cdot)\|_p \frac{y t^{n-1}}{\left(t^2 + y^2\right)^{(n+3)/2}}\, dt \right)^q \frac{dy}{y} \right)^{1/q} \\
&= M\,\omega_n \left(\int_0^\infty \left(\int_0^\infty \|\phi_t(\cdot)\|_p \frac{y t^{n-1}}{\left(t^2 + y^2\right)^{(n+3)/2}}\, dt \right)^q \frac{dy}{y^{1+q(\alpha-2)}} \right)^{1/q} \\
&\leq M\,\omega_n \left(\int_0^\infty \left(y^{-n-\alpha} \int_0^y t^{n-1} \|\phi_t(\cdot)\|_p\, dt \right)^q \frac{dy}{y} \right)^{1/q}
\end{aligned}$$

$$
\begin{aligned}
&+ M\,\omega_n \left(\int_0^\infty \left(y^{3-\alpha} \int_y^\infty t^{-4} \|\phi_t(\cdot)\|_p \, dt \right)^q \frac{dy}{y} \right)^{1/q} \\
&\le \frac{M\omega_n}{n+\alpha} \left(\int_0^\infty \left(y^{-n-\alpha+1} y^{n-1} \left\|\phi_y(\cdot)\right\|_p \right)^q \frac{dy}{y} \right)^{1/q} \\
&+ \frac{M\omega_n}{3-\alpha} \left(\int_0^\infty \left(y^{3-\alpha+1} y^{-4} \left\|\phi_y(\cdot)\right\|_p \right)^q \frac{dy}{y} \right)^{1/q} \\
&= M\,\omega_n \left(\frac{1}{n+\alpha} + \frac{1}{3-\alpha} \right) \left(\int_0^\infty \left(y^{-\alpha} \left\|\phi_y(\cdot)\right\|_p \right)^q \frac{dy}{y} \right)^{1/q} \\
&= M\,\omega_n \left(\frac{1}{n+\alpha} + \frac{1}{3-\alpha} \right) C.
\end{aligned}
$$

This proves that

$$E \le M_\alpha C,$$

with

$$M_\alpha := M\omega_n \left(\frac{1}{n+\alpha} + \frac{1}{3-\alpha} \right), \quad 0 < \alpha < 2.$$

Step 9: $H \le (1/\alpha)D$. If x is a point of the Lebesgue set of $f \in L^p(\mathbf{R}^n)$, it follows that

$$u(x, y) - f(x) = \int_0^y u_y(x, s)\, ds.$$

Hence, by applying Minkowski's inequality for integrals (Theorem 3.29) we obtain that

$$\|u(\cdot, y) - f(\cdot)\|_p \le \int_0^y \left\|u_y(\cdot, s)\right\|_p \, ds. \tag{6.24}$$

Step 9-a: The case $q = \infty$. Since we have the inequality

$$s^{1-\alpha} \left\|u_y(\cdot, s)\right\|_p \le D \quad \text{for all } s > 0,$$

it follows from inequality (6.24) that

$$\|u(\cdot, y) - f(\cdot)\|_p \le \int_0^y \left\|u_y(\cdot, s)\right\|_p \, ds \le D \int_0^y s^{\alpha-1}\, ds = \frac{D}{\alpha} y^\alpha.$$

This proves that

$$H = \sup_{y>0} \left\{ y^{-\alpha} \|u(\cdot, y) - f(\cdot)\|_p \right\} \le \frac{1}{\alpha} D.$$

Step 9-b: The case $1 \leq q < \infty$. In view of inequality (6.24), it follows from an application of Hardy's inequality (Theorem 3.31) that

$$\begin{aligned} H &= \left(\int_0^\infty \left(y^{-\alpha} \|u(\cdot, y) - f(\cdot)\|_p \right)^q \frac{dy}{y} \right)^{1/q} \\ &\leq \left(\int_0^\infty \left(y^{-\alpha} \int_0^y \left\| u_y(\cdot, s) \right\|_p ds \right)^q \frac{dy}{y} \right)^{1/q} \\ &\leq \frac{1}{\alpha} \left(\int_0^\infty \left(y^{1-\alpha} \left\| u_y(\cdot, y) \right\|_p \right)^q \frac{dy}{y} \right)^{1/q} \\ &= \frac{1}{\alpha} D, \quad 0 < \alpha < 1. \end{aligned}$$

Step 10: $D \leq M_\alpha H$ $(0 < \alpha < 1)$. We recall that

$$u_y(x, y) = u(x, y/2) * P_y(x, y/2),$$

and that

$$\left| P_y(x, y/2) \right| \leq \frac{n+1}{y} P(x, y/2) \quad \text{for all } y > 0.$$

Hence we have, by Young's inequality (Theorem 3.36),

$$\begin{aligned} \left\| u_y(\cdot, y) \right\|_p &\leq \|u(\cdot, y/2)\|_p \left\| P_y(\cdot, y/2) \right\|_1 \\ &\leq \frac{n+1}{y} \|P(\cdot, y/2)\|_1 \|u(\cdot, y/2)\|_p \\ &= \frac{n+1}{y} \|u(\cdot, y/2)\|_p = \frac{n+1}{y} \|P(\cdot, y/2) * f(\cdot)\|_p \\ &\leq \frac{n+1}{y} \|f\|_p \quad \text{for all } y > 0. \end{aligned}$$

This proves that, as $N \to \infty$,

$$\left\| u_y(\cdot, 2^N y) \right\|_p \longrightarrow 0 \quad \text{for each } y > 0.$$

Hence we have the assertion

$$\left\| u_y(\cdot, y) \right\|_p = \lim_{N \to \infty} \left\| u_y(\cdot, y) - u_y(\cdot, 2^N y) \right\|_p, \quad \text{for each } y > 0. \tag{6.25}$$

On the other hand, we have the following claim:

Claim 6.6 *There exists a positive constant M_α such that we have, for any integer $N > 1$,*

$$\left\| y^{1-\alpha}(u_y(\cdot, y) - u_y(\cdot, 2^N y)) \right\|_{pq} \le M_\alpha H, \quad 0 < \alpha < 1. \tag{6.26}$$

Proof By Minkowski's inequality (Theorem 3.25), we obtain that

$$\begin{aligned}
& y^{1-\alpha} \left\| u_y(\cdot, y) - u_y(\cdot, 2^N y) \right\|_p \\
& \le y^{1-\alpha} \sum_{k=1}^{N} \left\| u_y(\cdot, 2^{k-1} y) - u_y(\cdot, 2^k y) \right\|_p \\
& = y^{1-\alpha} \sum_{k=1}^{N} \left\| [f(\cdot) - u(\cdot, 2^{k-1} y)] * P_y(\cdot, 2^{k-1} y) \right\|_p \\
& \le y^{1-\alpha} \sum_{k=1}^{N} \left\| f(\cdot) - u(\cdot, 2^{k-1} y) \right\|_p \left\| P_y(\cdot, 2^{k-1} y) \right\|_1 \\
& \le y^{1-\alpha} \sum_{k=1}^{N} \frac{n+1}{y} 2^{1-k} \left\| f(\cdot) - u(\cdot, 2^{k-1} y) \right\|_p \\
& = (n+1) \sum_{k=1}^{N} (2^{\alpha-1})^{k-1} (2^{k-1} y)^{-\alpha} \left\| f(\cdot) - u(\cdot, 2^{k-1} y) \right\|_p .
\end{aligned} \tag{6.27}$$

However, we have, for $1 \le k \le N$,

$$\begin{aligned}
& \left(\int_0^\infty \left((2^{k-1} y)^{-\alpha} \left\| f(\cdot) - u(\cdot, 2^{k-1} y) \right\|_p \right)^q \frac{dy}{y} \right)^{1/q} \\
& = \left(\int_0^\infty \left(z^{-\alpha} \left\| f(\cdot) - u(\cdot, z) \right\|_p \right)^q \frac{dz}{z} \right)^{1/q} .
\end{aligned} \tag{6.28}$$

Hence it follows from inequalities (6.27) and (6.28) that

$$\begin{aligned}
& \left\| y^{1-\alpha}(u_y(\cdot, y) - u_y(\cdot, 2^N y)) \right\|_{pq} \\
& = \left(\int_0^\infty \left(y^{1-\alpha} \left\| u_y(\cdot, y) - u(\cdot, 2^N y) \right\|_p \right)^q \frac{dy}{y} \right)^{1/q} \\
& \le (n+1) \\
& \quad \times \left(\int_0^\infty \left(\sum_{k=1}^{N} (2^{\alpha-1})^{k-1} (2^{k-1} y)^{-\alpha} \left\| f(\cdot) - u(\cdot, 2^{k-1} y) \right\|_p \right)^q \frac{dy}{y} \right)^{1/q} \\
& = (n+1) \left(\int_0^\infty \left(\sum_{k=1}^{N} (2^{\alpha-1})^{k-1} \right)^q \left(z^{-\alpha} \left\| f(\cdot) - u(\cdot, z) \right\|_p \right)^q \frac{dz}{z} \right)^{1/q}
\end{aligned}$$

$$= (n+1)\left(\sum_{k=1}^{N}(2^{\alpha-1})^{k-1}\right)\left\|y^{-\alpha}(f(\cdot)-u(\cdot,y))\right\|_{pq}$$
$$\leq (n+1)\left(\sum_{k=1}^{\infty}\left(\frac{1}{2^{1-\alpha}}\right)^{k-1}\right)H, \quad 0<\alpha<1.$$

This proves the desired inequality (6.26), with

$$M_\alpha := \frac{n+1}{1-2^{\alpha-1}}, \quad 0<\alpha<1.$$

The proof of Claim 6.6 is complete. □

Therefore, by applying Fatou's lemma (Lemma 3.12) we obtain from formula (6.25) and inequality (6.26) that

$$\begin{aligned} D^q &= \left\|y^{1-\alpha}u_y(\cdot,y)\right\|_{pq}^q = \int_0^\infty \left(y^{1-\alpha}\left\|u_y(\cdot,y)\right\|_p\right)^q \frac{dy}{y} \\ &= \int_0^\infty \lim_{N\to\infty}\left(y^{1-\alpha}\left\|u_y(\cdot,y)-u_y(\cdot,2^N y)\right\|_p\right)^q \frac{dy}{y} \\ &\leq \liminf_{N\to\infty}\int_0^\infty \left(y^{1-\alpha}\left\|u_y(\cdot,y)-u_y(\cdot,2^N y)\right\|_p\right)^q \frac{dy}{y} \\ &= \liminf_{N\to\infty}\left\|y^{1-\alpha}(u_y(\cdot,y)-u_y(\cdot,2^N y))\right\|_{pq}^q \\ &\leq (M_\alpha H)^q. \end{aligned}$$

This proves that

$$D \leq M_\alpha H, \quad 0<\alpha<1.$$

The proof of Theorem 6.4 is now complete. □

Theorem 6.7 *Let $\alpha > 0$, $1 \leq p, q \leq \infty$ and let k be a positive integer less than α. Then the norm*

$$\|f\|_p + \sum_{|s|=k}\left\|D_x^s f\right\|_{\alpha-k;p,q}$$

is equivalent to the norm $\|f\|_{\alpha;p,q}$.

Proof The proof of Theorem 6.7 is divided into two steps.

Step (I): We assume that

$$\|f\|_{\alpha;p,q} = \|f\|_p + \left\|y^{\overline{\alpha}-\alpha}u_y^{(\overline{\alpha})}(\cdot,y)\right\|_{pq} < \infty,$$

where

$$u(x, y) = P(x, y) * f(x).$$

Then we have the following three assertions:

(a) The function $D_x^s u(x, y)$ is harmonic in the half-space $\mathbf{R}_+^{n+1}$ and bounded in each proper sub-half space of $\mathbf{R}_+^{n+1}$.
(b) $\left\| D_x^s u(\cdot, y) \right\|_p = \left\| D_x^s P(\cdot, y) * f(\cdot) \right\|_p \leq M_s \, \|f\|_p, \, y \geq 1.$
(c) It follows from a repeated application of Lemma 5.25 that

$$\begin{aligned}\left\| y^{\overline{\alpha}-(\alpha-k)} D_x^s (u_y^{(\overline{\alpha})}(\cdot, y)) \right\|_{pq} &\leq M_\alpha \left\| y^{\overline{\alpha}-\alpha} u_y^{(\overline{\alpha})}(\cdot, y) \right\|_{pq} \\ &\leq M_\alpha \, \|f\|_{\alpha; p,q}.\end{aligned}$$

Hence, by applying Lemma 6.5 to our situation we obtain that there exists a function $g \in B_{pq}^{\alpha-k}(\mathbf{R}^n)$ such that

$$D_x^s u(x, y) = P(x, y) * g(x), \quad |s| = k,$$

and that

$$\|g\|_{\alpha-k; p,q} \leq M \, \|f\|_{\alpha; p,q}.$$

However, it is easy to see that

$$g = D_x^s f.$$

Indeed, we have the formulas

$$P(x, y) * g(x) = D_x^s u(x, y) = D_x^s \left(P(x, y) * f(x) \right) = P(x, y) * D_x^s f(x).$$

Therefore, we obtain that

$$\begin{aligned}\left\| D_x^s f \right\|_{\alpha-k; p,q} &= \left\| D_x^s f \right\|_p + \left\| y^{\overline{\alpha}-(\alpha-k)} D_y^{\overline{\alpha}} (D_x^s u(\cdot, y)) \right\|_{pq} \\ &= \|g\|_p + \left\| y^{\overline{\alpha}-(\alpha-k)} D_y^{\overline{\alpha}} (P(\cdot, y) * g(\cdot)) \right\|_{pq} \\ &\leq \|g\|_{\alpha-k; p,q} \\ &\leq M \, \|f\|_{\alpha; p,q}.\end{aligned}$$

We are done with the proof in one direction:

$$\|f\|_p + \sum_{|s|=k} \left\| D_x^s f \right\|_{\alpha-k; p,q} \leq (M+1) \, \|f\|_{\alpha; p,q}.$$

Step (II): Conversely, we assume that

$$f \in L^p(\mathbf{R}^n),$$

and that

$$\sum_{|s|=k} \|D_x^s f\|_{\alpha-k;p,q} < \infty.$$

Then we have, for all $|s| = k$,

$$D_x^s u(x, y) = D_x^s \left(P(x, y) * f(x)\right) = P(x, y) * D_x^s f(x).$$

By applying Proposition 6.3 with $f(x) := D_x^s f(x)$, we obtain that

$$\begin{aligned} \left\| y^{k+\overline{\alpha}-\alpha} D_x^s (u_y^{(\overline{\alpha})}(\cdot, y)) \right\|_{pq} &= \left\| y^{\overline{\alpha}-(\alpha-k)} D_y^{\overline{\alpha}} (P(\cdot, y) * D_x^s f(\cdot)) \right\|_{pq} \\ &\leq \left\| D_x^s f \right\|_{\alpha-k;p,q}. \end{aligned} \tag{6.29}$$

Moreover, we have, by a repeated application of part (b) of Theorem 5.26,

$$\left\| y^{k+\overline{\alpha}-\alpha} u_y^{(k+\overline{\alpha})}(\cdot, y)) \right\|_{pq} \leq M_k \sum_{|s|=k} \left\| y^{k+\overline{\alpha}-\alpha} D_x^s (u_y^{(\overline{\alpha})}(\cdot, y)) \right\|_{pq}. \tag{6.30}$$

Hence, in view of Proposition 6.3, it follows from inequalities (6.29) and (6.30) that

$$\begin{aligned} \left\| y^{\overline{\alpha}-\alpha} u_y^{(\overline{\alpha})}(\cdot, y)) \right\|_{pq} &\leq M_k' \left\| y^{k+\overline{\alpha}-\alpha} u_y^{(k+\overline{\alpha})}(\cdot, y)) \right\|_{pq} \\ &\leq M_k' M_k \sum_{|s|=k} \left\| y^{k+\overline{\alpha}-\alpha} D_x^s (u_y^{(\overline{\alpha})}(\cdot, y)) \right\|_{pq} \\ &\leq M_k' M_k \sum_{|s|=k} \left\| D_x^s f \right\|_{\alpha-k;p,q}. \end{aligned} \tag{6.31}$$

Therefore, we obtain from inequality (6.31) that

$$\begin{aligned} \|f\|_{\alpha;p,q} &= \|f\|_p + \left\| y^{\overline{\alpha}-\alpha} u_y^{(\overline{\alpha})}(\cdot, y)) \right\|_{pq} \\ &\leq \left(1 + M_k' M_k\right) \left(\|f\|_p + \sum_{|s|=k} \left\| D_x^s f \right\|_{\alpha-k;p,q} \right). \end{aligned}$$

This completes the proof in the other direction.

The proof of Theorem 6.7 is complete. □

By combining Theorems 6.7 and 6.4, we obtain the following:

Theorem 6.8 *Let $\alpha > 0$, $1 \le p, q \le \infty$ and let k be a non-negative integer such that $k < \alpha \le k+1$. Then the norm*

$$\|f\|_p + \sum_{|s|=k} \left(\int_{\mathbf{R}^n} \frac{\|D^s f(\cdot + h) - 2D^s f(\cdot) + D^s f(\cdot - h)\|_p^q}{|h|^{n+(\alpha-k)q}} \, dh \right)^{1/q}$$

is equivalent to the norm $\|f\|_{\alpha;p,q}$ defined by formula (6.1)*:*

$$\|f\|_{\alpha;p,q} = \|f\|_p + \left(\int_0^\infty \left(y^{k+1-\alpha} \left\| u_y^{(k+1)}(\cdot, y) \right\|_p \right)^q \frac{dy}{y} \right)^{1/q}.$$

Chapter 7
Sobolev and Besov Spaces

This chapter is devoted to the precise definitions and statements of function spaces of L^p type with some detailed proofs. The function spaces we shall treat are the following:

(i) The generalized Sobolev spaces $W^{s,p}(\Omega)$ and $H^{s,p}(\Omega)$. When s is a positive integer, these spaces coincide with each other.
(ii) The Besov spaces $B^{s,p}(\partial\Omega)$ on the boundary $\partial\Omega$ of a Lipschitz domain Ω are function spaces defined in terms of the L^p modulus of continuity, and enter naturally in connection with boundary value problems in the framework of L^p Sobolev spaces.

In fact, we need to make sense of the restriction $u|_{\partial\Omega}$ to the boundary $\partial\Omega$ as an element of a Besov space on $\partial\Omega$ when u belongs to a Sobolev space on the domain Ω. In particular, we formulate an important trace theorem (Theorem 7.7) that will be used in the study of boundary value problems in Parts III and IV. It should be emphasized that the trace theorems remain valid on manifolds with boundary and bounded geometry (see [62, Theorem 4.10]).

Finally, in Sect. 7.4 we prove part (iv) of Examples 4.1 (Proposition 7.9):

$$W^{\theta,n/\theta}(\mathbf{R}^n) \subset \mathrm{VMO} \quad \text{for } 0 < \theta \leq 1.$$

For more thorough treatments of function spaces discussed here, the reader might refer to Adams–Fournier [2], Aronszajn–Smith [8], Bergh–Löfström [11], Friedman [54], Peetre [114], Sawano–Di Fazio–Hakim [125, 126], Stein [137], Taibleson [146] and Triebel [170].

K. Taira, *Real Analysis Methods for Markov Processes*,
https://doi.org/10.1007/978-981-97-3659-1_7

7.1 Sobolev Spaces

In this section we present a brief description of the basic concepts and results of L^p Sobolev spaces which will be used in subsequent chapters. Many problems in partial differential equations may be formulated in terms of abstract operators acting between suitable Sobolev spaces, and these operators are then analyzed by the methods of functional analysis.

7.1.1 *First Definition of Sobolev Spaces*

Let Ω be an open subset of $\mathbf{R}^n$. If $1 < p < \infty$ and if s is a non-negative integer, then we define the L^p Sobolev space

$$\begin{aligned} W^{s,p}(\Omega) := & \text{ the space of (equivalence classes of) functions} \\ & u \in L^p(\Omega) \text{whose derivatives} D^\alpha u, \, |\alpha| \leq s, \text{ in the} \\ & \text{sense of distributions are in} L^p(\Omega), \end{aligned}$$

with the norm

$$\|u\|_{W^{s,p}(\Omega)} = \left(\sum_{|\alpha| \leq s} \int_\Omega |D^\alpha u(x)|^p \, dx \right)^{1/p}. \tag{7.1}$$

If $1 < p < \infty$ and if

$$\boxed{s = m + \theta}$$

with a non-negative integer m and $0 < \theta < 1$, then the Sobolev space $W^{s,p}(\Omega)$ is defined to be the space of those functions $u \in W^{m,p}(\Omega)$ such that, for $|\alpha| = m$, the integral (Slobodeckiĭ seminorm)

$$\iint_{\Omega \times \Omega} \frac{|D^\alpha u(x) - D^\alpha u(y)|^p}{|x - y|^{n+p\theta}} \, dx \, dy \tag{7.2}$$

is finite. The norm $\|u\|_{W^{s,p}(\Omega)}$ of $W^{s,p}(\Omega)$ is defined by the formula

$$\begin{aligned} \|u\|_{W^{s,p}(\Omega)} = \Bigg(& \sum_{|\alpha| \leq m} \int_\Omega |D^\alpha u(x)|^p \, dx \\ & + \sum_{|\alpha| = m} \iint_{\Omega \times \Omega} \frac{|D^\alpha u(x) - D^\alpha u(y)|^p}{|x - y|^{n+p\theta}} \, dx \, dy \Bigg)^{1/p}. \end{aligned} \tag{7.3}$$

$$\begin{array}{ccc} u \in \mathcal{S}'(\mathbf{R}^n) & \xrightarrow{\mathcal{G}^s} & \mathcal{S}'(\mathbf{R}^n) \ni \mathcal{G}^s u \\ \mathcal{F}\big\downarrow & & \big\uparrow \mathcal{F}^* \\ \mathcal{F}u \in \mathcal{S}'(\mathbf{R}^n) & \xrightarrow[(1+|\xi|^2)^{-s/2}]{} & \mathcal{S}'(\mathbf{R}^n) \ni (1+|\xi|^2)^{-s/2}\mathcal{F}u \end{array}$$

Fig. 7.1 The mapping properties of the operator $\mathcal{G}^s$

7.1.2 Second Definition of Sobolev Spaces

Next, we introduce a second family of Sobolev spaces, by using the Fourier transform.

Let $\mathcal{S}(\mathbf{R}^n)$ be the *Schwartz space* or space of C^∞ functions on $\mathbf{R}^n$ rapidly decreasing at infinity. We recall that the (direct) Fourier transform $\mathcal{F}$ and the inverse Fourier transform $\mathcal{F}^*$ are isomorphisms of $\mathcal{S}(\mathbf{R}^n)$ onto itself. The dual space $\mathcal{S}'(\mathbf{R}^n)$ of $\mathcal{S}(\mathbf{R}^n)$ are called the space of *tempered distributions* on $\mathbf{R}^n$. Roughly speaking, the tempered distributions are those distributions which grow at most polynomially at infinity, since the functions in $\mathcal{S}(\mathbf{R}^n)$ die out faster than any power of x at infinity. The importance of tempered distributions lies in the fact that they have Fourier transforms. More precisely, if $u \in \mathcal{S}'(\mathbf{R}^n)$, we define its (direct) Fourier transform $\mathcal{F}u$ by the formula

$$\langle \mathcal{F}u, \varphi \rangle = \langle u, \mathcal{F}\varphi \rangle \quad \text{for all } \varphi \in \mathcal{S}(\mathbf{R}^n),$$

where $\langle \cdot, \cdot \rangle$ is the pairing of $\mathcal{S}'(\mathbf{R}^n)$ and $\mathcal{S}(\mathbf{R}^n)$. Similarly, if $v \in \mathcal{S}'(\mathbf{R}^n)$, we define its inverse Fourier transform $\mathcal{F}^*v$ by the formula

$$\langle \mathcal{F}^*v, \psi \rangle = \langle v, \mathcal{F}^*\psi \rangle \quad \text{for all } \psi \in \mathcal{S}(\mathbf{R}^n).$$

It should be emphasized that the Fourier transforms $\mathcal{F}$ and $\mathcal{F}^*$ are isomorphisms of $\mathcal{S}'(\mathbf{R}^n)$ onto itself.

If $s \in \mathbf{R}$, we define a linear map

$$\mathcal{G}^s = (I - \Delta)^{-s/2} \colon \mathcal{S}'(\mathbf{R}^n) \longrightarrow \mathcal{S}'(\mathbf{R}^n)$$

by the formula

$$\mathcal{G}^s u = \mathcal{F}^* \left((1 + |\xi|^2)^{-s/2} \mathcal{F}u \right), \quad u \in \mathcal{S}'(\mathbf{R}^n). \tag{7.4}$$

The operator $\mathcal{G}^s \colon \mathcal{S}'(\mathbf{R}^n) \to \mathcal{S}'(\mathbf{R}^n)$ can be visualized as in Fig. 7.1.

Then it is easy to see that the map $\mathcal{G}^s$ is an isomorphism of $\mathcal{S}'(\mathbf{R}^n)$ onto itself, and its inverse is the map $\mathcal{G}^{-s}$. The function $\mathcal{G}^s u$ is called the *Bessel potential* of order s of u.

We can calculate the convolution kernel $G_s(x)$ of the Bessel potential $\mathcal{G}^s u$ for all $s > 0$. More precisely, we have the following:

Theorem 7.1 *Let $s > 0$. (i) The inverse Fourier transform*

$$\mathcal{F}^* \left((1 + |\xi|^2)^{-s/2} \right)$$

is equal to the function

$$G_s(x) = \frac{1}{(4\pi)^{s/2}} \frac{1}{\Gamma(s/2)} \int_0^\infty e^{-\pi |x|^2/\delta} \, e^{-\delta/(4\pi)} \, \delta^{(s-n)/2} \, \frac{d\delta}{\delta}. \tag{7.5}$$

In other words, we have, by the Fourier inversion formula,

$$\mathcal{F}(G_s)(\xi) = (1 + |\xi|^2)^{-s/2}. \tag{7.6}$$

Moreover, we have the formula

$$\int_{\mathbf{R}^n} G_s(x) \, dx = 1,$$

and so

$$G_s \in L^1(\mathbf{R}^n).$$

(ii) Let $1 \leq p \leq \infty$. The Bessel potential $\mathcal{G}^s$ can be expressed as follows:

$$\mathcal{G}^s u(x) = G_s * u(x) = \int_{\mathbf{R}^n} G_s(x - y) u(y) \, dy, \quad u \in L^p(\mathbf{R}^n). \tag{7.7}$$

Furthermore, the Bessel potential $\mathcal{G}^s$ is bounded from $L^p(\mathbf{R}^n)$ into itself. More precisely, we have the inequality

$$\left\| \mathcal{G}^s u \right\|_{L^p(\mathbf{R}^n)} \leq \|u\|_{L^p(\mathbf{R}^n)}, \quad u \in L^p(\mathbf{R}^n). \tag{7.8}$$

One of the most important facts concerning Bessel potentials is that they can be used to define the generalized Sobolev spaces $H^{s,p}(\mathbf{R}^n)$ in the following way: If $s \in \mathbf{R}$ and $1 < p < \infty$, we let

$$\begin{aligned} H^{s,p}(\mathbf{R}^n) &= \text{the image of } L^p(\mathbf{R}^n) \text{ under the mapping } \mathcal{G}^s \\ &= \left\{ \mathcal{G}^s v : v \in L^p(\mathbf{R}^n) \right\}. \end{aligned}$$

We equip $H^{s,p}(\mathbf{R}^n)$ with the norm

$$\|u\|_{H^{s,p}(\mathbf{R}^n)} = \left\| \mathcal{G}^{-s} u \right\|_{L^p(\mathbf{R}^n)}, \quad u \in H^{s,p}(\mathbf{R}^n). \tag{7.9}$$

The space $H^{s,p}(\mathbf{R}^n)$ is called the Bessel-potential space of order s or the generalized *Sobolev space* of order s.

We list some basic topological properties of $H^{s,p}(\mathbf{R}^n)$:

(1) The Schwartz space $\mathcal{S}(\mathbf{R}^n)$ is dense in each $H^{s,p}(\mathbf{R}^n)$.
(2) The space $H^{-s,p'}(\mathbf{R}^n)$ is the dual space of $H^{s,p}(\mathbf{R}^n)$, where $p' = p/(p-1)$ is the exponent conjugate to p.
(3) If $s > t$, then we have the inclusions

$$\mathcal{S}(\mathbf{R}^n) \subset H^{s,p}(\mathbf{R}^n) \subset H^{t,p}(\mathbf{R}^n) \subset \mathcal{S}'(\mathbf{R}^n),$$

with continuous injections.
(4) If s is a non-negative integer, then the space $H^{s,p}(\mathbf{R}^n)$ is isomorphic to the Sobolev space $W^{s,p}(\mathbf{R}^n)$, and the norm (7.9) is equivalent to the norm (7.1).

7.1.3 Definition of General Sobolev Spaces

Now we define the generalized Sobolev spaces $H^{s,p}(\Omega)$ for general domains Ω.

For each $s \in \mathbf{R}$ and $1 < p < \infty$, we let

$$\begin{aligned} H^{s,p}(\Omega) := &\text{ the space of distributions } u \in \mathcal{D}'(\Omega) \text{ such that} \\ &\text{there exists a function } U \in H^{s,p}(\mathbf{R}^n) \text{ with } U|_{\Omega} = u. \end{aligned}$$

We equip the space $H^{s,p}(\Omega)$ with the norm

$$\|u\|_{H^{s,p}(\Omega)} = \inf\left\{\|U\|_{H^{s,p}(\mathbf{R}^n)} : U \in H^{s,p}(\mathbf{R}^n),\ U|_{\Omega} = u \text{ in } \Omega\right\}|1,$$

where the infimum is taken over all $U \in H^{s,p}(\mathbf{R}^n)$ that equal u in Ω. The space $H^{s,p}(\Omega)$ is a Banach space with respect to the norm $\|\cdot\|_{s,p}$. It should be noticed that

$$H^{0,p}(\Omega) = L^p(\Omega); \quad \|\cdot\|_{H^{0,p}(\Omega)} = \|\cdot\|_{L^p(\Omega)}.$$

Then we have the following important relationships between the Sobolev spaces $H^{s,p}(\Omega)$ and $W^{s,p}(\Omega)$ (see [104, Theorem], [2, Theorem 5.24]):

Theorem 7.2 (Meyers–Serrin) *Let Ω be a bounded, Lipschitz domain in Euclidean space $\mathbf{R}^n$. If $1 < p < \infty$ and if k is a non-negative integer, then we have the assertion*

$$H^{k,p}(\Omega) = W^{k,p}(\Omega).$$

7.1.4 Sobolev Imbedding Theorems

In Chap. 13, we shall need Sobolev's imbedding theorems (see [2, Theorem 4.12, Part I and Part II]; [155, Theorems 4.17, 4.19 and 4.23]).

Definition 7.3 An open subset Ω of $\mathbf{R}^n$ satisfies the *cone condition* if there exists a finite cone C such that each point $x \in \Omega$ is the vertex of a finite cone C_x contained in Ω and congruent to C.

We remark that the cone C_x need not be obtained from C by parallel translation, but simply by rigid motion.

Example 7.4 Any Lipschitz domain satisfies the cone condition (see [2, Paragraph 4.11]).

Theorem 7.5 (Sobolev) *Let $j \geq 0$ and $m \geq 1$ be integers, and let $1 \leq p < \infty$.*

***Part I**: Assume that Ω is an open subset of $\mathbf{R}^n$ which satisfies the* cone condition. *Then we have the imbeddings*

$$W^{m+j,p}(\Omega) \subset \begin{cases} W^{j,q}(\Omega) & \text{for all } p \leq q \leq np/(n-mp) \text{ if } 1 < p < n/m, \\ W^{j,q}(\Omega) & \text{for all } p \leq q < \infty \text{ if } p = n/m, \\ C_B^j(\Omega) & \text{if } n/m < p < \infty. \end{cases}$$

Here

$$C_B^j(\Omega) = \text{the space of those functions } u \in C^j(\Omega) \\ \text{for which } D^\alpha u \text{ is bounded for } 0 \leq |\alpha| \leq j,$$

with norm given by the formula

$$\|u\|_{C_B^j(\Omega)} = \max_{0 \leq |\alpha| \leq j} \sup_{x \in \Omega} |D^\alpha u(x)| .$$

***Part II**: Assume that Ω is a bounded,* Lipschitz domain *in $\mathbf{R}^n$. Then we have the imbeddings*

$$W^{m+j,p}(\Omega) \subset \begin{cases} C^{j+\lambda}(\overline{\Omega}) & \text{for all } 0 < \lambda \leq m - n/p \text{ if } (m-1)p < n < mp, \\ C^{j+\lambda}(\overline{\Omega}) & \text{for all } 0 < \lambda < 1 \text{ if } n = (m-1)p, \\ C^{j+1}(\overline{\Omega}) & \text{if } n = (m-1) \text{ and } p = 1. \end{cases}$$

7.1.5 The Rellich–Kondrachov Theorem

In Chap. 15 we shall need the following Rellich–Kondrachov theorem (see [2, Theorem 6.3, Parts I and II], [59, Section 7.12, Theorem 7.26]):

Table 7.1 A bird's-eye view of three compactness theorems in calculus

Subjects	Sequences	Compactness theorems
Theory of real numbers	Sequences of real numbers	The Bolzano–Weierstrass theorem
Calculus	Sequences of continuous functions	The Ascoli–Arzerà theorem
Theory of distributions	Sequences of distributions	The Rellich–Kondrachov theorem

Theorem 7.6 (Rellich–Kondrachov) *Let Ω be a bounded, open subset of* $\mathbf{R}^n$ *that satisfies the* cone condition. *Then the imbedding*

$$W^{j+1,p}(\Omega) \longrightarrow W^{j,q}(\Omega)$$

is compact *for any integer $j \geq 0$ if either $p \geq n$ and $1 \leq q < \infty$ or $1 \leq p < n$ and $1 \leq q < np/(n-p)$.*

The Rellich–Kondrachov theorem is an L^p Sobolev space version of the Bolzano–Weierstrass theorem and the Ascoli–Arzelà theorem in calculus (see Table 7.1).

7.2 Besov Spaces on the Boundary

In studying boundary value problems, we shall need to make sense of the restriction $u|_{\partial\Omega}$ as an element of a function space on the boundary $\partial\Omega$ when u belongs to an L^p Sobolev space on the Lipschitz domain Ω. In this way, the Besov spaces $B^{s,p}(\partial\Omega)$ on the boundary $\partial\Omega$ enter naturally in connection with boundary value problems. The Besov spaces $B^{s,p}(\partial\Omega)$ are defined to be locally the Besov spaces $B^{s,p}(\mathbf{R}^{n-1})$ on $\mathbf{R}^{n-1}$, upon using local coordinate systems flattening out $\partial\Omega$, together with a partition of unity.

An open set Ω in $\mathbf{R}^n$ is called a *Lipschitz hypograph* if its boundary $\partial\Omega$ can be represented as the graph of a Lipschitz continuous function. Namely, there exists a Lipschitz continuous function $\zeta : \mathbf{R}^{n-1} \to \mathbf{R}$ such that (see Fig. 5.1)

$$\Omega = \left\{x = (x', x_n) \in \mathbf{R}^n : x_n < \zeta(x'), \quad x' \in \mathbf{R}^{n-1}\right\}. \qquad (5.1)$$

We define Besov spaces $B^{s,p}(\partial\Omega)$ on the boundary $\partial\Omega$ of a Lipschitz domain Ω, upon using local coordinate systems flattening out $\partial\Omega$, together with a partition of unity, in the following way.

Step 1: If Ω is a Lipschitz hypograph defined by formula (5.1), then we remark that its boundary

$$\partial\Omega = \left\{x = (x', \zeta(x')) : x' \in \mathbf{R}^{n-1}\right\}$$

is an $(n-1)$-dimensional, $C^{0,1}$ submanifold of $\mathbf{R}^n$. Hence we find that $\partial\Omega$ has a surface measure $d\sigma$ and a unit outward normal $\boldsymbol{\nu}$ which exists $d\sigma$-almost everywhere in $\mathbf{R}^{n-1}$. Indeed, by Rademacher's theorem (Theorem 5.2) it follows that the function $\zeta(x')$ is Fréchet differentiable almost everywhere in $\mathbf{R}^{n-1}$ with

$$\|\nabla\zeta\|_{L^\infty(\mathbf{R}^{n-1})} \leq C,$$

where C is any Lipschitz constant for the function $\zeta(x')$. Then we have the following formulas for $d\sigma$ and $\boldsymbol{\nu}$:

$$d\sigma = \sqrt{1 + |\nabla\zeta(x')|^2}\, dx',$$
$$\boldsymbol{\nu} = \frac{(-\nabla\zeta(x'), 1)}{\sqrt{1 + |\nabla\zeta(x')|^2}}.$$

Step 1-1: Now we can define the Besov spaces $B^{s,p}(\partial\Omega)$ for $0 < s \leq 1$ in the following way: For any function $\varphi \in L^p(\partial\Omega) = L^p(\partial\Omega, d\sigma)$, we define a function

$$\varphi_\zeta(x') := \varphi(x', \zeta(x')), \quad x' \in \mathbf{R}^{n-1},$$

and let, for $0 < s < 1$,

$$B^{s,p}(\partial\Omega) = \left\{\varphi \in L^p(\partial\Omega) : \varphi_\zeta \in B^{s,p}(\mathbf{R}^{n-1})\right\}.$$

We equip this space with the norm (see the norm A in Theorem 6.4 with $n := n-1$, $\alpha := s$, $p = q$)

$$|\varphi|_{B^{s,p}(\partial\Omega)} = |\varphi_\zeta|_{B^{s,p}(\mathbf{R}^{n-1})} \tag{7.10}$$
$$= \left(\int_{\mathbf{R}^{n-1}} |\varphi_\zeta(x')|^p\, dx' + \iint_{\mathbf{R}^{n-1}\times\mathbf{R}^{n-1}} \frac{|\varphi_\zeta(x'+h') - \varphi_\zeta(x')|^p}{|h'|^{(n-1)+ps}}\, dx'\, dh'\right)^{1/p}.$$

For $s = 1$, we let

$B^{1,p}(\partial\Omega)$
= the space of (equivalence classes of) functions $\varphi \in L^p(\partial\Omega, d\sigma)$ for which

$$\iint_{\mathbf{R}^{n-1}\times\mathbf{R}^{n-1}} \frac{|\varphi_\zeta(x'+h') - 2\varphi_\zeta(x') + \varphi_\zeta(x'-h')|^p}{|h'|^{(n-1)+p}}\, dx'\, dh' < \infty.$$

The space $B^{1,p}(\partial\Omega)$ is a Banach space with respect to the norm (see the norm B in Theorem 6.4 with $n := n-1$, $\alpha := 1$, $p = q$)

$$
\begin{aligned}
|\varphi|_{B^{1,p}(\partial\Omega)} &= |\varphi_\zeta|_{B^{1,p}(\mathbf{R}^{n-1})} \\
&= \Bigg(\int_{\mathbf{R}^{n-1}} |\varphi_\zeta(x')|^p \, dx' \\
&+ \iint_{\mathbf{R}^{n-1}\times\mathbf{R}^{n-1}} \frac{|\varphi_\zeta(x'+h') - 2\varphi_\zeta(x') + \varphi_\zeta(x'-h')|^p}{|h'|^{(n-1)+p}} \, dx' \, dh' \Bigg)^{1/p} .
\end{aligned}
\tag{7.11}
$$

Step 1-2: If $\kappa(\Omega)$ is a Lipschitz hypograph for some rigid motion $\kappa\colon \mathbf{R}^n \to \mathbf{R}^n$, then we can define the Besov spaces $B^{s,p}(\partial\Omega)$ for $0 < s \leq 1$ in the same way except that

$$
\varphi_\zeta(x') := \varphi\left(\kappa^{-1}(x', \zeta(x'))\right), \quad x' \in \mathbf{R}^{n-1}.
$$

Step 2: We consider the general case where Ω is a bounded Lipschitz domain. By using the notation of Definition 5.1, we choose a partition of unity $\{\phi_j\}_{j=1}^J$ subordinate to the open covering $\{U_j\}_{j=1}^J$ of $\partial\Omega$ (see Fig. 5.4). Then we define the Besov spaces $B^{s,p}(\partial\Omega)$ for $0 < s \leq 1$ as follows:

$$
B^{s,p}(\partial\Omega) = \left\{\varphi \in L^p(\partial\Omega) : \phi_j\varphi \in B^{s,p}(\partial\Omega_j),\ 1 \leq j \leq J\right\},
$$

where the norm $|\varphi|_{B^{s,p}(\partial\Omega)}$ is defined by the formula

$$
|\varphi_{B^{s,p}(\partial\Omega)} = \sum_{j=1}^{J} |\phi_j\varphi|_{B^{s,p}(\partial\Omega_j)}.
$$

It should be emphasized that the Besov spaces $B^{s,p}(\partial\Omega)$ are independent of the open covering $\{U_j\}$ and the partition of unity $\{\phi_j\}$ used.

Step 3: Furthermore, we shall require Besov spaces $B^{s,p}(\partial\Omega)$ for $1 < s < 2$ defined on a bounded $C^{1,1}$ domain Ω.

Step 3-1: If Ω is a $C^{1,1}$ hypograph defined by formula (5.1) for some function $\zeta \in C^{1,1}(\mathbf{R}^{n-1})$, then we define the Besov spaces $B^{s,p}(\partial\Omega)$ for $s = 1 + \theta$ with $0 < \theta < 1$ in the same way by replacing the norm (7.10) by the norm

$$
\begin{aligned}
|\varphi|_{B^{s,p}(\partial\Omega)} &= |\varphi_\zeta|_{B^{s,p}(\mathbf{R}^{n-1})} \\
&= \Bigg(\sum_{|\alpha|\leq 1} \int_{\mathbf{R}^{n-1}} |D^\alpha\varphi_\zeta(x')|^p \, dx' \\
&+ \sum_{|\alpha|=1} \iint_{\mathbf{R}^{n-1}\times\mathbf{R}^{n-1}} \frac{|D^\alpha\varphi_\zeta(x'+h') - 2D^\alpha\varphi_\zeta(x') + D^\alpha\varphi_\zeta(x'-h')|^p}{|h'|^{(n-1)+p}} dx' dh' \Bigg)^{1/p} .
\end{aligned}
$$

Step 3-2: If Ω is a bounded $C^{1,1}$ domain, then the Besov spaces $B^{s,p}(\partial\Omega)$ for $1 < s < 2$ are defined to be locally the Besov spaces $B^{s,p}(\partial\Omega_j)$, $1 \leq j \leq J$, just as in Step 2. Here it should be emphasized that the boundary $\partial\Omega$ is an $(n-1)$-dimensional, $C^{1,1}$ submanifold of $\mathbf{R}^n$.

The norm of $B^{s,p}(\partial\Omega)$ for $0 \leq s < 2$ will be denoted by $|\cdot|_{s,p}$.

7.3 Trace Theorems

In this section we prove an important trace theorem which will be used in the study of boundary value problems in the framework of L^p Sobolev spaces (cf. [2, 11, 136, 146, 170]):

Theorem 7.7 *Let* $1 < p < \infty$. *For every function* $f \in H^{s,p}(\mathbf{R}^n)$ *with* $s > 1/p$, *the restriction*

$$g := Rf = f|_{\mathbf{R}^{n-1}}$$

is well defined almost everywhere in $\mathbf{R}^{n-1}$, *and* $g \in B^{s-1/p,p}(\mathbf{R}^{n-1})$. *Furthermore, the restriction mapping* R *so defined is continuous, that is, there exists a positive constant* C *such that*

$$\|Rf\|_{B^{s-1/p,p}(\mathbf{R}^{n-1})} \leq C\,\|f\|_{H^{s,p}(\mathbf{R}^n)} \quad \text{for all } f \in H^{s,p}(\mathbf{R}^n).$$

An elementary proof of Theorem 7.7 is given in [154, Theorem 6.6].

Under certain hypotheses on the domain Ω, functions in Sobolev spaces $H^{s,p}(\Omega)$ may be extended as functions in $H^{s,p}(\mathbf{R}^n)$. In this way, the trace theorem remains valid for $H^{s,p}(\Omega)$ and $B^{s-1/p,p}(\partial\Omega)$. More precisely, we have the following (see [2, Remarks 7.45]):

Theorem 7.8 (the trace theorem) *Let* Ω *be a bounded,* $C^{1,1}$ *domain of* $\mathbf{R}^n$. *If* $1 < p < \infty$, *then the trace map*

$$\gamma = (\gamma_0, \gamma_1) : \; W^{2,p}(\Omega) \longrightarrow B^{2-1/p,p}(\partial\Omega) \oplus B^{1-1/p,p}(\partial\Omega)$$
$$u \longmapsto (\gamma_0 u, \gamma_1 u) = \left(u|_{\partial\Omega}, \left.\frac{\partial u}{\partial \mathbf{n}}\right|_{\partial\Omega} \right)$$

is continuous and surjective. Here $\mathbf{n} = -\boldsymbol{\nu}$ *is the unit inward normal to the boundary* $\partial\Omega$ *(see Fig. 7.2).*

Indeed, it suffices to note that we have, by Theorem 7.2,

$$H^{2,p}(\Omega) = W^{2,p}(\Omega).$$

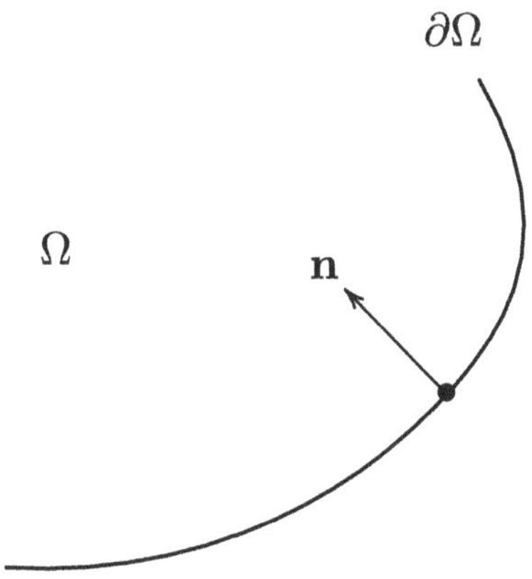

Fig. 7.2 The unit inward normal **n** to the boundary $\partial\Omega$

7.4 VMO Functions Revisited

In this last section we prove part (iv) of Examples 4.1. Namely, we prove the following:

Proposition 7.9 *We have, for* $0 < \theta \leq 1$,

$$W^{\theta,n/\theta}(\mathbf{R}^n) \subset \text{VMO}\,.$$

Proof The proof of Proposition 7.9 is divided into two steps.

Step (1): First, we consider the case $\theta = 1$. The proof is based on the following *Poincaré inequality* (see [59, Chap. 7, formula (7.45)], [97, Chap. 1, Theorem 1.51]):

Lemma 7.10 (Poincaré) *Let B be an open ball with diameter d. Let $1 \leq p < \infty$. Then we have, for all $u \in W^{1,p}(\mathbf{R}^n)$,*

$$\|u - u_B\|_p \leq 2^{n-1}\, d\, \|\nabla u\|_p\,, \tag{7.12}$$

where

$$u_B = \frac{1}{|B|}\int_B u(x)\,dx$$

is the integral average of u over B.

By using Poincaré's inequality (7.12) with $p := n$, we obtain that

$$\int_B |u(x) - u_B|^n\,dx \leq \left(2^{n-1}d\right)^n \int_B |\nabla u(x)|^n\,dx.$$

Hence it follows from an application of Hölder's inequality $p := n$ and $q := n/(n-1)$ (Theorem 3.24) that

$$\begin{aligned}\int_B |u(x) - u_B|\, dx &\le \left(\int_B |u(x) - u_B|^n\, dx\right)^{1/n} \left(\int_B dx\right)^{1-1/n} \\ &= \left(\int_B |u(x) - u_B|^n\, dx\right)^{1/n} |B|^{1-1/n} \\ &\le 2^{n-1}\, d \cdot \left(\int_B |\nabla u(x)|^n\, dx\right)^{1/n} \left(\frac{2^n n \Gamma(n/2)}{2d^n \pi^{n/2}}\right)^{1/n} |B|,\end{aligned}$$

so that

$$\frac{1}{|B|}\int_B |u(x) - u_B|\, dx \le \frac{2^n}{\sqrt{\pi}} \left(\frac{n}{2}\Gamma\left(\frac{n}{2}\right)\right)^{1/n} \left(\int_B |\nabla u(x)|^n\, dx\right)^{1/n}.$$

However, we find that the integral

$$\int_B |\nabla u(x)|^n\, dx$$

is *absolutely continuous*.

Therefore, we have proved that $u \in \mathrm{VMO}$ if $u \in W^{1,n}(\mathbf{R}^n)$.

Step (2): Secondly, we consider the case where $0 < \theta < 1$. By applying Hölder's inequality for $p := n/\theta$ and $q := n/(n - \theta)$, we obtain that

$$\begin{aligned}\int_B |u(x) - u_B|\, dx &\le \left(\int_B |u(x) - u_B|^{n/\theta}\, dx\right)^{\theta/n} \left(\int_B dx\right)^{1-\theta/n} \\ &= \left(\int_B |u(x) - u_B|^{n/\epsilon}\, dx\right)^{\theta/n} |B|^{1-\theta/n},\end{aligned}$$

so that

$$\frac{1}{|B|}\int_B |u(x) - u_B|\, dx \le \left(\frac{1}{|B|}\int_B |u(x) - u_B|^{n/\epsilon}\, dx\right)^{\theta/n}.$$

Moreover, we have, by Fubini's theorem (Theorem 3.20),

$$\begin{aligned}&\frac{1}{|B|}\int_B |u(x) - u_B|\, dx \le \left(\frac{1}{|B|}\int_B |u(x) - u_B|^{n/\epsilon}\, dx\right)^{\theta/n} \qquad (7.13) \\ &= \left(\frac{1}{|B|}\left(\int_B \frac{1}{|B|}\left|\int_B (u(x) - u(y))\, dy\right|\right)^{n/\theta} dx\right)^{\theta/n} \\ &\le \left(\frac{1}{|B|^{1+n/\theta}}\int_B \left[\int_B |u(x) - u(y)|\, dy\right]^{n/\theta} dx\right)^{\theta/n}\end{aligned}$$

$$\leq \left(\frac{1}{|B|^{1+n/\theta}} \int_B \left[\left(\int_B |u(x) - u(y)|^{n/\theta} \, dy \right)^{\theta/n} \left(\int_B dy \right)^{1-\theta/n} \right]^{n/\theta} dx \right)^{\theta/n}$$
$$= \left(\frac{1}{|B|^2} \int_B \int_B |u(x) - u(y)|^{n/\theta} \, dy \, dx \right)^{\theta/n}.$$

However, it is easy to see that we have, for the open ball B with diameter d,

$$|x - y|^{2n} \leq c(n) \, |B|^2 \quad \text{for all } x, y \in B,$$

where

$$c(n) = \frac{4^{n-1} n^2 \Gamma\left(\frac{n}{2}\right)^2}{\pi^n}.$$

Hence, by combining this inequality with inequality (7.13) we obtain that

$$\frac{1}{|B|} \int_B |u(x) - u_B| \, dx \leq \left(\frac{1}{|B|^2} \int_B \int_B |u(x) - u(y)|^{n/\theta} \, dy \, dx \right)^{\theta/n}$$
$$\leq c(n) \left(\int \int_{B \times B} \frac{|u(x) - u(y)|^{n/\theta}}{|x - y|^{2n}} \, dx \, dy \right)^{\theta/n}.$$

However, by formula (7.2) with $p := n/\theta$ we find that the integral

$$\int \int_{B \times B} \frac{|u(x) - u(y)|^{n/\theta}}{|x - y|^{2n}} \, dx \, dy$$

is *absolutely continuous*.

Therefore, we have proved that $u \in \mathrm{VMO}$ if $u \in W^{\theta, n/\theta}(\mathbf{R}^n)$ for $0 < \theta < 1$.

The proof of Proposition 7.9 is complete. □

Chapter 8
Maximum Principles in Sobolev Spaces

In this chapter we prove various maximum principles for second order, elliptic differential operators with discontinuous coefficients such as the weak and strong maximum principles (Theorems 8.1 and 8.9) and the Hopf boundary point lemma (Lemma 8.8) in the framework of L^p Sobolev spaces that will play an important role in the proof of uniqueness theorems for the Dirichlet problem in Part IV. The results here are adapted from Bony [13], Bony-Courrège--Priouret [15], Troianiello [172] and also Taira [148] and [154].

Let Ω be a bounded domain in Euclidean space $\mathbf{R}^N$, $N \geq 2$, with boundary $\partial\Omega$ of class $C^{1,1}$. We consider a second order, *uniformly elliptic* differential operator A with real *discontinuous* coefficients such that

$$Au(x) := \sum_{i,j=1}^{N} a^{ij}(x) \frac{\partial^2 u}{\partial x_i \partial x_j}(x) + \sum_{i=1}^{N} b^i(x) \frac{\partial u}{\partial x_i}(x) + c(x)u(x).$$

More precisely, we assume that the coefficients $a^{ij}(x)$, $b^i(x)$ and $c(x)$ of the differential operator A satisfy the following three conditions:

(1) $a^{ij} \in L^\infty(\Omega)$, $a^{ij}(x) = a^{ji}(x)$ for all $1 \leq i, j \leq N$ and for almost all $x \in \Omega$ and there exist a constant $\lambda > 0$ such that (the uniform ellipticity)

$$\frac{1}{\lambda} |\xi|^2 \leq \sum_{i,j=1}^{N} a^{ij}(x)\xi_i\xi_j \leq \lambda |\xi|^2 \tag{8.1}$$

$$\text{for almost all } x \in \Omega \text{ and all } \xi \in \mathbf{R}^N.$$

(2) $b^i \in L^\infty(\Omega)$ for all $1 \leq i \leq N$.
(3) $c \in L^\infty(\Omega)$ and $c(x) \leq 0$ for almost all $x \in \Omega$.

K. Taira, *Real Analysis Methods for Markov Processes*,
https://doi.org/10.1007/978-981-97-3659-1_8

8.1 Weak Maximum Principle

The purpose of this section is to prove a variant of the weak maximum principle in the framework of L^p Sobolev spaces, essentially due to Bony [13] (cf. [13, Théorème 2], [59, Sect. 9.1, Theorem 9.1], [172, Chap. 3, Lemma 3.25]):

Theorem 8.1 (the weak maximum principle) *Assume that a function $u \in W^{2,p}(\Omega)$, with $N < p < \infty$, satisfies the condition*

$$Au(x) \geq 0 \quad \text{for almost all } x \in \Omega. \tag{8.2}$$

Then we have the assertions

$$\begin{cases} \max_{x\in\overline{\Omega}} u(x) = \max_{x'\in\partial\Omega} u(x') & \text{if } c(x) = 0 \text{ for almost all } x \in \Omega, \\ \max_{x\in\overline{\Omega}} u(x) \leq \max_{x'\in\partial\Omega} u^+(x') & \text{if } c(x) \leq 0 \text{ for almost all } x \in \Omega, \end{cases}$$

where

$$u^+(x) = \max\{u(x), 0\} = \frac{1}{2}\{|u(x)| + u(x)\} \quad \text{for } x \in \overline{\Omega}.$$

Here it should be noticed that we have, by Sobolev's imbedding theorem (see Theorem 7.5),

$$W^{2,p}(\Omega) \subset C^1(\overline{\Omega}),$$

since $2 - N/p > 1$ for $N < p < \infty$.

The proof of Theorem 8.1 is divided into Step (I) and Step (II).

Step (I): The next lemma, due to Bony [13, Théorème 1], plays an essential role in the proof of Theorem 8.1 (cf. [172, Chap. 3, Lemma 3.24]):

Lemma 8.2 (Bony) *If a function $u \in W^{2,p}(\Omega)$, with $N < p < \infty$, attains a local maximum at a point $x_0 \in \Omega$, then we have the assertion*

$$\begin{aligned} &\lim \operatorname{ess\,inf}_{x\to x_0} \left(\sum_{i,j=1}^{N} a^{ij}(x) \frac{\partial^2 u}{\partial x_i \partial x_j}(x) \right) \\ &= \lim_{\rho\downarrow 0} \left\{ \operatorname{ess\,inf}_{B(x_0,\rho)} \left(\sum_{i,j=1}^{N} a^{ij}(x) \frac{\partial^2 u}{\partial x_i \partial x_j}(x) \right) \right\} \\ &\leq 0, \end{aligned}$$

where $B(x_0, \rho)$ denotes the open ball of radius ρ about x_0. More precisely, for every neighborhood $U(x_0)$ of x_0 there exists a subset M of $U(x_0)$, with positive measure, *such that*

$$\sum_{i,j=1}^{N} a^{ij}(x)\frac{\partial^2 u}{\partial x_i \partial x_j}(x) \leq 0 \quad \textit{for all } x \in M.$$

Proof The proof of Lemma 8.2 is divided into four steps.

Step 1: First, we may assume that $u \in W^{2,p}(\Omega)$ attains a *strict* local maximum at a point $x_0 \in \Omega$. Indeed, if we consider the function

$$\widetilde{u}(x) := u(x) - \varphi(x) = u(x) - |x - x_0|^4 \quad \text{for } x \in \Omega,$$

then it is easy to verify that $\widetilde{u} \in W^{2,p}(\Omega)$ attains a strict local maximum at a point x_0 and further that

$$\begin{aligned} &\lim \operatorname{ess\,inf}_{x \to x_0} \sum_{i,j=1}^{N} a^{ij}(x)\frac{\partial^2 \widetilde{u}}{\partial x_i \partial x_j}(x) \\ &= \lim \operatorname{ess\,inf}_{x \to x_0} \sum_{i,j=1}^{N} a^{ij}(x)\frac{\partial^2 u}{\partial x_i \partial x_j}(x), \end{aligned}$$

since we have the assertion

$$\lim \operatorname{ess\,inf}_{x \to x_0} \sum_{i,j=1}^{N} a^{ij}(x)\frac{\partial^2 \varphi}{\partial x_i \partial x_j}(x) = 0.$$

Step 2: Now, let $U(x_0)$ be an arbitrary open neighborhood of x_0 such that

$$u(x) < u(x_0) \quad \text{in } \overline{U(x_0)} \setminus \{x_0\}. \tag{8.3}$$

We define the *upper contact set* $\mathcal{Z}$ of u by the formula

$$\mathcal{Z} := \{x \in U(x_0) : u(y) \leq u(x) + \langle \nabla u(x), y - x\rangle \quad \text{for all } y \in U(x_0)\}.$$

By assumption (8.3), we can find a positive constant r such that we have, for each $\eta \in \mathbf{R}^N$ satisfying $|\eta| < r$,

$$u(x) < u(x_0) + \langle \eta, x - x_0\rangle \quad \text{for all } x \in \partial U(x_0). \tag{8.4}$$

If we let

$$\varphi(x, \eta) = u(x) - \langle x, \eta\rangle \quad \text{for } x \in U(x_0),$$

then inequality (8.4) can be rewritten in the form

$$\varphi(x, \eta) < \varphi(x_0, \eta) \quad \text{for all } x \in \partial U(x_0).$$

This implies that the function $\varphi(x, \eta)$ takes its maximum in the interior $U(x_0)$. Hence there exists a point $\widetilde{x} \in U(x_0)$, depending on η, such that

$$\varphi(y, \eta) \leq \varphi(\widetilde{x}, \eta) \quad \text{for all } x \in \partial U(x_0),$$

or equivalently, we have, for each $\eta \in \mathbf{R}^N$ satisfying $|\eta| < r$,

$$u(y) \leq u(\widetilde{x}) + \langle \eta, y - \widetilde{x} \rangle \quad \text{for all } y \in \overline{U(x_0)}. \tag{8.5}$$

In other words, we may translate vertically the plane

$$z = u(x) + \langle \eta, y - x \rangle$$

to the highest such position $\widetilde{x}$, that is, the surface

$$z = u(y) \quad \text{for } y \in \overline{U(x_0)},$$

lies below the plane.

More precisely, we have the following claim (see [13, Lemme 2]):

Claim 8.3 $\eta = \nabla u(\widetilde{x})$ *for* $\widetilde{x} \in U(x_0)$.

Proof If we let

$$\Phi(y) := u(\widetilde{x}) - u(y) + \langle \eta, y - \widetilde{x} \rangle \quad \text{for } y \in \overline{U(x_0)},$$

then it follows from inequality (8.5) that

$$\begin{aligned} &\Phi(y) \geq 0 \quad \text{for all } y \in \overline{U(x_0)}, \\ &\Phi(\widetilde{x}) = 0. \end{aligned}$$

This implies that $\nabla_y \Phi(\widetilde{x}) = 0$, so that $\eta = \nabla u(\widetilde{x})$.

The proof of Claim 8.3 is complete. □

By Claim 8.3, we can rewrite inequality (8.5) as follows:

$$u(y) \leq u(\widetilde{x}) + \langle \nabla u(\widetilde{x}), y - \widetilde{x} \rangle \quad \text{for all } y \in \overline{U(x_0)}.$$

This proves that $\widetilde{x} \in \mathcal{Z}$.

Summing up, we have proved the following claim:

Claim 8.4 *Let* $\mathcal{Z}$ *be the* upper contact set *of* u:

$$\mathcal{Z} = \{x \in U(x_0) : u(y) \leq u(x) + \langle \nabla u(x), y - x \rangle \quad \textit{for all } y \in U(x_0)\}.$$

Then we have, for some constant r,

$$B(0,r) \subset \{\nabla u(\widetilde{x}) : \widetilde{x} \in \mathcal{Z}\}.$$

Step 3: We prove that the measure of $\mathcal{Z}$ is *positive*. To do this, we introduce a mapping

$$\begin{aligned} F\colon\ U(x_0) &\longrightarrow \mathbf{R}^N \\ x &\longmapsto \nabla u(x), \end{aligned}$$

where $u \in W^{2,p}(\Omega)$ with $N < p < \infty$. Then we have, by Claim 8.4,

$$B(0,r) \subset F(\mathcal{Z}).$$

This proves that $\mu(F(\mathcal{Z})) > 0$, since the set $F(\mathcal{Z})$ contains the open ball $B(0,r)$.

Furthermore, the next lemma, due to Bony [13, Lemme 1], proves that $\mu(\mathcal{Z}) > 0$:

Lemma 8.5 (Bony) *Let $F = (f_1, f_2, \ldots, f_N)$ be a $W^{1,p}$ mapping of an open subset Ω of $\mathbf{R}^N$ into $\mathbf{R}^N$, where $N < p < \infty$. If S is a subset of measure zero, then the measure of the image $F(S)$ is equal to zero, that is,*

$$\mu(S) = 0 \implies \mu(F(S)) = 0.$$

Proof (1) If $\gamma > 0$, we let (see Fig. 8.1)

$$\begin{aligned} C_\gamma := &\text{ a compact cube with sides parallel to the axes} \\ &\text{ and non-empty interior, with side length } \gamma, \end{aligned}$$

and we estimate the measure $\mu(F(C_\gamma))$ of the image $F(C_\gamma)$:

$$F(C_\gamma) = \left\{(f_1(x), f_2(x), \ldots, f_N(x) : x \in C_\gamma\right\}.$$

To do this, we make use of a special case of Morrey's theorem (see [2, Lemma 4.28], [59, Theorem 7.17]; [155, Lemma 4.7]):

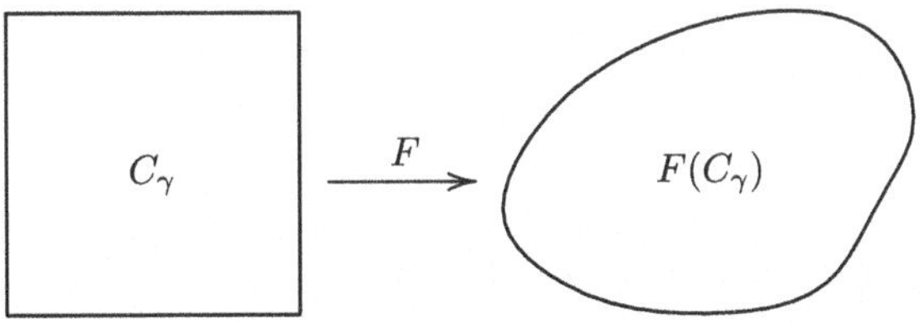

Fig. 8.1 The compact cube C_γ with sides parallel to the axes and the image $F(C_\gamma)$

Claim 8.6 (Morrey) *Let Ω be a cube in $\mathbf{R}^n$ and $n < p < \infty$. Then there exists a positive constant $C = C(p, n)$ such that we have, for all $u \in W^{1,p}(\Omega)$,*

$$\sup_{\substack{x,y\in\Omega \\ x\neq y}} \frac{|u(x) - u(y)|}{|x - y|^{1-n/p}} \leq C \, \|\nabla u\|_{L^p(\Omega)} . \tag{8.6}$$

Proof By a density argument, it suffices to show that we have, for all $u \in C^1(\overline{\Omega})$,

$$|u(x) - u(y)| \leq C \, \|\nabla u\|_{L^p(\Omega)} \, |x - y|^{1-n/p} \quad \text{for } x, y \in \Omega. \tag{8.6'}$$

Furthermore, by a non-singular linear transformation we may assume that Ω is a cube having unit side length.

For $0 < t < 1$, we denote by Q_t a subset of Ω which is a closed cube having side length t and faces parallel to those to Ω. If $x, y \in \Omega$ and

$$\sigma := |x - y| < 1,$$

we can find a cube Q_σ such that $x, y \in Q_\sigma$. Then we have, for all $z \in Q_\sigma$,

$$\begin{aligned} u(x) - u(z) &= -\int_0^1 \frac{d}{dt} u(x + t(z - x)) \, dt \\ &= -\int_0^1 \sum_{j=1}^n (z_j - x_j) \cdot \frac{\partial u}{\partial x_j}(x + t(z - x)) \, dt, \end{aligned}$$

and so, by Schwarz's inequality (Theorem 3.24 with $p = q := 2$),

$$|u(x) - u(z)| \leq \sqrt{n} \, \sigma \int_0^1 |\nabla u(x + t(z - x))| \, dt.$$

Hence it follows from an application of Hölder's inequality (Theorem 3.24) with $q := p/(p - 1)$ that

$$\begin{aligned} &\left| u(x) - \frac{1}{\sigma^n} \int_{Q_\sigma} u(z) \, dz \right| \\ &= \left| \frac{1}{\sigma^n} \int_{Q_\sigma} (u(x) - u(z)) \, dz \right| \\ &\leq \frac{1}{\sigma^n} \int_{Q_\sigma} |u(x) - u(z)| \, dz \\ &\leq \frac{\sqrt{n}}{\sigma^{n-1}} \int_{Q_\sigma} dz \int_0^1 |\nabla u(x + t(z - x))| \, dt \end{aligned} \tag{8.7}$$

$$
\begin{aligned}
&= \frac{\sqrt{n}}{\sigma^{n-1}} \int_0^1 \left(\int_{Q_\sigma} |\nabla u(x + t(z - x))| \, dz \right) dt \\
&= \frac{\sqrt{n}}{\sigma^{n-1}} \int_0^1 \left(\int_{Q_{t\sigma}+(1-t)x} |\nabla u(\zeta)| \, d\zeta \right) t^{-n} \, dt \\
&\leq \frac{\sqrt{n}}{\sigma^{n-1}} \left(\int_\Omega |\nabla u(z)|^p \, dz \right)^{1/p} \int_0^1 t^{-n} \left(|Q_{t\sigma} + (1-t)x| \right)^{1/q} dt \\
&= \frac{\sqrt{n}}{\sigma^{n-1}} \left(\int_\Omega |\nabla u(z)|^p \, dz \right)^{1/p} \int_0^1 t^{-n} \left((t\sigma)^n \right)^{1/q} dt \\
&= K \sigma^{1-n/p} \, \|\nabla u\|_{L^p(\Omega)} ,
\end{aligned}
$$

with

$$
K := K(n, p) = \sqrt{n} \int_0^1 t^{-n/p} \, dt = \frac{\sqrt{n}\, p}{p - n}.
$$

Similarly, we have, with y in place of x in inequality (8.7),

$$
\left| u(y) - \frac{1}{\sigma^n} \int_{Q_\sigma} u(z) \, dz \right| \leq K \sigma^{1-n/p} \, \|\nabla u\|_{L^p(\Omega)} . \tag{8.8}
$$

Therefore, by combining inequalities (8.7) and (8.8) we obtain that

$$
\begin{aligned}
|u(x) - u(y)| &\leq \left| u(x) - \frac{1}{\sigma^n} \int_{Q_\sigma} u(z) \, dz \right| + \left| \frac{1}{\sigma^n} \int_{Q_\sigma} u(z) \, dz - u(y) \right| \\
&\leq 2K \, \|\nabla u\|_{L^p(\Omega)} \, \sigma^{1-n/p} \\
&= 2K \, \|\nabla u\|_{L^p(\Omega)} \, |x - y|^{1-n/p} \quad \text{for all } x, y \in \Omega.
\end{aligned}
$$

This proves the desired inequality (8.6′) with $C := 2K$.

The proof of Claim 8.6 is complete. □

By applying Claim 8.6 with

$$
n := N, \quad \Omega := C_\gamma, \quad u := f_i, \ 1 \leq i \leq N,
$$

we obtain that, for some positive constant $K = K(p, N)$,

$$
\begin{aligned}
&\operatorname{osc} f_i := \sup_{x, y \in C_\gamma} |f_i(x) - f_i(y)| \leq K \gamma^{1-N/p} \, \|\nabla f_i\|_{L^p(C_\gamma)} , \\
&\qquad 1 \leq i \leq N.
\end{aligned} \tag{8.9}
$$

Since the measure $\mu\left(F\left(C_\gamma\right)\right)$ is estimated by $\operatorname{osc} f_1 \times \operatorname{osc} f_2 \times \cdots \times \operatorname{osc} f_N$, it follows from an application of inequality (8.9) that

$$
\mu\left(F\left(C_\gamma\right)\right) \tag{8.10}
$$

$$\leq K^N \gamma^{N(1-N/p)} \|\nabla f_1\|_{L^p(C_\gamma)} \times \|\nabla f_2\|_{L^p(C_\gamma)} \times \cdots \times \|\nabla f_N\|_{L^p(C_\gamma)}.$$

(2) Now we assume that

$$\mu(S) = 0.$$

Then, for any $\varepsilon > 0$ we can find a family of non-overlapping cubes C_α with side length γ_α, $\alpha \in \Lambda$, such that

$$S \subset \bigcup_{\alpha\in\Lambda} \overline{C_\alpha},$$

$$\sum_{\alpha\in\Lambda} \gamma_\alpha^N \leq \varepsilon.$$

By the subadditivity of the measure μ, it follows from inequality (8.10) with $C_\gamma := C_\alpha$ that

$$\begin{aligned} &\mu(F(S)) \\ &\leq \sum_{\alpha\in\Lambda} \mu(F(C_\alpha)) \\ &\leq K^N \sum_{\alpha\in\Lambda} \gamma_\alpha^{N(1-N/p)} \|\nabla f_1\|_{L^p(C_\alpha)} \times \|\nabla f_2\|_{L^p(C_\alpha)} \times \cdots \times \|\nabla f_N\|_{L^p(C_\alpha)} \\ &\leq K^N \sum_{\alpha\in\Lambda} \gamma_\alpha^{N(1-N/p)} \Big(\|\nabla f_1\|^p_{L^p(C_\alpha)} + \|\nabla f_2\|^p_{L^p(C_\alpha)} + \ldots \\ &\qquad + \|\nabla f_N\|^p_{L^p(C_\alpha)} \Big)^{N/p}. \end{aligned} \tag{8.11}$$

However, by applying a discrete version of Hölder's inequality (Theorem 3.24) to the last term of inequality (8.11) we obtain that

$$\begin{aligned} &\sum_{\alpha\in\Lambda} \gamma_\alpha^{N(1-N/p)} \\ &\times \left(\|\nabla f_1\|^p_{L^p(C_\alpha)} + \|\nabla f_2\|^p_{L^p(C_\alpha)} + \ldots + \|\nabla f_N\|^p_{L^p(C_\alpha)} \right)^{N/p} \\ &\leq \left(\sum_{\alpha\in\Lambda} \gamma_\alpha^N \right)^{1-N/p} \\ &\times \left(\sum_{\alpha\in\Lambda} \left(\|\nabla f_1\|^p_{L^p(C_\alpha)} + \|\nabla f_2\|^p_{L^p(C_\alpha)} + \ldots + \|\nabla f_N\|^p_{L^p(C_\alpha)} \right) \right)^{N/p} \\ &\leq \left(\sum_{\alpha\in\Lambda} \gamma_\alpha^N \right)^{1-N/p} \end{aligned} \tag{8.12}$$

$$\times \left(\|\nabla f_1\|^p_{L^p(\bigcup_{\alpha\in\Lambda} C_\alpha)} + \|\nabla f_2\|^p_{L^p(\bigcup_{\alpha\in\Lambda} C_\alpha)} + \|\nabla f_N\|^p_{L^p(\bigcup_{\alpha\in\Lambda} C_\alpha)} \right)^{N/p}$$

$$\leq \varepsilon^{1-N/p} \left(\|\nabla f_1\|^p_{L^p(\Omega)} + \|\nabla f_2\|^p_{L^p(\Omega)} + \ldots + \|\nabla f_N\|^p_{L^p(\Omega)} \right)^{N/p}.$$

Therefore, we have, by inequalities (8.11) and (8.12),

$$\mu(F(S)) \leq K^N \varepsilon^{1-N/p} \left(\|\nabla F\|^p_{L^p(\Omega)} \right)^N.$$

This implies that $\mu(F(S)) = 0$, since $\varepsilon > 0$ is arbitrary.

The proof of Lemma 8.5 is complete. □

By applying Lemma 8.5 to our situation, we obtain that $\mu(\mathcal{Z}) > 0$. In other words, for every neighborhood $U(x_0)$ of x_0 there exists a subset $\mathcal{Z}$ of $U(x_0)$, with positive measure, such that

$$u(y) \leq u(x) + \langle \nabla u(x), y - x \rangle \quad \text{for all } x \in \mathcal{Z} \text{ and all } y \in U(x_0). \tag{8.13}$$

Step 4: Moreover, we need the following claim (see [27, Theorem 12], [97, Theorem 1.72]):

Claim 8.7 (Calderón–Zygmund) *If $u \in W^{2,p}(\Omega)$ with $N < p < \infty$, then it is almost everywhere two times differentiable in the usual sense.*

By Claim 8.7, we may assume that $u(x)$ is two times differentiable almost everywhere in $\mathcal{Z}$.

Now we take an arbitrary vector ξ in the unit sphere Σ_{N-1} in $\mathbf{R}^N$, and let

$$\phi(t) := u(x + t\xi) \quad \text{for } x \in \widetilde{M} \text{ and } |t| < \operatorname{dist}(x, \partial U(x_0)).$$

By the change of variables, we may assume that, at almost all $x \in \mathcal{Z}$ each first partial derivative $\partial u/\partial x_i$, $1 \leq i \leq N$, has a directional derivative with respect to ξ in the usual sense. Then we have, by Taylor's formula,

$$\phi(t) = \phi(0) + t\, \frac{\partial \phi}{\partial t}(0) + \frac{t^2}{2!}\, \frac{\partial^2 \phi}{\partial t^2}(0) + o\left(t^2\right) \quad \text{as } t \to 0,$$

that is,

$$u(x + t\xi) = u(x) + \langle \nabla u(x), y - x \rangle + \frac{t^2}{2!} \sum_{i,j=1}^{N} \frac{\partial^2 u}{\partial x_i \partial x_j}(x)\xi_i\xi_j + o\left(t^2\right)$$
$$\text{as } t \to 0.$$

By inequality (8.13) with $y := x + t\xi$, we have, for any given vector $\xi \in \Sigma_{N-1}$,

$$\sum_{i,j=1}^{N} \frac{\partial^2 u}{\partial x_i \partial x_j}(x)\xi_i\,\xi_j \leq 0 \quad \text{for almost all } x \in W. \tag{8.14}$$

Let $\{\xi^{(n)}\}$ be a *countable* dense subset of Σ_{N-1}. By assertion (8.14), for every neighborhood $U(x_0)$ of x_0 we can find a subset $\widetilde{M}$ of $U(x_0)$, with $\mu(\widetilde{M}) > 0$, such that

$$\sum_{i,j=1}^{N} \frac{\partial^2 u}{\partial x_i \partial x_j}(x)\xi_i^{(n)}\,\xi_j^{(n)} \leq 0 \quad \text{for all } x \in \widetilde{M} \text{ and all } \xi^{(n)} \in \Sigma_{N-1}.$$

By passing to the limit in this inequality, we obtain that

$$\sum_{i,j=1}^{N} \frac{\partial^2 u}{\partial x_i \partial x_j}(x)\xi_i\xi_j \leq 0 \quad \text{for all } x \in \widetilde{M} \text{ and all } \xi \in \Sigma_{N-1}.$$

Summing up, we have proved [13, Proposition 1] that, for every neighborhood $U(x_0)$ of x_0 there exists a subset $\widetilde{M}$ of $U(x_0)$, with $\mu(\widetilde{M}) > 0$, such that the Hessian matrix

$$\left(\frac{\partial^2 u}{\partial x_i \partial x_j}(x)\right)$$

is negative semi-definite for all $x \in \widetilde{M}$.

On the other hand, condition (8.1) implies that the matrix $(a^{ij}(x))$ is positive definite for almost all $x \in \Omega$.

Therefore, for every neighborhood $U(x_0)$ of x_0 we can find a subset M of $U(x_0)$, with $\mu(M) > 0$, such that

$$\sum_{i,j=1}^{N} a^{ij}(x)\frac{\partial^2 u}{\partial x_i \partial x_j}(x) \leq 0 \quad \text{for all } x \in M.$$

The proof of Lemma 8.5 is complete. □

Step (II) (End of Proof of Theorem 8.1): We divide the proof of Theorem 8.1 into Case II-1 and Case II-2.

Case II-1: First, we consider the case where

$$A_0 = \sum_{i,j=1}^{N} a^{ij}(x)\frac{\partial^2}{\partial x_i \partial x_j} + \sum_{i=1}^{N} b^i(x)\frac{\partial}{\partial x_i}.$$

(1) We assume that a function $u \in W^{2,p}(\Omega)$, with $N < p < \infty$, satisfies the condition

$$A_0 u(x) \geq 0 \quad \text{for almost all } x \in \Omega, \tag{8.15}$$

and takes a local maximum at an interior point $x_0 \in \Omega$. Then we have, by Lemma 8.2,

$$\lim \operatorname{ess\,inf}_{x \to x_0} \sum_{i,j=1}^{N} a^{ij}(x) \frac{\partial^2 u}{\partial x_i \partial x_j}(x) \leq 0. \tag{8.16}$$

However, it should be noticed that

$$\begin{aligned} &b^i \in L^\infty(\Omega), \\ &\frac{\partial u}{\partial x_i}(x_0) = 0, \quad 1 \leq i \leq N. \end{aligned}$$

Therefore, we obtain from assertion (8.16) that

$$\lim \operatorname{ess\,inf}_{x \to x_0} A_0 u(x) \leq 0. \tag{8.17}$$

(2) We choose a positive constant γ so large that

$$\gamma > \lambda \left\| b^1 \right\|_{L^\infty(\Omega)}, \tag{8.18}$$

and let, for any $\varepsilon > 0$,

$$u_\varepsilon(x) := u(x) + \varepsilon e^{\gamma x_1}.$$

Then, by conditions (8.1), (8.15) and (8.18) we can find a positive constant $\eta = \eta(\varepsilon)$ such that

$$\begin{aligned} A_0 u_\varepsilon(x) &= A_0 u(x) + \varepsilon \left[a^{11}(x)\gamma^2 + b^1(x)\gamma \right] e^{\gamma x_1} \\ &\geq \varepsilon \left[a^{11}(x)\gamma^2 + b^1(x)\gamma \right] e^{\gamma x_1} > \varepsilon \left[\frac{\gamma^2}{\lambda} - \left\| b^1 \right\|_{L^\infty(\Omega)} \gamma \right] e^{\gamma x_1} \\ &\geq \eta(\varepsilon) > 0 \quad \text{for almost all } x \in \Omega. \end{aligned}$$

In view of inequality (8.17), this proves that the function $u_\varepsilon(x)$ does not take a maximum at any interior point of Ω. Hence we have the inequality

$$u_\varepsilon(x) = u(x) + \varepsilon e^{\gamma x_1} \leq \max_{y \in \partial\Omega} \left(u(y) \right) + \varepsilon e^{\gamma y_1} \Big) \quad \text{for all } x \in \overline{\Omega}. \tag{8.19}$$

Therefore, by letting $\varepsilon \downarrow 0$ in inequality (8.19) we obtain that

$$u(x) \leq \max_{y \in \partial\Omega} u(y) \quad \text{for all } x \in \overline{\Omega},$$

so that

$$\max_{\overline{\Omega}} u = \max_{\partial\Omega} u. \tag{8.20}$$

Here it should be emphasized that we do not make any regularity assumption on the boundary $\partial\Omega$.

Case II-2: Secondly, we consider the general case where

$$A = \sum_{i,j=1}^{N} a^{ij}(x) \frac{\partial^2}{\partial x_i \partial x_j} + \sum_{i=1}^{N} b^i(x) \frac{\partial}{\partial x_i} + c(x).$$

Here we recall that $c \in L^\infty(\Omega)$ and $c(x) \le 0$ for almost all $x \in \Omega$.

We assume that a function $u \in W^{2,p}(\Omega)$, with $N < p < \infty$, satisfies the condition

$$Au(x) \ge 0 \quad \text{for almost all } x \in \Omega.$$

(1) If $u(x) \le 0$ in Ω, then it follows that

$$\max_{\overline{\Omega}} u \le 0 \le \max_{\partial\Omega} u^+.$$

(2) We consider the case where the subdomain

$$\Omega^+ = \{x \in \Omega : u(x) > 0\}$$

is not empty. If we define a differential operator

$$A_0 := A - c(x) = \sum_{i,j=1}^{N} a^{ij}(x) \frac{\partial^2}{\partial x_i \partial x_j} + \sum_{i=1}^{N} b^i(x) \frac{\partial}{\partial x_i},$$

then it follows that

$$A_0 u(x) = Au(x) - c(x)u(x) \ge -c(x)u(x) \ge 0 \quad \text{for almost all } x \in \Omega^+.$$

Hence, by applying assertion (8.20) to our situation we obtain (see Fig. 8.2) that

$$\max_{\overline{\Omega}} u = \max_{\overline{\Omega^+}} u = \max_{\partial\Omega^+} u = \max_{\partial\Omega} u^+,$$

where

$$\partial\Omega^+ = \{x \in \Omega : u(x) = 0\} \cup \left\{x' \in \partial\Omega : u(x') \ge 0\right\}.$$

Summing up, we have proved that

$$\max_{x \in \overline{\Omega}} u \le \max_{x' \in \partial\Omega} u^+(x').$$

Now the proof of Theorem 8.1 is complete. □

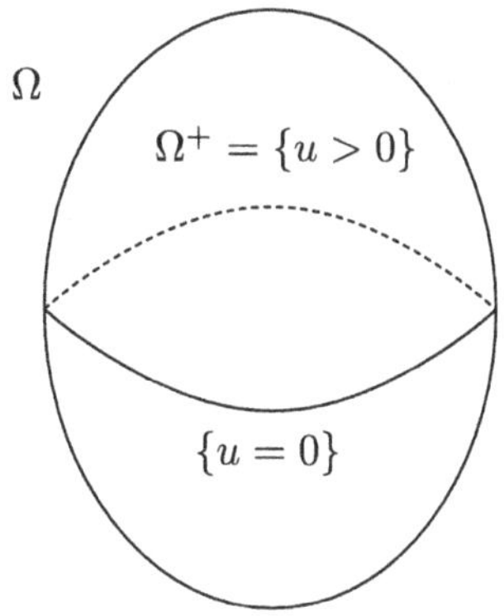

Fig. 8.2 The subdomain $\Omega^+ = \{u > 0\}$ of Ω with boundary $\partial\Omega^+$

8.2 Hopf's Boundary Point Lemma

In this section, we study the inward normal derivative $\partial u/\partial \mathbf{n}(x_0')$ at a boundary point x_0' where the function $u(x)$ takes its non-negative maximum.

The Hopf boundary point lemma reads as follows:

Lemma 8.8 (Hopf) *Assume that a function $u \in W^{2,p}(\Omega)$, with $N < p < \infty$, satisfies the condition*

$$Au(x) \geq 0 \quad \textit{for almost all } x \in \Omega. \tag{8.2}$$

If $u(x)$ attains a non-negative, strict local maximum at a point x_0' of $\partial\Omega$, then we have the inequality

$$\frac{\partial u}{\partial \mathbf{n}}(x_0') < 0, \tag{8.21}$$

where $\mathbf{n} = (n_1, n_2, \ldots, n_N)$ is the unit inward normal to the boundary $\partial\Omega$ (see Fig. 8.3).

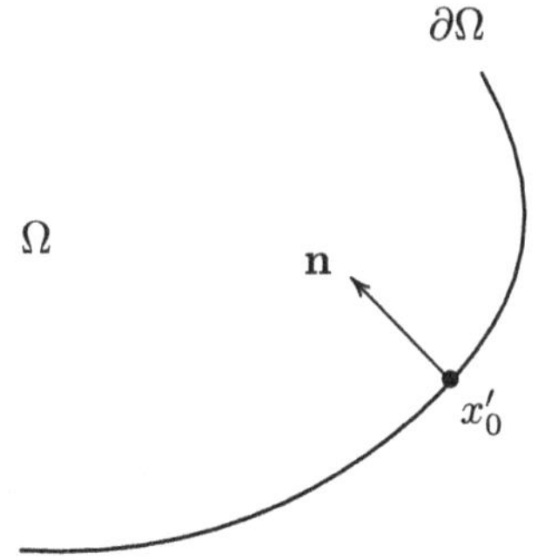

Fig. 8.3 The unit inward normal $\mathbf{n}$ to the boundary $\partial\Omega$ at x_0'

Proof By Theorem 8.1, it suffices to consider the case

$$\begin{cases} u(x_0') = m := \max_{x\in\overline{\Omega}} u(x) \geq 0, \\ u(y) < u(x_0') \quad \text{for all } y \in \Omega. \end{cases} \tag{8.22}$$

The proof of inequality (8.21) is divided into three steps.

Step 1: By condition (8.22), we can find an open ball $B(y, r)$ contained in the domain Ω, centered at y, such that (see Fig. 8.4):

(a) The point x_0' is on the boundary $S(y, r) = \{z \in \Omega : |z - y| = r\}$ of $B(y, r)$;
(b) $\mathbf{n} = s(y - x_0')$ for some $s > 0$.

Step 2: Near the boundary point x_0', we introduce local coordinate systems (x', x_N) such that $x' = (x_1, x_2, \ldots, x_{N-1})$ give local coordinates for the boundary $\partial\Omega$ and that (see Fig. 8.5)

$$\begin{aligned} \Omega &= \{(x', x_N) : x_N > 0\}, \\ \partial\Omega &= \{(x', x_N) : x_N = 0\}, \\ x_0' &= (0, \ldots, 0, 0), \\ y &= (0, \ldots, 0, r). \end{aligned}$$

Now we introduce a function $v(x)$ by the formula

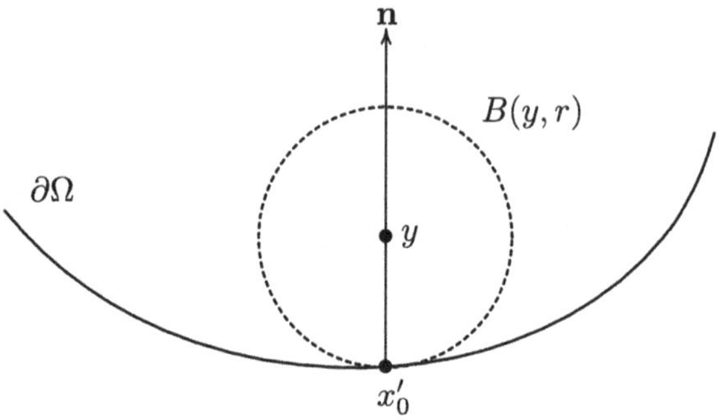

Fig. 8.4 The open ball $B(y, r)$ contained in the domain Ω centered at y

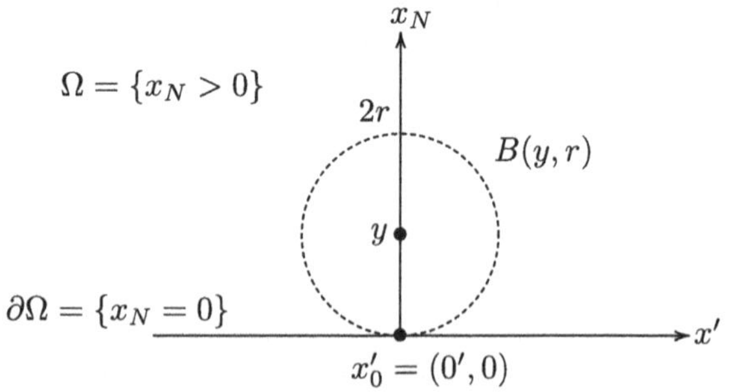

Fig. 8.5 The local coordinate systems (x', x_N) such that $\Omega = \{x_N > 0\}$ and $\partial\Omega = \{x_N = 0\}$

$$v(x) = v(x', x_N) = \exp\left[-\gamma|x-y|^2\right] - \exp\left[-\gamma R^2\right], \qquad (8.23)$$
$$0 < R < \frac{1}{2},$$

where γ is a positive constant to be chosen later on. Then it is easy to see that

$$\begin{aligned} Av(x) &= \sum_{i,j=1}^{N} a^{ij}(x)\frac{\partial^2 v}{\partial x_i \partial x_j}(x) + \sum_{i=1}^{N} b^i(x)\frac{\partial v}{\partial x_i}(x) + c(x)v(x) \qquad (8.24)\\ &= \exp\left[-\gamma|x-y|^2\right] \times \left[4\gamma^2\left(\sum_{i,j=1}^{N} a^{ij}(x)(x_i-y_i)(x_j-y_j)\right)\right.\\ &\quad \left. - 2\gamma\left(\sum_{i=1}^{N}\left(a^{ii}(x) + b^i(x)(x_i-y_i)\right)\right)\right] + c(x)v(x). \end{aligned}$$

However, we have, by formula (8.23) and condition (8.1),

$$v(x) \le \exp\left[-\gamma|x-y|^2\right] \quad \text{for all } x \in \Omega,$$

and

$$\sum_{i,j=1}^{N} a^{ij}(x)(x_i-y_i)(x_j-y_j) \ge \frac{1}{\lambda}|x-y|^2 \quad \text{for almost all } x \in \Omega.$$

Hence, it follows from formula (8.24) that

$$\begin{aligned} Av(x) &\ge \exp\left[-\gamma|x-y|^2\right] \qquad (8.25)\\ &\quad \times \left[\frac{4\gamma^2}{\lambda}|x-y|^2 - 2\gamma\left(\sum_{i=1}^{N}\left(a^{ii}(x) + |b^i(x)(x_i-y_i)|\right) + |c(x)|\right)\right]\\ &\quad \text{for almost all } x \in \Omega. \end{aligned}$$

Moreover, for any $\rho > 0$, we can choose a constant $\gamma = \gamma(\rho)$ so large that we have, for $\rho < |x-y| < R$,

$$\begin{aligned} &\frac{4\gamma^2}{\lambda}|x-y|^2 - 2\gamma\left(\sum_{i=1}^{N}\left(a^{ii}(x) + |b^i(x)(x_i-y_i)|\right) + |c(x)|\right) \qquad (8.26)\\ &\ge \frac{4\rho^2}{\lambda}\gamma^2 - 2\gamma\left(\sum_{i=1}^{N}\left(\|a^{ii}\|_{L^\infty(\Omega)} + |\,\|b^i\|_{L^\infty(\Omega)}\, R\right) + \|c\|_{L^\infty(\Omega)}\right)\\ &\ge 0 \quad \text{for almost all } x \in \Omega. \end{aligned}$$

Therefore, by combining inequalities (8.25) and (8.26) we obtain that, for any $\rho > 0$ there exists a positive constant $\gamma = \gamma(\rho)$ such that (see Fig. 8.6)

$$Av(x) \geq 0 \quad \text{for almost all } x \in \Gamma_{\rho R} := \{z \in \Omega : \rho < |z - y| < R\}. \tag{8.27}$$

Step 3: Without loss of generality, we may assume that, for $0 < R < 1/2$ sufficiently small,

$$u(x) < u(x_0) \quad \text{in } B(y, R). \tag{8.28}$$

If $\varepsilon > 0$, we let

$$w(x) := u(x) - u(x_0) + \varepsilon v(x).$$

(a) First, we have, by condition (8.28),

$$w(x) = u(x) - u(x_0) + \varepsilon v(x) \leq 0 \quad \text{on } S(y, \rho) = \{z \in \Omega : |z - y| = \rho\},$$

if $\varepsilon > 0$ is chosen sufficiently small.

(b) Secondly, it follows that

$$\begin{aligned} w(x) &= u(x) - u(x_0) + \varepsilon v(x) \\ &\leq 0 \quad \text{on } S(y, R) = \{z \in \Omega : |z - y| = R\}, \end{aligned}$$

since $v(x) = 0$ on $S(y, R)$.

Hence we have, by assertions (a) and (b),

$$w(x) \leq 0 \quad \text{on } \partial\Gamma_{\rho R} = S(y, \rho) \cup S(y, R).$$

On the other hand, by conditions (8.2) and (8.27) it follows that

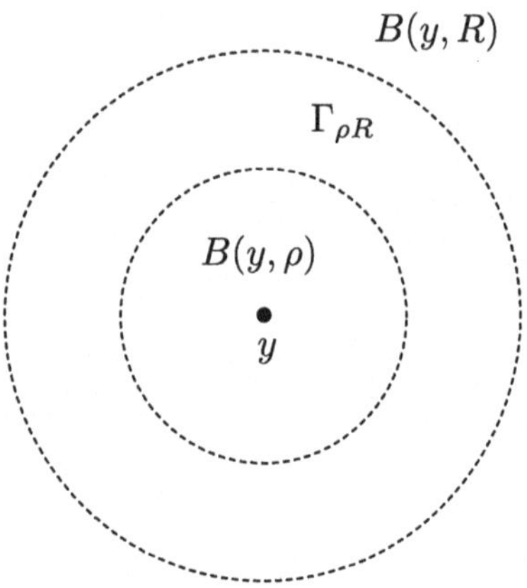

Fig. 8.6 The annular domain $\Gamma_{\rho R} = B(y, R) \setminus \overline{B(y, \rho)}$

$$\begin{aligned} Aw(x) &= Au(x) + \varepsilon Av(x) - c(x)u(x_0) \\ &\geq -c(x)u(x_0) \\ &\geq 0 \quad \text{for almost all } x \in \Gamma_{\rho R} = B(y, R) \setminus \overline{B(y, \rho)}. \end{aligned}$$

Therefore, by applying Theorem 8.1 with $\Omega := \Gamma_{\rho R}$ we obtain that

$$w(x) \leq 0 \quad \text{in } \Gamma_{\rho R} = B(y, R) \setminus \overline{B(y, \rho)}. \tag{8.29}$$

(c) On the other hand, we have, by formula (8.19),

$$w(x_0) = \varepsilon v(x_0) = 0. \tag{8.30}$$

Therefore, it follows from assertions (8.29) and (8.30) that

$$\frac{\partial w}{\partial \mathbf{n}}(x_0') = \frac{\partial u}{\partial \mathbf{n}}(x_0') + \varepsilon \frac{\partial v}{\partial \mathbf{n}}(x_0') \leq 0. \tag{8.31}$$

However, we have, by formula (8.23),

$$\frac{\partial v}{\partial \mathbf{n}}(x_0') = 2\gamma r e^{-\gamma r^2} > 0. \tag{8.32}$$

By combining inequalities (8.31) and (8.32), we obtain that

$$\frac{\partial u}{\partial \mathbf{n}}(x_0') \leq -\varepsilon \frac{\partial v}{\partial \mathbf{n}}(x_0') = -2\varepsilon\gamma r e^{-\gamma r^2} < 0.$$

Now the proof of Lemma 8.8 is complete. □

8.3 Strong Maximum Principle

Finally, we can prove the following strong maximum principle for the operator A [13, Théorème 2]:

Theorem 8.9 (the strong maximum principle) *Assume that a function $u \in W^{2,p}(\Omega)$, with $N < p < \infty$, satisfies the condition*

$$Au(x) \geq 0 \quad \textit{for almost all } x \in \Omega. \tag{8.2}$$

If $u(x)$ attains a non-negative maximum at an interior point of Ω, then it is a (non-negative) constant *function.*

Proof Our proof is based on a reduction to absurdity. We let

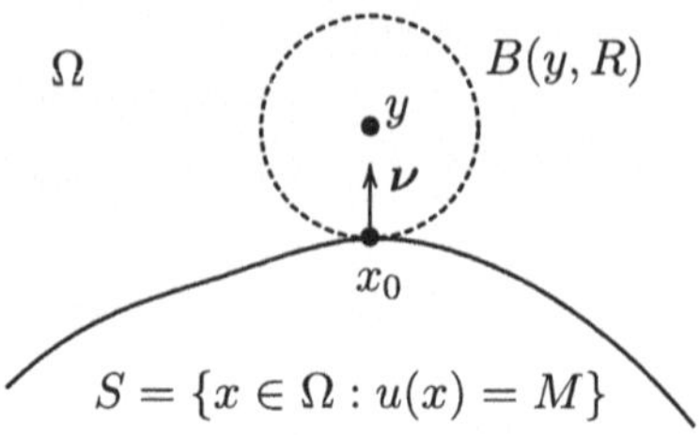

Fig. 8.7 The open ball $B(y, R)$ contained in the set $\Omega \setminus S$, centered at y

$$m := \max_{x\in\overline{\Omega}} u(x) \geq 0,$$
$$S := \{x \in \Omega : u(x) = m\},$$

and assume, to the contrary, that

$$S \subsetneqq \Omega.$$

Since S is closed in Ω, we can find a point x_0 of S and an open ball $B(y, R)$ contained in the set $\Omega \setminus S$, centered at y, such that (see Fig. 8.7):

(a) $B(y, R) \subset \Omega \setminus S$;
(b) x_0 is on the boundary $S(y, R) = \{z \in \Omega : |z - y| = R\}$ of $B(y, R)$.

By applying Hopf's boundary point lemma (Lemma 8.8) with $\Omega := B(y, R)$, we obtain that

$$\sum_{i=1}^{N} \nu_i \frac{\partial u}{\partial x_i}(x_0) < 0, \tag{8.33}$$

where

$$\nu = \frac{y - x_0}{|y - x_0|}.$$

However, since $u(x_0) = m$ for some interior point $x_0 \in \Omega$, it follows that

$$\frac{\partial u}{\partial x_i}(x_0) = 0 \quad \text{for } 1 \leq i \leq N.$$

Hence we have the assertion

$$\sum_{i=1}^{N} \nu_i \frac{\partial u}{\partial x_i}(x_0) = 0.$$

This contradicts condition (8.33).

The proof of Theorem 8.9 is complete. □

Part III
Theory of Singular Integral Operators

Chapter 9
Elements of Singular Integrals

The Calderón–Zygmund theory of singular integral operators is a very refined mathematical tool whose full power is yet to be exploited (see Stein [140], Meyer [102]). This chapter is devoted to a careful and accessible exposition of the most elementary part of the Calderón–Zygmund theory. We present a straightforward treatment of the Calderón–Zygmund theory necessary for the study of elliptic boundary value problems, assuming only basic knowledge of real analysis and functional analysis. The results discussed here are adapted from Calderón–Zygmund [22–25, 28] and Tanabe [166, 167].

In Sect. 9.1 we formulate precisely the notion of singular integrals of Alberto Pedro Calderón and Antoni Zygmund (Lemma 9.1). In Sect. 9.2 we study the case where the integral kernel $K(x)$ is a bounded function that satisfies Assumption 9.2 (Theorem 9.3). In Sect. 9.3 we study the case where the integral kernel $K(x)$ is a continuous function that satisfies Assumption 9.3 (Theorem 9.6). The proof of Theorem 9.6 is based on a version of the Calderón–Zygmund decomposition adapted to the present context (Lemma 9.4). In Sect. 9.5, we introduce a non-negative, measurable function $f^*(t)$ defined on the interval $[0, \infty)$ for a given non-negative, measurable function $f(x)$ on $\mathbf{R}^n$ such that

$$\int_{\mathbf{R}^n} f(x)^p \, dx = \int_0^\infty f^*(t)^p \, dt.$$

The function $f^*(t)$ is called an *equimeasurable function* of $f(x)$ (Lemmas 9.9 and 9.10).

Sections 9.4 and 9.6 are devoted to the basic theory of the Hilbert transform H that is a special case of the singular integral of a single independent variable (Theorems 9.7 and 9.15):

$$Hf(x) := \frac{1}{\pi} \lim_{\varepsilon \downarrow 0} \int_{|x-t|>\varepsilon} \frac{f(t)}{x-t} \, dt = \frac{1}{\pi} \lim_{\varepsilon \downarrow 0} \int_{|s|>\varepsilon} f(x-s) \, \frac{ds}{s}.$$

K. Taira, *Real Analysis Methods for Markov Processes*,
https://doi.org/10.1007/978-981-97-3659-1_9

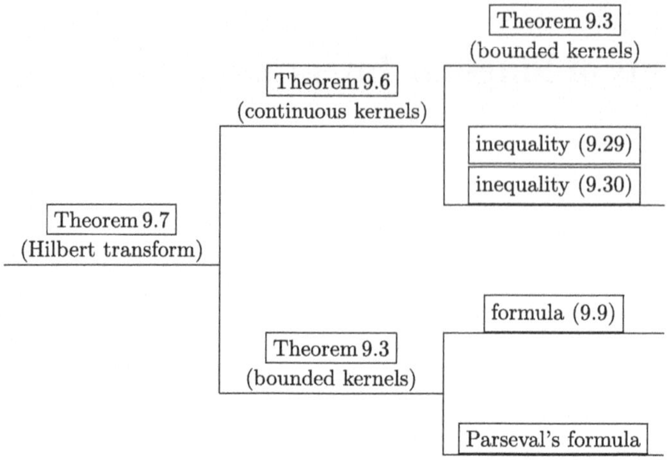

Fig. 9.1 A flowchart for the proof of Theorem 9.7

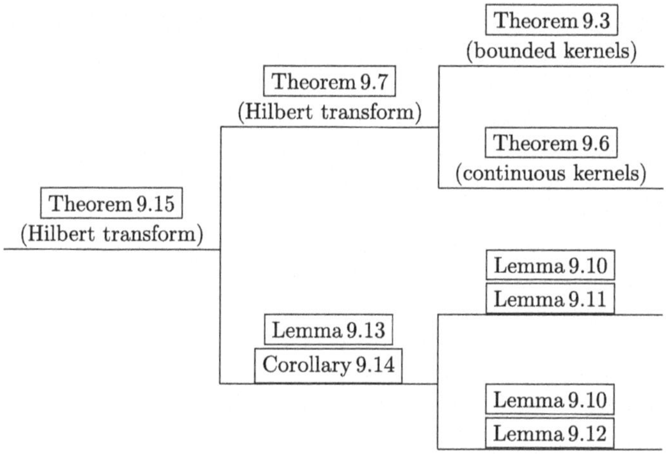

Fig. 9.2 A flowchart for the proof of Theorem 9.15

The proofs of Theorems 9.7 and 9.15 are flowcharted in Figs. 9.1 and 9.2 above.

In Sect. 9.7 we study the case where the integral kernel $K(x)$ is an *odd* function that satisfies Assumption 9.4 (Theorem 9.16). The proof of Theorem 9.16 can be reduced to the study of the Hilbert transform H (Theorem 9.15). More precisely, we have the following formula for the odd kernel $K(x)$ (see formula (9.86)):

$$
\begin{aligned}
K * f(x) &:= \lim_{\varepsilon \downarrow 0} \int_{|x-y|>\varepsilon} K(x-y)\, f(y)\, dy \\
&= \frac{1}{2} \int_{\Sigma_{n-1}} K(\sigma) \left(\lim_{\varepsilon \downarrow 0} \int_{|t|>\varepsilon} f(x - t\sigma) \frac{dt}{t} \right) d\sigma.
\end{aligned}
$$

Section 9.8 is devoted to the study of Riesz kernels (Theorem 9.17)

$$
R_j(x) = -\frac{\Gamma((n+1)/2)}{\pi^{(n+1)/2}} \frac{x_j}{|x|^{n+1}}, \quad 1 \leq j \leq n.
$$

It is easy to see that the Riesz kernels $R_j(x)$ satisfy Assumptions 9.3 and 9.4.

In Sect. 9.9 we study the case where the integral kernel $K(x)$ is an *even* function that satisfies Assumption 9.5 (Theorem 9.26). The case of even kernels can be reduced to the case of odd kernels. In fact, we have the following decomposition formula:

$$
K * f = -\sum_{j=1}^{n} R_j * (R_j * K) * f,
$$

where the operators R_j and $R_j * K$ have respectively the *odd kernels*

$$
\begin{aligned}
R_j(x) &= -\frac{\Gamma((n+1)/2)}{\pi^{(n+1)/2}} \frac{x_j}{|x|^{n+1}}, \\
R_j * K(x) &= -\frac{\Gamma((n+1)/2)}{\pi^{(n+1)/2}} \frac{x_j}{|x|^{n+1}} * K(x).
\end{aligned}
$$

This decomposition formula may be rephrased as follows:

$$
\{\text{Even kernels}\} = \{\text{Riesz kernels}\} * \{\text{Odd kernels}\}.
$$

Finally, in Sect. 9.10 we prove the existence of the singular integral in the general case (Theorem 9.27). The proof of Theorem 9.27 is flowcharted in Fig. 9.3 below.

9.1 Singular Integrals of Calderón and Zygmund

Let $K(x)$ be a real-valued, Lebesgue measurable function defined on $\mathbf{R}^n$. If $f \in L^p(\mathbf{R}^n)$ with $1 < p < \infty$, then it follows that the usual Lebesgue integral

$$
\int_{\mathbf{R}^n} K(x-y) f(y)\, dy \tag{9.1}
$$

does not exist. However, under suitable conditions on the integral kernel $K(x)$, we can prove that the principal value of the integral (9.1)

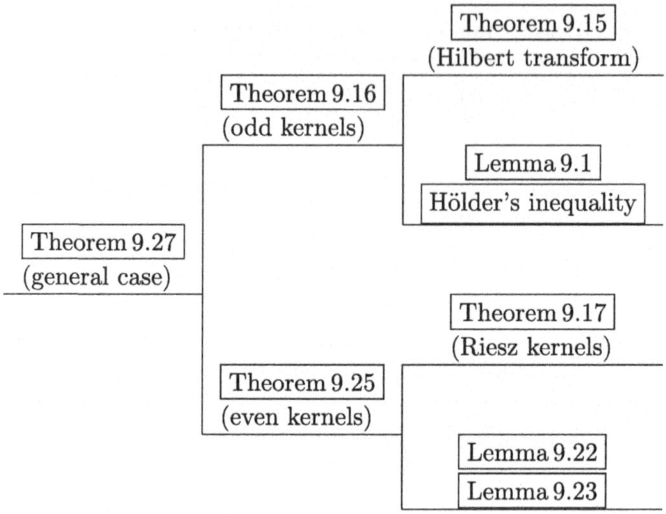

Fig. 9.3 A flowchart for the proof of Theorem 9.27

$$\text{v.p.} \int_{\mathbf{R}^n} K(x-y) f(y)\, dy = \lim_{\varepsilon \downarrow 0} \int_{|x-y|>\varepsilon} K(x-y)\, f(y)\, dy \tag{9.2}$$

exist. Here "v.p." stands for Cauchy's "valeur principale" in French. The integrals defined as formula (9.2) are called *singular integrals*. In what follows we shall use the notation

$$\boxed{K * f(x) = \lim_{\varepsilon \downarrow 0} \int_{|x-y|>\varepsilon} K(x-y)\, f(y)\, dy.}$$

Throughout this chapter, we consider the case where the integral kernel $K(x)$ in formula (9.2) satisfies the following assumption:

Assumption 9.1 The integral kernel $K(x)$ is positively homogeneous of degree $-n$, that is,

$$K(\lambda x) = \lambda^{-n} K(x) \quad \text{for all } \lambda > 0 \text{ and all } x \in \mathbf{R}^n, \tag{9.3}$$

and further it is integrable on the unit sphere

$$\Sigma_{n-1} = \{x \in \mathbf{R}^n : |x| = 1\}$$

and satisfies the *cancellation property*

$$\int_{\Sigma_{n-1}} K(\sigma)\, d\sigma = 0, \tag{9.4}$$

where $d\sigma$ is the surface measure on Σ_{n-1}.

In this chapter we shall denote the Lebesgue measure of a subset A of $\mathbf{R}^n$ by $|A|$ and the norm of the space $L^p(\mathbf{R}^n)$ by $\|\cdot\|_p$, respectively.

We begin by proving the following fundamental result:

Lemma 9.1 *Let $K(x)$ be an integral kernel satisfying Assumption 9.1. If $f \in L^p(\mathbf{R}^n)$ for $1 < p < \infty$ and $\varepsilon > 0$, then the Lebesgue integral*

$$\widetilde{f}_\varepsilon(x) := \int_{|x-y|>\varepsilon} K(x-y)\, f(y)\, dy \tag{9.5}$$

exists for almost all $x \in \mathbf{R}^n$.

Proof First, we have, by a suitable change of variables,

$$\begin{aligned}\widetilde{f}_\varepsilon(x) &= \int_{|y|>\varepsilon} K(y) f(x-y)\, dy \\ &= \int_{\Sigma_{n-1}} \left(\int_\varepsilon^\infty K(t\sigma) f(x-t\sigma) t^{n-1}\, dt \right) d\sigma \\ &= \int_{\Sigma_{n-1}} K(\sigma) \left(\int_\varepsilon^\infty f(x-t\sigma) \frac{dt}{t} \right) d\sigma.\end{aligned}$$

Hence it suffices to show that, for any bounded measurable subset S of $\mathbf{R}^n$, the integral

$$\int_S \left(\int_{\Sigma_{n-1}} |K(\sigma)| \left(\int_\varepsilon^\infty |f(x-t\sigma)| \frac{dt}{t} \right) d\sigma \right) dx$$

is finite. More precisely, we show that there exists a constant $C_{\varepsilon,S} > 0$ such that

$$\begin{aligned}&\int_S \left(\int_{\Sigma_{n-1}} |K(\sigma)| \left(\int_\varepsilon^\infty |f(x-t\sigma)| \frac{dt}{t} \right) d\sigma \right) dx \\ &\leq C_{\varepsilon,S} \, \|f\|_p \left(\int_{\Sigma_{n-1}} |K(\sigma)|\, d\sigma \right).\end{aligned} \tag{9.6}$$

By using Hölder's inequality (Theorem 3.24), we obtain that

$$\begin{aligned}&\int_S \left(\int_\varepsilon^\infty |f(x-t\sigma)| \frac{dt}{t} \right) dx \\ &\leq C_{\varepsilon,p} \int_S \left(\int_{-\infty}^\infty |f(x-t\sigma)|^p\, dt \right)^{1/p} dx \\ &\leq C_{\varepsilon,p} |S|^{1-1/p} \left(\int_S \int_{-\infty}^\infty |f(x-t\sigma)|^p\, dt\, dx \right)^{1/p},\end{aligned} \tag{9.7}$$

where $C_{\varepsilon,p}$ is a positive constant given by the formula

$$C_{\varepsilon,p} := (p-1)^{(p-1)/p} \varepsilon^{-1/p}.$$

For each $\sigma \in \Sigma_{n-1}$, we make a change of variables

$$x := z - s\,\sigma, \quad z \in \mathbf{R}^n, \; s \in \mathbf{R},$$

with

$$\langle z, \sigma \rangle = z_1\sigma_1 + \ldots + z_n\sigma_n = 0.$$

Moreover, we choose a subset S' of $\mathbf{R}^{n-1}$ and two real numbers a, b such that

$$S \subset \{x = z - s\sigma : z \in S', a < s < b\}.$$

Then the last term on inequality (9.7) can be estimated as follows:

$$\begin{aligned}
& C_{\varepsilon,p}|S|^{1-1/p} \left(\int_S \int_{-\infty}^{\infty} |f(x - t\,\sigma)|^p \, dt \, dx \right)^{1/p} \\
& \leq C_{\varepsilon,p}|S|^{1-1/p} \left(\int_a^b \int_{S'} \int_{-\infty}^{\infty} |f(z - (t+s)\sigma)|^p \, dt \, dz \, ds \right)^{1-p} \\
& \leq C_{\varepsilon,p}|S|^{1-1/p} \left((b-a) \int_{\mathbf{R}^n} |f(x)|^p \, dx \right)^{1/p} \\
& = C_{\varepsilon,p}\,|S|^{1-1/p}\,(b-a)^{1/p}\,\|f\|_p.
\end{aligned}$$

Here it should be noticed that the two numbers a, b can be chosen independent of $\sigma \in \Sigma_{n-1}$.

Therefore, we obtain that

$$\begin{aligned}
& \int_S \left(\int_{\Sigma_{n-1}} |K(\sigma)| \left(\int_{\varepsilon}^{\infty} |f(x - t\sigma)| \frac{dt}{t} \right) d\sigma \right) dx \\
& \leq C_{\varepsilon,p}\,|S|^{1-1/p}\,(b-a)^{1/p}\,\|f\|_p \left(\int_{\Sigma_{n-1}} |K(\sigma)| \, d\sigma \right).
\end{aligned}$$

This proves the desired inequality (9.6) with

$$C_{\varepsilon,S} := C_{\varepsilon,p}|S|^{1-1/p}(b-a)^{1/p}.$$

The proof of Lemma 9.1 is complete. □

Remark 9.2 We remark that the cancellation property (9.4) is not used in the proof of Lemma 9.1.

9.2 The Case of Bounded Kernels

In this section we consider the case where the integral kernel $K(x)$ satisfies the following assumption:

Assumption 9.2 The integral kernel $K(x)$ satisfies Assumption 9.1 and is *bounded* on the unit sphere Σ_{n-1}.

The purpose of this section is to prove the existence of the singular integral (9.2) in the space $L^2(\mathbf{R}^n)$ in the case of *bounded kernels*:

Theorem 9.3 *Let $K(x)$ be an integral kernel satisfying Assumption 9.2. If $f \in L^2(\mathbf{R}^n)$ and if $\varepsilon > 0$, we let*

$$\widetilde{f_\varepsilon}(x) := \int_{|x-y|>\varepsilon} K(x-y)\, f(y)\, dy.$$

Then we have the following four assertions:

(i) There exists a constant C, independent of ε, such that

$$\left\| \widetilde{f_\varepsilon} \right\|_2 \leq C \, \|f\|_2.$$

(ii) The sequence $\widetilde{f_\varepsilon}$ converges strongly to a function $K * f$ in the space $L^2(\mathbf{R}^n)$ as $\varepsilon \downarrow 0$. Namely, the singular integral

$$K * f(x) = \lim_{\varepsilon \downarrow 0} \widetilde{f_\varepsilon}(x) = \lim_{\varepsilon \downarrow 0} \int_{|x-y|>\varepsilon} K(x-y)\, f(y)\, dy$$

exists in the strong topology of $L^2(\mathbf{R}^n)$.

(iii) The mapping

$$\begin{aligned} L^2(\mathbf{R}^n) &\longrightarrow L^2(\mathbf{R}^n) \\ f &\longmapsto K * f \end{aligned}$$

is a bounded linear operator from $L^2(\mathbf{R}^n)$ into itself.

(iv) The Fourier transform $\mathcal{F}(K * f)$ of $K * f$ is given by the formula

$$\mathcal{F}(K * f)(\xi) = \mathcal{F}K(\xi) \cdot \mathcal{F}f(\xi), \tag{9.8}$$

where $\mathcal{F}K(\xi)$ is an essentially bounded, measurable function defined by the formula

$$\mathcal{F}K(\xi) := \int_{\Sigma_{n-1}} K(\sigma) \left(\int_0^\infty \left(e^{-is\frac{\sigma\cdot\xi}{|\xi|}} - e^{-s} \right) \frac{ds}{s} \right) d\sigma \quad \text{for } \xi \neq 0, \tag{9.9}$$

and is a positively homogeneous function of degree zero.

Proof For $0 < \varepsilon < \mu$, we introduce two functions

$$K_{\varepsilon,\mu}(x) := \begin{cases} K(x) & \text{if} \, \varepsilon < |x| < \mu, \\ 0 & \text{otherwise}, \end{cases} \tag{9.10}$$

and

$$K_{\varepsilon}(x) := \begin{cases} K(x) & \text{if} \, |x| > \varepsilon, \\ 0 & \text{if} |x| \leq \varepsilon. \end{cases} \tag{9.11}$$

Then it follows that

$$\widetilde{f}_{\varepsilon}(x) = \int_{\mathbf{R}^n} K_{\varepsilon}(x - y) \, f(y) \, dy = (K_{\varepsilon} * f)(x).$$

The proof of Theorem 9.3 is divided into four steps.

Step (I): First, we show that the sequence $K_{\varepsilon,\mu}$ converges strongly to the function K_{ε} in $L^2(\mathbf{R}^n)$ as $\mu \uparrow \infty$:

$$\lim_{\mu \uparrow \infty} K_{\varepsilon,\mu} = K_{\varepsilon} \quad \text{in} \, L^2(\mathbf{R}^n).$$

Indeed, it suffices to note that

$$\begin{aligned} \int_{\mathbf{R}^n} \left| K_{\varepsilon,\mu}(x) - K_{\varepsilon}(x) \right|^2 dx &= \int_{|x|>\mu} |K(x)|^2 \, dx \\ &= \int_{\mu}^{\infty} \int_{\Sigma_{n-1}} |K(r\sigma)|^2 \, r^{n-1} \, dr \, d\sigma \\ &= \left(\int_{\Sigma_{n-1}} |K(\sigma)|^2 \, d\sigma \right) \int_{\mu}^{\infty} \frac{1}{r^{n+1}} \, dr \\ &= \frac{1}{n} \left(\int_{\Sigma_{n-1}} |K(\sigma)|^2 \, d\sigma \right) \frac{1}{\mu^n} \longrightarrow 0 \quad \text{as} \, \mu \uparrow \infty. \end{aligned}$$

Therefore, it follows from an application of Parseval's formula that the sequence $\mathcal{F} K_{\varepsilon,\mu}$ converges strongly to the function $\mathcal{F} K_{\varepsilon}$ in $L^2(\mathbf{R}^n)$ as $\mu \uparrow \infty$:

$$\lim_{\mu \uparrow \infty} \mathcal{F} K_{\varepsilon,\mu} = \mathcal{F} K_{\varepsilon} \quad \text{in} \; L^2(\mathbf{R}^n). \tag{9.12}$$

Step (II): Secondly, we calculate explicitly the Fourier transform of $K_{\varepsilon,\mu}(x)$, which is an essential step in the proof of Theorem 9.3. To do this, we remark that

$$K_{\varepsilon,\mu} \in L^1(\mathbf{R}^n).$$

We write, for $x \in \mathbf{R}^n$ and $\xi \in \mathbf{R}^n$,

$$x = r\sigma, \quad |x| = r, \quad \sigma \in \Sigma_{n-1},$$
$$|\xi| = \rho, \quad x \cdot \xi = \langle x, \xi \rangle = r\rho \cos\phi, \quad \xi \neq 0.$$

Then, by using two conditions (9.3) and (9.4) (the cancellation property) we obtain from formula (9.10) that

$$\begin{aligned}
(\mathcal{F}K_{\varepsilon,\mu})(\xi) &= \int_{\mathbf{R}^n} K_{\varepsilon,\mu}(x)\, e^{-ix\cdot\xi}\, dx \\
&= \int_{\varepsilon<|x|<\mu} K(x)\, e^{-ix\cdot\xi}\, dx \\
&= \int_{\varepsilon}^{\mu} \left(\int_{\Sigma_{n-1}} K(\sigma)\, e^{-ir\rho\cos\phi}\, d\sigma \right) \frac{dr}{r} \\
&= \int_{\varepsilon\rho}^{\mu\rho} \left(\int_{\Sigma_{n-1}} K(\sigma)\, e^{-is\cos\phi}\, d\sigma \right) \frac{ds}{s} \\
&= \int_{\varepsilon\rho}^{\mu\rho} \left(\int_{\Sigma_{n-1}} K(\sigma) \left(e^{-is\cos\phi} - e^{-s} \right) d\sigma \right) \frac{ds}{s} \\
&= \int_{\Sigma_{n-1}} K(\sigma) \left(\int_{\varepsilon\rho}^{\mu\rho} \left(e^{-is\cos\phi} - e^{-s} \right) \frac{ds}{s} \right) d\sigma, \quad \rho = |\xi| \neq 0.
\end{aligned} \tag{9.13}$$

We decompose the integral

$$\int_{\varepsilon\rho}^{\mu\rho} \left(e^{-is\cos\phi} - e^{-s} \right) \frac{ds}{s} \quad (\rho \neq 0)$$

into the two terms:

$$\int_{\varepsilon\rho}^{\mu\rho} \left(e^{-is\cos\phi} - e^{-s} \right) \frac{ds}{s} := \mathrm{I}_1(\varepsilon, \mu) + i\, \mathrm{I}_2(\varepsilon, \mu), \tag{9.14}$$

where

$$\mathrm{I}_1(\varepsilon, \mu) := \int_{\varepsilon\rho}^{\mu\rho} \left(\cos(s\cos\phi) - e^{-s} \right) \frac{ds}{s},$$
$$\mathrm{I}_2(\varepsilon, \mu) := -\int_{\varepsilon\rho}^{\mu\rho} \sin(s\cos\phi)\, \frac{ds}{s}.$$

Step (1): First, we have, for $0 < a < b$,

$$\begin{aligned}\int_a^b \frac{\cos s}{s}\, ds &= \int_{a+\pi/2}^{b+\pi/2} \frac{\cos(s-\pi/2)}{s-\pi/2}\, ds \\ &= \int_{a+\pi/2}^{b+\pi/2} \frac{\sin s}{s-\pi/2}\, ds \\ &= \int_{a+\pi/2}^{b+\pi/2} \frac{\sin s}{s}\, ds + \int_{a+\pi/2}^{b+\pi/2} \left(\frac{1}{s-\pi/2} - \frac{1}{s}\right) \sin s\, ds \\ &= \int_{a+\pi/2}^{b+\pi/2} \frac{\sin s}{s}\, ds + \frac{\pi}{2} \int_a^b \frac{\cos t}{t(t+\pi/2)}\, dt.\end{aligned} \tag{9.15}$$

Since the improper integral

$$\lim_{b\uparrow\infty} \int_0^b \frac{\sin x}{x}\, dx$$

exists, it follows from formula (9.15) that the improper integral

$$\lim_{b\uparrow\infty} \int_a^b \frac{\cos s}{s}\, ds$$

also exists. Hence, we find that the improper integral

$$\lim_{\varepsilon\downarrow 0,\mu\uparrow\infty} \int_{\varepsilon\rho}^{\mu\rho} \left(e^{-is\cos\phi} - e^{-s}\right) \frac{ds}{s} \quad (\rho \neq 0)$$

exists for $\cos\phi \neq 0$.

Step (2): Secondly, we prove an estimate on integral (9.14) independent of ε and μ (see inequality (9.19) below).

To do this, we choose a constant $A > 0$ such that we have, for all $a > 0$,

$$\left| \int_0^a \frac{\sin x}{x}\, dx \right| \leq A. \tag{9.16}$$

For example, we may take

$$A := \sup_{a>0} \left| \int_0^a \frac{\sin x}{x}\, dx \right|.$$

Then it is easy to see that

$$\begin{aligned}|\mathrm{I}_2(\varepsilon,\mu)| &= \left| \int_{\varepsilon\rho}^{\mu\rho} \sin(s\cos\phi) \frac{ds}{s} \right| = \left| \int_{\varepsilon\rho\cos\phi}^{\mu\rho\cos\phi} \frac{\sin t}{t}\, dt \right| \\ &\leq 2A.\end{aligned}$$

On the other hand, by combining formula (9.15) and inequality (9.16) we obtain that

$$\begin{aligned}\left|\int_a^b \frac{\cos s}{s}\,ds\right| &\leq 2A + \int_{a+\pi/2}^{b+\pi/2}\left(\frac{1}{s-\pi/2} - \frac{1}{s}\right) ds \\ &= 2A + \log\frac{b(a+\pi/2)}{a(b+\pi/2)} = 2A + \log\left(1 + \frac{\pi(b-a)}{2a(b+\pi/2)}\right) \\ &\leq 2A + \log\left(1 + \frac{\pi}{2a}\right).\end{aligned} \tag{9.17}$$

Step (2-a): We consider the case where $\cos\phi > 0$. In this case, we have the inequality

$$\begin{aligned}&\int_0^1 \left|\cos(s\cos\phi) - e^{-s}\right| \frac{ds}{s} \\ &\leq \int_0^1 (1 - \cos(s\cos\phi))\frac{ds}{s} + \int_0^1 \left(1 - e^{-s}\right)\frac{ds}{s} \\ &\leq \int_0^1 (1 - \cos s)\frac{ds}{s} + \int_0^1 \left(1 - e^{-s}\right)\frac{ds}{s}.\end{aligned} \tag{9.18}$$

(i) If $\mu\rho < 1$, it follows that

$$|\mathrm{I}_1(\varepsilon,\mu)| = \left|\int_{\varepsilon\rho}^{\mu\rho}\left(\cos(s\cos\phi) - e^{-s}\right)\frac{ds}{s}\right| \leq B,$$

where

$$B := \int_0^1 (1 - \cos s)\frac{ds}{s} + \int_0^1 \left(1 - e^{-s}\right)\frac{ds}{s}.$$

(ii) If $\varepsilon\rho > 1$, it follows from inequality (9.17) that

$$\begin{aligned}|\mathrm{I}_1(\varepsilon,\mu)| &= \left|\int_{\varepsilon\rho\cos\phi}^{\mu\rho\cos\phi}\frac{\cos s}{s}\,ds - \int_{\varepsilon\rho}^{\mu\rho} e^{-s}\frac{ds}{s}\right| \\ &\leq 2A + \log\left(1 + \frac{\pi}{2\varepsilon\rho\cos\phi}\right) + \int_1^\infty e^{-s}\frac{ds}{s} \\ &\leq 2A + \log\left(1 + \frac{\pi}{2\cos\phi}\right) + \int_1^\infty e^{-s}\frac{ds}{s}.\end{aligned}$$

(iii) If $\varepsilon\rho < 1 < \mu\rho$, then, by combining two inequalities (9.17) and (9.18) we obtain that

$$\begin{aligned}
|\mathrm{I}_1(\varepsilon,\mu)| &= \left| \left(\int_{\varepsilon\rho}^{1} + \int_{1}^{\varepsilon\rho} \right) \left(\cos(s\cos\phi) - e^{-s} \right) \frac{ds}{s} \right| \\
&\le B + \left| \int_{\cos\phi}^{\mu\rho\cos\phi} \frac{\cos s}{s}\, ds \right| + \int_{1}^{\infty} e^{-s}\, \frac{ds}{s} \\
&\le B + 2A + \log\left(1 + \frac{\pi}{2\cos\phi}\right) + \int_{1}^{\infty} e^{-s}\, \frac{ds}{s}.
\end{aligned}$$

Step (2-b): The case where $\cos\phi < 0$ can be estimated similarly.
Summing up, we can find a constant $C > 0$, independent of ε and μ, such that

$$\begin{aligned}
\left| \int_{\varepsilon|\xi|}^{\mu|\xi|} \left(e^{-is\cos\phi} - e^{-s} \right) \frac{ds}{s} \right| &\le C + \log\left(\frac{1}{|\cos\phi|} \right) \\
&= C + \log\left(\frac{|\xi|}{|\sigma\cdot\xi|} \right) \quad \text{for all } \xi \neq 0.
\end{aligned} \tag{9.19}$$

Therefore, by combining formula (9.13) and inequality (9.19) we obtain that

$$\begin{aligned}
&\left| \left(\mathcal{F}K_{\varepsilon,\mu} \right)(\xi) \right| \\
&\le \int_{\Sigma_{n-1}} \left| K(\sigma) \int_{\varepsilon|\xi|}^{\mu|\xi|} \left(e^{-is\cos\phi} - e^{-s} \right) \frac{ds}{s} \right| d\sigma \\
&\le \left(\sup_{\sigma\in\Sigma_{n-1}} |K(\sigma)| \right) \int_{\Sigma_{n-1}} \left(C + \log\left(\frac{|\xi|}{|\sigma\cdot\xi|} \right) \right) d\sigma \quad \text{for all } \xi \neq 0.
\end{aligned}$$

This proves that there exists a constant $M > 0$, independent of ε anad μ, such that

$$\left| \left(\mathcal{F}K_{\varepsilon,\mu} \right)(\xi) \right| \le M \quad \text{for all} \xi \neq 0. \tag{9.20}$$

On the other hand, it follows from formula (9.13) and inequality (9.19) that we have, for all $\xi \neq 0$,

$$\lim_{\mu\uparrow\infty} \left(\mathcal{F}K_{\varepsilon,\mu} \right)(\xi) = \int_{\Sigma_{n-1}} K(\sigma) \left(\int_{\varepsilon|\xi|}^{\infty} \left(e^{-is\cos\phi} - e^{-s} \right) \frac{ds}{s} \right) d\sigma,$$

and

$$\left| \int_{\varepsilon|\xi|}^{\infty} \left(e^{-is\cos\phi} - e^{-s} \right) \frac{ds}{s} \right| \le C + \log\left(\frac{|\xi|}{|\sigma\cdot\xi|} \right) \quad \text{for all } \xi \neq 0. \tag{9.21}$$

By assertion (9.12), this proves that we have, for almost all $\xi \in \mathbf{R}^n \setminus \{0\}$,

$$\begin{aligned}(\mathcal{F}K_\varepsilon)(\xi) &= \lim_{\mu\uparrow\infty}(\mathcal{F}K_{\varepsilon,\mu})(\xi) \\ &= \int_{\Sigma_{n-1}} K(\sigma)\left(\int_{\varepsilon|\xi|}^{\infty}\left(e^{-is\frac{\sigma\cdot\xi}{|\xi|}} - e^{-s}\right)\frac{ds}{s}\right)d\sigma.\end{aligned} \tag{9.22}$$

Moreover, it follows from inequality (9.21) and formulas (9.22) and (9.9) that we have, for almost all $\xi \in \mathbf{R}^n \setminus \{0\}$,

$$\begin{aligned}\lim_{\varepsilon\downarrow 0}(\mathcal{F}K_\varepsilon)(\xi) &= \int_{\Sigma_{n-1}} K(\sigma)\left(\int_0^{\infty}\left(e^{-is\frac{\sigma\cdot\xi}{|\xi|}} - e^{-s}\right)\frac{ds}{s}\right)d\sigma \\ &= \mathcal{F}K(\xi).\end{aligned} \tag{9.23}$$

It should be noticed that the Fourier transform $\mathcal{F}K(\xi)$, defined by formula (9.23), is a positively homogeneous function of degree zero and further that we have, by inequality (9.18),

$$|(\mathcal{F}K_\varepsilon)(\xi)| \leq M \quad \text{for almost all } \xi \neq 0, \tag{9.24a}$$

$$|\mathcal{F}K(\xi)| \leq M \quad \text{for almost all } \xi \neq 0. \tag{9.24b}$$

Step (III): If $f \in L^2(\mathbf{R}^n)$, then it follows from an application of Young's inequality (Theorem 3.36) with $p := 1$ and $q = r := 2$ that

$$K_{\varepsilon,\mu} * f \in L^2(\mathbf{R}^n),$$

since $K_{\varepsilon,\mu} \in L^1(\mathbf{R}^n)$. However, we have, by assertion (9.22),

$$\mathcal{F}(K_{\varepsilon,\mu}(\xi) \longrightarrow \mathcal{F}K_\varepsilon(\xi) \quad \text{for almost all } \xi \in \mathbf{R}^n \setminus \{0\} \text{ as } \mu \uparrow \infty,$$

and also, by inequalities (9.20) and (9.24),

$$\left|\mathcal{F}\left(K_{\varepsilon,\mu}(\xi) - \mathcal{F}K_\varepsilon(\xi)\right)\mathcal{F}f(\xi)\right|^2 \leq 4\,M^2\,|\mathcal{F}f(\xi)|^2 \quad \text{for almost all } \xi \neq 0.$$

Therefore, by applying Lebesgue's dominated convergence theorem (Theorem 3.13) we obtain from Parseval's formula that

$$\begin{aligned}&\left\|K_{\varepsilon,\mu} * f - K_\varepsilon * f\right\|_2 \\ &= \frac{1}{(2\pi)^{n/2}}\left\|\mathcal{F}\left(K_{\varepsilon,\mu} * f\right) - \mathcal{F}\left(K_\varepsilon * f\right)\right\|_2 \\ &= \frac{1}{(2\pi)^{n/2}}\left\|\left(\mathcal{F}\left(K_{\varepsilon,\mu}\right) - \mathcal{F}\left(K_\varepsilon\right)\right)\mathcal{F}f\right\|_2 \\ &\longrightarrow 0 \quad \text{as } \mu \uparrow \infty.\end{aligned}$$

Step (IV): Moreover, by applying Lebesgue's dominated convergence theorem we obtain from assertion (9.23) and inequalities (9.24) that the sequence $\{K_\varepsilon * f\}$ is a Cauchy sequence in the space $L^2(\mathbf{R}^n)$ as $\varepsilon \downarrow 0$. This implies that $K_\varepsilon * f$ converges to some function g, denoted by $g := K * f$, in $L^2(\mathbf{R}^n)$ as $\varepsilon \downarrow 0$:

$$K * f := g = \lim_{\varepsilon \downarrow 0} K_\varepsilon * f \quad \text{in } L^2(\mathbf{R}^n).$$

Hence, by Parseval's formula and assertion (9.23) it follows that

$$\begin{aligned} \mathcal{F}(K * f)(\xi) &= \mathcal{F}g(\xi) \\ &= \lim_{\varepsilon \downarrow 0} \mathcal{F}(K_\varepsilon * f)(\xi) = \lim_{\varepsilon \downarrow 0} \mathcal{F}(K_\varepsilon)(\xi) \cdot \mathcal{F}f(\xi) \\ &= \mathcal{F}K(\xi) \cdot \mathcal{F}f(\xi) \quad \text{for almost all } \xi \neq 0. \end{aligned} \tag{9.25}$$

Therefore, by using Parseval's formula we obtain from formula (9.25) and inequality (9.24b) that

$$\begin{aligned} \|K * f\|_2 &= \frac{1}{(2\pi)^{n/2}} \|\mathcal{F}(K * f)\|_2 \\ &= \frac{1}{(2\pi)^{n/2}} \|\mathcal{F}K \cdot \mathcal{F}f\|_2 \\ &\leq \frac{M}{(2\pi)^{n/2}} \|\mathcal{F}f\|_2 \\ &= M \|f\|_2 \quad \text{for all } f \in L^2(\mathbf{R}^n). \end{aligned}$$

Now the proof of Theorem 9.3 is complete. □

9.3 The Case of Continuous Kernels

In this section, we make the following assumption:

Assumption 9.3 In addition to Assumption 9.2, the integral kernel $K(x)$ satisfies the following conditions:

(i) There exists a positively homogeneous function $\Omega(x)$ of degree zero defined on $\mathbf{R}^n \setminus \{0\}$ such that

$$K(x) = \Omega\left(\frac{x}{|x|}\right) \frac{1}{|x|^n} \quad \text{for all } x \in \mathbf{R}^n \setminus \{0\}. \tag{9.26}$$

(ii) The function $\Omega(x)$ satisfies the condition

$$|\Omega(x) - \Omega(y)| \leq \omega(|x - y|) \quad \text{for all } x, y \in \Sigma_{n-1}, \tag{9.27}$$

where $\omega(t)$ is a non-negative, increasing continuous function defined on the interval $[0, \infty)$ that satisfies the conditions

$$\int_0^1 \frac{\omega(t)}{t} \, dt < \infty, \tag{9.28a}$$

$$\frac{\omega(t)}{t} \geq c_0 \quad \text{for all } t \geq 0, \tag{9.28b}$$

with some constant $c_0 > 0$.

The next lemma is another version of the *Calderón–Zygmund decomposition* adapted to the present context (cf. Theorem 4.13):

Lemma 9.4 (the Calderón–Zygmund decomposition) *Let $f(x)$ be an arbitrary non-negative function in the space $L^p(\mathbf{R}^n)$ with $1 \leq p < \infty$. Then, for any positive number s, there exists a sequence of non-overlapping cubes $\{I_k\}$ such that*

$$s \leq \frac{1}{|I_k|} \int_{I_k} f(x) \, dx < 2^n s. \tag{9.29}$$

If we let

$$D_s := \bigcup_{k=1}^{\infty} I_k,$$

then it follows that

$$f(x) \leq s \quad \textit{almost everywhere outside the set } D_s,$$

and that

$$s \leq \frac{1}{|D_s|} \int_{D_s} f(x) \, dx < 2^n s. \tag{9.30}$$

Proof Let I be a cube. Since we have, as $|I| \to \infty$,

$$\frac{1}{|I|} \int_I f(x) \, dx \leq \frac{\|f\|_p}{|I|^{1/p}} \longrightarrow 0,$$

we can find a positive number ν such that, if $|I| \geq \nu$, we have the inequality

$$\frac{1}{|I|} \int_I f(x) \, dx < s.$$

We let

$$\mathbf{R}^n = \bigcup_{k=1}^{\infty} I_{0,k}$$

be the decomposition of $\mathbf{R}^n$ into a sum of cubes $I_{0,k}$ of volume ν. Then we have, for each k,

$$\frac{1}{|I_{0,k}|}\int_{I_{0,k}} f(x)\,dx < s.$$

First, we denote by $\{I_{1,k}\}$ the sequence of cubes obtained by dividing each $I_{0,k}$ into 2^n equal parts. We classify the cubes $\{I_{1,k}\}$ as follows:

$$\{I_{1,k}\} = \{I'_{1,k}\} \cup \{1''_{1,k}\},$$
$$\frac{1}{|I'_{1,k}|}\int_{I'_{1,k}} f(x)\,dx \geq s,$$
$$\frac{1}{|I''_{1,k}|}\int_{I''_{1,k}} f(x)\,dx < s.$$

Secondly, we denote by $\{I_{2,k}\}$ the sequence of cubes obtained by dividing each $I''_{1,k}$ into 2^n equal parts, and we classify the cubes $\{I_{2,k}\}$ as follows:

$$\{I_{2,k}\} = \{I'_{2,k}\} \cup \{I''_{2,k}\},$$
$$\frac{1}{|I'_{2,k}|}\int_{I'_{2,k}} f(x)\,dx \geq s,$$
$$\frac{1}{|I''_{2,k}|}\int_{I''_{2,k}} f(x)\,dx < s.$$

Repeating this process, we obtain a sequence of cubes $\{I_{m,k}\}$ such that we have, for each integer m,

- $\{I_{m,k}\} = \{I'_{m,k}\} \cup \{I''_{m,k}\},$
- $\dfrac{1}{|I'_{m,k}|}\displaystyle\int_{I'_{m,k}} f(x)\,dx \geq s,$
- $\dfrac{1}{|I''_{m,k}|}\displaystyle\int_{I''_{m,k}} f(x)\,dx < s.$

Then we have the decomposition

$$\mathbf{R}^n = \left(\cup I'_{1,k}\right) \cup \left(\cup I''_{1,k}\right) \cup \cdots \cup \left(\cup I'_{m,k}\right) \cup \left(\cup I''_{m,k}\right).$$

Moreover, if $m \geq 1$, then each cube $I'_{m,k}$ is obtained by dividing some $I''_{m-1,l}$, where $I''_{0,l} = I_{0,l}$. This implies that

$$|I'_{m,k}| = \frac{1}{2^n}|I''_{m-1,l}|.$$

Hence we have the inequalities

$$\frac{1}{|I'_{m,k}|}\int_{I'_{m,k}} f(x)\,dx \leq \frac{2^n}{|I''_{m-1,l}|}\int_{I''_{m-1,l}} f(x)\,dx < 2^n s.$$

Now, we let $\{I_k\}$ be the totality of the cubes $I'_{m,k}$, $m = 1, 2, \ldots$, $k = 1, 2, \ldots$, and let

$$D_s = \bigcup_{k=1}^{\infty} I_k.$$

Then it follows that inequalities (9.29) hold true for all k:

$$s \leq \frac{1}{|I_k|}\int_{I_k} f(x)\,dx < 2^n s.$$

Since we have, by Hölder's inequality (Theorem 3.24),

$$|I_k| \leq \frac{1}{s}\int_{I_k} f(x)\,dx \leq \frac{1}{s}|I_k|^{1-1/p}\left(\int_{I_k} f(x)^p\,dx\right)^{1/p},$$

we obtain that

$$|I_k| \leq \frac{1}{s^p}\int_{I_k} f(x)^p\,dx,$$

so that

$$|D_s| = \sum_{k=1}^{\infty} |I_k| \leq \frac{1}{s^p}\int_{D_s} f(x)^p\,dx < \infty.$$

Moreover, it is easy to verify that inequalities (9.30) hold true. Indeed, we have, by inequalities (9.29),

$$|D_s| = \sum_{k=1}^{\infty} |I_k| \leq \frac{1}{s}\sum_{k=1}^{\infty}\int_{I_k} f(x)\,dx = \frac{1}{s}\int_{D_s} f(x)\,dx.$$

Similarly, we have the inequality

$$\int_{D_s} f(x)\,dx = \sum_{k=1}^{\infty}\int_{I_k} f(x)\,dx \leq \sum_{k=1}^{\infty} |I_k|\,2^n\,s = 2^n\,s\,|D_s|.$$

Finally, let x_0 be an arbitrary Lebesgue point of the function $f(x)$ which does not belong to the set D_s (see Definition 4.1). Then, for any integer $m \geq 1$, there exists an integer $k(m)$ such that $x_0 \in I''_{m,k(m)}$. Since $\{I''_{m,k(m)}\}$ is a sequence of regular closed sets converging to x_0, it follows from an application of Corollary 4.8 that

$$s > \frac{1}{|I''_{m,k}(m)|} \int_{I''_{m,k(m)}} f(x)\,dx \longrightarrow f(x_0),$$

so that

$$f(x_0) \leq s.$$

Therefore, we have proved that $f(x) \leq s$ almost everywhere outside the set D_s.

The proof of Lemma 9.4 is complete. □

The next lemma is an essential step for the proof of existence of the singular integral (9.2) in the space $L^p(\mathbf{R}^n)$ for $1 < p < \infty$:

Lemma 9.5 *Let $K(x)$ be an integral kernel satisfying Assumption 9.3. Assume that $f(x)$ is a non-negative function in $L^p(\mathbf{R}^n)$ for $1 \leq p \leq 2$. Let $K_\varepsilon(x)$ be the function defined by formula* (9.11) *and let*

$$\widetilde{f}_\varepsilon(x) := K_\varepsilon * f(x)$$

be the function defined by formula (9.5). *Moreover, if s is a positive number, we let*

$$E_s := \left\{x : |\widetilde{f}_\varepsilon(x)| > s\right\}.$$

Then we can find two positive constants C_1 and C_2, independent of ε and s, such that

$$|E_s| \leq \frac{C_1}{s^2} \int_{\mathbf{R}^n} [f(x)]_s^2\,dx + C_2|D_s|, \tag{9.31}$$

where D_s is the set as in Lemma 9.4, and $[f(x)]_s$ is a function defined by the formula

$$[f(x)]_s := \begin{cases} f(x) & \text{if } f(x) \leq s, \\ s & \text{if } f(x) > s. \end{cases}$$

Proof In the following we shall denote by C a generic positive constant, independent of ε and s.

Let $\{I_k\}$ be the sequence of cubes as in Lemma 9.4, and let $g(x)$ and $h(x)$ be functions defined respectively by the formulas

$$h(x) := \begin{cases} \frac{1}{|I_k|} \int_{I_k} f(y)\,dy & \text{if } x \in I_k, \\ f(x) & \text{if } x \notin D_s, \end{cases}$$

and

$$g(x) := f(x) - h(x).$$

Then it follows that

$$s \leq h(x) \leq 2^n s \quad \text{in } D_s, \tag{9.32a}$$

$$h(x) = f(x) = [f(x)]_s \quad \text{outside } D_s, \tag{9.32b}$$

$$g(x) = 0 \quad \text{outside } D_s. \tag{9.32c}$$

We remark that we have, for each $k \in \mathbf{N}$,

$$\int_{I_k} g(y)\, dy = \int_{I_k} f(y)\, dy - \int_{I_k} h(y)\, dy = 0. \tag{9.33}$$

If we let

$$\widetilde{h}_\varepsilon(x) := K_\varepsilon * h(x),$$
$$\widetilde{g}_\varepsilon(x) := K_\varepsilon * g(x),$$

and

$$E_1 := \left\{x : \widetilde{h}_\varepsilon(x) \geq \frac{s}{2}\right\},$$
$$E_2 := \left\{x : \widetilde{g}_\varepsilon(x) \geq \frac{s}{2}\right\},$$

then we have the inclusion

$$E_s \subset E_1 \cup E_2. \tag{9.34}$$

However, by arguing just as in the proof of Theorem 9.3 we obtain from Parseval's formula and inequality (9.24a) that

$$\begin{aligned}
\int_{\mathbf{R}^n} |\widetilde{h}_\varepsilon(x)|^2\, dx &= \frac{1}{(2\pi)^{n/2}} \int_{\mathbf{R}^n} |(\mathcal{F}(K_\varepsilon * h)(\xi)|^2\, d\xi \\
&= \frac{1}{(2\pi)^{n/2}} \int_{\mathbf{R}^n} |(\mathcal{F}K_\varepsilon)(\xi) \cdot (\mathcal{F}h)(\xi)|^2\, d\xi \\
&\leq \frac{M^2}{(2\pi)^{n/2}} \int_{\mathbf{R}^n} |(\mathcal{F}h)(\xi)|^2\, d\xi \\
&= M^2 \int_{\mathbf{R}^n} h(x)^2\, dx.
\end{aligned}$$

This proves that

$$\begin{aligned}\frac{s^2}{4}|E_1| &\leq \int_{E_1} |\widetilde{h}_\varepsilon(x)|^2\,dx \\ &\leq \int_{\mathbf{R}^n} |\widetilde{h}_\varepsilon(x)|^2\,dx \\ &\leq M^2 \int_{\mathbf{R}^n} h(x)^2\,dx.\end{aligned} \tag{9.35}$$

On the other hand, it follows from inequalities (9.32a) and formula (9.32b) that

$$\begin{aligned}\int_{\mathbf{R}^n} h(x)^2\,dx &= \int_{D_s} h(x)^2\,dx + \int_{\mathbf{R}^n \setminus D_s} h(x)^2\,dx \\ &\leq 2^{2n}s^2|D_s| + \int_{\mathbf{R}^n} [f(x)]_s^2\,dx.\end{aligned}$$

Therefore, by combining this inequality with inequality (9.35) we obtain that

$$|E_1| \leq 2^{2n+2}M^2|D_s| + \frac{4M^2}{s^2}\int_{\mathbf{R}^n} [f(x)]_s^2\,dx. \tag{9.36}$$

If S_k is the ball centered at the center of I_k and of radius the diameter of I_k, we let

$$\widetilde{D}_s := \bigcup_{k=1}^{\infty} S_k,$$

and also

$$\begin{aligned}\widetilde{D}_s^c &:= \mathbf{R}^n \setminus \widetilde{D}_s, \\ S_k^c &:= \mathbf{R}^n \setminus S_k.\end{aligned}$$

Then it follows that

$$|\widetilde{D}_s| \leq C|D_s|. \tag{9.37}$$

Since we have, by assertion (9.32c),

$$\begin{aligned}\widetilde{g}_\varepsilon(x) &= \int_{D_s} K_\varepsilon(x-y)g(y)\,dy \\ &= \sum_{k=1}^{\infty}\int_{I_k} K_\varepsilon(x-y)g(y)\,dy,\end{aligned}$$

we obtain that

$$\int_{\widetilde{D}_s^c} |\widetilde{g}_\varepsilon(x)|\, dx \leq \sum_{k=1}^{\infty} \int_{\widetilde{D}_s^c} \left| \int_{I_k} K_\varepsilon(x-y) g(y)\, dy \right| dx \qquad (9.38)$$
$$\leq \sum_{k=1}^{\infty} \int_{S_k^c} \left| \int_{I_k} K_\varepsilon(x-y) g(y)\, dy \right| dx.$$

Now, let x be a fixed point of $S_k^c = \mathbf{R}^n \setminus S_k$, and denote the center of I_k by y_k. We consider the following two cases:

(I) The set $\{y : |y-x| \leq \varepsilon\} \cap I_k$ is empty.
(II) The set $\{y : |y-x| \leq \varepsilon\} \cap I_k$ is not empty.

Case I: First, we consider the case where

$$\{y : |y-x| \leq \varepsilon\} \cap I_k = \emptyset.$$

In this case, it follows from formula (9.33) that

$$\int_{I_k} K_\varepsilon(x-y) g(y)\, dy = \int_{I_k} K(x-y) g(y)\, dy \qquad (9.39)$$
$$= \int_{I_k} \left(K(x-y) - K(x-y_k)\right) g(y)\, dy.$$

However, we have, by formula (9.26),

$$|K(x-y) - K(x-y_k)| \qquad (9.40)$$
$$= \left| \frac{1}{|x-y|^n} \Omega\left(\frac{x-y}{|x-y|}\right) - \frac{1}{|x-y_k|^n} \Omega\left(\frac{x-y_k}{|x-y_k|}\right) \right|$$
$$\leq \left| \frac{1}{|x-y|^n} - \frac{1}{|x-y_k|^n} \right| \left| \Omega\left(\frac{x-y}{|x-y|}\right) \right|$$
$$+ \frac{1}{|x-y_k|^n} \left| \Omega\left(\frac{x-y}{|x-y|}\right) - \Omega\left(\frac{x-y_k}{|x-y_k|}\right) \right|.$$

We estimate each term on the right-hand side of inequality (9.40).

(a) Since we have the inequality

$$|y-y_k| \leq \frac{|x-y_k|}{2} \quad \text{for all } y \in I_k,$$

it follows that

$$\frac{|x-y_k|}{2} \leq |x-y| \leq \frac{3|x-y_k|}{2}.$$

Hence we obtain that

$$\begin{aligned}
&\left|\frac{1}{|x-y|^n}-\frac{1}{|x-y_k|^n}\right| \\
&=\left|\frac{1}{|x-y|}-\frac{1}{|x-y_k|}\right|\sum_{i=0}^{n-1}\frac{1}{|x-y|^{n-1-i}|x-y_k|^i} \\
&\le \frac{|y-y_k|}{|x-y|\,|x-y_k|}\sum_{i=0}^{n-1}\frac{1}{|x-y|^{n-1-i}\,|x-y_k|^i} \\
&\le \frac{C\,|y-y_k|}{|x-y_k|^{n+1}} \le \frac{C\,|I_k|^{1/n}}{|x-y_k|^{n+1}}.
\end{aligned} \tag{9.41}$$

(b) On the other hand, since we have the inequality

$$\begin{aligned}
&\left|\frac{x-y}{|x-y|}-\frac{x-y_k}{|x-y_k|}\right| \\
&\le \left|\frac{x-y}{|x-y|}-\frac{x-y}{|x-y_k|}\right|+\left|\frac{x-y}{|x-y_k|}-\frac{x-y_k}{|x-y_k|}\right| \\
&\le \frac{2|y-y_k|}{|x-y_k|} \le \frac{C\,|I_k|^{1/n}}{|x-y_k|},
\end{aligned}$$

it follows from condition (9.27) that

$$\left|\Omega\left(\frac{x-y}{|x-y|}\right)-\Omega\left(\frac{x-y_k}{|x-y_k|}\right)\right| \le \omega\left(\frac{C\,|I_k|^{1/n}}{|x-y_k|}\right). \tag{9.42}$$

By combining three inequalities (9.40), (9.41) and (9.42) and by using condition (9.28b), we obtain that

$$\begin{aligned}
&|K(x-y)-K(x-y_k)| \\
&\le \frac{C\,|I_k|^{1/n}}{|x-y_k|^{n+1}}+\frac{1}{|x-y_k|^n}\,\omega\left(\frac{C\,|I_k|^{1/n}}{|x-y_k|}\right) \\
&\le \frac{1}{|x-y_k|^n}\,c_0\,\omega\left(\frac{C\,|I_k|^{1/n}}{|x-y_k|}\right)+\frac{1}{|x-y_k|^n}\,\omega\left(\frac{C\,|I_k|^{1/n}}{|x-y_k|}\right) \\
&\le \frac{C}{|x-y_k|^n}\,\omega\left(\frac{C\,|I_k|^{1/n}}{|x-y_k|}\right).
\end{aligned}$$

Therefore, it follows from this inequality and formula (9.39) that

$$\begin{aligned}
&\left|\int_{I_k} K_\varepsilon(x-y)g(y)\,dy\right| \\
&\leq \int_{I_k} |K(x-y)-K(x-y_k)|\,g(y)\,dy \\
&\leq \frac{C}{|x-y_k|^n}\,\omega\left(\frac{C\,|I_k|^{1/n}}{|x-y_k|}\right)\int_{I_k}|g(y)|\,dy.
\end{aligned} \tag{9.43}$$

Case II: Secondly, we consider the case where

$$\{y : |y-x| \leq \varepsilon\} \cap I_k \neq \emptyset.$$

It should be noticed that, in this case, we have the inclusion

$$I_k \subset \{y : |y-x| \leq 3\varepsilon\}. \tag{9.44}$$

Let $\gamma(t)$ be the characteristic function of the interval $[0, 3]$:

$$\gamma(t) = \begin{cases} 1 & \text{if}\, 0 \leq t \leq 3, \\ 0 & \text{otherwise.} \end{cases}$$

If $y \in I_k$, then it follows from condition (9.44) that

$$\gamma\left(\frac{|x-y|}{\varepsilon}\right) = 1.$$

Moreover, since we have the inequality

$$|K_\varepsilon(x-y)| \leq \frac{C}{\varepsilon^n},$$

it follows that

$$\left|\int_{I_k} K_\varepsilon(x-y)g(y)\,dy\right| \leq \frac{C}{\varepsilon^n}\int_{I_k}\gamma\left(\frac{|x-y|}{\varepsilon}\right)|g(y)|\,dy. \tag{9.45}$$

Therefore, by combining the inequalities (9.43) and (9.45) we obtain that, for $x \in S_k^c = \mathbf{R}^n \setminus S_k$,

$$\begin{aligned}
\left|\int_{I_k} K_\varepsilon(x-y)g(y)\,dy\right| &\leq \frac{C}{|x-y_k|^n}\,\omega\left(\frac{C\,|I_k|^{1/n}}{|x-y_k|}\right)\int_{I_k}|g(y)|\,dy \\
&\quad + \frac{C}{\varepsilon^n}\int_{I_k}\gamma\left(\frac{|x-y|}{\varepsilon}\right)|g(y)|\,dy.
\end{aligned} \tag{9.46}$$

By inequalities (9.38) and (9.46), it follows that

$$\begin{aligned}\int_{\widetilde{D}_s^c} |\widetilde{g}_\varepsilon(x)|\,dx &\le \sum_{k=1}^{\infty} \int_{S_k^c} \frac{C}{|x-y_k|^n}\,\omega\left(\frac{C\,|I_k|^{1/n}}{|x-y_k|}\right)\left(\int_{I_k} |g(y)|\,dy\right)dx \\ &\quad + \sum_{k=1}^{\infty} \int_{S_k^c} \frac{C}{\varepsilon^n}\left(\int_{I_k} \gamma\left(\frac{|x-y|}{\varepsilon}\right)|g(y)|\,dy\right)dx \\ &= \sum_{k=1}^{\infty} \left(\int_{S_k^c} \frac{C}{|x-y_k|^n}\,\omega\left(\frac{C\,|I_k|^{1/n}}{|x-y_k|}\right)dx\right)\int_{I_k} |g(y)|\,dy \\ &\quad + C\sum_{k=1}^{\infty} \int_{I_k} |g(y)|\left(\frac{1}{\varepsilon^n}\int_{S_k^c} \gamma\left(\frac{|x-y|}{\varepsilon}\right)dx\right)dy.\end{aligned} \tag{9.47}$$

However, since we have, by condition (9.28a),

$$\begin{aligned}&\int_{S_k^c} \frac{C}{|x-y_k|^n}\,\omega\left(\frac{C\,|I_k|^{1/n}}{|x-y_k|}\right)dx \\ &= \int_{\Sigma_{n-1}} \left(\int_{C\,|I_k|^{1/n}}^{\infty} \frac{C}{r^n}\,\omega\left(\frac{C\,|I_k|^{1/n}}{r}\right)r^{n-1}\,dr\right)d\sigma \\ &= \int_{\Sigma_{n-1}} \left(\int_{C\,|I_k|^{1/n}}^{\infty} \frac{C}{r}\,\omega\left(\frac{C\,|I_k|^{1/n}}{r}\right)dr\right)d\sigma \\ &= C\int_{\Sigma_{n-1}} d\sigma \cdot \int_0^1 \frac{\omega(t)}{t}\,dt \\ &= C\,\omega_n \int_0^1 \frac{\omega(t)}{t}\,dt < \infty,\end{aligned}$$

and the inequality

$$\begin{aligned}\frac{1}{\varepsilon^n}\int_{S_k^c} \gamma\left(\frac{|x-y|}{\varepsilon}\right)dx &\le \frac{1}{\varepsilon^n}\int_{\mathbf{R}^n} \gamma\left(\frac{|x|}{\varepsilon}\right)dx \\ &= \int_{\mathbf{R}^n} \gamma(|x|)\,dx \\ &= \frac{3^n\,\omega_n}{n},\end{aligned}$$

it follows from inequality (9.47) that

$$\begin{aligned}\int_{\widetilde{D}_s^c} |\widetilde{g}_\varepsilon(x)|\,dx &\le C\,\omega_n\left(\int_0^1 \frac{\omega(t)}{t}\,dt + \frac{3^n}{n}\right)\sum_{k=1}^{\infty} \int_{I_k} |g(y)|\,dy \\ &\le C\int_{D_s} |g(y)|\,dy.\end{aligned} \tag{9.48}$$

Here

$$\omega_n = |\Sigma_{n-1}| = \frac{2\pi^{n/2}}{\Gamma(n/2)}$$

is the surface area of the unit sphere Σ_{n-1} in $\mathbf{R}^n$.

Moreover, we have the inequality

$$\begin{aligned}
\int_{D_s} |g(y)|\, dy &\leq \int_{D_s} f(y)\, dy + \int_{D_s} h(y)\, dy \\
&= 2 \int_{D_s} f(y)\, dy \\
&= 2 \sum_{k=1}^{\infty} \int_{I_k} f(x)\, dx \\
&\leq 2 \left(\sum_{k=1}^{\infty} |I_k| \right) 2^n\, s \\
&= 2^{n+1}\, s\, |D_s|\,.
\end{aligned}$$

By combining this inequality with inequality (9.48), we obtain that

$$\int_{\widetilde{D}_s^c} |\widetilde{g}_\varepsilon(x)|\, dx \leq C \int_{D_s} |g(y)|\, dy \leq C\, s\, |D_s|\,.$$

This implies that

$$\begin{aligned}
\frac{s}{2} \left| \widetilde{D}_s^c \cap E_2 \right| &\leq \int_{\widetilde{D}_s^c \cap E_2} |\widetilde{g}_\varepsilon(x)|\, dx \\
&\leq \int_{\widetilde{D}_s^c} |\widetilde{g}_\varepsilon(x)|\, dx \\
&\leq C\, s\, |D_s|\,,
\end{aligned}$$

so that

$$\left| \widetilde{D}_s^c \cap E_2 \right| \leq C |D_s|.$$

Hence, we have, by this inequality and inequality (9.37),

$$\begin{aligned}
|E_2| &\leq |\widetilde{D}_s^c \cap E_2| + |\widetilde{D}_s \cap E_2| \\
&\leq C\, |D_s|\,.
\end{aligned} \tag{9.49}$$

Therefore, by combining two inequalities (9.36) and (9.49) we obtain from inclusion (9.34) that

$$
\begin{aligned}
|E_s| &\le |E_1| + |E_2| \\
&\le 2^{2n+2} M^2 |D_s| + \frac{4M^2}{s^2} \int_{\mathbf{R}^n} [f(x)]_s^2 \, dx + C|D_s| \\
&= \frac{4M^2}{s^2} \int_{\mathbf{R}^n} [f(x)]_s^2 \, dx + \left(2^{2n+2} M^2 + C\right) |D_s|.
\end{aligned}
$$

This proves the desired inequality (9.31) with

$$
C_1 := 4M^2, \qquad C_2 := 2^{2n+2} M^2 + C.
$$

Now the proof of Lemma 9.5 is complete. □

The next theorem asserts the existence of the singular integral (9.2) in the space $L^p(\mathbf{R}^n)$ for $1 < p < \infty$ in the case of *continuous kernels*:

Theorem 9.6 *Assume that $K(x)$ is an integral kernel satisfying Assumption 9.3. Let $f \in L^p(\mathbf{R}^n)$ for $1 < p < \infty$. If $\varepsilon > 0$, we let $\widetilde{f_\varepsilon}(x)$ be the function defined by formula* (9.5)

$$
\widetilde{f_\varepsilon}(x) := \int_{|x-y|>\varepsilon} K(x-y) \, f(y) \, dy.
$$

Then we have the following three assertions:

(i) There exists a constant $C_p > 0$, independent of ε, such that

$$
\left\| \widetilde{f_\varepsilon} \right\|_p \le C_p \, \|f\|_p \, . \tag{9.50}
$$

(ii) The sequence $\widetilde{f_\varepsilon}$ converges strongly to a function $K * f$ in the space $L^p(\mathbf{R}^n)$ as $\varepsilon \downarrow 0$. Namely, the singular integral

$$
K * f(x) = \lim_{\varepsilon \downarrow 0} \widetilde{f_\varepsilon}(x) = \lim_{\varepsilon \downarrow 0} \int_{|x-y|>\varepsilon} K(x-y) \, f(y) \, dy \tag{9.51}
$$

exists in the strong topology of $L^p(\mathbf{R}^n)$.

(iii) The mapping

$$
\begin{aligned}
L^p(\mathbf{R}^n) &\longrightarrow L^p(\mathbf{R}^n) \\
f &\longmapsto K * f
\end{aligned}
$$

is a bounded linear operator from $L^p(\mathbf{R}^n)$ into itself. More precisely, we have the inequality

$$
\|K * f\|_p \le C_p \, \|f\|_p \quad \textit{for all } f \in L^p(\mathbf{R}^n). \tag{9.52}
$$

Proof In the proof, we write

$$T_\varepsilon f(x) := \widetilde{f}_\varepsilon(x) = \int_{|x-y|>\varepsilon} K(x-y)\, f(y)\, dy \quad \text{for } \varepsilon > 0.$$

By applying Theorem 9.3 to our situation, we can find a constant $C > 0$, independent of ε, such that

$$\|T_\varepsilon f\|_2 \leq C\, \|f\|_2 \quad \text{for all } f \in L^2(\mathbf{R}^n).$$

This proves that the operator T_ε is of type $(2, 2)$ uniformly with respect to ε.

(1) If $f(x)$ is a non-negative function in $L^1(\mathbf{R}^n)$, then we have the inequality

$$[f(x)]_s^2 \leq s\, f(x) \quad \text{for all } s > 0.$$

Hence it follows from two inequalities (9.30) and (9.31) that

$$|\{x : |(T_\varepsilon f)(x)| > s\}| \leq \frac{C\, \|f\|_1}{s} \quad \text{for all } s > 0.$$

If $f(x)$ is an arbitrary function in $L^1(\mathbf{R}^n)$, then, by decomposing it into the positive part $f_1(x)$ and the negative part $f_2(x)$ as

$$\begin{aligned} f(x) &= f_1(x) - f_2(x), \\ f_1(x) &:= \max\{f(x), 0\} = \frac{1}{2}\left(f(x) + |f(x)|\right), \\ f_2(x) &:= \max\{-f(x), 0\} = \frac{1}{2}\left(|f(x)| - f(x)\right), \end{aligned}$$

we obtain that

$$\begin{aligned} &|\{x : |(T_\varepsilon f)(x)| > s\}| \\ &\leq |\{x : |(T_\varepsilon f_1)(x)| > s/2\}| + |\{x : |(T_\varepsilon f_2)(x)| > s/2\}| \\ &\leq \frac{2C\, \|f_1\|_1}{s} + \frac{2C\, \|f_2\|_1}{s} \\ &= \frac{2C\, \|f\|_1}{s} \quad \text{for all } s > 0. \end{aligned}$$

This proves that the operator T_ε is of weak type $(1, 1)$ uniformly in ε. Therefore, by arguing just as in the proof of Marcinkiewicz's interpolation theorem (Theorem 3.47) we can find a positive constant C_p such that the desired inequality (9.50) holds true for $1 < p \leq 2$:

$$\|T_\varepsilon f\|_p \leq C_p\, \|f\|_p \quad (1 < p \leq 2).$$

Furthermore, by passing to the adjoint operator T_ε^* we obtain that the desired inequality (9.49) holds true for $2 < p < \infty$:

$$\|T_\varepsilon f\|_p \le C_p \|f\|_p \quad (2 < p < \infty).$$

(2) Secondly, let $f \in L^p(\mathbf{R}^n)$ for $1 < p < \infty$. If $g \in C_0^1(\mathbf{R}^n)$, i.e., if $g(x)$ is a continuously differentiable function with compact support in $\mathbf{R}^n$ and if $\varepsilon, \delta > 0$, then we have, by inequality (9.50),

$$\begin{aligned} &\|T_\varepsilon f - T_\delta f\|_p \\ &\le \|T_\varepsilon f - T_\varepsilon g\|_p + \|T_\delta g - T_\delta f\|_p + \|T_\varepsilon g - T_\delta g\|_p \\ &\le 2C_p \|f - g\|_p + \|T_\varepsilon g - T_\delta g\|_p \quad (1 < p < \infty). \end{aligned} \tag{9.53}$$

If $0 < \varepsilon < 1$, it follows from the cancellation property (9.4) that

$$\begin{aligned} (T_\varepsilon g)(x) &= \int_{|x-y|>1} K(x-y)\, g(y)\, dy \\ &\quad + \int_{\varepsilon<|x-y|<1} K(x-y)\, (g(y) - g(x))\, dy. \end{aligned} \tag{9.54}$$

Since $K \in L^p(\{x : |x| > 1\})$ and $g \in L^1(\mathbf{R}^n)$, it follows from an application of Young's inequality (Theorem 3.36) with $q := 1$ and $r = p$ that the first term on the right-hand side of formula (9.53) belongs to $L^p(\mathbf{R}^n)$:

$$\int_{|x-y|>1} K(x-y)\, g(y)\, dy \in L^p(\mathbf{R}^n).$$

On the other hand, we remark that the second term on the right-hand side of formula (9.53) is a function with support contained in a fixed compact set, so that it converges uniformly as $\varepsilon \downarrow 0$, since $g(x)$ is uniformly Lipschitz continuous. Hence we find that the second term on the right-hand side of formula (9.54) converges strongly in the space $L^p(\mathbf{R}^n)$:

$$\lim_{\varepsilon \downarrow 0} \int_{\varepsilon<|x-y|<1} K(x-y)\, (g(y) - g(x))\, dy \in L^p(\mathbf{R}^n).$$

This implies that

$$\|T_\varepsilon g - T_\delta g\|_p \longrightarrow 0 \quad \text{as } \varepsilon,\ \delta \downarrow 0.$$

However, Corollary 3.43 tells us that the space $C_0^1(\mathbf{R}^n)$ is dense in $L^p(\mathbf{R}^n)$. Hence, we obtain from inequality (9.53) that the sequence $T_\varepsilon f = \widetilde{f_\varepsilon}$ converges strongly to some function $K * f$ in the space $L^p(\mathbf{R}^n)$ as $\varepsilon \downarrow 0$:

$$\lim_{\varepsilon \downarrow 0} T_\varepsilon f = \lim_{\varepsilon \downarrow 0} \widetilde{f_\varepsilon} = K * f \in L^p(\mathbf{R}^n).$$

Therefore, by letting $\varepsilon \downarrow 0$ in inequality (9.50) we obtain from Lebesgue's dominated convergence theorem (Theorem 3.13) that

$$\|K * f\|_p \leq C_p \|f\|_p \quad \text{for all } f \in L^p(\mathbf{R}^n).$$

This proves that the mapping $f \mapsto K * f$ is bounded from $L^p(\mathbf{R}^n)$ into itself.

The proof of Theorem 9.6 is complete. □

9.4 The Hilbert Transform

In the case where $n = 1$, we can apply Theorem 9.6 with

$$K(x) := \frac{1}{\pi x}$$

to obtain that the singular integral $K * f$ reduces to the following formula:

$$K * f(x) = \frac{1}{\pi} \lim_{\varepsilon \downarrow 0} \int_{|x-t|>\varepsilon} \frac{f(t)}{x-t}\, dt.$$

Moreover, it should be noticed that this formula is also expressed as

$$\frac{1}{\pi}\left(\text{v.p.}\,\frac{1}{x}\right) * f,$$

where v.p.$(1/x)$ is a distribution defined by the formula

$$\left\langle \text{v.p.}\,\frac{1}{x}, \varphi \right\rangle = \lim_{\varepsilon \downarrow 0} \int_{|x|>\varepsilon} \frac{\varphi(x)}{x}\, dx \quad \text{for all } \varphi \in C_0^1(\mathbf{R}).$$

The function $Hf(x)$, defined by the formula

$$\begin{aligned} Hf(x) &= \frac{1}{\pi}\left(\text{v.p.}\,\frac{1}{x}\right) * f \\ &= \frac{1}{\pi} \lim_{\varepsilon \downarrow 0} \int_{|x-t|>\varepsilon} \frac{f(t)}{x-t}\, dt \\ &= \frac{1}{\pi} \lim_{\varepsilon \downarrow 0} \int_{|s|>\varepsilon} f(x-s)\, \frac{ds}{s}, \end{aligned} \tag{9.55}$$

is called the *Hilbert transform* of $f(x)$.

The purpose of this section is to prove the following basic results for the Hilbert transform:

Theorem 9.7 *Let H be the Hilbert transform defined by formula* (9.55). *Then we have the following three assertions:*

(i) The Hilbert transform

$$H\colon L^p(\mathbf{R}) \longrightarrow L^p(\mathbf{R})$$
$$f \longmapsto Hf = \frac{1}{\pi} \lim_{\varepsilon \downarrow 0} \int_{|x-t|>\varepsilon} \frac{f(t)}{x-t}\, dt$$

is a bounded linear operator from $L^p(\mathbf{R})$ into itself for $1 < p < \infty$.

(ii) If $g = Hf$ is the Hilbert transform of $f \in L^p(\mathbf{R})$ for $1 < p < \infty$, then its inverse transform is given by the formula

$$f(x) = -h * g(x) = -\frac{1}{\pi} \lim_{\varepsilon \downarrow 0} \int_{|x-t|>\varepsilon} \frac{g(t)}{x-t}\, dt. \tag{9.56}$$

Furthermore, the singular integrals (9.55) *and* (9.56) *exist in the strong topology of $L^p(\mathbf{R})$.*

(iii) If $f \in L^p(\mathbf{R})$ and $\widehat{f} \in L^{p'}(\mathbf{R})$ with $p' = p/(p-1)$ and if $g = Hf$ and $\widehat{g} = H\widehat{f}$ are the Hilbert transforms of f and $\widehat{f}$, respectively, then the formula

$$\begin{aligned}\int_{-\infty}^{\infty} f(x) \cdot \widehat{f}(x)\, dx &= \int_{-\infty}^{\infty} Hf(x) \cdot H\widehat{f}(x)\, dx \\ &= \int_{-\infty}^{\infty} g(x) \cdot \widehat{g}(x)\, dx\end{aligned} \tag{9.57}$$

holds true. In particular, the Hilbert transform H is a unitary operator *in the Hilbert space $L^2(\mathbf{R})$, and the formula*

$$\|g\|_2 = \|Hf\|_2 = \|f\|_2 \tag{9.58}$$

holds true if $g = Hf$ is the Hilbert transform of $f \in L^2(\mathbf{R})$.

Proof The proof of Theorem 9.7 is divided into four steps.

Step (I): First, we consider the case where $p = 2$. Following the proof of Theorem 9.3, we calculate explicitly the Fourier transform of the distribution (formula (9.60) below)

$$h(x) := \frac{1}{\pi} \,\text{v.p.}\, \frac{1}{x}.$$

To do this, we let, for $0 < \varepsilon < \mu$,

$$h_{\varepsilon,\mu}(x) := \begin{cases} \frac{1}{\pi x} & \text{if } \varepsilon < |x| < \mu, \\ 0 & \text{otherwise,} \end{cases}$$

and

$$h_\varepsilon(x) := \begin{cases} \frac{1}{\pi x} & \text{if } |x| > \varepsilon, \\ 0 & \text{if } |x| \leq \varepsilon. \end{cases}$$

Then it follows that

$$\begin{aligned}
(\mathcal{F}h_{\varepsilon,\mu})(\xi) &= \int_{\varepsilon<|x|<\mu} \frac{e^{-ix\cdot\xi}}{\pi x}\,dx \\
&= -\frac{2i}{\pi}\int_\varepsilon^\mu \frac{\sin(x\cdot\xi)}{x}\,dx \\
&= \begin{cases} -\frac{2i}{\pi}\int_{\varepsilon\xi}^{\mu\xi} \frac{\sin x}{x}\,dx & \text{if } \xi > 0, \\ \frac{2i}{\pi}\int_{-\varepsilon\xi}^{-\mu\xi} \frac{\sin x}{x}\,dx & \text{if } \xi < 0 \end{cases} \\
&= -\frac{2i}{\pi}\int_{\varepsilon|\xi|}^{\mu|\xi|} \frac{\sin x}{x}\,dx \cdot \operatorname{sign}\xi \quad \text{for } \xi \neq 0,
\end{aligned}$$

so that we have, as $\mu \uparrow \infty$,

$$(\mathcal{F}h_\varepsilon)(\xi) = -\frac{2i}{\pi}\int_{\varepsilon|\xi|}^{\infty} \frac{\sin x}{x}\,dx \cdot \operatorname{sign}\xi \quad \text{for } \xi \neq 0. \tag{9.59}$$

Therefore, by letting $\varepsilon \downarrow 0$ in formula (9.59) we obtain that

$$\begin{aligned}
(\mathcal{F}h)(\xi) &= -\frac{2i}{\pi}\left(\int_0^\infty \frac{\sin x}{x}\,dx\right)\operatorname{sign}\xi \\
&= -\frac{2i}{\pi}\,\frac{\pi}{2}\operatorname{sign}\xi \\
&= -i\operatorname{sign}\xi \quad \text{for } \xi \neq 0.
\end{aligned} \tag{9.60}$$

If $f \in L^2(\mathbf{R})$, then we have, by formula (9.59) and inequality (9.16),

$$|(\mathcal{F}h_\varepsilon)(\xi)| \leq \frac{4A}{\pi},$$

and so, by Parseval's formula,

$$\begin{aligned}
\|h_\varepsilon * f\|_2 &= \frac{1}{(2\pi)^{1/2}}\|\mathcal{F}(h_\varepsilon * f)\|_2 \\
&= \frac{1}{(2\pi)^{1/2}}\|\mathcal{F}h_\varepsilon \cdot \mathcal{F}f\|_2 \\
&\leq \frac{1}{(2\pi)^{1/2}}\,\frac{4A}{\pi}\|\mathcal{F}f\|_2 \\
&= \frac{4A}{\pi}\|f\|_2\,.
\end{aligned}$$

Step (II): On the other hand, Theorem 9.3 asserts that the singular integral

$$\begin{aligned} g(x) = Hf(x) &= \frac{1}{\pi}\left(\text{v.p.}\,\frac{1}{x}\right) * f \\ &= \frac{1}{\pi}\lim_{\varepsilon\downarrow 0}\int_{|x-t|>\varepsilon}\frac{f(t)}{x-t}\,dt \end{aligned} \tag{9.61}$$

exists in the strong topology of $L^2(\mathbf{R})$ ($p = 2$):

$$H\colon L^2(\mathbf{R}) \longrightarrow L^2(\mathbf{R}).$$

By combining two formulas (9.8) and (9.60), we obtain that

$$\begin{aligned} (\mathcal{F}g)(\xi) &= (\mathcal{F}h)(\xi)\cdot(\mathcal{F}f)(\xi) \\ &= -i\,\operatorname{sign}\xi\cdot(\mathcal{F}f)(\xi) \quad \text{for } \xi \neq 0. \end{aligned} \tag{9.62}$$

Therefore, by Parseval's formula we have the desired formula (9.58)

$$\|g\|_2 = \|Hf\|_2 = \|f\|_2\,,$$

and also the formula

$$\begin{aligned} (\mathcal{F}f)(\xi) &= i\,\operatorname{sign}\xi\cdot(\mathcal{F}g)(\xi) \\ &= -(\mathcal{F}h)(\xi)\cdot(\mathcal{F}g)(\xi) \quad \text{for } \xi \neq 0. \end{aligned} \tag{9.63}$$

By formulas (9.61) and (9.63), it follows from an application of the Fourier inversion formula that the singular integral (9.56) exists in the strong topology of $L^2(\mathbf{R})$ ($p = 2$).

Step (III): We consider the general case where $1 < p < \infty$. By applying Theorem 9.6, we obtain that the Hilbert transform H is a bounded linear operator from $L^p(\mathbf{R})$ into itself.

Let $f \in L^p(\mathbf{R})$ and let $g = Hf$ be its Hilbert transform. If we take a sequence $f_n \in L^p(\mathbf{R}) \cap L^2(\mathbf{R})$ such that

$$f_n \longrightarrow f \quad \text{in } L^p(\mathbf{R}) \text{ as } n \to \infty,$$

then it follows that the Hilbert transform $g_n = Hf_n$ of f_n belongs to $L^p(\mathbf{R}) \cap L^2(\mathbf{R})$ and further that

$$g_n = Hf_n \longrightarrow g = Hf \quad \text{in } L^p(\mathbf{R}) \text{ as } n \to \infty.$$

By assertion (9.56), we remark that

$$-\,f_n(x) = \frac{1}{\pi} \lim_{\varepsilon \downarrow 0} \int_{|x-t|>\varepsilon} \frac{g_n(t)}{x-t}\, dt$$

in the strong topology of $L^2(\mathbf{R})$ and further that the right-hand side of this formula is the Hilbert transform of g_n. Hence it follows from an application of Theorem 9.6 that the above limit exists also in the strong topology of $L^p(\mathbf{R})$. Moreover, by letting $n \to \infty$ we obtain from Theorem 9.6 that formula (9.56) holds true also in the strong topology of $L^p(\mathbf{R})$:

$$f(x) = -h * g(x) = -\frac{1}{\pi} \lim_{\varepsilon \downarrow 0} \int_{|x-t|>\varepsilon} \frac{g(t)}{x-t}\, dt \quad (1 < p < \infty).$$

Step (IV): It remains to prove formula (9.57).

Now, let $f \in L^p(\mathbf{R})$ and $\widehat{f} \in L^{p'}(\mathbf{R})$ with $p' = p/(p-1)$, and let $g = Hf$ and $\widehat{g} = H\widehat{f}$ be the Hilbert transforms of f and $\widehat{f}$, respectively. By using Friedrichs' mollifiers (Sect. 3.7.2), we can choose two sequences $f_n \in L^p(\mathbf{R}) \cap L^2(\mathbf{R})$ and $\widehat{f_n} \in L^{p'}(\mathbf{R}) \cap L^2(\mathbf{R})$ such that

$$\begin{aligned} f_n &\longrightarrow f \quad \text{in } L^p(\mathbf{R}) \text{ as } n \to \infty, \\ \widehat{f_n} &\longrightarrow \widehat{f} \quad \text{in } L^{p'}(\mathbf{R}) \text{as } n \to \infty. \end{aligned}$$

Then we have the assertions

$$\begin{aligned} g_n &= Hf_n \in L^p(\mathbf{R}) \cap L^2(\mathbf{R}), \\ \widehat{g_n} &= H\widehat{f_n} \in L^{p'}(\mathbf{R}) \cap L^2(\mathbf{R}), \end{aligned}$$

and

$$\begin{aligned} g_n = Hf_n &\longrightarrow g = Hf \quad \text{in } L^p(\mathbf{R}), \\ \widehat{g_n} = H\widehat{f_n} &\longrightarrow \widehat{g} = H\widehat{f} \quad \text{in } L^{p'}(\mathbf{R}). \end{aligned}$$

However, it follows from formula (9.58) that the Hilbert transform H is a unitary operator in $L^2(\mathbf{R})$. Hence we have the formulas

$$\begin{aligned} \int_{-\infty}^{\infty} f_n(x) \cdot \widehat{f_n}(x)\, dx &= \int_{-\infty}^{\infty} Hf_n(x) \cdot H\widehat{f_n}(x)\, dx \\ &= \int_{-\infty}^{\infty} g_n(x) \cdot \widehat{g_n}(x)\, dx. \end{aligned}$$

Therefore, by letting $n \to \infty$ in these formulas we obtain the desired formula (9.57)

$$\int_{-\infty}^{\infty} f(x) \cdot \widehat{f}(x) \, dx = \int_{-\infty}^{\infty} Hf(x) \cdot H\widehat{f}(x) \, dx$$
$$= \int_{-\infty}^{\infty} g(x) \cdot \widehat{g}(x) \, dx.$$

The proof of Theorem 9.7 is now complete. □

9.5 Equimeasurable Functions

Let $f(x)$ be a non-negative, measurable function defined on $\mathbf{R}^n$. For each $\tau \geq 0$, we let

$$m(\tau) := \left|\left\{x \in \mathbf{R}^n : f(x) > \tau\right\}\right|. \tag{9.64}$$

In this section we consider only functions such that $m(\tau) < \infty$ for any $\tau > 0$ and that

$$\lim_{\tau \uparrow \infty} m(\tau) = 0.$$

It should be emphasized that non-zero functions $f(x)$ in $L^p(\mathbf{R}^n)$ with $1 \leq p < \infty$ satisfy this condition.

Let $f(x)$ be a non-negative, measurable function defined on $\mathbf{R}^n$. A non-negative, measurable function $f^*(t)$ defined on the interval $[0, \infty)$ is called an *equimeasurable function* of $f(x)$ if it satisfies the condition

$$m^*(\tau) := \left|\left\{t : f^*(t) > \tau\right\}\right| = m(\tau) \quad \text{for all } \tau \geq 0. \tag{9.65}$$

We begin by proving the following elementary result:

Lemma 9.8 *The function $m(\tau)$ defined by formula* (9.64) *is monotone decreasing and right-continuous.*

Proof It is clear that $m(\tau)$ is monotone decreasing. If $\tau_k \downarrow \tau$, then it follows that

$$\{x : f(x) > \tau\} = \bigcup_{k=1}^{\infty} \{x : f(x) > \tau_k\},$$

so that $m(\tau) = \lim_{k \to \infty} m(\tau_k)$. This proves the right-continuity of $m(\tau)$.

The proof of Lemma 9.8 is complete. □

We let

$$t_0 := m(0), \quad \tau_0 := \sup\{\tau : m(\tau) > 0\}.$$

It is easy to see that $0 < t_0 \leq \infty$ and $0 < \tau_0 \leq \infty$.

The next lemma proves the existence and uniqueness of the equimeasurable function:

Lemma 9.9 *The function $f^*(t)$, defined by the formula*

$$f^*(t) := \inf\{\tau : t \geq m(\tau)\}, \quad f^*(0) := \tau_0,$$

is the unique monotone decreasing, right-continuous equimeasurable function of $f(x)$.

Proof The proof of Lemma 9.9 is divided into two steps.

Step (1): We prove that the non-negative function $f^*(t)$ is a monotone decreasing, right-continuous equimeasurable function of $f(x)$.

(a) First, we show that

$$\tau \geq f^*(t) \Longleftrightarrow t \geq m(\tau).$$

Indeed, we remark that $t \geq m(\tau)$ implies that $\tau \geq f^*(t)$.

Conversely, if $\tau > f^*(t)$, then, by the definition of f^* it follows that $t \geq m(\tau)$. Assume that $\tau = f^*(t)$. If $\tau_k \downarrow \tau$, then it follows that $t \geq m(\tau_k)$. Since the function $m(\tau)$ is right-continuous, by letting $k \to \infty$ we obtain that

$$t \geq m(\tau).$$

Therefore, we have, for all $\tau > 0$,

$$\left\{t : f^*(t) > \tau\right\} = [0, m(\tau)).$$

Hence, if $m^*(\tau)$ is the function defined by formula (9.65), then it follows that $m^*(\tau) = m(\tau)$. Namely, the function $f^*(t)$ satisfies the condition (9.65).

(b) Secondly, it is clear that $f^*(t)$ is *monotone decreasing*.

(c) Thirdly, we show that $f^*(t)$ is *right-continuous*. Since either $t_0 = m(0) \geq m(\tau)$ or $\tau \geq f^*(t_0)$ for all $\tau > 0$, it follows that $f^*(t_0) = 0$. If $t_0 < \infty$, then $f^*(t) = 0$ for all $t \geq t_0$.

We show that $f^*(t) > 0$ for all $0 < t < t_0 = m(0)$. Indeed, there exists a number $\tau > 0$ such that $t < m(\tau)$. This proves that $0 < \tau < f^*(t)$.

Let $\{t_k\}$ be an arbitrary sequence such that $t_k \downarrow t$. For any given number $\tau < f^*(t)$, it follows that $t < m(\tau)$. In particular, we have the assertion

$$t_k < m(\tau) \quad \text{if } k \text{ is sufficiently large},$$

or equivalently

$$\tau < f^*(t_k) \leq f^*(t) \quad \text{if } k \text{ is sufficiently large}.$$

This proves that

$$f^*(t) = \lim_{k \to \infty} f^*(t_k),$$

since $\tau < f^*(t)$ is arbitrary.

Therefore, we have proved the right-continuity of $f^*(t)$.

Step (2): We prove the *uniqueness* of the equimeasurable function.

Let $g^*(t)$ be another monotone decreasing, right-continuous equimeasurable function of $f(x)$. If t_1 is a positive number, then we let $\tau := f^*(t_1)$ and

$$t_2 := \min\left\{t : f^*(t) = \tau\right\} = \inf\left\{t : f^*(t) = \tau\right\}.$$

We recall that $f^*(t)$ is right-continuous. Since we have the formula

$$\left\{t : f^*(t) > \tau\right\} = [0, t_2),$$

it follows that

$$\left|\left\{t : g^*(t) > \tau\right\}\right| = m(\tau) = \left|\left\{t : f^*(t) > \tau\right\}\right| = t_2.$$

Moreover, since $g^*(t)$ is monotone decreasing and right-continuous, it follows that $\{t : g^*(t) > \tau\} = [0, t_2)$. This proves that $g^*(t_2) \leq \tau = f^*(t_1)$. Hence we have, for $t_2 \leq t_1$,

$$g^*(t_1) \leq g^*(t_2) \leq \tau = f^*(t_1).$$

Similarly, we can prove that $f^*(t_1) \leq g^*(t_1)$ for any positive number t_1.

Summing up, we obtain that $f^*(t_1) = g^*(t_1)$ for any positive number t_1. This proves the uniqueness of the equimeasurable function of $f(x)$.

The proof of Lemma 9.9 is now complete. □

Lemma 9.10 *Let $f(x)$ be a non-negative, measurable function on $\mathbf{R}^n$. If $f^*(t)$ is an equimeasurable function of $f(x)$, then we have, for $1 \leq p < \infty$,*

$$\int_{\mathbf{R}^n} f(x)^p\,dx = \int_0^\infty f^*(t)^p\,dt.$$

Proof By applying formula (3.24) with $\sigma(\tau) := m(\tau)$ and $\sigma(\tau) := m^*(\tau)$, we obtain from formulas (9.64) and (9.65) that

$$\begin{aligned}\int_{\mathbf{R}^n} f(x)^p\,dx &= -\int_0^\infty \tau^p\,dm(\tau) = -\int_0^\infty \tau^p\,dm^*(\tau)\\ &= \int_0^\infty f^*(t)^p\,dt.\end{aligned}$$

The proof of Lemma 9.10 is complete. □

Lemma 9.11 *Let $f(x)$ be a non-negative, measurable function on $\mathbf{R}^n$. If A is a measurable subset of $\mathbf{R}^n$ of finite measure, then we have the inequality*

$$\int_A f(x)\,dx \le \int_0^{|A|} f^*(t)\,dt.$$

Proof We let

$$f_1(x) := \begin{cases} f(x) & \text{if } x \in A, \\ 0 & \text{if } x \notin A. \end{cases}$$

Since $f_1(x) \le f(x)$, it follows that

$$m_1(\tau) := |\{x : f_1(x) > \tau\}| \le |\{x : f(x) > \tau\}| = m(\tau).$$

Hence, if $f_1^*(t)$ is the monotone decreasing, right-continuous equimeasurable function of $f_1(x)$, then we have the inequality

$$f_1^*(t) = \inf\{\tau : t \ge m_1(\tau)\} \le f^*(t).$$

Since we have, for all $\tau > 0$,

$$\{x : f_1(x) > \tau\} \subset \{x : f_1(x) > 0\} \subset A,$$

it follows that $m_1(\tau) \le |A|$. This proves that

$$f_1^*(t) = 0 \quad \text{for } t \ge |A|.$$

Therefore, we obtain from Lemma 9.10 with $p := 1$ that

$$\begin{aligned} \int_A f(x)\,dx &= \int_{\mathbf{R}^n} f_1(x)\,dx = \int_0^\infty f_1^*(t)\,dt \\ &\le \int_0^{|A|} f_1^*(t)\,dt \\ &\le \int_0^{|A|} f^*(t)\,dt. \end{aligned}$$

The proof of Lemma 9.11 is complete. □

From now on, we assume that $f(x)$ is a non-negative function in $L^p(\mathbf{R}^n)$ for $1 \le p < \infty$, and let $f^*(t)$ be the monotone decreasing, right-continuous equimeasurable function of $f(x)$. If $t > 0$, we let

$$\beta_f(t) := \frac{1}{t}\int_0^t f^*(s)\,ds. \tag{9.66}$$

By Lemma 9.10, it follows that $f^* \in L^p(0,\infty)$ for $1 \le p < \infty$. Hence it follows from an application of Hölder's inequality (Theorem 3.24) that the right-hand side of formula (9.66) is finite. Moreover, we remark that

$$\frac{d}{dt}\beta_f(t) = \frac{1}{t^2}\int_0^t \left(f^*(t) - f^*(s)\right) ds \leq 0.$$

If $\beta'_f(t_1) = 0$ for some $t_1 > 0$, then it follows that $f^*(t) = f^*(t_1)$ in the interval $(0, t_1)$, and so $\beta_f(t) = f^*(t_1)$ there. Hence it is easy to see that $f^*(t) \to 0$ as $t \uparrow \infty$. This proves that $\beta_{f(t)} \to 0$ as $t \uparrow \infty$.

Therefore, we obtain that either

(a) $\beta_f(t)$ is strictly decreasing in $(0, \infty)$

or

(b) there exists a positive number t_1 such that $\beta_f(t)$ is constant in $(0, t_1]$ and strictly decreasing in $[t_1, \infty)$.

In the case (a), we let $t_1 := 0$, and let $\beta_f(0) := s_1$ if $s_1 = \lim_{t\to 0}\beta_f(t)$. Then it is easy to see that $0 < s_1 \leq \infty$ and that $s_1 = \beta_f(t_1) = \beta_f(0)$.

We denote the inverse function of $s = \beta_f(t)$ by $t = \beta^f(s)$.

(1) If $t_1 = 0$ and $s_1 = \infty$, then it follows that $\beta^f(s)$ is uniquely determined in $0 < s < \infty$.
(2) If $t_1 = 0$ and $s_1 < \infty$, then we let $\beta^f(s) := 0$ for $s > s_1$.
(3) If $t_1 > 0$, then we let $\beta^f(s_1) := t_1$ and $\beta^f(s) := 0$ for $s > s_1$.

Then we have the following lemma:

Lemma 9.12 *If $f \in L^p(\mathbf{R}^n)$ for $1 < p < \infty$, then we have the inequality*

$$\left(\int_0^\infty \beta_f(t)^p\, dt\right)^{1/p} \leq \frac{p}{p-1}\left(\int_{\mathbf{R}^n} f(x)^p\, dx\right)^{1/p}. \tag{9.67}$$

Proof By Lemma 9.10, it follows that $f^*) \in L^p(0, \infty)$. Let $0 < a < b$. Then we have, by integration by parts,

$$\begin{aligned}\int_a^b \beta_f(t)^p\, dt &= \int_a^b t^{-p}\left(\int_0^t f^*(s)\, ds\right)^p dt \\ &\leq \frac{a^{1-p}}{p-1}\left(\int_0^a f^*(s)\, ds\right)^p \\ &\quad + \frac{p}{p-1}\int_a^b t^{1-p} f^*(t)\left(\int_0^t f^*(s)\, ds\right)^{p-1} dt.\end{aligned} \tag{9.68}$$

However, we obtain that the first term on the right-hand side of inequality (9.68) tends to 0 as $a \downarrow 0$. Indeed, it suffices to note that we have, by Hölder's inequality (Theorem 3.24),

$$\frac{a^{1-p}}{p-1}\left(\int_0^a f^*(s)\, ds\right)^p \leq \frac{1}{p-1}\int_0^a f^*(s)^p\, ds.$$

Therefore, by letting $a \downarrow 0$ and $b \uparrow \infty$ in inequality (9.68) and by using Hölder's inequality (Theorem 3.24) we obtain that

$$\begin{aligned}\int_0^\infty \beta_f(t)^p\,dt &\le \frac{p}{p-1}\int_0^\infty f^*(t)\,\beta_f(t)^{p-1}\,dt\\ &\le \frac{p}{p-1}\left(\int_0^\infty f^*(t)^p\,dt\right)^{1/p}\left(\int_0^\infty \beta_f(t)^p\,dt\right)^{1-1/p}.\end{aligned}$$

This proves the desired inequality (9.66).

The proof of Lemma 9.12 is complete. □

The main purpose of this section is to prove the following lemma:

Lemma 9.13 *Assume that $f(x)$ is a non-negative function in $L^p(\mathbf{R})$ for $1 < p < \infty$. If $\varepsilon > 0$, we let*

$$F_\varepsilon(x) := \frac{1}{\varepsilon}\int_0^\varepsilon f(x+y)\,dy, \quad G(x) := \sup_{\varepsilon>0} F_\varepsilon(x). \tag{9.69}$$

Then we have the inequality

$$\left(\int_{-\infty}^\infty G(x)^p\,dx\right)^{1/p} \le \frac{p}{p-1}\left(\int_{-\infty}^\infty f(x)^p\,dx\right)^{1/p}. \tag{9.70}$$

Proof First, we remark that the function $G(x)$ is measurable, since the supremum of the right-hand side of the second equality of formula (9.69) may be taken only over the rational $\varepsilon \in \mathbf{Q}$. If $\tau > 0$, we let

$$H(\tau) := \{x : G(x) > \tau\}.$$

Since we have the formula

$$G(x) = \sup_{y>x}\frac{1}{y-x}\int_x^y f(t)\,dt,$$

it is easy to see that $x \in H(\tau)$ if and only if the inequality

$$\int_0^y f(t)\,dt - y\tau > \int_0^x f(t)\,dt - x\tau$$

holds true for some $y > x$. Hence, if we let

$$F(x) := \int_0^x f(t)\,dt - x\tau,$$

then we have the formula

$$H(\tau) = \{x : F(y) > F(x) \quad \text{for some } y > x\}.$$

This proves that the set $H(\tau)$ is open, so that it is the sum of disjoint open intervals, that is,

$$H(\tau) = \bigcup_{k=1}^{\infty} (a_k, b_k) .$$

We have, by Hölder's inequality (Theorem 3.24),

$$\lim_{x \to \pm\infty} F(x) = \mp\infty. \tag{9.71}$$

Step (1): We show that, for any $k \in \mathbf{N}$,

$$-\infty < a_k < b_k < \infty.$$

Our proof is based on a reduction to absurdity. We assume, to the contrary, that $a_k = -\infty$ for some k. Then we have the inclusion

$$(-\infty, b_k) \subset H(\tau).$$

By assertion (9.71), it follows that $F(c) > F(b_k)$ for some $c < b_k$. Let c_1 be a real number in the interval $[c, b_k]$ such that

$$F(c_1) = \max_{c \leq x \leq b_k} F(x),$$

Then it follows that $c \leq c_1 < b_k$, since we have the inequalities

$$F(c_1) \geq F(c) > F(b_k).$$

This implies that $c_1 \in (-\infty, b_k) \subset H(\tau)$. Furthermore, it follows that

$$F(x) \leq F(c_1) \quad \text{for } c_1 < x \leq b_k.$$

On the other hand, since we have the assertion $b_k \notin H(\tau)$, it follows that

$$F(x) \leq F(b_k) < F(c_1) \quad \text{for all } x > b_k.$$

Hence we have proved that

$$F(x) \leq F(c_1) \quad \text{for all } x > c_1.$$

This contradicts the condition that $c_1 \in H(\tau)$.

Next, we assume, to the contrary, that $a_k > -\infty$ and $b_k = \infty$. Then we have the inclusion

$$(a_k, \infty) \subset H(\tau).$$

Let c be an arbitrary number satisfying the condition $c > a_k$. By assertion (9.71), it follows that

$$F(c_1) = \max_{x \geq c} F(x) \quad \text{for some } c_1 \geq c.$$

This implies that $c_1 \in (a_k, \infty) \subset H(\tau)$. However, we have the inequality

$$F(x) \leq F(c_1) \quad \text{for all } x > c_1.$$

This is also a contradiction.

Summing up, we have proved that $-\infty < a_k < b_k < \infty$.

Step (2): Secondly, we show that $F(a_k) = F(b_k)$ for all k.

Since $a_k \notin H(\tau)$ and $b_k > a_k$, it follows that

$$F(b_k) \leq F(a_k).$$

Our proof is based on a reduction to absurdity. Assume, to the contrary, that $F(b_k) < F(a_k)$ for some k. Then it follows that

$$F(b_k) < F(c) \quad \text{for some } c \in (a_k, b_k).$$

Let c_1 be a real number in the interval $[c, b_k]$ such that

$$F(c_1) = \max_{c \leq x \leq b_k} F(x).$$

Since $F(c_1) \geq F(c) > F(b_k)$, it follows that $c \leq c_1 < b_k$, so that

$$c_1 \in (a_k, b_k) \subset H(\tau).$$

Hence we have the inequality

$$F(x) \leq F(c_1) \quad \text{for all } c_1 < x \leq b_k.$$

On the other hand, since $b_k \notin H(\tau)$, it follows that

$$F(x) \leq F(c_1) \quad \text{for all } x > c_1.$$

This contradicts the condition that $c_1 \in H(\tau)$.

Summing up, we have proved that $F(a_k) = F(b_k)$ for all k.

Step (3) Therefore, we have, by the definition of $F(x)$ and by Hölder's inequality (Theorem 3.24),

$$(b_k - a_k)\tau = \int_{a_k}^{b_k} f(t)\,dt \leq (b_k - a_k)^{1-1/p} \left(\int_{a_k}^{b_k} f(t)^p\,dt \right)^{1/p},$$

and so

$$(b_k - a_k)^{1/p}\tau \le \left(\int_{a_k}^{b_k} f(t)^p\,dt\right)^{1/p}.$$

Hence, it follows that

$$|H(\tau)|\tau^p = \sum_{k=1}^{\infty}(b_k - a_k)\tau^p \le \sum_{k=1}^{\infty}\int_{a_k}^{b_k} f(t)^p\,dt < \infty.$$

By applying Lemma 9.11 to our situation, we obtain that

$$\begin{aligned}|H(\tau)|\tau = \sum_{k=1}^{\infty}(b_k - a_k)\tau &= \sum_{k=1}^{\infty}\int_{a_k}^{b_k} f(t)\,dt \\ &= \int_{H(\tau)} f(t)\,dt \\ &\le \int_0^{|H(\tau)|} f^*(t)\,dt,\end{aligned}$$

so that

$$\tau \le \frac{1}{|H(\tau)|}\int_0^{|H(\tau)|} f^*(t)\,dt = \beta_f(|H(\tau)|).$$

This proves that

$$|H(\tau)| \le \beta^f(\tau).$$

Hence, we have, by this inequality and Theorem 3.45,

$$\begin{aligned}\int_{-\infty}^{\infty} G(x)^p\,dx &= p\int_0^{\infty} \tau^{p-1}|H(\tau)|\,d\tau \\ &\le p\int_0^{\infty} \tau^{p-1}\beta^f(\tau)\,d\tau \\ &= \int_0^{s_1} \beta_f(\tau)\,d\tau^p.\end{aligned} \tag{9.72}$$

Here we recall that $s_1 = \lim_{t\to 0}\beta_f(t)$.

However, just as in the proof of Lemma 9.12 it follows that

$$\lim_{t\to 0} t\beta_f(t)^p = \lim_{t\to 0} t^{1-p}\left(\int_0^t f^*(s)\,ds\right)^p = 0.$$

Hence, by making the change of the independent variable $t := \beta^f(\tau)$ or $\tau := \beta_f(t)$ in the last integral of (9.72) we obtain that

$$\begin{aligned}
\int_{-\infty}^{\infty} G(x)^p \, dx &\le \int_0^{s_1} \beta_f(\tau) \, d\tau^p \\
&= -\int_{t_1}^{\infty} t \, d\beta_f(t)^p \\
&= -\left[t\beta_f(t)\right]_{t_1}^{\infty} + \int_{t_1}^{\infty} \beta_f(t)^p \, dt \\
&\le t_1\beta_f(t_1)^p + \int_{t_1}^{\infty} \beta_f(t)^p \, dt = t_1 \, s_1^p + \int_{t_1}^{\infty} \beta_f(t)^p \, dt \\
&= \int_0^{t_1} \beta_f(t)^p \, dt + \int_{t_1}^{\infty} \beta_f(t)^p \, dt \\
&= \int_0^{\infty} \beta_f(t)^p \, dt.
\end{aligned}$$

Therefore, we have, by this inequality and Lemma 9.12,

$$\begin{aligned}
\int_{-\infty}^{\infty} G(x)^p \, dx &\le \int_0^{\infty} \beta_f(t)^p \, dt \\
&\le \left(\frac{p}{p-1}\right)^p \int_{\mathbf{R}^n} f(x)^p \, dx.
\end{aligned}$$

This proves the desired inequality (9.70).

Now the proof of Lemma 9.13 is complete. □

Corollary 9.14 *Under the assumptions of Lemma 9.13, we let*

$$\overline{f}(x) := \sup_{\varepsilon>0} \frac{1}{2\varepsilon} \int_{-\varepsilon}^{\varepsilon} f(x+y) \, dy.$$

Then we have the inequality

$$\left(\int_{-\infty}^{\infty} \overline{f}(x)^p \, dx\right)^{1/p} \le \frac{2^{1/p} p}{p-1} \left(\int_{-\infty}^{\infty} f(x)^p \, dx\right)^{1/p}. \tag{9.73}$$

Proof If $\tau > 0$, we let

$$\overline{E}(\tau) := \left\{x : \overline{f}(x) > \tau\right\},$$

and

$$F_\varepsilon^-(x) := \frac{1}{\varepsilon} \int_0^{\varepsilon} f(x-y) \, dy, \qquad G^-(x) := \sup_{\varepsilon>0} F_\varepsilon^-(x).$$

Moreover, we let

$$H^-(\tau) := \left\{x : G^-(x) > \tau\right\}.$$

Recall that

$$F_\varepsilon(x) = \frac{1}{\varepsilon}\int_0^\varepsilon f(x+y)\,dy, \quad G(x) = \sup_{\varepsilon>0} F_\varepsilon(x),$$

and that

$$H(\tau) = \{x : G(x) > \tau\}.$$

Then, just as in the proof of Lemma 9.13 we obtain that

$$|H^-(\tau)| \leq \beta^f(\tau).$$

On the other hand, since $\overline{f}(x) \leq (G(x) + G^-(x))/2$, it follows that

$$\overline{E}(\tau) \subset H(\tau) \cup H^-(\tau).$$

Hence we have the inequality

$$|\overline{E}(\tau)| \leq |H(\tau)| + |H^-(\tau)| \leq 2\beta^f(\tau).$$

Therefore, just as in the proof of Lemma 9.13 we can obtain the desired inequality (9.73).

The proof of Corollary 9.14 is complete. □

9.6 The Hilbert Transform (Continued)

Let $\phi(x)$ be an even function in $C^1(\mathbf{R})$ such that

$$\phi(x) = \begin{cases} 0 & \text{if } |x| \leq 1/4, \\ 1 & \text{if } |x| \geq 3/4, \end{cases}$$

and that $0 \leq \phi(x) \leq 1$ on $\mathbf{R}$. We denote by $\psi(x)$ the Hilbert transform of $\phi(x)/x$:

$$\psi(x) = \lim_{\varepsilon\downarrow 0} \frac{1}{\pi}\int_{|x-t|>\varepsilon} \frac{\phi(t)}{t}\frac{dt}{x-t}.$$

First, since we have the formula

$$\begin{aligned}
\frac{d}{dx}\left(\int_{|x-t|>\varepsilon} \frac{\phi(t)}{t}\frac{dt}{x-t}\right) &= \frac{d}{dx}\left(\int_{|t|>\varepsilon} \frac{\phi(x-t)}{x-t}\frac{dt}{t}\right) \\
&= \int_{|t|>\varepsilon} \frac{\partial}{\partial x}\left(\frac{\phi(x-t)}{x-t}\right)\frac{dt}{t} \\
&= \int_{|x-t|>\varepsilon} \frac{d}{dt}\frac{\phi(t)}{t} \cdot \frac{dt}{x-t},
\end{aligned}$$

it follows that that the derivative

$$\frac{d}{dx}\psi(x) = \lim_{\varepsilon \downarrow 0} \frac{1}{\pi} \int_{|x-t|>\varepsilon} \frac{d}{dt}\frac{\phi(t)}{t} \cdot \frac{dt}{x-t}$$

exists and is continuous, that is, $\psi(x)$ is of class C^1. It is easy to see that $\psi(x)$ is an even function.

Secondly, we show that, as $x \to \pm\infty$,

$$\psi(x) = O(x^{-2}), \quad \psi'(x) = O(|x|^{-3}). \tag{9.74}$$

If $x > 1$, we let

$$\pi\psi(x) := \lim_{N\to\infty, \varepsilon\downarrow 0} \left(\int_{-N}^{-1} + \int_{-1}^{0} + \int_{0}^{1} + \int_{1}^{x-\varepsilon} + \int_{x+\varepsilon}^{N} \right) \frac{\phi(t)}{t} \frac{dt}{x-t}.$$

Then, since we have the four formulas

- $\displaystyle\int_{-N}^{-1} \frac{\phi(t)}{t} \frac{dt}{x-t} = \int_{-N}^{-1} \frac{dt}{t(x-t)} = -\frac{1}{x}\left[\log\frac{N}{x+N} + \log(x+1)\right],$
- $\displaystyle\int_{-1}^{0} \frac{\phi(t)}{t} \frac{dt}{x-t} = -\int_{0}^{1} \frac{\phi(t)}{t} \frac{dt}{x+t},$
- $\displaystyle\int_{1}^{x-\varepsilon} \frac{\phi(t)}{t} \frac{dt}{x-t} = \int_{1}^{x-\varepsilon} \frac{dt}{t(x-t)} = \frac{1}{x}\left[\log\frac{x-\varepsilon}{\varepsilon} + \log(x-1)\right],$
- $\displaystyle\int_{x+\varepsilon}^{N} \frac{\phi(t)}{t} \frac{dt}{x-t} = \int_{x+\varepsilon}^{N} \frac{dt}{t(x-t)} = \frac{1}{x}\left(\log\frac{N}{N-x} - \log\frac{x+\varepsilon}{\varepsilon}\right),$

it follows that we have, for $x > 1$,

$$\pi\psi(x) = \frac{1}{x}\log\frac{x-1}{x+1} + \int_0^1 \frac{2\phi(t)}{(x-t)(x+t)}\, dt.$$

Therefore, the desired assertions (9.74) follow immediately from this formula, by using de L'Hôpital's rule.

The next theorem asserts the existence of the Hilbert transform (9.55) in the space $L^p(\mathbf{R}^n)$ for $1 < p < \infty$, refining Theorem 9.7:

Theorem 9.15 *Assume that $f \in L^p(\mathbf{R})$ for $1 < p < \infty$. If $\varepsilon > 0$, we let*

$$\widetilde{f_\varepsilon}(x) := \frac{1}{\pi} \int_{|x-t|>\varepsilon} \frac{f(t)}{x-t}\, dt.$$

Then we have the following three assertions:

(i) *There exists a constant $C_p > 0$, independent of ε, such that*

$$\left(\int_{-\infty}^{\infty} \sup_{\varepsilon>0} \left|\widetilde{f_\varepsilon}(x)\right|^p \, dx\right)^{1/p} \leq C_p \left(\int_{-\infty}^{\infty} |f(x)|^p \, dx\right)^{1/p}. \tag{9.75}$$

(ii) *The sequence $\widetilde{f_\varepsilon}$ converges almost everywhere in $\mathbf{R}$ and in the strong topology of $L^p(\mathbf{R})$ as $\varepsilon \downarrow 0$. Namely, the singular integral*

$$Hf(x) = \lim_{\varepsilon\downarrow 0} \widetilde{f_\varepsilon}(x) = \frac{1}{\pi} \lim_{\varepsilon\downarrow 0} \int_{|x-t|>\varepsilon} \frac{f(t)}{x-t} \, dt$$

exists for almost all $x \in \mathbf{R}$ and in the strong topology of $L^p(\mathbf{R})$.

(iii) *The Hilbert transform*

$$\begin{aligned} H\colon L^2(\mathbf{R}^n) &\longrightarrow L^2(\mathbf{R}^n) \\ f &\longmapsto Hf \end{aligned}$$

is a bounded linear operator from $L^p(\mathbf{R})$ into itself. More precisely, we have the inequality

$$\|Hf\|_p \leq C_p \|f\|_p \quad \text{for all } f \in L^p(\mathbf{R}).$$

Proof Let $f(x)$ be an arbitrary function in $L^p(\mathbf{R})$ with $1 < p < \infty$. The proof of Theorem 9.15 is divided into three steps.

Step (1): First, we prove inequality (9.75). If $\phi(x)$ and $\psi(x)$ are the two functions defined as above, then we have the formula

$$\begin{aligned} &\widetilde{f_\varepsilon}(x) \\ &= \frac{1}{\pi} \int_{|x-t|>\varepsilon} \phi\left(\frac{x-t}{\varepsilon}\right) \frac{f(t)}{x-t} \, dt \\ &= \frac{1}{\pi} \int_{-\infty}^{\infty} \phi\left(\frac{x-t}{t}\right) \frac{f(t)}{x-t} \, dt - \frac{1}{\pi} \int_{|x-t|<\varepsilon} \phi\left(\frac{x-t}{\varepsilon}\right) \frac{f(t)}{x-t} \, dt \\ &:= \widetilde{f_{1,\varepsilon}}(x) + \widetilde{f_{2,\varepsilon}}(x). \end{aligned} \tag{9.76}$$

The Hilbert transform of the function

$$t \longmapsto \phi\left(\frac{x-t}{\varepsilon}\right) \frac{1}{x-t} \qquad \text{(for a fixed } x\text{)}$$

is equal to the following:

$$\frac{1}{\varepsilon} \psi\left(\frac{x-t}{\varepsilon}\right).$$

Indeed, by the change of variables

$$s = \frac{x-\sigma}{\varepsilon},$$

we find that

$$\begin{aligned}&\frac{1}{\pi}\,\text{v.p.}\int_{\mathbf{R}} \phi\left(\frac{x-\sigma}{\varepsilon}\right)\frac{1}{x-\sigma}\,\frac{1}{t-\sigma}\,d\sigma \\ &= \frac{1}{\varepsilon}\,\frac{1}{\pi}\,\text{v.p.}\int_{\mathbf{R}} \frac{\phi(s)}{s}\,\frac{1}{(x-t)/\varepsilon - s}\,ds \\ &= \frac{1}{\varepsilon}\,\psi\left(\frac{x-t}{\varepsilon}\right).\end{aligned}$$

Hence, if we denote the Hilbert transform of f by $g = Hf$, then it follows from an application of Theorem 9.7 (formula (9.57)) that

$$\begin{aligned}\widetilde{f}_{1,\varepsilon}(x) &= \frac{1}{\pi}\int_{-\infty}^{\infty} \phi\left(\frac{x-t}{t}\right)\frac{1}{x-t}\cdot f(t)\,dt \\ &= \frac{1}{\pi\varepsilon}\int_{-\infty}^{\infty} \psi\left(\frac{x-t}{\varepsilon}\right)\cdot g(t)\,dt \\ &= \frac{1}{\pi\varepsilon}\int_{-\infty}^{\infty} \psi\left(\frac{s}{\varepsilon}\right) g(x-s)\,ds \\ &= \frac{1}{\pi\varepsilon}\left(\int_{0}^{\infty} \psi\left(\frac{s}{\varepsilon}\right) g(x-s)\,ds + \int_{-\infty}^{0} \psi\left(\frac{s}{\varepsilon}\right) g(x-s)\,ds\right) \\ &= \frac{1}{\pi\varepsilon}\int_{0}^{\infty} \psi\left(\frac{s}{\varepsilon}\right)(g(x-t)+g(x+t))\,dt.\end{aligned} \tag{9.77}$$

Here we have used the fact that the function $\psi(x)$ is even.

If $t > 0$ and $-\infty < x < \infty$, we let

$$\begin{aligned}I(x:t) &:= \int_{-t}^{t} g(x+y)\,dy, \\ \overline{g}(x) &:= \sup_{\varepsilon>0}\frac{1}{2\varepsilon}\int_{-\varepsilon}^{\varepsilon} |g(x+y)|\,dy.\end{aligned}$$

Then we have, by Corollary 9.14,

$$\overline{g} \in L^p(\mathbf{R}).$$

Let x be a point such that $\overline{g}(x) < \infty$. We remark that

$$|I(t:x)| \leq \int_{-t}^{t} |g(x+y)|\, dy \leq 2t\,\overline{g}(x) \quad \text{for all } t > 0, \tag{9.78}$$

and further that

$$dI(t:x) = (g(x-t) + g(x+t))\, dt. \tag{9.79}$$

By combining two formulas (9.77) and (9.79), we obtain that

$$\begin{aligned}\widetilde{f_{1,\varepsilon}}(x) &= \frac{1}{\pi\varepsilon} \int_0^\infty \psi\left(\frac{t}{\varepsilon}\right) dI(t:x) \\ &= \frac{1}{\pi\varepsilon}\left[\psi\left(\frac{t}{\varepsilon}\right) I(x:t)\right]_{t=0}^{t=\infty} - \frac{1}{\pi\varepsilon^2}\int_0^\infty \psi'\left(\frac{t}{\varepsilon}\right) I(t:x)\, dt.\end{aligned}$$

However, it follows from assertions (9.74) that

$$\begin{aligned}\psi\left(\frac{t}{\varepsilon}\right) &= O\left(\frac{\varepsilon^2}{t^2}\right) \quad \text{as } t \to \infty, \\ \psi'\left(\frac{t}{\varepsilon}\right) &= O\left(\frac{\varepsilon^3}{t^3}\right) \quad \text{as } t \to \infty.\end{aligned}$$

Hence we find from inequality (9.78) that the first term on the right-hand side equals 0, while the second term on the right-hand side converges. Summing up, we have proved the formula

$$\widetilde{f_{1,\varepsilon}}(x) = -\frac{1}{\pi\varepsilon^2}\int_0^\infty \psi'\left(\frac{t}{\varepsilon}\right) I(x:t)\, dt. \tag{9.80}$$

Therefore, by combining inequality (9.78) and formula (9.80) we obtain from formula (9.77) that

$$\begin{aligned}|\widetilde{f_{1,\varepsilon}}(x)| &\leq \frac{1}{\pi\varepsilon^2}\int_0^\infty \left|\psi'\left(\frac{t}{\varepsilon}\right)\right| 2t\,\overline{g}(x)\, dt \\ &= \frac{1}{\pi\varepsilon^2}\int_0^\infty |\psi'(s)|\, 2(\varepsilon\, s)\, \varepsilon\, ds \cdot \overline{g}(x) \\ &= \frac{2}{\pi}\int_0^\infty t\,\left|\psi'(t)\right|\, dt \cdot \overline{g}(x).\end{aligned} \tag{9.81}$$

On the other hand, we have the inequality

$$
\begin{aligned}
|\widetilde{f_{2,\varepsilon}}(x)| &\leq \frac{1}{\pi} \int_{\varepsilon/4<|x-t|<\varepsilon} \phi\left(\frac{x-t}{\varepsilon}\right) \left|\frac{f(t)}{x-t}\right| dt \\
&\leq \frac{1}{\pi} \int_{\varepsilon/4<|x-t|<\varepsilon} \left|\frac{f(t)}{x-t}\right| dt \\
&\leq \frac{4}{\pi\,\varepsilon} \int_{|x-t|<\varepsilon} |f(t)|\, dt = \frac{4}{\pi\varepsilon} \int_{x-\varepsilon}^{x+\varepsilon} |f(t)|\, dt \\
&= \frac{4}{\pi\varepsilon} \int_{-\varepsilon}^{\varepsilon} |f(x+y)|\, dy.
\end{aligned}
$$

If we let

$$
\overline{f}(x) := \sup_{\varepsilon>0} \frac{1}{2\varepsilon} \int_{-\varepsilon}^{\varepsilon} |f(x+y)|\, dy,
$$

then it follows that

$$
|\widetilde{f_{2,\varepsilon}}(x)| \leq \frac{8}{\pi}\, \overline{f}(x). \tag{9.82}
$$

Therefore, we obtain from formula (9.76) and two inequalities (9.81) and (9.82) that

$$
\begin{aligned}
\sup_{\varepsilon>0} |\widetilde{f_\varepsilon}(x)| &\leq |\widetilde{f_{1,\varepsilon}}(x)| + |\widetilde{f_{2,\varepsilon}}(x)| \\
&\leq \frac{2}{\pi} \int_0^\infty t|\psi'(t)|\, dt \cdot \overline{g}(x) + \frac{8}{\pi}\, \overline{f}(x).
\end{aligned}
$$

The desired inequality (9.75) follows by applying Theorem 9.7 and Corollary 9.14 to the right-hand side of the above inequality:

$$
\begin{aligned}
\int_{-\infty}^{\infty} \sup_{\varepsilon>0} |\widetilde{f_\varepsilon}(x)|^p\, dx &\leq D_p \left(\int_{-\infty}^{\infty} |\overline{g}(x)|^p\, dx + \int_{-\infty}^{\infty} |\overline{f}(x)|^p\, dx \right) \\
&\leq D_p' \left(\int_{-\infty}^{\infty} |Hf(x)|^p\, dx + \int_{-\infty}^{\infty} |f(x)|^p\, dx \right) \\
&\leq D_p'' \int_{-\infty}^{\infty} |f(x)|^p\, dx.
\end{aligned}
$$

Here D_p, D_p', D_p'' are positive constants independent of ε.

Step (2): We have already proved that, as $\varepsilon \downarrow 0$, the sequence $\widetilde{f_\varepsilon}$ converges in the strong topology of $L^p(\mathbf{R})$.

Step (3): Finally, we prove that the sequence $\widetilde{f_\varepsilon}$ converges almost everywhere in $\mathbf{R}$, as $\varepsilon \downarrow 0$.

To do this, we take a function $g(x)$ in $C_0^1(\mathbf{R})$ and let

$$h(x) := f(x) - g(x).$$

If we define two functions $\widetilde{g}_\varepsilon(x)$ and $\widetilde{h}_\varepsilon(x)$ as we defined the function $\widetilde{f}_\varepsilon(x)$, then, by arguing just as in the proof of Theorem 9.6 we obtain that the sequence $\widetilde{g}_\varepsilon$ converges uniformly as $\varepsilon \downarrow 0$. Hence we have the inequality

$$\begin{aligned}
&\limsup_{\varepsilon\downarrow 0} \widetilde{f}_\varepsilon(x) - \liminf_{\varepsilon\downarrow 0} \widetilde{f}_\varepsilon(x) \\
&= \limsup_{\varepsilon\downarrow 0} \left(\widetilde{h}_\varepsilon(x) + \widetilde{g}_\varepsilon(x)\right) - \liminf_{\varepsilon\downarrow 0} \left(\widetilde{h}_\varepsilon(x) + \widetilde{g}_\varepsilon(x)\right) \\
&= \limsup_{\varepsilon\downarrow 0} \widetilde{h}_\varepsilon(x) - \liminf_{\varepsilon\downarrow 0} \widetilde{h}_\varepsilon(x) \\
&\leq 2 \sup_{\varepsilon>0} |\widetilde{h}_\varepsilon(x)|.
\end{aligned}$$

By combining this inequality and inequality (9.75) with $f(x) := h(x)$, we obtain that

$$\begin{aligned}
&\int_{-\infty}^{\infty} \left| \limsup_{\varepsilon\downarrow 0} \widetilde{f}_\varepsilon(x) - \liminf_{\varepsilon\downarrow 0} \widetilde{f}_\varepsilon(x) \right|^p dx \leq 2^p \int_{-\infty}^{\infty} |\widetilde{h}_\varepsilon(x)|^p \, dx \qquad (9.83) \\
&\leq 2^p \, C_p^p \int_{-\infty}^{\infty} |h(x)|^p \, dx,
\end{aligned}$$

where

$$h = f - g.$$

However, Corollary 3.43 tells us that $C_0^1(\mathbf{R})$ is dense in $L^p(\mathbf{R})$. Hence, the right-hand side of inequality (9.83) can be made arbitrarily small if we choose the function $g(x)$ appropriately. This proves that the integrand of the left hand side of inequality (9.83) vanishes almost everywhere in $\mathbf{R}$:

$$\limsup_{\varepsilon\downarrow 0} \widetilde{f}_\varepsilon(x) = \liminf_{\varepsilon\downarrow 0} \widetilde{f}_\varepsilon(x).$$

Namely, the sequence $\widetilde{f}_\varepsilon$ converges almost everywhere in $\mathbf{R}$ as $\varepsilon \downarrow 0$.

Now the proof of Theorem 9.15 is complete. □

9.7 The Case of Odd Kernels

In this section we consider the case where the integral kernel $K(x)$ is an odd function, and make the following assumption:

Assumption 9.4 The integral kernel $K(x)$ is an *odd function* satisfying Assumption 9.1.

It should be noticed that the odd integral kernel $K(x)$ satisfies condition (9.2).

In the case of *odd kernels*, the existence of the singular integral (9.2) in the space $L^p(\mathbf{R}^n)$ for $1 < p < \infty$ can be reduced to the study of the Hilbert transform (Theorem 9.15). In fact, we can prove the following theorem:

Theorem 9.16 *Assume that the integral kernel $K(x)$ satisfies Assumption 9.4. If $f \in L^p(\mathbf{R}^n)$ for $1 < p < \infty$ and $\varepsilon > 0$, we let*

$$\widetilde{f_\varepsilon}(x) := \int_{|x-y|>\varepsilon} K(x-y)\, f(y)\, dy.$$

Then we have the following three assertions:

(i) The inequality

$$\left(\int_{\mathbf{R}^n} \sup_{\varepsilon>0} \left| \widetilde{f_\varepsilon}(x) \right|^p dx \right)^{1/p} \qquad (9.84)$$
$$\leq \frac{\pi C_p}{2} \left(\int_{\Sigma_{n-1}} |K(\sigma)|\, d\sigma \right) \left(\int_{\mathbf{R}^n} |f(x)|^p\, dx \right)^{1/p}$$

holds true where C_p is the positive constant given in Theorem 9.15.

(ii) The sequence $\widetilde{f_\varepsilon}$ converges almost everywhere in $\mathbf{R}^n$ and in the strong topology of $L^p(\mathbf{R}^n)$ as $\varepsilon \downarrow 0$. Namely, the singular integral

$$K * f(x) = \lim_{\varepsilon \downarrow 0} \widetilde{f_\varepsilon}(x) = \lim_{\varepsilon \downarrow 0} \int_{|x-y|>\varepsilon} K(x-y)\, f(y)\, dy$$

exists for almost all $x \in \mathbf{R}^n$ and in the strong topology of $L^p(\mathbf{R}^n)$.

(iii) The mapping

$$L^p(\mathbf{R}^n) \longrightarrow L^p(\mathbf{R}^n)$$
$$f \longmapsto K * f$$

is a bounded linear operator from $L^p(\mathbf{R}^n)$ into itself. More precisely, we have the inequality

$$\|K * f\|_p \leq \frac{\pi C_p}{2} \left(\int_{\Sigma_{n-1}} |K(\sigma)|\, d\sigma \right) \|f\|_p \quad \text{for all } f \in L^p(\mathbf{R}^n). \qquad (9.85)$$

Proof The proof of Theorem 9.16 is divided into three steps.

Step (1): As is shown in the proof of Lemma 9.1, we have the formula

$$\widetilde{f}_\varepsilon(x) = \int_{|y|>\varepsilon} K(y) f(x-y)\,dy = \int_{\Sigma_{n-1}} K(\sigma)\left(\int_\varepsilon^\infty f(x-t\sigma)\frac{dt}{t}\right) d\sigma.$$

However, since the kernel $K(x)$ is odd, it follows that

$$\begin{aligned}
&\int_{\Sigma_{n-1}} K(\sigma)\left(\int_\varepsilon^\infty f(x-t\sigma)\,\frac{dt}{t}\right) d\sigma \\
&= -\int_{\Sigma_{n-1}} K(\sigma)\left(\int_\varepsilon^\infty f(x+t\sigma)\,\frac{dt}{t}\right) d\sigma \\
&= \int_{\Sigma_{n-1}} K(\sigma)\left(\int_{-\infty}^{-\varepsilon} f(x-t\sigma)\,\frac{dt}{t}\right) d\sigma.
\end{aligned}$$

Hence we have the formula

$$\widetilde{f}_\varepsilon(x) = \frac{1}{2}\int_{\Sigma_{n-1}} K(\sigma)\left(\int_{|t|>\varepsilon} f(x-t\sigma)\frac{dt}{t}\right) d\sigma. \tag{9.86}$$

By using Hölder's inequality (Theorem 3.24), we obtain that

$$\begin{aligned}
&|\widetilde{f}_\varepsilon(x)| \\
&\le \frac{1}{2}\int_{\Sigma_{n-1}} |K(\sigma)|^{1-1/p}|K(\sigma)|^{1/p}\left|\int_{|t|>\varepsilon} f(x-t\sigma)\frac{dt}{t}\right| d\sigma \\
&\le \frac{1}{2}\left(\int_{\Sigma_{n-1}} |K(\sigma)|\,d\sigma\right)^{1-1/p}\left(\int_{\Sigma_{n-1}} |K(\sigma)|\left|\int_{|t|>\varepsilon} f(x-t\sigma)\frac{dt}{t}\right|^p d\sigma\right)^{1/p}.
\end{aligned}$$

Hence we have the inequality

$$\begin{aligned}
&\sup_{\varepsilon>0} |\widetilde{f}_\varepsilon(x)|^p \\
&\le \frac{1}{2^p}\left(\int_{\Sigma_{n-1}} |K(\sigma)|\,d\sigma\right)^{p-1}\int_{\Sigma_{n-1}} |K(\sigma)| \sup_{\varepsilon>0}\left|\int_{|t|>\varepsilon} f(x-t\sigma)\frac{dt}{t}\right|^p d\sigma.
\end{aligned}$$

Moreover, by integrating the both sides of this inequality over $\mathbf{R}^n$ we obtain that

$$\begin{aligned}
&\int_{\mathbf{R}^n} \sup_{\varepsilon>0} |\widetilde{f}_\varepsilon(x)|^p\,dx \\
&\le \frac{1}{2^p}\left(\int_{\Sigma_{n-1}} |K(\sigma)|\,d\sigma\right)^{p-1} \\
&\quad \times \int_{\Sigma_{n-1}} |K(\sigma)|\left(\int_{\mathbf{R}^n} \sup_{\varepsilon>0}\left|\int_{|t|>\varepsilon} f(x-t\sigma)\frac{dt}{t}\right|^p dx\right) d\sigma.
\end{aligned} \tag{9.87}$$

For each $\sigma \in \Sigma_{n-1}$, we make a change of variables

$$x := y + s\,\sigma, \quad y \in \mathbf{R}^n, \; s \in \mathbf{R},$$

with

$$\langle y, \sigma \rangle = y_1\sigma_1 + \ldots + y_n\sigma_n = 0.$$

Then we have the formula

$$\begin{aligned}
&\int_{\mathbf{R}^n} \sup_{\varepsilon>0} \left| \int_{|t|>\varepsilon} f(x - t\,\sigma) \frac{dt}{t} \right|^p dx \\
&= \int_{\mathbf{R}^{n-1}} \left(\int_{-\infty}^{\infty} \sup_{\varepsilon>0} \left| \int_{|t|>\varepsilon} f(y + (s-t)\sigma) \frac{dt}{t} \right|^p ds \right) dy.
\end{aligned}$$

Here it should be noticed that

$$\int_{|t|>\varepsilon} f(y + (s-t)\sigma) \frac{dt}{t} = - \int_{|s-\tau|>\varepsilon} \frac{f(y + \tau\sigma)}{s - \tau} \, d\tau.$$

Hence, by applying Theorem 9.15 with

$$f(t) := f(y + t\sigma) \quad \text{for } t \in \mathbf{R},$$

we obtain that

$$\begin{aligned}
&\int_{\mathbf{R}^n} \sup_{\varepsilon>0} \left| \int_{|t|>\varepsilon} f(x - t\sigma) \frac{dt}{t} \right|^p dx \\
&= \pi^p \int_{\mathbf{R}^{n-1}} \left(\int_{-\infty}^{\infty} \sup_{\varepsilon>0} \left| \frac{1}{\pi} \int_{|s-\tau|>\varepsilon} \frac{f(y + \tau\sigma)}{s - \tau} \, d\tau \right|^p ds \right) dy \\
&\le (\pi C_p)^p \int_{\mathbf{R}^{n-1}} \left(\int_{-\infty}^{\infty} |f(y + s\sigma)|^p \, ds \right) dy \\
&= (\pi C_p)^p \int_{\mathbf{R}^n} |f(x)|^p \, dx.
\end{aligned}$$

By combining this inequality with inequality (9.87), we have the inequality

$$\begin{aligned}
&\int_{\mathbf{R}^n} \sup_{\varepsilon>0} |\widetilde{f}_\varepsilon(x)|^p \, dx \\
&\le \frac{1}{2^p} \left(\int_{\Sigma_{n-1}} |K(\sigma)| \, d\sigma \right)^{p-1} \left(\int_{\Sigma_{n-1}} |K(\sigma)| \left((\pi C_p)^p \int_{\mathbf{R}^n} |f(x)|^p \, dx \right) d\sigma \right) \\
&= \left(\frac{\pi C_p}{2} \right)^p \left(\int_{\Sigma_{n-1}} |K(\sigma)| \, d\sigma \right)^p \int_{\mathbf{R}^n} |f(x)|^p \, dx.
\end{aligned}$$

This proves the desired inequality (9.84).

Step (2): We consider the case where $f(x)$ is an arbitrary function in the space $C_0^1(\mathbf{R}^n)$.

We remark that if $f(x) = 0$ for $|x| > M$ and $0 < \varepsilon < 1$, then it is easy to see that the function $\widetilde{f_\varepsilon}(x)$ is independent of ε, for $|x| \geq M + 1$. If $|x| < M + 1$, then we have the formula

$$\begin{aligned}
&\widetilde{f_\varepsilon}(x) \\
&= \int_{\substack{|x-y|>\varepsilon \\ |x|<M+1, |y|<M}} K(x-y) f(y)\, dy = \int_{\varepsilon<|x-y|<2M+1} K(x-y) f(y)\, dy \\
&= \int_{\varepsilon<|x-y|<2M+1} K(x-y)(f(y) - f(x))\, dy.
\end{aligned}$$

However, by letting

$$L := \max_{x \in \mathbf{R}^n} |\nabla f(x)|,$$

we obtain that

$$\begin{aligned}
&\int_{|x-y|<\varepsilon} |K(x-y)(f(y) - f(x))|\, dy \\
&\leq L \int_{|x-y|<\varepsilon} |K(x-y)|\, |x-y|\, dy \\
&\leq \varepsilon L \int_{|x-y|>\varepsilon} |K(x-y)|\, dy \\
&= \varepsilon L \int_{\Sigma_{n-1}} |K(\sigma)|\, d\sigma \quad \text{for all } |x| < M + 1.
\end{aligned}$$

This proves that the sequence $\widetilde{f_\varepsilon}$ converges uniformly to some function $\widetilde{f}$ in the open ball $B(0, M + 1) = \{|x| < M + 1\}$, as $\varepsilon \downarrow 0$. Hence we have, by inequality (9.84),

$$\begin{aligned}
\int_{\mathbf{R}^n} |\widetilde{f}(x)|^p\, dx &= \lim_{\varepsilon \downarrow 0} \int_{\mathbf{R}^n} |\widetilde{f_\varepsilon}(x)|^p\, dx \leq \int_{\mathbf{R}^n} \sup_{\varepsilon>0} |\widetilde{f_\varepsilon}(x)|^p\, dx \\
&\leq \left(\frac{\pi C_p}{2}\right)^p \left(\int_{\Sigma_{n-1}} |K(\sigma)|\, d\sigma\right)^p \cdot \int_{\mathbf{R}^n} |f(x)|^p\, dx,
\end{aligned}$$

so that

$$\widetilde{f} \in L^p(\mathbf{R}^n).$$

Therefore, by applying Lebesgue's dominated convergence theorem (Theorem 3.13) we obtain that

$$\|\widetilde{f_\varepsilon} - \widetilde{f}\|_p^p = \int_{|x|<M+1} |\widetilde{f_\varepsilon}(x) - \widetilde{f}(x)|^p\, dx \longrightarrow 0 \quad \text{as } \varepsilon \downarrow 0.$$

Step (3): We consider the general case where $f(x)$ is an arbitrary function in the space $L^p(\mathbf{R}^n)$.

We take a function $g \in C_0^1(\mathbf{R}^n)$ and let

$$h(x) := f(x) - g(x).$$

Just as in the proof of Theorem 9.15, we can prove that the sequence $\widetilde{f_\varepsilon}$ converges almost everywhere in $\mathbf{R}^n$ to some function $K * f$ as $\varepsilon \downarrow 0$:

$$\widetilde{f_\varepsilon} \longrightarrow K * f \quad \text{almost everywhere in } \mathbf{R}^n \text{ as } \varepsilon \downarrow 0.$$

Moreover, we show that

$$\widetilde{f_\varepsilon} \longrightarrow K * f \quad \text{in } L^p(\mathbf{R}^n) \text{ as } \varepsilon \downarrow 0.$$

To do this, we remark that

$$\begin{aligned} \left\| \widetilde{f_\varepsilon} - \widetilde{f_\delta} \right\|_p &= \left\| \left(\widetilde{h_\varepsilon} + \widetilde{g_\varepsilon}\right) - \left(\widetilde{h_\delta} + \widetilde{g_\varepsilon}\right) \right\|_p \\ &\leq \|\widetilde{g_\varepsilon} - \widetilde{g_\delta}\|_p + \left\| \widetilde{h_\varepsilon} - \widetilde{h_\delta} \right\|_p. \end{aligned}$$

Since g is uniformly Lipschitz continuous, it follows that the first term on the right-hand side of the inequality tends to 0 as ε, $\delta \downarrow 0$:

$$\lim_{\varepsilon,\delta\downarrow 0} \|\widetilde{g_\varepsilon} - \widetilde{g_\delta}\|_p = 0.$$

On the other hand, we have, by inequality (9.84),

$$\begin{aligned} \left\| \widetilde{h_\varepsilon} - \widetilde{h_\delta} \right\|_p &\leq 2 \left(\int_{\mathbf{R}^n} \sup_{\varepsilon>0} |\widetilde{h_\varepsilon}(x)|^p \, dx \right)^{1/p} \\ &\leq \pi C_p \left(\int_{\Sigma_{n-1}} |K(\sigma)| \, d\sigma \right) \|h\|_p. \end{aligned}$$

However, we remark that the last term on this inequality can be made arbitrarily small if we choose the function $g(x)$ appropriately.

In this way, we find that the sequence $\widetilde{f_\varepsilon}$ converges strongly to the function $\widetilde{f}$ in $L^p(\mathbf{R}^n)$ as $\varepsilon \downarrow 0$:

$$\widetilde{f_\varepsilon} \longrightarrow \widetilde{f} \quad \text{in } L^p(\mathbf{R}^n) \text{ as } \varepsilon \downarrow 0.$$

This proves that

$$K * f = \widetilde{f} \in L^p(\mathbf{R}^n),$$

so that

$$\widetilde{f_\varepsilon} \longrightarrow K * f \quad \text{in } L^p(\mathbf{R}^n) \text{ as } \varepsilon \downarrow 0.$$

Finally, by letting $\varepsilon \downarrow 0$ in inequality (9.84) we obtain from Lebesgue's dominated convergence theorem (Theorem 3.13) that

$$\|K * f\|_p \leq \frac{\pi C_p}{2}\left(\int_{\Sigma_{n-1}} |K(\sigma)|\, d\sigma\right)\|f\|_p \quad \text{for all } f \in L^p(\mathbf{R}^n).$$

This proves that the mapping $f \mapsto K * f$ is a bounded linear operator from $L^p(\mathbf{R}^n)$ into itself.

Now the proof of Theorem 9.16 is complete. □

9.8 Riesz Kernels

The *Riesz kernels* are functions defined by the formulas

$$R_j(x) = -\frac{\Gamma((n+1)/2)}{\pi^{(n+1)/2}} \frac{x_j}{|x|^{n+1}} \quad \text{for } 1 \leq j \leq n. \tag{9.88}$$

It is easy to see that the Riesz kernels $R_j(x)$ satisfy Assumption 9.3 and Assumption 9.4 with $\omega(t) := t$. Moreover, it should be emphasized that the Riesz kernels transform even kernels to *odd* kernels. Therefore, by making use of Theorem 9.16 we can prove the existence of singular integral operators with even kernel. In fact, this will be carried out in Sect. 9.9 below.

In this section we calculate explicitly the Fourier transforms $(\mathcal{F}R_j)(\xi)$ of the Riesz kernels $R_j(x)$ (see formula (9.100)).

To do this, we let $0 < \varepsilon < 1 < \mu$. Then we have, by the divergence theorem (Theorem 5.3),

$$\begin{aligned}
&(1-n)\int_{\varepsilon<|x|<1} e^{-ix\cdot\xi} \frac{x_j}{|x|^{n+1}}\, dx \\
&= \int_{\varepsilon<|x|<1} e^{-ix\cdot\xi} \frac{\partial}{\partial x_j}\left(|x|^{1-n}\right) dx \\
&= \int_{|x|=1} e^{-ix\cdot\xi} \frac{x_j}{|x|^n}\, d\sigma - \int_{|x|=\varepsilon} e^{-ix\cdot\xi} \frac{x_j}{|x|^n}\, dS \\
&= \int_{|x|=1} e^{-ix\cdot\xi} \frac{x_j}{|x|^n}\, d\sigma + \int_{\varepsilon<|x|<1} i\xi j e^{-ix\cdot\xi} |x|^{1-n}\, dx \\
&= \int_{|x|=1} x_j e^{-ix\cdot\xi}\, d\sigma - \frac{1}{\varepsilon^n}\int_{|x|=\varepsilon} x_j e^{-ix\cdot\xi}\, dS \\
&= \int_{|x|=1} x_j e^{-ix\cdot\xi}\, d\sigma + i\xi_j \int_{\varepsilon<|x|<1} e^{-ix\cdot\xi} |x|^{1-n}\, dx.
\end{aligned} \tag{9.89}$$

However, it follows that

$$\int_{|x|=1} x_j e^{-ix\cdot\xi}\, d\sigma = i\,\frac{\partial}{\partial \xi_j}\left(\int_{|x|=1} e^{-ix\cdot\xi}\, d\sigma\right)$$
$$= i\,\frac{\partial}{\partial \xi_j}\left(\int_{|x|=1} \cos(x\cdot\xi)\, d\sigma\right).$$

If we make an orthogonal transformation which maps ξ to $(|\xi|, 0, \ldots, 0)$, then we obtain that

$$\int_{|x|=1} x_j e^{-ix\cdot\xi}\, d\sigma = i\,\frac{\partial}{\partial \xi_j}\left(\int_{|y|=1} \cos(|\xi| y_1)\, d\sigma\right) \tag{9.90}$$
$$= -i\,\frac{\xi_j}{|\xi|}\int_{|y|=1} y_1 \sin(|\xi| y_1)\, d\sigma.$$

Moreover, it is easy to verify that

$$\lim_{\varepsilon\downarrow 0}\frac{1}{\varepsilon^n}\int_{|x|=\varepsilon} x_j e^{-ix\cdot\xi}\, dS = \lim_{\varepsilon\downarrow 0}\int_{|x|=1} \sigma_j e^{-i\varepsilon\sigma\xi}\, d\sigma \tag{9.91}$$
$$= \int_{\Sigma_{n-1}} \sigma_j\, d\sigma = 0,$$

and further that

$$\lim_{\varepsilon\downarrow 0}\int_{\varepsilon<|x|<1} e^{-ix\cdot\xi}|x|^{1-n}\, dx = \int_{|x|<1} e^{-ix\cdot\xi}|x|^{1-n}\, dx \tag{9.92}$$
$$= \int_{|x|<1} |x|^{1-n}\cos(x\cdot\xi)\, dx$$
$$= \int_{|y|<1} |y|^{1-n}\cos(|\xi| y_1)\, dy.$$

By combining four formulas (9.89), (9.90), (9.91) and (9.92), we obtain that

$$\lim_{\varepsilon\downarrow 0}\int_{\varepsilon<|x|<1} e^{-ix\cdot\xi}\frac{x_j}{|x|^{n+1}}\, dx$$
$$= \frac{1}{1-n}\left(-i\,\frac{\xi_j}{|\xi|}\int_{|y|=1} y_1 \sin(|\xi| y_1)\, d\sigma + i\,\xi_j \int_{|y|<1} |y|^{1-n}\cos(|\xi| y_1)\, dy\right).$$

Hence, if we let

$$c_1 := \frac{i}{1-n}\left(-\int_{|y|=1} y_1 \sin y_1\, d\sigma + \int_{|y|<1} |y|^{1-n}\cos y_1\, dy\right),$$

then we have, for $|\xi| = 1$,

$$\lim_{\varepsilon \downarrow 0} \int_{\varepsilon < |x| < 1} e^{-ix\cdot\xi} \frac{x_j}{|x|^{n+1}}\, dx = c_1 \xi_j. \tag{9.93}$$

Similarly, we have the formula

$$\int_{1<|x|<\mu} e^{-ix\cdot\xi} \frac{x_j}{|x|^{n+1}}\, dx = i \frac{\partial}{\partial \xi_j} \left(\int_{1<|x|<\mu} \frac{e^{-ix\cdot\xi}}{|x|^{n+1}}\, dx \right). \tag{9.94}$$

It is easy to see that the limit

$$\lim_{\mu \uparrow \infty} \int_{1<|x|<\mu} \frac{e^{-ix\cdot\xi}}{|x|^{n+1}}\, dx = \int_{|x|>1} \frac{e^{-ix\cdot\xi}}{|x|^{n+1}}\, dx \tag{9.95}$$

exists uniformly in $\mathbf{R}^n$. If $\xi_j \neq 0$, then it follows from an application of the divergence theorem (Theorem 5.3) that

$$\begin{aligned}
&\int_{1<|x|<\mu} e^{-ix\cdot\xi} \frac{x_j}{|x|^{n+1}}\, dx \\
&= \frac{i}{\xi_j} \int_{1<|x|<\mu} \frac{x_j}{|x|^{n+1}} \frac{\partial}{\partial x_j} \left(e^{-ix\cdot\xi}\right) dx \\
&= \frac{i}{\xi_j} \left(\int_{|x|=\mu} e^{-ix\cdot\xi} \frac{x_j}{|x|^{n+1}} \frac{x_j}{|x|}\, dS + \int_{|x|=1} e^{-ix\cdot\xi} \frac{x_j}{|x|^{n+1}} \frac{-x_j}{|x|}\, d\sigma \right. \\
&\quad \left. - \int_{1<|x|<\mu} e^{-ix\cdot\xi} \frac{\partial}{\partial x_j} \left(\frac{x_j}{|x|^{n+1}} \right) dx \right) \\
&= \frac{i}{\xi_j} \left(\int_{|x|=\mu} e^{-ix\cdot\xi} \frac{x_j^2}{|x|^{n+2}}\, dS - \int_{|x|=1} e^{-ix\cdot\xi} \frac{x_j^2}{|x|^{n+2}}\, d\sigma \right. \\
&\quad \left. - \int_{1<|x|<\mu} e^{-ix\cdot\xi} \frac{\partial}{\partial x_j} \left(\frac{x_j}{|x|^{n+1}} \right) dx \right) \\
&= \frac{i}{\xi_j} \left(\frac{1}{\mu} \int_{|x|=1} x_j^2 e^{-i\mu x\cdot\xi}\, d\sigma - \int_{|x|=1} x_j^2 e^{-ix\cdot\xi}\, d\sigma \right. \\
&\quad \left. - \int_{1<|x|<\mu} e^{-ix\cdot\xi} \frac{\partial}{\partial x_j} \left(\frac{x_j}{|x|^{n+1}} \right) dx \right),
\end{aligned}$$

where $dS = \mu^{n-1}\, d\sigma$ is the surface measure on the sphere $\{|x| = \mu\}$.

Hence we find that the limit

$$\begin{aligned}
&\lim_{\mu \uparrow \infty} \int_{1<|x|<\mu} e^{-ix\cdot\xi} \frac{x_j}{|x|^{x+1}}\, dx \qquad (9.96) \\
&= -\frac{i}{\xi_j} \left(\int_{|x|=1} x_j^2 e^{-ix\cdot\xi}\, d\sigma + \int_{|x|>1} e^{-ix\cdot\xi} \frac{\partial}{\partial x_j} \left(\frac{x_j}{|x|^{n+1}} \right) dx \right)
\end{aligned}$$

exists uniformly on any compact subset of the space $\{\xi \in \mathbf{R}^n : \xi_j \neq 0\}$.

By combining three formulas (9.94), (9.95) and (9.96), we have proved that

$$\int_{|x|>1} \frac{e^{-ix\cdot\xi}}{|x|^{n+1}}\,dx \in C^1(\{\xi \in \mathbf{R}^n : \xi_1 \cdots \xi_n \neq 0\}) \quad \text{for all } 1 \leq j \leq n, \tag{9.97}$$

and further that

$$\begin{aligned} & i\,\frac{\partial}{\partial \xi_j}\left(\int_{|x|>1} \frac{e^{-ix\cdot\xi}}{|x|^{n+1}}\,dx\right) = i \lim_{\mu\uparrow\infty} \frac{\partial}{\partial \xi_j}\left(\int_{1<|x|<\mu} \frac{e^{-ix\cdot\xi}}{|x|^{n+1}}\,dx\right) \\ &= \lim_{\mu\uparrow\infty} \int_{1<|x|<\mu} e^{-ix\cdot\xi}\frac{x_j}{|x|^{n+1}}\,dx \\ &= -\frac{i}{\xi_j}\left(\int_{|x|=1} x_j^2 e^{-ix\cdot\xi}\,d\sigma + \int_{|x|>1} e^{-ix\cdot\xi}\frac{\partial}{\partial x_j}\left(\frac{x_j}{|x|^{n+1}}\right)dx\right) \quad \text{for } \xi_j \neq 0. \end{aligned} \tag{9.98}$$

If we let (cf. formula (9.90))

$$\begin{aligned} f(|\xi|) &:= \int_{|x|>1} \frac{e^{-ix\cdot\xi}}{|x|^{n+1}}\,dx = \int_{|x|>1} \frac{\cos(x\cdot\xi)}{|x|^{n+1}}\,dx - i\int_{|x|>1} \frac{\sin(x\cdot\xi)}{|x|^{n+1}}\,dx \\ &= \int_{|y|>1} \frac{\cos(y_1|\xi|)}{|y|^{n+1}}\,dy, \end{aligned}$$

then it follows from assertion (9.97) that the function $f(|\xi|)$ is of class C^1 in the space $\{\xi \in \mathbf{R}^n : \xi_1 \cdots \xi_n \neq 0\}$. Hence we obtain that

$$f(t) = \int_{|y|>1} \frac{\cos(y_1 t)}{|y|^{n+1}}\,dy \in C^1(0,\infty),$$

and further that $f(|\xi|)$ is continuously differentiable in the space $\mathbf{R}^n \setminus \{0\}$. By formula (9.98), it follows that

$$\lim_{\mu\uparrow\infty} \int_{1<|x|<\mu} e^{-ix\cdot\xi}\frac{x_j}{|x|^{n+1}}\,dx = i\,\frac{\partial}{\partial \xi_j}(f(|\xi|)) = i\,\frac{\xi_j}{|\xi|}f'(|\xi|).$$

Hence, if we let

$$c_2 := if'(1) = -i\int_{|y|>1} \frac{y_1 \sin y_1}{|x|^{n+1}}\,dy,$$

then we have, for $|\xi| = 1$,

$$\lim_{\mu\uparrow\infty} \int_{1<|x|<\mu} e^{-ix\cdot\xi}\frac{x_j}{|x|^{n+1}}\,dx = i\xi_j\,f'(1) = c_2\,\xi_j. \tag{9.99}$$

By two formulas (9.93) and (9.99), it follows that we have, for $|\xi| = 1$,

$$\lim_{\varepsilon\downarrow 0,\mu\uparrow\infty}\int_{\varepsilon<|x|<\mu} e^{-ix\cdot\xi}\frac{x_j}{|x|^{n+1}}\,dx = c_1\,\xi_j + c_2\,\xi_j := c_3\,\xi_j, \tag{9.100}$$

where

$$c_3 := c_1 + c_2.$$

However, we remark that, since the left-hand side of formula (9.100) is a positively homogeneous function of ξ degree zero, we have, for all $\xi \neq 0$,

$$\lim_{\varepsilon\downarrow 0,\mu\uparrow\infty}\int_{\varepsilon<|x|<\mu} e^{-ix\cdot\xi}\frac{x_j}{|x|^{n+1}}\,dx = c_3\,\frac{\xi_j}{|\xi|}. \tag{9.101}$$

We calculate explicitly the value of the constant c_3. To do this, we may choose $j = 1$ and $\xi = (1, 0, \ldots, 0)$. Then it follows from formula (9.101) that

$$\lim_{\varepsilon\downarrow 0,\mu\uparrow\infty}\int_{\varepsilon<|x|<\mu} e^{-ix_1}\,\frac{x_1}{|x|^{n+1}}\,dx = c_3, \tag{9.102}$$

and further that

$$\begin{aligned}
&\int_{\varepsilon<|x|<\mu} e^{-ix_1}\frac{x_1}{|x|^{n+1}}\,dx \\
&= \int_{\varepsilon<|x|<\mu} \frac{x_1\cos x_1}{|x|^{n+1}}\,dx - i\int_{\varepsilon<|x|<\mu}\frac{x_1\sin x_1}{|x|^{n+1}}\,dx \\
&= -i\int_{\Sigma_{n-1}}\sigma_1\left(\int_{\varepsilon}^{\mu}\frac{\sin(r\sigma_1)}{r}\,dr\right)d\sigma.
\end{aligned} \tag{9.103}$$

However, since we have the formula

$$\lim_{\varepsilon\downarrow 0,\mu\uparrow\infty}\int_{\varepsilon}^{\mu}\frac{\sin(r\sigma_1)}{r}\,dr = \frac{\pi}{2}\,\operatorname{sign}\sigma_1$$

and the inequality (see inequality (9.16))

$$\left|\int_{\varepsilon}^{\mu}\frac{\sin(r\sigma_1)}{r}\,dr\right| \leq 2A,$$

by applying Lebesgue's dominated convergence theorem (Theorem 3.13) we obtain from two formulas (9.102) and (9.103) that

$$\begin{aligned}
c_3 &= \lim_{\varepsilon \downarrow 0, \mu \uparrow \infty} \int_{\varepsilon<|x|<\mu} e^{-ix_1} \frac{x_1}{|x|^{n+1}}\, dx \\
&= -i \lim_{\varepsilon \downarrow 0, \mu \uparrow \infty} \int_{\Sigma_{n-1}} \sigma_1 \left(\int_\varepsilon^\mu \frac{\sin(r\sigma_1)}{r}\, dr \right) d\sigma \\
&= -\frac{\pi}{2} i \left(\int_{\Sigma_{n-1}} \sigma_1 \operatorname{sign} \sigma_1 \, d\sigma \right) = -\pi i \int_{\sigma_1>0} \sigma_1 \, d\sigma.
\end{aligned}$$

Moreover, we have the formula

$$\begin{aligned}
\int_{\sigma_1>0} \sigma_1 \, d\sigma &= \omega_{n-1} \int_0^1 t(1-t^2)^{(n-3)/2}\, dt = \frac{1}{2}\omega_{n-1} \int_0^1 (1-s)^{(n-3)/2}\, ds \\
&= \frac{\omega_{n-1}}{n-1},
\end{aligned}$$

where

$$\omega_{n-1} := |\Sigma_{n-2}| = \frac{2\pi^{(n-1)/2}}{\Gamma((n-1)/2)}$$

is the surface area of the unit sphere Σ_{n-2} in $\mathbf{R}^{n-1}$.

Therefore, we obtain that

$$c_3 = -\pi i \frac{\omega_n}{n-1} = -\frac{\pi i}{n-1} \frac{2\pi^{(n-1)/2}}{\Gamma((n-1)/2)} = -\frac{i\pi^{(n+1)/2}}{\Gamma((n+1)/2)}.$$

Summing up, we find from formula (9.101) that the Fourier transforms $(\mathcal{F}R_j)(\xi)$ of the Riesz kernels $R_j(x)$ are given by the formulas

$$\begin{aligned}
(\mathcal{F}R_j)(\xi) &= \lim_{\varepsilon \downarrow 0, \mu \uparrow \infty} \int_{\varepsilon<|x|<\mu} e^{-ix\cdot\xi} R_j(x)\, dx \qquad (9.104) \\
&= -\frac{\Gamma((n+1)/2)}{\pi^{(n+1)/2}} \lim_{\varepsilon \downarrow 0, \mu \uparrow \infty} \int_{\varepsilon<|x|<\mu} e^{-ix\cdot\xi} \frac{x_j}{|x|^{n+1}}\, dx \\
&= -\frac{\Gamma((n+1)/2)}{\pi^{(n+1)/2}} c_3 \frac{\xi_j}{|\xi|} \\
&= \frac{\Gamma((n+1)/2)}{\pi^{(n+1)/2}} \frac{i\pi^{(n+1)/2}}{\Gamma((n+1)/2)} \frac{\xi_j}{|\xi|} \\
&= i \frac{\xi_j}{|\xi|} \quad \text{for } 1 \leq j \leq n.
\end{aligned}$$

Now, we let $f \in L^2(\mathbf{R}^n)$ and

$$
\begin{aligned}
g_j(x) &:= -\lim_{\varepsilon\downarrow 0}\int_{|x-y|>\varepsilon} R_j(x-y)f(y)\,dy \\
&= -\left(R_j * f\right)(x).
\end{aligned}
$$

Then we have, by formula (9.104),

$$
\begin{aligned}
&-\mathcal{F}\left(\sum_{j=1}^{n} R_j * \left(R_j * f\right)\right)(\xi) = \mathcal{F}\left(\sum_{j=1}^{n} R_j * g_j\right)(\xi) \\
&= \sum_{j=1}^{n}(\mathcal{F}R_j)(\xi)\cdot(\mathcal{F}g_j)(\xi) \\
&= -\sum_{j=1}^{n}\left(\left(\mathcal{F}R_j\right)(\xi)\right)^2\cdot\left(\mathcal{F}f\right)(\xi) = -\sum_{j=1}^{n}\frac{(-\xi_j^2)}{|\xi|^2}\cdot(\mathcal{F}f)(\xi) \\
&= \left(\mathcal{F}f\right)(\xi).
\end{aligned}
$$

By the Fourier inversion formula, this proves that we have, for all $f \in L^2(\mathbf{R}^n)$,

$$
-\sum_{j=1}^{n} R_j * \left(R_j * f\right) = f. \tag{9.105}
$$

Moreover, by arguing just as in the proof of Theorem 9.6 we obtain that formula (9.105) holds true for all $f \in L^p(\mathbf{R}^n)$ with $1 < p < \infty$.

Therefore, we have proved the following fundamental theorem for the *Riesz kernels*:

Theorem 9.17 *The Fourier transforms* $(\mathcal{F}R_j)(\xi)$ *of the Riesz kernels* $R_j(x)$ *are given by formulas* (9.104). *Moreover, formula* (9.105) *holds true for all* $f \in L^p(\mathbf{R}^n)$ *with* $1 < p < \infty$.

Remark 9.18 By virtue of formula (9.104), we have, for all $1 \leq j, k \leq n$,

$$
\frac{\partial^2 f}{\partial x_j \partial x_k} = -R_j * (R_k * (\Delta f)) \quad \text{for all } f \in C_0^\infty(\mathbf{R}^n).
$$

Therefore, by applying Theorem 9.16 (inequality (9.85)) we find that

$$
\left\|\frac{\partial^2 f}{\partial x_j \partial x_k}\right\|_{L^p(\mathbf{R}^n)} \leq C_p' \left\|\Delta f\right\|_{L^p(\mathbf{R}^n)} \quad \text{for all } 1 \leq j, k \leq n,
$$

where $C_p' > 0$ is a constant. In other words, the Laplacian Δ controls all second order partial derivatives in the L^p-norm for $1 < p < \infty$.

9.9 The Case of Even Kernels

In this section we consider the case where the integral kernel $K(x)$ is an even function, and make the following assumption:

Assumption 9.5 The integral kernel $K(x)$ is an even function satisfying Assumption 9.1 and the condition

$$\int_{\Sigma_{n-1}} |K(\sigma)| \log^+ |K(\sigma)| \, d\sigma < \infty. \tag{9.106}$$

Here $\log^+ x$ is a function defined by the formula

$$\log^+ x := \max\{\log x, 0\} = \begin{cases} \log x & \text{if } x > 1, \\ 0 & \text{if } 0 < x \leq 1. \end{cases}$$

The case of even kernels can be reduced to the case of odd kernels. Indeed, we have, by formula (9.105) with $f := K * f$,

$$K * f = -\sum_{j=1}^{n} R_j * (R_j * K) * f.$$

However, it should be noticed that the operators R_j and $R_j * K$ have respectively the odd kernels

$$\begin{aligned} R_j(x) &= -\frac{\Gamma((n+1)/2)}{\pi^{(n+1)/2}} \frac{x_j}{|x|^{n+1}}, \\ R_j * K(x) &= -\frac{\Gamma((n+1)/2)}{\pi^{(n+1)/2}} \frac{x_j}{|x|^{n+1}} * K(x) \end{aligned}$$

of positively homogeneous of degree $-n$. In other words, a singular integral operator with even kernel can be expressed as a finite sum of products of singular integral operators with odd kernel.

We begin by proving the following elementary result:

Lemma 9.19 *For all $a > 0$ and $b > 0$, we have the inequality*

$$ab \leq a \log^+ a + e^{b-1}. \tag{9.107}$$

Proof Let $\varphi(x)$ and $\psi(y)$ be two functions defined respectively by the formulas

$$\varphi(x) := \begin{cases} \log x + 1 & \text{if } x > 1, \\ 0 & \text{if } 0 \leq x \leq 1, \end{cases}$$

and

$$\psi(y) := \begin{cases} e^{y-1} & \text{if } y > 1, \\ 1 & \text{if } 0 \leq y \leq 1x. \end{cases}$$

Then it follows that $\psi(y)$ is the inverse function of $\varphi(x)$ in the region $\{y > 1\}$ (or $\{x > 1\}$). Hence we obtain that

$$ab \leq \int_0^a \varphi(x)\, dx + \int_0^b \psi(y)\, dy. \tag{9.108}$$

However, we have the formulas

$$\int_0^a \varphi(x)\, dx = \begin{cases} 0 & \text{if } 0 \leq a \leq 1, \\ a \log a & \text{if } a > 1, \end{cases}$$

and

$$\int_0^b \psi(y)\, dy = \begin{cases} 0 & \text{if } 0 \leq b \leq 1, \\ e^{b-1} - 1 & \text{if } b > 1. \end{cases}$$

Therefore, the desired inequality (9.107) follows immediately from inequality (9.108).

The proof of Lemma 9.19 is complete. □

Lemma 9.20 *Let $K(x)$ be an integral kernel satisfying Assumption 9.3. Assume that $f(x)$ is a measurable function with compact support in $\mathbf{R}^n$ which satisfies the condition*

$$\int_{\mathbf{R}^n} |f(x)| \log^+ |f(x)|\, dx < \infty. \tag{9.109}$$

If S is a measurable set of $\mathbf{R}^n$ with $|\text{supp} f| < |S| < \infty$, where $\text{supp} f$ denotes the support of $f(x)$, then there exist positive constants C_3 and C_4 such that we have, for all $\varepsilon > 0$,

$$\int_S \left|\widetilde{f_\varepsilon}(x)\right| dx \leq C_3 \int_{\mathbf{R}^n} |f(x)| \log^+ |f(x)|\, dx + C_4 |S|, \tag{9.110}$$

where $\widetilde{f_\varepsilon}(x)$ is the function defined by formula (9.5)*:*

$$\widetilde{f_\varepsilon}(x) = \int_{|x-y|>\varepsilon} K(x-y)\, f(y)\, dy.$$

Proof Since we have the inequality

$$\begin{aligned} \int_{\mathbf{R}^n} |f(x)|\, dx &= \int_{|f|>e} |f(x)|\, dx + \int_{|f|\leq e} |f(x)|\, dx \\ &\leq \int_{\mathbf{R}^n} |f(x)| \log^+ |f(x)|\, dx + e\, |\text{supp} f|, \end{aligned} \tag{9.111}$$

it follows from condition (9.109) that

$$f \in L^1(\mathbf{R}^n).$$

Step (1): First, we consider the case where $f(x)$ is non-negative: We let

$$E_s := \left\{x : \left|\widetilde{f_\varepsilon}(x)\right| > s\right\},$$
$$E_s' := E_s \cap S.$$

Let $\beta_f(t)$ and $\beta^f(t)$ be the two functions defined in Sect. 9.5, and $s_0 := \beta_f(|S|)$. Then we have, by Theorem 3.45,

$$\int_S \left|\widetilde{f_\varepsilon}(x)\right| dx = \int_0^\infty \left|E_s'\right| ds \leq |S|s_0 + \int_{s_0}^\infty |E_s|\, ds. \tag{9.112}$$

However, by Lemma 9.10 it follows that

$$|S|s_0 = |S|\beta_f(|S|) = \int_0^{|S|} f^*(s)\, ds \leq \int_{\mathbf{R}^n} f(x)\, dx. \tag{9.113}$$

Moreover, by combining Theorem 3.45 and Lemma 9.11 we obtain that

$$|D_s| \leq \beta^f(s).$$

Hence, we have, by Lemma 9.5,

$$\int_{s_0}^\infty |E_s|\, ds \leq C_1 \int_0^\infty \frac{1}{s^2} \int_{\mathbf{R}^n} [f(x)]_s^2\, dx\, ds + C_2 \int_{s_0}^\infty \beta^f(s)\, ds, \tag{9.114}$$

where

$$[f(x)]_s = \begin{cases} f(x) & \text{if } f(x) \leq s, \\ s & \text{if } f(x) > s. \end{cases}$$

Hence we have, by Fubini's theorem (Theorem 3.20),

$$\int_0^\infty \frac{1}{s^2}\left(\int_{\mathbf{R}^n} [f(x)]_s^2\, dx\right) ds = \int_{\mathbf{R}^n}\left(\int_0^\infty \frac{1}{s^2}[f(x)]_s^2\, ds\right) dx \tag{9.115}$$
$$= \int_{\mathbf{R}^n}\left[\int_0^{f(x)} \frac{1}{s^2}\cdot s^2\, ds + \int_{f(x)}^\infty \frac{1}{s^2}\, f(x)^2\, ds\right] dx$$
$$= \int_{\mathbf{R}^n}\left[\int_0^{f(x)} ds + f(x)^2 \cdot \frac{1}{f(x)}\right] dx$$
$$= 2\int_{\mathbf{R}^n} f(x)\, dx.$$

Let $s_1 = \beta_f(t_1)$ be the number defined in Sect. 9.5. If we make the change of the variable $t = \beta^f(s)$ or $s = \beta_f(t)$, then it is easy to see that $\beta^f(s_0) = |S|$ and $\beta^f(s_1) = t_1$. Hence we have the formula

$$\begin{aligned}
\int_{s_0}^{\infty} \beta^f(s)\,ds &= \int_{s_0}^{s_1} \beta^f(s)\,ds = \int_{|S|}^{t_1} t\,d\beta_f(t) \\
&= \left[t\beta_f(t)\right]_{|S|}^{t_1} + \int_{t_1}^{|S|} \beta_f(t)\,dt = \left[\int_0^t f^*(s)\,ds\right]_{|S|}^{t_1} + \int_{t_1}^{|S|} \beta_f(t)\,dt \\
&= \int_0^{t_1} f^*(s)\,ds - \int_0^{|S|} f^*(s)\,ds + \int_{t_1}^{|S|} \beta_f(t)\,dt \\
&:= A.
\end{aligned}$$

Here it should be noticed that $t_1 \leq |S|$, since $|S| > |\mathrm{supp}\, f|$. Moreover, since $f^*(s) = \beta_f(s) = \beta_f(t_1)$ for $0 < s \leq t_1$, we can estimate the last term on the above formula as follows:

$$\begin{aligned}
A &:= \int_0^{t_1} f^*(s)\,ds - \int_0^{|S|} f^*(s)\,ds + \int_{t_1}^{|S|} \beta_f(t)\,dt \\
&= \int_0^{|S|} \beta_f(t)\,dt - \int_0^{|S|} f^*(s)\,ds \\
&\leq \int_0^{|S|} \beta_f(t)\,dt = \int_0^{|S|} \frac{1}{t} \int_0^t f^*(s)\,ds\,dt \\
&= \int_0^{|S|} f^*(s) \left(\int_s^{|S|} \frac{dt}{t} \right) ds = \int_0^{|S|} f^*(t) \log \frac{|S|}{t}\,dt \\
&= 2 \int_0^{|S|} f^*(t) \log \left(\frac{|S|}{t} \right)^{1/2} dt.
\end{aligned}$$

However, by applying Lemma 9.19 with

$$\begin{aligned}
a &:= f^*(t), \\
b &:= \log \left(\frac{|S|}{t} \right)^{1/2},
\end{aligned}$$

we obtain that

$$\begin{aligned}
\int_{s_0}^{\infty} \beta^f(s)\,ds &= A \leq 2 \int_0^{|S|} f^*(t) \log \left(\frac{|S|}{t} \right)^{1/2} dt \qquad (9.116) \\
&\leq 2 \int_0^{|S|} \left[f^*(t) \log^+ f^*(t) + \exp \left(\log \left(\frac{|S|}{t} \right)^{1/2} - 1 \right) \right] dt
\end{aligned}$$

$$
\begin{aligned}
&\leq 2\int_0^{\infty} f^*(t)\,\log^+ f^*(t)\,dt + \frac{2C_2}{e}\int_0^{|S|}\left(\frac{|S|}{t}\right)^{1/2} dt \\
&= 2\int_{\mathbf{R}^n} f(x)\,\log^+ f(x)\,dx + \frac{4C_2}{e}|S|.
\end{aligned}
$$

Therefore, by combining inequality (9.112) with inequalities (9.113) and (9.114) and formulas (9.115) and (9.116) and then using inequality (9.111), we obtain that

$$
\begin{aligned}
&\int_S \left|\widetilde{f_\varepsilon}(x)\right| dx \\
&\leq |S|s_0 + \int_{s_0}^{\infty} |E_s|\,ds \\
&\leq \int_{\mathbf{R}^n} f(x)\,dx + C_1\int_0^{\infty}\frac{1}{s^2}\left(\int_{\mathbf{R}^n}[f(x)]_s^2\,dx\right)ds + C_2\int_{s_0}^{\infty}\beta^f(s)\,ds \\
&\leq (1+2C_1)\int_{\mathbf{R}^n} f(x)\,dx + C_2\int_{s_0}^{\infty}\beta^f(s)\,ds \\
&\leq (1+2C_1)\left(\int_{\mathbf{R}^n} f(x)\,\log^+ f(x)\,dx + e\,|S|\right) \\
&\quad + 2C_2\int_{\mathbf{R}^n} f(x)\log^+ f(x)\,dx + \frac{4C_2}{e}|S| \\
&= (1+2C_1+2C_2)\int_{\mathbf{R}^n} f(x)\,\log^+ f(x)\,dx + \left((1+2C_1)e + \frac{4C_2}{e}\right)|S|.
\end{aligned}
$$

This proves the desired inequality (9.110) in the case where $f(x)$ is non-negative, with

$$
C_3 := 1 + 2C_1 + 2C_2, \qquad C_4 := (1+2C_1)e + \frac{4C_2}{e}.
$$

Step (2): In the general case, by decomposing a function $f(x)$ into its positive part $f^+(x)$ and its negative part $f^-(x)$ as

$$
\begin{aligned}
&f(x) = f^+(x) - f^-(x), \\
&f^+(x) := \max\{f(x), 0\} = \frac{1}{2}\left(f(x) + |f(x)|\right), \\
&f^-(x) := \max\{-f(x), 0\} = \frac{1}{2}\left(-f(x) + |f(x)|\right),
\end{aligned}
$$

we can easily prove that the desired inequality (9.110) holds true with

$$
C_3 := 1 + 2C_1 + 2C_2, \qquad C_4 := 2(1+2C_1)e + \frac{8C_2}{e}.
$$

The proof of Lemma 9.20 is now complete. □

We need the following elementary inequalities for the function $\log^+ x$:

Lemma 9.21 *For all $\alpha > 0$ and $\beta > 0$, we have the two inequalities*

$$\log^+(\alpha\beta) \leq \log^+ \alpha + \log^+ \beta, \tag{9.117}$$

and

$$\frac{\alpha+\beta}{2} \log^+ \left(\frac{\alpha+\beta}{2}\right) \leq \frac{1}{2}\left(\alpha \log^+ \alpha + \beta \log^+ \beta\right). \tag{9.118}$$

Proof (1) If $\alpha\beta \leq 1$, it follows that $\log^+(\alpha\beta) = 0$. If $\alpha\beta > 1$, then it follows that

$$\log^+(\alpha\beta) = \log(\alpha\beta) = \log\alpha + \log\beta \leq \log^+ \alpha + \log^+ \beta.$$

so that the first inequality (9.117) follows.

(2) Next, we remark that the function $f(x) := x \log^+ x$ is convex for $x > 1$ and $f(x) = 0$ for $0 < x \leq 1$. Hence we have the inequality

$$f\left(\frac{\alpha+\beta}{2}\right) \leq \frac{1}{2}\left(f(\alpha) + f(\beta)\right).$$

This proves the second inequality (9.118).

The proof of Lemma 9.21 is complete. □

Now we can prove the following fundamental result:

Lemma 9.22 *Assume that the integral kernel $K(x)$ satisfies Assumption 9.3. If $f(x)$ is a measurable function with compact support in $\mathbf{R}^n$ that satisfies condition (9.109), then the sequence $\widetilde{f}_\varepsilon$ converges in the space $L^1_{\mathrm{loc}}(\mathbf{R}^n)$ as $\varepsilon \downarrow 0$.*

Proof Let S be an arbitrary bounded measurable set such that $|\mathrm{supp}\, f| < |S|$. It suffices to show that

$$\lim_{\varepsilon,\varepsilon' \downarrow 0} \int_S \left|\widetilde{f}_\varepsilon(x) - \widetilde{f}_{\varepsilon'}(x)\right| dx = 0.$$

If k is a positive integer, we let

$$[f(x)]_k := \begin{cases} f(x) & \text{if } |f(x)| \leq k, \\ k\, \frac{f(x)}{|f(x)|} & \text{if } |f(x)| > k. \end{cases}$$

For any given $0 < \delta < 2$, we can choose the integer k so large that

$$\int_{\mathbf{R}^n} |f(x) - [f(x)]_k| \cdot \log^+ |f(x) - [f(x)]_k| \, dx < \frac{\delta}{4}, \tag{9.119}$$

and that

$$\int_{\mathbf{R}^n} |f(x) - [f(x)]_k| \, dx < \frac{\delta}{4} \frac{1}{\log^+(2/\delta)}. \tag{9.120}$$

Indeed, we have, by condition (9.109),

$$\begin{aligned}
&\lim_{k \to \infty} \int_{\mathbf{R}^n} |f(x) - [f(x)]_k| \cdot \log^+ |f(x) - [f(x)]_k| \, dx \\
&= \lim_{k \to \infty} \int_{|f(x)| > k+1} (|f(x)| - k) \cdot \log^+ (|f(x)| - k) \, dx \\
&\leq \lim_{k \to \infty} \int_{|f(x)| > k+1} |f(x)| \cdot \log^+ |f(x)| \, dx = 0.
\end{aligned}$$

Similarly, since $f \in L^1(\mathbf{R}^n)$, it follows that

$$\begin{aligned}
\lim_{k \to \infty} \int_{\mathbf{R}^n} |f(x) - [f(x)]_k| \, dx &= \lim_{k \to \infty} \int_{|f(x)| > k} |f(x) - k| \, dx \\
&\leq \lim_{k \to \infty} \int_{|f(x)| > k} |f(x)| \, dx = 0.
\end{aligned}$$

By applying inequality (9.117) with

$$\begin{aligned}
\alpha &:= |f(x) - [f(x)]_k|, \\
\beta &:= \frac{2}{\delta},
\end{aligned}$$

we obtain from two inequalities (9.119) and (9.120) that

$$\begin{aligned}
&\int_{\mathbf{R}^n} \frac{|f(x) - [f(x)]_k|}{\delta/2} \log^+ \left(\frac{|f(x) - [f(x)]_k|}{\delta/2} \right) dx \\
&\leq \int_{\mathbf{R}^n} \frac{|f(x) - [f(x)]_k|}{\delta/2} \log^+ (|f(x) - [f(x)]_k|) \, dx \\
&\quad + \int_{\mathbf{R}^n} \frac{|f(x) - [f(x)]_k|}{\delta/2} \log^+ \left(\frac{2}{\delta} \right) dx \\
&< 1.
\end{aligned} \tag{9.121}$$

We take a function $g(x)$ in $C_0^1(\mathbf{R}^n)$ that satisfies the conditions

$$\begin{aligned}
&|\mathrm{supp} g| < |S|, \\
&\int_{\mathbf{R}^n} |g(x) - [f(x)]_k|^2 \, dx < \left(\frac{\delta}{2} \right)^2.
\end{aligned}$$

Then, since we have the elementary inequality

$$\alpha \log^+ \alpha \leq \alpha^2 \quad \text{for all } \alpha > 0,$$

it follows that

$$\begin{aligned} &\int_{\mathbf{R}^n} \frac{|g(x) - [f(x)]_k|}{\delta/2} \log^+ \left(\frac{|g(x) - [f(x)]_k|}{\delta/2} \right) dx \\ &\leq \int_{\mathbf{R}^n} \left(\frac{|g(x) - [f(x)]_k|}{\delta/2} \right)^2 dx \\ &< 1. \end{aligned} \tag{9.122}$$

Moreover, by letting

$$\alpha := \frac{|f(x) - [f(x)]_k|}{\delta/2},$$
$$\beta := \frac{|[f(x)]_k - g(x)|}{\delta/2},$$

we have the inequality

$$\begin{aligned} \frac{|f(x) - g(x)|}{\delta/2} &\leq \frac{|f(x) - [f(x)]_k|}{\delta/2} + \frac{|[f(x)]_k - g(x)|}{\delta/2} \\ &= \alpha + \beta. \end{aligned}$$

Hence it follows from an application of inequality (9.118) that

$$\begin{aligned} &2\,\frac{|f(x) - g(x)|}{\delta} \log^+ \left(\frac{|f(x) - g(x)|}{\delta} \right) \\ &\leq (\alpha + \beta) \log^+ \left(\frac{\alpha + \beta}{2} \right) \leq \alpha \log^+ \alpha + \beta \log^+ \beta \\ &= 2\,\frac{|f(x) - [f(x)]_k|}{\delta} \log^+ \left(\frac{|f(x) - [f(x)]_k|}{\delta/2} \right) \\ &\quad + 2\,\frac{|[f(x)]_k - g(x)|}{\delta} \log^+ \left(\frac{|[f(x)]_k - g(x)|}{\delta/2} \right). \end{aligned} \tag{9.123}$$

If we let

$$h(x) := f(x) - g(x),$$

then we obtain from inequalities (9.121), (9.122) and (9.123) that

$$\begin{aligned}
&\int_{\mathbf{R}^n} \frac{|h(x)|}{\delta} \log^+\left(\frac{|h(x)|}{\delta}\right) dx \qquad (9.124)\\
&\le \int_{\mathbf{R}^n} \frac{|f(x)-[f(x)]_k|}{\delta} \log^+\left(\frac{|f(x)-[f(x)]_k|}{\delta/2}\right) dx\\
&+ \int_{\mathbf{R}^n} \frac{|g(x)-[f(x)]_k|}{\delta} \log^+\left(\frac{|g(x)-[f(x)]_k|}{\delta/2}\right) dx\\
&< \frac{1}{2}+\frac{1}{2}=1.
\end{aligned}$$

On the other hand, since $|\text{supp} h| < |S|$, by applying inequality (9.110) (Lemma 9.20) with

$$f(x) := \frac{h(x)}{\delta},$$

we obtain from inequality (9.124) that

$$\begin{aligned}
\int_S \frac{|\widetilde{h}_\varepsilon(x)|}{\delta}\, dx &\le C_3 \int_{\mathbf{R}^n} \frac{|h(x)|}{\delta} \log^+\left(\frac{|h(x)|}{\delta}\right) dx + C_4|S|\\
&\le C_3 + C_4|S|.
\end{aligned}$$

Hence we have, for all $\varepsilon > 0$ and $\varepsilon' > 0$,

$$\begin{aligned}
&\int_S \left|\widetilde{f}_\varepsilon(x) - \widetilde{f}_{\varepsilon'}(x)\right| dx\\
&= \int_S \left|\left(\widetilde{g}_\varepsilon(x) - \widetilde{h}_\varepsilon(x)\right) - \left(\widetilde{g}_{\varepsilon'}(x) - \widetilde{h}_{\varepsilon'}(x)\right)\right| dx\\
&\le \int_S |\widetilde{g}_\varepsilon(x) - \widetilde{g}_{\varepsilon'}(x)|\, dx + \int_S \left|\widetilde{h}_\varepsilon(x)\right| dx + \int_S \left|\widetilde{h}_{\varepsilon'}(x)\right| dx\\
&\le \int_S |\widetilde{g}_\varepsilon(x) - \widetilde{g}_{\varepsilon'}(x)|\, dx + 2\delta(C_3 + C_4|S|).
\end{aligned}$$

However, the sequence $\widetilde{g}_\varepsilon(x)$ converges uniformly as $\varepsilon \downarrow 0$, as is shown in the proof of Theorem 9.6. Therefore, we obtain that

$$\limsup_{\varepsilon,\varepsilon' \downarrow 0} \int_S \left|\widetilde{f}_\varepsilon(x) - \widetilde{f}_{\varepsilon'}(x)\right| dx \le 2\delta(C_3 + C_4|S|).$$

This proves the desired assertion

$$\lim_{\varepsilon,\varepsilon' \downarrow 0} \int_S \left|\widetilde{f}_\varepsilon(x) - \widetilde{f}_{\varepsilon'}(x)\right| dx = 0,$$

since $\delta > 0$ is arbitrary.

The proof of Lemma 9.22 is complete. □

Now we take a function $\phi(t)$ in $C^1(\mathbf{R})$ such that

$$\phi(t) = \begin{cases} 0 & \text{if} 0 \leq t \leq 1/4, \\ 1 & \text{if} t > 3/4. \end{cases}$$

If the $R_j(x)$ are Riesz kernels

$$R_j(x) = -\frac{\Gamma((n+1)/2)}{\pi^{(n+1)/2}} \frac{x_j}{|x|^{n+1}} \quad \text{for } 1 \leq j \leq n,$$

we define integral kernels

$$K_j^{(1)}(x) := \lim_{\varepsilon \downarrow 0} \lim_{\delta \downarrow 0} \int_{|x-y|>\varepsilon, |y|>\delta} R_j(x-y)\, K(y)\, dy \quad \text{for } 1 \leq j \leq n, \tag{9.125}$$

and

$$K_j^{(2)}(x) := \lim_{\varepsilon \downarrow 0} \int_{|x-y|>\varepsilon} R_j(x-y)\, K(y)\, \phi(|y|)\, dy \quad \text{for } 1 \leq j \leq n. \tag{9.126}$$

The first purpose of this section is to prove the following lemma:

Lemma 9.23 *(i) The right-hand side of formula* (9.125) *converges in the space* $L^1_{\mathrm{loc}}(\mathbf{R}^n \setminus \{0\})$

$$K_j^{(1)} \in L^1_{\mathrm{loc}}(\mathbf{R}^n \setminus \{0\}) \quad \textit{for } 1 \leq j \leq n,$$

while the right-hand side of formula (9.126) *converges in the space* $L^1_{\mathrm{loc}}(\mathbf{R}^n)$

$$K_j^{(2)} \in L^1_{\mathrm{loc}}(\mathbf{R}^n) \quad \textit{for } 1 \leq j \leq n.$$

The integral kernels $K_j^{(1)}(x)$ *and* $K_j^{(2)}(x)$ *are both* odd *functions, and* $K_1(x)$ *is positively homogeneous of degree* $-n$.

(ii) If $K \in L^q(\Sigma_{n-1})$ *for some* $1 < q < \infty$, *then it follows that*

$$K_j^{(1)} \in L^q(\Sigma_{n-1}) \quad \textit{for } 1 \leq j \leq n,$$

and further that we have, for some constant $C_q > 0$,

$$\left(\int_{\Sigma_{n-1}} \left| K_j^{(1)}(\sigma) \right|^q d\sigma \right)^{1/q} \tag{9.127}$$
$$\leq C_q \left(\int_{\Sigma_{n-1}} |K(\sigma)|^q \, d\sigma \right)^{1/q} \quad \textit{for } 1 \leq j \leq n.$$

Proof The proof of Lemma 9.23 is divided into four steps. In the following we shall denote by C a generic positive constant.

Step (1): First, we assume that $1/2 \leq |x| \leq 1$ and that $0 < \varepsilon < 1/4$. Then we have the formula

$$\int_{|x-y|>\varepsilon} R_j(x-y)\,K(y)\,dy = \lim_{\delta\downarrow 0} \int_{|x-y|>\varepsilon, |y|>\delta} R_j(x-y)\,K(y)\,dy \tag{9.128}$$
$$= \lim_{\delta\downarrow 0} \int_{\delta<|y|<1/4} R_j(x-y)\,K(y)\,dy$$
$$+ \int_{|x-y|>\varepsilon, 1/4<|y|<2} R_j(x-y)\,K(y)\,dy + \int_{|y|>2} R_j(x-y)\,K(y)\,dy$$
$$:= \mathrm{I}_j^{(1)}(x) + \mathrm{I}_j^{(2)}(x;\varepsilon) + \mathrm{I}_j^{3}(x) \quad \text{for } 1 \leq j \leq n.$$

Here it should be noticed that

$$K_j^{(1)}(x) = \mathrm{I}_j^{(1)}(x) + \lim_{\varepsilon\downarrow 0} \mathrm{I}_j^{(2)}(x;\varepsilon) + \mathrm{I}_j^{(3)}(x) \quad \text{for } 1 \leq j \leq n. \tag{9.129}$$

We estimate the three terms $\mathrm{I}_j^{(1)}(x)$, $\mathrm{I}_j^{(2)}(x;\varepsilon)$ and $\mathrm{I}_j^{(3)}(x)$ on the right-hand side of formula (9.128). In the following we shall denote by C a generic positive constant.

Step (1-a): We recall that the Riesz kernels $R_j(x)$ satisfy Assumption 9.3 with $\omega(t) := t$. By using inequality (9.41) with $y_k := 0$, we can find a constant $C > 0$ such that we have, for all $|x| \geq 1/2$ and $|y| < 1/4$,

$$|R_j(x-y) - R_j(x)| \leq \frac{C}{|x|^{n+1}}\,|y|.$$

Hence we have the inequality

$$\left|\mathrm{I}_j^{(1)}(x)\right| = \left|\lim_{\delta\downarrow 0} \int_{\delta<|y|<1/4} (R_j(x-y) - R_j(x))K(y)\,dy\right| \tag{9.130}$$
$$\leq \frac{C}{|x|^{n+1}} \int_{0<|y|<1/4} |y|\,|K(y)|\,dy$$
$$= \frac{C}{|x|^{n+1}} \int_{\Sigma_{n-1}} \left(\int_0^{1/4} |t\sigma|\,|K(t\sigma)|\,t^{n-1}\,dt\right) d\sigma$$
$$= \frac{C}{4\,|x|^{n+1}} \int_{\Sigma_{n-1}} |K(\sigma)|\,d\sigma \quad \text{for all } 1/2 \leq |x| \leq 1 \quad \text{for } 1 \leq j \leq n.$$

This proves that

$$\mathrm{I}_j^{(1)} \in L^1(\{1/2 \leq |x| \leq 1\}) \quad \text{for } 1 \leq j \leq n.$$

Step (1-b): If $1/2 \leq |x| \leq 1$ and $|y| > 2$, then it follows that

$$|x| \leq 1 < \frac{|y|}{2},$$

so that

$$|x-y| \geq |y| - |x| > \frac{|y|}{2}.$$

Hence we have the inequality

$$\begin{aligned}\left|\mathrm{I}_j^{(3)}(x)\right| &= \left|\int_{|y|>2} R_j(x-y)\,K(y)\,dy\right| \\ &\leq \int_{|y|>2} \frac{C}{|x-y|^n}|K(y)|\,dy \leq 2^n\,C \int_{|y|>2} \frac{|K(y)|}{|y|^n}\,dy \\ &= 2^n\,C \int_{\Sigma_{n-1}} \left(\int_2^\infty \frac{1}{|t\sigma|^n}\,|K(t\sigma)|\,t^{n-1}\,dt\right) d\sigma \\ &= \frac{C}{n}\int_{\Sigma_{n-1}} |K(\sigma)|\,d\sigma \quad \text{for all } 1/2 \leq |x| \leq 1.\end{aligned} \tag{9.131}$$

This proves that

$$\mathrm{I}_j^{(3)} \in L^1\left(\{1/2 \leq |x| \leq 1\}\right) \quad \text{for } 1 \leq j \leq n.$$

Step (1-c): In order to estimate the middle term $\mathrm{I}_j^{(2)}(x;\varepsilon)$, we introduce a function $f(x)$ by the formula

$$f(x) := \begin{cases} K(x) & \text{if } 1/4 < |x| < 2, \\ 0 & \text{otherwise.}\end{cases}$$

Then we have the inequality

$$\begin{aligned}&\int_{\mathbf{R}^n} |f(x)|\log^+|f(x)|\,dx \\ &= \int_{1/4<|x|<2} |K(x)|\log^+|K(x)|\,dx \\ &= \int_{\Sigma_{n-1}} \left(\int_{1/4}^2 |K(t\sigma)|\log^+|K(t\sigma)|\,t^{n-1}\,dt\right) d\sigma \\ &= \int_{\Sigma_{n-1}} \left(\int_{1/4}^2 |K(\sigma)|\log^+(t^{-n}|K(\sigma)|)\,\frac{dt}{t}\right) d\sigma \\ &\leq \int_{\Sigma_{n-1}} \left(\int_{1/4}^2 |K(\sigma)|\left(\log^+|K(\sigma) + \log^+(t^{-n})\right)\frac{dt}{t}\right) d\sigma \\ &= \int_{\Sigma_{n-1}} |K(\sigma)|\log^+|K(\sigma)|\,d\sigma \cdot \int_{1/4}^2 \frac{dt}{t}\end{aligned}$$

$$
\begin{aligned}
&+\int_{\Sigma_{n-1}} |K(\sigma)|\,d\sigma \cdot \int_{1/4}^{2} \log^{+}(t^{-n})\,\frac{dt}{t} \\
&\leq C\left(\int_{\Sigma_{n-1}} |K(\sigma)| \log^{+}|K(\sigma)|\,d\sigma + \int_{\Sigma_{n-1}} |K(\sigma)|\,d\sigma\right) \\
&< \infty.
\end{aligned}
$$

This proves that the function $f(x)$ satisfies condition (9.109). Hence, by applying Lemma 9.22 with

$$
K(x) := R_j(x) = -\frac{\Gamma((n+1)/2)}{\pi^{(n+1)/2}} \frac{x_j}{|x|^{n+1}} \quad \text{for } 1 \leq j \leq n,
$$

we find that the term

$$
\begin{aligned}
\mathrm{I}_j^{(2)}(x;\varepsilon) &= \int_{|x-y|>\varepsilon,\,1/4<|y|<2} R_j(x-y)\,K(y)\,dy \\
&= \int_{|x-y|>\varepsilon} R_j(x-y)\,f(y)\,dy \quad \text{for } 1 \leq j \leq n,
\end{aligned}
$$

converges in the space $L^1_{\mathrm{loc}}(\mathbf{R}^n)$ as $\varepsilon \downarrow 0$.

Therefore, we have proved that formula (9.126) converges in the space $L^1(\{1/2 \leq |x| \leq 1\})$.

Step (2): We assume that $1 \leq |x| \leq 2$. Then we have the formula

$$
\begin{aligned}
\int_{|x-y|>\varepsilon} R_j(x-y)\,K(y)\,dy &= \int_{|x-2z|>\varepsilon/2} R_j\,(x-2z)\,K(2z)\,2^n\,dz \\
&= \frac{1}{2^n} \int_{|x/2-z|>\varepsilon/2} R_j\left(\frac{x}{2}-z\right) K(z)\,dz.
\end{aligned}
$$

Hence it follows that formula (9.129) converges in the space $L^1(\{1 \leq |x| \leq 2\})$.

Similarly, we have, for $1/2^2 \leq |x| \leq 1/2$,

$$
\begin{aligned}
\int_{|x-y|>\varepsilon} R_j(x-y)\,K(y)\,dy &= \int_{|x-z/2|>\varepsilon/2} R_j\left(x-\frac{z}{2}\right) K\left(\frac{z}{2}\right) \frac{1}{2^n}\,dz \\
&= 2^n \int_{|2x-z|>2\varepsilon} R_j\,(2x-z)\,K(z)\,dz, \quad 1 \leq j \leq n.
\end{aligned}
$$

Hence it follows that formula (9.129) converges in the space $L^1(\{1/2^2 \leq |x| \leq 1/2\})$.

Repeating this process, we obtain that formula (9.129) (and hence formula (9.125)) converges in the space $L^1_{\mathrm{loc}}(\mathbf{R}^n \setminus \{0\})$:

$$
K_j^{(1)}(x) = \mathrm{I}_j^{(1)}(x) + \lim_{\varepsilon \downarrow 0} \mathrm{I}_j^{(2)}(x;\varepsilon) + \mathrm{I}_j^{(3)}(x) \quad \text{for } 1 \leq j \leq n.
$$

Moreover, it is easy to see that the function $K_j^{(1)}(x)$ is positively homogeneous of degree $-n$.

Step (3): Now we assume that $K \in L^q(\Sigma_{n-1})$ for $1 < q < \infty$. By applying Hölder's inequality (Theorem 3.24), we obtain from inequalities (9.130) and (9.131) that

$$\int_{1/2<|x|<1} \left| \mathrm{I}_j^{(1)}(x) \right|^q dx \leq C \int_{\Sigma_{n-1}} |K(\sigma)|^q \, d\sigma \quad \text{for } 1 \leq j \leq n, \tag{9.132a}$$

$$\int_{1/2<|x|<1} \left| \mathrm{I}_j^{(3)}(x) \right|^q dx \leq C \int_{\Sigma_{n-1}} |K(\sigma)|^q \, d\sigma \quad \text{for } 1 \leq j \leq n. \tag{9.132b}$$

We let

$$g(y) := \begin{cases} K(y) & \text{if } 1/4 < || < 2, \\ 0 & \text{otherwise.} \end{cases}$$

Then, by applying Theorem 9.16 with

$$K(x) := R_j(x) = -\frac{\Gamma((n+1)/2)}{\pi^{(n+1)/2}} \frac{x_j}{|x|^{n+1}} \quad \text{for } 1 \leq j \leq n,$$

we find that

$$\begin{aligned} \int_{\mathbf{R}^n} \left| \mathrm{I}_j^{(2)}(x;\varepsilon) \right|^q dx &\leq \left(\frac{\pi C_q}{2} \right)^q \left(\int_{\Sigma_{n-1}} \left| R_j(\sigma) \right| d\sigma \right)^q \int_{1/4<|x|<2} |K(x)|^q \, dx \\ &\leq C \int_{1/4<|x|<2} |K(x)|^q \, dx \\ &= C \int_{\Sigma_{n-1}} \left(\int_{1/4}^2 |K(t\,\sigma)|^q \, t^{n-1} \, dt \right) d\sigma \\ &\leq C \, \frac{4^{n(q-1)}}{n(q-1)} \int_{\Sigma_{n-1}} |K(\sigma)|^q \, d\sigma. \end{aligned}$$

By combining this inequality with inequalities (9.132), we obtain that

$$\int_{1/2<|x|<1} |K_j^{(1)}(x)|^q \, dx \leq C \int_{\Sigma_{n-1}} |K(\sigma)|^q \, d\sigma \quad \text{for } 1 \leq j \leq n. \tag{9.133}$$

Therefore, the desired inequality (9.127) follows from inequality (9.133) and the homogeneity of $K_j^{(1)}(x)$. Indeed, it suffices to note that

$$\int_{1/2<|x|<1} \left|K_j^{(1)}(x)\right|^q dx = \int_{\Sigma_{n-1}} \left(\int_{1/2}^1 \left|K_j^{(1)}(t\,\sigma)\right|^q t^{n-1}\,dt\right) d\sigma$$
$$= \frac{1}{n(q-1)}\left(2^{n(q-1)}-1\right)\int_{\Sigma_{n-1}} \left|K_j^{(1)}(\sigma)\right|^q d\sigma.$$

Step (4): Finally, we consider the convergence of formula (9.126). Let N be an arbitrary positive integer. If $|x| < N$ and $0 < \varepsilon < 1$, it follows that

$$\begin{aligned}
&\int_{|x-y|>\varepsilon} R_j(x-y)\,K(y)\,\phi(|y|)\,dy \\
&= \int_{|x-y|>\varepsilon,\,|y|<N+1} R_j(x-y)\,K(y)\,\phi(|y|)\,dy \\
&\quad + \int_{|y|>N+1} R_j(x-y)\,K(y)\,dy \\
&:= \mathrm{I}_j^{(4)}(x;\varepsilon) + \mathrm{I}_j^{(5)}(x).
\end{aligned}$$

Here it should be noticed that

$$K_j^{(2)}(x) = \lim_{\varepsilon\downarrow 0} \mathrm{I}_j^{(4)}(x;\varepsilon) + \mathrm{I}_j^{(5)}(x;\varepsilon) \quad \text{for } 1 \le j \le n. \tag{9.134}$$

We study the convergence of the two terms $\mathrm{I}_j^{(4)}(x;\varepsilon)$ and $\mathrm{I}_j^{(5)}(x)$ on the right-hand side of the above formula in the space $L^1_{\mathrm{loc}}(\mathbf{R}^n)$.

Step (4-a): Since the function $h(y)$, defined by the formula

$$\begin{aligned}
h(y) &= \begin{cases} K(y)\phi(|y|) & \text{if } |y| < N+1, \\ 0 & \text{if } |y| > N+1 \end{cases} \\
&= \begin{cases} K(y)\phi(|y|) & \text{if } 1/4 < |y| < N+1, \\ 0 & \text{otherwise}, \end{cases}
\end{aligned}$$

satisfies condition (9.105) (see **Step 1**), we can apply Lemma 9.22 to obtain that the term

$$\begin{aligned}
\mathrm{I}_j^{(4)}(x;\varepsilon) &= \int_{|x-y|>\varepsilon,\ 1/4<|y|<N+1} R_j(x-y)\,K(y)\,\phi(|y|)\,dy \\
&= \int_{|x-y|>\varepsilon} R_j(x-y)\,h(y)\,dy \quad \text{for } 1 \le j \le n,
\end{aligned}$$

converges in the space $L^1_{\mathrm{loc}}(\mathbf{R}^n)$ as $\varepsilon \downarrow 0$.

Step (4-b): On the other hand, we have the inequality

$$\begin{aligned}
\left|I_j^{(5)}(x)\right| &= \int_{|y|>N+1} R_j(x-y)\,K(y)\,dy \\
&\le C \int_{|y|>N+1} \frac{1}{|x-y|^n}\,|K(y)|\,dy \\
&\le C \int_{|y|>N+1} \frac{1}{(|y|-N)^n}\,|K(y)|\,dy \\
&\le C \int_{|y|>N+1} \frac{(N+1)^n}{|y|^n}\,|K(y)|\,dy \\
&= C(N+1)^n \left(\int_{N+1}^{\infty} \frac{1}{|t\sigma|^n} \int_{\Sigma_{n-1}} |K(t\,\sigma)|\, t^{n-1}\,dt \right) d\sigma \\
&= \frac{C}{n} \int_{\Sigma_{n-1}} |K(\sigma)|\,d\sigma.
\end{aligned}$$

This proves that

$$I_j^{(5)} \in L^1_{\mathrm{loc}}(\mathbf{R}^n) \quad \text{for } 1 \le j \le n.$$

Therefore, we find that formula (9.134) (and hence formula (9.126)) converges in the space $L^1_{\mathrm{loc}}(\mathbf{R}^n)$:

$$K_j^{(2)}(x) = \lim_{\varepsilon \downarrow 0} I_j^{(4)}(x;\varepsilon) + I_j^{(5)}(x) \quad \text{for } 1 \le j \le n.$$

The proof of Lemma 9.23 is now complete. □

The second purpose of this section is to prove the following lemma:

Lemma 9.24 *We can find positive constants C, C_1 and a non-negative, positively homogeneous function $G_j(x)$ of degree zero such that*

$$|K_j^{(1)}(x) - K_j^{(2)}(x)| \le \frac{C}{|x|^{n+1}} \int_{\Sigma_{n-1}} |K(\sigma)|\,d\sigma \quad \textit{for all } |x| \ge 1 \tag{9.135}$$

and

$$\int_{\Sigma_{n-1}} G_j(\sigma)\,d\sigma < \infty,$$

and further that

$$|K_j^{(2)}(x)| \le G_j(x) \quad \textit{for all } |x| \le 1. \tag{9.136}$$

Moreover, if $K \in L^q(\Sigma_{n-1})$ for $1 < q < \infty$, then we have, for some constant $C_q > 0$,

$$\left(\int_{\Sigma_{n-1}} \left| K_j^{(2)}(\sigma) \right|^q d\sigma \right)^{1/q} \leq \left(\int_{\Sigma_{n-1}} G_j(\sigma)^q \, d\sigma \right)^{1/q} \tag{9.137}$$
$$\leq C_q \left(\int_{\Sigma_{n-1}} |K(\sigma)|^q \, d\sigma \right)^{1/q} \quad \textit{for } 1 \leq j \leq n.$$

This inequality implies that the expression formula (9.126)

$$K_j^{(2)}(x) = \lim_{\varepsilon \downarrow 0} \int_{|x-y|>\varepsilon} R_j(x-y)\, K(y)\, \phi(|y|)\, dy \quad \textit{for } 1 \leq j \leq n,$$

holds true in the strong topology of $L^q(\mathbf{R}^n)$.

Proof By Lemma 9.23, it follows that

$$K_j^{(1)} \in L^1_{\text{loc}}(\mathbf{R}^n \setminus \{0\}) \quad \text{for } 1 \leq j \leq n,$$
$$K_j^{(2)} \in L^1_{\text{loc}}(\mathbf{R}^n) \quad \text{for } 1 \leq j \leq n.$$

and further that we have, in the space $L^1_{\text{loc}}(\mathbf{R}^n \setminus \{0\})$,

$$K_j^{(1)}(x) - K_j^{(2)}(x) \tag{9.138}$$
$$= \int_{|x-y|>1/4} R_j(x-y)\, K(y)\, (1 - \phi(|y|))\, dy \quad \text{for } 1 \leq j \leq n.$$

In the following we shall denote by C a generic positive constant.

Step (1): We may assume that $0 < \varepsilon < 1/16$. If $|x| \geq 1$ and $|y| < 3/4$, we have, by inequality (9.41) with $y_k := 0$,

$$|R_j(x-y) - R_j(x)| \leq \frac{C}{|x|^{n+1}}\, |y|, \quad |x-y| > \frac{1}{4}.$$

Hence we obtain from formula (9.138) that

$$\begin{aligned}
\left| K_j^{(1)}(x) - K_j^{(2)}(x) \right| &= \left| \int_{\mathbf{R}^n} R_j(x-y)\, K(y)\, (1 - \phi(|y|))\, dy \right| \\
&= \left| \int_{\mathbf{R}^n} \big(R_j(x-y) - R_j(x) \big)\, K(y)\, (1 - \phi(|y|))\, dy \right| \\
&\leq \frac{C}{|x|^{n+1}} \int_{|y|<3/4} |y|\, |K(y)|\, dy \\
&= \frac{C}{|x|^{n+1}} \int_{\Sigma_{n-1}} \left(\int_0^{3/4} |t\sigma|\, |K(t\sigma)|\, t^{n-1}\, dt \right) d\sigma \\
&\leq \frac{3C}{4|x|^{n+1}} \int_{\Sigma_{n-1}} |K(\sigma)|\, d\sigma \quad \text{for all } |x| \geq 1.
\end{aligned}$$

This proves the desired inequality (9.135).

Step (2): If $|x| \leq 1/8$ and $|y| > 1/4$, it follows that

$$|y| > \frac{1}{4} \geq 2|x|,$$

so that

$$|x - y| \geq |y| - |x| > \frac{|y|}{2}.$$

Hence we have the inequality

$$\begin{aligned} \left|K_j^{(2)}(x)\right| &= \left|\int_{|y|>1/4} R_j(x-y)\, K(y)\, \phi(|y|)\, dy\right| \\ &\leq C \int_{|y|>1/4} \frac{|K(y)|}{|x-y|^n}\, dy \leq 2^n\, C \int_{|y|>1/4} \frac{|K(y)|}{|y|^n}\, dy \\ &= 2^n\, C \int_{\Sigma_{n-1}} \left(\int_{1/4}^{\infty} \frac{1}{|t\sigma|^n} |K(t\sigma)|\, t^{n-1}\, dt \right) d\sigma \\ &\leq \frac{8^n\, C}{n} \int_{\Sigma_{n-1}} |K(\sigma)|\, d\sigma \quad \text{for all } |x| \leq 1/8. \end{aligned} \tag{9.139}$$

Step (3): If $1/8 \leq |x| \leq 1$, we consider the characteristic function $\chi(t)$ of the interval $[0, 1/16]$:

$$\chi(t) = \begin{cases} 1 & \text{if } 0 \leq t \leq \frac{1}{16}, \\ 0 & \text{otherwise.} \end{cases}$$

Then we have the formula

$$\begin{aligned} &\int_{|x-y|>\varepsilon} R_j(x-y)\, K(y)\, \phi(|y|)\, dy \\ &= \phi(|x|) \int_{|x-y|>\varepsilon} R_j(x-y)\, K(y)\, dy \\ &\quad + \int_{|x-y|>\varepsilon} R_j(x-y)\, K(y)\, (\phi(|y|) - \phi(|x|))\, dy \\ &= \phi(|x|) \int_{|x-y|>\varepsilon} R_j(x-y)\, K(y)\, dy \\ &\quad + \int_{|x-y|>\varepsilon} \big(R_j(x-y) - \chi(|y|) R_j(x)\big)\, K(y)\, (\phi(|y|) - \phi(|x|))\, dy \\ &\quad + R_j(x) \int_{|x-y|>\varepsilon} \chi(|y|)\, K(y)\, (\phi(|y|) - \phi(|x|))\, dy \quad \text{for } 1 \leq j \leq n. \end{aligned} \tag{9.140}$$

However, since $\phi(t) = 0$ for $t \leq 1/4$ it follows from the cancellation property (9.4) that the last integral of formula (9.140) tends to zero as $\varepsilon \downarrow 0$:

$$\begin{aligned}
&\int_{|x-y|>1/16>\varepsilon,\, |y|<1/16} \chi(|y|)\, K(y)\, (\phi(|y|) - \phi(|x|))\, dy \\
&= -\int_{0<|y|<1/16} K(y)\phi(|x|)\, dy = -\int_{0<|y|<1/16} K(y)\, dy \cdot \phi(|x|) \\
&= 0.
\end{aligned}$$

Hence, by letting $\varepsilon \downarrow 0$ in formula (9.140) we obtain that

$$\begin{aligned}
K_j^{(2)}(x) &= \phi(|x|)\, K_j^{(1)}(x) \\
&\quad + \int_{\mathbf{R}^n} \big(R_j(x-y) - \chi(|y|) R_j(x)\big)\, K(y)\, (\phi(|y|) - \phi(|x|))\, dy \\
&\quad \text{for } 1/8 \leq |x| \leq 1.
\end{aligned} \tag{9.141}$$

Here it should be noticed that the function $\phi(|x|)$ is uniformly Lipschitz continuous.

Step (3-a): If $1/8 \leq |x| \leq 1$ and if $|y| \leq 1/16$, then it follows that

$$2\,|x-y| \leq 2\,|x| + 2\,|y| \leq 2\,|x| + \frac{1}{16} \leq 3\,|x|.$$

Hence we have the inequality

$$\bullet\ \big|R_j(x-y) - \chi(|y|)\, R_j(x)\big| = \big|R_j(x-y) - R_j(x)\big| \leq \frac{C}{|x-y|^n} + \frac{C}{|x|^n} \leq \left(1 + \left(\frac{3}{2}\right)^n\right) \frac{C}{|x-y|^n}.$$

On the other hand, we have, by inequality (9.41) with $y_k := 0$,

$$\bullet\ \big|R_j(x-y) - \chi(|y|) R_j(x)\big| = \big|R_j(x-y) - R_j(x)\big| \leq \frac{C}{|x-y|^{n+1}}\, |y|.$$

Therefore, by combining these two inequalities we obtain that

$$\begin{aligned}
&\big|R_j(x-y) - \chi(|y|)\, R_j(x)\big| \\
&\leq C \left(\frac{|y|}{|x-y|^{n+1}}\right)^{1/2} \cdot \left(\frac{1}{|x-y|^n}\right)^{1/2} \\
&= \frac{C}{|x-y|^{n+1/2}}\, |y|^{1/2} \quad \text{for all } 1/8 \leq |x| \leq 1 \text{ and } |y| \leq 1/16.
\end{aligned} \tag{9.142}$$

Step (3-b): If $1/8 \leq |x| \leq 1$ and if $|y| > 1/16$, then it follows that

$$\begin{aligned}|x-y| &\leq |x| + |y| \leq 1 + |y| \leq 16\,|y| + |y| \\ &= 17\,|y|.\end{aligned}$$

Hence we have the inequality

$$\begin{aligned}&\left|R_j(x-y) - \chi(|y|)\,R_j(x)\right| = \left|R_j(x-y)\right| \qquad (9.143)\\ &\leq \frac{C}{|x-y|^n} \leq \left(\frac{17\,|y|}{|x-y|}\right)^{1/2} \frac{C}{|x-y|^n}\\ &= \frac{C}{|x-y|^{n+1/2}}\,|y|^{1/2} \quad \text{for all } 1/8 \leq |x| \leq 1 \text{ and } |y| > 1/16.\end{aligned}$$

By using two inequalities (9.142) and (9.143), we have, for $1/8 \leq |x| \leq 1$,

$$|R_j(x-y) - \chi(|y|)\,R_j(x)| \leq \frac{C}{|x-y|^{n+1/2}}\,|y|^{1/2} \quad \text{on } \mathbf{R}^n.$$

Therefore, by carrying this inequality into formula (9.141) we obtain that

$$\begin{aligned}&\left|K_j^{(2)}(x)\right| \qquad (9.144)\\ &\leq |K_j^{(1)}(x)| + C \int_{\mathbf{R}^n} |y|^{1/2}\,|K(y)|\,|x-y|^{-n+1/2}\,dy\\ &\leq 8^n |x|^n |K_1(x)| + 8^{n-1} C |x|^{n-1} \int_{\mathbf{R}^n} |y|^{1/2}\,|K(y)|\,|x-y|^{-n+1/2}\,dy\\ &\quad \text{for all } 1/8 \leq |x| \leq 1.\end{aligned}$$

Step (4): By combining two inequalities (9.139) and (9.144), we can choose a constant $C_0 > 0$ such that the function $G_j(x)$, defined by the formula

$$\begin{aligned}G_j(x) := C_0\Bigg(&\int_{\Sigma_{n-1}} |K(\sigma)|\,d\sigma + |x|^n \left|K_j^{(1)}(x)\right| \qquad (9.145)\\ &+ |x|^{n-1} \int_{\mathbf{R}^n} |y|^{1/2}\,|K(y)|\,|x-y|^{-n+1/2}\,dy\Bigg) \quad \text{for } 1 \leq j \leq n,\end{aligned}$$

satisfies the inequality

$$\left|K_j^{(2)}(x)\right| \leq G_j(x) \quad \text{for all } |x| \leq 1.$$

Here it should be noticed that the functions $G_j(x)$ are positively homogeneous of degree zero.

Next, we show that

$$G_j \in L^1(\Sigma_{n-1}) \quad \text{for } 1 \le j \le n.$$

To do this, since we have the formula

$$\begin{aligned}\int_{1/2<|x|<3/2} G_j(x)\,dx &= \int_{\Sigma_{n-1}} \int_{1/2}^{3/2} G_j(t\,\sigma)\,t^{n-1}\,dt\,d\sigma \\ &= \frac{1}{n}\left(\left(\frac{3}{2}\right)^n - \left(\frac{1}{2}\right)^n\right) \int_{\Sigma_{n-1}} G_j(\sigma)\,d\sigma,\end{aligned}$$

it suffices to prove that

$$\int_{1/2<|x|<3/2} G_j(x)\,dx < \infty. \tag{9.146}$$

However, we remark that

$$\int_{1/2<|x|<3/2} |x|^n \left|K_j^{(1)}(x)\right| dx \le \left(\frac{3}{2}\right)^n \int_{1/2<|x|<3/2} \left|K_j^{(1)}(x)\right| dx < \infty,$$

since the functions $K_j^{(1)}(x)$ are locally integrable on $\mathbf{R}^n \setminus \{0\}$.

Therefore, we have only to show that

$$\begin{aligned}&\int_{1/2<|x|<3/2} \left(\int_{\mathbf{R}^n} |y|^{1/2} |K(y)|\,|x-y|^{-n+1/2}\,dy \right) dx \\ &\le C \int_{\Sigma_{n-1}} |K(\sigma)|\,d\sigma.\end{aligned} \tag{9.147}$$

To do this, we note the following three inequalities:

$$\begin{aligned}\bullet\ &\int_{1/2<|x|<3/2} \left(\int_{1/4<|y|<3} |y|^{1/2} |K(y)|\,|x-y|^{-n+1/2}\,dy \right) dx \\ &\le \int_{1/4<|y|<3} |y|^{1/2} |K(y)| \left(\int_{|x-y|<9/2} |x-y|^{-n+1/2}\,dx \right) dy \\ &= 3\sqrt{2}\omega_n \int_{1/4<|y|<3} |y|^{1/2} |K(y)|\,dy \\ &= C \int_{\Sigma_{n-1}} |K(\sigma)|\,d\sigma,\end{aligned} \tag{9.148}$$

where

$$\omega_n = |\Sigma_{n-1}| = \frac{2\pi^{n/2}}{\Gamma(n/2)}$$

is the surface area of the unit sphere Σ_{n-1} in $\mathbf{R}^n$.

$$\bullet \int_{1/2<|x|<3/2} \left(\int_{0<|y|<1/4} |y|^{1/2} |K(y)| \, |x-y|^{-n+1/2} \, dy \right) dx \tag{9.149}$$
$$\leq 4^{n-1/2} \int_{1/2<|x|<3/2} \left(\int_{|y|<1/4} |y|^{1/2} |K(y)| \, dy \right) dx$$
$$= C \int_{\Sigma_{n-1}} |K(\sigma)| \, d\sigma.$$

$$\bullet \int_{1/2<|x|<3/2} \left(\int_{|y|>3} |y|^{1/2} |K(y)| \, |x-y|^{-n+1/2} \, dy \right) dx \tag{9.150}$$
$$\leq \left(\frac{2}{3}\right)^{n-1/2} \int_{1/2<|x|<3/2} \left(\int_{|y|>3} |y|^{1-n} |K(y)| \, dy \right) dx$$
$$= C \int_{\Sigma_{n-1}} |K(\sigma)| \, d\sigma.$$

The desired inequality (9.147) follows by combining inequalities (9.148), (9.149) and (9.150).

Furthermore, we obtain from inequality (9.147) that the last integral of inequality (9.144) is finite for almost all x:

$$\int_{\mathbf{R}^n} |y|^{1/2} \, |K(y)| \, |x-y|^{-n+1/2} \, dy < \infty \quad \text{for almost all } x \in \mathbf{R}^n.$$

Step (5): Finally, we consider the case where $K \in L^q(\Sigma_{n-1})$ for $1 < q < \infty$. The first two terms in formula (9.145) can be estimated as follows:

(1) We have, by Hölder's inequality (Theorem 3.24),

$$\int_{\Sigma_{n-1}} |K(\sigma)| \, d\sigma \leq \omega_n^{1-1/q} \left(\int_{\Sigma_{n-1}} |K(\sigma)|^q \, d\sigma \right)^{1/q}.$$

(2) We have already proved inequality (9.127):

$$\left(\int_{\Sigma_{n-1}} |K_1(\sigma)|^q \, d\sigma \right)^{1/q} \leq C_q \left(\int_{\Sigma_{n-1}} |K(\sigma)|^q \, d\sigma \right)^{1/q}.$$

Therefore, in order to prove inequality (9.137) it remains to prove the following inequality for the last term in formula (9.145):

$$\int_{1/2<|x|<3/2} \left(\int_{\mathbf{R}^n} |y|^{1/2} |K(y)| \, |x-y|^{-n+1/2} \, dy \right)^q dx \tag{9.151}$$
$$\leq C \int_{\Sigma_{n-1}} |K(\sigma)|^q \, d\sigma.$$

Step (5-a): First, by Hölder's inequality (Theorem 3.24) it follows that

$$\begin{aligned}
&\left(\int_{|y|<1/4} |y|^{1/2} |K(y)| \, |x-y|^{-n+1/2} \, dy \right)^q \\
&= \left(\int_{|y|<1/4} \left(|y|^{1/2} |K(y)| \right)^{1-1/q} \left(|y|^{1/2} |K(y)| \right)^{1/q} |x-y|^{-n+1/2} \, dy \right)^q \\
&\leq \left(\int_{|y|<1/4} |y|^{1/2} |K(y)| \, dy \right)^{q-1} \\
&\quad \times \left(\int_{|y|<1/4} |y|^{1/2} |K(y)| \, |x-y|^{(-n+1/2)q} \, dy \right).
\end{aligned}$$

Hence, by integrating the both sides of this inequality over the annular region

$$\left\{ x \in \mathbf{R}^n : \frac{1}{2} < |x| < \frac{3}{2} \right\},$$

we obtain from Fubini's theorem (Theorem 3.20) and Hölder's inequality (Theorem 3.24) that

$$\begin{aligned}
&\int_{1/2<|x|<3/2} \left(\int_{|y|<1/4} |y|^{1/2} |K(y)| \, |x-y|^{-n+1/2} \, dy \right)^q dx \qquad (9.152) \\
&\leq \left(\int_{|y|<1/4} |y|^{1/2} |K(y)| \, dy \right)^{q-1} \\
&\quad \times \int_{1/2<|x|<3/2} \left(\int_{|y|<1/4} |y|^{1/2} |K(y)| \, |x-y|^{(-n+1/2)q} \, dy \right) dx \\
&= \left(\int_{|y|<1/4} |y|^{1/2} |K(y)| \, dy \right)^{q-1} \\
&\quad \times \int_{|y|<1/4} |y|^{1/2} |K(y)| \left(\int_{1/2<|x|<3/2} |x-y|^{(-n+1/2)q} \, dx \right) dy \\
&\leq \left(\int_{|y|<1/4} |y|^{1/2} \, |K(y)| \, dy \right)^q \left(\int_{1/4<|z|<7/4} |z|^{(-n+1/2)q} \, dz \right) \\
&\leq C \left(\int_{\Sigma_{n-1}} |K(\sigma)| \, d\sigma \right)^q \\
&\leq C \, \omega_n^{q-1} \int_{\Sigma_{n-1}} |K(\sigma)|^q \, d\sigma.
\end{aligned}$$

Step (5-b): Secondly, since we have, for $1/2 < |x| < 3/2$ and $1/4 < |y| < 3$,

$$\begin{aligned}
&\left(\int_{1/4<|y|<3} |y|^{1/2}|K(y)|\,|x-y|^{-n+1/2}\,dy\right)^q \\
&\le \int_{1/4<|y|<3} \left(|y|^{1/2}|K(y)|\right)^q |x-y|^{-n+1/2}\,dy \\
&\quad \times \left(\int_{1/4<|y|<3} |x-y|^{-n+1/2}\,dy\right)^{q-1} \\
&\le \left(\int_{1/4<|y|<3} \left(|y|^{1/2}|K(y)|\right)^q |x-y|^{-n+1/2}\,dy\right) \\
&\quad \times \left(\int_{|z|<9/2} |z|^{-n+1/2}\,dz\right)^{q-1},
\end{aligned}$$

it follows from an application of Fubini's theorem (Theorem 3.20) that we have, for $1/4 < |y| < 3$,

$$\begin{aligned}
&\int_{1/2<|x|<3/2} \left(\int_{1/4<|y|<3} |y|^{1/2}|K(y)|\,|x-y|^{-n+1/2}\,dy\right)^q dx \qquad (9.153)\\
&\le \int_{1/2<|x|<3/2} \left(\int_{1/4<|y|<3} \left(|y|^{1/2}|K(y)|\right)^q |x-y|^{-n+1/2}\,dy\right) dx \\
&\quad \times \left(\int_{|z|<9/2} |z|^{-n+1/2}\,dz\right)^{q-1} \\
&\le \left(\int_{1/4<|y|<3} \left(|y|^{1/2}|K(y)|\right)^q\,dy\right)\left(\int_{|y|<9/2} |y|^{-n+1/2}\,dy\right)^q \\
&\le C \int_{\Sigma_{n-1}} |K(\sigma)|^q\,d\sigma.
\end{aligned}$$

Step (5-c): Thirdly, by Hölder's inequality (Theorem 3.24) it follows that

$$\begin{aligned}
&\int_{1/2<|x|<3/2} \left(\int_{|y|>3} |y|^{1/2}|K(y)|\,|x-y|^{-n+1/2}\,dy\right)^q dx \qquad (9.154)\\
&\le C \int_{1/2<|x|<3/2} \left(\int_{|y|>3} |y|^{1-n}|K(y)|\,dy\right)^q dx \\
&\le C \left(\int_{\Sigma_{n-1}} |K(\sigma)|\,d\sigma\right)^q \\
&\le C\,\omega_n^{q-1} \int_{\Sigma_{n-1}} |K(\sigma)|^q\,d\sigma.
\end{aligned}$$

Therefore, by combining inequalities (9.152), (9.153) and (9.154) we can obtain inequality (9.151) and hence the desired inequality (9.137).

The proof of Lemma 9.24 is now complete. □

The third purpose of this section is to prove the following lemma:

Lemma 9.25 *If $f(x)$ is a non-negative function in $L^p(\mathbf{R})$ for $1 < p < \infty$, then we have the inequalities*

$$\left(\int_{-\infty}^{\infty}\left(\sup_{\varepsilon>0}\frac{1}{\varepsilon^n}\int_0^{\varepsilon} f(t+s)s^{n-1}\,ds\right)^p dt\right)^{1/p} \tag{9.155}$$
$$\leq \frac{p}{p-1}\left(\int_{-\infty}^{\infty} f(t)^p\,dt\right)^{1/p},$$

and

$$\left(\int_{-\infty}^{\infty}\left(\sup_{\varepsilon>0}\varepsilon\int_{\varepsilon}^{\infty} f(t+s)s^{-2}\,ds\right)^p dt\right)^{1/p} \tag{9.156}$$
$$\leq \frac{2p}{p-1}\left(\int_{-\infty}^{\infty} f(t)^p\,dt\right)^{1/p}.$$

Proof (1) If $\varepsilon > 0$, we define two functions $F_\varepsilon(t)$ and $G(t)$ by formulas (9.69):

$$F_\varepsilon(t) := \frac{1}{\varepsilon}\int_0^{\varepsilon} f(t+s)\,ds,$$
$$G(t) := \sup_{\varepsilon>0} F_\varepsilon(t).$$

Then it follows from inequality (9.70) (Lemma 9.13) that

$$\int_{-\infty}^{\infty} G(t)^p\,dt \leq \left(\frac{p}{p-1}\right)^p \int_{-\infty}^{\infty} f(t)^p\,dt. \tag{9.157}$$

However, we have the inequality

$$\sup_{\varepsilon>0}\left(\frac{1}{\varepsilon^n}\int_0^{\varepsilon} f(t+s)\,s^{n-1}\,ds\right)$$
$$= \sup_{\varepsilon>0}\left(\frac{1}{\varepsilon}\int_0^{\varepsilon} f(t+s)\left(\frac{s}{\varepsilon}\right)^{n-1} ds\right)$$
$$\leq \sup_{\varepsilon>0}\left(\frac{1}{\varepsilon}\int_0^{\varepsilon} f(t+s)\,ds\right) = \sup_{\varepsilon>0} F_\varepsilon(t)$$
$$= G(t).$$

Therefore, the desired inequality (9.155) follows from inequality (9.157).

(2) First, we have, by Hölder's inequality (Theorem 3.24),

$$\begin{aligned}\varepsilon F_\varepsilon(t) &= \int_0^\varepsilon f(t+s)\,ds \\ &\leq \left(\int_0^\varepsilon ds\right)^{1-1/p}\left(\int_0^\varepsilon f(t+s)^p\,ds\right)^{1/p} \\ &\leq \varepsilon^{1-1/p}\,\|f\|_p,\end{aligned}$$

and so

$$\lim_{\varepsilon\uparrow\infty}\frac{F_\varepsilon(t)}{\varepsilon} = 0.$$

Secondly, we have the formula

$$\frac{d}{ds}\,(s\,F_s(t)) = \frac{d}{ds}\left(\int_0^s f(t+\sigma)\,d\sigma\right) = f(t+s).$$

Hence, by integration by parts it follows that

$$\begin{aligned}\varepsilon\int_\varepsilon^\infty f(t+s)s^{-2}\,ds &= \varepsilon\int_\varepsilon^\infty \frac{d}{ds}\,(s\,F_s(t))\,s^{-2}\,ds \qquad (9.158)\\ &= \varepsilon\left[s^{-1}F_s(t)\right]_{s=\varepsilon}^{s=\infty} + 2\varepsilon\int_\varepsilon^\infty s^{-2}F_s(t)\,ds \\ &= -F_\varepsilon(t) + 2\varepsilon\int_\varepsilon^\infty s^{-2}F_s(t)\,ds \\ &\leq 2\varepsilon\int_\varepsilon^\infty s^{-2}\,F_s(t)\,ds \\ &\leq 2\varepsilon\int_\varepsilon^\infty s^{-2}\,ds\cdot G(t) \\ &= 2G(t).\end{aligned}$$

Therefore, the desired inequality (9.156) follows by combining inequalities (9.158) and (9.157).

The proof of Lemma 9.25 is complete. □

Now, by using Lemmas 9.23, 9.24 and 9.25 we can prove the existence of the singular integral (9.2) in the space $L^p(\mathbf{R}^n)$ for $1 < p < \infty$ in the case of *even kernels*:

Theorem 9.26 *Assume that the integral kernel $K(x)$ satisfies Assumption 9.5. If $f \in L^p(\mathbf{R}^n)$ for $1 < p < \infty$, we let*

$$\widetilde{f_\varepsilon}(x) := \int_{|x-y|>\varepsilon} K(x-y)\,f(y)\,dy.$$

Then we have the following three assertions:

(i) There exists a constant $C_p > 0$, independent of ε, such that

$$\left(\int_{\mathbf{R}^n} \sup_{\varepsilon>0} |\widetilde{f_\varepsilon}(x)|^p \, dx\right)^{1/p} \leq C_p \left(\int_{\mathbf{R}^n} |f(x)|^p \, dx\right)^{1/p}. \tag{9.159}$$

(ii) *The sequence* $\widetilde{f_\varepsilon}$ *converges almost everywhere in* $\mathbf{R}^n$ *and in the strong topology of* $L^p(\mathbf{R}^n)$ *as* $\varepsilon \downarrow 0$. *Namely, the singular integral*

$$K * f(x) = \lim_{\varepsilon \downarrow 0} \widetilde{f_\varepsilon}(x) = \lim_{\varepsilon \downarrow 0} \int_{|x-y|>\varepsilon} K(x-y)\, f(y)\, dy$$

exists for almost all $x \in \mathbf{R}^n$ *and in the strong topology of* $L^p(\mathbf{R}^n)$.

(iii) *The mapping*

$$\begin{aligned} L^p(\mathbf{R}^n) &\longrightarrow L^p(\mathbf{R}^n) \\ f &\longmapsto K * f \end{aligned}$$

is a bounded linear operator from $L^p(\mathbf{R}^n)$ *into itself. More precisely, we have the inequality*

$$\|K * f\|_p \leq C_p \, \|f\|_p \quad \text{for all } f \in L^p(\mathbf{R}^n).$$

Proof Let $K_j^{(1)}(x)$ and $K_j^{(2)}(x)$ be the functions defined by formulas (9.125) and (9.126), respectively. If we let

$$\begin{aligned} g_j(x) &:= -\left(R_j * f\right)(x) \\ &= -\lim_{\varepsilon \downarrow 0} \int_{|x-y|>\varepsilon} R_j(x-y)\, f(y)\, dy \quad \text{for } 1 \leq j \leq n, \end{aligned}$$

by applying Theorems 9.16 and 9.17 we obtain from formula (9.105) that the function

$$\begin{aligned} f(x) &= \sum_{j=1}^{n} (R_j * g_j)(x) \\ &= \lim_{\varepsilon \downarrow 0} \sum_{j=1}^{n} \int_{|x-y|>\varepsilon} R_j(x-y)\, g_j(y)\, dy \\ &= -\sum_{j=1}^{n} R_j * \left(R_j * f\right) \end{aligned}$$

exists for almost all $x \in \mathbf{R}^n$ and in the strong topology of $L^p(\mathbf{R}^n)$.

The proof of Theorem 9.26 is divided into three steps.

Step (1): Now we show that

$$\begin{aligned}&\int_{\mathbf{R}^n} K(x-y)\phi\left(\frac{|x-y|}{\varepsilon}\right) f(y)\,dy \\ &= \frac{1}{\varepsilon^n}\sum_{j=1}^{n}\int_{\mathbf{R}^n} K_j^{(2)}\left(\frac{x-y}{\varepsilon}\right) g_j(y)\,dy \\ &= -\frac{1}{\varepsilon^n}\sum_{j=1}^{n}\int_{\mathbf{R}^n} K_j^{(2)}\left(\frac{x-y}{\varepsilon}\right)\left(R_j * f\right)(y)\,dy.\end{aligned} \tag{9.160}$$

Here $\phi \in C^1(\mathbf{R})$ is the function used in the definition of the functions $K_j^{(2)}(x)$:

$$\phi(t) = \begin{cases} 0 & \text{if } 0 \leq t \leq 1/4, \\ 1 & \text{if } t > 3/4. \end{cases}$$

Step (1-a): First, we assume that

$$g_j \in C_0^1\left(\mathbf{R}^n\right) \quad \text{for } 1 \leq j \leq n,$$

and that

$$\operatorname{supp} g_j \subset B(0, N) := \left\{x \in \mathbf{R}^n : |x| \leq N\right\}$$

for some positive integer N. If $|y| \leq N+1$, then we find that the integral

$$\sum_{j=1}^{n}\int_{|y-z|>\delta} R_j(y-z)\, g_j(z)\,dz = \sum_{j=1}^{n}\int_{|y-z|>\delta} R_j(y-z)\left(g_j(z) - g_j(y)\right)dz$$

converges to the function

$$f(y) = \sum_{j=1}^{n}(R_j * g_j)(y)$$

uniformly in $y \in B(0, N)$, as $\delta \downarrow 0$. Indeed, it suffices to note that the function $g(x)$ is uniformly Lipschitz continuous.

If $|y| > N+1$, then it follows that

$$\begin{aligned}2(N+1)\,|y-z| &= (N+1)|y-z| + (N+1)|y-z| \\ &\geq |y-z| + (N+1) \\ &\geq |y| + (N-|z|) + 1 \\ &\geq |y| + 1 \quad \text{for all } |z| \leq N.\end{aligned}$$

Hence we have, for $0 < \delta < 1$,

$$\begin{aligned}\left|\int_{|y-z|>\delta} R_j(y-z)\, g_j(z)\, dz\right| &\leq \int_{|z|<N} \frac{C|g_j(z)|}{|y-z|^n}\, dz \\ &\leq 2^n\,(N+1)^n\, C\, \left\|g_j\right\|_1 \frac{1}{(|y|+1)^n}.\end{aligned}$$

Moreover, since we have the inequality

$$\begin{aligned}2\,(|x-z|+1) &\geq (|z|-|x|)+(|z|-|x|) \\ &\geq 2|x|-2|x|+|z| \\ &= |z| \quad \text{for all } |z|>2|x|,\end{aligned}$$

it follows that

$$\begin{aligned}&\int_{\mathbf{R}^n} |K(x-y)|\,\phi\left(\frac{|x-y|}{\varepsilon}\right)\frac{dy}{(|y|+1)^n} \\ &= \int_{\mathbf{R}^n} |K(z)|\phi\left(\frac{|z|}{\varepsilon}\right)\frac{dz}{(|x-z|+1)^n} \\ &\leq \int_{|z|>\varepsilon/4} |K(z)|\,\phi\left(\frac{|z|}{\varepsilon}\right)\frac{dz}{(|x-z|+1)^n} \\ &\leq \int_{\varepsilon/4<|z|<2|x|} |K(y)|\,dz + 2^n \int_{|z|>2|x|} \frac{|K(z)|}{|z|^n}\,dz \\ &= \left(\log\left(\frac{8|x|}{\varepsilon}\right)+\frac{1}{n\,|x|^n}\right)\int_{\Sigma_{n-1}} |K(\sigma)|\,d\sigma \\ &< \infty \quad \text{for all } x \neq 0.\end{aligned}$$

Hence it follows from an application of Lebesgue's dominated convergence theorem (Theorem 3.13) that

$$\begin{aligned}&\lim_{\delta\downarrow 0}\int_{\mathbf{R}^n} K(x-y)\phi\left(\frac{|x-y|}{\varepsilon}\right)\left(\sum_{j=1}^n \int_{|y-z|>\delta} R_j(y-z)\,g_j(z)\,dz\right)dy \qquad (9.161)\\ &= \int_{\mathbf{R}^n} K(x-y)\,\phi\left(\frac{|x-y|}{\varepsilon}\right)\left(\sum_{j=1}^n (R_j * g_j)(y)\right)dy \\ &= \int_{\mathbf{R}^n} K(x-y)\,\phi\left(\frac{|x-y|}{\varepsilon}\right) f(y)\,dy.\end{aligned}$$

Step (1-b): If $|y-z|>\delta$, then it follows that

$$\begin{aligned}2\,|y-z| &= |y-z| + |y-z| \\ &\geq |y-z| + \delta \\ &\geq \delta\,(|y-z|+1)\,.\end{aligned}$$

Hence we have the inequality

$$\begin{aligned}&\left| K(x-y)\phi\left(\frac{|x-y|}{\varepsilon}\right) R_j(y-z)\, g_j(z)\right| \qquad (9.162)\\ &\leq \frac{2^n\,C}{\delta^n}\,|K(x-y)|\,\phi\left(\frac{|x-y|}{\varepsilon}\right)\frac{|g_j(z)|}{(|y-z|+1)^n} \\ &\leq \frac{2^n\,C(N+1)^n}{\delta^n}\,|K(x-y)|\,\phi\left(\frac{|x-y|}{\varepsilon}\right)\frac{|g_j(z)|}{(|y|+1)^n} \quad \text{for all } |z| \leq N.\end{aligned}$$

However, we find that this term is integrable in $(y, z) \in \mathbf{R}^n \times \mathbf{R}^n$ for $x \neq 0$. Indeed, it suffices to note that

$$\begin{aligned}&\int_{\mathbf{R}^n} |K(x-y)|\phi\left(\frac{|x-y|}{\varepsilon}\right)\frac{dy}{(|y|+1)^n} \\ &= \int_{\mathbf{R}^n} |K(z)|\phi\left(\frac{|z|}{\varepsilon}\right)\frac{dz}{(|x-z|+1)^n} \\ &\leq \int_{\varepsilon/4<|z|<2|x|} |K(y)|\,dz + 2^n \int_{|z|>2|x|} \frac{|K(z)|}{|z|^n}\,dz \\ &= \left(\log\left(\frac{8|x|}{\varepsilon}\right) + \frac{1}{n\,|x|^n}\right)\int_{\Sigma_{n-1}} |K(\sigma)|\,d\sigma \\ &< \infty \quad \text{for all } x \neq 0.\end{aligned}$$

Therefore, by using Fubini's theorem (Theorem 3.20) and then formula (9.126) we obtain that the left-hand side of formula (9.161) is equal to the following:

$$\begin{aligned}&\int_{\mathbf{R}^n} K(x-y)\,\phi\left(\frac{|x-y|}{\varepsilon}\right) f(y)\,dy \\ &= \lim_{\delta\downarrow 0}\int_{\mathbf{R}^n} K(x-y)\phi\left(\frac{|x-y|}{\varepsilon}\right)\left(\sum_{j=1}^n \int_{|y-z|>\delta} R_j(y-z)\,g_j(z)\,dz\right)dy \\ &= \lim_{\delta\downarrow 0}\sum_{j=1}^n \int_{\mathbf{R}^n}\left(\int_{|y-z|>\delta} K(x-y)\phi\left(\frac{|x-y|}{\varepsilon}\right) R_j(y-z)\,dy\right) g_j(z)\,dz \\ &= \lim_{\delta\downarrow 0}\int_{\mathbf{R}^n}\sum_{j=1}^n\left(\int_{|x-\varepsilon w-z|>\delta} K(\varepsilon w)\phi(|w|)R_j(x-\varepsilon w-z)\varepsilon^n\,dw\right) g_j(z)\,dz\end{aligned}$$

$$
\begin{aligned}
&= \frac{1}{\varepsilon^n} \lim_{\delta \downarrow 0} \int_{\mathbf{R}^n} \sum_{j=1}^n \left(\int_{|(x-z)/\varepsilon - w| > \delta/\varepsilon} K(w)\phi(|w|) R_j \left(\frac{x-z}{\varepsilon} - w \right) dw \right) g_j(z)\, dz \\
&= \frac{1}{\varepsilon^n} \sum_{j=1}^n \int_{\mathbf{R}^n} K_j^{(2)} \left(\frac{x-z}{\varepsilon} \right) g_j(z)\, dz.
\end{aligned}
$$

Namely, the desired formula (9.160) holds true for all $f = \sum_{j=1}^n R_j * g_j$ with $g_j \in C_0^1(\mathbf{R}^n)$:

$$
\begin{aligned}
&\int_{\mathbf{R}^n} K(x-y)\, \phi\left(\frac{|x-y|}{\varepsilon} \right) f(y)\, dy \\
&= -\frac{1}{\varepsilon^n} \sum_{j=1}^n \int_{\mathbf{R}^n} K_j^{(2)} \left(\frac{x-y}{\varepsilon} \right) \left(R_j * f \right)(y)\, dy.
\end{aligned}
$$

Step (1-c): Now we consider the general case where $f \in L^p(\mathbf{R}^n)$ for $1 < p < \infty$. We let

$$
g_j(x) := -\left(R_j * f \right)(x) \in L^p(\mathbf{R}^n), \quad 1 \leq j \leq n.
$$

Then we choose a sequence $\{g_j^k\}$ in the space $C_0^1(\mathbf{R}^n)$ such that (see Corollary 3.43)

$$
\begin{aligned}
&g_j^k - g_j \longrightarrow 0 \quad \text{in } L^p(\mathbf{R}^n) \ \text{as}\, k \to \infty, \\
&\sum_{k=1}^\infty \left\| g_j^{k+1} - g_j^k \right\|_p < \frac{1}{2^k},
\end{aligned}
$$

and define a sequence

$$
f_k := \sum_{j=1}^n R_j * g_j^k.
$$

By applying Theorem 9.6 or Theorem 9.16 to our situation, we obtain that

$$
\begin{aligned}
\|f_k - f\|_p &\leq \left\| \sum_{j=1}^n R_j * \left(g_j^k - g_j \right) \right\|_p \\
&\leq C \sum_{j=1}^n \left\| g_j^k - g_j) \right\|_p \longrightarrow 0 \quad \text{as}\, k \to \infty,
\end{aligned}
$$

and

$$\begin{aligned}\sum_{k=1}^{\infty} \|f_{k+1} - f_k\|_p &= \sum_{k=1}^{\infty} \left\| \sum_{j=1}^{n} R_j * \left(g_j^{k+1} - g_j^k\right) \right\|_p \\ &\le \sum_{k=1}^{\infty} \sum_{j=1}^{n} \left\| R_j * \left(g_j^{k+1} - g_j^k\right) \right\|_p \\ &\le C \sum_{k=1}^{\infty} \sum_{j=1}^{n} \left\| g_j^{k+1} - g_j^k \right\|_p = C \sum_{j=1}^{n} \left(\sum_{k=1}^{\infty} \left\| g_j^{k+1} - g_j^k \right\|_p \right) \\ &< \frac{nC}{2^k}.\end{aligned}$$

Moreover, we find that the series

$$\overline{g}_j(x) := \left|g_j^1(x)\right| + \sum_{k=1}^{\infty} \left| g_j^{k+1}(x) - g_j^k(x) \right| \quad \text{for } 1 \le j \le n,$$

and that the series

$$\overline{f}(x) := |f_1(x)| + \sum_{k=1}^{\infty} |f_{k+1}(x) - f_k(x)|$$

converge for almost all $x \in \mathbf{R}^n$, and satisfy the following three conditions:

(i) $\overline{g}_j \in L^p(\mathbf{R}^n)$ and $\overline{f} \in L^p(\mathbf{R}^n)$.
(ii) $\left|g_j^k(x)\right| \le \overline{g}_j(x)$ for $1 \le j \le n$ and $|f_k(x)| \le \overline{f}(x)$ almost everywhere in $\mathbf{R}^n$.
(iii) $g_j^k(x) \to g_j(x)$ and $f_k(x) \to f(x)$ in $\mathbf{R}$ for almost all $x \in \mathbf{R}^n$, as $k \to \infty$.

Moreover, since $\overline{f} \in L^p(\mathbf{R}^n)$, it follows from an application of Lemma 9.1 that the integral

$$\int_{\mathbf{R}^n} |K(x-y)|\, \phi\left(\frac{|x-y|}{\varepsilon}\right) \overline{f}(y)\, dy \le \int_{|x-y|>\varepsilon/4} |K(x-y)|\, \overline{f}(y)\, dy$$

exists for almost all $x \in \mathbf{R}^n$, as $\varepsilon \downarrow 0$.

On the other hand, we have, by inequality (9.135),

$$\begin{aligned}&\sum_{j=1}^{n} \int_{\mathbf{R}^n} \left| K_j^{(2)}\left(\frac{x-y}{\varepsilon}\right) \right| \overline{g}_j(y)\, dy \\ &= \sum_{j=1}^{n} \int_{|x-y|<\varepsilon} \left| K_j^{(2)}\left(\frac{x-y}{\varepsilon}\right) \right| \overline{g}_j(y)\, dy + \sum_{j=1}^{n} \int_{|x-y|>\varepsilon} \left| K_j^{(2)}\left(\frac{x-y}{\varepsilon}\right) \right| \overline{g}_j(y)\, dy\end{aligned}$$

$$\leq \sum_{j=1}^{n} \int_{|x-y|<\varepsilon} \left| K_j^{(2)} \left(\frac{x-y}{\varepsilon} \right) \right| \overline{g}_j(y)\, dy$$

$$+ \sum_{j=1}^{n} \int_{|x-y|>\varepsilon} \left| K_j^{(2)} \left(\frac{x-y}{\varepsilon} \right) - K_j^{(1)} \left(\frac{x-y}{\varepsilon} \right) \right| \overline{g}_j(y)\, dy$$

$$+ \sum_{j=1}^{n} \int_{|x-y|>\varepsilon} \left| K_j^{(1)} \left(\frac{x-y}{\varepsilon} \right) \right| \overline{g}_j(y)\, dy$$

$$\leq \sum_{j=1}^{n} \int_{|x-y|<\varepsilon} \left| K_j^{(2)} (\frac{x-y}{\varepsilon}) \right| \overline{g}_j(y)\, dy + \sum_{j=1}^{n} \int_{|x-y|>\varepsilon} \left| K_j^{(1)} (\frac{x-y}{\varepsilon}) \right| \overline{g}_j(y)\, dy$$

$$+ C\varepsilon^{n+1} \sum_{j=1}^{n} \int_{|x-y|>\varepsilon} \frac{\overline{g}_j(y)}{|x-y|^{n+1}}\, dy.$$

By applying Theorem 3.36 with $p := 1$ and $q = r := p$, we obtain that the first term on the right-hand side of this inequality is finite for almost every $x \in \mathbf{R}^n$. Indeed, we have, by inequality (9.136),

$$\int_{|x|<\varepsilon} \left| K_j^{(2)} \left(\frac{x}{\varepsilon} \right) \right| dx$$
$$= \varepsilon^n \int_{|y|<1} |K_j^{(2)}(y)|\, dy$$
$$\leq \varepsilon^n \int_{|y|<1} G_j(y)\, dy = \varepsilon^n \int_{\Sigma_{n-1}} \left(\int_0^1 G_j(t\sigma)\, t^{n-1}\, dt \right) d\sigma$$
$$= \frac{\varepsilon^n}{n} \int_{\Sigma_{n-1}} G_j(\sigma)\, d\sigma \quad \text{for } 1 \leq j \leq n.$$

Similarly, the third term is also finite for almost every $x \in \mathbf{R}^n$ if we apply Theorem 3.36 with $p := 1$ and $q = r := p$. Indeed, it suffices to note the inequality

$$\varepsilon^{n+1} \int_{|x|>\varepsilon} \frac{1}{|x|^{n+1}}\, dx = \varepsilon^n \int_{|y|>1} \frac{1}{|y|^{n+1}}\, dy$$
$$= \varepsilon^n \int_{\Sigma_{n-1}} \left(\int_1^{\infty} \frac{1}{t^{n+1}}\, t^{n-1}\, dt \right) d\sigma = \varepsilon^n\, \omega_n.$$

The second term is finite for almost every $x \in \mathbf{R}^n$ if we apply Lemma 9.1.

Therefore, by applying formula (9.160) to the sequences f_k and g_j^k and by letting $k \to \infty$ we obtain from Lebesgue's dominated convergence theorem (Theorem 3.13) that the desired formula (9.160) holds true for all $f \in L^p(\mathbf{R}^n)$ with $1 < p < \infty$:

$$\begin{aligned}
&\int_{\mathbf{R}^n} K(x-y)\,\phi\left(\frac{|x-y|}{\varepsilon}\right) f(y)\,dy \\
&= \lim_{k\to\infty}\int_{\mathbf{R}^n} K(x-y)\,\phi\left(\frac{|x-y|}{\varepsilon}\right) f_k(y)\,dy \\
&= \lim_{k\to\infty}\frac{1}{\varepsilon^n}\sum_{j=1}^{n}\int_{\mathbf{R}^n} K_j^{(2)}\left(\frac{x-y}{\varepsilon}\right) g_j^k(y)\,dy \\
&= \frac{1}{\varepsilon^n}\sum_{j=1}^{n}\int_{\mathbf{R}^n} K_j^{(2)}\left(\frac{x-y}{\varepsilon}\right) g_j(y)\,dy \\
&= -\frac{1}{\varepsilon^n}\sum_{j=1}^{n}\int_{\mathbf{R}^n} K_j^{(2)}\left(\frac{x-y}{\varepsilon}\right) (R_j * f)(y)\,dy.
\end{aligned}$$

Step (2): By formula (9.160), it follows that

$$\begin{aligned}
\widetilde{f}_\varepsilon(x) &= \int_{|x-y|>\varepsilon} K(x-y)\,f(y)\,dy \qquad (9.163)\\
&= \int_{|x-y|>\varepsilon} K(x-y)\phi\left(\frac{|x-y|}{\varepsilon}\right) f(y)\,dy \\
&= \int_{\mathbf{R}^n} K(x-y)\phi\left(\frac{|x-y|}{\varepsilon}\right) f(y)\,dy \\
&\quad - \int_{|x-y|<\varepsilon} K(x-y)\phi\left(\frac{|x-y|}{\varepsilon}\right) f(y)\,dy \\
&= \frac{1}{\varepsilon^n}\sum_{j=1}^{\infty}\int_{\mathbf{R}^n} K_j^{(2)}\left(\frac{x-y}{\varepsilon}\right) g_j(y)\,dy \\
&\quad - \int_{|x-y|<\varepsilon} K(x-y)\phi\left(\frac{|x-y|}{\varepsilon}\right) f(y)\,dy \\
&= \frac{1}{\varepsilon^n}\sum_{j=1}^{\infty}\int_{|x-y|>\varepsilon} K_j^{(1)}\left(\frac{x-y}{\varepsilon}\right) g_j(y)\,dy \\
&\quad + \frac{1}{\varepsilon^n}\sum_{j=1}^{\infty}\int_{|x-y|>\varepsilon} \left(K_j^{(2)}\left(\frac{x-y}{\varepsilon}\right) - K_j^{(1)}\left(\frac{x-y}{\varepsilon}\right)\right) g_j(y)\,dy \\
&\quad + \frac{1}{\varepsilon^n}\sum_{j=1}^{\infty}\int_{|x-y|<\varepsilon} K_j^{(2)}\left(\frac{x-y}{\varepsilon}\right) g_j(y)\,dy \\
&\quad - \int_{|x-y|<\varepsilon} K(x-y)\phi\left(\frac{|x-y|}{\varepsilon}\right) f(y)\,dy \\
&:= \mathrm{I}_1(x;\varepsilon) + \mathrm{I}_2(x;\varepsilon) + \mathrm{I}_3(x;\varepsilon) + \mathrm{I}_4(x;\varepsilon).
\end{aligned}$$

We estimate the four terms $\mathrm{I}_1(x;\varepsilon)$ through $\mathrm{I}_4(x;\varepsilon)$ on the right-hand side of formula (9.163).

Step (2-a): First, since $K_j^{(1)}(x)$ are positively homogeneous of degree $-n$, we have the formula

$$\begin{aligned}\mathrm{I}_1(x;\varepsilon) &= \sum_{j=1}^{\infty} \int_{|x-y|>\varepsilon} K_j^{(1)}(x-y)\, g_j(y)\, dy \\ &= -\sum_{j=1}^{\infty} \int_{|x-y|>\varepsilon} K_j^{(1)}(x-y)\, (R_j * f)(y)\, dy.\end{aligned}$$

Hence, by applying inequality (9.85) with

$$K(x) := K_j^{(1)}(x), \quad f(x) := R_j * f(x) \quad \text{for } 1 \le j \le n,$$

we obtain that

$$\begin{aligned}&\bullet \int_{\mathbf{R}^n} \sup_{\varepsilon>0} |\mathrm{I}_1(x;\varepsilon)|^p\, dx \\ &\le \left(\frac{\pi C_p}{2}\right)^p \left(\sum_{j=1}^{n} \int_{\Sigma_{n-1}} \left|K_j^{(1)}(\sigma)\right| d\sigma\right)^p \int_{\mathbf{R}^n} |g_j(x)|^p\, dx \\ &= \left(\frac{\pi C_p}{2}\right)^p \left(\int_{\Sigma_{n-1}} \left|K_j^{(1)}(\sigma)\right| d\sigma\right)^p \sum_{j=1}^{n} \|R_j * f\|_p^p \\ &\le C \left(\sum_{j=1}^{n} \int_{\Sigma_{n-1}} \left|K_j^{(1)}(\sigma)\right| d\sigma\right)^p \|f\|_p^p.\end{aligned} \tag{9.164}$$

Here we recall that the functions $K_j^{(1)}(x)$ are integrable on Σ_{n-1}, since they are positively homogeneous of degree $-n$ and belong to the space $L^1_{\mathrm{loc}}(\mathbf{R}^n \setminus \{0\})$.

Step (2-b): Secondly, by inequality (9.135) with

$$x := \frac{x-y}{\varepsilon},$$

it follows that

$$\begin{aligned}&|\mathrm{I}_2(x;\varepsilon)| \\ &= \left|\frac{1}{\varepsilon^n} \int_{|x-y|>\varepsilon} \left(K_j^{(2)}\left(\frac{x-y}{\varepsilon}\right) - K_j^{(1)}\left(\frac{x-y}{\varepsilon}\right)\right) g_j(y)\, dy\right| \\ &\le \frac{1}{\varepsilon^n} \int_{|x-y|>\varepsilon} \left|\left(K_j^{(2)}\left(\frac{x-y}{\varepsilon}\right) - K_j^{(1)}\left(\frac{x-y}{\varepsilon}\right)\right)\right| |g_j(y)|\, dy\end{aligned}$$

$$\leq C \frac{1}{\varepsilon^n} \int_{|x-y|>\varepsilon} \frac{\varepsilon^{n+1}}{|x-y|^{n+1}} |g_j(y)|\, dy \cdot \int_{\Sigma_{n-1}} |K(\sigma)|\, d\sigma$$
$$\leq C \left(\varepsilon \int_{|x-y|>\varepsilon} \frac{|g_j(y)|}{|x-y|^{n+1}}\, dy \right) \int_{\Sigma_{n-1}} |K(\sigma)|\, d\sigma.$$

However, we have, by Hölder's inequality (Theorem 3.24),

$$\varepsilon \int_{|x-y|>\varepsilon} \frac{|g_j(y)|}{|x-y|^{n+1}}\, dy = \varepsilon \int_{|z|>\varepsilon} \frac{|g_j(x-z)|}{|z|^{n+1}}\, dz$$
$$= \varepsilon \int_{\Sigma_{n-1}} \left(\int_\varepsilon^\infty \frac{|g_j(x-s\sigma)|}{|s\sigma|^{n+1}} s^{n-1}\, ds \right) dz$$
$$= \varepsilon \int_{\Sigma_{n-1}} \left(\int_\varepsilon^\infty s^{-2} |g_j(x-s\sigma)|\, ds \right) d\sigma$$
$$\leq \left(\int_{\Sigma_{n-1}} d\sigma \right)^{1-1/p} \left(\int_{\Sigma_{n-1}} \left(\varepsilon \int_\varepsilon^\infty s^{-2} |g_j(x-s\sigma)|\, ds \right)^p d\sigma \right)^{1/p}$$
$$= \omega_n^{1-1/p} \left(\int_{\Sigma_{n-1}} \left(\varepsilon \int_\varepsilon^\infty s^{-2} |g_j(x-s\sigma)|\, ds \right)^p d\sigma \right)^{1/p}.$$

Hence it follows from an application of Fubini's theorem (Theorem 3.20) that

$$\int_{\mathbf{R}^n} \left(\sup_{\varepsilon>0} \varepsilon \int_{|x-y|>\varepsilon} \frac{|g_j(y)|}{|x-y|^{n+1}}\, dy \right)^p dx$$
$$\leq \omega_n^{p-1} \int_{\Sigma_{n-1}} \left(\int_{\mathbf{R}^n} \left(\sup_{\varepsilon>0} \varepsilon \int_\varepsilon^\infty s^{-2} |g_j(x-s\sigma)|\, ds \right)^p dx \right) d\sigma.$$

For each $\sigma \in \Sigma_{n-1}$, we make the change of the variables

$$x := y - t\,\sigma, \quad y \in \mathbf{R}^n, \ t \in \mathbf{R},$$

with

$$\langle y, \sigma \rangle = y_1\sigma_1 + \ldots + y_n\sigma_n = 0,$$

as in the proof of Theorem 9.6. Then, by applying inequality (9.156) with

$$f(t) := |g_j(y - t\,\sigma)|,$$

we obtain that

$$\begin{aligned}
&\int_{\mathbf{R}^n} \left(\sup_{\varepsilon>0} \varepsilon \int_{\varepsilon}^{\infty} s^{-2} |g_j(x - s\,\sigma)|\, ds \right)^p dx \\
&= \int_{\mathbf{R}^{n-1}} \left(\int_{-\infty}^{\infty} \left(\sup_{\varepsilon>0} \varepsilon \int_{\varepsilon}^{\infty} s^{-2} |g_j(y - (t+s)\sigma)|\, ds \right)^p dt \right) dy \\
&\leq \left(\frac{2p}{p-1} \right)^p \int_{\mathbf{R}^{n-1}} \left(\int_{-\infty}^{\infty} |g_j(y - t\sigma)|^p\, dt \right) dy \\
&= \left(\frac{2p}{p-1} \right)^p \int_{\mathbf{R}^n} |g_j(x)|^p\, dx \quad \text{for every } \sigma \in \Sigma_{n-1}.
\end{aligned}$$

This proves that

$$\begin{aligned}
&\int_{\mathbf{R}^n} \left(\sup_{\varepsilon>0} \varepsilon \int_{|x-y|>\varepsilon} \frac{|g_j(y)|}{|x-y|^{n+1}}\, dy \right)^p dx \\
&\leq \omega_n^{p-1} \left(\int_{\Sigma_{n-1}} \left(\frac{2p}{p-1} \right)^p \int_{\mathbf{R}^n} |g_j(x)|^p\, dx \right) d\sigma \\
&= \left(\frac{2p}{p-1} \right)^p \omega_n^{p-1} \left(\int_{\mathbf{R}^n} |g_j(x)|^p\, dx \right) \int_{\Sigma_{n-1}} d\sigma \\
&= \left(\frac{2p\omega_n}{p-1} \right)^p \int_{\mathbf{R}^n} |g_j(x)|^p\, dx.
\end{aligned}$$

Summing up, we have proved that

$$\begin{aligned}
\bullet\ &\int_{\mathbf{R}^n} \sup_{\varepsilon>0} |\mathrm{I}_2(x;\varepsilon)|^p\, dx \qquad (9.165) \\
&\leq \left(\frac{2pC\omega_n}{p-1} \right)^p \left(\sum_{j=1}^{n} \int_{\Sigma_{n-1}} |K(\sigma)|\, d\sigma \right)^p \int_{\mathbf{R}^n} |g_j(x)|^p\, dx \\
&= \left(\frac{2pC\omega_n}{p-1} \right)^p \left(\int_{\Sigma_{n-1}} |K(\sigma)|\, d\sigma \right)^p \sum_{j=1}^{n} \|R_j * f\|_p^p \\
&\leq C \left(\int_{\Sigma_{n-1}} |K(\sigma)|\, d\sigma \right)^p \|f\|_p^p.
\end{aligned}$$

Step (2-c): Thirdly, we have, by inequality (9.136) and Hölder's inequality (Theorem 3.24),

$$\begin{aligned}
&|\mathrm{I}_3(x;\varepsilon)| \\
&= \left| \frac{1}{\varepsilon^n} \sum_{j=1}^{n} \int_{|x-y|<\varepsilon} K_2\left(\frac{x-y}{\varepsilon}\right) g_j(y)\,dy \right| \\
&\le \frac{1}{\varepsilon^n} \sum_{j=1}^{n} \int_{|(x-y)/\varepsilon|<1} \left| K_j^{(2)}\left(\frac{x-y}{\varepsilon}\right)\right| g_j(y)\,dy \\
&\le \frac{1}{\varepsilon^n} \sum_{j=1}^{n} \int_{|x-y|<\varepsilon} G_j(x-y)\,|g_j(y)|\,dy \\
&= \frac{1}{\varepsilon^n} \sum_{j=1}^{n} \int_{|z|<\varepsilon} G_j(z)|g_j(x-z)|\,dz \\
&= \int_{\Sigma_{n-1}} G_j(\sigma) \left(\frac{1}{\varepsilon^n} \int_0^{\varepsilon} |g_j(x-s\sigma)| s^{n-1}\,ds \right) d\sigma \\
&\le \sum_{j=1}^{n} \left(\int_{\Sigma_{n-1}} G_j(\sigma)\,d\sigma \right)^{1-1/p} \\
&\quad \times \left(\int_{\Sigma_{n-1}} G_j(\sigma) \left(\frac{1}{\varepsilon^n} \int_0^{\varepsilon} |g_j(x-s\sigma)| s^{n-1}\,ds \right)^p d\sigma \right)^{1/p}.
\end{aligned}$$

However, by applying inequality (9.155) with

$$f(t) := |g_j(y - t\,\sigma)| \quad \text{for } t \in \mathbf{R},$$

we obtain that

$$\begin{aligned}
&\int_{\mathbf{R}^n} \left(\sup_{\varepsilon>0} \frac{1}{\varepsilon^n} \int_0^{\varepsilon} |g_j(x-s\sigma)| s^{n-1}\,ds \right)^p dx \\
&= \int_{\mathbf{R}^{n-1}} \left(\int_{-\infty}^{\infty} \left(\sup_{\varepsilon>0} \frac{1}{\varepsilon^n} \int_0^{\varepsilon} \left|g_j(y-(t+s)\sigma)\right| s^{n-1}\,ds \right)^p dt \right) dy \\
&\le \int_{\mathbf{R}^{n-1}} \left(\frac{p}{p-1} \right)^p \left(\int_{-\infty}^{\infty} |g_j(y-t\sigma)|^p\,dt \right) dy \\
&= \left(\frac{p}{p-1} \right)^p \int_{\mathbf{R}^{n-1}} \int_{-\infty}^{\infty} |g_j(y-t\sigma)|^p\,dt\,dy \\
&= \left(\frac{p}{p-1} \right)^p \int_{\mathbf{R}^n} |g_j(x)|^p\,dx.
\end{aligned}$$

Therefore, by Fubini's theorem (Theorem 3.20) it follows that

$$
\begin{aligned}
\bullet & \int_{\mathbf{R}^n} \sup_{\varepsilon>0} |\mathrm{I}_3(x;\varepsilon)|^p\,dx \qquad (9.166)\\
&\le \sum_{j=1}^n \left(\int_{\Sigma_{n-1}} G_j(\sigma)\,d\sigma\right)^{p-1}\\
&\quad\times \int_{\mathbf{R}^n}\left(\int_{\Sigma_{n-1}} G_j(\sigma)\sup_{\varepsilon>0}\left(\frac{1}{\varepsilon^n}\int_0^\varepsilon \left|g_j(x-s\,\sigma)\right| s^{n-1}\,ds\right)^p d\sigma\right)dx\\
&= \sum_{j=1}^n \left(\int_{\Sigma_{n-1}} G_j(\sigma)\,d\sigma\right)^{p-1}\\
&\quad\times \int_{\Sigma_{n-1}} G_j(\sigma)\left(\int_{\mathbf{R}^n}\sup_{\varepsilon>0}\left(\frac{1}{\varepsilon^n}\int_0^\varepsilon \left|g_j(x-s\,\sigma)\right| s^{n-1}\,ds\right)^p dx\right)d\sigma\\
&\le \sum_{j=1}^n \left(\int_{\Sigma_{n-1}} G_j(\sigma)\,d\sigma\right)^{p-1}\left(\frac{p}{p-1}\right)^p \int_{\mathbf{R}^n}|g_j(x)|^p\,dx\cdot\int_{\Sigma_{n-1}} G_j(\sigma)\,d\sigma\\
&= \sum_{j=1}^n \left(\frac{p}{p-1}\right)^p\left(\int_{\Sigma_{n-1}} G_j(\sigma)\,d\sigma\right)^p\int_{\mathbf{R}^n}|g_j(x)|^p\,dx\\
&\le \sum_{j=1}^n \left(\frac{p}{p-1}\right)^p\left(\int_{\Sigma_{n-1}} G_j(\sigma)\,d\sigma\right)^p \|R_j * f\|_p^p\\
&\le C\sum_{j=1}^n \left(\int_{\Sigma_{n-1}} G_j(\sigma)\,d\sigma\right)^p \|f\|_p^p.
\end{aligned}
$$

Here we recall that the functions $G_j(\sigma)$ are integrable on Σ_{n-1}.

Step (2-d): Finally, we have, by Hölder's inequality (Theorem 3.24),

$$
\begin{aligned}
&|\mathrm{I}_4(x;\varepsilon)|\\
&= \left|\int_{\varepsilon/4<|x-y|<\varepsilon} |x-y|^{-n} K\left(\frac{x-y}{|x-y|}\right)\phi\left(\frac{|x-y|}{\varepsilon}\right) f(y)\,dy\right|\\
&\le \left(\frac{4}{\varepsilon}\right)^n \int_{|x-y|<\varepsilon}\left|K\left(\frac{x-y}{|x-y|}\right)\right| |f(y)|\,dy\\
&= \left(\frac{4}{\varepsilon}\right)^n \int_{\Sigma_{n-1}} |K(\sigma)|^{1-1/p}\,|K(\sigma)|^{1/p}\left(\int_0^\varepsilon |f(x-s\sigma)|s^{n-1}\,ds\right)d\sigma\\
&\le 4^n\left(\int_{\Sigma_{n-1}}|K(\sigma)|\,d\sigma\right)^{1-1/p}\\
&\quad\times\left(\int_{\Sigma_{n-1}}|K(\sigma)|\left(\frac{1}{\varepsilon^n}\int_0^\varepsilon |f(x-s\sigma)|s^{n-1}\,ds\right)^p d\sigma\right)^{1/p}.
\end{aligned}
$$

Hence, by applying Fubini's theorem (Theorem 3.20) and inequality (9.144) with

$$f(t) := |f(y - t\,\sigma)| \quad \text{for } t \in \mathbf{R},$$

we obtain that

$$\begin{aligned}
&\bullet \int_{\mathbf{R}^n} \sup_{\varepsilon>0} |\mathrm{I}_4(x;\varepsilon)|^p\,dx \qquad (9.167)\\
&\leq 4^{np} \left(\int_{\Sigma_{n-1}} |K(\sigma)|\,d\sigma \right)^{p-1} \\
&\quad \times \int_{\Sigma_{n-1}} |K(\sigma)| \left(\int_{\mathbf{R}^n} \left(\sup_{\varepsilon>0} \frac{1}{\varepsilon^n} \int_0^{\varepsilon} |f(x - s\sigma)| s^{n-1}\,ds \right)^p dx \right) d\sigma \\
&\leq 4^{np} \left(\int_{\Sigma_{n-1}} |K(\sigma)|\,d\sigma \right)^{p-1} \\
&\quad \times \left(\frac{p}{p-1} \right)^p \int_{\Sigma_{n-1}} |K(\sigma)| \left(\int_{\mathbf{R}^{n-1}} \int_{-\infty}^{\infty} |f(y - t\sigma)|^p\,dt\,dy \right) d\sigma \\
&= \left(\frac{4^n p}{p-1} \right)^p \left(\int_{\Sigma_{n-1}} |K(\sigma)|\,d\sigma \right)^p \int_{\mathbf{R}^n} |f(x)|^p\,dx.
\end{aligned}$$

Therefore, the desired inequality (9.159) follows from four inequalities (9.164), (9.165), (9.166) and (9.167), if we recall the formula

$$\widetilde{f}_{\varepsilon}(x) = \mathrm{I}_1(x;\varepsilon) + \mathrm{I}_2(x;\varepsilon) + \mathrm{I}_3(x;\varepsilon) + \mathrm{I}_4(x;\varepsilon). \qquad (9.163)$$

Step (3): The proof of the remaining part is analogous to that of Theorem 9.16. Now the proof of Theorem 9.26 is complete. □

9.10 The General Case

The next theorem asserts the existence of the singular integral (9.2) in the space $L^p(\mathbf{R}^n)$ for $1 < p < \infty$ in the general case:

Theorem 9.27 *Assume that the integral kernel $K(x)$ is a measurable, positively homogeneous function of degree $-n$ defined on $\mathbf{R}^n$ and satisfies the two conditions*

$$\int_{\Sigma_{n-1}} |K(\sigma)| \log^+ |K(\sigma)|\,d\sigma < \infty, \qquad (9.168a)$$

$$\int_{\Sigma_{n-1}} K(\sigma)\,d\sigma = 0 \quad \textit{(the cancellation property)}, \qquad (9.168b)$$

where Σ_{n-1} is the unit sphere in $\mathbf{R}^n$ and $d\sigma$ is the surface measure on Σ_{n-1}. If $f \in L^p(\mathbf{R}^n)$ with $1 < p < \infty$ and $\varepsilon > 0$, we let

$$\widetilde{f_\varepsilon}(x) := \int_{|x-y|>\varepsilon} K(x-y)\, f(y)\, dy.$$

Then we have the following three assertions:

(i) There exists a constant $C_p > 0$*, independent of* ε*, such that*

$$\left(\int_{\mathbf{R}^n} \sup_{\varepsilon>0} |\widetilde{f_\varepsilon}(x)|^p \, dx \right)^{1/p} \le C_p \left(\int_{\mathbf{R}^n} |f(x)|^p \, dx \right)^{1/p}.$$

(ii) The sequence $\widetilde{f_\varepsilon}$ *converges almost everywhere in* $\mathbf{R}$ *and in the strong topology of* $L^p(\mathbf{R})$ *as* $\varepsilon \downarrow 0$*. Namely, the singular integral*

$$K * f(x) = \lim_{\varepsilon\downarrow 0} \widetilde{f_\varepsilon}(x) = \lim_{\varepsilon\downarrow 0} \int_{|x-y|>\varepsilon} K(x-y)\, f(y)\, dy$$

exists for almost all $x \in \mathbf{R}^n$ *and in the strong topology of* $L^p(\mathbf{R}^n)$*.*

(iii) The mapping

$$\begin{aligned} L^p(\mathbf{R}^n) &\longrightarrow L^p(\mathbf{R}^n) \\ f &\longmapsto K * f \end{aligned}$$

is a bounded linear operator from $L^p(\mathbf{R}^n)$ *into itself. More precisely, we have the inequality*

$$\|K * f\|_p \le C_p \, \|f\|_p \quad \text{for all } f \in L^p(\mathbf{R}^n).$$

Proof It suffices to apply Theorems 9.16 and 9.26 to the odd part

$$K_1(x) := \frac{K(x) - K(-x)}{2}$$

and the even part

$$K_2(x) := \frac{K(x) + K(-x)}{2}$$

of the integral kernel $K(x)$, respectively.

Indeed, it is easy to see that

$$\int_{\Sigma_{n-1}} K_1(\sigma)\, d\sigma = \frac{1}{2} \left(\int_{\Sigma_{n-1}} K(\sigma)\, d\sigma - \int_{\Sigma_{n-1}} K(-\sigma)\, d\sigma \right) = 0,$$

This proves that the odd part $K_1(x)$ satisfies the cancellation property (9.4).

On the other hand, by inequality (9.118) with

$$\alpha := |K(\sigma)|,$$
$$\beta := |K(-\sigma)|,$$

it follows that

$$\begin{aligned} |K_2(\sigma)| \log^+ |K_2(\sigma)| &\leq \frac{|K(\sigma)| + |K(-\sigma)|}{2} \log^+ \left(\frac{|K(\sigma)| + |K(-\sigma)|}{2} \right) \\ &\leq \frac{1}{2} \left(|K(\sigma)| \log^+ |K(\sigma)| + |K(-\sigma)| \log^+ |K(-\sigma)| \right). \end{aligned}$$

Hence we have, by conditions (9.168),

$$\begin{aligned} &\int_{\Sigma_{n-1}} |K_2(\sigma)| \log^+ |K_2(\sigma)| \, d\sigma \\ &\leq \frac{1}{2} \left(\int_{\Sigma_{n-1}} |K(\sigma)| \log^+ |K(\sigma)| \, d\sigma + \int_{\Sigma_{n-1}} |K(-\sigma)| \log^+ |K(-\sigma)| \, d\sigma \right) \\ &= \frac{1}{2} \left(\int_{\Sigma_{n-1}} |K(\sigma)| \log^+ |K(\sigma)| \, d\sigma + \int_{\Sigma_{n-1}} |K(\sigma)| \log^+ |K(\sigma)| \, d\sigma \right) \\ &= \int_{\Sigma_{n-1}} |K(\sigma)| \log^+ |K(\sigma)| \, d\sigma \\ &< \infty. \end{aligned}$$

This proves that the even part $K_2(x)$ satisfies condition (9.106).

Now the proof of Theorem 9.27 is complete. □

Chapter 10
Calderón–Zygmund Kernels and Their Commutators

This chapter and Chap. 11 are the heart of the subject. The Calderón–Zygmund theory of singular integrals continues to be one of the most influential works in modern history of analysis. The first main result (Theorem 10.2) asserts the existence of singular integral operators and the second main result (Theorem 10.3) concerns commutators of bounded mean oscillation functions (BMO) and singular integral operators. It should be emphasized that singular integral operators with non-smooth kernels provide a powerful tool to deal with smoothness of solutions of partial differential equations, with minimal assumptions of regularity on the coefficients (see [26, 28, 107]). The results discussed here are adapted from Coifman–Rochberg–Weiss [39] and Bramanti–Cerutti [17].

10.1 Calderón–Zygmund Kernels

Let $k(x)$ be a real-valued function defined on $\mathbf{R}^n \setminus \{0\}$. We say that $k(x)$ is a *Calderón–Zygmund kernel* on $\mathbf{R}^n$ if it satisfies the following three conditions:

(i) $k \in C^\infty(\mathbf{R}^n \setminus \{0\})$.
(ii) $k(x)$ is positively homogeneous of degree $-n$, that is, $k(tx) = t^{-n}k(x)$ for all $t > 0$.
(iii) $\int_{\Sigma_{n-1}} k(\sigma)\, d\sigma = 0$ where Σ_{n-1} is the unit sphere in $\mathbf{R}^n$ and $d\sigma$ is the surface measure on Σ_{n-1}.

Example 10.1 Let $h(x)$ be a function in $C^\infty(\mathbf{R}^n \setminus \{0\})$ which is positively homogeneous of degree $1 - n$. Then the derivatives

$$\frac{\partial h}{\partial x_j} \quad \text{for } 1 \leq j \leq n,$$

are Calderón–Zygmund kernels.

K. Taira, *Real Analysis Methods for Markov Processes*,
https://doi.org/10.1007/978-981-97-3659-1_10

Proof It is easy to see that $\partial h/\partial x_j$ is positively homogeneous of degree $-n$. Hence we have only to verify that $\partial h/\partial x_j$ has the *cancellation property*

$$\int_{\Sigma_{n-1}} \frac{\partial h}{\partial x_j}(\sigma)\, d\sigma = 0. \tag{10.1}$$

Let $\rho(t)$ be a non-negative, smooth function on $\mathbf{R}$ such that

$$\operatorname{supp} \rho \subset [1, 2], \tag{10.2a}$$

$$\int_0^\infty \frac{\rho(t)}{t}\, dt = 1. \tag{10.2b}$$

First, by integration by parts it follows that

$$\int_{\mathbf{R}^n} \frac{\partial h}{\partial x_j}(x) \cdot \rho(|x|)\, dx = -\int_{\mathbf{R}^n} h(x)\rho'(|x|)\frac{x_j}{|x|}\, dx. \tag{10.3}$$

Since $h(x)$ is positively homogeneous of degree $1 - n$ and since $\partial h/\partial x_j$ is positively homogeneous of degree $-n$, by introducing polar coordinates

$$\begin{aligned} x &= r\,\sigma, \\ r &= |x|, \\ \sigma &\in \Sigma_{n-1}, \end{aligned}$$

we obtain from formula (10.3) that

$$\begin{aligned} &\int_0^\infty \int_{\Sigma_{n-1}} r^{-n} \frac{\partial h}{\partial x_j}(\sigma)\rho(r) r^{n-1}\, d\sigma\, dr \\ &= -\int_0^\infty \int_{\Sigma_{n-1}} r^{1-n} h(\sigma)\rho'(r)\sigma_j r^{n-1}\, d\sigma\, dr, \end{aligned}$$

so that

$$\begin{aligned} &\left(\int_0^\infty \frac{\rho(r)}{r}\, dr\right) \int_{\Sigma_{n-1}} \frac{\partial h}{\partial x_j}(\sigma)\, d\sigma \\ &= -\left(\int_0^\infty \rho'(r)\, dr\right) \int_{\Sigma_{n-1}} h(\sigma)\sigma_j\, d\sigma. \end{aligned} \tag{10.4}$$

However, we have, by conditions (10.2),

$$\begin{aligned} &\int_0^\infty \frac{\rho(r)}{r}\, dr = 1, \\ &\int_0^\infty \rho'(r)\, dr = 0. \end{aligned}$$

Therefore, the desired assertion (10.1) follows from formula (10.4). □

The first main result—the most important property of Calderón–Zygmund kernels—asserts the existence of singular integral operators [23]. The next theorem is a special case of Theorem 9.6 (cf. [34, Theorem 2.5]):

Theorem 10.2 *Let $k(x)$ be a Calderón–Zygmund kernel on $\mathbf{R}^n$. If $\varepsilon > 0$ and $f \in L^p(\mathbf{R}^n)$ for $1 < p < \infty$, we let*

$$K_\varepsilon f(x) := \int_{|x-y|>\varepsilon} k(x-y)\, f(y)\, dy. \tag{10.5}$$

Then there exists a function $Kf \in L^p(\mathbf{R}^n)$ such that

$$\lim_{\varepsilon \downarrow 0} \|K_\varepsilon f - Kf\|_{L^p(\mathbf{R}^n)} = 0. \tag{10.6}$$

Moreover, the mapping

$$\begin{aligned} K \colon L^p(\mathbf{R}^n) &\longrightarrow L^p(\mathbf{R}^n) \\ f &\longmapsto Kf = \lim_{\varepsilon \downarrow 0} \int_{|x-y|>\varepsilon} k(x-y)\, f(y)\, dy \end{aligned} \tag{10.7}$$

is a bounded linear operator from $L^p(\mathbf{R}^n)$ into itself. More precisely, there exists a constant $c_1 = c_1\left(n, p, \|k\|_{L^2(\Sigma_{n-1})}\right) > 0$ such that

$$\|Kf\|_{L^p(\mathbf{R}^n)} \leq c_1 \|f\|_{L^p(\mathbf{R}^n)} \quad \text{for all } f \in L^p(\mathbf{R}^n). \tag{10.8}$$

The operator K is called a *Calderón–Zygmund singular integral operator*. In what follows we shall use the notation

$$\begin{aligned} Kf(x) &= (\text{v.p.}\, k) * f(x) = \text{v.p.} \int_{\mathbf{R}^n} k(x-y)\, f(y)\, dy \\ &= \lim_{\varepsilon \downarrow 0} \int_{|x-y|>\varepsilon} k(x-y)\, f(y)\, dy. \end{aligned}$$

10.2 Commutators of Calderón–Zygmund Kernels

Now we are in a position to state the second main result concerning commutators of singular integrals. If $k(x)$ is a Calderón–Zygmund kernel on $\mathbf{R}^n$, we define a singular integral operator K by formula (10.7)

$$Kf = \text{v.p.} \int_{\mathbf{R}^n} k(x-y)\, f(y)\, dy \quad \text{for } f \in L^p(\mathbf{R}^n) \text{ with } 1 < p < \infty.$$

Let $\varphi \in \mathrm{BMO}$. Then we define the commutator $C[\varphi, K]$ of φ and K as the principal value

$$\begin{aligned} C[\varphi, K] f &:= \varphi(Kf) - K(\varphi f) \\ &= \text{v.p.} \int_{\mathbf{R}^n} k(x-y)\{\varphi(x) - \varphi(y)\} f(y)\, dy \\ &= \lim_{\varepsilon \downarrow 0} \int_{|x-y|>\varepsilon} \int_{\mathbf{R}^n} k(x-y)\{\varphi(x) - \varphi(y)\} f(y)\, dy. \end{aligned} \tag{10.9}$$

Then we have the following theorem (cf. [39, Theorem I], [17, Theorem 2.6], [34, Theorem 2.7]):

Theorem 10.3 *If* $\varphi \in \mathrm{BMO}$, *then the commutator of singular integrals*

$$C[\varphi, K] f = \varphi(Kf) - K(\varphi f)$$

is well-defined. Moreover, the commutator $C[\varphi, K]$, *defined by the formula* (10.9), *is bounded on the space* $L^p(\mathbf{R}^n)$ *for* $1 < p < \infty$. *More precisely, there exists a constant* $c_2 = c_2\left(n, p, \|k\|_{L^2(\Sigma_{n-1})}\right) > 0$ *such that*

$$\|C[\varphi, K] f\|_{L^p(\mathbf{R}^n)} \leq c_2 \|\varphi\|_* \|f\|_{L^p(\mathbf{R}^n)} \quad \textit{for all } f \in L^p(\mathbf{R}^n). \tag{10.10}$$

Proof The idea of our proof is due to Strömberg (cf. [168, pp. 417–419]). The proof is divided into three steps.

Step 1: Let Q be a cube with sides parallel to the coordinate axes. If $f \in L^1_{\mathrm{loc}}(\mathbf{R}^n)$, then we define the Hardy–Littlewood maximal function (see Sect. 4.4)

$$Mf(x) := \sup_{Q \ni x} \frac{1}{|Q|} \int_Q |f(y)|\, dy, \tag{10.11}$$

and the sharp function (see Sect. 4.6)

$$f^\sharp(x) := \sup_{Q \ni x} \frac{1}{|Q|} \int_Q |f(y) - f_Q|\, dy, \tag{10.12}$$

where the supremum is taken over all cubes containing x and f_Q is the integral average of f over Q:

$$f_Q := \frac{1}{|Q|} \int_Q f(z)\, dz. \tag{10.13}$$

For the sake of simplicity, we write

$$Tf(x) := C[\varphi, K] f(x) \quad \text{for } x \in \mathbf{R}^n. \tag{10.14}$$

For each $r \in (1, p)$, we shall prove a pointwise estimate

$$
\begin{aligned}
&(Tf)^\sharp(x) &\quad (10.15)\\
&\leq c(n,r)\,\|\varphi\|_* \left(\left(M(|Kf|^r)(x)\right)^{1/r} + \left(M(|f|^r)(x)\right)^{1/r} \right) \quad \text{for } x \in \mathbf{R}^n,
\end{aligned}
$$

with a positive constant $c(n, r)$.

Step 2: Assuming estimate (10.15) for the moment, we shall prove Theorem 10.3. The proof of estimate (10.15) will be given in Sect. 10.3 below, due to its length.

By applying Corollary 4.21 and Remark 4.22 and estimate (10.15), we obtain that, for $1 < r < p$,

$$
\begin{aligned}
&\|Tf\|_{L^p(\mathbf{R}^n)} \leq c_p \left\|(Tf)^\sharp\right\|_{L^p(\mathbf{R}^n)} &\quad (10.16)\\
&\leq 2^{(p-1)/p} c_p c(n,r)\,\|\varphi\|_*\\
&\times \left\{ \left(\int_{\mathbf{R}^n} (M(|Kf|^r)(x))^{p/r}\,dx \right)^{1/p} + \left(\int_{\mathbf{R}^n} (M(|f|^r)(x))^{p/r}\,dx \right)^{1/p} \right\}.
\end{aligned}
$$

Step 2-1: However, it should be noticed that

$$
\left(\int_{\mathbf{R}^n} (M(|f|^r)(x))^{p/r}\,dx \right)^{1/p} = \left(\left\| M(|f|^r) \right\|_{L^{p/r}(\mathbf{R}^n)} \right)^{1/r},
$$

so that we have, by Theorem 4.10 with $p := p/r$ and Remark 4.12,

$$
\left\| M(|f|^r) \right\|_{L^{p/r}(\mathbf{R}^n)} \leq C(p,r) \left\| |f|^r \right\|_{L^{p/r}(\mathbf{R}^n)} = C(p,r)\,\|f\|^r_{L^p(\mathbf{R}^n)}.
$$

Hence, we have the inequality

$$
\left(\int_{\mathbf{R}^n} (M(|f|^r)(x))^{p/r}\,dx \right)^{1/p} \leq C(p,r)^{1/r}\,\|f\|_{L^p(\mathbf{R}^n)}. \quad (10.17)
$$

Step 2-2: Similarly, we have the formula

$$
\left(\int_{\mathbf{R}^n} (M(|Kf|^r)(x))^{p/r}\,dx \right)^{1/p} = \left(\left\| M(|Kf|^r) \right\|_{L^{p/r}(\mathbf{R}^n)} \right)^{1/r},
$$

and also we have, by Theorem 4.10 with $p := p/r$,

$$
\left\| M(|Kf|^r) \right\|_{L^{p/r}(\mathbf{R}^n)} \leq C(p,r) \left\| |Kf|^r \right\|_{L^{p/r}(\mathbf{R}^n)} = C(p,r)\,\|Kf\|^r_{L^p(\mathbf{R}^n)}.
$$

Hence, it follows from an application of Theorem 10.2 that

$$\left(\int_{\mathbf{R}^n} (M(|Kf|^r)(x))^{p/r} \, dx \right)^{1/p} \leq C(p,r)^{1/r} \, \|Kf\|_{L^p(\mathbf{R}^n)} \tag{10.18}$$
$$\leq c_1(n,p) C(p,r)^{1/r} \, \|f\|_{L^p(\mathbf{R}^n)}.$$

Step 3: Therefore, the desired estimate (10.10) follows by combining estimates (10.16), (10.17) and (10.18), with the constant

$$c_2(n,p) := 2^{(p-1)/p} c_p c(n,r) \, (c_1(n,p) + 1) \, C(p,r)^{1/r}.$$

The proof of Theorem 10.3 is now complete, apart from the proof of estimate (10.15). □

10.3 Proof of Estimate 10.15

The purpose of this section is to prove estimate (10.15). If Q is a cube, then we denote by δ_Q its side length and by x_Q its center, respectively. For each $j \in \mathbf{N}$, we denote by $2^j Q$ the cube centered at x_Q with side length $2^j \delta_Q$ (see Fig. 4.3). Let Q be an arbitrary cube containing x. We write

$$Tf(x) := C\,[\varphi, K]\, f(x) \tag{10.14}$$

in the form

$$\begin{aligned} Tf(x) &= K\left((\varphi(\cdot) - \varphi_Q) f(\cdot) \chi_{2Q}\right)(x) \\ &\quad + K\left((\varphi(\cdot) - \varphi_Q) f(\cdot) \chi_{\mathbf{R}^n \setminus 2Q}\right)(x) - \left(\varphi(x) - \varphi_Q\right) Kf(x) \\ &:= A(x) + B(x) - C(x), \end{aligned}$$

where $\chi_A(x)$ is the characteristic function of the set A:

$$\chi_A(x) = \begin{cases} 1 & \text{if } x \in A, \\ 0 & \text{if } x \in \mathbf{R}^n \setminus A. \end{cases}$$

The proof is divided into four steps.
Step 1: The estimate of the term $A(x)$. We prove that, for $1 < r < p$,

$$\frac{1}{|Q|} \int_Q |A(x) - A_Q| \, dx \leq c_1(n,r) \, \|\varphi\|_* \left(M(|f|^r)(y)\right)^{1/r} \tag{10.19}$$
$$\text{for } y \in Q.$$

First, we have the inequality

$$\begin{aligned}\frac{1}{|Q|}\int_Q |A(x)-A_Q|\,dx &\le \frac{1}{|Q|}\int_Q |A(x)|\,dx + |A_Q| \\ &= \frac{1}{|Q|}\int_Q |A(x)|\,dx + \frac{1}{|Q|}\left|\int_Q A(y)\,dy\right| \\ &\le \frac{2}{|Q|}\int_Q |A(x)|\,dx \\ &= \frac{2}{|Q|}\int_Q \left|K\left((\varphi(\cdot)-\varphi_Q)f(\cdot)\chi_{2Q}\right)(x)\right|\,dx.\end{aligned} \tag{10.20}$$

If q is a number such that $1 < q < r$, then it follows that we have, by Hölder's inequality (Theorem 3.24),

$$\begin{aligned}&\frac{2}{|Q|}\int_Q \left|K\left((\varphi(\cdot)-\varphi_Q)f(\cdot)\chi_{2Q}\right)(x)\right|\,dx \\ &\le \frac{2}{|Q|}\left(\int_Q \left|K\left((\varphi(\cdot)-\varphi_Q)f(\cdot)\chi_{2Q}\right)(x)\right|^q\,dx\right)^{1/q}\left(\int_Q dx\right)^{1-1/q} \\ &\le \frac{2}{|Q|^{1/q}}\left(\int_{\mathbf{R}^n} \left|K\left((\varphi(\cdot)-\varphi_Q)f(\cdot)\chi_{2Q}\right)(x)\right|^q\,dx\right)^{1/q}.\end{aligned} \tag{10.21}$$

However, we have, by Theorem 10.2 and Hölder's inequality (Theorem 3.24),

$$\begin{aligned}&\frac{2}{|Q|^{1/q}}\left(\int_{\mathbf{R}^n} \left|K\left((\varphi(\cdot)-\varphi_Q)f(\cdot)\chi_{2Q}\right)(x)\right|^q\,dx\right)^{1/q} \\ &\le c(n,q)\frac{2}{|Q|^{1/q}}\left(\int_{\mathbf{R}^n} \left|\left(\varphi(x)-\varphi_Q\right)f(x)\chi_{2Q}(x)\right|^q\,dx\right)^{1/q} \\ &= 2c(n,q)\left(\frac{1}{|Q|}\int_{2Q} |\varphi(x)-\varphi_Q|^q\,|f(x)|^q\,dx\right)^{1/q} \\ &\le 2c(n,q) \\ &\quad\times\left(\frac{1}{|Q|}\left(\int_{2Q}|f(x)|^r\,dx\right)^{q/r}\left(\int_{2Q}|\varphi(x)-\varphi_Q|^{rq/(r-q)}\,dx\right)^{(r-q)/r}\right)^{1/q}.\end{aligned} \tag{10.22}$$

Moreover, we have the following claim:

Claim 10.4 *There exists a constant $c(n,q,r) > 0$ such that we have, for $1 < q < r < p$,*

$$\int_{2Q} \left|\varphi(x)-\varphi_Q\right|^{rq/(r-q)}\,dx \le c(n,q,r)\,\|\varphi\|_*^{rq/(r-q)}\,|2Q|. \tag{10.23}$$

Proof First, by Lemma 4.3 with $f := \varphi$ and $j := 1$ it follows that

$$\begin{aligned}|\varphi(x) - \varphi_Q| &\leq |\varphi(x) - \varphi_{2Q}| + |\varphi_{2Q} - \varphi_Q| \\ &\leq |\varphi(x) - \varphi_{2Q}| + c(n)\, \|\varphi\|_*,\end{aligned}$$

so that

$$\begin{aligned}&\int_{2Q} |\varphi(x) - \varphi_Q|^{rq/(r-q)}\, dx \qquad (10.24)\\ &\leq 2^{C(q,r)} \left\{ \int_{2Q} |\varphi(x) - \varphi_{2Q}|^{rq/(r-q)}\, dx + \int_{2Q} \left(c(n)\, \|\varphi\|_*\right)^{rq/(r-q)}\, dx \right\} \\ &= 2^{C(q,r)} \left\{ \int_{2Q} |\varphi(x) - \varphi_{2Q}|^{rq/(r-q)}\, dx + c(n)^{rq/(r-q)}\, \|\varphi\|_*^{rq/(r-q)}\, |2Q| \right\},\end{aligned}$$

where

$$C(q,r) := \frac{rq}{r-q} - 1 \quad \text{for } 1 < q < r.$$

However, we have, by Theorem 4.14 with $p := rq/(r-q)$ and Remark 4.15,

$$\int_Q |\varphi(x) - \varphi_{2Q}|^{rq/(r-q)}\, dx \leq c_1(q,r)\, \|\varphi\|_*^{rq/(r-q)}\, |2Q|. \qquad (10.25)$$

Therefore, by combining two estimates (10.24) and (10.25) we obtain that

$$\begin{aligned}&\int_{2Q} |\varphi(x) - \varphi_Q|^{rq/(r-q)}\, dx \\ &\leq 2^{C(q,r)} \left(c_1(q,r) + c(n)^{rq/(r-q)}\right) \|\varphi\|_*^{rq/(r-q)}\, |2Q|.\end{aligned}$$

This proves the desired estimate (10.23), with the constant

$$c(n,q,r) := 2^{C(q,r)} \left(c_1(q,r) + c(n)^{rq/(r-q)}\right).$$

The proof of Claim 10.4 is complete. □

Therefore, by combining four estimates (10.20), (10.21), (10.22) and (10.23) we obtain that

$$\begin{aligned}&\frac{1}{|Q|} \int_Q |A(x) - A_Q|\, dx \qquad (10.26)\\ &\leq 2c(n,q) \\ &\times \left(\frac{1}{|Q|} \left(\int_{2Q} |f(x)|^r\, dx \right)^{q/r} \left(\int_{2Q} |\varphi(x) - \varphi_Q|^{rq/(r-q)}\, dx \right)^{(r-q)/r} \right)^{1/q}\end{aligned}$$

$$= 2^{1+n/r} c(n,q)c(n,q,r)\frac{1}{|Q|^{1/q}}\left(\frac{1}{|2Q|}\int_{2Q}|f(x)|^r\,dx\right)^{1/r}|Q|^{1/r}$$
$$\times\left(\|\varphi\|_*^{rq/(r-q)}\,|2Q|\right)^{(r-q)/rq}$$
$$= 2^{1+n/r} c(n,q)c(n,q,r)\,\|\varphi\|_*\left(\frac{1}{|2Q|}\int_{2Q}|f(x)|^r\,dx\right)^{1/r}$$
$$\leq 2^{1+n/r} c(n,q)c(n,q,r)\,\|\varphi\|_*\left(M(|f|^r)(y)\right)^{1/r} \quad \text{for } y \in Q,$$

since we have the estimate

$$\frac{1}{|2Q|}\int_{2Q}|f(x)|^r\,dx \leq M(|f|^r)(y) \quad \text{for } y \in Q.$$

The desired estimate (10.19) follows from estimate (10.26), with the constant

$$c_1(n,r) := 2^{1+n/r} c(n,q)c(n,q,r).$$

Step 2: The estimate of the term $C(x)$. Similarly, we prove that, for $1 < r < p$,

$$\frac{1}{|Q|}\int_Q\left|C(x)-C_Q\right|dx \leq c_2(n,r)\,\|\varphi\|_*\left(M(|Kf|^r)(y)\right)^{1/r} \tag{10.27}$$
$$\text{for } y \in Q.$$

First, we have the inequality

$$\frac{1}{|Q|}\int_Q\left|C(x)-C_Q\right|dx \leq \frac{1}{|Q|}\int_Q|C(x)|\,dx + |C_Q| \tag{10.28}$$
$$= \frac{1}{|Q|}\int_Q|C(x)|\,dx + \frac{1}{|Q|}\left|\int_Q C(y)\,dy\right|$$
$$\leq \frac{2}{|Q|}\int_Q|C(x)|\,dx.$$

If q is a number such that $1 < q < r$, then it follows from Hölder's inequality (Theorem 3.24) that we have, for $r' = r/(r-1)$,

$$\frac{2}{|Q|}\int_Q|C(x)|\,dx$$
$$= \frac{2}{|Q|}\int_Q\left|\left(\varphi(x)-\varphi_Q\right)Kf(x)\right|\,dx$$
$$\leq 2\left(\frac{1}{|Q|}\int_Q\left|\varphi(x)-\varphi_Q\right|^{r'}\,dx\right)^{1/r'}\left(\frac{1}{|Q|}\int_Q|Kf(x)|^r\,dx\right)^{1/r}.$$

However, we have, by Theorem 4.14 with $p := r'$ and Remark 4.15,

$$\frac{1}{|Q|}\int_Q \left|\varphi(x) - \varphi_Q\right|^{r'} dx \leq c_1(r')\, \|\varphi\|_*^{r'},$$

and also

$$\frac{1}{|Q|}\int_Q |Kf(x)|^r\, dx \leq M(|Kf|^r)(y) \quad \text{for } y \in Q.$$

Hence it follows that

$$\begin{aligned}\frac{2}{|Q|}\int_Q |C(x)|\, dx &\leq 2\left(c_1(r')\, \|\varphi\|_*^{r'}\right)^{1/r'} \left(M(|Kf|^r)(y)\right)^{1/r} \qquad (10.29)\\ &= 2c_1(r')^{1/r'}\, \|\varphi\|_* \left(M(|Kf|^r)(y)\right)^{1/r} \quad \text{for } y \in Q.\end{aligned}$$

Therefore, the desired estimate (10.27) follows by combining estimates (10.28) and (10.29), with the constant

$$c_2(n, r) := 2c_1(r')^{1/r'}.$$

Step 3: The estimate of the term $B(x)$. Thirdly, we prove that, for $1 < r < p$,

$$\frac{1}{|Q|}\int_Q \left|B(x) - B_Q\right| dx \leq c_3(n, r)\, \|\varphi\|_* \left(M(|f|^r)(y)\right)^{1/r} \qquad (10.30)$$
$$\text{for } y \in Q.$$

The proof of estimate (10.30) is divided into four steps.

Step 3-1: We begin with the *pointwise Hörmander condition* for the Calderón–Zygmund kernels $k(x)$:

Lemma 10.5 *Let $k(x)$ be a Calderón–Zygmund kernel on $\mathbf{R}^n$. If Q is a cube with center x_Q, then we have, for all $x \in Q$ and $y \notin (2Q)$ (see Fig. 10.1),*

$$\left|k(x - y) - k\left(x_Q - y\right)\right| \leq c\frac{|x - x_Q|}{|x_Q - y|^{n+1}}, \qquad (10.31)$$

with a positive constant c.

Proof (1) Since $k(x)$ is positively homogeneous of degree $-n$, it follows that, for $h \neq 0$,

$$\frac{k(x + h\mathbf{e}_j) - k(x)}{h} = \frac{k\left(\frac{x}{|x|} + \frac{h}{|x|}\mathbf{e}_j\right) - k\left(\frac{x}{|x|}\right)}{h} \cdot \frac{1}{|x|^n} \quad \text{for } x \in \mathbf{R}^n \setminus \{0\},$$

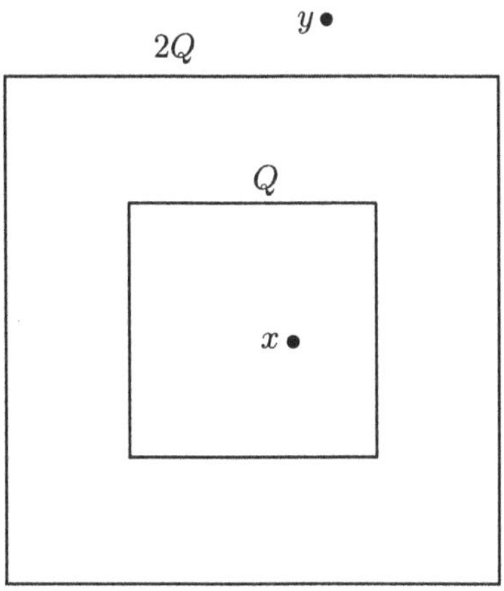

Fig. 10.1 The points $x \in Q$ and $y \in \mathbf{R}^n \setminus (2Q)$

so that

$$\frac{k(x + h\mathbf{e}_j) - k(x)}{h} = \frac{k(\overline{x} + \delta\mathbf{e}_j) - k(\overline{x})}{\delta} \cdot \frac{1}{|x|^{n+1}}, \tag{10.32}$$

with

$$\overline{x} = \frac{x}{|x|} \in \Sigma_{n-1}, \quad \delta = \frac{h}{|x|}.$$

Hence, by letting $h \to 0$ in formula (10.32) we obtain that

$$|\nabla k(x)| \leq \frac{C}{|x|^{n+1}} \quad \text{for all } x \in \mathbf{R}^n \setminus \{0\}, \tag{10.33}$$

where C is a positive constant given by the formula

$$C := \max_{|z|=1} |\nabla k(z)| .$$

(2) If we let

$$x_Q = (x_Q^1, x_Q^2, \ldots, x_Q^n),$$

then it follows from the mean value theorem that we have, for $0 < \theta < 1$,

$$\begin{aligned}
&\left|k(x-y) - k\left(x_Q - y\right)\right| \qquad (10.34)\\
&= \left|k\left((x_Q - y) + (x - x_Q)\right) - k\left(x_Q - y\right)\right| \\
&\leq \left|\frac{\partial k}{\partial x_1}(x_Q^1 - y_1 + \theta(x_1 - x_Q^1), x_2 - y_2, \ldots, x_n - y_n)\right| \left|x_1 - x_Q^1\right| \\
&\quad + \left|\frac{\partial k}{\partial x_2}(x_Q^1 - y_1, x_Q^2 - y_2 + \theta(x_2 - x_Q^2), x_3 - y_3, \ldots, x_n - y_n)\right|
\end{aligned}$$

$$
\begin{aligned}
&\times \left|x_2 - x_Q^2\right| \\
&+ \ldots + \left|\frac{\partial k}{\partial x_n}(x_Q^1 - y_1, x_Q^2 - y_2, \ldots, x_Q^n - y_n + \theta(x_n - x_Q^n))\right| \\
&\times \left|x_n - x_Q^n\right| \\
\leq &\left|\frac{\partial k}{\partial x_1}(x_Q^1 - y_1 + \theta(x_1 - x_Q^1), x_2 - y_2, \ldots, x_n - y_n)\right| \left|x - x_Q\right| \\
&+ \left|\frac{\partial k}{\partial x_2}(x_Q^1 - y_1, x_Q^2 - y_2 + \theta(x_2 - x_Q^2), x_3 - y_3, \ldots, x_n - y_n)\right| \\
&\times \left|x - x_Q\right| \\
&+ \ldots + \left|\frac{\partial k}{\partial x_n}(x_Q^1 - y_1, x_Q^2 - y_2, \ldots, x_Q^n - y_n + \theta(x_n - x_Q^n))\right| \\
&\times \left|x - x_Q\right|.
\end{aligned}
$$

However, we have, for all $x \in Q$ and $y \notin 2Q$,

$$
\begin{aligned}
\left|x_Q^j - y_j + \theta(x_j - x_Q^j)\right| &\geq \left|x_Q^j - y_j\right| - \left|x_j - x_Q^j\right| \\
&\geq \frac{1}{2}\left|y_j - x_Q^j\right| \quad \text{for } 0 < \theta < 1,
\end{aligned}
$$

and also

$$
\begin{aligned}
\left|x_j - y_j\right| = \left|(x_Q^j - y_j) + (x_j - x_Q^j)\right| &\geq \left|x_Q^j - y_j\right| - \left|x_j - x_Q^j\right| \\
&\geq \frac{1}{2}\left|y_j - x_Q^j\right|.
\end{aligned}
$$

In particular, it follows that, for $1 \leq j \leq n$,

$$
\left|(x_Q^1 - y_1, \ldots, x_Q^j - y_j + \theta(x_j - x_Q^j), \ldots, x_n - y_n)\right| \geq \frac{1}{2}\left|y - x_Q\right|. \quad (10.35)
$$

Therefore, by using inequality (10.33) we obtain from inequalities (10.34) and (10.35) that

$$
\begin{aligned}
&\left|k(x - y) - k\left(x_Q - y\right)\right| \\
\leq &\left(\sum_{j=1}^{n}\left|\frac{\partial k}{\partial x_j}(x_Q^1 - y_1, \ldots, x_Q^j - y_j + \theta(x_j - x_Q^j), \ldots, x_Q^n - y_n)\right|^2\right)^{1/2} \\
&\times \left|x - x_Q\right|
\end{aligned}
$$

$$\leq \frac{C}{\sum_j (|x_Q^1 - y_1|^2 + \ldots + |x_Q^j - y_j + \theta(x_j - x_Q^j)|^2 + \ldots + |x_Q^n - y_n|^2)^{(n+1)/2}}$$
$$\times \left|x - x_Q\right|$$
$$\leq 2^{n+1}\, C\, \frac{|x - x_Q|}{|y - x_Q|^{n+1}}.$$

This proves the desired inequality (10.31), with the constant

$$c := 2^{n+1}\, C = 2^{n+1} \max_{|z|=1} |\nabla k(z)|.$$

The proof of Lemma 10.5 is complete. □

Step 3-2: First, it follows that

$$\begin{aligned}
&B(x) - B(x_Q) \\
&= K\left((\varphi(\cdot) - \varphi_Q) f(\cdot) \chi_{\mathbf{R}^n \setminus 2Q}\right)(x) - K\left((\varphi(\cdot) - \varphi_Q) f(\cdot) \chi_{\mathbf{R}^n \setminus 2Q}\right)(x_Q) \\
&= \int_{\mathbf{R}^n \setminus (2Q)} k(x - y) \left\{\varphi(y) - \varphi_Q\right\} f(y)\, dy \\
&\quad - \int_{\mathbf{R}^n \setminus (2Q)} k\left(x_Q - y\right) \left\{\varphi(y) - \varphi_Q\right\} f(y)\, dy,
\end{aligned}$$

so that

$$\left|B(x) - B(x_Q)\right| \leq \int_{\mathbf{R}^n \setminus (2Q)} \left|k(x - y) - k\left(x_Q - y\right)\right| \left|\varphi(y) - \varphi_Q\right| |f(y)|\, dy.$$

However, by Lemma 10.5 we can find a positive constant c such that

$$\left|k(x - y) - k\left(x_Q - y\right)\right| \leq c \frac{|x - x_Q|}{|x_Q - y|^{n+1}} \quad \text{for } x \in Q \text{ and } y \notin 2Q.$$

Hence we have, by Hölder's inequality (Theorem 3.24),

$$\begin{aligned}
&\left|B(x) - B(x_Q)\right| && (10.36)\\
&\leq \int_{\mathbf{R}^n \setminus (2Q)} \left|k(x - y) - k\left(x_Q - y\right)\right| \left|\varphi(y) - \varphi_Q\right| |f(y)|\, dy \\
&\leq c \int_{\mathbf{R}^n \setminus (2Q)} \frac{|x - x_Q|}{|x_Q - y|^{n+1}} \left|\varphi(x) - \varphi_Q\right| |f(y)|\, dy \\
&\leq c' \delta_Q \int_{\mathbf{R}^n \setminus (2Q)} \frac{|f(y)|}{|x_Q - y|^{(n+1)/r}} \frac{|\varphi(x) - \varphi_Q|}{|x_Q - y|^{(n+1)/r'}}\, dy \\
&\leq c' \delta_Q \left(\int_{\mathbf{R}^n \setminus (2Q)} \frac{|f(y)|^r}{|x_Q - y|^{n+1}}\, dy\right)^{1/r}
\end{aligned}$$

$$\times \left(\int_{\mathbf{R}^n \setminus (2Q)} \frac{|\varphi(x) - \varphi_Q|^{r'}}{|x_Q - y|^{n+1}} \, dy \right)^{1/r'}.$$

Here δ_Q is the side length of Q.

Step 3-3: Now we prove the following two estimates:

$$I(x) := \int_{\mathbf{R}^n \setminus (2Q)} \frac{|f(z)|^r}{|x_Q - z|^{n+1}} \, dz \leq \frac{C_1}{\delta_Q} M(|f|^r)(y) \quad \text{for } y \in Q. \tag{10.37}$$

$$II(x) := \int_{\mathbf{R}^n \setminus (2Q)} \frac{|\varphi(x) - \varphi_Q|^{r'}}{|x_Q - y|^{n+1}} \, dy \leq \frac{C_2}{\delta_Q} \|\varphi\|_*^{r'}. \tag{10.38}$$

Proof *(of Estimate* (10.37)*)* Indeed, we have, for all $z \in 2^j Q$ with $j \in \mathbf{N}$,

$$|x_Q - z|^{n+1} \geq (2^{j-1}\delta_Q)^{n+1} = 2^{(n+1)j} \delta_Q^{n+1} \frac{1}{2^{n+1}},$$

and also

$$2^j \delta_Q \left|2^j Q\right| = 2^{(n+1)j} \delta_Q^{n+1}.$$

Hence it follows that

$$\begin{aligned} \int_{\mathbf{R}^n \setminus (2Q)} \frac{|f(z)|^r}{|x_Q - z|^{n+1}} \, dz &= \sum_{j=2}^{\infty} \int_{2^j Q \setminus (2^{j-1} Q)} \frac{|f(z)|^r}{|x_Q - z|^{n+1}} \, dz \\ &\leq C \sum_{j=2}^{\infty} \frac{1}{2^j \delta_Q} \left(\frac{1}{|2^j Q|} \int_{2^j Q} |f(z)|^r \, dz \right) \\ &\leq \frac{C}{2\delta_Q} M(|f|^r)(y) \quad \text{for } y \in Q. \end{aligned}$$

This proves the desired estimate (10.37), with the constant

$$C_1 := \frac{C}{2}.$$

The proof of estimate (10.37) is complete. □

Proof *(of Estimate* (10.38)*)* Similarly, we have, for $r' = r/(r-1)$,

$$\begin{aligned} \int_{\mathbf{R}^n \setminus (2Q)} \frac{|\varphi(z) - \varphi_Q|^{r'}}{|x_Q - z|^{n+1}} \, dz &= \sum_{j=2}^{\infty} \int_{2^j Q \setminus (2^{j-1} Q)} \frac{|\varphi(z) - \varphi_Q|^{r'}}{|x_Q - z|^{n+1}} \, dz \\ &\leq C \sum_{j=2}^{\infty} \frac{1}{2^j \delta_Q} \left(\frac{1}{|2^j Q|} \int_{2^j Q} |\varphi(z) - \varphi_Q|^{r'} \, dz \right). \end{aligned} \tag{10.39}$$

However, by Lemma 4.3 with $f := \varphi$ it follows that

$$\begin{aligned}|\varphi(z) - \varphi_Q| &\leq left|\varphi(z) - \varphi_{2^j Q} right| + \left|\varphi_{2^j Q} - \varphi_Q\right| \\ &\leq \left|\varphi(z) - \varphi_{2^j Q}\right| + c(n) j \, \|\varphi\|_*,\end{aligned}$$

so that

$$\begin{aligned}&\frac{1}{|2^j Q|} \int_{2^j Q} \left|\varphi(z) - \varphi_Q\right|^{r'} dz \\ &\leq 2^{r'-1} \left(\frac{1}{|2^j Q|} \int_{2^j Q} \left|\varphi(z) - \varphi_{2^j Q}\right|^{r'} dz \right) \\ &\quad + \frac{1}{|2^j Q|} \int_{2^j Q} \left(c(n) j \, \|\varphi\|_*\right)^{r'} dz \\ &\leq 2^{r'-1} \left(\frac{1}{|2^j Q|} \int_{2^j Q} \left|\varphi(z) - \varphi_{2^j Q}\right|^{r'} dz \right) + 2^{r'-1} \left(c(n) j \, \|\varphi\|_*\right)^{r'}.\end{aligned} \tag{10.40}$$

However, we have, by Theorem 4.14 with $p := r'$ and Remark 4.15,

$$\frac{1}{|2^j Q|} \int_{2^j Q} \left|\varphi(z) - \varphi_{2^j Q}\right|^{r'} dz \leq c_1(r') \, \|\varphi\|_*^{r'}. \tag{10.41}$$

Hence, it follows from two estimates (10.40) and (10.41) that

$$\frac{1}{|2^j Q|} \int_{2^j Q} \left|\varphi(z) - \varphi_Q\right|^{r'} dz \leq 2^{r'-1} \left(c(n)^{r'} j^{r'} + c_1(r') \right) \|\varphi\|_*^{r'}. \tag{10.42}$$

Therefore, by combining two estimates (10.39) and (10.42) we obtain that

$$\begin{aligned}&\int_{\mathbf{R}^n \setminus (2Q)} \frac{|\varphi(z) - \varphi_Q|^{r'}}{|x_Q - z|^{n+1}} \, dz \\ &= \sum_{j=2}^{\infty} \int_{2^j Q \setminus (2^{j-1} Q)} \frac{|\varphi(z) - \varphi_Q|^{r'}}{|x_Q - z|^{n+1}} \, dz \\ &\leq 2^{r'-1} C \sum_{j=2}^{\infty} \frac{1}{2^j \delta_Q} \left(c(n)^{r'} j^{r'} + c_1(r') \right) \|\varphi\|_*^{r'} \\ &= 2^{r'-1} \frac{C}{\delta_Q} \left\{ c_1(r') \left(\sum_{j=2}^{\infty} \frac{1}{2^j} \right) + c(n)^{r'} \left(\sum_{j=2}^{\infty} \frac{j^{r'}}{2^j} \right) \right\} \|\varphi\|_*^{r'}.\end{aligned}$$

This proves the desired estimate (10.38), with the constant

$$C_2 := 2^{r'-1} C \left\{ c_1(r') \left(\sum_{j=2}^{\infty} \frac{1}{2^j} \right) + c(n)^{r'} \left(\sum_{j=2}^{\infty} \frac{j^{r'}}{2^j} \right) \right\}.$$

The proof of estimate (10.38) is complete. □

Step 3-4: Therefore, by combining three estimates (10.36), (10.37) and (10.38) we obtain that

$$\begin{aligned}
&\frac{1}{|Q|} \int_Q \left| B(x) - B_Q \right| dx \leq \frac{2}{|Q|} \int_Q \left| B(x) - B(x_Q) \right| dx \\
&\leq \frac{2}{|Q|} c(n, r) \, \|\varphi\|_* \left(M(|f|^r)(y) \right)^{1/r} \int_Q dx \\
&= 2c(n, r) \, \|\varphi\|_* \left(M(|f|^r)(y) \right)^{1/r} \quad \text{for all } y \in Q.
\end{aligned}$$

This proves the desired estimate (10.30), with the constant

$$c_3(n, r) := 2c(n, r).$$

The proof of estimate (10.30) is complete. □

Step 4: Finally, if we let

$$c(n, r) := \max\{c_1(n, r), c_2(n, r), c_3(n, r)\},$$

then it follows from three estimates (10.19), (10.27) and (10.30) that

$$\begin{aligned}
&\frac{1}{|Q|} \int_Q \left| Tf(x) - (Tf)_Q \right| dx \\
&\leq \frac{1}{|Q|} \int_Q \left| A(x) - A_Q \right| dx + \frac{1}{|Q|} \int_Q \left| B(x) - B_Q \right| dx \\
&\quad + \frac{1}{|Q|} \int_Q \left| C(x) - C_Q \right| dx \\
&\leq c(n, r) \, \|\varphi\|_* \left(\left(M(|Kf|^r)(y) \right)^{1/r} + \left(M(|f|^r)(y) \right)^{1/r} \right) \quad \text{for all } y \in Q.
\end{aligned}$$

This proves that

$$\begin{aligned}
&(Tf)^{\sharp} f(y) \\
&\leq c(n, r) \, \|\varphi\|_* \left(\left(M(|Kf|^r)(y) \right)^{1/r} + \left(M(|f|^r)(y) \right)^{1/r} \right) \quad \text{for all } y \in \mathbf{R}^n,
\end{aligned}$$

since Q is arbitrary.

Now the proof of estimate (10.15) (and hence that of Theorem 10.3) is complete. □

Chapter 11
Calderón–Zygmund Variable Kernels and Their Commutators

In this chapter we consider singular integrals with kernels depending on a *parameter*, and prove theorems about singular integrals and commutators of L^∞ functions and singular integral operators (Theorems 11.2 and 11.3), generalizing Theorems 10.2 and 10.3 in Chap. 10. The main idea of proof is to reduce the variable kernel case to the constant kernel case. This is done by expanding the kernel into a series of spherical harmonics (Theorem 4.41), each term defining a constant kernel operator treated in Chap. 10. Theorems about singular integrals and commutators are usually formulated in the whole space $\mathbf{R}^n$. However, our application to the theory of elliptic equations with discontinuous coefficients will require a local version of Theorems 11.2 and 11.3 (Theorems 11.4 and 11.5 and Corollary 11.7).

The results here are adapted from Chiarenza–Frasca–Longo [34] and Calderón–Zygmund [28].

11.1 Operators Having Calderón–Zygmund Variable Kernels

Let $k(x, z)$ be a real-valued function defined on $\mathbf{R}^n \times (\mathbf{R}^n \setminus \{0\})$ with $n \geq 3$ that satisfies the following two conditions:

(i) $k(x, \cdot)$ is a Calderón–Zygmund kernel on $\mathbf{R}^n$ for almost all $x \in \mathbf{R}^n$.
(ii) The quantity

$$M := \max_{|\alpha| \leq 2n} \left\| \frac{\partial^\alpha k}{\partial z^\alpha}(\cdot, \cdot) \right\|_{L^\infty(\mathbf{R}^n \times \Sigma_{n-1})}$$

is finite. Here Σ_{n-1} is the unit sphere in $\mathbf{R}^n$.

It should be emphasized that the lack of regularity in the first variable x of $k(x, z)$ prevents the application of many recent results on singular integral operators.

K. Taira, *Real Analysis Methods for Markov Processes*,
https://doi.org/10.1007/978-981-97-3659-1_11

However, the good regularity in the second variable z of $k(x, z)$ allows us to use an old argument employed by Giraud, Calderón and Zygmund based on an expansion into spherical harmonics (Theorem 4.41).

Example 11.1 Assume that the functions $a^{ij} \in \mathrm{VMO} \cap L^\infty(\mathbf{R}^n)$ satisfy the following two conditions:

(1) $a^{ij}(x) = a^{ji}(x)$ for all $1 \leq i, j \leq n$ and for almost all $x \in \Omega$.
(2) There exists a positive constant λ such that

$$\frac{1}{\lambda}|\xi|^2 \leq \sum_{i,j=1}^{n} a^{ij}(x)\xi_i\xi_j \leq \lambda |\xi|^2$$

for almost all $x \in \Omega$ and for all $\xi \in \mathbf{R}^n$.

Let $\widetilde{\Omega}$ be the subset of Ω where conditions (1) and (2) hold true, and let

$$\Gamma(x, z) = \frac{1}{(2-n)\omega_n} \frac{1}{\sqrt{\det(a^{ij}(x))}} \left(\sum_{i,j=1}^{n} A_{ij}(x) z_i z_j \right)^{(2-n)/2}$$

for all $x \in \widetilde{\Omega}$ and all $z \in \mathbf{R}^n \setminus \{0\}$.

Here:

$$\big(A_{ij}(x)\big) := \text{ the inverse matrix of } \big(a^{ij}(x)\big),$$

$$\omega_n := |\Sigma_{n-1}| = \frac{2\pi^{n/2}}{\Gamma(n/2)} \quad \text{(the surface area of the unit sphere } \Sigma_{n-1} \text{ in } \mathbf{R}^n\text{)}.$$

Then it follows that the functions

$$\begin{aligned}\Gamma_i(x, z) &= \frac{\partial \Gamma}{\partial z_i}(x, z) \\ &= \frac{1}{\omega_n} \frac{1}{\sqrt{\det(a^{ij}(x))}} \left(\sum_{i,j=1}^{n} A_{ij}(x) z_i z_j \right)^{-n/2} \left(\sum_{j=1}^{n} A_{ij}(x) z_j \right)\end{aligned}$$

for all $x \in \widetilde{\Omega}$ and all $z \in \mathbf{R}^n \setminus \{0\}$,

are positively homogeneous of degree $1 - n$ with respect to the variable z ($1 \leq i \leq n$). Therefore, by applying Example 10.1 to the functions $\Gamma_i(x, z)$ we obtain that the functions

$$\Gamma_{ij}(x, z) = \frac{\partial^2 \Gamma}{\partial z_i \partial z_j}(x, z), \quad 1 \leq i, j \leq n,$$

are Calderón–Zygmund kernels in the z variable.

Then we have the following existence theorem of singular integrals based on Theorems 9.6 and 3.27 (see [34, Theorem 2.10]):

Theorem 11.2 *Let $k(x, z)$ be a real-valued function defined on $\mathbf{R}^n \times (\mathbf{R}^n \setminus \{0\})$ that satisfies conditions (i) and (ii). Then we have the following two assertions for all $f \in L^p(\mathbf{R}^n)$ with $1 < p < \infty$:*

(I) If $\varepsilon > 0$, we let

$$K_\varepsilon f(x) := \int_{|x-y|>\varepsilon} k(x, x-y)\, f(y)\, dy, \tag{11.1}$$

then there exists a function $Kf \in L^p(\mathbf{R}^n)$ such that

$$\lim_{\varepsilon \downarrow 0} \|K_\varepsilon f - Kf\|_{L^p(\mathbf{R}^n)} = 0. \tag{11.2}$$

(II) The mapping

$$\begin{aligned} K\colon L^p(\mathbf{R}^n) &\longrightarrow L^p(\mathbf{R}^n) \\ f &\longmapsto Kf = \lim_{\varepsilon \downarrow 0} \int_{|x-y|>\varepsilon} k(x, x-y)\, f(y)\, dy \end{aligned} \tag{11.3}$$

is a bounded linear operator from $L^p(\mathbf{R}^n)$ into itself. More precisely, there exists a constant $c_3\,(n, p, M) > 0$ such that

$$\|Kf\|_{L^p(\mathbf{R}^n)} \leq c_3\,(n, p, M)\, \|f\|_{L^p(\mathbf{R}^n)} \quad \text{for all } f \in L^p(\mathbf{R}^n). \tag{11.4}$$

Proof By a density argument, it suffices to prove Theorem 11.2 for all $f \in C_0^\infty(\mathbf{R}^n)$. The proof is divided into five steps.

An idea of the proof of Theorem 11.2 can be visualized as follows:

$$\begin{array}{ccc} K_{N;\varepsilon} f & \xrightarrow{\varepsilon \downarrow 0} & K_N f \\ {\scriptstyle N\to\infty}\big\downarrow & & \big\downarrow{\scriptstyle N\to\infty} \\ K_\varepsilon f & \xrightarrow[\varepsilon \downarrow 0]{} & Kf \end{array}$$

Step 1: First, it should be noticed that the function

$$z \longmapsto |z|^n k(x, z)$$

belongs to $C^\infty\,(\mathbf{R}^n \setminus \{0\})$ for almost all $x \in \mathbf{R}^n$, and it is positively homogeneous of degree zero and enjoys the cancellation property

$$\int_{\Sigma_{n-1}} k(x, z)\, dz = 0.$$

For each $m = 1, 2, \dots$ and $k = 1, 2, \dots, d(m)$, we let

$$a_{km}(x) := \int_{\Sigma_{n-1}} k(x, z)\, Y_{km}(z)\, dz,$$

then, by the completeness of the spherical harmonics $\{Y_{km}\}$ in $L^2(\Sigma_{n-1})$ it follows that

$$|z|^n k(x, z) = \sum_{m=1}^{\infty} \sum_{k=1}^{d(m)} a_{km}(x)\, Y_{km}(z). \tag{11.5}$$

Moreover, we have, by assertion (4.76) with $r := n$ and assertion (4.75) with $\alpha := 0$ of Theorem 4.41,

$$\|a_{km}\|_{L^\infty(\mathbf{R}^n)} \leq \frac{c_1(n)}{m^{2n}} M, \tag{11.6a}$$

$$\|Y_{km}\|_{L^\infty(\Sigma_{n-1})} \leq c_2(n)\, m^{(n-2)/2}, \tag{11.6b}$$

and, by assertion (4.74) of Theorem 4.41,

$$d(m) = \dim \mathcal{H}_m \leq c_3(n)\, m^{n-2}. \tag{11.6c}$$

Step 2: Secondly, we let

$$R_{km;\varepsilon} f(x) := \int_{|x-y|>\varepsilon} \frac{Y_{km}(x-y)}{|x-y|^n} f(y)\, dy \quad \text{for } f \in L^p(\mathbf{R}^n).$$

It should be noticed that the function

$$\frac{Y_{km}(z)}{|z|^n}$$

is a Calderón–Zygmund kernel and that

$$\|Y_{km}\|_{L^2(\Sigma_{n-1})} = 1.$$

Therefore, by applying Theorem 10.2 or Theorem 9.6 we can find a constant $c_1(n, p) > 0$, independent of ε, such that (see inequality (9.50))

$$\left\| R_{km;\varepsilon} f \right\|_{L^p(\mathbf{R}^n)} \leq c_1(n, p)\, \|f\|_{L^p(\mathbf{R}^n)} \quad \text{for all } f \in L^p(\mathbf{R}^n). \tag{11.7}$$

Here it should be emphasized that the constant $c_1(n, p)$ depends neither k nor m.

Moreover, there exists a limit function

$$R_{km} f = \lim_{\varepsilon \downarrow 0} R_{km;\varepsilon} f \in L^p(\mathbf{R}^n)$$

such that (see assertion (9.51))

$$\lim_{\varepsilon \downarrow 0} \left\| R_{km;\varepsilon} f - R_{km} f \right\|_{L^p(\mathbf{R}^n)} = 0 \quad \text{for all } f \in L^p(\mathbf{R}^n), \tag{11.8}$$

and further that (see inequality (9.52))

$$\| R_{km} f \|_{L^p(\mathbf{R}^n)} \leq c_1 (n, p) \| f \|_{L^p(\mathbf{R}^n)} \quad \text{for all } f \in L^p(\mathbf{R}^n). \tag{11.9}$$

Step 3: Thirdly, if $\varepsilon > 0$, by using the spherical expansion (11.5) with $z := x - y$ we have the formula

$$\begin{aligned} K_\varepsilon f(x) &:= \int_{|x-y|>\varepsilon} k(x, x-y)\, f(y)\, dy \qquad (11.10) \\ &= \sum_{m=1}^{\infty} \sum_{k=1}^{d(m)} a_{km}(x) \int_{|x-y|>\varepsilon} \frac{Y_{km}(x-y)}{|x-y|^n} f(y)\, dy \\ &= \lim_{N \to \infty} \sum_{m=1}^{N} \int_{|x-y|>\varepsilon} \sum_{k=1}^{d(m)} a_{km}(x) \frac{Y_{km}(x-y)}{|x-y|^n} f(y)\, dy \\ &= \lim_{N \to \infty} \sum_{m=1}^{N} \sum_{k=1}^{d(m)} a_{km}(x)\, R_{km;\varepsilon} f(x) \quad \text{in } L^p(\mathbf{R}^n). \end{aligned}$$

Indeed, for a given positive integer N we let

$$K_{N;\varepsilon} f(x) := \sum_{m=1}^{N} \sum_{k=1}^{d(m)} a_{km}(x)\, R_{km;\varepsilon} f(x) \quad \text{for } f \in L^p(\mathbf{R}^n). \tag{11.11}$$

Then it follows from estimates (11.6a), (11.6b) and (11.6c) and inequality (11.7) that

$$\begin{aligned} &\left\| K_{N;\varepsilon} f \right\|_{L^p(\mathbf{R}^n)} \qquad (11.12) \\ &\leq \sum_{m=1}^{N} \sum_{k=1}^{d(m)} \left\| a_{km}\, R_{km;\varepsilon} f \right\|_{L^p(\mathbf{R}^n)} \\ &\leq \sum_{m=1}^{N} \sum_{k=1}^{d(m)} \| a_{km} \|_{L^\infty(\mathbf{R}^n)} \left\| R_{km;\varepsilon} f \right\|_{L^p(\mathbf{R}^n)} \\ &\leq \sum_{m=1}^{N} \sum_{k=1}^{d(m)} \left(\frac{c_1(n)}{m^{2n}} M\, c_1(n, p) \right) \| f \|_{L^p(\mathbf{R}^n)} \end{aligned}$$

$$
\begin{aligned}
&= \sum_{m=1}^{N} d(m) \left(\frac{c_1(n)}{m^{2n}} M\, c_1(n, p) \right) \|f\|_{L^p(\mathbf{R}^n)} \\
&\leq (c_1(n)\, c_1(n, p)\, c_3(n)\, M) \sum_{m=1}^{N} \frac{1}{m^{2n}} m^{n-2} \|f\|_{L^p(\mathbf{R}^n)} \\
&\leq (c_1(n)\, c_1(n, p)\, c_3(n)\, M) \left(\sum_{m=1}^{\infty} \frac{1}{m^{n+2}} \right) \|f\|_{L^p(\mathbf{R}^n)} \\
&= C\,(n, p, M) \|f\|_{L^p(\mathbf{R}^n)} \quad \text{for all } f \in L^p(\mathbf{R}^n).
\end{aligned}
$$

with

$$
C\,(n, p, M) := (c_1(n)\, c_1(n, p)\, c_3(n)\, M) \left(\sum_{m=1}^{\infty} \frac{1}{m^{n+2}} \right).
$$

Therefore, by applying Theorem 3.27 to our situation we obtain from inequality (11.12) that the following two assertions hold true:

(A) The limit

$$
\begin{aligned}
K_\varepsilon f(x) &= \lim_{N \to \infty} K_{N;\varepsilon} f(x) \\
&= \lim_{N \to \infty} \sum_{m=1}^{N} \sum_{k=1}^{d(m)} a_{km}(x)\, R_{km;\varepsilon} f(x) \\
&= \sum_{m=1}^{\infty} \sum_{k=1}^{d(m)} a_{km}(x)\, R_{km;\varepsilon} f(x)
\end{aligned}
\tag{11.13}
$$

converges absolutely in the space $L^p(\mathbf{R}^n)$, for every $\varepsilon > 0$. More precisely, we have the assertion

$$
\lim_{N \to \infty} \left\| K_{N;\varepsilon} f - K_\varepsilon f \right\|_{L^p(\mathbf{R}^n)} = 0 \quad \text{for all } f \in L^p(\mathbf{R}^n). \tag{11.14}
$$

(B) There exists a constant $C\,(n, p, M) > 0$, independent of ε, such that

$$
\|K_\varepsilon f\|_{L^p(\mathbf{R}^n)} \leq C\,(n, p, M) \|f\|_{L^p(\mathbf{R}^n)} \quad \text{for all } f \in L^p(\mathbf{R}^n). \tag{11.15}
$$

Step 4: We show that the series of operators

$$
\begin{aligned}
Kf(x) &= \sum_{m=1}^{\infty} \sum_{k=1}^{d(m)} a_{km}(x)\, R_{km} f(x) \\
&= \lim_{N \to \infty} \sum_{m=1}^{N} \sum_{k=1}^{d(m)} a_{km}(x)\, R_{km} f(x)
\end{aligned}
\tag{11.16}
$$

converges absolutely in the L^p space. More precisely, we have the following two assertions for the mapping (11.3):

(C) $Kf \in L^p(\mathbf{R}^n)$ for all $f \in L^p(\mathbf{R}^n)$.
(D) We have the inequality

$$\|Kf\|_{L^p(\mathbf{R}^n)} \leq C\,(n, p, M)\,\|f\|_{L^p(\mathbf{R}^n)} \quad \text{for all } f \in L^p(\mathbf{R}^n). \tag{11.4}$$

Indeed,, for a given positive integer N we let

$$K_N f(x) := \sum_{m=1}^{N} \sum_{k=1}^{d(m)} a_{km}(x)\, R_{km} f(x). \tag{11.17}$$

Then we have, by estimates (11.6a) and inequality (11.9),

$$\begin{aligned}
&\|K_N f\|_{L^p(\mathbf{R}^n)} \\
&\leq \sum_{m=1}^{N} \sum_{k=1}^{d(m)} \|a_{km}\, R_{km} f\|_{L^p(\mathbf{R}^n)} \\
&\leq \sum_{m=1}^{N} \sum_{k=1}^{d(m)} \|a_{km}\|_{L^\infty(\mathbf{R}^n)}\, \|R_{km} f\|_{L^p(\mathbf{R}^n)} \\
&\leq \sum_{m=1}^{N} \sum_{k=1}^{d(m)} \frac{c_1(n)\, M}{m^{2n}}\, c_1(n, p)\, \|f\|_{L^p(\mathbf{R}^n)} \\
&\leq \sum_{m=1}^{N} \frac{c_1(n)\, M}{m^{2n}}\, c_1(n, p)\, c_3(n)\, m^{n-2}\, \|f\|_{L^p(\mathbf{R}^n)} \\
&= \sum_{m=1}^{N} \frac{c_1(n)\, M}{m^{n+2}}\, c_1(n, p)\, c_3(n)\, \|f\|_{L^p(\mathbf{R}^n)} \\
&\leq \left(\sum_{m=1}^{\infty} \frac{1}{m^{n+2}}\right) c_1(n)\, c_1(n, p)\, c_3(n)\, M\, \|f\|_{L^p(\mathbf{R}^n)} \\
&= C\,(n, p, M)\, \|f\|_{L^p(\mathbf{R}^n)}\,.
\end{aligned} \tag{11.18}$$

By applying again Theorem 3.27 to our situation, we obtain from inequality (11.18) that the desired two assertions (C) and (D) hold true.

Step 5: Finally, we prove assertion (11.1)

$$\lim_{\varepsilon \downarrow 0} \|K_\varepsilon f - Kf\|_{L^p(\mathbf{R}^n)} = 0.$$

Indeed, we have, by formulas (11.10) and (11.16) and inequalities (11.7) and (11.9),

$$
\begin{aligned}
&\|K_\varepsilon f - Kf\|_{L^p(\mathbf{R}^n)} \qquad (11.19)\\
&= \left\| \sum_{m=1}^{\infty} \sum_{k=1}^{d(m)} a_{km}\, R_{km;\varepsilon} f - \sum_{m=1}^{\infty} \sum_{k=1}^{d(m)} a_{km}\, R_{km} f \right\|_{L^p(\mathbf{R}^n)} \\
&= \left\| \sum_{m=1}^{\infty} \sum_{k=1}^{d(m)} \left(a_{km}\, R_{km;\varepsilon} f - a_{km}\, R_{km} f \right) \right\|_{L^p(\mathbf{R}^n)} \\
&\le \sum_{m=1}^{\infty} \sum_{k=1}^{d(m)} \|a_{km}\|_{L^\infty(\mathbf{R}^n)} \left\| R_{km;\varepsilon} f - R_{km} f \right\|_{L^p(\mathbf{R}^n)} \\
&\le \sum_{m=1}^{\infty} \sum_{k=1}^{d(m)} \frac{c_1(n)\, M}{m^{2n}} \left\| R_{km;\varepsilon} f - R_{km} f \right\|_{L^p(\mathbf{R}^n)} \\
&\le \sum_{m=1}^{\infty} \frac{c_1(n)\, M}{m^{2n}}\, c_3(n)\, m^{n-2} \left\| R_{km;\varepsilon} f - R_{km} f \right\|_{L^p(\mathbf{R}^n)} \\
&= c_1(n)\, c_3(n)\, M \left(\sum_{m=1}^{\infty} \frac{1}{m^{n+2}} \right) \left\| R_{km;\varepsilon} f - R_{km} f \right\|_{L^p(\mathbf{R}^n)} \\
&\le c_1(n)\, c_3(n)\, M \left(\sum_{m=1}^{N} \frac{1}{m^{n+2}} \right) \left\| R_{km;\varepsilon} f - R_{km} f \right\|_{L^p(\mathbf{R}^n)} \\
&\quad + c_1(n)\, c_3(n)\, M \left(\sum_{m=N+1}^{\infty} \frac{1}{m^{n+2}} \right) \left(\left\| R_{km;\varepsilon} f \right\|_{L^p(\mathbf{R}^n)} + \|R_{km} f\|_{L^p(\mathbf{R}^n)} \right) \\
&\le c_1(n)\, c_3(n)\, M \left(\sum_{m=1}^{N} \frac{1}{m^{n+2}} \right) \left\| R_{km;\varepsilon} f - R_{km} f \right\|_{L^p(\mathbf{R}^n)} \\
&\quad + 2\, c_1(n)\, c_3(n)\, M\, c_1\,(n,\, p) \left(\sum_{m=N+1}^{\infty} \frac{1}{m^{n+2}} \right) \|f\|_{L^p(\mathbf{R}^n)}\,,
\end{aligned}
$$

where N is arbitrary.

Hence, by passing to $\varepsilon \downarrow \infty$ in inequality (11.19) we obtain from assertion (11.8) that

$$
\begin{aligned}
&\limsup_{\varepsilon \downarrow 0} \|K_\varepsilon f - Kf\|_{L^p(\mathbf{R}^n)} \qquad (11.20)\\
&\le 2\, c_1(n)\, c_3(n)\, M\, c_1\,(n,\, p) \left(\sum_{m=N+1}^{\infty} \frac{1}{m^{n+2}} \right) \|f\|_{L^p(\mathbf{R}^n)}
\end{aligned}
$$

for any positive integer N.

Therefore, the desired assertion (11.4) follows by letting $N \to \infty$ in inequality (11.20).

Now the proof of Theorem 11.2 is complete. □

Let $\varphi \in L^\infty(\mathbf{R}^n)$. If $\varepsilon > 0$ and $f \in L^p(\mathbf{R}^n)$ for $1 < p < \infty$, we define the commutator $C[\varphi, K_\varepsilon]$ by the formula

$$\begin{aligned} C[\varphi, K_\varepsilon] f &:= \varphi(K_\varepsilon f) - K_\varepsilon(\varphi f) \\ &= \int_{|x-y|>\varepsilon} k(x, x-y) \{\varphi(x) - \varphi(y)\} f(y)\, dy. \end{aligned}$$

Then we have the following theorem (see [34, Theorem 2.10]):

Theorem 11.3 *Let $f \in L^p(\mathbf{R}^n)$ for $1 < p < \infty$. If $\varphi \in L^\infty(\mathbf{R}^n)$, then there exists a function $C[\varphi, K] f \in L^p(\mathbf{R}^n)$ such that*

$$\lim_{\varepsilon \downarrow 0} \|C[\varphi, K_\varepsilon] f - C[\varphi, K] f\|_{L^p(\mathbf{R}^n)} = 0. \tag{11.21}$$

Furthermore, the commutator $C[\varphi, K]$, defined by the formula

$$C[\varphi, K]f = \lim_{\varepsilon \downarrow 0} C[\varphi, K_\varepsilon] f \quad \text{in } L^p(\mathbf{R}^n), \tag{11.22}$$

is bounded on $L^p(\mathbf{R}^n)$. More precisely, there exists a constant $c_4(n, p, M) > 0$ such that

$$\|C[\varphi, K] f\|_{L^p(\mathbf{R}^n)} \leq c_4(n, p, M) \|\varphi\|_* \|f\|_{L^p(\mathbf{R}^n)} \quad \text{for all } f \in L^p(\mathbf{R}^n). \tag{11.23}$$

The proof of Theorem 11.3 is essentially the same as that of Theorem 11.2 if we use Theorem 10.3 (instead of Theorem 10.2).

11.2 Local Version of Theorems 11.2 and 11.3

Theorems about singular integrals and commutators are usually formulated in the whole space $\mathbf{R}^n$. However, our application to the theory of elliptic equations with discontinuous coefficients will require a local version of Theorems 11.2 and 11.3.

Let Ω be an open subset of $\mathbf{R}^n$, and let $k(x, z)$ be a real-valued function defined on $\Omega \times (\mathbf{R}^n \setminus \{0\})$ that satisfies the following two conditions:

(i) $k(x, \cdot)$ is a Calderón–Zygmund kernel on $\mathbf{R}^n$ for almost all $x \in \Omega$.
(ii) The quantity

$$M := \max_{|\alpha|\leq 2n} \left\| \frac{\partial^\alpha k}{\partial z^\alpha}(\cdot, \cdot) \right\|_{L^\infty(\Omega\times\Sigma_{n-1})}$$

is finite.

The next existence theorem of singular integrals is a local version of Theorem 11.2 (cf. [34, Theorem 2.11]):

Theorem 11.4 *Let $k(x, z)$ be a real-valued function defined on $\mathbf{R}^n \times (\mathbf{R}^n \setminus \{0\})$ that satisfies conditions (i) and (ii). Then we have the following two assertions for all $f \in L^p(\Omega)$ with $1 < p < \infty$:*

(I) If $\varepsilon > 0$, we let

$$K_\varepsilon f(x) := \int_{\substack{y\in\Omega \\ |x-y|>\varepsilon}} k(x, x-y)\, f(y)\, dy,$$

then there exists a function $Kf \in L^p(\Omega)$ such that

$$\lim_{\varepsilon\downarrow 0} \|K_\varepsilon f - Kf\|_{L^p(\Omega)} = 0 \quad \text{for every } f \in L^p(\Omega).$$

(II) The operator K, defined by the formula

$$Kf = \lim_{\varepsilon\downarrow 0} K_\varepsilon f \quad \text{in } L^p(\Omega),$$

is bounded on $L^p(\Omega)$. More precisely, there exists a positive constant $c_5(n, p, M)$ such that

$$\|Kf\|_{L^p(\Omega)} \leq c_5(n, p, M)\, \|f\|_{L^p(\Omega)} \quad \text{for all } f \in L^p(\Omega).$$

Proof First, we remark that the function

$$\widetilde{k}(x, z) := \begin{cases} k(x, z) & \text{if } x \in \Omega \text{ and } z \in \mathbf{R}^n \setminus \{0\}, \\ 0 & \text{if } x \notin \Omega \text{ and } z \in \mathbf{R}^n \setminus \{0\} \end{cases}$$

satisfies both the conditions (i) and (ii) of Theorem 11.2.

(i) $\widetilde{k}(x, \cdot)$ is a Calderón–Zygmund kernel on $\mathbf{R}^n$ for almost all $x \in \mathbf{R}^n$.
(ii) The quantity

$$\begin{aligned} M &:= \max_{|\alpha|\leq 2n} \left\| \frac{\partial^\alpha \widetilde{k}}{\partial z^\alpha}(\cdot, \cdot) \right\|_{L^\infty(\mathbf{R}^n\times\Sigma_{n-1})} \\ &= \max_{|\alpha|\leq 2n} \left\| \frac{\partial^\alpha k}{\partial z^\alpha}(\cdot, \cdot) \right\|_{L^\infty(\Omega\times\Sigma_{n-1})} \end{aligned}$$

is finite.

Moreover, if $f \in L^p(\Omega)$, we let

$$\widetilde{f}(x) := \begin{cases} f(x) & \text{if } x \in \Omega, \\ 0 & \text{if } x \notin \Omega. \end{cases}$$

Then it follows that

$$\|\widetilde{f}\|_{L^p(\mathbf{R}^n)} = \|f\|_{L^p(\Omega)}.$$

By applying Theorem 11.2, we obtain that there exists a function

$$\widetilde{K}\widetilde{f} \in L^p(\mathbf{R}^n)$$

such that

$$\lim_{\varepsilon \downarrow 0} \|\widetilde{K}_\varepsilon \widetilde{f} - \widetilde{K}\widetilde{f}\|_{L^p(\mathbf{R}^n)} = 0,$$

and further that

$$\|\widetilde{K}\widetilde{f}\|_{L^p(\mathbf{R}^n)} \leq c_5\,(n, p, M)\,\|\widetilde{f}\|_{L^p(\mathbf{R}^n)} = c_5\,(n, p, M)\,\|f\|_{L^p(\Omega)},$$

with a positive constant $c_5\,(n, p, M)$.

However, it should be noticed that, for almost all $x \in \Omega$,

$$\begin{aligned} \widetilde{K}_\varepsilon \widetilde{f}(x) &= \int_{|x-y|>\varepsilon} \widetilde{k}(x, x-y)\widetilde{f}(y)\,dy = \int_{\substack{|x-y|>\varepsilon \\ y \in \Omega}} \widetilde{k}(x, x-y) f(y)\,dy \\ &= \int_{\substack{|x-y|>\varepsilon \\ y \in \Omega}} k(x, x-y)\, f(y)\,dy \\ &= K_\varepsilon f(x). \end{aligned}$$

Therefore, if we let

$$Kf := \widetilde{K}\widetilde{f}|_\Omega \in L^p(\Omega),$$

then we have the inequality

$$\|K_\varepsilon f - Kf\|_{L^p(\Omega)} \leq \|\widetilde{K}_\varepsilon \widetilde{f} - \widetilde{K}\widetilde{f}\|_{L^p(\mathbf{R}^n)},$$

and hence

$$\lim_{\varepsilon \downarrow 0} \|K_\varepsilon f - Kf\|_{L^p(\Omega)} = 0 \quad \text{for all } f \in L^p(\Omega).$$

Finally, we obtain that

$$\|Kf\|_{L^p(\Omega)} \leq \left\|\widetilde{K}\widetilde{f}\right\|_{L^p(\mathbf{R}^n)} \leq c_5(n, p, M)\left\|\widetilde{f}\right\|_{L^p(\mathbf{R}^n)}$$
$$= c_5(n, p, M)\|f\|_{L^p(\Omega)} \quad \text{for all } f \in L^p(\Omega).$$

The proof of Theorem 11.4 is complete. □

The next theorem is a local version of Theorem 11.3 (cf. [34, Theorem 2.11]):

Theorem 11.5 *(I) If $\varphi \in L^\infty(\mathbf{R}^n)$, then the commutator*

$$C[\varphi, K]f := \varphi(Kf) - K(\varphi f) = \lim_{\varepsilon\downarrow 0} C[\varphi, K_\varepsilon]f$$
$$= \lim_{\varepsilon\downarrow 0}\int_{\substack{y\in\Omega\\ |x-y|>\varepsilon}} k(x, x-y)\{\varphi(x) - \varphi(y)\} f(y)\,dy,$$

is well-defined for all $f \in L^p(\Omega)$ with $1 < p < \infty$.

(II) Moreover, the commutator $C[\varphi, K]$, defined by the formula

$$C[\varphi, K]f = \lim_{\varepsilon\downarrow 0} C[\varphi, K_\varepsilon]f \quad \text{in } L^p(\Omega),$$

is bounded on $L^p(\Omega)$. More precisely, there exists a constant $c_6(n, p, M) > 0$ such that

$$\|C[\varphi, K]f\|_{L^p(\Omega)} \leq c_6(n, p, M)\|\varphi\|_*\|f\|_{L^p(\Omega)} \quad \text{for all } f \in L^p(\Omega).$$

The proof of Theorem 11.5 is essentially the same as that of Theorem 11.4 if we use Theorem 11.3 (instead of Theorem 11.2).

Remark 11.6 Let $\varphi, \psi \in L^\infty(\mathbf{R}^n)$ such that $\varphi(x) = \psi(x)$ almost everywhere in Ω. Then we have, for all $f \in L^p(\Omega)$,

$$\int_{\substack{y\in\Omega\\ |x-y|>\varepsilon}} k(x, x-y)\{\varphi(x) - \varphi(y)\} f(y)\,dy$$
$$= \int_{\substack{y\in\Omega\\ |x-y|>\varepsilon}} k(x, x-y)\{\psi(x) - \psi(y)\} f(y)\,dy \quad \text{almost everywhere in } \Omega.$$

Therefore, we have, for all $f \in L^p(\Omega)$,

$$C[\varphi, K]f = C[\psi, K]f \quad \text{almost everywhere in } \Omega.$$

Now we are in a position to do the last job in this chapter, which asserts that the norm of singular commutators can be made small if $\varphi \in \text{VMO}$ (see [34, Theorem 2.13]):

Corollary 11.7 *Let $\varphi \in \mathrm{VMO} \cap L^\infty(\mathbf{R}^n)$ and η the* VMO *modulus of φ. Then, for each $\varepsilon > 0$, there exists a constant $\rho_0 = \rho_0(\varepsilon, \eta) > 0$ such that, for any ball B_r of radius r, $0 < r < \rho_0$, contained in Ω, we have the inequality*

$$\|C[\varphi, K] f\|_{L^p(B_r)} \leq c_7(n, p, M)\, \varepsilon \|f\|_{L^p(B_r)} \quad \text{for all } f \in L^p(B_r), \tag{11.24}$$

with a positive constant $c_7(n, p, M)$.

Proof The proof of Corollary 11.7 is divided into three steps.

Step 1: By using Theorem 4.5, for each $\varepsilon > 0$ we can find a bounded, uniformly continuous function $a(x)$ on $\mathbf{R}^n$ such that

$$\|a - \varphi\|_* < \frac{\varepsilon}{2}. \tag{11.25}$$

Let $\omega_a(r)$ be the modulus of uniform continuity of $a(x)$ defined by the formula

$$\omega_a(r) = \sup_{|x-y| \leq r} |a(x) - a(y)|,$$

and choose a constant $\rho_0 = \rho_0(\varepsilon, \eta) > 0$ such that

$$\omega_a(\rho_0) < \frac{\varepsilon}{2}. \tag{11.26}$$

If $B_r = B_r(x_0)$ is a ball of radius r about x_0, we let

$$b(x) := \begin{cases} a(x) & \text{if } x \in B_r(x_0), \\ a\left(x_0 + r\frac{x - x_0}{|x - x_0|}\right) & \text{if } x \in \mathbf{R}^n \setminus B_r(x_0). \end{cases}$$

It should be noticed that the function $b(x)$ is uniformly continuous on $\mathbf{R}^n$ and that the oscillation of $b(x)$ in $\mathbf{R}^n$ equals the oscillation of $a(x)$ in B_r.

Step 2: Now we remark that

$$\begin{aligned} C[\varphi, K] f &= \varphi(Kf) - K(\varphi f) \\ &= (\varphi - a) Kf - K((\varphi - a) f) + a(Kf) - K(af) \\ &= C[\varphi - a, K] f + C[a, K] f, \end{aligned}$$

so that

$$\|C[\varphi, K] f\|_{L^p(B_r)} \leq \|C[\varphi - a, K] f\|_{L^p(B_r)} + \|C[a, K] f\|_{L^p(B_r)}.$$

However, we have, by Theorem 11.5 with $\Omega := B_r$,

$$\|C[\varphi - a, K] f\|_{L^p(B_r)} \leq c_7(n, p, M)\, \|\varphi - a\|_*\, \|f\|_{L^p(B_r)}, \tag{11.27}$$

and also, by Remark 11.6,

$$\begin{aligned}\|C[a,K]f\|_{L^p(B_r)} &= \|C[b,K]f\|_{L^p(B_r)} \\ &\le c_7(n,p,M)\,\|b\|_*\,\|f\|_{L^p(B_r)}\,.\end{aligned} \tag{11.28}$$

Moreover, it is easy to see that

$$\|b\|_* \le \omega_b(r) = \omega_a(r). \tag{11.29}$$

Hence, by combining inequalities (11.28) and (11.29) we obtain that

$$\begin{aligned}\|C[a,K]f\|_{L^p(B_r)} &\le c_7(n,p,M)\,\omega_a(r)\,\|f\|_{L^p(B_r)} \\ &\quad \text{for all } f \in L^p(B_r).\end{aligned} \tag{11.30}$$

Step 3: Therefore, we obtain from inequalities (11.27) and (11.30) that, for all $0 < r < \rho_0$,

$$\begin{aligned}&\|C[\varphi,K]f\|_{L^p(B_r)} \\ &\le \|C[\varphi-a,K]f\|_{L^p(B_r)} + \|C[a,K]f\|_{L^p(B_r)} \\ &\le c_7(n,p,M)\,\|\varphi-a\|_*\,\|f\|_{L^p(B_r)} + c_7(n,p,M)\,\omega_a(r)\,\|f\|_{L^p(B_r)} \\ &\le c_7(n,p,M)\left(\|\varphi-a\|_* + \omega_a(\rho_0)\right)\|f\|_{L^p(B_r)}\,.\end{aligned} \tag{11.31}$$

In view of inequalities (11.25) and (11.26), the last inequality (11.31) proves the desired inequality (11.24).

The proof of Corollary 11.7 is complete. □

Remark 11.8 As a *notational convenience*, we express inequality (11.24) as follows:

$$\begin{aligned}\|C[\varphi,K]f\|_{L^p(B_r)} &\le c_8(n,p,M)\,\eta(r)\,\|f\|_{L^p(B_r)} \\ &\quad \text{for all } f \in L^p(B_r),\end{aligned} \tag{11.24'}$$

with a positive constant $c_8(n,p,M)$. Here

$$\eta(r) = \sup_{\rho \le r} \frac{1}{|B|} \int_B |\varphi(x) - \varphi_B|\,dx$$

is the VMO modulus of φ (see Sect. 4.2).

Part IV
Dirichlet Problems for Elliptic Differential Equations with Discontinuous Coefficients

Chapter 12
Dirichlet Problems in Sobolev Spaces

The purpose of this chapter is to formulate the *homogeneous* Dirichlet problem in the framework of L^p Sobolev spaces. We state interior and global *a priori* estimates for the Dirichlet problem (Theorems 12.1 and 12.2) that will play an essential role in the proof of the unique solvability theorem for the homogeneous Dirichlet problem (Theorem 15.1) in Chap. 15. The results discussed here are adapted from Chiarenza–Frasca–Longo [34, 35] and Chiarenza [33]. Our approach can be traced back to the pioneering work of J. Schauder [127] and [128] on the Dirichlet problem for second order, elliptic differential operators.

12.1 Formulation of the Dirichlet Problem

Let Ω be a bounded domain in Euclidean space $\mathbf{R}^n$ with $n \geq 3$, with boundary $\partial\Omega$ of class $C^{1,1}$. If $1 < p < \infty$ and if $k = 1$ or $k = 2$, then we define the Sobolev space as follows:

$$\begin{aligned} W^{k,p}(\Omega) = &\text{ the space of (equivalence classes of) functions} \\ &u \in L^p(\Omega) \text{ whose derivatives } D^\alpha u, |\alpha| \leq k, \text{ in the} \\ &\text{sense of distributions are in } L^p(\Omega). \end{aligned}$$

Its norm $\|u\|_{W^{k,p}(\Omega)}$ is defined by the formula

$$\|u\|_{W^{k,p}(\Omega)} = \left(\sum_{|\alpha| \leq k} \int_\Omega |D^\alpha u(x)|^p dx \right)^{1/p}.$$

For a bounded $C^{1,1}$ domain Ω, we can define the boundary space of traces as follows:

K. Taira, *Real Analysis Methods for Markov Processes*,
https://doi.org/10.1007/978-981-97-3659-1_12

$$B^{k-1/p,p}(\partial\Omega) = \text{the space of distributions } \varphi \in \mathcal{D}'(\partial\Omega) \text{ such that there exists a function } u \in W^{k,p}(\Omega) \text{ with } \gamma_0 u = \varphi.$$

In the space $B^{k-1/p,p}(\partial\Omega)$, we introduce a norm

$$|\varphi|_{B^{k-1/p,p}(\partial\Omega)} = \inf\left\{\|u\|_{W^{k,p}(\Omega)} : u \in W^{k,p}(\Omega),\ \gamma_0 u = \varphi \text{ on } \partial\Omega\right\},$$

where the infimum is taken over all functions $u \in W^{k,p}(\Omega)$ that equal φ on the boundary $\partial\Omega$. The space $B^{k-1/p,p}(\partial\Omega)$ is a Banach space with respect to this norm. We recall that the space $B^{k-1/p,p}(\partial\Omega)$ is a Besov space (see the trace theorem (Theorem 7.8)).

Moreover, it should be emphasized (see [2, Theorem 5.37]) that the closure $W_0^{1,p}(\Omega)$ of $C_0^\infty(\Omega)$ in $W^{1,p}(\Omega)$ can be characterized as follows:

$$W_0^{1,p}(\Omega) = \left\{u \in W^{1,p}(\Omega) : \gamma_0 u = 0 \text{ on } \partial\Omega\right\}.$$

Now we consider a second order, elliptic differential operator $\mathcal{L}$ with real *discontinuous* coefficients of the form

$$\mathcal{L}u := \sum_{i,j=1}^{n} a^{ij}(x)\frac{\partial^2 u}{\partial x_i \partial x_j}.$$

More precisely, we assume that the coefficients $a^{ij}(x)$ are the components of a symmetric contravariant tensor of type $\binom{2}{0}$ in $\mathbf{R}^n$ and satisfy the following three conditions:

(i) $a^{ij} \in \mathrm{VMO} \cap L^\infty(\mathbf{R}^n)$ for all $1 \leq i, j \leq n$.
(ii) $a^{ij}(x) = a^{ji}(x)$ for all $1 \leq i, j \leq n$ and for almost all $x \in \Omega$.
(iii) There exists a positive constant λ such that (the uniform ellipticity)

$$\frac{1}{\lambda}|\xi|^2 \leq \sum_{i,j=1}^{n} a^{ij}(x)\xi_i\xi_j \leq \lambda|\xi|^2 \quad \text{for almost all } x \in \Omega \text{ and for all } \xi \in \mathbf{R}^n.$$

In this chapter we study the following *homogeneous* Dirichlet problem in the framework of L^p Sobolev spaces:

$$\begin{cases} \mathcal{L}u = \sum_{i,j=1}^{n} a^{ij}(x)\frac{\partial^2 u}{\partial x_i \partial x_j} = f & \text{in } \Omega, \\ \gamma_0 u = 0 & \text{on } \partial\Omega. \end{cases} \tag{12.1}$$

More precisely, for a given function $f \in L^p(\Omega)$ we find a function $u \in W^{2,p}(\Omega) \cap W_0^{1,p}(\Omega)$ that satisfies the equation $\mathcal{L}u = f$ almost everywhere in Ω.

If $\widetilde{\Omega}$ is the subset of Ω where conditions (ii) and (iii) hold true, then we let

$$\Gamma(x, z) := \frac{1}{(2-n)\omega_n} \frac{1}{\sqrt{\det(a^{ij}(x))}} \left(\sum_{i,j=1}^{n} A_{ij}(x) z_i z_j \right)^{(2-n)/2}$$

$$\text{for } x \in \widetilde{\Omega} \text{ and } z \in \mathbf{R}^n \setminus \{0\}.$$

Here:

$$(A_{ij}(x)) = \text{the inverse matrix of } (a^{ij}(x)),$$

$$\omega_n := |\Sigma_{n-1}| = \frac{2\pi^{n/2}}{\Gamma(n/2)} \quad \text{(the surface area of the unit sphere} \Sigma_{n-1} \text{ in } \mathbf{R}^n).$$

Moreover, we let

$$\Gamma_i(x, z) := \frac{\partial \Gamma}{\partial z_i}(x, z) \quad \text{for } x \in \widetilde{\Omega}, z \in \mathbf{R}^n \setminus \{0\} \text{and } 1 \leq i \leq n,$$

$$\Gamma_{ij}(x, z) := \frac{\partial^2 \Gamma}{\partial z_i \partial z_j}(x, z) \quad \text{for } x \in \widetilde{\Omega}, z \in \mathbf{R}^n \setminus \{0\} \text{ and } 1 \leq i, j \leq n,$$

and

$$M := \max_{1 \leq i,j \leq n} \max_{|\alpha| \leq 2n} \left\| \frac{\partial^\alpha}{\partial z^\alpha} \Gamma_{ij}(\cdot, \cdot) \right\|_{L^\infty(\Omega \times \Sigma_{n-1})}.$$

Finally, if $\eta^{ij}(r)$ are the VMO moduli of $a^{ij}(x)$, we let

$$\eta(r) := \left(\sum_{i,j=1}^{n} \eta^{ij}(r)^2 \right)^{1/2}.$$

12.2 Statement of Main Results (Theorems 12.1 and 12.2)

The next *interior* $W^{2,p}$ estimate plays an important role in the proof of the existence theorem for the Dirichlet problem (12.1) in Chap. 15 (see [34, Theorem 4.2]):

Theorem 12.1 (the interior regularity theorem) *Let* $1 < q < p < \infty$ *and* $f \in L^p(\Omega)$. *If a function* $u \in W^{2,q}_{\text{loc}}(\Omega)$ *satisfies the equation*

$$\mathcal{L}u = f \quad \text{in } \Omega,$$

then it follows that $u \in W^{2,p}_{\text{loc}}(\Omega)$. *Moreover, for any open subsets* $\Omega' \Subset \Omega'' \Subset \Omega$ *(see Fig. 12.1), we have the interior* a priori *estimate*

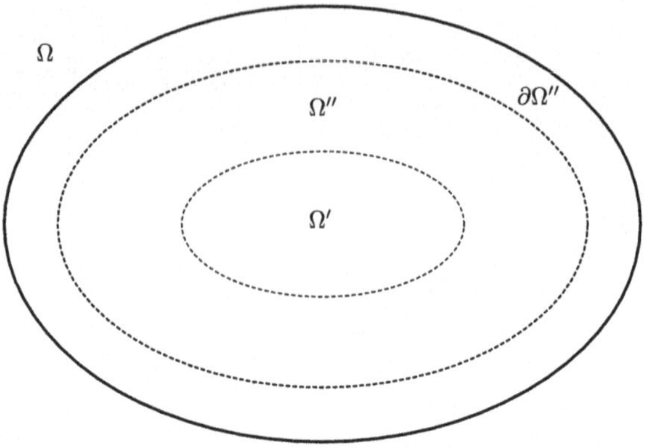

Fig. 12.1 The subsets Ω', Ω'' of Ω such that $\Omega' \Subset \Omega'' \Subset \Omega$

$$\|u\|_{W^{2,p}(\Omega')} \leq C_1 \left(\|u\|_{L^p(\Omega'')} + \|f\|_{L^p(\Omega)} \right), \tag{12.2}$$

with a positive constant $C_1 = C_1(n, p, M, \text{dist}(\Omega', \partial\Omega''), \lambda, \eta)$.

Theorem 12.1 is proved in Sect. 13.3 of Chap. 13.

The next *global* $W^{2,p}$ estimate plays an essential role in the proof of the existence theorem for the Dirichlet problem (12.1) in Chap. 14 (see [35, Theorem 4.2]).

Theorem 12.2 (the global regularity theorem) *Let* $1 < q < p < \infty$ *and* $f \in L^p(\Omega)$. *If a function* $u \in W^{2,q}(\Omega) \cap W_0^{1,q}(\Omega)$ *satisfies the equation*

$$\mathcal{L}u = f \quad \text{in } \Omega,$$

then it follows that $u \in W^{2,p}(\Omega)$. *Moreover, we have the global* a priori *estimate*

$$\|u\|_{W^{2,p}(\Omega)} \leq C_2 \left(\|u\|_{L^p(\Omega)} + \|f\|_{L^p(\Omega)} \right), \tag{12.3}$$

with a positive constant $C_2 = C_2(n, p, M, \partial\Omega, \lambda, \eta)$.

Theorem 12.2 is proved in Sect. 14.5 of Chap. 14.

The desired interior *a priori* estimate (12.2) and global *a priori* estimate (12.3) are consequences of the following two facts:

(I) The explicit representation formulas for the solutions of the homogeneous Dirichlet problem (Chaps. 13 and 14).

(II) An L^p boundedness of some singular integral operators appearing in those formulas (Chap. 14).

The proofs of Theorems 12.1 and 12.2 can be visualized as in Figs. 12.2 and 12.3 below.

Finally, some remarks are in order (see [1, Proposition 1.3]; [74, Theorem 1]; [169, Theorem 5.1]).

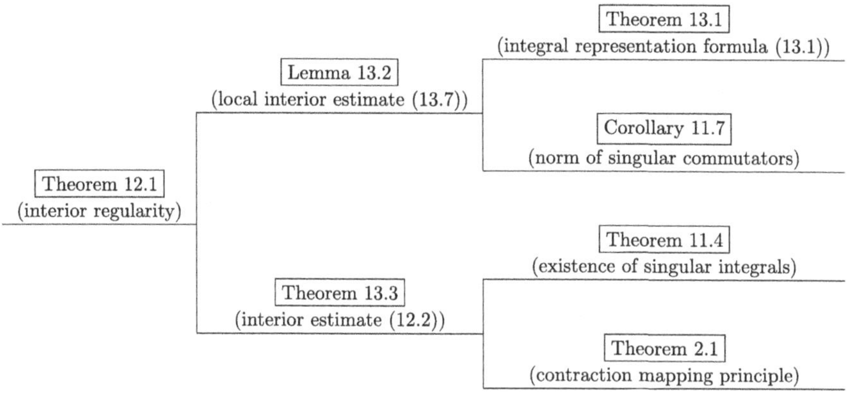

Fig. 12.2 A flowchart for the proof of Theorem 12.1

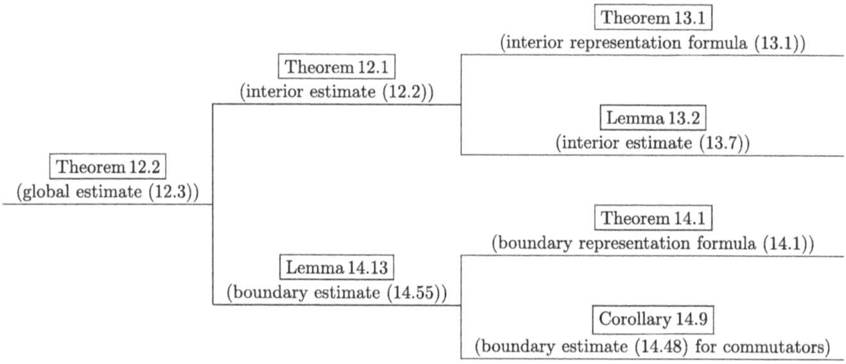

Fig. 12.3 A flowchart for the proof of Theorem 12.2

Remark 12.3 1° It is worthwhile pointing out here that VMO functions are invariant under $C^{1,1}$ diffeomorphisms.

2° Let Ω be a $C^{1,1}$ domain with boundary $\partial\Omega$. We recall that the definitions of the classes BMO(Ω) and VMO(Ω) are given if we replace the ball B by the intersection $B \cap \Omega$ in Sects. 4.1 and 4.2, respectively. For a given function in the class BMO(Ω) (resp. VMO(Ω)), we can extend it to the whole $\mathbf{R}^n$ preserving its BMO (resp. VMO) character.

Chapter 13
Calderón–Zygmund Kernels and Interior Estimates

This chapter is devoted to the proof of Theorem 12.1 (Theorem 13.3) that is based on some local interior *a priori* estimates for the solutions of the homogeneous Dirichlet problem (Lemma 13.2). The main idea of proof may be considered as an integral perturbation about the constant coefficient case, which goes back to Eugenio Elia Levi [89] (Theorem 13.1). The vanishing mean oscillation (VMO) assumption on the coefficients is of the greatest relevance in the study of an error term expressed by singular commutators (Corollary 11.7). The desired interior *a priori* estimate (12.2) follows in a standard way from Lemma 13.2 by a covering argument if we make use of the following three theorems:

(1) Sobolev's imbedding theorem (Theorem 7.5).
(2) The contraction mapping principle (Theorem 2.1).
(3) The interpolation inequality (Theorem 13.4).

The results here are adapted from Chiarenza–Frasca–Longo [34].

13.1 Interior Representation Formula for Solutions

In this chapter we consider a second order, elliptic differential operator with variable coefficients

$$\mathcal{L} = \sum_{i,j=1}^{n} a^{ij}(x) \frac{\partial^2}{\partial x_i \partial x_j}.$$

We assume that the coefficients a^{ij} satisfy the following three conditions: Let B be an open ball of $\mathbf{R}^n$ with $n \geq 3$.

(i) $a^{ij} \in \mathrm{VMO} \cap L^\infty(\mathbf{R}^n)$ for all $1 \leq i, j \leq n$.
(ii) $a^{ij}(x) = a^{ji}(x)$ for all $1 \leq i, j \leq n$ and for almost all $x \in B$.

K. Taira, *Real Analysis Methods for Markov Processes*,
https://doi.org/10.1007/978-981-97-3659-1_13

(iii) There exists a positive constant λ such that (the uniform ellipticity)

$$\frac{1}{\lambda}|\xi|^2 \le \sum_{i,j=1}^{n} a^{ij}(x)\xi_i\xi_j \le \lambda|\xi|^2 \quad \text{for almost all } x \in B \text{ and for all } \xi \in \mathbf{R}^n.$$

If $\widetilde{B}$ is the subset of B where conditions (ii) and (iii) hold true, then we let

$$\Gamma(x,z) := \frac{1}{(2-n)\omega_n}\frac{1}{\sqrt{\det(a^{ij}(x))}}\left(\sum_{i,j=1}^{n} A_{ij}(x)z_iz_j\right)^{(2-n)/2}$$
$$\text{for all } x \in \widetilde{B} \text{ and all } z \in \mathbf{R}^n \setminus \{0\}.$$

Here:

$$(A_{ij}(x)) = \text{the inverse matrix of } (a^{ij}(x)),$$
$$\omega_n := |\Sigma_{n-1}| = \frac{2\pi^{n/2}}{\Gamma(n/2)} \quad \text{(the surface area of the unit sphere } \Sigma_{n-1} \text{ in } \mathbf{R}^n\text{)}.$$

It should be emphasized that the function $\Gamma(x_0, z)$, $x_0 \in \widetilde{B}$, is a fundamental solution for the constant coefficients elliptic differential operator

$$\mathcal{L}_0 = \sum_{i,j=1}^{n} a^{ij}(x_0)\frac{\partial^2}{\partial x_i \partial x_j}.$$

This is Korn's freezing trick.

Moreover, we recall that the functions

$$\begin{aligned}\Gamma_i(x,z) &= \frac{\partial \Gamma}{\partial z_i}(x,z)\\ &= \frac{1}{\omega_n}\frac{1}{\sqrt{\det(a^{ij}(x))}}\left(\sum_{i,j=1}^{n} A_{ij}(x)z_iz_j\right)^{-n/2}\left(\sum_{j=1}^{n} A_{ij}(x)z_j\right)\end{aligned}$$
$$\text{for } x \in \widetilde{B},\ z \in \mathbf{R}^n \setminus \{0\} \text{ and } 1 \le i \le n$$

are positively homogeneous of degree $1-n$ with respect to the variable z, so that the functions

$$\Gamma_{ij}(x,z) = \frac{\partial^2 \Gamma}{\partial z_i \partial z_j}(x,z) \quad \text{for } x \in \widetilde{B},\ z \in \mathbf{R}^n \setminus \{0\} \text{ and } 1 \le i, j \le n$$

are Calderón–Zygmund kernels in the z variable (see Example 10.1).

The next theorem gives *integral representation formulas* for the second derivatives of solutions of the homogeneous Dirichlet problem for the elliptic differential operator $\mathcal{L}$ with variable coefficients (see [34, Theorem 3.1]):

Theorem 13.1 (the integral representation formula) *Let B be an open ball of $\mathbf{R}^n$ and assume that $u \in W_0^{2,p}(B)$ for $1 < p < \infty$. If $\widetilde{B}$ is the subset of B where conditions (ii) and (iii) hold true, then we have, for all $x \in \widetilde{B}$,*

$$\begin{aligned}\frac{\partial^2 u}{\partial x_i \partial x_j}(x) &= \text{v.p.} \int_B \Gamma_{ij}(x, x-y) \\ &\quad \times \left(\sum_{h,k=1}^{n} \left\{ a^{hk}(x) - a^{hk}(y) \right\} \frac{\partial^2 u}{\partial x_h \partial x_k}(y) + \mathcal{L}u(y) \right) dy \\ &\quad + \mathcal{L}u(x) \left(\int_{|t|=1} \Gamma_i(x, t) t_j \, d\sigma_t \right) \quad \text{for } 1 \le i, j \le n.\end{aligned} \tag{13.1}$$

Proof By a *density argument*, it suffices to prove formula (13.1) for all $u \in C_0^\infty(B)$. Indeed, the general case can be proved by using Theorems 11.4 and 11.5. The proof is divided into three steps.

Step 1: If x_0 is an arbitrary point of $\widetilde{B}$, we consider the constant coefficients elliptic differential operator

$$\mathcal{L}_0 = \sum_{i,j=1}^{n} a^{ij}(x_0) \frac{\partial^2}{\partial x_i \partial x_j}.$$

Then it follows that

$$\mathcal{L}_0 u(x) = \sum_{i,j=1}^{n} a^{ij}(x_0) \frac{\partial^2 u}{\partial x_i \partial x_j}(x) \in C_0^\infty(B).$$

Moreover, we have the formula

$$\begin{aligned}\mathcal{L}_0 u(y) &= \sum_{i,j=1}^{n} \left\{ a^{ij}(x_0) - a^{ij}(y) \right\} \frac{\partial^2 u}{\partial x_i \partial x_j}(y) + \sum_{i,j=1}^{n} a^{ij}(y) \frac{\partial^2 u}{\partial x_i \partial x_j}(y) \\ &= (\mathcal{L}_0 - \mathcal{L})u(y) + \mathcal{L}u(y) \quad \text{for } y \in B.\end{aligned}$$

This implies that

$$\begin{aligned}u(x) &= \text{v.p.} \int_B \Gamma(x_0, x-y) \mathcal{L}_0 u(y) \, dy \\ &= \text{v.p.} \int_B \Gamma(x_0, x-y)\end{aligned} \tag{13.2}$$

$$\times \left(\sum_{h,k=1}^{n} \left\{ a^{hk}(x_0) - a^{hk}(y) \right\} \frac{\partial^2 u}{\partial x_h \partial x_k}(y) + \mathcal{L}u(y) \right) dy.$$

We introduce a function $g \in C_0^\infty(\mathbf{R}^n)$ by the formula

$$g(y) = \begin{cases} \mathcal{L}_0 u(y) = \sum_{i,j=1}^n a^{ij}(x_0) \frac{\partial^2 u}{\partial x_i \partial x_j}(y) & \text{if } y \in B, \\ 0 & \text{if } y \notin B. \end{cases}$$

Then it follows that

$$u(x) = \text{v.p.} \int_{\mathbf{R}^n} \Gamma(x_0, x-y) g(y)\, dy = (\Gamma * g)(x) \quad \text{for } x \in \mathbf{R}^n.$$

Hence we have, by the Fourier transform,

$$\widehat{u}(\xi) = \widehat{\Gamma}(x_0, \xi) \widehat{g}(\xi) \quad \text{for } \xi \in \mathbf{R}^n,$$

and, for $1 \leq i, j \leq n$,

$$\xi_i \xi_j \widehat{u}(\xi) = \xi_i \xi_j \widehat{\Gamma}(x_0, \xi) \widehat{g}(\xi) \quad \text{for } \xi \in \mathbf{R}^n. \tag{13.3}$$

Step 2: Secondly, we prove that

$$-\xi_i \xi_j \widehat{\Gamma}(x_0, \xi) = \widehat{\text{v.p. } \Gamma_{ij}}(x_0, \xi) + c_{ij}(x_0) \widehat{\delta}(\xi) \quad \text{for } 1 \leq i, j \leq n, \tag{13.4}$$

where $\delta(x)$ is the Dirac measure at 0 on $\mathbf{R}^n$ and

$$c_{ij}(x_0) := \int_{|y|=1} \Gamma_i(x_0, y) y_j \, d\sigma_y.$$

Indeed, it follows that we have, for all $\varphi \in \mathcal{S}(\mathbf{R}^n)$,

$$\begin{aligned} &- \langle \xi_i \xi_j \widehat{\Gamma}(x_0, \xi), \varphi(\xi) \rangle = -\langle \Gamma(x_0, t), \widehat{\xi_i \xi_j \varphi}(t) \rangle \qquad (13.5) \\ &= \left\langle \Gamma(x_0, t), \frac{\partial^2}{\partial t_i \partial t_j} \widehat{\varphi}(t) \right\rangle \\ &= \left\langle \frac{\partial^2}{\partial t_i \partial t_j} \Gamma(x_0, t), \widehat{\varphi}(t) \right\rangle \\ &= \langle \Gamma_{ij}(x_0, t), \widehat{\varphi}(t) \rangle \quad \text{for } 1 \leq i, j \leq n. \end{aligned}$$

On the other hand, we have, for all $\varphi \in \mathcal{S}(\mathbf{R}^n)$,

$$
\begin{aligned}
&\langle \Gamma_{ij}(x_0, t), \widehat{\varphi}(t)\rangle \\
&= \left\langle \frac{\partial^2}{\partial t_i \partial t_j} \Gamma(x_0, t), \psi(t) \right\rangle = \left\langle \frac{\partial}{\partial t_j} \Gamma_i(x_0, t), \widehat{\varphi}(t) \right\rangle \\
&= -\left\langle \Gamma_i(x_0, t), \frac{\partial \widehat{\varphi}}{\partial t_j}(t) \right\rangle = -\int_{\mathbf{R}^n} \Gamma_i(x_0, t) \frac{\partial \widehat{\varphi}}{\partial t_j}(t)\, dt \\
&= -\lim_{\varepsilon \downarrow 0} \int_{|t|>\varepsilon} \Gamma_i(x_0, t) \frac{\partial \widehat{\varphi}}{\partial t_j}(t)\, dt \\
&= \lim_{\varepsilon \downarrow 0} \left\{ \int_{|t|>\varepsilon} \frac{\partial}{\partial t_j} \Gamma_i(x_0, t) \widehat{\varphi}(t)\, dt + \int_{|t|=\varepsilon} \Gamma_i(x_0, t) \frac{t_j}{|t|} \widehat{\varphi}(t)\, d\sigma_t \right\} \\
&= \langle \text{v.p. } \Gamma_{ij}(x_0, t), \widehat{\varphi}\rangle + \lim_{\varepsilon \downarrow 0} \int_{|y|=1} \Gamma_i(x_0, y) y_j \widehat{\varphi}(\varepsilon y)\, d\sigma_y \\
&= \langle \text{v.p. } \Gamma_{ij}(x_0, t), \widehat{\varphi}\rangle + \left(\int_{|y|=1} \Gamma_i(x_0, y) y_j\, d\sigma_y \right) \widehat{\varphi}(0),
\end{aligned}
$$

since the $\Gamma_i(x, z)$ are positively homogeneous of degree $1 - n$ with respect to the variable z.

Hence we have the formula

$$
\Gamma_{ij}(x_0, t) = \text{v.p. } \Gamma_{ij}(x_0, t) + c_{ij}(x_0)\delta(t) \quad \text{for } 1 \leq i, j \leq n. \tag{13.6}
$$

Therefore, by combining formulas (13.5) and (13.6) we obtain that

$$
\begin{aligned}
&- \langle \xi_i \xi_j \widehat{\Gamma}(x_0, \xi), \varphi(\xi)\rangle = \langle \Gamma_{ij}(x_0, t), \widehat{\varphi}(t)\rangle \\
&= \langle \text{v.p. } \Gamma_{ij}(x_0, t) + c_{ij}(x_0)\delta(t), \widehat{\varphi}(t)\rangle \\
&= \langle \widehat{\text{v.p. } \Gamma}_{ij}(x_0, \xi) + c_{ij}(x_0)\widehat{\delta}(\xi), \varphi(\xi)\rangle \quad \text{for all } \varphi \in \mathcal{S}(\mathbf{R}^n).
\end{aligned}
$$

This proves the desired formula (13.4).

Step 3: By formulas (13.3) and (13.4), it follows that

$$
\begin{aligned}
\widehat{\frac{\partial^2 u}{\partial x_i \partial x_j}} &= -\xi_i \xi_j \widehat{u}(\xi) = -\xi_i \xi_j \widehat{\Gamma}(x_0, \xi) \widehat{g}(\xi) \\
&= \left(\widehat{\text{v.p. } \Gamma_{ij}}(x_0, \xi) + c_{ij}(x_0)\widehat{\delta}(\xi) \right) \widehat{g}(\xi) \\
&= \mathcal{F}\left(\text{v.p. } \Gamma_{ij}(x_0, \cdot) * g\right)(\xi) + c_{ij}(x_0)\widehat{\delta * g}(\xi) \\
&= \mathcal{F}\left(\text{v.p. } \Gamma_{ij}(x_0, \cdot) * g\right)(\xi) + c_{ij}(x_0)\widehat{g}(\xi).
\end{aligned}
$$

Hence we have, by the Fourier inversion formula,

$$
\frac{\partial^2 u}{\partial x_i \partial x_j}(x) = \lim_{\varepsilon \downarrow 0} \int_{|x-y|>\varepsilon} \Gamma_{ij}(x_0, x - y) g(y)\, dy + c_{ij}(x_0) g(x) \quad \text{for } x \in \mathbf{R}^n.
$$

In particular, by taking

$$x = x_0 \in \widetilde{B} \subset B,$$

we obtain that

$$\begin{aligned}\frac{\partial^2 u}{\partial x_i \partial x_j}(x_0) &= \text{v.p.} \int_B \Gamma_{ij}(x_0, x_0 - y)\mathcal{L}_0 u(y)\, dy + c_{ij}(x_0)\mathcal{L}_0 u(x_0) \\ &= \text{v.p.} \int_B \Gamma_{ij}(x_0, x_0 - y) \\ &\quad \times \left\{ \sum_{i,j=1}^n \left\{a^{ij}(x_0) - a^{ij}(y)\right\} \frac{\partial^2 u}{\partial x_i \partial x_j}(y) + \mathcal{L}u(y) \right\} dy \\ &\quad + c_{ij}(x_0)\mathcal{L}u(x_0) \quad \text{for } 1 \le i, j \le n.\end{aligned}$$

This proves the representation formula (13.1) with $x := x_0$.

The proof of Theorem 13.1 is complete. □

13.2 Local Interior Estimates

Let Ω be an open subset of $\mathbf{R}^n$ with $n \ge 3$, and we assume that the functions $a^{ij} \in \text{VMO} \cap L^\infty(\mathbf{R}^n)$, $1 \le i, j \le n$, satisfy the following three conditions:

(i) $a^{ij} \in \text{VMO} \cap L^\infty(\mathbf{R}^n)$ for all $1 \le i, j \le n$.
(ii) $a^{ij}(x) = a^{ji}(x)$ for all $1 \le i, j \le n$ and for almost all $x \in \Omega$.
(iii) There exist a positive constant λ such that

$$\frac{1}{\lambda}|\xi|^2 \le \sum_{i,j=1}^n a^{ij}(x)\xi_i\xi_j \le \lambda|\xi|^2 \quad \text{for almost all } x \in \Omega \text{ and all } \xi \in \mathbf{R}^n.$$

If $\widetilde{\Omega}$ is the subset of Ω where conditions (ii) and (iii) hold true, then we let

$$\Gamma(x, z) := \frac{1}{(2-n)\omega_n} \frac{1}{\sqrt{\det(a^{ij}(x))}} \left(\sum_{i,j=1}^n A_{ij}(x) z_i z_j \right)^{(2-n)/2}$$
$$\text{for all } x \in \widetilde{\Omega} \text{ and all } z \in \mathbf{R}^n \setminus \{0\},$$

and

$$M := \max_{1 \le i,j \le n} \max_{|\alpha| \le 2n} \left\| \frac{\partial^\alpha}{\partial z^\alpha} \Gamma_{ij}(\cdot, \cdot) \right\|_{L^\infty(\Omega \times \Sigma_{n-1})}.$$

Moreover, if $\eta^{ij}(r)$ is the VMO modulus of $a^{ij}(x)$, we let

$$\eta(r) := \left(\sum_{i,j=1}^{n} \eta^{ij}(r)^2 \right)^{1/2}.$$

Then we can prove local interior *a priori* estimates for the second derivatives of solutions of the homogeneous Dirichlet problem for the elliptic differential operator

$$\mathcal{L} = \sum_{i,j=1}^{n} a^{ij}(x) \frac{\partial^2}{\partial x_i \partial x_j}$$

with variable coefficients (cf. [34, Lemma 4.1]):

Lemma 13.2 *Let* $1 < p < \infty$ *and assume that*

$$M = \max_{1 \leq i,j \leq n} \max_{|\alpha| \leq 2n} \left\| \frac{\partial^\alpha}{\partial z^\alpha} \Gamma_{ij}(\cdot, \cdot) \right\|_{L^\infty(\Omega \times \Sigma_{n-1})} < \infty.$$

Then there exist positive constants $C = C(n, p, M)$ *and* $\rho_0 = \rho_0(C, \eta)$ *such that, for any* $u \in W_0^{2,p}(B_r)$ *in an open ball* B_r *of radius* r *with* $0 < r < \rho_0$, *contained in* Ω *(see Fig. 13.1), we have the interior a priori estimate*

$$\left\| \frac{\partial^2 u}{\partial x_i \partial x_j} \right\|_{L^p(B_r)} \leq C \, \|\mathcal{L}u\|_{L^p(B_r)} \quad \text{for all } 1 \leq i, j \leq n. \tag{13.7}$$

In other words, the second order, elliptic differential operator $\mathcal{L}$ *controls all second order partial derivatives in the* L^p*-norm for* $1 < p < \infty$ *in the open ball* B_r *(cf. Remark 9.18).*

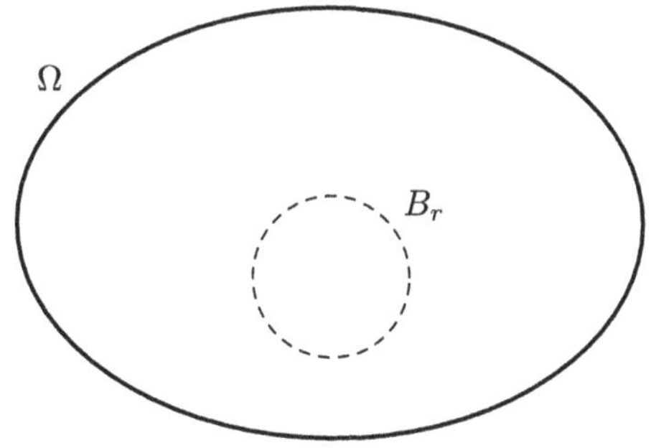

Fig. 13.1 The open ball B_r of radius r contained in Ω

Proof By Theorem 13.1, we can represent the second derivatives $\partial^2 u/\partial x_i \partial x_j$ of u in the following form:

$$\begin{aligned}\frac{\partial^2 u}{\partial x_i \partial x_j}(x) &= \text{v.p.} \int_{B_r} \Gamma_{ij}(x, x-y) \\ &\quad \times \left(\sum_{h,k=1}^{n} \left\{ a^{hk}(x) - a^{hk}(y) \right\} \frac{\partial^2 u}{\partial x_h \partial x_k}(y) + \mathcal{L}u(y) \right) dy \\ &\quad + \left(\int_{|t|=1} \Gamma_i(x,t) t_j \, d\sigma_t \right) \mathcal{L}u(x) \\ &:= \sum_{h,k=1}^{n} C\left\{a^{hk}, K_{ij}\right\} \left(\frac{\partial^2 u}{\partial x_h \partial x_k} \right)(x) + K_{ij}(\mathcal{L}u)(x) \\ &\quad + c_{ij}(x)\, \mathcal{L}u(x) \quad \text{for } 1 \le i, j \le n,\end{aligned} \tag{13.8}$$

where

$$K_{ij} f(x) = \text{v.p.} \int_{B_r} \Gamma_{ij}(x, x-y) f(y)\, dy,$$

$$C\left\{\varphi, K_{ij}\right\} = \text{v.p.} \int_{B_r} \Gamma_{ij}(x, x-y) [\varphi(x) - \varphi(y)] f(y)\, dy,$$

and

$$c_{ij}(x) = \int_{|t|=1} \Gamma_i(x,t) t_j \, d\sigma_t.$$

We can estimate the three terms on the right-hand side of formula (13.8) as follows:

(1) First, we have, for some positive constant c_1,

$$\|\Gamma_i(\cdot,\cdot)\|_{L^\infty(\Omega\times\Sigma_{n-1})} \le c_1 \quad \text{for } 1 \le i \le n,$$

and hence

$$\left\|c_{ij}\, \mathcal{L}u\right\|_{L^p(B_r)} \le C_1 \|\mathcal{L}u\|_{L^p(B_r)} \quad \text{for } 1 \le i, j \le n,$$

for some positive constant C_1.

(2) Secondly, by applying Theorem 11.4 with

$$\Omega := B_r, \quad k(x,z) := \Gamma_{ij}(x,z),$$

we obtain that

$$\left\|K_{ij}(\mathcal{L}u)\right\|_{L^p(B_r)} \le C_2 \|\mathcal{L}u\|_{L^p(B_r)} \quad \text{for } 1 \le i, j \le n.$$

(3) Thirdly, we show that the norm of singular commutators can be made small if the coefficients have a small integral oscillation.
Indeed, by applying Corollary 11.7 with

$$k(x, z) := \Gamma_{ij}(x, z), \quad \varphi(x) := a^{hk}(x),$$

we obtain from Remark 11.8 that

$$\left\| C\left\{a^{hk}, K_{ij}\right\} \frac{\partial^2 u}{\partial x_h \partial x_k} \right\|_{L^p(B_r)} \leq c_2\, \eta^{hk}(r) \left\| \frac{\partial^2 u}{\partial x_h \partial x_k} \right\|_{L^p(B_r)}.$$

Hence it follows that

$$\left\| C\left\{a^{hk}, K_{ij}\right\} \frac{\partial^2 u}{\partial x_h \partial x_k} \right\|_{L^p(B_r)} \leq c_3\, \eta(r) \left\| \nabla^2 u \right\|_{L^p(B_r)} \quad \text{for } 1 \leq h, k \leq n.$$

Summing up, we have proved that, for all $1 \leq i, j \leq n$,

$$\left\| \frac{\partial^2 u}{\partial x_i \partial x_j} \right\|_{L^p(B_r)} \leq c_3\, \eta(r) \left\| \nabla^2 u \right\|_{L^p(B_r)} + C_3 \left\| \mathcal{L} u \right\|_{L^p(B_r)}$$
$$\text{for } u \in W_0^{2,p}(B_r).$$

This implies that

$$\begin{aligned} &\left\| \nabla^2 u \right\|_{L^p(B_r)} \\ &\leq n^2 c_3\, \eta(r) \left\| \nabla^2 u \right\|_{L^p(B_r)} + n^2\, C_3 \left\| \mathcal{L} u \right\|_{L^p(B_r)} \quad \text{for all } u \in W_0^{2,p}(B_r). \end{aligned} \tag{13.9}$$

Therefore, the desired interior *a priori* estimate (13.7) follows from estimate (13.9) if we take $\rho_0 > 0$ sufficiently small so that

$$n^2\, c_3\, \eta(r) < \frac{1}{2} \quad \text{for all } 0 < r < \rho_0.$$

The proof of Lemma 13.2 is complete. □

13.3 Proof of Theorem 12.1

This section is devoted to the proof of Theorem 12.1. In fact, the next theorem proves Theorem 12.1 (see [34, Theorem 4.2]):

Theorem 13.3 *Let $1 < q < p < \infty$ and $f \in L^p_{\text{loc}}(\Omega)$. If a function $u \in W^{2,q}_{\text{loc}}(\Omega)$ satisfies the equation*

$$\mathcal{L}u = \sum_{i,j=1}^{n} a^{ij}(x)\frac{\partial^2 u}{\partial x_i \partial x_j} = f \quad in\ \Omega,$$

then it follows that $u \in W^{2,p}_{\mathrm{loc}}(\Omega)$. *Moreover, for any open subsets* $\Omega' \Subset \Omega'' \Subset \Omega$ *(see Fig. 12.1), we have the interior a priori estimate*

$$\|u\|_{W^{2,p}(\Omega')} \leq C\left(\|u\|_{L^p(\Omega'')} + \|f\|_{L^p(\Omega)}\right), \tag{12.2}$$

with a positive constant $C = C(n, p, M, \mathrm{dist}(\Omega', \partial\Omega''), \lambda, \eta)$.

Proof The proof is divided into two steps.

Step 1: Let B_ρ be an open ball such that $B_\rho \Subset \Omega$, and let $f \in L^r(B_\rho)$ for some $r \in (1, \infty)$. For $i, j, h, k = 1, 2, \ldots, n$, we introduce a linear bounded operator S_{ijhk} on $L^r(B_\rho)$ by the formula

$$S_{ijhk}(f) = \text{v.p.} \int_{B_\rho} \Gamma_{ij}(x, x-y)\left\{a^{hk}(x) - a^{hk}(y)\right\} f(y)\, dy.$$

By applying Corollary 11.7 with

$$k(x, z) := \Gamma_{ij}(x, z), \quad a(x) := a^{hk}(x),$$

we can find a constant $\rho_0 > 0$ such that

$$\sum_{i,j,h,k=1}^{n} \left\|S_{ijhk}\right\| < \frac{1}{2} \quad \text{for all } 0 < \rho < \rho_0. \tag{13.10}$$

Here the norm of the operators S_{ijhk} is the norm in the space of bounded linear operators on $L^r(B_\rho)$ for $0 < \rho < \rho_0$.

Step 1-1: Let B be an open ball with radius less than ρ_0 such that $B \Subset \Omega$, and take a function $\beta \in C_0^\infty(B)$ such that $\beta(x) = 1$ on some ball $B' \Subset B$ (see Fig. 13.2).

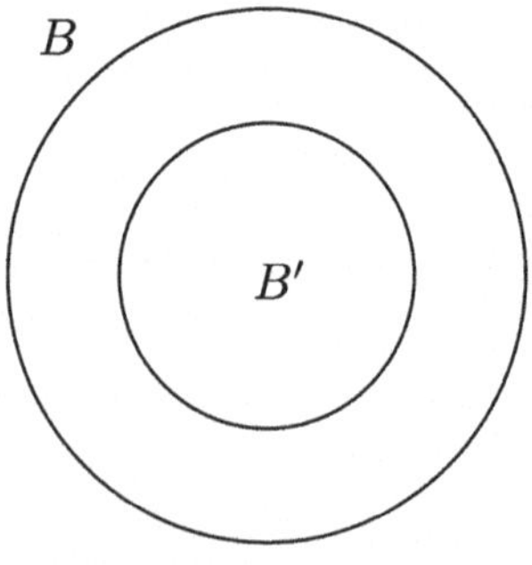

Fig. 13.2 The open balls B and B' such that $B \Subset \Omega$

Since we have the assertions

$$\begin{cases} u \in W^{2,q}_{\mathrm{loc}}(\Omega), \\ \mathcal{L}u = f \in L^p_{\mathrm{loc}}(\Omega), \end{cases}$$

it follows that

$$v = \beta u \in W^{2,q}_0(B), \tag{13.11}$$

and that

$$\mathcal{L}u \in L^p(B). \tag{13.12}$$

Step 1-2: In order to estimate the term $\mathcal{L}v$, we need Sobolev's imbedding theorem (Theorem 7.5).

Since we have the assertions

$$\frac{\partial \beta}{\partial x_i}\frac{\partial u}{\partial x_j} \in W^{1,q}(B) \quad \text{for } 1 \le i, j \le n,$$
$$\frac{\partial^2 \beta}{\partial x_i \partial x_j} u \in W^{2,q}(B) \quad \text{for } 1 \le i, j \le n,$$

by applying Sobolev's imbedding theorem (Theorem 7.5) to our situation we obtain that

$$\sum_{i,j=1}^{n} a^{ij}(x)\frac{\partial \beta}{\partial x_i}\frac{\partial u}{\partial x_j} \in L^{q_1}(B), \tag{13.13}$$

$$(\mathcal{L}\beta)u = \sum_{i,j=1}^{n} a^{ij}(x)\frac{\partial^2 \beta}{\partial x_i \partial x_j} u \in L^{q_1}(B), \tag{13.14}$$

for some $q_1 \in (q, p]$.

Therefore, by assertions (13.12), (13.13) and (13.14) it follows that

$$\begin{aligned} \mathcal{L}v &= \sum_{i,j=1}^{n} a^{ij}(x)\frac{\partial^2}{\partial x_i \partial x_j}(\beta u) \\ &= \beta(\mathcal{L}u) + 2\sum_{i,j=1}^{n} a^{ij}(x)\frac{\partial \beta}{\partial x_i}\frac{\partial u}{\partial x_j} + (\mathcal{L}\beta)u \in L^{q_1}(B). \end{aligned}$$

Step 1-3: Now we have, by formula (13.1),

$$\frac{\partial^2 v}{\partial x_i \partial x_j}(x) \tag{13.15}$$

$$
\begin{aligned}
&= \text{v.p.} \int_B \Gamma_{ij}(x, x-y) \\
&\quad \times \left\{ \sum_{h,k=1}^{n} \left\{ a^{hk}(x) - a^{hk}(y) \right\} \frac{\partial^2 v}{\partial x_h \partial x_k}(y) + \mathcal{L}v(y) \right\} dy \\
&\quad + c_{ij}(x)\mathcal{L}v(x) \quad \text{for } 1 \leq i, j \leq n,
\end{aligned}
$$

with

$$c_{ij}(x) := \int_{|y|=1} \Gamma_i(x, y) y_j \, d\sigma_y \in L^\infty(\Omega).$$

If we let

$$h_{ij}(x) := \text{v.p.} \int_B \Gamma_{ij}(x, x-y)\mathcal{L}v(y)\, dy + c_{ij}(x)\mathcal{L}v(x) \quad \text{for } 1 \leq i, j \leq n,$$

then, since $\mathcal{L}v \in L^{q_1}(B)$, it follows from an application of Theorem 11.4 with

$$\Omega := B, \quad k(x, z) := \Gamma_{ij}(x, z), \quad p := q_1$$

that

$$h_{ij} \in L^{q_1}(B).$$

Therefore, if $r \in [q, q_1]$, we can introduce a mapping

$$T : \left(L^r(B)\right)^{n^2} \longrightarrow \left(L^r(B)\right)^{n^2}$$

by the formula

$$
\begin{cases}
T\mathbf{w} = \left(\sum_{h,k=1}^{n} S_{ijhk}(w_{hk}) + h_{ij}(x)\right)_{1 \leq i, j \leq n}, \\
\mathbf{w} = \left(w_{ij}\right)_{1 \leq i, j \leq n}.
\end{cases}
$$

Then it follows that T is a *contraction mapping* (see condition (2.1)). Indeed, we have, by condition (13.10),

$$
\begin{aligned}
&\left\| T\mathbf{w}^{(1)} - T\mathbf{w}^{(2)} \right\| \\
&= \sum_{i,j=1}^{n} \left\| \sum_{h,k=1}^{n} S_{ijhk}\left(w_{hk}^{(1)}\right) - \sum_{h,k=1}^{n} S_{ijhk}\left(w_{hk}^{(2)}\right) \right\|_{L^r(B)} \\
&\leq \sum_{i,j,h,k=1}^{n} \left\| S_{ijhk} \right\| \left\| w_{hk}^{(1)} - w_{hk}^{(2)} \right\|_{L^r(B)} \\
&\leq \sum_{h,k=1}^{n} \left(\sum_{i,j=1}^{n} \left\| S_{ijhk} \right\| \right) \left\| w_{hk}^{(1)} - w_{hk}^{(2)} \right\|_{L^r(B)}
\end{aligned}
$$

$$\leq \frac{1}{2} \sum_{h,k=1}^{n} \left\| w_{hk}^{(1)} - w_{hk}^{(2)} \right\|_{L^r(B)}$$
$$= \frac{1}{2} \left\| \mathbf{w}^{(1)} - \mathbf{w}^{(2)} \right\| .$$

Therefore, by applying the *contraction mapping principle* (Theorem 2.1) we obtain that the mapping T has a unique fixed point

$$\mathbf{w} = \left(w_{ij}\right) \in \left(L^{q_1}(B)\right)^{n^2},$$
$$T\mathbf{w} = \mathbf{w}.$$

This implies that

$$\sum_{h,k=1}^{n} S_{ijhk}\left(w_{hk}\right) + h_{ij}(x)$$
$$= \text{v.p.} \int_B \Gamma_{ij}(x, x-y) \left\{ \sum_{h,k=1}^{n} \left\{a^{hk}(x) - a^{hk}(y)\right\} w_{hk}(y) + \mathcal{L}v(y) \right\} dy$$
$$+ c_{ij}(x)\, \mathcal{L}v(x)$$
$$= w_{ij}(x) \quad \text{for } 1 \leq i, j \leq n.$$

On the other hand, we obtain from formula (13.15) and assertion (13.11) that the Hessian matrix

$$\nabla^2 v = \left(\frac{\partial^2 v}{\partial x_i \partial x_j}\right) \in \left(L^q(B)\right)^{n^2}$$

is also a fixed point of T.

Therefore, since we have, for $q_1 > q$,

$$\left(L^{q_1}(B)\right)^{n^2} \subset \left(L^q(B)\right)^{n^2},$$

by the uniqueness of fixed points of T it follows that

$$\frac{\partial^2(\beta u)}{\partial x_i \partial x_j} = \frac{\partial^2 v}{\partial x_i \partial x_j} = w_{ij} \in L^{q_1}(B) \quad \text{for } 1 \leq i, j \leq n.$$

If $q_1 = p$, then we obtain that

$$u \in W^{2,p}_{\text{loc}}(\Omega),$$

since $B \Subset \Omega$ and $B' \Subset B$ are arbitrary.

If $q_1 < p$, by continuing this procedure we obtain, after a finite number of steps, that

$$u \in W^{2,p}_{\mathrm{loc}}(\Omega).$$

Step 2: The interior *a priori* estimate (12.2) follows in a standard way from Lemma 13.2 by a covering argument.

Step 2-1: Let Ω' and Ω'' be any open subsets such that $\Omega' \Subset \Omega'' \Subset \Omega$ (see Fig. 12.1). First, we cover the closure $\overline{\Omega'}$ of Ω' by a finite number of open balls $\{B_r(x_k)\}_{k=1}^N$ in each of which inequality (13.7) holds true (see Fig. 13.3), and we take a partition of unity $\{\alpha_k\}_{k=1}^N$

subordinate to the open covering $\{B_r(x_k)\}_{k=1}^N$. Moreover, we may assume that

$$\operatorname{supp} \alpha_k \subset \Omega'' \quad \text{for } 1 \leq k \leq N.$$

If $u \in W^{2,p}_{\mathrm{loc}}(\Omega)$, then it follows that

$$\|u\|_{W^{2,p}(\Omega')} \leq \sum_{k=1}^{N} \|\alpha_k u\|_{W^{2,p}(\Omega)} \tag{13.16}$$

$$\leq \sum_{k=1}^{N} \left(\left\|\nabla^2(\alpha_k u)\right\|_{L^p(\Omega)} + \|\nabla(\alpha_k u)\|_{L^p(\Omega)} + \|\alpha_k u\|_{L^p(\Omega)} \right)$$

$$\leq \sum_{k=1}^{N} \left(\left\|\nabla^2(\alpha_k u)\right\|_{L^p(B_r(x_k))} + \|\nabla(\alpha_k u)\|_{L^p(B_r(x_k))} + \|u\|_{L^p(\Omega'')} \right).$$

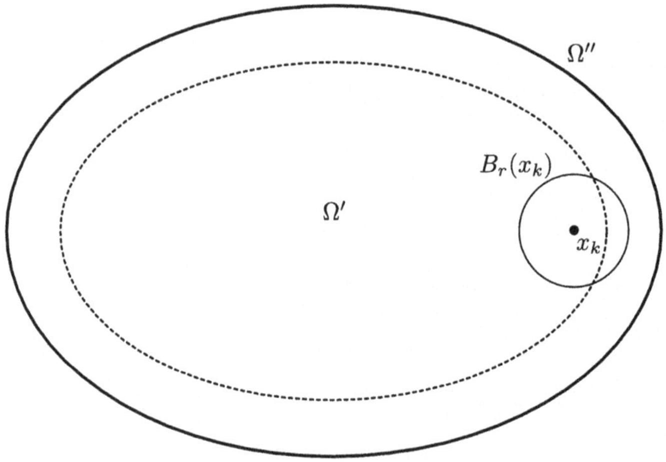

Fig. 13.3 The open subsets Ω', Ω'' such that $\Omega' \Subset \Omega''$ and the open ball $B_r(x_k)$

Step 2-2: In order to estimate the terms $\nabla(\alpha_k u)$, we need the following *interpolation inequality* (see [2, Theorem 5.2]; [59, Theorem 7.28]; [155, Theorem 4.14]):

Theorem 13.4 (the interpolation inequality) *Let Ω be a $C^{1,1}$ domain in $\mathbf{R}^n$, and $1 \leq p < \infty$. Then there exists a positive constant $C = C(\Omega, p)$ such that, for any $\varepsilon > 0$ we have the inequality*

$$\|\nabla v\|_{L^p(\Omega)} \leq \varepsilon \left\|\nabla^2 v\right\|_{L^p(\Omega)} + \frac{C}{\varepsilon} \|v\|_{L^p(\Omega)} \quad \text{for all } v \in W^{2,p}(\Omega). \tag{13.17}$$

Proof Since we have the formula

$$\nabla(\alpha_k u) = \alpha_k \nabla u + u(\nabla \alpha_k),$$

by applying inequality (13.17) to the function $\alpha_k u$ we obtain that, for some positive constants C_1 and C_2,

$$\begin{aligned}
&\|\nabla(\alpha_k u)\|_{L^p(B_r(x_k))} \\
&\leq \|\alpha_k(\nabla u)\|_{L^p(B_r(x_k))} + \|u(\nabla \alpha_k)\|_{L^p(B_r(x_k))} \\
&\leq \|\nabla u\|_{L^p(\Omega'')} + C_1 \|u\|_{L^p(\Omega'')} \\
&\leq \varepsilon \left\|\nabla^2 u\right\|_{L^p(\Omega'')} + \frac{C_2}{\varepsilon} \|u\|_{L^p(\Omega'')} + C_1 \|u\|_{L^p(\Omega'')} \quad \text{for } 1 \leq k \leq N.
\end{aligned} \tag{13.18}$$

Step 2-3: On the other hand, by applying inequality (13.7) to the functions $\alpha_k u$ we obtain that

$$\left\|\nabla^2(\alpha_k u)\right\|_{L^p(B_r(x_k))} \leq C \|\mathcal{L}(\alpha_k u)\|_{L^p(B_r(x_k))} \quad \text{for } 1 \leq k \leq N, \tag{13.19}$$

and also that

$$\mathcal{L}(\alpha_k u) = \alpha_k f + u\,\mathcal{L}(\alpha_k) + 2 \sum_{i,j=1}^{n} a^{ij}(x) \frac{\partial \alpha_k}{\partial x_i} \frac{\partial u}{\partial x_j} \quad \text{for } 1 \leq k \leq N. \tag{13.20}$$

Therefore, it follows from inequality (13.19) and formula (13.20) that

$$\begin{aligned}
&\left\|\nabla^2(\alpha_k u)\right\|_{L^p(B_r(x_k))} \leq C \|\mathcal{L}(\alpha_k u)\|_{L^p(B_r(x_k))} \\
&\leq C\Bigg(\|\alpha_k f\|_{L^p(B_r(x_k))} + \|u\,\mathcal{L}(\alpha_k)\|_{L^p(B_r(x_k))} \\
&\quad + 2 \sum_{i,j=1}^{n} \left\| a^{ij} \frac{\partial \alpha_k}{\partial x_i} \frac{\partial u}{\partial x_j} \right\|_{L^p(B_r(x_k))} \Bigg) \\
&\leq C_3 \left(|f|_{L^p(\Omega)} + \|u\|_{L^p(\Omega'')} + |\nabla u\|_{L^p(\Omega'')}\right) \\
&\leq C_3 \left(\|f\|_{L^p(\Omega)} + \|u\|_{L^p(\Omega'')}\right) + \varepsilon \left\|\nabla^2 u\right\|_{L^p(\Omega'')} + \frac{C_4}{\varepsilon} \|u\|_{L^p(\Omega'')}.
\end{aligned} \tag{13.21}$$

By combining inequalities (13.16), (13.18) and (13.21), we obtain that

$$\|u\|_{W^{2,p}(\Omega')} \leq 2N\varepsilon \left\|\nabla^2 u\right\|_{L^p(\Omega')} + C_5 \left(\|f\|_{L^p(\Omega)} + \|u\|_{L^p(\Omega'')}\right) + \frac{C_6}{\varepsilon} \|u\|_{L^p(\Omega'')}.$$

This proves the desired interior *a priori* estimate (12.2)

$$\|u\|_{W^{2,p}(\Omega')} \leq C_1 \left(\|u\|_{L^p(\Omega'')} + \|f\|_{L^p(\Omega)}\right),$$

if we take

$$\varepsilon := \frac{1}{4N}.$$

The proof of Theorem 13.3 (and hence that of Theorem 12.1) is complete. □

Chapter 14
Calderón–Zygmund Kernels and Boundary Estimates

The purpose of this chapter is to prove the global *a priori* estimate (12.3) for the homogeneous Dirichlet problem stated in Theorem 12.2 (see [35, Theorem 4.2]). The desired global *a priori* estimate (12.3) is a consequence of the following two facts:

(I) The explicit boundary representation formula (14.2) for the solutions of the homogeneous Dirichlet problem, which is obtained from the half space Green function, involves the same integral operators as in the interior case.
(II) An L^p boundedness of some singular integral operators and boundary commutators appearing in boundary representation formula (14.2) (Theorems 14.2 and 14.5).

The results here can be visualized as in Fig. 14.1 below.

This chapter is adapted from Chiarenza–Frasca–Longo [35] and Bramanti–Cerutti [17].

14.1 Boundary Representation Formula for Solutions

In this section we give a boundary representation formula for solutions of the homogeneous Dirichlet problem, by using the half space Green function for the uniformly elliptic differential operator

$$\mathcal{L} = \sum_{i,j=1}^{n} a^{ij}(x) \frac{\partial^2}{\partial x_i \partial x_j}.$$

Let B_σ be an open ball of radius σ in Euclidean space $\mathbf{R}^n$ with $n \geq 3$, and assume that the functions a^{ij} satisfy the following three conditions:

K. Taira, *Real Analysis Methods for Markov Processes*,
https://doi.org/10.1007/978-981-97-3659-1_14

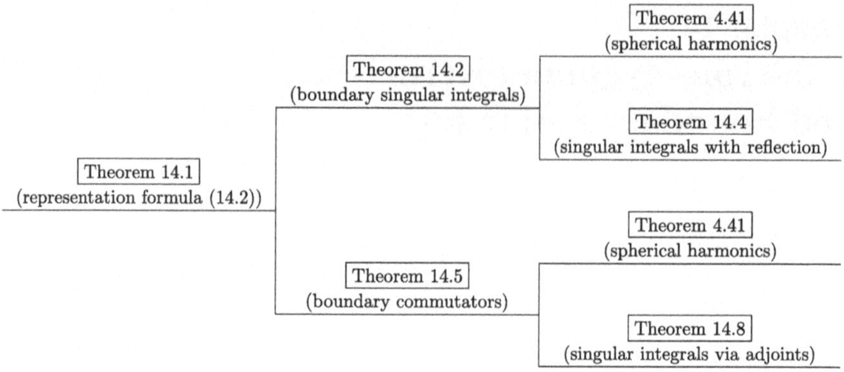

Fig. 14.1 A flowchart for the proof of Theorem 14.1

(i) $a^{ij} \in \mathrm{VMO} \cap L^\infty(\mathbf{R}^n)$ for all $1 \le i, j \le n$.
(ii) $a^{ij}(x) = a^{ji}(x)$ for all $1 \le i, j \le n$ and for almost all $x \in B_\sigma$.
(iii) There exists a positive constant λ such that (the uniform ellipticity)

$$\frac{1}{\lambda}|\xi|^2 \le \sum_{i,j=1}^{n} a^{ij}(x)\xi_i\xi_j \le \lambda|\xi|^2 \tag{14.1}$$

for almost all$x \in B_\sigma$ and for all $\xi \in \mathbf{R}^n$.

If $\widetilde{B}_\sigma$ is the subset of B_σ where conditions (ii) and (iii) hold true, then we let (see Sect. 12.1)

$$\Gamma(x,z) := \frac{1}{(2-n)\omega_n} \frac{1}{\sqrt{\det(a^{ij}(x))}} \left(\sum_{i,j=1}^{n} A_{ij}(x) z_i z_j \right)^{(2-n)/2}$$

for $x \in \widetilde{B}_\sigma$ and $z \in \mathbf{R}^n \setminus \{0\}$.

Here:

$$(A_{ij}(x)) = \text{ the inverse matrix of } (a^{ij}(x)),$$

$$\omega_n := |\Sigma_{n-1}| = \frac{2\pi^{n/2}}{\Gamma(n/2)} \quad \text{(the surface area of the unit sphere } \Sigma_{n-1} \text{ in } \mathbf{R}^n).$$

Moreover, we let

$$\Gamma_i(x,z) := \frac{\partial \Gamma}{\partial z_i}(x,z) \quad \text{for } x \in \widetilde{B}_\sigma,\ z \in \mathbf{R}^n \setminus \{0\} \text{ and } 1 \le i \le n,$$

$$\Gamma_{ij}(x,z) := \frac{\partial^2 \Gamma}{\partial z_i \partial z_j}(x,z) \quad \text{for } x \in \widetilde{B}_\sigma,\, z \in \mathbf{R}^n \setminus \{0\} \text{ and } 1 \leq i, j \leq n.$$

Now we let (see Fig. 14.2)

$$B_\sigma^+ := B_\sigma \cap \mathbf{R}_+^n = B_\sigma \cap \{x_n > 0\},$$

and introduce two function spaces

$$C_{\gamma_0} := \left\{u|_{B_\sigma^+} : u \in C_0^\infty(B_\sigma),\ u = 0 \text{ on } B_\sigma \cap \{x_n = 0\}\right\},$$

and

$$W_{\gamma_0}^{2,p}(B_\sigma^+) := \text{the closure of } C_{\gamma_0} \text{in the Sobolev space } W^{2,p}(B_\sigma^+).$$

Then it is easy to verify that $u \in W_0^{2,p}(B_\sigma)$ belongs to $W_{\gamma_0}^{2,p}(B_\sigma^+)$ if and only if it vanishes on $B_\sigma \cap \{x_n = 0\}$ (see [2, Theorem 5.37]).

The next theorem gives *integral representation formulas* for the second derivatives of solutions of the homogeneous Dirichlet problem for the elliptic differential operator

$$\mathcal{L} = \sum_{i,j=1}^n a^{ij}(x) \frac{\partial^2}{\partial x_i \partial x_j}$$

with variable coefficients (see [35, Theorem 3.2]):

Theorem 14.1 *(the integral representation formula) Let $\widetilde{B}_\sigma$ be the subset of B_σ where conditions (ii) and (iii) hold true, and let $u \in W_{\gamma_0}^{2,p}(B_\sigma^+)$ for $1 < p < \infty$. Then we have, for all $x \in \widetilde{B}_\sigma^+ = \widetilde{B}_\sigma \cap \mathbf{R}_+^n$,*

$$\frac{\partial^2 u}{\partial x_i \partial x_j}(x) = \text{v.p.} \int_{B_\sigma^+} \Gamma_{ij}(x, x-y) \tag{14.2}$$

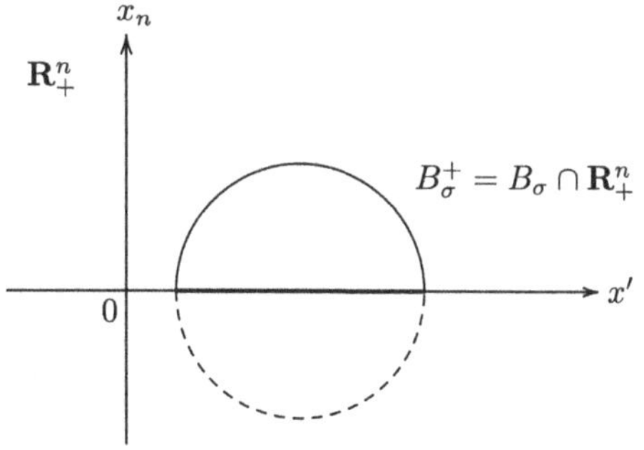

Fig. 14.2 The semi-ball B_σ^+ in the half-space $\mathbf{R}_+^n$

$$\times \left(\sum_{h,k=1}^{n} \left\{ a^{hk}(x) - a^{hk}(y) \right\} \frac{\partial^2 u}{\partial x_h \partial x_k}(y) + \mathcal{L}u(y) \right) dy$$

$$+ \mathcal{L}u(x) \left(\int_{|t|=1} \Gamma_i(x,t) t_j \, d\sigma_t \right) - I_{ij}(x;x) \quad \mathit{for} 1 \leq i, j \leq n.$$

Here:

$$I_{ij}(x;z) := \int_{B_\sigma^+} \Gamma_{ij}(z, T(x;z) - y)$$
$$\times \left\{ \sum_{h,k=1}^{n} \left\{ a^{hk}(z) - a^{hk}(y) \right\} \frac{\partial^2 u}{\partial x_h \partial x_k}(y) + \mathcal{L}u(y) \right\} dy$$
$$\mathit{for}\ 1 \leq i, j \leq n-1;$$

$$I_{in}(x;z) = I_{ni}(x;z)$$
$$:= \int_{B_\sigma^+} \sum_{\ell=1}^{n} \Gamma_{i\ell}(z, T(x;z) - y) A_\ell(z)$$
$$\times \left\{ \sum_{h,k=1}^{n} \left\{ a^{hk}(z) - a^{hk}(y) \right\} \frac{\partial^2 u}{\partial x_h \partial x_k}(y) + \mathcal{L}u(y) \right\} dy$$
$$\mathit{for}\ 1 \leq i \leq n-1;$$

and

$$I_{nn}(x;z) := \int_{B_\sigma^+} \sum_{\ell,m=1}^{n} \Gamma_{\ell m}(z, T(x;z) - y) A_\ell(z) A_m(z)$$
$$\times \left\{ \sum_{h,k=1}^{n} \left\{ a^{hk}(z) - a^{hk}(y) \right\} \frac{\partial^2 u}{\partial x_h \partial x_k}(y) + \mathcal{L}u(y) \right\} dy.$$

Moreover, the generalized reflection map $T(x;z)$ *is defined by the formula*

$$T(x;z) = \begin{pmatrix} x_1 - 2x_n \frac{a^{1n}(z)}{a^{nn}(z)} \\ x_2 - 2x_n \frac{a^{2n}(z)}{a^{nn}(z)} \\ \vdots \\ -x_n \end{pmatrix}, \quad x = \begin{pmatrix} x_1 \\ x_2 \\ \vdots \\ x_n \end{pmatrix},$$

and the vector $A(x)$ is defined by the formula

$$A(x) = \begin{pmatrix} A_1(x) \\ A_2(x) \\ \vdots \\ A_n(x) \end{pmatrix} := \frac{\partial T}{\partial x_n}(x; x_0)\Big|_{x_0=x} = \begin{pmatrix} -2\frac{a^{1n}(x)}{a^{nn}(x)} \\ -2\frac{a^{2n}(x)}{a^{nn}(x)} \\ \vdots \\ -1 \end{pmatrix}.$$

Proof By a *density argument*, it suffices to prove boundary representation formula (14.2) for all $u \in C_{\gamma_0}$. Indeed, the general case can be obtained by using Theorems 11.4 and 11.5, and Theorems 14.2 and 14.5 which will be proved in Sects. 14.2 and 14.3, respectively. More precisely, we shall prove the following two assertions:

(i) If we let

$$\widetilde{K}_{ij} f(x) := \int_{\mathbf{R}^n_+} \Gamma_{ij}(x, T(x) - y) f(y)\, dy \quad \text{for } 1 \leq i, j \leq n,$$

where

$$T(x) := T(x; x) = \begin{pmatrix} x_1 - 2x_n \frac{a^{1n}(x)}{a^{nn}(x)} \\ x_2 - 2x_n \frac{a^{2n}(x)}{a^{nn}(x)} \\ \vdots \\ -x_n \end{pmatrix}, \quad x = \begin{pmatrix} x_1 \\ x_2 \\ \vdots \\ x_n \end{pmatrix},$$

then there exists a positive constant $C_1 = C_1(n, p, \lambda, M)$ such that (Theorem 14.2)

$$\left\| \widetilde{K}_{ij} f \right\|_{L^p(\mathbf{R}^n_+)} \leq C_1 \, \| f \|_{L^p(\mathbf{R}^n_+)} \quad \text{for all } f \in L^p(\mathbf{R}^n_+).$$

(ii) Let $\varphi \in L^\infty(\mathbf{R}^n)$. If we let

$$\begin{aligned} &\widetilde{C}\left[\varphi, K_{ij}\right] f(x) \\ &:= \int_{\mathbf{R}^n_+} \Gamma_{ij}(x, T(x) - y) \{\varphi(x) - \varphi(y)\} f(y)\, dy \quad \text{for } 1 \leq i, j \leq n, \end{aligned}$$

then there exists a positive constant $C_2 = C_2(n, p, \lambda, M)$ such that (Theorem 14.5)

$$\left\| \widetilde{C}\left[\varphi, K_{ij}\right] f \right\|_{L^p(\mathbf{R}^n_+)} \leq C_2 \, \|\varphi\|_* \, \| f \|_{L^p(\mathbf{R}^n_+)} \quad \text{for all } f \in L^p(\mathbf{R}^n_+).$$

Let x_0 be an arbitrary point of $\widetilde{B}_\sigma$. First, it is easy to verify that the *half space Green function* $G(x_0, x, y)$ for the elliptic differential operator

$$\mathcal{L}_0 = \sum_{i,j=1}^{n} a^{ij}(x_0) \frac{\partial^2}{\partial x_i \partial x_j}$$

with *constant coefficients* is given by the formula (cf. [59, Sect. 4.4, formula (4.28)])

$$G(x_0, x, y) = \Gamma(x_0, x - y) - \Gamma(x_0, T(x; x_0) - y).$$

Then we have the representation formula

$$\begin{aligned} u(x) &= \int_{B_\sigma^+} G(x_0, x, y)\mathcal{L}_0 u(y)\, dy \qquad (14.3)\\ &= \int_{B_\sigma^+} \Gamma(x_0, x - y)\mathcal{L}_0 u(y)\, dy - \int_{B_\sigma^+} \Gamma(x_0, T(x; x_0) - y)\mathcal{L}_0 u(y)\, dy \\ &:= J(x; x_0) - I(x; x_0). \end{aligned}$$

We can differentiate the first term $J(x; x_0)$ in formula (14.3) twice to obtain that

$$\begin{aligned} &\frac{\partial^2 J}{\partial x_i \partial x_j}(x; x_0) = \frac{\partial^2}{\partial x_i \partial x_j}\left(\int_{B_\sigma^+} \Gamma(x_0, x - y)\mathcal{L}_0 u(y)\, dy\right) \qquad (14.4)\\ &= \text{v.p.} \int_{B_\sigma^+} \Gamma_{ij}(x_0, x - y) \\ &\quad \times \left(\sum_{h,k=1}^{n} \left\{a^{hk}(x_0) - a^{hk}(y)\right\} \frac{\partial^2 u}{\partial x_h \partial x_k}(y) + \mathcal{L}u(y)\right) dy \\ &\quad + \mathcal{L}u(x)\left(\int_{|t|=1} \Gamma_i(x_0, t) t_j\, d\sigma_t\right) \\ &= J_{ij}(x; x_0) \quad \text{for } 1 \le i, j \le n, \end{aligned}$$

since we have the formula

$$\begin{aligned} \mathcal{L}_0 u(y) &= (\mathcal{L}_0 - \mathcal{L})u(y) + \mathcal{L}u(y) \\ &= \sum_{i,j=1}^{n} \left\{a^{ij}(x_0) - a^{ij}(y)\right\} \frac{\partial^2 u}{\partial x_i \partial x_j}(y) + \sum_{i,j=1}^{n} a^{ij}(y) \frac{\partial^2 u}{\partial x_i \partial x_j}(y). \end{aligned}$$

As for the second term $I(x; x_0)$ in formula (14.3), we can differentiate it under the integral sign to obtain that

$$\begin{aligned} &\frac{\partial^2 I}{\partial x_i \partial x_j}(x; x_0) = \frac{\partial^2}{\partial x_i \partial x_j}\left(\int_{B_\sigma^+} \Gamma(x_0, T(x; x_0) - y)\mathcal{L}_0 u(y)\, dy\right) \qquad (14.5a)\\ &= \int_{B_\sigma^+} \Gamma_{ij}(x_0, T(x; x_0) - y) \\ &\quad \times \left\{\sum_{h,k=1}^{n} \left\{a^{hk}(x_0) - a^{hk}(y)\right\} \frac{\partial^2 u}{\partial x_h \partial x_k}(y) + \mathcal{L}u(y)\right\} dy \qquad (14.5b)\\ &= I_{ij}(x; x_0) \quad \text{for } 1 \le i, j \le n - 1; \end{aligned}$$

$$
\begin{aligned}
\frac{\partial^2 I}{\partial x_i \partial x_n}(x; x_0) &= \frac{\partial^2}{\partial x_i \partial x_n}\left(\int_{B_\sigma^+} \Gamma(x_0, T(x; x_0) - y)\mathcal{L}_0 u(y)\, dy\right) \qquad (14.5c) \\
&= \int_{B_\sigma^+} \sum_{\ell=1}^{n} \Gamma_{i\ell}(x_0, T(x; x_0) - y) \frac{\partial T_\ell}{\partial x_n}(x; x_0) \\
&\quad \times \left\{ \sum_{h,k=1}^{n} \left\{a^{hk}(x_0) - a^{hk}(y)\right\} \frac{\partial^2 u}{\partial x_h \partial x_k}(y) + \mathcal{L}u(y) \right\} dy \\
&= I_{in}(x; x_0) \quad \text{for } 1 \le i \le n-1;
\end{aligned}
$$

and

$$
\begin{aligned}
\frac{\partial^2 I}{\partial x_n^2}(x; x_0) &= \frac{\partial^2}{\partial x_n^2}\left(\int_{B_\sigma^+} \Gamma(x_0, T(x; x_0) - y)\mathcal{L}_0 u(y)\, dy\right) \qquad (14.5d) \\
&= \int_{B_\sigma^+} \sum_{\ell,m=1}^{n} \Gamma_{\ell m}(x_0, T(x; x_0) - y) \frac{\partial T_\ell}{\partial x_n}(x; x_0) \frac{\partial T_m}{\partial x_n}(x; x_0) \\
&\quad \times \left\{ \sum_{h,k=1}^{n} \left\{a^{hk}(x_0) - a^{hk}(y)\right\} \frac{\partial^2 u}{\partial x_h \partial x_k}(y) + \mathcal{L}u(y) \right\} dy \\
&= I_{nn}(x; x_0).
\end{aligned}
$$

Therefore, the desired boundary representation formula (14.2) for $x := x_0$ follows by setting $x = x_0$ in formulas (14.4) and 14.5.

The proof of Theorem 14.1 is complete. □

14.2 L^p Boundedness of Boundary Singular Integral Operators

This section is devoted to the proof of L^p boundedness of boundary singular integral operators appearing in boundary representation formula (14.2).

14.2.1 Boundary Singular Integral Operators

Let $k(x, z)$ be a real-valued function defined on $\mathbf{R}^n \times (\mathbf{R}^n \setminus \{0\})$ that satisfies the following two conditions:

(i) $k(x, \cdot)$ is a Calderón–Zygmund kernel on $\mathbf{R}^n$ for almost all $x \in \mathbf{R}^n$.
(ii) The quantity

$$M := \max_{|\alpha| \leq 2n} \left\| \frac{\partial^\alpha k}{\partial z^\alpha}(\cdot, \cdot) \right\|_{L^\infty(\mathbf{R}^n \times \Sigma_{n-1})}$$

is finite, where Σ_{n-1} is the unit sphere in $\mathbf{R}^n$. Moreover, we recall that

$$T(x) := T(x; x) = \begin{pmatrix} x_1 - 2x_n \frac{a^{1n}(x)}{a^{nn}(x)} \\ x_2 - 2x_n \frac{a^{2n}(x)}{a^{nn}(x)} \\ \vdots \\ -x_n \end{pmatrix}, \quad x = \begin{pmatrix} x_1 \\ x_2 \\ \vdots \\ x_n \end{pmatrix},$$

and

$$A(x) := \begin{pmatrix} A_1(x) \\ A_2(x) \\ \vdots \\ A_n(x) \end{pmatrix} = \begin{pmatrix} -2\frac{a^{1n}(x)}{a^{nn}(x)} \\ -2\frac{a^{2n}(x)}{a^{nn}(x)} \\ \vdots \\ -1 \end{pmatrix}.$$

Then we have the following L^p boundedness of boundary singular integral operators (cf. [17, Theorem 3.1]):

Theorem 14.2 *If $f \in L^p(\mathbf{R}^n_+)$ for $1 < p < \infty$, we let*

$$\widetilde{K} f(x) := \int_{\mathbf{R}^n_+} k(x, T(x) - y) f(y)\, dy.$$

Then there exists a positive constant $c = c(n, p, \lambda, M)$ such that

$$\|\widetilde{K} f\|_{L^p(\mathbf{R}^n_+)} \leq c\, \|f\|_{L^p(\mathbf{R}^n_+)} \quad \text{for all } f \in L^p(\mathbf{R}^n_+). \tag{14.6}$$

The proof of Theorem 14.2 will be given in Sect. 14.2.2, due to its length.

Before the proof of Theorem 14.2, we need the following geometric properties of the map $T(x) = T(x; x)$ (cf. [35, Lemma 3.1], [17, Lemma 3.2]):

Lemma 14.3 *There exist positive constants c_1, c_2 such that we have, for all $y \in \mathbf{R}^n_+$ and all $x \in \mathbf{R}^n_+$ for which $T(x)$ is defined,*

$$c_1 |\widetilde{x} - y| \leq |T(x) - y| \leq c_2\, |\widetilde{x} - y|, \tag{14.7}$$

where $\widetilde{x} = \left(x', -x_n\right)$ for $x = \left(x', x_n\right) \in \mathbf{R}^n_+$.

Proof (1) First, we have, for all $y = \left(y', y_n\right) = (y_1, \ldots, y_{n-1}, y_n) \in \mathbf{R}^n_+$,

$$T(x) - y = \begin{pmatrix} T_1(x) - y_1 \\ T_2(x) - y_2 \\ \vdots \\ T_{n-1}(x) - y_{n-1} \\ -x_n - y_n \end{pmatrix} := \begin{pmatrix} T(x)' - y' \\ -x_n - y_n \end{pmatrix}, \tag{14.8}$$

and so

$$|T(x) - y| \geq x_n + y_n \geq x_n.$$

Hence it follows from (14.8) that

$$\begin{aligned} \frac{|T(x) - \widetilde{x}|}{|T(x) - y|} &\leq \frac{|T(x) - \widetilde{x}|}{x_n} \\ &= \frac{1}{x_n} \left|(x', 0) + x_n A(x) - (x', -x_n)\right| \\ &= \left|(0', 1) + A(x)\right| \\ &= \frac{2}{a^{nn}(x)} \sqrt{\left(a^{1n}(x)\right)^2 + \ldots + \left(a^{n-1\,n}(x)\right)^2} \end{aligned}$$

for all $y \in \mathbf{R}^n_+$ and all $x \in \mathbf{R}^n_+$ for which $T(x)$ is defined.

However, since the functions $a^{ij} \in L^\infty(\mathbf{R}^n)$ satisfy condition (14.1), we have the inequality

$$\left|(0', 1) + A(x)\right| \leq C_1(n, \lambda),$$

where

$$C_1(n\lambda) = 2\,\lambda \left(\left\|a^{1n}\right\|_\infty + \ldots + \left\|a^{n-1\,n}\right\|_\infty\right).$$

Hence we have, for all $y \in \mathbf{R}^n_+$ and all $x \in \mathbf{R}^n_+$ for which $T(x)$ is defined,

$$\frac{|T(x) - \widetilde{x}|}{|T(x) - y|} \leq C_1(n, \lambda). \tag{14.9}$$

Therefore, we obtain from inequality (14.9) that, for all $y \in \mathbf{R}^n_+$ and for all $x \in \mathbf{R}^n_+$ for which $T(x)$ is defined,

$$\begin{aligned} |\widetilde{x} - y| &\leq |T(x) - \widetilde{x}| + |T(x) - y| \\ &\leq |T(x) - y| \left(1 + \frac{|T(x) - \widetilde{x}|}{|T(x) - y|}\right) \\ &\leq (1 + C_1(n, \lambda)|T(x) - y|. \end{aligned}$$

This proves the desired inequality (14.7) with

$$c_1 := \frac{1}{1 + C_1(n, \lambda)}.$$

(2) On the other hand, it follows that, for all $y = \left(y', y_n\right) \in \mathbf{R}^n_+$ and for all $x = \left(x', x_n\right) \in \mathbf{R}^n_+$,

$$\begin{aligned}
|\widetilde{x} - y| &= \left|\left(x' - y', -x_n - y_n\right)\right| = \sqrt{|x' - y'|^2 + (x_n + y_n)^2}, \\
|T(x) - y| &= \sqrt{|T(x)' - y'|^2 + (x_n + y_n)^2},
\end{aligned}$$

where

$$T(x)' - y' = \begin{pmatrix} x_1 - y_1 - 2x_n \frac{a^{1n}(x)}{a^{nn}(x)} \\ x_2 - y_2 - 2x_n \frac{a^{2n}(x)}{a^{nn}(x)} \\ \vdots \\ x_{n-1} - y_{n-1} - 2x_n \frac{a^{n-1n}(x)}{a^{nn}(x)} \end{pmatrix}.$$

We remark that

$$\begin{aligned}
\left|T(x)' - y'\right| &\leq \left|x' - y'\right| + C_1(n, \lambda)\, x_n \\
&\leq \left|x' - y'\right| + C_1(n, \mu)\, (x_n + y_n).
\end{aligned}$$

Hence we have, for all $y \in \mathbf{R}^n_+$ and for all $x \in \mathbf{R}^n_+$ for which $T(x)$ is defined,

$$\begin{aligned}
|T(x) - y| &\leq \left|T(x)' - y'\right| + (x_n + y_n) \\
&\leq |x' - y'| + (1 + C_1(n, \lambda))\, (x_n + y_n) \\
&\leq (C_1(n, \lambda) + 1) \left(|x' - y| + (x_n + y_n)\right) \\
&\leq \sqrt{2}\, (C_1(n, \lambda) + 1) \sqrt{|x' - y|^2 + (x_n + y_n)^2} \\
&= \sqrt{2}\, (C_1(n, \lambda) + 1)\, |\widetilde{x} - y|.
\end{aligned}$$

This proves the desired inequality (14.7) with

$$c_2 := \sqrt{2}\, (C_1(n, \lambda) + 1).$$

The proof of Lemma 14.3 is complete. □

Moreover, we have the following L^p boundedness theorem for singular intgrals with reflection (cf. [35, Theorem 2.5], [17, Lemma 3.3]):

Theorem 14.4 *If $f \in L^p(\mathbf{R}^n_+)$ for $1 < p < \infty$, we let*

$$\widetilde{K} f(x) := \int_{\mathbf{R}^n_+} \frac{f(y)}{|\widetilde{x} - y|^n}\, dy,$$

where $\widetilde{x} = (x', -x_n)$. *Then there exists a positive constant* $c(n, p)$ *such that*

$$\|\widetilde{K} f\|_{L^p(\mathbf{R}^n_+)} \leq c(n, p)\, \|f\|_{L^p(\mathbf{R}^n_+)} \quad \text{for all } f \in L^p(\mathbf{R}^n_+). \tag{14.10}$$

For example, we may take

$$c(n, p) := B\left(\frac{1}{p}, 1 - \frac{1}{p}\right) \int_{\mathbf{R}^{n-1}} \frac{1}{(|z'|^2 + 1)^{n/2}}\, dz'.$$

Proof The proof of Theorem 14.4 is divided into three steps.

Step (1): First, we have the inequality

$$\begin{aligned}\|\widetilde{K} f(x)\|^p_{L^p(\mathbf{R}^n_+)} &\leq \int_{\mathbf{R}^n_+} \left(\int_{\mathbf{R}^n} \frac{|f(y)|}{|\widetilde{x} - y|^n}\, dy\right)^p dx \\ &= \int_0^\infty \int_{\mathbf{R}^{n-1}} \left(\int_{\mathbf{R}^n_+} \frac{|f(y)|}{|\widetilde{x} - y|^n}\, dy\right)^p dx'dx_n \\ &= \int_0^\infty I(x_n)\, dx_n,\end{aligned} \tag{14.11}$$

where

$$I(x_n) := \int_{\mathbf{R}^{n-1}} \left(\int_0^\infty \int_{\mathbf{R}^{n-1}} \frac{|f(y', y_n)|}{\left(|x' - y'|^2 + (x_n + y_n)^2\right)^{n/2}}\, dy'dy_n\right)^p dx'.$$

Thus we are reduced to the estimate of the function $I(x_n)$.

Step (2): We prove that

$$I(x_n)^{1/p} \leq C_n \int_0^\infty \frac{1}{x_n + y_n} \left(\int_{\mathbf{R}^{n-1}} |f(y', y_n)|^p\, dy'\right)^{1/p} dy_n, \tag{14.12}$$

where C_n is a positive constant given by the formula

$$C_n := \int_{\mathbf{R}^{n-1}} \frac{1}{(|z'|^2 + 1)^{n/2}}\, dz'.$$

By applying Minkowski's inequality for integrals (Theorem 3.29)

$$\left\|\int_0^\infty F(\cdot, y_n)\, dy_n\right\|_{L^p(\mathbf{R}^{n-1})} \leq \int_0^\infty \|F(\cdot, y_n)\|_{L^p(\mathbf{R}^{n-1})}\, dy_n$$

to the function

$$F(x', y_n) = \int_0^\infty \int_{\mathbf{R}^{n-1}} \frac{|f(y', y_n)|}{\left(|x' - y'|^2 + (x_n + y_n)^2\right)^{n/2}}\, dy',$$

we obtain that

$$
\begin{aligned}
& I(x_n)^{1/p} \\
& \leq \int_0^\infty \left(\int_{\mathbf{R}^{n-1}} \left(\int_{\mathbf{R}^{n-1}} \frac{|f(y', y_n)|}{\left(|x' - y'|^2 + (x_n + y_n)^2\right)^{n/2}} \, dy' \right)^p dx' \right)^{1/p} dy_n \\
& = \int_0^\infty \|G * f(\cdot, x_n + y_n)\|_{L^p(\mathbf{R}^{n-1})} \, dy_n,
\end{aligned}
$$

where

$$
G(x', x_n + y_n) := \frac{1}{\left(|x'|^2 + (x_n + y_n)^2\right)^{n/2}},
$$

and

$$
(G * f)(x', x_n + y_n) := \int_{\mathbf{R}^{n-1}} \frac{|f(y', y_n)|}{\left(|x' - y'|^2 + (x_n + y_n)^2\right)^{n/2}} \, dy'.
$$

Moreover, by Young's inequality (Theorem 3.36) it follows that

$$
\begin{aligned}
& \|G * f(\cdot, x_n + y_n)\|_{L^p(\mathbf{R}^{n-1})} \\
& \leq \|G(\cdot, x_n + y_n)\|_{L^1(\mathbf{R}^{n-1})} \, \|f(\cdot, y_n)\|_{L^p(\mathbf{R}^{n-1})}.
\end{aligned} \tag{14.13}
$$

However, we have, for the above positive constant C_n,

$$
\begin{aligned}
\|G(\cdot, x_n + y_n)\|_{L^1(\mathbf{R}^{n-1})} &= \int_{\mathbf{R}^{n-1}} \frac{1}{\left(|y'|^2 + (x_n + y_n)^2\right)^{n/2}} dy' \\
&= \frac{1}{x_n + y_n} \int_{\mathbf{R}^{n-1}} \frac{1}{(|z'|^2 + 1)^{n/2}} \, dz' \\
&= \frac{C_n}{x_n + y_n}.
\end{aligned} \tag{14.14}
$$

Therefore, the desired inequality (14.12) follows by combining inequality (14.13) and formula (14.14).

Step (3): By inequality (14.12), it follows that

$$
\begin{aligned}
& \int_0^\infty I(x_n) \, dx_n \\
& \leq C_n^p \int_0^\infty \left(\int_0^\infty \frac{\left(\int_{\mathbf{R}^{n-1}} |f(y', y_n)|^p \, dy'\right)^{1/p}}{x_n + y_n} \, dy_n \right)^p dx_n \\
& = C_n^p \int_0^\infty \left(\int_0^\infty \frac{\left(\int_{\mathbf{R}^{n-1}} |f(y', \lambda x_n)|^p \, dy'\right)^{1/p}}{1 + \lambda} \, d\lambda \right)^p dx_n
\end{aligned} \tag{14.15}
$$

$$= C_n^p \int_0^\infty \left(\int_0^\infty \frac{\|f(\cdot, \lambda x_n)\|_{L^p(\mathbf{R}^{n-1})}}{1+\lambda} \, d\lambda \right)^p dx_n.$$

However, by applying Minkowski's inequality for integrals (Theorem 3.29)

$$\left\| \int_0^\infty G(\cdot, \lambda) \, d\lambda \right\|_{L^p(0,\infty)} \leq \int_0^\infty \|G(\cdot, \lambda)\|_{L^p(0,\infty)} \, d\lambda$$

to the function

$$G(x_n, \lambda) = \frac{\|f(\cdot, \lambda x_n)\|_{L^p(\mathbf{R}^{n-1})}}{1+\lambda},$$

we obtain that

$$\begin{aligned} &\int_0^\infty \left(\int_0^\infty \frac{\|f(\cdot, \lambda x_n)\|_{L^p(\mathbf{R}^{n-1})}}{1+\lambda} \, d\lambda \right)^p dx_n \qquad (14.16) \\ &\leq \left(\int_0^\infty \left(\int_0^\infty \left(\frac{\|f(\cdot, \lambda x_n)\|_{L^p(\mathbf{R}^{n-1})}}{1+\lambda} \right)^p dx_n \right)^{1/p} d\lambda \right)^p \\ &= \left(\int_0^\infty \frac{1}{1+\lambda} \left(\int_0^\infty \|f(\cdot, z_n)\|_{L^p(\mathbf{R}^{n-1})}^p \frac{dz_n}{\lambda} \right)^{1/p} d\lambda \right)^p \\ &= \left(\int_0^\infty \frac{1}{(1+\lambda)\lambda^{1/p}} \, d\lambda \right)^p \left(\int_0^\infty \int_{\mathbf{R}^{n-1}} |f(z', z_n)|^p \, dz' dz_n \right) \\ &= D_p^p \, \|f\|_{L^p(\mathbf{R}_+^n)}^p, \end{aligned}$$

where D_p is a positive constant given by the formula

$$D_p := \int_0^\infty \frac{1}{(1+\lambda)\lambda^{1/p}} \, d\lambda = B\left(\frac{1}{p}, 1 - \frac{1}{p} \right).$$

Therefore, we have, by inequalities (14.11), (14.15) and (14.16),

$$\begin{aligned} &\|\widetilde{K} f(x)\|_{L^p(\mathbf{R}_+^n)}^p \leq \int_0^\infty I(x_n) \, dx_n \\ &\leq C_n^p \int_0^\infty \left(\int_0^\infty \frac{\|f(\cdot, \lambda x_n)\|_{L^p(\mathbf{R}^{n-1})}}{1+\lambda} \, d\lambda \right)^p dx_n \\ &\leq C_n^p \, D_p^p \, \|f\|_{L^p(\mathbf{R}_+^n)}^p \quad \text{for all } f \in L^p(\mathbf{R}_+^n). \end{aligned}$$

This proves the desired inequality (14.10) with

$$c(n, p) := C_n \, D_p.$$

The proof of Theorem 14.4 is complete. □

14.2.2 Proof of Theorem 14.2

This subsection is devoted to the proof of Theorem 14.2. By a *density argument*, it suffices to prove the theorem for all $f \in C_0^\infty(\mathbf{R}_+^n)$. The proof is divided into three steps.

Step 1: First, it should be noticed that the function

$$z \longmapsto |z|^n k(x, z)$$

belongs to $C^\infty(\mathbf{R}^n \setminus \{0\})$ for almost all $x \in \mathbf{R}_+^n$, and it is positively homogeneous of degree zero and satisfies the condition

$$\int_{\Sigma_{n-1}} k(x, z)\, d\sigma_z = 0.$$

For each $m = 1, 2, \ldots$ and $k = 1, 2, \ldots, d(m)$, we let

$$a_{km}(x) := \int_{\Sigma_{n-1}} k(x, z)\, Y_{km}(z)\, dz,$$

then, by the completeness of the spherical harmonics $\{Y_{km}\}$ in $L^2(\Sigma_{n-1})$ it follows that

$$\begin{aligned} &|T(x) - y|^n k(x, T(x) - y) \\ &= \sum_{m=1}^{\infty} \sum_{k=1}^{d(m)} a_{km}(x)\, Y_{km}(T(x) - y) \quad \text{for } x \in \mathbf{R}_+^n \text{ and } y \in \mathbf{R}^n \setminus \{0\}. \end{aligned} \tag{14.17}$$

Moreover, we have, by assertion (4.76) with $r := n$ and assertion (4.75) with $\alpha := 0$ of Theorem 4.41,

$$\|a_{km}\|_{L^\infty(\mathbf{R}_+^n)} \leq \frac{c_1(n)}{m^{2n}} M, \tag{14.18a}$$

$$\|Y_{km}\|_{L^\infty(\Sigma_{n-1})} \leq c_2(n)\, m^{(n-2)/2}, \tag{14.18b}$$

and, by assertion (4.74) of Theorem 4.41,

$$d(m) \leq c_3(n)\, m^{n-2}. \tag{14.18c}$$

Step 2: By using the expansion (14.17), we obtain from inequality (14.18a) that

$$\|\widetilde{K} f\|_{L^p(\mathbf{R}_+^n)} = \left\| \int_{\mathbf{R}_+^n} k(\cdot, T(\cdot) - y) f(y)\, dy \right\|_{L^p(\mathbf{R}_+^n)} \tag{14.19}$$

$$
\begin{aligned}
&= \left\| \sum_{m=1}^{\infty} \sum_{k=1}^{d(m)} a_{km}(\cdot) \int_{\mathbf{R}_+^n} \frac{Y_{km}(T(\cdot)-y)}{|T(\cdot)-y|^n} f(y)\, dy \right\|_{L^p(\mathbf{R}_+^n)} \\
&\le \sum_{m=1}^{\infty} \sum_{k=1}^{d(m)} \|a_{km}\|_{L^\infty(\mathbf{R}_+^n)} \left\| \int_{\mathbf{R}_+^n} \frac{Y_{km}(T(\cdot)-y)}{|T(\cdot)-y|^n} f(y)\, dy \right\|_{L^p(\mathbf{R}_+^n)} \\
&\le c_1(n)\, M \sum_{m=1}^{\infty} \sum_{k=1}^{d(m)} \frac{1}{m^{2n}} \left\| \int_{\mathbf{R}_+^n} \frac{Y_{km}(T(\cdot)-y)}{|T(\cdot)-y|^n} f(y)\, dy \right\|_{L^p(\mathbf{R}_+^n)} .
\end{aligned}
$$

However, by inequality (14.18b) it follows from an application of Lemma 14.3 and Theorem 14.4 that

$$
\begin{aligned}
&\left\| \int_{\mathbf{R}_+^n} \frac{Y_{km}(T(\cdot)-y)}{|T(\cdot)-y|^n} f(y)\, dy \right\|_{L^p(\mathbf{R}_+^n)} \\
&\le c(n,p)\, c_2(n)\, m^{(n-2)/2} \|f\|_{L^p(\mathbf{R}_+^n)} .
\end{aligned}
\tag{14.20}
$$

Indeed, we have, by Lemma 14.3,

$$
\begin{aligned}
&\left(\int_{\mathbf{R}_+^n} \left| \int_{\mathbf{R}_+^n} \frac{|Y_{km}(T(x)-y)|}{|T(x)-y|^n} f(y)\, dy \right|^p dx \right)^{1/p} \\
&\le \left(\int_{\mathbf{R}_+^n} \left| \int_{\mathbf{R}_+^n} \frac{\|Y_{km}\|_{L^\infty(\Sigma_{n-1})}}{|T(x)-y|^n} |f(y)|\, dy \right|^p dx \right)^{1/p} \\
&\le c_2(n)\, m^{(n-2)/2} \left(\int_{\mathbf{R}_+^n} \left| \int_{\mathbf{R}_+^n} \frac{|f(y)|}{|\widetilde{x}-y|^n}\, dy \right|^p dx \right)^{1/p} \\
&= c_2(n)\, m^{(n-2)/2} \left\| \widetilde{K}(|f|) \right\|_{L^p(\mathbf{R}_+^n)} \\
&\le c(n,p)\, c_2(n)\, m^{(n-2)/2} \|f\|_{L^p(\mathbf{R}_+^n)} .
\end{aligned}
$$

Step 3: Finally, the desired inequality (14.6) follows by combining inequalities (14.19), (14.20) and (14.18c):

$$
\begin{aligned}
&\|\widetilde{K} f\|_{L^p(\mathbf{R}_+^n)} \\
&\le c_1(n)\, c(n,p)\, c_2(n)\, M \sum_{m=1}^{\infty} \sum_{k=1}^{d(m)} \frac{1}{m^{2n}} m^{(n-2)/2} \|f\|_{L^p(\mathbf{R}_+^n)} \\
&\le c_1(n)\, c_2(n)\, c_3(n)\, c(n,p) M \left(\sum_{m=1}^{\infty} \frac{1}{m^{2n}} m^{(n-2)/2} m^{n-2} \right) \|f\|_{L^p(\mathbf{R}_+^n)}
\end{aligned}
$$

$$= c_1(n)\, c_2(n)\, c_3(n)\, c(n, p)\, M \left(\sum_{m=1}^{\infty} \frac{1}{m^{n/2+3}} \right) \|f\|_{L^p(\mathbf{R}^n_+)}.$$

The proof of Theorem 14.2 is now complete. □

14.3 L^p Boundedness of Boundary Commutators

The next theorem concerns the L^p boundedness of boundary commutators appearing in boundary representation formula (14.2) (cf. [17, Theorem 3.4]):

Theorem 14.5 *Assume that $k(x, z)$ is a real-valued function defined on $\mathbf{R}^n \times (\mathbf{R}^n \setminus \{0\})$ which satisfies conditions (i) and (ii). If $\varphi \in L^\infty(\mathbf{R}^n)$ and $f \in L^p(\mathbf{R}^n_+)$ for $1 < p < \infty$, we define the boundary commutator $\widetilde{C}[\varphi, K]$ by the formula*

$$\begin{aligned} \widetilde{C}\,[\varphi, K]\, f(x) &:= \int_{\mathbf{R}^n_+} k\,(x, T(x) - y)\, \{\varphi(x) - \varphi(y)\}\, f(y)\, dy \\ &= \varphi(x)\, K f(x) - K(\varphi f)(x). \end{aligned}$$

Then there exists a positive constant $c_5 = c_5(n, p, \lambda, M)$ such that

$$\left\| \widetilde{C}\,[\varphi, K]\, f \right\|_{L^p(\mathbf{R}^n_+)} \leq c_5\, \|\varphi\|_*\, \|f\|_{L^p(\mathbf{R}^n_+)} \quad \text{for all } f \in L^p(\mathbf{R}^n_+). \tag{14.21}$$

The proof of Theorem 14.5 will be given in Sect. 14.3.2, due to its length.

Before the proof of Theorem 14.5, we need one more result concerning constant kernel boundary commutators. However, for lack of regularity of $T(x)$, we cannot apply Theorem 11.5. This difficulty may be overcome by taking into consideration the *adjoint operators* $\widetilde{K}^*$ and $\widetilde{C}^*[\varphi, K]$, just as in Bramanti–Cerutti [17].

More precisely, we can prove the following $L^{p'}$ boundedness theorem (cf. [17, Theorem 3.5]):

Theorem 14.6 *Let $k(x)$ be a Calderón–Zygmund kernel on $\mathbf{R}^n$. Then we let*

$$\widetilde{K}^* g(x) := \int_{\mathbf{R}^n_+} k\,(T(y) - x)\, g(y)\, dy.$$

If $\varphi \in L^\infty(\mathbf{R}^n)$ and $g \in L^{p'}(\mathbf{R}^n_+)$ for $p' = p/(p-1)$, we define the boundary commutator $\widetilde{C}^*[\varphi, K]$ by the formula

$$\begin{aligned} \widetilde{C}^*\,[\varphi, K]\, g(x) &:= \int_{\mathbf{R}^n_+} k(T(y) - x)\, \{\varphi(x) - \varphi(y)\}\, g(y)\, dy \\ &= \varphi(x)\, \widetilde{K}^* g(x) - \widetilde{K}^*(\varphi g)(x). \end{aligned}$$

Then there exists a positive constant $c_6 = c_6(n, p', \lambda, M)$ *such that*

$$\left\| \widetilde{C}^* [\varphi, K] g \right\|_{L^{p'}(\mathbf{R}^n_+)} \leq c_6 \, \|\varphi\|_* \, \|g\|_{L^{p'}(\mathbf{R}^n_+)} \quad \textit{for all } g \in L^{p'}(\mathbf{R}^n_+). \tag{14.22}$$

Proof The proof is divided into three steps.

Step 1: First, by Lemma 14.3 it follows that

$$|k(T(y) - x)| \leq \frac{C}{|T(y) - x|^n} \leq \frac{C'}{|x - \widetilde{y}|^n} \quad \text{for } x, y \in \mathbf{R}^n_+.$$

However, we have the formula

$$|x - \widetilde{y}|^2 = |x' - y'|^2 + (x_n + y_n)^2 = |\widetilde{x} - y|^2.$$

Hence it follows that

$$|k(T(y) - x)| \leq \frac{C'}{|\widetilde{x} - y|^n} \quad \text{for } x, y \in \mathbf{R}^n_+.$$

Therefore, by applying Theorem 14.4 to the operator $\widetilde{K}^*$ we obtain that $\widetilde{K}^*$ is bounded on $L^{p'}(\mathbf{R}^n)$; more precisely, there exists a constant $c^*(n, p') > 0$ such that

$$\left\| \widetilde{K}^* g \right\|_{L^{p'}(\mathbf{R}^n_+)} \leq c^*(n, p') \, \|g\|_{L^{p'}(\mathbf{R}^n_+)} \quad \text{for all } g \in L^{p'}(\mathbf{R}^n_+). \tag{14.23}$$

Step 2: Now let Q be a cube with sides parallel to the coordinate axes contained in $\mathbf{R}^n_+$. If $f \in L^1_{\mathrm{loc}}(\mathbf{R}^n)$, then we define the Hardy–Littlewood maximal function (see Sect. 4.4)

$$M_+ f(x) := \sup_{Q \ni x} \frac{1}{|Q|} \int_Q |f(y)| \, dy,$$

and the sharp function (see Sect. 4.6)

$$f_+^\sharp(x) := \sup_{Q \ni x} \frac{1}{|Q|} \int_Q |f(y) - f_Q| \, dy,$$

where the supremum is taken over all cubes containing x and f_Q is the integral average of f over Q

$$f_Q := \frac{1}{|Q|} \int_Q f(z) \, dz.$$

Here it should be emphasized that we have, for all $f \in L^p(\mathbf{R}^n_+)$ with $1 < p < \infty$ (Theorem 4.10 and Remark 4.12),

$$\|M_+ f\|_{L^p(\mathbf{R}^n_+)} \leq c \, \|f\|_{L^p(\mathbf{R}^n_+)},$$

and that we have, for all $f \in L^p(\mathbf{R}^n_+)$ with $1 < p < \infty$ (Corollary 4.21 and Remark 4.22),

$$\|f\|_{L^p(\mathbf{R}^n_+)} \leq c \left\| f^\sharp_+ \right\|_{L^p(\mathbf{R}^n_+)}.$$

Now we consider the operator

$$Sg(x) := \widetilde{C}^* \left[\varphi, \widetilde{K}^*\right] g(x) = \int_{\mathbf{R}^n_+} k(T(y) - x) \left\{\varphi(x) - \varphi(y)\right\} g(y)\, dy,$$

and, for each $r \in (1, p')$, we shall prove a pointwise estimate

$$\begin{aligned} &(Sg)^\sharp_+(x) \qquad\qquad (14.24)\\ &\leq c(n,r)\, \|\varphi\|_* \left(\left(M_+(|\widetilde{K}^* g|^r)(x)\right)^{1/r} + \left(M_+(|g|^r)(x)\right)^{1/r} \right) \quad \text{for all } x \in \mathbf{R}^n_+, \end{aligned}$$

with a constant $c(n, r) > 0$.

Step 3: Assuming estimate (14.24) for the moment, we shall prove Theorem 14.6. The proof of estimate (14.24) will be given in the next Sect. 14.3.1, due to its length.

By applying Corollary 4.21 with $p := p'$ and Remark 4.22 and the pointwise estimate (14.24), we obtain that

$$\begin{aligned} &\|Sg\|_{L^{p'}(\mathbf{R}^n_+)} \qquad\qquad (14.25)\\ &\leq c_{p'} \left\| (Sg)^\sharp_+ \right\|_{L^{p'}(\mathbf{R}^n_+)} \\ &\leq c_{p'}\, c(n,r)\, \|\varphi\|_* \left\{ \left(\int_{\mathbf{R}^n_+} \left(M_+(|\widetilde{K}^* g|^r)(x)\right)^{p'/r} dx \right)^{1/p'} \right. \\ &\qquad \left. + \left(\int_{\mathbf{R}^n_+} \left(M_+(|g|^r)(x)\right)^{p'/r} dx \right)^{1/p'} \right\}. \end{aligned}$$

Step 3-1: However, it should be noticed that

$$\left(\int_{\mathbf{R}^n_+} \left(M_+(|g|^r)(x)\right)^{p'/r} dx \right)^{1/p'} = \left(\left\| M_+(|g|^r) \right\|_{L^{p'/r}(\mathbf{R}^n)} \right)^{1/r},$$

so that, by Theorem 4.10 with $p := p'/r$ and Remark 4.12,

$$\left\| M_+(|g|^r) \right\|_{L^{p'/r}(\mathbf{R}^n_+)} \leq C(p', r) \left\| |g|^r \right\|_{L^{p'/r}(\mathbf{R}^n_+)} = C(p', r)\, \|g\|^r_{L^{p'}(\mathbf{R}^n_+)}.$$

Hence we have the estimate

$$\left(\int_{\mathbf{R}^n_+} (M_+(|g|^r)(x))^{p'/r}\, dx\right)^{1/p'} \leq C(p',r)^{1/r}\, \|g\|_{L^{p'}(\mathbf{R}^n_+)}. \tag{14.26}$$

Step 3-2: Similarly, we have the formula

$$\left(\int_{\mathbf{R}^n_+} (M_+(|\widetilde{K}^* g|^r)(x))^{p'/r}\, dx\right)^{1/p'} = \left(\left\| M_+(|\widetilde{K}^* g|^r)\right\|_{L^{p'/r}(\mathbf{R}^n_+)}\right)^{1/r},$$

and, by Theorem 4.10 with $p := p'/r$ and Remark 4.12,

$$\begin{aligned}
&\left\| M_+(|\widetilde{K}^* g|^r)\right\|_{L^{p'/r}(\mathbf{R}^n_+)} \\
&\leq C(p',r) \left\| |\widetilde{K}^* g|^r \right\|_{L^{p/r}(\mathbf{R}^n_+)} \\
&= C(p',r) \left\| \widetilde{K}^* g\right\|^r_{L^{p'}(\mathbf{R}^n_+)}.
\end{aligned}$$

Hence it follows from an application of Theorem 14.4 that

$$\begin{aligned}
&\int_{\mathbf{R}^n_+} \left(M_+(|\widetilde{K}^* g|^r)(x))^{p'/r} dx\right)^{1/p'} \qquad (14.27)\\
&\leq C(p',r)^{1/r} \left\| \widetilde{K}^* g\right\|_{L^{p'}(\mathbf{R}^n_+)} \\
&\leq c(n,p') C(p',r)^{1/r} c^*(n,p')\, \|g\|_{L^{p'}(\mathbf{R}^n_+)}.
\end{aligned}$$

Therefore, the desired estimate (14.22) follows by combining estimates (14.25), (14.26) and (14.27), with

$$c_6(n,p') = c_{p'}\, c(n,r) \left(c(n,p')c^*(n,p') + 1\right) C(p',r)^{1/r} \quad \text{for } 1 < r < p'.$$

The proof of Theorem 14.6 is now complete, apart from the proof of estimate (14.24). □

14.3.1 End of Proof of Theorem 14.6

This subsection is devoted to the proof of estimate (14.24). If Q is a cube contained in $\mathbf{R}^n_+$, then we denote by δ_Q its side length and by x_Q its center, respectively. For each $j \in \mathbf{N}$, we denote by $2^j Q$ the cube centered at x_Q with side length $2^j \delta_Q$. For any cube Q containing x, we write $Sg(x) := \widetilde{C}^* g(x)$ in the form

$$
\begin{aligned}
Sg(x) &= \int_{\mathbf{R}^n_+} k(T(y)-x)\left(\varphi(x)-\varphi_Q\right) g(y)\,dy \\
&\quad + \int_{\mathbf{R}^n_+} k(T(y)-x)\left(\varphi_Q-\varphi(y)\right) g(y)\,dy \\
&= \left(\varphi(x)-\varphi_Q\right)\int_{\mathbf{R}^n_+} k(T(y)-x) g(y)\,dy \\
&\quad + \int_{\mathbf{R}^n_+\cap(2Q)} k(T(y)-x)\left(\varphi_Q-\varphi(y)\right) g(y)\,dy \\
&\quad + \int_{\mathbf{R}^n_+\setminus(2Q)} k(T(y)-x)\left(\varphi_Q-\varphi(y)\right) g(y)\,dy \\
&:= A(x)+B(x)+C(x).
\end{aligned}
$$

The proof of estimate (14.24) is divided into four steps.

Step I: The estimate of the term $A(x)$. We prove that, for $1 < r < p' = p/(p-1)$,

$$
\begin{aligned}
&\frac{1}{|Q|}\int_Q |A(x)-A_Q|\,dx \\
&\le c_1(n,r)\,\|\varphi\|_*\left(M_+\left(\left|\widetilde{K}^* g\right|^r\right)(y)\right)^{1/r} \quad \text{for all } y\in Q.
\end{aligned}
\tag{14.28}
$$

First, we have the inequality

$$
\begin{aligned}
&\frac{1}{|Q|}\int_Q \left|A(x)-A_Q\right|\,dx \\
&\le \frac{1}{|Q|}\int_Q |A(x)|\,dx + \left|A_Q\right| \\
&= \frac{1}{|Q|}\int_Q |A(x)|\,dx + \frac{1}{|Q|}\left|\int_Q A(y)\,dy\right| \\
&\le \frac{2}{|Q|}\int_Q |A(x)|\,dx.
\end{aligned}
\tag{14.29}
$$

If r is a number such that $1 < r < p' = p/(p-1)$, then, by Hölder's inequality (Theorem 3.24) it follows that, for $r' = r/(r-1)$,

$$
\begin{aligned}
&\frac{2}{|Q|}\int_Q |A(x)|\,dx \\
&= \frac{2}{|Q|}\int_Q \left|\left(\varphi(x)-\varphi_Q\right)\int_{\mathbf{R}^n_+} k(T(y)-x) g(y)\,dy\right|\,dx
\end{aligned}
\tag{14.30}
$$

$$
\begin{aligned}
&\leq \frac{2}{|Q|} \int_Q \left|\varphi(x) - \varphi_Q\right| \left|\widetilde{K}^* g(x)\right| dx \\
&\leq \frac{2}{|Q|} \left(\int_Q \left|\varphi(x) - \varphi_Q\right|^{r'} dx \right)^{1/r'} \left(\int_Q \left|\widetilde{K}^* g(x)\right|^r dx \right)^{1/r}. \\
&= 2 \left(\frac{1}{|Q|} \int_Q \left|\varphi(x) - \varphi_Q\right|^{r'} dx \right)^{1/r'} \left(\frac{1}{|Q|} \int_Q \left|\widetilde{K}^* g(x)\right|^r dx \right)^{1/r}.
\end{aligned}
$$

However, we have, by Theorem 4.14 with $p := r'$ and Remark 4.15,

$$
\frac{1}{|Q|} \int_Q \left|\varphi(y) - \varphi_Q\right|^{r'} dy \leq c_1(r') \, \|\varphi\|_*^{r'}. \tag{14.31}
$$

and, by the definition of the maximal functions,

$$
\begin{aligned}
&\left(\frac{1}{|Q|} \int_Q \left|\widetilde{K}^* g(x)\right|^r dx \right)^{1/r} \\
&\leq \left(M_+ \left(\left|\widetilde{K}^* g\right|^r \right)(y) \right)^{1/r} \quad \text{for all } y \in Q.
\end{aligned} \tag{14.32}
$$

Therefore, the desired estimate (14.28) follows by combining estimates (14.29), (14.30), (14.31) and (14.32) with

$$
c_1(n, r) = 2 \, c_1(r')^{1/r'}.
$$

Step II: The estimate of the term $B(x)$. Secondly, we prove that, for $1 < r < p' = p/(p-1)$,

$$
\begin{aligned}
&\frac{1}{|Q|} \int_Q \left|B(x) - B_Q\right| dx \\
&\leq c_2(n, q, r) \, \|\varphi\|_* \left(M_+ \left(|g|^r \right)(y) \right)^{1/r} \quad \text{for all } y \in Q.
\end{aligned} \tag{14.33}
$$

Here

$$
1 < q < r < \frac{p}{p-1}.
$$

First, we remark that

$$
\begin{aligned}
B(x) &= \int_{\mathbf{R}^n_+ \cap (2Q)} k(T(y) - x) \left(\varphi_Q - \varphi(y) \right) g(y) \, dy \\
&= \int_{\mathbf{R}^n_+} k(T(y) - x) \left(\varphi_Q - \varphi(y) \right) \chi_{2Q}(y) \, g(y) \, dy \\
&= -\widetilde{K}^* \left(\left(\varphi - \varphi_Q \right) \chi_{2Q} \, g \right)(x),
\end{aligned}
$$

where χ_{2Q} is the characteristic function of $2Q$:

$$\chi_{2Q}(x) = \begin{cases} 1 & \text{if } x \in 2Q, \\ 0 & \text{if } x \in \mathbf{R}^n \setminus (2Q). \end{cases}$$

Then we have the inequality

$$\begin{aligned} &\frac{1}{|Q|}\int_Q \left|B(x) - B_Q\right| dx \leq \frac{1}{|Q|}\int_Q |B(x)|\, dx + \left|B_Q\right| \qquad (14.34) \\ &= \frac{1}{|Q|}\int_Q |B(x)|\, dx + \frac{1}{|Q|}\left|\int_Q B(y)\, dy\right| \\ &\leq \frac{2}{|Q|}\int_Q |B(x)|\, dx \\ &= \frac{2}{|Q|}\int_Q \left|\widetilde{K}^*\left(\left(\varphi - \varphi_Q\right)\chi_{2Q}\, g\right)(x)\right| dx. \end{aligned}$$

If q is a number such that $1 < q < r$, then, by Hölder's inequality (Theorem 3.24) it follows that, for $q' = q/(q-1)$,

$$\begin{aligned} &\frac{2}{|Q|}\int_Q \left|\widetilde{K}^*\left(\left(\varphi - \varphi_Q\right)\chi_{2Q}\, g\right)(x)\right| dx \qquad (14.35) \\ &\leq 2\frac{1}{|Q|}\left(\int_Q \left|\widetilde{K}^*\left(\left(\varphi - \varphi_Q\right)\chi_{2Q}\, g\right)\right|^q dx\right)^{1/q}\left(\int_Q dx\right)^{1/q'} \\ &= \frac{2}{|Q|^{1/q}}\left(\int_Q \left|\widetilde{K}^*\left(\left(\varphi - \varphi_Q\right)\chi_{2Q}\, g\right)\right|^q dx\right)^{1/q}. \end{aligned}$$

However, we have, by inequality (14.23) with $p' := q$ and then by Hölder's inequality for the exponent r/q,

$$\begin{aligned} &\left(\int_Q \left|\widetilde{K}^*\left(\left(\varphi - \varphi_Q\right)\chi_{2Q}\, g\right)\right|^q dx\right)^{1/q} \qquad (14.36) \\ &\leq \left(\int_{\mathbf{R}^n_+} \left|\widetilde{K}^*\left(\left(\varphi - \varphi_Q\right)\chi_{2Q}\, g\right)\right|^q dx\right)^{1/q} \\ &\leq c^*(n,q)\left\|(\varphi - \varphi_Q)\chi_{2Q}\, g\right\|_{L^q(\mathbf{R}^n_+)} \\ &= c^*(n,q)\left(\int_{2Q} \left|\varphi(x) - \varphi_Q\right|^q \cdot |g(x)|^q\, dx\right)^{1/q} \\ &\leq c^*(n,q)\left(\int_{2Q} |g(x)|^r\, dx\right)^{1/r}\left(\int_{2Q} \left|\varphi(x) - \varphi_Q\right|^{qr/(r-q)}\, dx\right)^{(r-q)/qr}. \end{aligned}$$

Therefore, by combining inequalities (14.34), (14.35) and (14.36) we obtain that

$$\begin{aligned}
&\frac{1}{|Q|}\int_Q \left|B(x)-B_Q\right|\,dx \qquad (14.37)\\
&\leq \frac{2}{|Q|}\int_Q \left|\widetilde{K}^*\left(\left(\varphi-\varphi_Q\right)\chi_{2Q}\,g\right)(x)\right|\,dx\\
&\leq \frac{2}{|Q|^{1/q}}\left(\int_Q \left|\widetilde{K}^*\left(\left(\varphi-\varphi_Q\right)\chi_{2Q}\,g\right)\right|^q\,dx\right)^{1/q}\\
&\leq \frac{2\,c^*(n,q)}{|Q|^{1/q}}\left(\int_{2Q}|g(x)|^r\,dx\right)^{1/r}\left(\int_{2Q}\left|\varphi(x)-\varphi_Q\right|^{qr/(r-q)}\,dx\right)^{(r-q)/qr}.
\end{aligned}$$

Now we recall the following inequality (see Claim 10.4): There exists a constant $c(n,q,r)>0$ such that we have, for $1<q<r$,

$$\int_{2Q}|\varphi(x)-\varphi_Q|^{rq/(r-q)}\,dx \leq c(n,q,r)\,\|\varphi\|_*^{rq/(r-q)}\,|2Q|\,. \tag{14.38}$$

Therefore, it follows from inequalities (14.37) and (14.38) that

$$\begin{aligned}
&\frac{1}{|Q|}\int_Q \left|B(x)-B_Q\right|\,dx\\
&\leq \frac{2\,c^*(n,q)}{|Q|^{1/q}}\left(\int_{2Q}|g(x)|^r\,dx\right)^{1/r}\left(c(n,q,r)\,\|\varphi\|_*^{rq/(r-q)}\,|2Q|\right)^{(r-q)/qr}\\
&= 2^{1+(n/q)}\,c^*(n,q)\,c(n,q,r)^{(r-q)/qr}\,\|\varphi\|_*\left(\frac{1}{|2Q|}\int_{2Q}|g(x)|^r\,dx\right)^{1/r}\\
&\leq 2^{1+(n/q)}\,c^*(n,q)\,c(n,q,r)^{(r-q)/qr}\,\|\varphi\|_*\left(M_+\left(|g|^r\right)(y)\right)^{1/r} \quad \text{for all } y\in Q.
\end{aligned}$$

This proves the desired estimate (14.33) with

$$c_2(n,q,r) := 2^{1+(n/q)}\,c^*(n,q)\,c(n,q,r)^{(r-q)/qr} \quad \text{for } 1<q<r.$$

Step III: The estimate of the term $C(x)$. Thirdly, we prove that, for $1<r<p'=p/(p-1)$,

$$\begin{aligned}
&\frac{1}{|Q|}\int_Q \left|C(x)-C_Q\right|\,dx \qquad (14.39)\\
&\leq c_3(n,r)\,\|\varphi\|_*\left(M_+(|g|^r)(y)\right)^{1/r} \quad \text{for all } y\in Q.
\end{aligned}$$

The proof of estimate (14.39) is divided into four steps.

Step III-1: We begin with the *pointwise Hörmander condition* for adjoint kernels $k\,(x-T(y))$.

Lemma 14.7 *Let $k(x)$ be a Calderón–Zygmund kernel on $\mathbf{R}^n$. If Q is a cube with center x_Q contained in $\mathbf{R}^n_+$, then we have, for all $x \in Q$, $y \in \mathbf{R}^n_+ \setminus (2Q)$ (see Fig. 14.3),*

$$\left|k\left(x - T(y)\right) - k\left(x_Q - T(y)\right)\right| \leq c \frac{|x - x_Q|}{|x_Q - \widetilde{y}|^{n+1}}, \tag{14.40}$$

with a positive constant c. Here recall that the map $T(y)$ is defined by the formula

$$T(y) := T(y; y) = \begin{pmatrix} y_1 - 2y_n \frac{a^{1n}(y)}{a^{nn}(y)} \\ y_2 - 2y_n \frac{a^{2n}(y)}{a^{nn}(y)} \\ \vdots \\ -y_n \end{pmatrix}, \quad y = \begin{pmatrix} y_1 \\ y_2 \\ \vdots \\ y_n \end{pmatrix}.$$

Proof The proof of Lemma 14.7 is essentially the same as that of Lemma 10.5 if we use Lemma 14.3.

If Q is a cube with center x_Q contained in $\mathbf{R}^n_+$ and if $y \in \mathbf{R}^n_+ \setminus (2Q)$, we let

$$x = (x_1, x_2, \ldots, x_n) = \left(x', x_n\right) \in Q,$$
$$x_Q = \left(x^1_Q, x^2_Q, \ldots, x^n_Q\right) = \left(x'_Q, x^n_Q\right) \in Q,$$

and

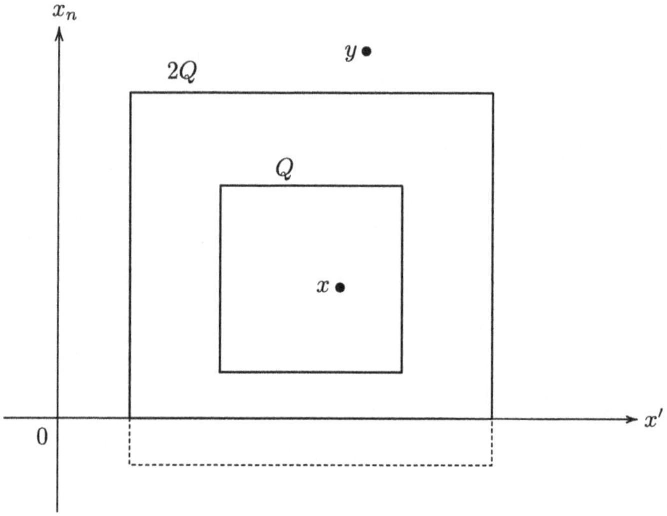

Fig. 14.3 The points $x \in Q$ and $y \in \mathbf{R}^n_+ \setminus (2Q)$

$$T(y) = (T_1(y), T_2(y), \dots, T_n(y)) \in \mathbf{R}^n_-,$$
$$\widetilde{y} = (y_1, y_2, \dots, y_{n-1}, -y_n) \in \mathbf{R}^n_-.$$

Then, by the mean value theorem it follows that we have, for $0 < \theta < 1$,

$$\begin{aligned}
&|k(x - T(y)) - k(x_Q - T(y))| \qquad (14.41)\\
&= |k((x_Q - T(y)) + (x - x_Q)) - k(x_Q - T(y))|\\
&\le \left|\frac{\partial k}{\partial x_1}\left(x_Q^1 - T_1(y) + \theta(x_1 - x_Q^1), x_2 - T_2(y), \dots, x_n - T_n(y)\right)\right| \times \left|x_1 - x_Q^1\right|\\
&+ \left|\frac{\partial k}{\partial x_2}(x_Q^1 - T_1(y), x_Q^2 - T_2(y) + \theta(x_2 - x_Q^2), x_3 - T_3(y), \dots, x_n - T_n(y))\right|\\
&\times \left|x_2 - x_Q^2\right|\\
&+ \ldots + \left|\frac{\partial k}{\partial x_n}\left(x_Q^1 - T_1(y), x_Q^2 - T_2(y), \dots, x_Q^n - T_n(y) + \theta(x_n - x_Q^n)\right)\right|\\
&\times \left|x_n - x_Q^n\right|\\
&\le \left|\frac{\partial k}{\partial x_1}(x_Q^1 - T_1(y) + \theta(x_1 - x_Q^1), x_2 - T_2(y), \dots, x_n - T_n(y))\right|\\
&\times \left|x - x_Q\right|\\
&+ \left|\frac{\partial k}{\partial x_2}(x_Q^1 - T_1(y), x_Q^2 - T_2(y) + \theta(x_2 - x_Q^2), x_3 - T_3(y), \dots, x_n - T_n(y))\right|\\
&\times \left|x - x_Q\right|\\
&+ \ldots + \left|\frac{\partial k}{\partial x_n}(x_Q^1 - T_1(y), x_Q^2 - T_2(y), \dots, x_Q^n - T_n(y) + \theta(x_n - x_Q^n))\right|\\
&\times \left|x - x_Q\right|.
\end{aligned}$$

However, we obtain from Lemma 14.3 that

$$\begin{aligned}
&c_1 \left|x_Q^j - \widetilde{y}_j + \theta(x_j - x_Q^j)\right| \le \left|x_Q^j - T_j(y) + \theta(x_j - x_Q^j)\right|\\
&\le c_2 \left|x_Q^j - \widetilde{y}_j + \theta(x_j - x_Q^j)\right| \quad \text{for } 1 \le j \le n,
\end{aligned}$$

and further that, for all $x \in Q$, $y \in \mathbf{R}^n_+ \setminus (2Q)$,

$$\begin{aligned}
&\left|x_Q^j - \widetilde{y}_j + \theta(x_j - x_Q^j)\right| \ge \left|x_Q^j - \widetilde{y}_j\right| - \left|x_j - x_Q^j\right|\\
&\ge \frac{1}{2}\left|x_Q^j - \widetilde{y}_j\right| \quad \text{for } 1 \le j \le n.
\end{aligned}$$

In particular, we have, for $1 \leq j \leq n$,

$$\begin{aligned}&\left|\left(x_Q^1 - T_1(y), \ldots, x_Q^j - T_j(y) + \theta(x_j - x_Q^j), \ldots, x_n - T_n(y)\right)\right| \\ &\geq \frac{c_1}{2}\left|x_Q - \widetilde{y}\right|.\end{aligned} \tag{14.42}$$

Therefore, by using inequality (10.33) we obtain from inequalities (14.41) and (14.42) that

$$|k(x - T(y)) - k(x_Q - T(y))| \leq \frac{2^{n+1}}{c_1^{n+1}}\, n\, C\, \frac{|x - x_Q|}{|x_Q - \widetilde{y}|^{n+1}} \quad \text{for all } x \in Q,\ y \notin 2Q.$$

This proves the desired inequality (14.40), with

$$c := \frac{2^{n+1}}{c_1^{n+1}}\, n\, C = \frac{2^{n+1}}{c_1^{n+1}}\, n \max_{|z|=1} |\nabla k(z)|.$$

The proof of Lemma 14.7 is complete. □

Step III-2: First, it follows that

$$\begin{aligned}&\frac{1}{|Q|}\int_Q \left|C(x) - C_Q\right| dx \\ &\leq \frac{1}{|Q|}\int_Q \left|C(x) - C(x_Q)\right| dx + \frac{1}{|Q|}\int_Q \left|C(x_Q) - C_Q\right| dx.\end{aligned}$$

However, we have the estimate

$$\begin{aligned}|C(x) - C_Q| &= \frac{1}{|Q|}\left|\int_Q (C(x) - C(x_Q))\, dx\right| \\ &\leq \frac{1}{|Q|}\int_Q |C(x) - C(x_Q)|\, dx.\end{aligned}$$

Hence we obtain that

$$\frac{1}{|Q|}\int_Q \left|C(x) - C_Q\right| dx \leq \frac{2}{|Q|}\int_Q \left|C(x) - C(x_Q)\right| dx. \tag{14.43}$$

Moreover, by Lemma 14.7 we can find a constant $c > 0$ such that

$$\left|k(x - T(y)) - k(x_Q - T(y))\right| \leq c\, \frac{|x - x_Q|}{|x_Q - \widetilde{y}|^{n+1}}, \quad \text{for all } x \in Q \text{ and } y \notin 2Q.$$

Therefore, we have, by Hölder's inequality (Theorem 3.24),

$$
\begin{aligned}
&\left|C(x) - C(x_Q)\right| \qquad (14.44)\\
&= \left| \int_{\mathbf{R}^n_+ \setminus (2Q)} \big(k(x - T(y)) - k(x_Q - T(y))\big) \big(\varphi_Q - \varphi(y)\big)\, g(y)\, dy \right| \\
&\le \int_{\mathbf{R}^n_+ \setminus (2Q)} \left|k(x - T(y)) - k(x_Q - T(y))\right| |\varphi(y) - \varphi_Q| |g(y)|\, dy \\
&\le c \int_{\mathbf{R}^n_+ \setminus (2Q)} \frac{|x - x_Q|}{|x_Q - \widetilde{y}|^{n+1}} |\varphi(x) - \varphi_Q| |g(y)|\, dy \\
&\le c'\, \delta_Q \int_{\mathbf{R}^n_+ \setminus (2Q)} \frac{|g(y)|}{|x_Q - \widetilde{y}|^{(n+1)/r}} \frac{|\varphi(x) - \varphi_Q|}{|x_Q - \widetilde{y}|^{(n+1)/r'}}\, dy \\
&\le c'\, \delta_Q \left(\int_{\mathbf{R}^n_+ \setminus (2Q)} \frac{|g(z)|^r}{|x_Q - \widetilde{z}|^{n+1}}\, dz \right)^{1/r} \left(\int_{\mathbf{R}^n_+ \setminus (2Q)} \frac{|\varphi(z) - \varphi_Q|^{r'}}{|x_Q - \widetilde{z}|^{n+1}}\, dz \right)^{1/r'}.
\end{aligned}
$$

Here δ_Q is the side length of Q.

Step III-3: Now we prove the following two estimates:

$$
I(x) := \int_{\mathbf{R}^n_+ \setminus (2Q)} \frac{|g(z)|^r}{|x_Q - \widetilde{z}|^{n+1}}\, dz \le \frac{C_1}{\delta_Q} M_+(|g|^r)(y) \quad \text{for all } y \in Q. \qquad (14.45)
$$

$$
II(x) := \int_{\mathbf{R}^n_+ \setminus (2Q)} \frac{|\varphi(x) - \varphi_Q|^{r'}}{|x_Q - \widetilde{z}|^{n+1}}\, dz \le \frac{C_2}{\delta_Q} \|\varphi\|_*^{r'}. \qquad (14.46)
$$

Proof (of Estimate (14.45)) Indeed, we have, for all $x_Q^n \ge 0$, $z_n \ge 0$,

$$
\begin{aligned}
|x_Q - \widetilde{z}|^2 &= |x_Q' - z'|^2 + (x_Q^n + z_n)^2 \\
&\ge |x_Q' - z'|^2 + (x_Q^n - z_n)^2 \\
&= |x_Q - z|^2.
\end{aligned}
$$

If Q_j' is the cube in $\mathbf{R}^n_+$ containing $2^j Q \cap \mathbf{R}^n_+$ and having the same measure as $2^j Q$, then we obtain that

$$
\begin{aligned}
&\int_{\mathbf{R}^n_+ \setminus (2Q)} \frac{|g(z)|^r}{|x_Q - \widetilde{z}|^{n+1}}\, dz \\
&\le \int_{\mathbf{R}^n_+ \setminus (2Q)} \frac{|g(z)|^r}{|x_Q - z|^{n+1}}\, dz \\
&= \sum_{j=2}^{\infty} \int_{\mathbf{R}^n_+ \cap (2^j Q \setminus (2^{j-1} Q))} \frac{|g(z)|^r}{|x_Q - z|^{n+1}}\, dz \\
&\le \frac{C}{\delta_Q} \sum_{j=2}^{\infty} \frac{1}{2^j} \left(\frac{1}{|Q_j'|} \int_{Q_j'} |g(z)|^r\, dz \right)
\end{aligned}
$$

$$\leq \frac{C}{\delta_Q} \left(\sum_{j=2}^{\infty} \frac{1}{2^j} \right) M_+(|g|^r)(y) \quad \text{for all } y \in Q.$$

This proves the desired estimate (14.45), with

$$C_1 := C \sum_{j=2}^{\infty} \frac{1}{2^j} = \frac{C}{2}.$$

The proof of Estimate (14.45) is complete. □

Proof (of Estimate (14.46)) Similarly, we have the estimate

$$\begin{aligned} &\int_{\mathbf{R}^n_+ \setminus (2Q)} \frac{|\varphi(z) - \varphi_Q|^{r'}}{|x_Q - \widetilde{z}|^{n+1}}\, dz \qquad (14.47)\\ &\leq \sum_{j=2}^{\infty} \int_{2^j Q \setminus (2^{j-1} Q)} \frac{1}{2^j \delta_Q} \frac{|\varphi(z) - \varphi_Q|^{r'}}{|x_Q - z|^{n+1}}\, dz \\ &\leq \frac{C}{\delta_Q} \sum_{j=2}^{\infty} \left(\frac{1}{|2^j Q|} \int_{2^j Q} |\varphi(z) - \varphi_Q|^{r'}\, dz \right). \end{aligned}$$

However, by Lemma 4.3 with $f := \varphi$ it follows that

$$\begin{aligned} |\varphi(z) - \varphi_Q| &\leq |\varphi(z) - \varphi_{2^j Q}| + |\varphi_{2^j Q} - \varphi_Q| \\ &\leq |\varphi(z) - \varphi_{2^j Q}| + c(n) j \, \|\varphi\|_*, \end{aligned}$$

so that

$$\begin{aligned} &\frac{1}{|2^j Q|} \int_{2^j Q} |\varphi(z) - \varphi_Q|^{r'}\, dz \qquad (14.48)\\ &\leq 2^{r'-1} \left(\frac{1}{|2^j Q|} \int_{2^j Q} |\varphi(z) - \varphi_{2^j Q}|^{r'}\, dz \right) + \frac{1}{|2^j Q|} \int_{2^j Q} (c(n) j \, \|\varphi\|_*)^{r'}\, dz \\ &\leq 2^{r'-1} \left(\frac{1}{|2^j Q|} \int_{2^j Q} |\varphi(z) - \varphi_{2^j Q}|^{r'}\, dz \right) + 2^{r'-1} \, (c(n) j \, \|\varphi\|_*)^{r'}. \end{aligned}$$

However, we have, by Theorem 4.14 with $p := r'$ and Remark 4.15,

$$\frac{1}{|2^j Q|} \int_{2^j Q} |\varphi(z) - \varphi_{2^j Q}|^{r'}\, dz \leq c_1(r') \, \|\varphi\|_*^{r'}. \qquad (14.49)$$

Hence, it follows from estimates (14.48) and (14.49) that

$$\frac{1}{|2^j Q|} \int_{2^j Q} |\varphi(z) - \varphi_Q|^{r'}\, dz \leq 2^{r'-1} \left(c(n)^{r'} j^{r'} + c_1(r') \right) \|\varphi\|_*^{r'}. \qquad (14.50)$$

Therefore, by combining estimates (14.47) and (14.50) we obtain that

$$
\begin{aligned}
&\int_{\mathbf{R}^n_+\setminus(2Q)} \frac{|\varphi(z)-\varphi_Q|^{r'}}{|x_Q-z|^{n+1}}\,dz \\
&= \sum_{j=2}^{\infty} \int_{2^jQ\setminus(2^{j-1}Q)} \frac{|\varphi(z)-\varphi_Q|^{r'}}{|x_Q-z|^{n+1}}\,dz \\
&\leq 2^{r'-1}\,C \sum_{j=2}^{\infty} \frac{1}{2^j\delta_Q}\left(c(n)^{r'}\,j^{r'} + c_1(r')\right)\|\varphi\|_*^{r'} \\
&= 2^{r'-1}\frac{C}{\delta_Q}\left\{c_1(r')\left(\sum_{j=2}^{\infty}\frac{1}{2^j}\right) + c(n)^{r'}\left(\sum_{j=2}^{\infty}\frac{j^{r'}}{2^j}\right)\right\}\|\varphi\|_*^{r'}.
\end{aligned}
$$

This proves the desired estimate (14.46), with

$$
C_2 := 2^{r'-1}\,C\left\{c_1(r')\left(\sum_{j=2}^{\infty}\frac{1}{2^j}\right) + c(n)^{r'}\left(\sum_{j=2}^{\infty}\frac{j^{r'}}{2^j}\right)\right\}.
$$

The proof of estimate (14.46) is complete. □

Step III-4: Therefore, by combining estimates (14.43), (14.44), (14.45) and (14.46) we obtain that

$$
\begin{aligned}
&\frac{1}{|Q|}\int_Q |C(x)-C_Q|\,dx \\
&\leq \frac{2}{|Q|}\int_Q |C(x)-C(x_Q)|\,dx \\
&\leq \frac{2}{|Q|}\,c'\,C_1^{1/r}\,C_2^{1/r'}\,\|\varphi\|_*\left(M_+(|g|^r)(y)\right)^{1/r}\int_Q dx \\
&= 2\,c'\,C_1^{1/r}\,C_2^{1/r'}\,\|\varphi\|_*\left(M_+(|g|^r)(y)\right)^{1/r} \quad \text{for all } y \in Q.
\end{aligned}
$$

This proves the desired estimate (14.39), with

$$
c_3(n,r) := 2\,c'\,C_1^{1/r}\,C_2^{1/r'} \quad \text{for } r' = \frac{r}{r-1}.
$$

Step IV: If we let

$$
c(n,r) := \max\{c_1(n,r), c_2(n,r), c_3(n,r)\},
$$

then, by combining estimates (14.28), (14.33) and (14.39) we obtain that

$$\begin{aligned}
&\frac{1}{|Q|}\int_Q |Sg(x) - (Sg)_Q|\,dx \\
&\le \frac{1}{|Q|}\int_Q |A(x) - A_Q|\,dx + \frac{1}{|Q|}\int_Q |B(x) - B_Q|\,dx \\
&\quad + \frac{1}{|Q|}\int_Q |C(x) - C_Q|\,dx \\
&\le c(n,r)\,\|\varphi\|_* \left(\left(M_+(|\widetilde{K}^* g|^r)(y)\right)^{1/r} + \left(M_+(|g|^r)(y)\right)^{1/r} \right) \quad \text{for all } y \in Q.
\end{aligned}$$

This proves that

$$\begin{aligned}
&(Sg)^\sharp_+(y) \\
&\le c(n,r)\,\|\varphi\|_* \left(\left(M_+(|\widetilde{K}^* g|^r)(y)\right)^{1/r} + \left(M_+(|g|^r)(y)\right)^{1/r} \right) \quad \text{for all } y \in \mathbf{R}^n,
\end{aligned}$$

since Q is arbitrary.

The proof of estimate (14.24) (and hence that of Theorem 14.6) is now complete. □

14.3.2 Proof of Theorem 14.5

This subsection is devoted to the proof of Theorem 14.5. By a *density argument*, it suffices to prove the theorem for all $f \in C_0^\infty(\mathbf{R}^n_+)$. The proof is divided into four steps.

Step 1: First, by passing to adjoint operators we obtain from Theorem 14.6 the following:

Theorem 14.8 *Let $k(x)$ be a Calderón–Zygmund kernel on $\mathbf{R}^n$. Then we let*

$$\widetilde{K} f(x) := \int_{\mathbf{R}^n_+} k\,(T(x) - y)\, f(y)\,dy \quad \textit{for all } f \in L^p(\mathbf{R}^n_+).$$

If $\varphi \in L^\infty(\mathbf{R}^n)$, we define the boundary commutator $\widetilde{C}[\varphi, \widetilde{K}]$ by the formula

$$\begin{aligned}
\widetilde{C}\left[\varphi, \widetilde{K}\right] f(x) &:= \int_{\mathbf{R}^n_+} k\,(T(x) - y)\,\{\varphi(x) - \varphi(y)\}\, f(y)\,dy \\
&= \varphi(x)\,\widetilde{K} f(x) - \widetilde{K}(\varphi f)(x)
\end{aligned}$$

for all $f \in L^p(\mathbf{R}^n_+)$ with $1 < p < \infty$. Then there exists a positive constant $c_7 = c_7(n, p, \lambda, M)$ such that

$$\left\|\widetilde{C}\,[\varphi, K]\, f\right\|_{L^p(\mathbf{R}^n_+)} \le c_7\,\|\varphi\|_*\,\|f\|_{L^p(\mathbf{R}^n_+)} \quad \textit{for all } f \in L^p(\mathbf{R}^n_+). \tag{14.51}$$

Proof First, we prove the L^p boundedness of the operator $\widetilde{K}$. For any $f \in L^p(\mathbf{R}^n_+)$ and any $g \in L^{p'}(\mathbf{R}^n_+)$, we have, by Theorem 14.6 and Fubini's theorem (Theorem 3.20),

$$\begin{aligned}\langle f, \widetilde{K}^* g\rangle &= \int_{\mathbf{R}^n_+} f(y) \left(\int_{\mathbf{R}^n_+} k(T(x) - y) g(x)\, dx \right) dy \\ &= \int_{\mathbf{R}^n_+} \left(\int_{\mathbf{R}^n_+} k(T(x) - y) f(y)\, dy \right) g(x)\, dx \\ &= \langle \widetilde{K} f, g\rangle.\end{aligned}$$

Therefore, by passing to the adjoint operator we obtain from inequality (14.23) that

$$\widetilde{K} f \in L^p(\mathbf{R}^n_+),$$

and further that

$$\begin{aligned}\|\widetilde{K} f\|_{L^p(\mathbf{R}^n_+)} &= \sup_{\|g\|_{p'} \le 1} |\langle \widetilde{K} f, g\rangle| = \sup_{\|g\|_{p'} \le 1} |\langle f, \widetilde{K}^* g\rangle| \\ &\le c^*(n, p')\, \|f\|_{L^p(\mathbf{R}^n_+)} .\end{aligned}$$

The operators $\widetilde{K} : L^p(\mathbf{R}^n_+) \to L^p(\mathbf{R}^n_+)$ and $\widetilde{K}^* : L^{p'}(\mathbf{R}^n_+) \to L^{p'}(\mathbf{R}^n_+)$ can be visualized as follows:

$$\boxed{\begin{array}{ccc} f \in L^p(\mathbf{R}^n_+) & \xrightarrow{\ \widetilde{K}\ } & L^p(\mathbf{R}^n_+) \ni \widetilde{K} f \\ \updownarrow & & \updownarrow \\ \widetilde{K}^* g \in L^{p'}(\mathbf{R}^n_+) & \xleftarrow{\ \widetilde{K}^*\ } & L^{p'}(\mathbf{R}^n_+) \ni g \end{array}}$$

Secondly, we prove inequality (14.51). For any $f \in L^p(\mathbf{R}^n_+)$ and any $g \in L^{p'}(\mathbf{R}^n_+)$, we have, by Theorem 14.6 and Fubini's theorem (Theorem 3.20),

$$\begin{aligned}\langle f, \widetilde{C}^* [\varphi, K]\, g\rangle &= \int_{\mathbf{R}^n_+} f(y) \left(\int_{\mathbf{R}^n_+} k(T(x) - y) \{\varphi(x) - \varphi(y)\}\, g(x)\, dx \right) dy \\ &= \int_{\mathbf{R}^n_+} \left(\int_{\mathbf{R}^n_+} k(T(x) - y) \{\varphi(x) - \varphi(y)\}\, f(y)\, dy \right) g(x)\, dx \\ &= \langle \widetilde{C} [\varphi, K]\, f, g\rangle.\end{aligned}$$

Therefore, by passing to the adjoint operator we obtain from inequality (14.22) that

$$\widetilde{C}\,[\varphi, K]\, f \in L^p(\mathbf{R}^n_+),$$

and further that

$$\begin{aligned}\left\| \widetilde{C}\,[\varphi, K]\, f \right\|_{L^p(\mathbf{R}^n_+)} &= \sup_{\|g\|_{p'} \leq 1} \left| \langle \widetilde{C}\,[\varphi, K]\, f, g \rangle \right| = \sup_{\|g\|_{p'} \leq 1} \left| \langle f, \widetilde{C}^*\,[\varphi, K]\, g \rangle \right| \\ &\leq c_6\, \|\varphi\|_*\, \|f\|_{L^p(\mathbf{R}^n_+)}\,.\end{aligned}$$

The operators

$$\widetilde{C}\,[\varphi, K] : L^p(\mathbf{R}^n_+) \longrightarrow L^p(\mathbf{R}^n_+)$$

and

$$\widetilde{C}^*\,[\varphi, K] : L^{p'}(\mathbf{R}^n_+) \longrightarrow L^{p'}(\mathbf{R}^n_+)$$

can be visualized as follows:

$$\boxed{\begin{array}{ccc} f \in L^p(\mathbf{R}^n_+) & \xrightarrow{\widetilde{C}[\varphi, K]} & L^p(\mathbf{R}^n_+) \ni \widetilde{C}\,[\varphi, K]\, f \\ \updownarrow & & \updownarrow \\ \widetilde{C}^*\,[\varphi, K]\, g \in L^{p'}(\mathbf{R}^n_+) & \xleftarrow{\widetilde{C}^*[\varphi, K]} & L^{p'}(\mathbf{R}^n_+) \ni g \end{array}}$$

The proof of Theorem 14.8 is complete. □

Step 2: Now let $k(x, z)$ be a real-valued function on $\mathbf{R}^n \times (\mathbf{R}^n \setminus \{0\})$ that satisfies conditions (i) and (ii). It should be noticed that the function

$$z \longmapsto |z|^n k(x, z)$$

belongs to $C^\infty(\mathbf{R}^n \setminus \{0\})$ for almost all $x \in \mathbf{R}^n_+$, and it is positively homogeneous of degree zero and satisfies the condition

$$\int_{\Sigma_{n-1}} k(x, z)\, d\sigma_z = 0.$$

For each $m = 1, 2, \ldots$ and $k = 1, 2, \ldots, d(m)$, we let

$$a_{km}(x) := \int_{\Sigma_{n-1}} k(x, z) Y_{km}(z)\, dz,$$

then, by the completeness of the spherical harmonics $\{Y_{km}\}$ in $L^2(\Sigma_{n-1})$ it follows that

$$|T(x) - y|^n k(x, T(x) - y) \tag{14.52}$$

$$= \sum_{m=1}^{\infty} \sum_{k=1}^{d(m)} a_{km}(x) Y_{km}(T(x) - y), \quad x \in \mathbf{R}^n_+, \ y \in \mathbf{R}^n \setminus \{0\}.$$

Moreover, we have, by Theorem 4.41,

$$\|a_{km}\|_{L^\infty(\mathbf{R}^n_+)} \leq \frac{c_1(n)}{m^{2n}} M, \tag{14.18a}$$

$$\|Y_{km}\|_{L^\infty(\Sigma_{n-1})} \leq c_2(n)\, m^{(n-2)/2}, \tag{14.18b}$$

and

$$d(m) \leq c_3(n)\, m^{n-2}. \tag{14.18c}$$

Step 3: By using the expansion (14.52), we obtain that

$$\begin{aligned}
\widetilde{K} f(x) &= \int_{\mathbf{R}^n_+} k(x, T(x) - y) f(y)\, dy \\
&= \sum_{m=1}^{\infty} \sum_{k=1}^{d(m)} a_{km}(x) \int_{\mathbf{R}^n_+} \frac{Y_{km}(T(x) - y)}{|T(x) - y|^n} f(y)\, dy.
\end{aligned}$$

Hence it follows that

$$\begin{aligned}
&\widetilde{C}\,[\varphi, K]\, f(x) \\
&= \int_{\mathbf{R}^n_+} k(x, T(x) - y) \{\varphi(x) - \varphi(y)\}\, f(y)\, dy \\
&= \sum_{m=1}^{\infty} \sum_{k=1}^{d(m)} a_{km}(x) \int_{\mathbf{R}^n_+} \{\varphi(x) - \varphi(y)\} \frac{Y_{km}(T(x) - y)}{|T(x) - y|^n} f(y)\, dy,
\end{aligned}$$

and so

$$\widetilde{C}\,[\varphi, K]\, f(x) = \sum_{m=1}^{\infty} \sum_{k=1}^{d(m)} a_{km}(x) \widetilde{C}_{km}\,[\varphi, K]\, f(x),$$

where

$$\widetilde{C}_{km}\,[\varphi, K]\, f(x) := \int_{\mathbf{R}^n_+} \{\varphi(x) - \varphi(y)\} \frac{Y_{km}(T(x) - y)}{|T(x) - y|^n} f(y)\, dy.$$

Therefore, we have, by inequality (14.18a),

$$\begin{aligned}
&\left\|\widetilde{C}\,[\varphi, K]\,f\right\|_{L^p(\mathbf{R}^n_+)} \qquad (14.53)\\
&= \left\|\sum_{m=1}^{\infty}\sum_{k=1}^{d(m)} a_{km}\widetilde{C}_{km}\,[\varphi, K]\,f\right\|_{L^p(\mathbf{R}^n_+)}\\
&\le \sum_{m=1}^{\infty}\sum_{k=1}^{d(m)} \|a_{km}\|_{L^\infty(\mathbf{R}^n_+)} \left\|\int_{\mathbf{R}^n_+} \widetilde{C}_{km}\,[\varphi, K]\,f\right\|_{L^p(\mathbf{R}^n_+)}\\
&\le c_1(n) M \sum_{m=1}^{\infty}\sum_{k=1}^{d(m)} \frac{1}{m^{2n}} \left\|\int_{\mathbf{R}^n_+} \widetilde{C}_{km}\,[\varphi, K]\,f\right\|_{L^p(\mathbf{R}^n_+)}.
\end{aligned}$$

However, by inequality (14.18b) it follows from an application of Theorem 14.8 that

$$\left\|\widetilde{C}_{km}\,[\varphi, K]\,f\right\|_{L^p(\mathbf{R}^n_+)} \le c_7 c_2(n)\, m^{(n-2)/2}\, \|\varphi\|_*\, \|f\|_{L^p(\mathbf{R}^n_+)}\,. \qquad (14.54)$$

Step 4: Finally, the desired inequality (14.21) follows by combining inequalities (14.53), (14.54) and (14.18c):

$$\begin{aligned}
&\left\|\widetilde{C}\,[\varphi, K]\,f\right\|_{L^p(\mathbf{R}^n_+)}\\
&\le c_1(n)\, M \sum_{m=1}^{\infty}\sum_{k=1}^{d(m)} \frac{1}{m^{2n}} c_2(n)\, m^{(n-2)/2}\, \|\varphi\|_*\, \|f\|_{L^p(\mathbf{R}^n_+)}\\
&\le c_1(n)\, c_2(n)\, c_3(n)\, c_7 M \left(\sum_{m=1}^{\infty} \frac{1}{m^{2n}} m^{(n-2)/2} m^{n-2}\right) \|\varphi\|_*\, \|f\|_{L^p(\mathbf{R}^n_+)}\\
&= c_1(n)\, c_2(n)\, c_3(n)\, c_7 M \left(\sum_{m=1}^{\infty} \frac{1}{m^{n/2+3}}\right) \|\varphi\|_*\, \|f\|_{L^p(\mathbf{R}^n_+)}.
\end{aligned}$$

Now the proof of Theorem 14.5 is complete. □

The next result is a local version of Theorem 14.5 (see [35, Theorem 2.7]):

Corollary 14.9 *Let* $\varphi \in \mathrm{VMO} \cap L^\infty(\mathbf{R}^n)$ *and* η *the* VMO *modulus of* φ. *Then, for each* $\varepsilon > 0$, *there exists a positive constant* $\rho_0 = \rho_0(\varepsilon, \eta)$ *such that, for any* $r \in (0, \rho_0)$, *we have the inequality*

$$\left\|\widetilde{C}\,[\varphi, K]\,f\right\|_{L^p(B_r^+)} \le c_5\, \varepsilon\, \|f\|_{L^p(B_r^+)} \quad \textit{for all } f \in L^p(B_r^+), \qquad (14.55)$$

where (see Fig. 14.4)

$$B_r^+ := B_r \cap \mathbf{R}^n_+.$$

Proof The proof of Corollary 14.9 is divided into three steps.

Fig. 14.4 The semi-ball B_r^+ in the half-space $\mathbf{R}_+^n$

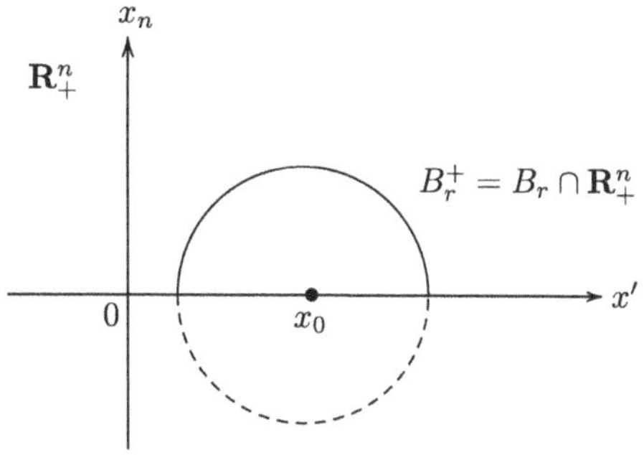

Step (1): By using Theorem 4.5, for each $\varepsilon > 0$ we can find a bounded, uniformly continuous function $a(x)$ on $\mathbf{R}^n$ such that

$$\|a - \varphi\|_* < \frac{\varepsilon}{2}. \tag{14.56}$$

Let $\omega_a(r)$ be the modulus of uniform continuity of $a(x)$ defined by the formula

$$\omega_a(r) = \sup_{|x-y|\leq r} |a(x) - a(y)|,$$

and choose a constant $\rho_0 = \rho_0(\varepsilon, \eta) > 0$ such that

$$\omega_a(\rho_0) < \frac{\varepsilon}{2}. \tag{14.57}$$

If $B_r = B_r(x_0)$ is a ball of radius r about x_0, we let

$$b(x) := \begin{cases} a(x) & \text{if } x \in B_r(x_0), \\ a\left(x_0 + r\frac{x-x_0}{|x-x_0|}\right) & \text{if } x \in \mathbf{R}^n \setminus B_r(x_0). \end{cases}$$

It should be noticed that the function $b(x)$ is uniformly continuous on $\mathbf{R}^n$ and that the oscillation of $b(x)$ in $\mathbf{R}^n$ equals the oscillation of $a(x)$ in B_r.

Step (2): Now we remark that

$$\begin{aligned} \widetilde{C}[\varphi, K] f &= \varphi(Kf) - K(\varphi f) \\ &= (\varphi - a) Kf - K((\varphi - a) f) + a(Kf) - K(af) \\ &= \widetilde{C}[\varphi - a, K] f + \widetilde{C}[a, K] f, \end{aligned}$$

so that

$$\left\|\widetilde{C}[\varphi, K] f\right\|_{L^p(B_r^+)} \leq \left\|\widetilde{C}[\varphi - a, K] f\right\|_{L^p(B_r^+)} + \left\|\widetilde{C}[a, K] f\right\|_{L^p(B_r^+)}.$$

However, we have, by Theorem 14.5,

$$\left\|\widetilde{C}\,[\varphi - a, K]\, f\right\|_{L^p(B_r^+)} \leq c_5\, \|\varphi - a\|_*\, \|f\|_{L^p(B_r^+)}, \tag{14.58}$$

and also

$$\begin{aligned} \left\|\widetilde{C}\,[a, K]\, f\right\|_{L^p(B_r^+)} &= \left\|\widetilde{C}\,[b, K]\, f\right\|_{L^p(B_r^+)} \\ &\leq c_5\, \|b\|_*\, \|f\|_{L^p(B_r^+)}. \end{aligned} \tag{14.59}$$

Moreover, it is easy to see that

$$\|b\|_* \leq \omega_b(r) = \omega_a(r). \tag{14.60}$$

Hence, by combining inequalities (14.59) and (14.60) we obtain that

$$\left\|\widetilde{C}\,[a, K]\, f\right\|_{L^p(B_r^+)} \leq c_5\, \omega_a(r)\, \|f\|_{L^p(B_r^+)}. \tag{14.61}$$

Step (3): Therefore, it follows from inequalities (14.58) and (14.61) that we have, for all $0 < r < \rho_0$,

$$\begin{aligned} &\left\|\widetilde{C}\,[\varphi, K]\, f\right\|_{L^p(B_r^+)} \\ &\leq \left\|\widetilde{C}\,[\varphi - a, K]\, f\right\|_{L^p(B_r^+)} + \left\|\widetilde{C}\,[a, K]\, f\right\|_{L^p(B_r^+)} \\ &\leq c_5\, \|\varphi - a\|_*\, \|f\|_{L^p(B_r^+)} + c_5\, \omega_a(r)\, \|f\|_{L^p(B_r^+)} \\ &\leq c_5\, (\|\varphi - a\|_* + \omega_a(\rho_0))\, \|f\|_{L^p(B_r^+)}. \end{aligned}$$

By inequalities (14.56) and (14.57), this inequality proves the desired inequality (14.55).

The proof of Corollary 14.9 is complete. □

Remark 14.10 Roughly speaking, inequality (14.55) may be expressed as follows:

$$\left\|\widetilde{C}\,[\varphi, K]\, f\right\|_{L^p(B_r^+)} \leq c_5\, \eta(r)\, \|f\|_{L^p(B_r^+)} \quad \text{for } f \in L^p\left(B_r^+\right). \tag{14.55'}$$

Here

$$\eta(r) = \sup_{\rho \leq r} \frac{1}{|B|} \int_B |\varphi(x) - \varphi_B|\, dx$$

is the VMO modulus of φ (see Sect. 4.2).

14.3.3 L^p Boundedness of Integral Operators with Positive Kernel

Just as in the proof of Theorem 14.5, we can prove an L^p boundedness of integral operators which have *positive kernels* depending on the difference $|\varphi(x) - \varphi(y)|$. In fact, we can obtain the following theorem ([16, Theorem 0.1]):

Theorem 14.11 *(Bramanti) If $\varphi \in \mathrm{BMO}$ and $f \in L^p(\mathbf{R}^n_+)$ for $1 < p < \infty$, we define the commutator $\widetilde{C}_\varphi$ by the formula*

$$\widetilde{C}_\varphi f(x) := \int_{\mathbf{R}^n_+} \frac{|\varphi(x) - \varphi(y)|}{|T(x) - y|^n} f(y)\, dy$$
$$= \varphi(x)$$

Then there exists a positive constant $c = c(n, p)$ such that

$$\left\| \widetilde{C}_\varphi f \right\|_{L^p(\mathbf{R}^n_+)} \leq c \, \|\varphi\|_* \, \|f\|_{L^p(\mathbf{R}^n_+)} \quad \text{for all } f \in L^p(\mathbf{R}^n_+).$$

Remark 14.12 It should be emphasized that the results in Sect. 14.3 remain valid if the function $|T(x) - y|$ is replaced by an equivalent function.

14.4 Local Boundary Estimates

First, we recall (see Sect. 14.1) that

$$C_{\gamma_0} = \left\{ u|_{B_r^+} : u \in C_0^\infty(B_r), \ u = 0 \text{ on } B_r \cap \{x_n = 0\} \right\},$$

and that $W^{2,p}_{\gamma_0}(B_r^+)$ is the closure of C_{γ_0} in the Sobolev space $W^{2,p}(B_r^+)$. In particular, it should be emphasized that $u \in W^{2,p}_0(B_r)$ belongs to $W^{2,p}_{\gamma_0}(B_r^+)$ if and only if it vanishes on $B_r \cap \{x_n = 0\}$ (see [2, Theorem 5.37]).

Now we can prove local boundary *a priori* estimates for the second derivatives of solutions of the homogeneous Dirichlet problem, similar to Lemma 13.2 (see [35, Theorem 3.3]):

Lemma 14.13 *Let $1 < q < p < \infty$, and let*

$$M := \max_{1 \leq i, j \leq n} \max_{|\alpha| \leq 2n} \left\| \frac{\partial^\alpha}{\partial z^\alpha} \Gamma_{ij}(\cdot, \cdot) \right\|_{L^\infty(B_\sigma^+ \times \Sigma_{n-1})}.$$

Then there exists a constant $\rho_0 = \rho_0(n, p, q, M, \lambda, \eta)$, with $0 < \rho_0 < \sigma$, such that if $u \in W^{2,q}_{\gamma_0}(B_r^+)$ with $\mathcal{L}u \in L^p(B_r^+)$ for $0 < r < \rho_0$, it follows that $u \in W^{2,p}(B_r^+)$. Moreover, we have the boundary a priori estimate

$$\left\| \frac{\partial^2 u}{\partial x_i \partial x_j} \right\|_{L^p(B_r^+)} \leq C \, \|\mathcal{L}u\|_{L^p(B_r^+)} \quad \text{for } 1 \leq i, j \leq n, \tag{14.62}$$

with a positive constant $C = C(n, p, M, \lambda, \eta)$. *Here (see Fig. 14.5)*

$$B_r^+ = B_r \cap \mathbf{R}_+^n \quad \text{for } 0 < r < \rho_0.$$

In other words, the second order, elliptic differential operator $\mathcal{L}$ *controls all second order partial derivatives in the* L^p*-norm for* $1 < p < \infty$ *in the semi-ball* B_r^+ *(cf. Lemma 13.2).*

Proof Let $\nu \in [q, p]$. For $1 \leq i, j, h, k \leq n$, we introduce linear operators S_{ijhk} and $\widetilde{S}_{ijhk}$ on $L^\nu(B_r^+)$, $0 < r \leq \sigma$, as follows:

(1) If $1 \leq i, j, h, k \leq n$, we let

$$S_{ijhk}(f) := \text{v.p.} \int_{B_r^+} \Gamma_{ij}(x, x - y) \left\{ a^{hk}(x) - a^{hk}(y) \right\} f(y) \, dy.$$

(2) If $1 \leq i, j \leq n - 1$ and $1 \leq h, k \leq n$, we let

$$\widetilde{S}_{ijhk}(f) := \int_{B_r^+} \Gamma_{ij}(x, T(x) - y) \left\{ a^{hk}(x) - a^{hk}(y) \right\} f(y) \, dy.$$

(3) If $1 \leq i \leq n - 1$, $j = n$ and $1 \leq h, k \leq n$, we let

$$\widetilde{S}_{inhk}(f) := \int_{B_r^+} \left(\sum_{j=1}^{n} \Gamma_{ij}(x, T(x) - y) A_j(x) \right) \left\{ a^{hk}(x) - a^{hk}(y) \right\} f(y) \, dy.$$

(4) If $i = n$, $j = n$ and $1 \leq h, k \leq n$, we let

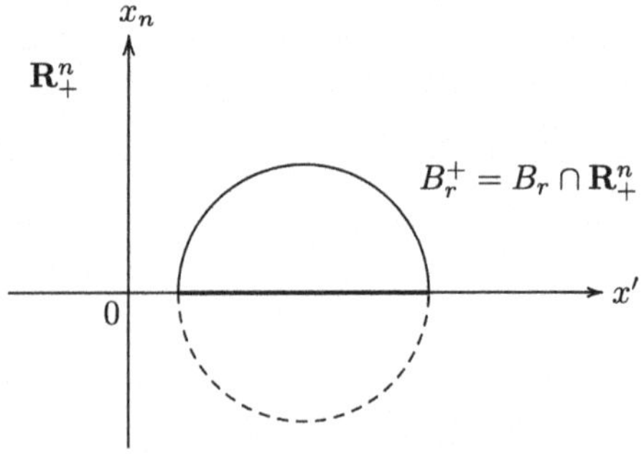

Fig. 14.5 The semi-ball B_r^+ in the half-space $\mathbf{R}_+^n$

$$\widetilde{S}_{nnhk}(f) := \int_{B_r^+} \left(\sum_{i,j=1}^{n} \Gamma_{ij}(x, T(x) - y) A_i(x) A_j(x) \right) \times \left\{ a^{hk}(x) - a^{hk}(y) \right\} f(y)\, dy.$$

Here we recall that the map $T(x)$ is defined by the formula

$$T(x) := T(x; x) = \begin{pmatrix} x_1 - 2x_n \frac{a^{1n}(x)}{a^{nn}(x)} \\ x_2 - 2x_n \frac{a^{2n}(x)}{a^{nn}(x)} \\ \vdots \\ -x_n \end{pmatrix}, \quad x = \begin{pmatrix} x_1 \\ x_2 \\ \vdots \\ x_n \end{pmatrix},$$

and that the vector $A(x)$ is defined by the formula

$$A(x) = \begin{pmatrix} A_1(x) \\ A_2(x) \\ \vdots \\ A_n(x) \end{pmatrix} := \frac{\partial T}{\partial x_n}(x; z)\Big|_{z=x} = \begin{pmatrix} -2\frac{a^{1n}(x)}{a^{nn}(x)} \\ -2\frac{a^{2n}(x)}{a^{nn}(x)} \\ \vdots \\ -1 \end{pmatrix}.$$

By using Corollaries 11.7 and 14.9, we can find a constant $\rho_0 > 0$ such that

$$\sum_{i,j,h,k=1}^{n} \left\| S_{ijhk} + \widetilde{S}_{ijhk} \right\| < \frac{1}{2} \quad \text{for } 0 < \rho < \rho_0. \tag{14.63}$$

Here the norm of the operators $S_{ijhk} + \widetilde{S}_{ijhk}$ is the norm in the space of bounded, linear operators on $L^\nu(B_r^+)$ for $0 < r < \rho_0$ and $q \leq \nu \leq p$.

Let $u \in W^{2,p}_{\gamma_0}(B_r^+)$ such that $\mathcal{L}u \in L^p(B_r^+)$, $0 < r < \rho_0$, and let

$$h_{ij}(x) = \text{v.p.} \int_{B_r^+} \Gamma_{ij}(x, x - y) \mathcal{L}u(y)\, dy + c_{ij}(x)\, \mathcal{L}u(x) - \widetilde{I}_{ij}(x) \quad \text{for } 1 \leq i, j \leq n.$$

Here the terms $\widetilde{I}_{ij}(x)$ are defined respectively as follows:

$$\widetilde{I}_{ij}(x) := \int_{B_r^+} \Gamma_{ij}(x, T(x) - y) \mathcal{L}u(y)\, dy \quad \text{for } 1 \leq i, j \leq n - 1;$$

$$\widetilde{I}_{in}(x) := \widetilde{I}_{ni}(x) = \int_{B_r^+} \left(\sum_{\ell=1}^{n} \Gamma_{ij}(x, T(x) - y) A_\ell(y) \right) \mathcal{L}u(y)\, dy$$

for $1 \leq i \leq n-1$,

$$\widetilde{I}_{nn}(x) := \int_{B_r^+} \left(\sum_{\ell,m=1}^{n} \Gamma_{ij}(x, T(x)-y) A_\ell(x) A_m(x) \right) \mathcal{L}u(y)\, dy.$$

Then, since $\mathcal{L}u \in L^p(B_r^+)$, it follows from an application of Theorem 14.2 that

$$h_{ij} \in L^p(B_r^+).$$

Therefore, if we introduce a mapping

$$T\colon \left(L^p(B_r^+)\right)^{n^2} \longrightarrow \left(L^p(B_r^+)\right)^{n^2}$$

by the formula

$$T\mathbf{w} = \left(\sum_{h,k=1}^{n} \left(S_{ijhk} + \widetilde{S}_{ijhk}\right)(w_{hk}) + h_{ij}(x) \right)_{1 \leq i,j \leq n},$$
$$\mathbf{w} = (w_{ij})_{1 \leq i,j \leq n},$$

then we find that T is a *contraction mapping* (see condition (2.1)). Indeed, we have, by condition (14.63),

$$\begin{aligned}
\left\| T\mathbf{w}^{(1)} - T\mathbf{w}^{(2)} \right\| &= \sum_{i,j=1}^{n} \left\| \sum_{h,k=1}^{n} \left(S_{ijhk} + \widetilde{S}_{ijhk}\right)\left(w_{hk}^{(1)} - w_{hk}^{(2)}\right) \right\|_{L^r(B_r^+)} \\
&\leq \sum_{i,j,h,k=1}^{n} \left\| S_{ijhk} + \widetilde{S}_{ijhk} \right\| \left\| w_{hk}^{(1)} - w_{hk}^{(2)} \right\|_{L^r(B_r^+)} \\
&\leq \sum_{h,k=1}^{n} \left(\sum_{i,j=1}^{n} \left\| S_{ijhk} + \widetilde{S}_{ijhk} \right\| \right) \left\| w_{hk}^{(1)} - w_{hk}^{(2)} \right\|_{L^r(B_r^+)} \\
&\leq \frac{1}{2} \sum_{h,k=1}^{n} \left\| w_{hk}^{(1)} - w_{hk}^{(2)} \right\|_{L^r(B_r^+)} \\
&= \frac{1}{2} \left\| \mathbf{w}^{(1)} - \mathbf{w}^{(2)} \right\|.
\end{aligned}$$

Hence it follows an application of the *contraction mapping principle* (Theorem 2.1) that the mapping T has a unique fixed point

$$\mathbf{w} = \left(w_{ij}\right) \in \left(L^{q_1}(B_r^+)\right)^{n^2},$$
$$T\mathbf{w} = \mathbf{w}.$$

On the other hand, boundary representation formula (14.2) implies that the Hessian matrix

$$\nabla^2 u = \left(\frac{\partial^2 u}{\partial x_i \partial x_j}\right) \in \left(L^q(B_r^+)\right)^{n^2}$$

is also a fixed point of T. However, since we have the inclusion

$$\left(L^p(B_r^+)\right)^{n^2} \subset \left(L^q(B_r^+)\right)^{n^2},$$

by the uniqueness of fixed points of T, it follows that

$$\frac{\partial^2 u}{\partial x_i \partial x_j} = w_{ij} \in L^p(B_r^+) \quad \text{for } 1 \leq i, j \leq n.$$

Therefore, we obtain that

$$u \in W^{2,p}(B_r^+).$$

Furthermore, by Theorem 14.1 we can write the second derivatives

$$\frac{\partial^2 u}{\partial x_i \partial x_j} \quad \text{for } 1 \leq i, j \leq n,$$

in the form

$$\begin{aligned}
&\frac{\partial^2 u}{\partial x_i \partial x_j}(x) \\
&= \sum_{h,k=1}^{n} C\left\{a^{hk}, K_{ij}\right\}\left(\frac{\partial^2 u}{\partial x_h \partial x_k}\right)(x) \\
&\quad + K_{ij}(\mathcal{L}u)(x) + \left(\int_{|t|=1} \Gamma_i(x,t) t_j \, d\sigma_t\right) \mathcal{L}u(x) - I_{ij}(x).
\end{aligned}$$

Here the singular integral operators K_{ij} and boundary commutators $C\left[\varphi, K_{ij}\right]$ are defined respectively as follows:

$$\begin{aligned}
K_{ij} f(x) &:= \text{v.p.} \int_{B_r^+} \Gamma_{ij}(x, x-y) f(y) \, dy, \\
C\left[\varphi, K_{ij}\right] f(x) &:= \text{v.p.} \int_{B_r^+} \Gamma_{ij}(x, x-y) \left\{\varphi(x) - \varphi(y)\right\} f(y) \, dy,
\end{aligned}$$

and the terms $I_{ij}(x)$ are defined respectively as follows:

$$I_{ij}(x) := \sum_{h,k=1}^{n} \widetilde{C}\left[a^{hk}, K_{ij}\right]\left(\frac{\partial^2 u}{\partial x_h \partial x_k}\right)(x) + \widetilde{K}_{ij}(\mathcal{L}u)(x)$$

for $1 \leq i, j \leq n-1$;

$$I_{in}(x) = I_{in}(x) := \sum_{j=1}^{n} A_j(x) \times \left(\sum_{h,k=1}^{n} \widetilde{C}\left[a^{hk}, K_{ij}\right]\left(\frac{\partial^2 u}{\partial x_h \partial x_k}\right)(x) + \widetilde{K}_{ij}(\mathcal{L}u)(x)\right)$$

for $1 \leq i \leq n-1$;

$$I_{nn}(x) := \sum_{i,j=1}^{n} A_i(x) A_j(x) \times \left(\sum_{h,k=1}^{n} \widetilde{C}\left[a^{hk}, K_{ij}\right]\left(\frac{\partial^2 u}{\partial x_h \partial x_k}\right)(x) + \widetilde{K}_{ij}(\mathcal{L}u)(x)\right),$$

and the singular integral operators $\widetilde{K}_{ij}$ and boundary commutators $\widetilde{C}\left[\varphi, K_{ij}\right]$ are defined respectively as follows:

$$\widetilde{K}_{ij} f(x) := \int_{B_r^+} \Gamma_{ij}(x, T(x) - y) f(y)\, dy,$$

$$\widetilde{C}\left[\varphi, K_{ij}\right] f(x) := \int_{B_r^+} \Gamma_{ij}(x, T(x) - y)\, \{\varphi(x) - \varphi(y)\}\, f(y)\, dy.$$

Therefore, the desired boundary *a priori* estimate (14.62) follows from an application of Theorems 11.4, 14.2, 14.5 and Corollary 14.9 with

$$\begin{cases} k(x, z) := \Gamma_{ij}(x, z), \\ \varphi(x) := a^{hk}(x). \end{cases}$$

The proof of Lemma 14.13 is complete. □

14.5 Proof of Theorem 12.2

The purpose of this section is to prove Theorem 12.2. The proof of Theorem 12.2 is divided into two steps.

Step 1: Since the boundary $\partial\Omega$ is of class $C^{1,1}$, for each point $x_0 \in \partial\Omega$ we can find a neighborhood $U(x_0)$ and a $C^{1,1}$ diffeomorphism

$$G = (G_1, G_2, \ldots, G_n),$$

which maps $U(x_0) \cap \Omega$ onto B_r^+. Then it is easy to see that the coefficients $a^{ij}(x)$ of $\mathcal{L}$ are transformed into the functions

$$b^{ij}(y) = \sum_{h,k=1}^{n} a^{hk}(x)\frac{\partial G_i}{\partial x_h}(x)\frac{\partial G_j}{\partial x_k}(x) \quad \text{for } x = G^{-1}(y) \text{ and } y \in B_r^+,$$

where

$$G^{-1}\colon B_r^+ \longrightarrow U(x_0) \cap \Omega$$

is the inverse of G. Moreover, without loss of generality, we may assume that the functions $b^{ij}(y)$ are defined on the whole space $\mathbf{R}^n$ (see [1, Proposition 1.3]; [74, Theorem 1]; [169, Theorem 5.1]). Then we can verify that the functions $b^{ij}(y)$ are in VMO and further that their VMO moduli are estimated in terms of the VMO moduli of a^{hk}, the $C^{1,1}$-norm of G, the uniform continuity moduli of the derivatives $\partial G_i/\partial x_h$.

Step 2: The desired global *a priori* estimate (12.3) follows in a standard way from Theorem 12.1 and Lemma 14.13 by a covering argument, by flattening the boundary $\partial\Omega$ and by using interpolation inequalities (Theorem 13.4).

Step 2-1: Now we choose a finite covering $\{U_j\}_{j=1}^N$ of the boundary $\partial\Omega$ by open subsets of $\mathbf{R}^n$ and $C^{1,1}$ diffeomorphisms G_j of $U_j \cap \Omega$ onto B_r^+ in each of which inequality (14.62) holds true (see Fig. 14.6). Furthermore, we choose an open subset U_0 of Ω, bounded away from $\partial\Omega$, such that (see Fig. 14.7).

$$\Omega \subset U_0 \cup \left(\bigcup_{j=1}^{N} U_j\right),$$

and take a partition of unity $\{\alpha_k\}_{k=0}^N$ subordinate to the open covering $\{U_k\}_{k=0}^N$ of Ω.

Step 2-2: Now we assume that a function $u \in W^{2,p}(\Omega) \cap W_0^{1,p}(\Omega)$ satisfies the equation

$$\mathcal{L}u = f \quad \text{in } \Omega.$$

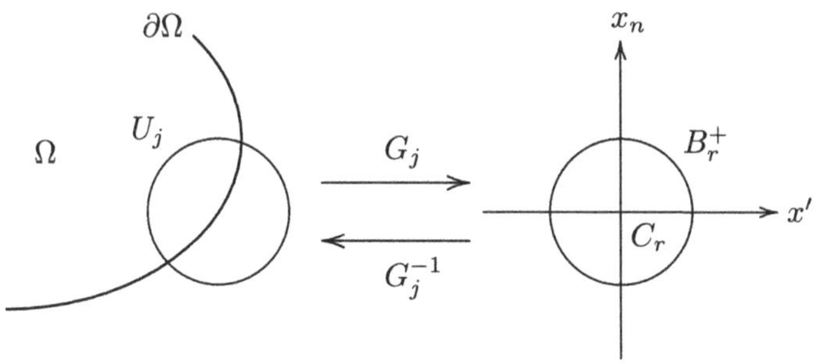

Fig. 14.6 The covering $\{U_j\}$ of $\partial\Omega$ and the $C^{1,1}$ diffeomorphism G_j of $U_j \cap \Omega$ onto B_r^+

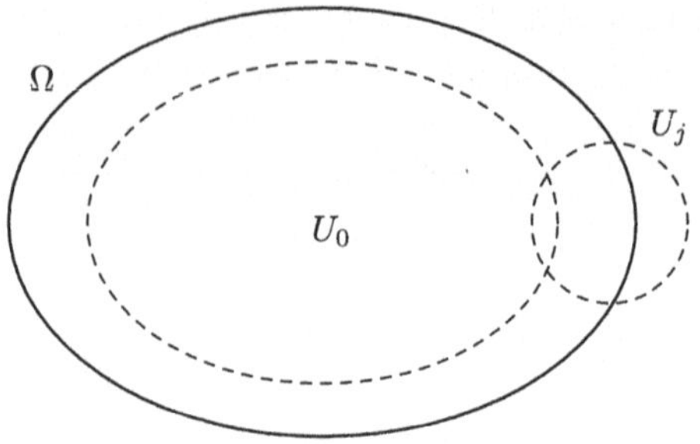

Fig. 14.7 The open covering $\{U_j\}$ of $\partial\Omega$ and the open set U_0 bounded away from $\partial\Omega$

Then, by applying the interior *a priori* estimate (12.2) to the function $\alpha_0 u$ we obtain that

$$
\begin{aligned}
\|u\|_{W^{2,p}(\Omega)} &\le \sum_{k=0}^{N} \|\alpha_k u\|_{W^{2,p}(\Omega)} \qquad (14.64)\\
&\le \|\alpha_0 u\|_{W^{2,p}(\Omega)} + \sum_{k=1}^{N}\left(\left\|\nabla^2(\alpha_k u)\right\|_{L^p(\Omega)} + \|\nabla(\alpha_k u)\|_{L^p(\Omega)} + \|\alpha_k u\|_{L^p(\Omega)}\right)\\
&\le C_0\left(\|u\|_{L^p(\Omega)} + \|f\|_{L^p(\Omega)}\right)\\
&\quad + \sum_{k=1}^{N}\left(\left\|\nabla^2(\alpha_k u)\right\|_{L^p(\Omega\cap U_k)} + \|\nabla(\alpha_k u)\|_{L^p(\Omega\cap U_k)} + \|u\|_{L^p(\Omega)}\right).
\end{aligned}
$$

(a) In order to estimate the terms $\nabla(\alpha_k u)$, $1 \le k \le N$, we recall the interpolation inequality (Theorem 13.4)

$$
\|\nabla v\|_{L^p(\Omega)} \le \varepsilon \left\|\nabla^2 v\right\|_{L^p(\Omega)} + \frac{C}{\varepsilon}\|v\|_{L^p(\Omega)} \quad \text{for } v \in W^{2,p}(\Omega). \qquad (13.17)
$$

Since we have the formula

$$
\nabla(\alpha_k u) = \alpha_k \nabla u + u(\nabla\alpha_k),
$$

by applying inequality (13.17) to the function $u \in W^{2,p}(\Omega)$, we obtain that, for some positive constants C_1 and C_2,

$$
\begin{aligned}
&\|\nabla(\alpha_k u)\|_{L^p(\Omega\cap U_k)} \qquad (14.65)\\
&\le \|\alpha_k(\nabla u)\|_{L^p(\Omega\cap U_k)} + \|u(\nabla\alpha_k)\|_{L^p(\Omega\cap U_k)}\\
&\le \|\nabla u\|_{L^p(\Omega)} + C_1\|u\|_{L^p(\Omega)}\\
&\le \varepsilon\left\|\nabla^2 u\right\|_{L^p(\Omega)} + \frac{C_2}{\varepsilon}\|u\|_{L^p(\Omega)} + C_1\|u\|_{L^p(\Omega)} \quad \text{for } 1 \le k \le N.
\end{aligned}
$$

(b) Moreover, by applying inequality (14.62) to the functions $\alpha_k u \in W^{2,p}_{\gamma_0}(\Omega \cap U_k)$, $1 \leq k \leq N$, we obtain that

$$\left\| \nabla^2(\alpha_k u) \right\|_{L^p(\Omega\cap U_k)} \leq C \left\| \mathcal{L}(\alpha_k u) \right\|_{L^p(\Omega\cap U_k)} . \tag{14.66}$$

However, we have the formula

$$\mathcal{L}(\alpha_k u) = \alpha_k f + u\,\mathcal{L}(\alpha_k) + 2\sum_{i,j=1}^{n} a^{ij}(x) \frac{\partial \alpha_k}{\partial x_i}\frac{\partial u}{\partial x_j}. \tag{14.67}$$

Hence, it follows from inequality (14.66), formula (14.67) and inequality (13.17) that

$$\begin{aligned}
&\left\| \nabla^2(\alpha_k u) \right\|_{L^p(\Omega\cap U_k)} \leq C \left\| \mathcal{L}(\alpha_k u) \right\|_{L^p(\Omega\cap U_k)} \qquad (14.68)\\
&\leq C\Bigg(\|\alpha_k f\|_{L^p(\Omega\cap U_k)} + \|u\,\mathcal{L}(\alpha_k)\|_{L^p(\Omega\cap U_k)}\\
&\quad + 2\sum_{i,j=1}^{n} \left\| a^{ij}\frac{\partial \alpha_k}{\partial x_i}\frac{\partial u}{\partial x_j} \right\|_{L^p(\Omega\cap U_k)} \Bigg)\\
&\leq C_3 \left(\|f\|_{L^p(\Omega)} + \|u\|_{L^p(\Omega)} + \|\nabla u\|_{L^p(\Omega)} \right)\\
&\leq C_3 \left(\|f\|_{L^p(\Omega)} + \|u\|_{L^p(\Omega)} \right) + \varepsilon \left\| \nabla^2 u \right\|_{L^p(\Omega)} + \frac{C_4}{\varepsilon}\|u\|_{L^p(\Omega)}\\
&\quad \text{for } 1 \leq k \leq N.
\end{aligned}$$

(c) Therefore, by combining inequalities (14.64), (14.65) and (14.68) we obtain that

$$\begin{aligned}
\|u\|_{W^{2,p}(\Omega)} &\leq 2N\varepsilon \left\| \nabla^2 u \right\|_{L^p(\Omega)} + C_5 \left(\|f\|_{L^p(\Omega)} + \|u\|_{L^p(\Omega)} \right)\\
&\quad + \frac{C_6}{\varepsilon}\|u\|_{L^p(\Omega)}.
\end{aligned}$$

This proves the desired global *a priori* estimate (12.3)

$$\|u\|_{W^{2,p}(\Omega)} \leq C_2 \left(\|u\|_{L^p(\Omega)} + \|f\|_{L^p(\Omega)} \right) \quad \text{for all } u \in W^{2,p}(\Omega) \cap W_0^{1,p}(\Omega),$$

if we take

$$\varepsilon := \frac{1}{4N}.$$

The proof of Theorem 12.2 is now complete. □

Chapter 15
Unique Solvability of the Homogeneous Dirichlet Problem

This chapter is devoted to the study of the *homogeneous* Dirichlet problem for a second order, uniformly elliptic differential operator with vanishing mean oscillation (VMO) coefficients in the framework of Sobolev spaces of L^p style. We prove an existence and uniqueness theorem for the Dirichlet problem (Theorem 15.1). Our proof is based on some interior and boundary *a priori* estimates for the solutions of problem (15.2) (Theorems 12.1 and 12.2). Both the interior and boundary *a priori* estimates are consequences of explicit representation formulas (13.1) and (14.1) for the solutions of problem (15.2) (Theorems 13.1 and 14.1) and also of the L^p-boundedness of Calderón–Zygmund singular integral operators and boundary commutators appearing in those representation formulas (Theorems 14.2 and 14.5). It should be emphasized that the VMO assumption on the coefficients a^{ij} is of the greatest relevance in the study of singular commutators. The uniqueness result in Theorem 15.1 follows from a variant of the Bakel'man–Aleksandrov maximum principle in the framework of Sobolev spaces due to Bony [13] (Theorem 8.1). Moreover, in order to prove the existence theorem for the homogeneous Dirichlet problem, we make good use of the method of continuity (Theorem 2.16).

Now, let Ω be a bounded domain in Euclidean space $\mathbf{R}^n$ ($n \geq 3$) with boundary $\partial\Omega$ of class $C^{1,1}$, and we consider a second order, uniformly elliptic differential operator $\mathcal{L}$ with real *discontinuous* coefficients of the form

$$\mathcal{L}u := \sum_{i,j=1}^{n} a^{ij}(x) \frac{\partial^2 u}{\partial x_i \partial x_j}.$$

Here the coefficients a^{ij} satisfy the following three conditions:

(i) $a^{ij} \in \mathrm{VMO} \cap L^\infty(\mathbf{R}^n)$ for all $1 \leq i, j \leq n$.
(ii) $a^{ij}(x) = a^{ji}(x)$ for all $1 \leq i, j \leq n$ and for almost all $x \in \Omega$.
(iii) There exists a positive constant λ such that (uniform ellipticity)

K. Taira, *Real Analysis Methods for Markov Processes*,
https://doi.org/10.1007/978-981-97-3659-1_15

$$\frac{1}{\lambda}|\xi|^2 \leq \sum_{i,j=1}^{n} a^{ij}(x)\xi_i\xi_j \leq \lambda\,|\xi|^2 \tag{15.1}$$

for almost all $x \in \Omega$ and for all $\xi \in \mathbf{R}^n$.

The purpose of this chapter is devoted to the study of the following *homogeneous* Dirichlet problem:

$$\begin{cases} \mathcal{L}u(x) = \sum_{i,j=1}^{n} a^{ij}(x)\frac{\partial^2 u}{\partial x_i \partial x_j}(x) = f(x) & \text{in } \Omega, \\ \gamma_0 u = 0 & \text{on } \partial\Omega. \end{cases} \tag{15.2}$$

It should be emphasized that a local $C^{1,1}$ diffeomorphism that straightens the boundary $\partial\Omega$ preserves the VMO character of the $a^{ij}(x)$ (see [1, Proposition 1.3], [74, Theorem 1], [169, Theorem 5.1]).

The next existence and uniqueness theorem for problem (15.2) is essentially due to Chiarenza–Frasca–Longo [35, Theorems 4.3 and 4.4]:

Theorem 15.1 (the existence and uniqueness theorem) *Let* $1 < p < \infty$. *Then, for any given* $f \in L^p(\Omega)$ *there exists a unique solution* $u \in W^{2,p}(\Omega) \cap W_0^{1,p}(\Omega)$ *of problem* (15.2). *Moreover, we have the a priori estimate*

$$\|u\|_{W^{2,p}(\Omega)} \leq C\,\|f\|_{L^p(\Omega)}, \tag{15.3}$$

with a positive constant $C = C(n, p, \lambda, \eta, M, \partial\Omega)$.

The proof of Theorem 15.1 can be visualized as in Fig. 15.1 below.

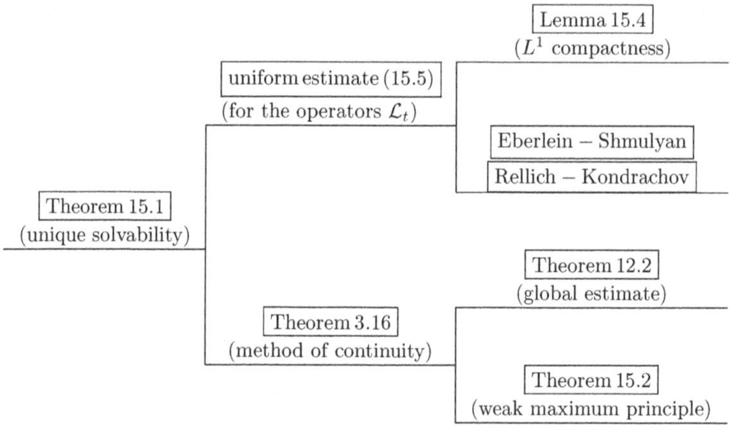

Fig. 15.1 A flowchart for the proof of Theorem 15.1

15.1 VMO Functions and Friedrichs' Mollifiers

The VMO assumption means a kind of continuity in the average sense, not in the pointwise sense. This property guarantees that VMO functions may be approximated by smooth functions. In fact, we can prove the following (cf. condition (iii) of Theorem 4.7):

Proposition 15.2 *Let* $f \in \mathrm{VMO}$. *For any* $\varepsilon > 0$, *there exists a smooth function* $g_\varepsilon \in \mathrm{VMO}$ *such that*

$$\|f - g_\varepsilon\|_* < \varepsilon, \tag{15.4}$$

and further that

$$\eta_\varepsilon(r) \leq \eta(r),$$

where $\eta(r)$ *and* $\eta_\varepsilon(r)$ *are the* VMO *moduli of* f *and* g_ε, *respectively. In other words,* VMO *functions can be approximated by smooth functions.*

Proof Just as in Sect. 3.7.2, we take a bell-shaped function $\varphi(x)$ on $\mathbf{R}^n$ that satisfies the following four conditions:

$$\varphi \in C_0^\infty(\mathbf{R}^n).$$
$$\varphi(x) \geq 0 \quad \text{on } \mathbf{R}^n.$$
$$\operatorname{supp} \varphi \subset B(0,1) := \left\{x \in \mathbf{R}^n : |x| \leq 1\right\}.$$
$$\int_{\mathbf{R}^n} \varphi(x)\, dx = 1.$$

For each $\varepsilon > 0$, we let

$$\varphi_\varepsilon(x) := \frac{1}{\varepsilon^n} \varphi\left(\frac{x}{\varepsilon}\right).$$

Then it is easy to verify the following four conditions:

$$\varphi_\varepsilon \in C_0^\infty(\mathbf{R}^n).$$
$$\varphi_\varepsilon(x) \geq 0 \quad \text{on } \mathbf{R}^n.$$
$$\operatorname{supp} \varphi_\varepsilon \subset B(0,\varepsilon) := \left\{x \in \mathbf{R}^n : |x| \leq \varepsilon\right\},$$
$$\int_{\mathbf{R}^n} \varphi_\varepsilon(x)\, dx = 1.$$

We recall that the functions $\{\varphi_\varepsilon\}$ are *Friedrichs' mollifiers*.

If we let

$$\begin{aligned} f_\varepsilon(x) &:= (f * \varphi_\varepsilon)(x) \\ &= \frac{1}{\varepsilon^n} \int_{\mathbf{R}^n} \varphi\left(\frac{x-y}{\varepsilon}\right) f(y)\, dy \end{aligned}$$

$$= \int_{|z|\leq 1} \varphi(z)\, f(x - \varepsilon z)\, dz,$$

then we have the following assertions:

(1) $f_\varepsilon \in C^\infty(\mathbf{R}^n)$.
(2) By applying Minkowski's inequality for integrals (Theorem 3.29), we obtain that

$$\|f_\varepsilon - f\|_* = \left\| \int_{|z|\leq 1} \varphi(z)\,(f(\cdot - \varepsilon z) - f(\cdot))\, dz \right\|_* \leq \int_{|z|\leq 1} \varphi(z)\, \|f(\cdot - \varepsilon z) - f(\cdot)\|_*\, dz. \tag{15.5}$$

However, by Theorem 4.5, (iii) it follows that there exists a positive constant C, independent of ε, such that

$$\|f(\cdot - \varepsilon z) - f(\cdot)\|_* \leq C\, \eta(\varepsilon) \quad \text{for } |z| \leq 1. \tag{15.6}$$

By combining inequalities (15.5) and (15.6), we find that

$$\begin{aligned} \|f_\varepsilon - f\|_* &\leq \int_{|z|\leq 1} \varphi(z)\, \|f(\cdot - \varepsilon z) - f(\cdot)\|_*\, dz \\ &\leq C \int_{|z|\leq 1} \varphi(z)\, dz \cdot \eta(\varepsilon) \\ &= C\, \eta(\varepsilon). \end{aligned}$$

This proves the desired inequality (15.4) with $g_\varepsilon := f_\varepsilon$.
(3) Moreover, we have, by Theorem 3.29,

$$\begin{aligned} \|f_\varepsilon\|_* &= \left\| \int_{|z|\leq 1} \varphi(z)\, f(\cdot - \varepsilon z)\, dz \right\|_* \\ &\leq \int_{|z|\leq 1} \varphi(z)\, \|f(\cdot - \varepsilon z)\|_*\, dz \\ &= \int_{|z|\leq 1} \varphi(z)\, dz \cdot \|f\|_* \\ &= \|f\|_*\,. \end{aligned}$$

This proves that

$$\begin{aligned} &g_\varepsilon = f_\varepsilon \in \mathrm{VMO} \cap C^\infty(\mathbf{R}^n), \\ &\eta_\varepsilon(r) \leq \eta(r). \end{aligned}$$

The proof of Proposition 15.2 is complete. □

15.2 Proof of Theorem 15.1

The proof of Theorem 15.1 is divided into four steps (see [34, 35, 99]).

Step 1: First, by applying Theorem 12.2 we obtain the following global *a priori* estimate:

$$\|u\|_{W^{2,p}(\Omega)} \leq c_1 \left(\|\mathcal{L}u\|_{L^p(\Omega)} + \|u\|_{L^p(\Omega)}\right) \quad (15.7)$$
$$\text{for } u \in W^{2,p}(\Omega) \cap W_0^{1,p}(\Omega),$$

where

$$W_0^{1,p}(\Omega) = \left\{u \in W^{1,p}(\Omega) : \gamma_0 u = 0 \text{ on } \partial\Omega\right\}.$$

Here $c_1 > 0$ is a constant depending on the coefficients a^{ij} only through the ellipticity constant λ, the bound on the norms $\|a^{ij}\|_{L^p(\Omega)}$ and the VMO moduli of the a^{ij}.

Step 2: The *uniqueness result* of problem (15.2) follows from an application of the global regularity theorem (Theorem 12.2) and the weak maximum principle (Theorem 8.1) due to Bony [13]:

Theorem 15.3 (the weak maximum principle) *Assume that a function u in* $W^{2,p}(\Omega)$ *with* $n < p < \infty$ *satisfies the condition*

$$\mathcal{L}u(x) \geq 0 \quad \textit{for almost all } x \in \Omega.$$

Then it follows that

$$\max_{\overline{\Omega}} u \leq \max_{\partial\Omega} u^+,$$

where

$$u^+(x) := \max\{u(x), 0\} = \frac{1}{2}\{|u(x)| + u(x)\} \quad \textit{for } x \in \overline{\Omega} = \Omega \cup \partial\Omega.$$

Remark 15.4 In fact, Bony proved this maximum principle under the weaker condition that $a^{ij} \in L^\infty(\Omega)$ (see the proof of Theorem 8.1 in Chap. 8).

In light of the global regularity theorem (Theorem 12.2), we may assume that

$$n < p < \infty.$$

Hence we have, by Sobolev's imbedding theorem (see Theorem 7.5),

$$W^{2,p}(\Omega) \subset C^1(\overline{\Omega}),$$

since $2 - n/p > 1$ for $n < p < \infty$. By applying the weak maximum principle (Theorem 15.3) to the functions $\pm u(x)$, we find that

$$\begin{cases} \mathcal{L}u = 0 & \text{almost everywhere in } \Omega, \\ \gamma_0 u = 0 & \text{on } \partial\Omega \end{cases} \Longrightarrow u = 0 \quad \text{in } \Omega.$$

This proves the uniqueness result of problem (15.2).

Step 3: In order to prove the *existence result* of problem (15.2), we make use of the *method of continuity* (Theorem 2.16).

We shall apply Theorem 2.16 with

$$\begin{aligned} \mathcal{B} &:= W^{2,p}(\Omega) \cap W_0^{1,p}(\Omega), \\ \mathcal{V} &:= L^p(\Omega), \\ \mathcal{L}_1 &:= \mathcal{L} = \sum_{i,j=1}^n a^{ij}(x) \frac{\partial^2}{\partial x_i \partial x_j}, \\ \mathcal{L}_0 &:= \Delta. \end{aligned}$$

Here it should be noticed that

$$W^{2,p}(\Omega) \cap W_0^{1,p}(\Omega) = \left\{ u \in W^{2,p}(\Omega) : \gamma_0 u = 0 \text{ on } \partial\Omega \right\}$$

is a closed subspace of $W^{2,p}(\Omega)$, since the trace theorem (Theorem 7.8) asserts that the trace $\gamma_0 \colon W^{2,p}(\Omega) \to B^{2-1/p,p}(\partial\Omega)$ is continuous.

Substep 3-1: The essential step in our proof is how to show the *a priori* estimate (corresponding to inequality (2.9)) for a family of elliptic differential operators

$$\begin{aligned} \mathcal{L}_t &:= t\mathcal{L} + (1-t)\Delta \\ &= \sum_{i,j=1}^n \left(t\, a^{ij}(x) + (1-t)\delta^{ij} \right) \frac{\partial^2}{\partial x_i \partial x_j} \quad \text{for } 0 \le t \le 1. \end{aligned}$$

More precisely, we consider, instead of the original homogeneous Dirichlet problem $(t = 1)$

$$\begin{cases} \mathcal{L}u = f & \text{in } \Omega, \\ \gamma_0 u = 0 & \text{on } \partial\Omega, \end{cases} \tag{15.2}$$

a family of homogeneous Dirichlet problems

$$\begin{cases} \mathcal{L}_t u = f & \text{in } \Omega, \\ \gamma_0 u = 0 & \text{on } \partial\Omega. \end{cases}$$

By the uniqueness result of problem (15.2), we can get rid of the term $\|u\|_{L^p(\Omega)}$ on the right-hand side of estimate (15.4). Namely, we shall prove that the *a priori* estimate

$$\|u\|_{W^{2,p}(\Omega)} \le c_2 \, \|\mathcal{L}_t u\|_{L^p(\Omega)} \quad \text{for } u \in \mathfrak{B} \tag{15.8}$$

holds true for the elliptic differential operators $\mathcal{L}_t$, $0 \leq t \leq 1$. Here $c_2 > 0$ is a structure constant depending on the coefficients a^{ij} only through the ellipticity constant λ, the bound on the norms $\|a^{ij}\|_{L^p(\Omega)}$ and the VMO moduli of the a^{ij}.

Now it should be noticed that the coefficients

$$a^{ij}_{(t)}(x) := t\, a^{ij}(x) + (1-t)\,\delta^{ij} \quad \text{for } 0 \leq t \leq 1,$$

satisfy the following three conditions:

(i) $a^{ij}_{(t)} \in \mathrm{VMO} \cap L^\infty(\mathbf{R}^n)$ for all $1 \leq i, j \leq n$.

Indeed, we have, for all $0 \leq t \leq 1$,

$$\left\| a^{ij}_{(t)} \right\|_{L^\infty(\mathbf{R}^n)} \leq \left\| a^{ij} \right\|_{L^\infty(\mathbf{R}^n)} + 1,$$

and

$$\left\| a^{ij}_{(t)} \right\|_* \leq t \left\| a^{ij} \right\|_* \leq \left\| a^{ij} \right\|_* .$$

(ii) $a^{ij}_{(t)}(x) = a^{ji}_{(t)}(x)$ for almost all $x \in \Omega$.

(iii) We have, for almost all $x \in \Omega$ and for all $\xi \in \mathbf{R}^n$,

$$\frac{1}{\lambda + 1} |\xi|^2 \leq \sum_{i,j=1}^n a^{ij}_{(t)}(x) \xi_i \xi_j \leq (\lambda + 1)\, |\xi|^2,$$

where λ is the same constant as in condition (15.1).

Our proof of the *a priori* estimate (15.8) is based on a reduction to absurdity. We assume, to the contrary, that the *a priori* estimate (15.8) does not hold true. Then we can find a sequence of elliptic differential operators

$$\mathcal{L}^{(m)} = \sum_{i,j=1}^n a^{ij}_{(m)}(x) \frac{\partial^2}{\partial x_i \partial x_j},$$

and a sequence of functions $\{u^{(m)}\}$ such that the coefficients $a^{ij}_{(m)}(x)$ and the functions $\{u^{(m)}\}$ satisfy the following four conditions:

(i) $a^{ij}_{(m)} \in \mathrm{VMO} \cap L^\infty(\mathbf{R}^n)$ for all $1 \leq i, j \leq n$ and

$$\left\| a^{ij}_{(m)} \right\|_{L^\infty(\mathbf{R}^n)} \leq \left\| a^{ij} \right\|_{L^\infty(\mathbf{R}^n)} + 1, \tag{15.9a}$$

$$\eta^{ij}_{(m)}(r) \leq \eta^{ij}(r). \tag{15.9b}$$

Here we recall that

$$\eta^{ij}_{(m)}(r) = \sup_{\rho \leq r} \frac{1}{|B|} \int_B \left| a^{ij}_{(m)}(y) - \left(a^{ij}_{(m)}\right)_B \right| dy,$$

$$\eta^{ij}(r) = \sup_{\rho \leq r} \frac{1}{|B|} \int_B \left| a^{ij}(y) - \left(a^{ij}\right)_B \right| dy,$$

where the supremum is taken over all balls B with radius $\rho \leq r$.

(ii) $a^{ij}_{(m)}(x) = a^{ji}_{(m)}(x)$ for almost all $x \in \Omega$.

(iii) We have, for almost all $x \in \Omega$ and for all $\xi \in \mathbf{R}^n$,

$$\frac{1}{\lambda + 1} |\xi|^2 \leq \sum_{i,j=1}^{n} a^{ij}_{(m)}(x) \xi_i \xi_j \leq (\lambda + 1) |\xi|^2. \tag{15.10}$$

(iv) $u^{(m)} \in W^{2,p}(\Omega) \cap W^{1,p}_0(\Omega)$ and

$$\left\| u^{(m)} \right\|_{W^{2,p}(\Omega)} = 1, \tag{15.11a}$$

$$\left\| \mathcal{L}^{(m)} u^{(m)} \right\|_{L^p(\Omega)} \longrightarrow 0. \tag{15.11b}$$

(3-1a): First, it follows that the sequence $\{(a^{ij}_{(m)})_B\}$ is bounded for every $1 \leq i, j \leq n$. Indeed, we have, by inequality (15.9a),

$$\left| \left(a^{ij}_{(m)}\right)_B \right| \leq \left\| a^{ij}_{(m)} \right\|_{L^\infty(\mathbf{R}^n)} \leq \left\| a^{ij} \right\|_{L^\infty(\mathbf{R}^n)} + 1.$$

Moreover, we have the following lemma:

Lemma 15.5 *For any ball B in $\mathbf{R}^n$, every sequence*

$$\left\{ a^{ij}_{(m)}(x) - \left(a^{ij}_{(m)}\right)_B \right\} \quad \textit{for } 1 \leq i, j \leq n,$$

is compact *in the space* $L^1(B)$.

Proof The proof of Lemma 15.5 is divided into two steps (1) and (2).

Step (1): For simplicity, we write

$$a_{(m)}(x) := a^{ij}_{(m)}(x),$$

and let

$$f_{(m)}(x) := a_{(m)}(x) - \left(a_{(m)}\right)_B, \tag{15.12a}$$

$$f^{\varepsilon}_{(m)}(x) := a_{(m)} * \varphi_\varepsilon(x) - \left(a_{(m)} * \varphi_\varepsilon\right)_B, \tag{15.12b}$$

where

$$a_{(m)} * \varphi_\varepsilon(x) = \int_{\mathbf{R}^n} a_{(m)}(x-y)\varphi_\varepsilon(y)\, dy$$

is a *mollification* of $a_{(m)}$.

In light of the mollification, the VMO functions may be approximated by smooth functions. In fact, we can prove the following two Claims 15.6 and 15.7:

Claim 15.6 *The sequence $\{f_{(m)}^\varepsilon\}$ is uniformly bounded and equicontinuous in B, for each $\varepsilon > 0$. Hence $\{f_{(m)}^\varepsilon\}$ has a subsequence that is uniformly convergent in B to a continuous function, for each $\varepsilon > 0$.*

Proof We have only to prove the first statement. Indeed, the second statement follows from an application of the Ascoli–Arzelà theorem (Corollary 2.71).

(a) The *uniform boundedness* of $\{f_{(m)}^\varepsilon\}$: First, we have, for all $x \in B$,

$$\begin{aligned}|a_{(m)} * \varphi_\varepsilon(x)| &\le \int_{\mathbf{R}^n} |a_{(m)}(x-y)|\, \varphi_\varepsilon(y)\, dy \\ &\le \left\|a_{(m)}\right\|_{L^\infty(\mathbf{R}^n)} \\ &\le \|a\|_{L^\infty(\mathbf{R}^n)} + 1,\end{aligned}$$

and also

$$\begin{aligned}\left| \left(a_{(m)} * \varphi_\varepsilon\right)_B \right| &\le \frac{1}{|B|}\int_{\mathbf{R}^n} |a_{(m)}(x-y)|\, \varphi_\varepsilon(y)\, dy \\ &\le \left\|a_{(m)}\right\|_{L^\infty(\mathbf{R}^n)} \\ &\le \|a\|_{L^\infty(\mathbf{R}^n)} + 1.\end{aligned}$$

Hence it follows that

$$\begin{aligned}|f_{(m)}^\varepsilon(x)| &\le |a_{(m)} * \varphi_\varepsilon(x)| + \left| \left(a_{(m)} * \varphi_\varepsilon\right)_B \right| \\ &\le 2\left(\|a\|_{L^\infty(\mathbf{R}^n)} + 1\right) \quad \text{for all } x \in B.\end{aligned}$$

(b) The *equicontinuity* of $\{f_{(m)}^\varepsilon\}$: By the mean value theorem, it suffices to show that the gradient $\{\nabla f_{(m)}^\varepsilon\}$ is uniformly bounded in B, for each $\varepsilon > 0$.

We have, for all $x \in B$,

$$\begin{aligned}\nabla_x f_{(m)}^\varepsilon(x) &= \nabla_x \left(\int_{B(0,\varepsilon)} a_{(m)}(x-y)\varphi_\varepsilon(y)\, dy - \left(a_{(m)} * \varphi_\varepsilon\right)_{B(0,\varepsilon)} \right) \\ &= \nabla_x \left(\int_{B(x,\varepsilon)} \varphi_\varepsilon(x-y) a_{(m)}(y)\, dy \right) \\ &= \frac{1}{\varepsilon^n} \nabla_x \left(\int_{B(x,\varepsilon)} \varphi\left(\frac{x-y}{\varepsilon}\right) a_{(m)}(y)\, dy \right).\end{aligned}$$

However, it should be noticed that

$$\nabla_x \left(\varphi\left(\frac{x-y}{\varepsilon}\right)\right) = -\nabla_y \left(\varphi\left(\frac{x-y}{\varepsilon}\right)\right) = \frac{1}{\varepsilon}(\nabla\varphi)\left(\frac{x-y}{\varepsilon}\right),$$

and further that

$$\int_{B(x,\varepsilon)} \nabla\varphi\left(\frac{x-y}{\varepsilon}\right) dy = 0.$$

Hence we can express the function $\nabla_x f^{\varepsilon}_{(m)}(x)$ as follows:

$$\begin{aligned}\nabla_x f^{\varepsilon}_{(m)}(x) &= \frac{1}{\varepsilon^{n+1}} \int_{B(x,\varepsilon)} \nabla\varphi\left(\frac{x-y}{\varepsilon}\right) a_{(m)}(y)\, dy \\ &= \frac{1}{\varepsilon^{n+1}} \int_{B(x,\varepsilon)} \nabla\varphi\left(\frac{x-y}{\varepsilon}\right) \left(a_{(m)}(y) - \left(a_{(m)}\right)_{B(x,\varepsilon)}\right) dy.\end{aligned}$$

Therefore, we have, for all $x \in B$,

$$\begin{aligned}&\left|\nabla f^{\varepsilon}_{(m)}(x)\right| \qquad (15.13)\\ &\leq \frac{1}{\varepsilon^{n+1}} \int_{B(x,\varepsilon)} \left|a_{(m)}(y) - \left(a_{(m)}\right)_{B(x,\varepsilon)}\right| \cdot \left|\nabla\varphi\left(\frac{x-y}{\varepsilon}\right)\right| dy \\ &\leq \frac{1}{\varepsilon^{n+1}} \left(\sup_{\mathbf{R^n}} |\nabla\varphi|\right) \int_{B(x,\varepsilon)} \left|a_{(m)}(y) - \left(a_{(m)}\right)_{B(x,\varepsilon)}\right| dy \\ &\leq \frac{1}{\varepsilon^{n+1}} \left(\sup_{\mathbf{R^n}} |\nabla\varphi|\right) |B(x,\varepsilon)|\, \eta_{(m)}(\varepsilon) \\ &= \frac{1}{\varepsilon^{n+1}} \left(\sup_{\mathbf{R^n}} |\nabla\varphi|\right) \left(\varepsilon^n \omega_n\right) \eta_{(m)}(\varepsilon) \\ &\leq \frac{\omega_n}{\varepsilon} \left(\sup_{\mathbf{R^n}} |\nabla\varphi|\right) \eta(\varepsilon) \quad \text{for }, \varepsilon > 0.\end{aligned}$$

Here we recall that

$$\begin{aligned}\omega_n &= \frac{2\pi^{n/2}}{\Gamma(n/2)} \quad \text{(the surface area of the unit sphere } \Sigma_{n-1} \text{ in } \mathbf{R}^n\text{)}, \\ \eta_{(m)}(r) &:= \eta^{ij}_{(m)}(r) = \sup_{\rho\leq r} \frac{1}{|B|} \int_B \left|a^{ij}_{(m)}(y) - \left(a^{ij}_{(m)}\right)_B\right| dy, \\ \eta(r) &:= \eta^{ij}(r) = \sup_{\rho\leq r} \frac{1}{|B|} \int_B \left|a^{ij}(y) - \left(a^{ij}\right)_B\right| dy.\end{aligned}$$

By inequality (15.13), it follows that $\{\nabla f^{\varepsilon}_{(m)}\}$ is uniformly bounded in B, for each $\varepsilon > 0$.

The proof of Claim 15.6 is complete. □

Claim 15.7 *There exists a positive constant C, independent of ε, such that*

$$\frac{1}{|B|}\int_B \left|f^{\varepsilon}_{(m)}(x) - f_{(m)}(x)\right| dx \leq C\,\eta(\varepsilon) \quad \textit{for } \varepsilon > 0. \tag{15.14}$$

In particular, we have the assertion

$$\left\| f^{\varepsilon}_{(m)} - f_{(m)} \right\|_* \leq C\,\eta(\varepsilon) \quad \textit{for } \varepsilon > 0.$$

Proof First, by Fubini's theorem (Theorem 3.20) it follows that

$$\begin{aligned}
\left(a_{(m)} * \varphi_\varepsilon\right)_B &= \frac{1}{|B|}\int_B a_{(m)} * \varphi_\varepsilon(x)\,dx \qquad (15.15)\\
&= \frac{1}{|B|}\int_B \left(\int_{B(0,\varepsilon)} a_{(m)}(x-y)\varphi_\varepsilon(y)\,dy\right) dx\\
&= \int_{B(0,\varepsilon)} \left(\frac{1}{|B|}\int_B a_{(m)}(x-y)\,dx\right)\varphi_\varepsilon(y)\,dy\\
&= \int_{B(0,\varepsilon)} \left(\frac{1}{|B-y|}\int_{B-y} a_{(m)}(z)\,dz\right)\varphi_\varepsilon(y)\,dy\\
&= \int_{B(0,\varepsilon)} \left(a_{(m)}\right)_{B-y}\varphi_\varepsilon(y)\,dy,
\end{aligned}$$

where $B - y$ is the translation of the ball B by y-units

$$B - y = \{x - y : x \in B\}.$$

Hence we have, by formulas (15.12) and (15.15),4

$$\begin{aligned}
&\int_B \left|f^{\varepsilon}_{(m)}(x) - f_{(m)}(x)\right| dx \qquad (15.16)\\
&= \int_B \left|a_{(m)} * \varphi_\varepsilon(x) - \left(a_{(m)} * \varphi_\varepsilon\right)_B - \left(a_{(m)}(x) - \left(a_{(m)}\right)_B\right)\right| dx\\
&= \int_B \left|\int_{B(0,\varepsilon)} \left(a_{(m)}(x-y) - a_{(m)}(x)\right)\right.\\
&\qquad \left. - \left(\left(a_{(m)}\right)_{B-y} - \left(a_{(m)}\right)_B\right)\varphi_\varepsilon(y)\,dy\right| dx\\
&\leq \int_B \int_{B(0,\varepsilon)} \left|\left(a_{(m)}(x-y) - a_{(m)}(x)\right)\right.\\
&\qquad \left. - \left(\left(a_{(m)}\right)_{B-y} - \left(a_{(m)}\right)_B\right)\right| \varphi_\varepsilon(y)\,dy\,dx
\end{aligned}$$

$$\leq \int_{B(0,\varepsilon)} \varphi_\varepsilon(y) \left(\int_B \left| \left(a_{(m)}(x-y) - a_{(m)}(x)\right) \right.\right.$$
$$\left.\left. - \left(\left(a_{(m)}\right)_{B-y} - \left(a_{(m)}\right)_B \right) \right| dx \right) dy.$$

However, we have the formula

$$\begin{aligned}\left(a_{(m)}\right)_{B-y} &= \frac{1}{|B-y|} \int_{B-y} a_{(m)}(w)\, dw = \frac{1}{|B|} \int_B a_{(m)}(z-y)\, dz \\ &= a_{(m)}(\cdot - y)_B.\end{aligned}$$

Hence it follows from an application of inequality (4.9) with $f := a_{(m)}$ that

$$\begin{aligned}&\frac{1}{|B|} \int_B \left| \left(a_{(m)}(x-y) - a_{(m)}(x)\right) - \left(\left(a_{(m)}\right)_{B-y} - \left(a_{(m)}\right)_B \right) \right| dx \qquad (15.17)\\ &= \frac{1}{|B|} \int_B \left| \left(a_{(m)}(x-y) - a_{(m)}(x)\right) - \left(a_{(m)}(\cdot - y) - a_{(m)}\right)_B \right| dx \\ &\leq \left\| a_{(m)}(\cdot - y) - a_{(m)}(\cdot) \right\|_* \\ &\leq C\, \eta_{(m)}(\varepsilon) \quad \text{for } |y| < \varepsilon.\end{aligned}$$

Therefore, by combining inequalities (15.16) and (15.17) we obtain that

$$\begin{aligned}\int_B \left| f^\varepsilon_{(m)}(x) - f_{(m)}(x) \right| dx &\leq C\, \eta_{(m)}(\varepsilon) \left(\int_{B(0,\varepsilon)} \varphi_\varepsilon(y)\, dy \right) |B| \\ &= C\, \eta_{(m)}(\varepsilon)\, |B| \\ &\leq C\, |B|\, \eta(\varepsilon).\end{aligned}$$

This proves the desired inequality (15.14).

The proof of Claim 15.7 is complete. □

Step (2): By combining Claims 15.6 and 15.7, we find from Theorems 2.4 and 2.6 that the sequence $\{f_{(m)}\}$ is compact in the space $L^1(B)$. Indeed, it suffices to apply Example 2.5 with $X := L^1(B)$, $A := \{f_{(m)}\}$ and $A_h := \{f^h_{(m)}\}$.

The proof of Lemma 15.5 is complete. □

(3-1b): By using Lemma 15.5 and the Bolzano–Weierstrass theorem, we can choose a subsequence of the sequence

$$a^{ij}_{(m)}(x) = \left(a^{ij}_{(m)}(x) - \left(a^{ij}_{(m)}\right)_B \right) + \left(a^{ij}_{(m)}\right)_B$$

that converges almost everywhere in B. Therefore, by considering an exhaustive sequence of balls of $\mathbf{R}^n$ we can choose a subsequence of $a^{ij}_{(m)}$, denoted again by $a^{ij}_{(m)}$, that converges to a function α^{ij} almost everywhere in $\mathbf{R}^n$, as $m \to \infty$. Then it is easy to verify the following three assertions (i), (ii) and (iii):

(i) $\alpha^{ij} \in \mathrm{VMO} \cap L^\infty(\mathbf{R}^n)$ for all $1 \leq i, j \leq n$.
Indeed, since we have the inequality

$$\left\| a^{ij}_{(m)} \right\|_{L^\infty(\mathbf{R}^n)} \leq \left\| a^{ij} \right\|_{L^\infty(\mathbf{R}^n)} + 1,$$

it follows that

$$\left\| \alpha^{ij} \right\|_{L^\infty(\mathbf{R}^n)} \leq \left\| a^{ij} \right\|_{L^\infty(\mathbf{R}^n)} + 1.$$

Moreover, since we have, for all balls B with radius $\rho \leq r$,

$$\begin{aligned} \frac{1}{|B|} \int_B |a^{ij}_{(m)}(x) - \left(a^{ij}_{(m)}\right)_B | \, dx &\leq \eta^{ij}_{(m)}(r) \\ &\leq \eta^{ij}(r), \end{aligned}$$

and since we have, by the Lebesgue dominated convergence theorem (Theorem 3.13),

$$\begin{aligned} (a^{ij}_{(m)})_B &= \frac{1}{|B|} \int_B a^{ij}_{(m)}(y) \, dy \\ &\longrightarrow \frac{1}{|B|} \int_B \alpha^{ij}(y) \, dy = \left(\alpha^{ij}\right)_B \quad \text{as } m \to \infty, \end{aligned}$$

it follows that we have, for all balls B with radius $\rho \leq r$,

$$\begin{aligned} \frac{1}{|B|} \int_B |\alpha^{ij}(x) - \left(\alpha^{ij}\right)_B | \, dx &= \lim_{m \to \infty} \frac{1}{|B|} \int_B |a^{ij}_{(m)}(x) - \left(a^{ij}_{(m)}\right)_B | \, dx \\ &\leq \eta^{ij}(r). \end{aligned}$$

This proves that the VMO modulus of α^{ij} is dominated by $\eta^{ij}(r)$:

$$\sup_{\rho \leq r} \frac{1}{|B|} \int_B \left|\alpha^{ij}(x) - \left(\alpha^{ij}\right)_B\right| \, dx \leq \eta^{ij}(r).$$

(ii) $\alpha^{ij}(x) = \alpha^{ji}(x)$ for almost all $x \in \Omega$.
Indeed, we have, for almost all $x \in \Omega$,

$$\alpha^{ij}(x) = \lim_{m \to \infty} a^{ij}_{(m)}(x) = \lim_{m \to \infty} a^{ji}_{(m)}(x) = \alpha^{ji}(x).$$

(iii) We have, for almost all $x \in \Omega$ and for all $\xi \in \mathbf{R}^n$,

$$\frac{1}{\lambda+1}|\xi|^2 \leq \sum_{i,j=1}^{n} \alpha^{ij}(x)\xi_i\xi_j \leq (\lambda+1)|\xi|^2.$$

These inequalities can be obtained by passing to the limit in the inequalities (15.10).

We introduce a new second order, uniformly elliptic differential operator $\mathcal{A}$ by the formula

$$\mathcal{A} := \sum_{i,j=1}^{n} \alpha^{ij}(x)\frac{\partial^2}{\partial x_i \partial x_j}.$$

(3-1c): Now it is known (see [2, Theorem 3.6]) that the L^p Sobolev space $W^{2,p}(\Omega)$ is reflexive if $1 < p < \infty$. By applying the Eberlein–Shmulyan theorem (Theorem 2.32) and the Rellich–Kondrachov theorem (Theorem 7.6) to our situation, we can obtain the following two assertions:

(A) Let $\{w_m\}$ be any sequence that is norm bounded in the Sobolev space $W^{2,p}(\Omega)$ for $1 < p < \infty$. Then there exists a subsequence $\{w_{m'}\}$ that converges *weakly* to an element of $W^{2,p}(\Omega)$.
(B) The injection $W^{2,p}(\Omega) \to W^{1,p}(\Omega)$ is *compact* for $1 < p < \infty$ ($j := 1$ and $p = q$ in Theorem 7.6).

Therefore, by using condition (15.11a) we can find a subsequence of the sequence

$$\{u^{(m)}\} \subset W^{2,p}(\Omega) \cap W_0^{1,p}(\Omega),$$

denoted again by $\{u^{(m)}\}$, that converges *weakly* to a function $v \in W^{2,p}(\Omega)$ in the space $W^{2,p}(\Omega)$ and also converges *strongly* to v in the space $L^p(\Omega)$:

$$\begin{aligned} u^{(m)} &\longrightarrow v \quad \text{weakly in } W^{2,p}(\Omega) \text{ as } m \to \infty, \\ u^{(m)} &\longrightarrow v \quad \text{in } L^p(\Omega) \text{ as } m \to \infty. \end{aligned}$$

We shall prove that $v = 0$, that is,

$$u^{(m)} \longrightarrow 0 \quad \text{weakly in } W^{2,p}(\Omega) \text{ as } m \to \infty, \tag{15.18a}$$

$$u^{(m)} \longrightarrow 0 \quad \text{in } L^p(\Omega) \text{ as } m \to \infty. \tag{15.18b}$$

First, we show that the function $v \in W^{2,p}(\Omega)$ satisfies the boundary condition

$$v = 0 \quad \text{on } \partial\Omega,$$

that is,

$$v \in W^{2,p}(\Omega) \cap W_0^{1,p}(\Omega). \tag{15.19}$$

To do this, it should be noticed that the subspace

$$\mathfrak{B} := W^{2,p}(\Omega) \cap W_0^{1,p}(\Omega)$$

is balanced, convex and *strongly closed* in $W^{2,p}(\Omega)$. Indeed, it suffices to note that if $\{u_j\}$ is any sequence in $\mathfrak{B}$ that converges strongly to a function u in $W^{2,p}(\Omega)$, then we have, by the trace theorem (Theorem 7.8),

$$\gamma_0 u = \lim_{j\to\infty} \gamma_0(u_j) = 0 \quad \text{in } B^{2-1/p,p}(\partial\Omega).$$

Hence, we find that the subspace $\mathfrak{B}$ is *weakly closed* in $W^{2,p}(\Omega)$, by applying Mazur's theorem (or Theorem 2.31) with

$$X := W^{2,p}(\Omega), \quad M := \mathfrak{B}.$$

Therefore, we obtain the desired assertion (15.19), since the sequence $\{u^{(m)}\}$ in $W^{2,p}(\Omega) \cap W_0^{1,p}(\Omega)$ converges weakly to v in $W^{2,p}(\Omega)$.

To prove assertions (15.18), let φ be an arbitrary function in $L^q(\Omega)$ with $q = p/(p-1)$. Then we have, by Hölder's inequality (Theorem 3.24) and condition (15.18a),

$$\begin{aligned}
&\left|\int_\Omega \left(\mathcal{L}^{(m)}u^{(m)} - \mathcal{A}v\right)\varphi\,dx\right| \qquad (15.20)\\
&= \left|\sum_{i,j=1}^n \int_\Omega \left(a^{ij}_{(m)}(x)\frac{\partial^2 u^{(m)}}{\partial x_i \partial x_j} - \alpha^{ij}(x)\frac{\partial^2 v}{\partial x_i \partial x_j}\right)\varphi\,dx\right| \\
&\le \sum_{i,j=1}^n \left|\int_\Omega \left(a^{ij}_{(m)}(x)\frac{\partial^2 u^{(m)}}{\partial x_i \partial x_j} - \alpha^{ij}(x)\frac{\partial^2 v}{\partial x_i \partial x_j}\right)\varphi\,dx\right| \\
&\le \sum_{i,j=1}^n \left|\int_\Omega \left(a^{ij}_{(m)}(x) - \alpha^{ij}(x)\right)\frac{\partial^2 u^{(m)}}{\partial x_i \partial x_j}\,\varphi\,dx\right| \\
&\quad + \sum_{i,j=1}^n \left|\int_\Omega \alpha^{ij}(x)\left(\frac{\partial^2 u^{(m)}}{\partial x_i \partial x_j} - \frac{\partial^2 v}{\partial x_i \partial x_j}\right)\varphi\,dx\right| \\
&\le \sum_{i,j=1}^n \left\|\nabla^2 u^{(m)}\right\|_{L^p(\Omega)} \left\|\left(a^{ij}_{(m)} - \alpha^{ij}\right)\varphi\right\|_{L^q(\Omega)} \\
&\quad + \sum_{i,j=1}^n \left|\int_\Omega \alpha^{ij}(x)\varphi(x)\cdot\left(\frac{\partial^2 u^{(m)}}{\partial x_i \partial x_j} - \frac{\partial^2 v}{\partial x_i \partial x_j}\right)dx\right| \\
&\le \sum_{i,j=1}^n \left\|\left(a^{ij}_{(m)} - \alpha^{ij}\right)\varphi\right\|_{L^q(\Omega)}
\end{aligned}$$

$$+ \sum_{i,j=1}^{n} \left| \int_{\Omega} \alpha^{ij}(x)\varphi(x) \cdot \left(\frac{\partial^2 u^{(m)}}{\partial x_i \partial x_j} - \frac{\partial^2 v}{\partial x_i \partial x_j} \right) dx \right| .$$

However, by condition (15.9a) it follows from an application of the Lebesgue dominated convergence theorem (Theorem 3.13) that the first term on the last inequality (15.20) tends to zero as $m \to \infty$:

$$\sum_{i,j=1}^{n} \left\| \left(a^{ij}_{(m)} - \alpha^{ij} \right) \varphi \right\|_{L^q(\Omega)} \longrightarrow 0.$$

Moreover, we recall that $\{u^{(m)}\}$ converges weakly to a function $v \in W^{2,p}(\Omega) \cap W_0^{1,p}(\Omega)$. Since $\alpha^{ij}(x)\varphi \in L^q(\Omega)$, we find that the second term on the last inequality (15.20) tends to zero as $m \to \infty$:

$$\sum_{i,j=1}^{n} \left| \int_{\Omega} \alpha^{ij}(x)\varphi(x) \cdot \left(\frac{\partial^2 u^{(m)}}{\partial x_i \partial x_j} - \frac{\partial^2 v}{\partial x_i \partial x_j} \right) dx \right| \longrightarrow 0.$$

Hence we have, by inequality (15.20),

$$\int_{\Omega} \mathcal{L}^{(m)} u^{(m)} \cdot \varphi \, dx \longrightarrow \int_{\Omega} \mathcal{A}v \cdot \varphi \, dx \quad \text{as } m \to \infty.$$

On the other hand, by Hölder's inequality (Theorem 3.24) and condition (15.11b) it follows that

$$\left| \int_{\Omega} \mathcal{L}^{(m)} u^{(m)} \cdot \varphi \, dx \right| \leq \left\| \mathcal{L}^{(m)} u^{(m)} \right\|_{L^p(\Omega)} \cdot \|\varphi\|_{L^q(\Omega)} \longrightarrow 0 \quad \text{as } m \to \infty.$$

This proves that

$$\int_{\Omega} \mathcal{A}v \cdot \varphi \, dx = \lim_{n \to \infty} \int_{\Omega} \mathcal{L}^{(m)} u^{(m)} \cdot \varphi \, dx = 0 \quad \text{for all } \varphi \in L^q(\Omega).$$

Summing up, we have proved that

$$\begin{aligned} & v \in W^{2,p}(\Omega) \cap W_0^{1,p}(\Omega), \\ & \mathcal{A}v = 0 \quad \text{almost everywhere in } \Omega. \end{aligned}$$

By applying Theorem 15.3 to the operator $\mathcal{A}$, we obtain that

$$v = 0 \quad \text{in } \Omega.$$

This proves the desired assertions (15.18).

(3-1d): By combining assertions (15.11b) and (15.18b), we have proved that, as $m \to \infty$,

$$\mathcal{L}^{(m)} u^{(m)} \longrightarrow 0 \quad \text{in } L^p(\Omega),$$
$$u^{(m)} \longrightarrow 0 \quad \text{in } L^p(\Omega).$$

Therefore, by applying estimate (15.7) to the operators $\{\mathcal{L}^{(m)}\}$ we obtain that

$$\left\| u^{(m)} \right\|_{W^{2,p}(\Omega)} \leq c_1 \left(\left\| \mathcal{L}^{(m)} u^{(m)} \right\|_{L^p(\Omega)} + \left\| u^{(m)} \right\|_{L^p(\Omega)} \right),$$

so that

$$u^{(m)} \longrightarrow 0 \quad \text{in } W^{2,p}(\Omega) \text{ as } m \to \infty.$$

However, this assertion contradicts condition (15.11a):

$$\left\| u^{(m)} \right\|_{W^{2,p}(\Omega)} = 1.$$

This contradiction proves the desired *a priori* estimate (15.8).

Substep 3-2: By virtue of estimate (15.8), we can apply the *method of continuity* (Theorem 2.16) to obtain that the Dirichlet problem is uniquely solvable for the operator $\mathcal{L}_0$ if and only if it is uniquely solvable for the operator $\mathcal{L}_1$. However, it is known (see [59, Theorem 13.15]) that the homogeneous Dirichlet problem is uniquely solvable for the Laplace operator $\mathcal{L}_0 = \Delta$: More precisely, for any $f \in L^p(\Omega)$ there exists a unique solution $u \in W^{2,p}(\Omega) \cap W_0^{1,p}(\Omega)$ of the Dirichlet problem

$$\begin{cases} \Delta u = f & \text{in } \Omega, \\ \gamma_0 u = 0 & \text{on } \partial\Omega. \end{cases}$$

In other words, the operator $\mathcal{L}_0 := \Delta$ maps $W^{2,p}(\Omega) \cap W_0^{1,p}(\Omega)$ onto $L^p(\Omega)$.

Therefore, it follows from an application of Theorem 2.16 that the operator $\mathcal{L}_1 := \mathcal{L}$ maps $W^{2,p}(\Omega) \cap W_0^{1,p}(\Omega)$ onto $L^p(\Omega)$. Namely, for any $f \in L^p(\Omega)$, there exists a unique solution $u \in W^{2,p}(\Omega) \cap W_0^{1,p}(\Omega)$ of problem (15.2).

Finally, the desired *a priori* estimate (15.3) follows from estimate (15.8) with $t := 1$.

Now the proof of Theorem 15.1 is complete. □

Part V
Oblique Derivative Problems for Elliptic Differential Equations with Discontinuous Coefficients

Chapter 16
Regular Oblique Derivative Problems in Sobolev Spaces

This chapter is devoted to the study of the *regular* oblique derivative problem for a second order, uniformly elliptic differential operator with discontinuous coefficients in the framework of L^p Sobolev spaces. More precisely, we consider a second order, uniformly elliptic differential operator with vanishing mean oscillation (VMO) coefficients and an oblique derivative boundary operator that is *nowhere* tangential to the boundary. We state global regularizing property of the oblique derivative problem in the framework of L^p Sobolev spaces (Theorem 16.1). Furthermore, we state an existence and uniqueness theorem for the oblique derivative problem in the framework of L^p Sobolev spaces (Theorem 16.2). The results discussed here are adapted from Di Fazio–Palagachev [40] and Maugeri–Palagachev–Softova [99].

16.1 Formulation of the Oblique Derivative Problem

Let Ω be a bounded domain of $\mathbf{R}^n$, $n \geq 3$, with boundary $\partial\Omega$ of class $C^{1,1}$. In the interior Ω, we consider a second order, *uniformly elliptic* differential operator $\mathcal{L}$ with real *discontinuous* coefficients of the form

$$\mathcal{L}u(x) := \sum_{i,j=1}^{n} a^{ij}(x)\frac{\partial^2 u}{\partial x_i \partial x_j}(x) \quad \text{for } x \in \Omega.$$

More precisely, we assume that the coefficients $a^{ij}(x)$ are the components of a symmetric contravariant tensor of type $\binom{2}{0}$ in $\mathbf{R}^n$ and satisfy the following three conditions:

(1) $a^{ij} \in \mathrm{VMO} \cap L^\infty(\Omega)$.
(2) $a^{ij}(x) = a^{ji}(x)$ for almost all $x \in \Omega$.
(3) There exists a positive constant λ such that (the uniform ellipticity condition)

K. Taira, *Real Analysis Methods for Markov Processes*,
https://doi.org/10.1007/978-981-97-3659-1_16

$$\frac{1}{\lambda}\,|\xi|^2 \leq \sum_{i,j=1}^{n} a^{ij}(x)\xi_i\xi_j \leq \lambda\,|\xi|^2 \tag{16.1}$$

for almost all $x \in \Omega$ and for all $\xi \in \mathbf{R}^n$.

If $\eta^{ij}(r)$ is the VMO modulus of $a^{ij}(x)$, then we let

$$\eta(r) := \left(\sum_{i,j=1}^{n} \eta^{ij}(r)^2\right)^{1/2}.$$

On the boundary $\partial\Omega$, we consider a first order boundary operator $\mathcal{B}$ with real continuous coefficients of the from

$$\begin{aligned}\mathcal{B}u(x') &:= \frac{\partial u}{\partial \ell}(x') + \sigma(x')u(x') \\ &= \sum_{i=1}^{n} \ell^i(x')\frac{\partial u}{\partial x_i}(x') + \sigma(x')u(x') \quad \text{for } x' \in \partial\Omega.\end{aligned} \tag{16.2}$$

Concerning the boundary operator $\mathcal{B}$, we assume that the following three conditions are satisfied:

$$\ell^i(x') \text{ and } \sigma(x') \text{ are Lipschitz continuous functions on } \partial\Omega. \tag{16.3a}$$

$$\big\langle \ell(x'), \mathbf{n}(x')\big\rangle := \sum_{i=1}^{n} \ell^i(x')n_i(x') > 0 \quad \text{on } \partial\Omega. \tag{16.3b}$$

$$\sigma(x') < 0 \quad \text{on } \partial\Omega. \tag{16.3c}$$

Here $\mathbf{n}(x') = \big(n_1(x'), n_2(x'), \ldots, n_n(x')\big)$ is the unit *inward normal* to $\partial\Omega$ and $\ell(x') = (\ell^1(x'), \ell^2(x'), \ldots, \ell^n(x'))$ is a vector field on $\partial\Omega$. It should be emphasized that the boundary operator $\mathcal{B}$ is given by a directional derivative with respect to the vector field $\ell(x')$ on $\partial\Omega$. The simple geometric meaning of conditions (16.3b) is that the vector field $\ell(x')$ is *nowhere* tangential to the boundary $\partial\Omega$ (see Fig. 16.1).

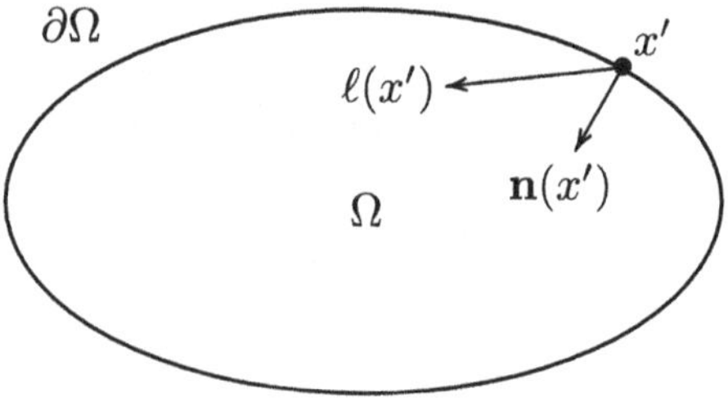

Fig. 16.1 the unit inward normal $\mathbf{n}(x')$ and the oblique vector field $\ell(x')$ at x'

To interpret the boundary condition (16.2) in the sense of traces on $\partial\Omega$, we recall some definitions and useful notations. If $1 < p < \infty$ and if k is a non-negative integer, then the Sobolev space $W^{k,p}(\Omega)$ is defined to be the space of those functions $u \in L^p(\Omega)$ such that $D^\alpha u \in L^p(\Omega)$ for $|\alpha| \leq k$, and the norm $\|u\|_{W^{k,p}(\Omega)}$ is defined by the formula

$$\|u\|_{W^{k,p}(\Omega)} = \left(\sum_{|\alpha| \leq k} \int_\Omega |D^\alpha u(x)|^p \, dx \right)^{1/p}.$$

We define the boundary space of traces

$$B^{k-1/p,p}(\partial\Omega) := \text{the space of distributions } \varphi \in \mathcal{D}'(\partial\Omega) \text{ such that there exists a function } u \in W^{k,p}(\Omega) \text{ with } \gamma_0 u = \varphi.$$

In the space $B^{k-1/p,p}(\partial\Omega)$, we introduce a norm

$$|\varphi|_{B^{k-1/p,p}(\partial\Omega)} = \inf \left\{ \|u\|_{W^{k,p}(\Omega)} : u \in W^{k,p}(\Omega), \ \gamma_0 u = \varphi \text{ on } \partial\Omega \right\},$$

where the infimum is taken over all functions $u \in W^{k,p}(\Omega)$ that equal φ on the boundary $\partial\Omega$. The space $B^{k-1/p,p}(\partial\Omega)$ is a Banach space with respect to this norm. We recall that the space $B^{k-1/p,p}(\partial\Omega)$ is a Besov space (see the trace theorem (Theorem 7.6)).

The purpose of this chapter is to study global regularity and solvability in the framework of Sobolev spaces of the following non-homogeneous oblique derivative problem:

$$\begin{cases} \mathcal{L}u(x) = f(x) & \text{for almost all } x \in \Omega, \\ \mathcal{B}u(x') = \varphi(x') & \text{in the sense of traces on } \partial\Omega. \end{cases} \tag{16.4}$$

It should be emphasized that the boundary value problem (16.4) is a *regular* oblique derivative problem, since the vector field $\ell(x')$ is nowhere tangential to the boundary $\partial\Omega$.

The interest in the study of oblique derivative problems for uniformly elliptic operators with VMO coefficients increased significantly in the last twenty years. This is mainly due to the fact that VMO contains as a proper subspace $C(\overline{\Omega})$ which ensures the extension of the L^p Schauder theory of operators with continuous coefficients to *discontinuous* coefficients (see [59, 85]). On the other hand, the Sobolev spaces $W^{1,n}(\Omega)$ and $W^{\theta,n/\theta}(\Omega)$, $0 < \theta < 1$, are also contained in VMO; hence the VMO discontinuity of the $a^{ij}(x)$ becomes more general than those studied before (see [103, 105, 165]).

The crucial point of our investigations is the local boundary Sobolev regularity of the solutions of problem (16.4). Our approach is based on explicit integral representation formulas (18.7) for the second derivatives of solutions of problem (16.4) with constant coefficients operators and homogeneous boundary conditions (near the boundary), in terms of singular integral operators with Calderón–Zygmund kernels and their commutators and operators with positive kernels (Theorem 18.3). This method has been already used in the study of the Dirichlet problem in Part III of this book. In order to deal with non-homogeneous oblique derivative boundary conditions with variable coefficients, we introduce a special auxiliary function which, roughly speaking, absorbs the right-hand side of the boundary condition (16.4) (Lemma 17.1). Moreover, we make use of special non-dimensional norms (17.1) and (17.6) to estimate effectively the Sobolev norms of the second derivatives of solutions of the oblique derivative problem (16.4).

16.2 Statement of Main Results (Theorems 16.1 and 16.2)

The first main result of this chapter is stated as follows (see [99, Chap. 2, Theorem 2.2.1]):

Theorem 16.1 (the regularity theorem) *Let* $1 < p < \infty$, *and assume that conditions* (16.1) *and* (16.3) *are satisfied. If a function* $u \in W^{2,q}(\Omega)$, $1 < q < p < \infty$, *is a solution of problem* (16.4) *with* $f \in L^p(\Omega)$ *and* $\varphi \in B^{1-1/p,p}(\partial\Omega)$, *then it follows that* $u \in W^{2,p}(\Omega)$. *Moreover, we have the global* a priori *estimate*

$$\|u\|_{W^{2,p}(\Omega)} \leq C_1 \left(\|u\|_{L^p(\Omega)} + \|f\|_{L^p(\Omega)} + \|\varphi\|_{B^{1-1/p,p}(\partial\Omega)}\right), \tag{16.5}$$

with a positive constant $C_1 = C_1(n, p, \lambda, \eta, \ell, \sigma, \partial\Omega)$.

The proof of Theorem 16.1 can be visualized as in Fig. 16.2 below.

The regularizing property of the couple $(\mathcal{L}, \mathcal{B})$ implies the well-posedness of problem (16.4) in the framework of L^p Sobolev spaces. More precisely, the second main result of this chapter is stated as follows (see [99, Chap. 2, Theorem 2.2.2]):

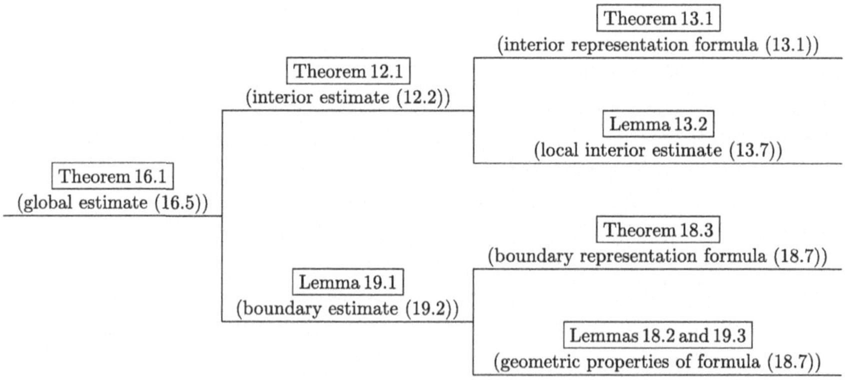

Fig. 16.2 A flowchart for the proof of Theorem 16.1

Theorem 16.2 (the existence and uniqueness theorem) *Let* $1 < p < \infty$, *and assume that conditions* (16.1) *and* (16.3) *are satisfied. Then, for any* $f \in L^p(\Omega)$ *and any* $\varphi \in B^{1-1/p,p}(\partial\Omega)$ *there exists a unique solution of problem* (16.4). *Moreover, we have the global* a priori *estimate*

$$\|u\|_{W^{2,p}(\Omega)} \leq C_2 \left(\|f\|_{L^p(\Omega)} + \|\varphi\|_{B^{1-1/p,p}(\partial\Omega)}\right), \tag{16.6}$$

with a positive constant $C_2 = C_2(n, p, \lambda, \eta, \ell, \sigma, \partial\Omega)$.

Remark 16.3 Some remarks are in order:

1° The results presented here can be applied to the study of Sobolev regularity of the solutions of problem (16.4) for a general second order uniformly elliptic operator

$$Au(x) = \sum_{i,j=1}^{n} a^{ij}(x) \frac{\partial^2 u}{\partial x_i \partial x_j}(x) + \sum_{i=1}^{n} b^i(x) \frac{\partial u}{\partial x_i}(x) + c(x)u(x),$$

where the lower order coefficients $b^i(x)$ and $c(x)$ satisfy suitable Lebesgue integrability conditions (see [98]) such as the case where

$$b^i, \ c \in L^\infty(\Omega),$$

and

$$c(x) \leq 0 \quad \text{almost everywhere in} \Omega.$$

2° It should be emphasized that VMO functions are invariant under $C^{1,1}$ diffeomorphisms (see [1, Proposition 1.3]; [74, Theorem 1]; [169, Theorem 5.1]).

16.3 The Oblique Derivative Problem (16.4) and the Distance Function

In this section, we show that the condition

$$\sigma(x') < 0 \quad \text{on } \partial\Omega \tag{16.3c}$$

is not necessary for the study of the oblique derivative problem (16.4). More precisely, we assume that

$$\sigma(x') \leq 0 \quad \text{on } \partial\Omega. \tag{16.7}$$

First, we introduce the *distance function* $d(x)$ by the formula

$$d(x) = \text{dist}\,(x, \partial\Omega) \quad \text{for } x \in \overline{\Omega}.$$

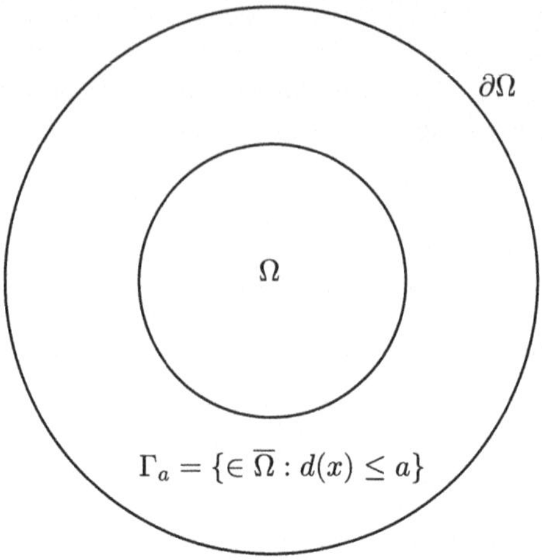

Fig. 16.3 The distance function $d(x)$ and a tubular neighborhood Γ_a of the boundary

It is easy to see that $d(x)$ is uniformly Lipschitz continuous. In fact, we have, by the triangle inequality,

$$|d(x) - d(y)| \leq |x - y| \quad \text{for all } x, \ y \in \overline{\Omega}.$$

Furthermore, for $a > 0$ we define a tubular neighborhood Γ_a of $\partial\Omega$ by the formula (see Fig. 16.3)

$$\Gamma_a := \left\{x \in \overline{\Omega} : d(x) \leq a\right\}.$$

Then we find from the proof of [59, Lemma 14.16] that

$$d \in C^{1,1}\left(\Gamma_{a_0}\right) \quad \text{for some } a_0 > 0.$$

It should be noticed that the points $x \in \Gamma_{a_0}$ and $y' \in \partial\Omega$ are related as follows:

$$x = y' + d(x)\,\mathbf{n}(y'),$$

where $\mathbf{n}(y')$ is the unit inward normal to $\partial\Omega$ at y'.

By using [97, Lemma 1.67], we can extend the distance function $d(x)$ in such a way that

$$d \in C^{1,1}\left(\overline{\Omega}\right).$$

Hence, it follows from an application of Sobolev imbedding theorem [2, Theorem 4.12, Part II] and Rademacher's theorem [97, Corollary 1.73] that

$$d \in W^{2,\infty}(\Omega) = C^{1,1}\left(\overline{\Omega}\right). \tag{16.8}$$

Finally, we have the formula

$$\operatorname{grad} d(x') = \mathbf{n}(x') \quad \text{at every point } x' \in \partial\Omega. \tag{16.9}$$

Now we are in a position to study the non-homogeneous oblique derivative problem via the distance function (cf. [98, Theorem 4.1], [99, Theorem 2.3.5], [160, Theorem 5.1]):

Theorem 16.4 *Let $n \geq 3$ and $n < p < \infty$. Assume that the differential operator*

$$\mathcal{L}u = \sum_{i,j=1}^{n} a^{ij}(x) \frac{\partial^2 u}{\partial x_i \partial x_j}$$

satisfies the uniform elliptic condition (16.1), *and further that the boundary condition*

$$\mathcal{B}u = \frac{\partial u}{\partial \ell} + \sigma(x')u$$

satisfies the non-tangential condition (16.3b) *and condition* (16.7):

$$\langle \ell(x'), \mathbf{n}(x') \rangle = \sum_{i=1}^{n} \ell^i(x')\, n_i(x') > 0 \quad on\ \partial\Omega. \tag{16.3b}$$

$$\sigma(x') \leq 0 \quad on\ \partial\Omega. \tag{16.7}$$

Then there exists a constant $\alpha_0 > 0$ (see formula (16.14) *below) such that the non-homogeneous oblique derivative problem*

$$\begin{cases} (\mathcal{L} - \alpha_0)\, u = f & in\ \Omega, \\ \mathcal{B}u = \varphi & on\ \partial\Omega \end{cases} \tag{16.10}$$

has a unique solution $u \in W^{2,p}(\Omega)$ for any function $f \in L^p(\Omega)$ and any function $\varphi \in B^{1-1/p,p}(\partial\Omega)$.

Proof The proof of Theorem 16.4 is divided into three steps.

Step (1): If we let

$$u(x) := v(x)\, e^{-d(x)} \quad \text{for } x \in \overline{\Omega},$$

then we have, by a direct calculation (cf. [99, pp. 138–139]),

$$(\mathcal{L} - \alpha)\, u = f \quad \text{in } \Omega \iff \left(\widetilde{\mathcal{L}} - \alpha\right) v = f\, e^{d(x)} \quad \text{in } \Omega. \tag{16.11}$$

Here the differential operator $\widetilde{\mathcal{L}}$ is given by the formula

$$
\begin{aligned}
\widetilde{\mathcal{L}}v &= \sum_{i,j=1}^{n} a^{ij}(x)\frac{\partial^2 v}{\partial x_i \partial x_j} + \sum_{i=1}^{n} \widetilde{b^i}(x)\frac{\partial v}{\partial x_i} + \widetilde{c}(x)\, v(x) \\
&:= \sum_{i,j=1}^{n} a^{ij}(x)\frac{\partial^2 v}{\partial x_i \partial x_j} - 2\sum_{i=1}^{n}\left(\sum_{j=1}^{n} a^{ij}(x)\frac{\partial d}{\partial x_j}\right)\frac{\partial v}{\partial x_i} \\
&\quad + \left(\sum_{i,j=1}^{n} a^{ij}(x)\frac{\partial d}{\partial x_i}\frac{\partial d}{\partial x_j} - \sum_{i,j=1}^{n} a^{ij}(x)\frac{\partial^2 d}{\partial x_i \partial x_j}\right).
\end{aligned}
$$

By assertion (16.8), we remark that the coefficients $\widetilde{b^i}(x)$ and $\widetilde{c}(x)$ satisfy the following regularity conditions:

- $$\widetilde{b^i}(x) = -2\sum_{j=1}^{n} a^{ij}(x)\frac{\partial d}{\partial x_j} \in L^\infty(\Omega) \quad \text{for } 1 \le i \le n, \tag{16.12}$$
- $$\widetilde{c}(x) = \sum_{i,j=1}^{n} a^{ij}(x)\frac{\partial d}{\partial x_i}\frac{\partial d}{\partial x_j} - \sum_{i,j=1}^{n} a^{ij}(x)\frac{\partial^2 d}{\partial x_i \partial x_j} \in L^\infty(\Omega). \tag{16.13}$$

Therefore, if we let

$$\alpha_0 := \max_{x \in \overline{\Omega}} \left| \sum_{i,j=1}^{n} a^{ij}(x)\left(\frac{\partial d}{\partial x_i}\frac{\partial d}{\partial x_j} - \frac{\partial^2 d}{\partial x_i \partial x_j}\right)\right|, \tag{16.14}$$

then we have the inequality

$$\widetilde{c}(x) - \alpha_0 \le 0 \quad \text{on } \overline{\Omega}. \tag{16.15}$$

Step (2): On the other hand, by using assertion (16.8) and formula (16.9) we have the formula

$$\mathcal{B}u = \frac{\partial u}{\partial \ell} + \sigma(x')u = \varphi \quad \text{on } \partial\Omega \tag{16.16}$$

$$\Longleftrightarrow$$

$$\widetilde{\mathcal{B}}v := \frac{\partial v}{\partial \ell} + \widetilde{\sigma}(x')v = \varphi \quad \text{on } \partial\Omega,$$

where

$$\widetilde{\sigma}(x') = \sigma(x') - \langle \ell(x'), \mathbf{n}(x') \rangle \quad \text{on } \partial\Omega. \tag{16.17}$$

Indeed, it suffices to note the following:

$$\frac{\partial}{\partial \ell}\left(e^{d(x)}\right) = \left\langle \ell, \operatorname{grad} d(x')\right\rangle e^{d(x')} \quad \text{on } \partial\Omega.$$

Step (3): By combining formulas (16.11) and (16.16), we have the *equivalence* of the two oblique derivative problems (16.10) and (16.18):

$$\begin{cases} (\mathcal{L} - \alpha_0)\, u = f & \text{in } \Omega, \\ \mathcal{B}u = \varphi & \text{on } \partial\Omega \end{cases} \tag{16.10}$$

$$\Longleftrightarrow$$

$$\begin{cases} \left(\widetilde{\mathcal{L}} - \alpha_0\right) v = f\, e^{d(x)} & \text{in } \Omega, \\ \widetilde{\mathcal{B}}v = \varphi & \text{on } \partial\Omega. \end{cases} \tag{16.18}$$

By assertion (16.8) and inequality (16.15), it should be noticed that the coefficients $\widetilde{b^i}(x)$ and $\widetilde{c}(x)$ of the differential operator $\widetilde{\mathcal{L}}$ satisfies the following two conditions:

(2) $\widetilde{b^i} \in L^\infty(\Omega)$ for $1 \leq i \leq n$.
(3) $\widetilde{c} \in L^\infty(\Omega)$ and $\widetilde{c}(x) - \alpha_0 \leq 0$ for almost all $x \in \Omega$.

Moreover, by formula (16.17) it follows that the coefficients of the modified boundary condition $\widetilde{\mathcal{B}}$ satisfies the following two conditions:

$$\left\langle \ell(x'), \mathbf{n}(x')\right\rangle = \sum_{i=1}^{n} \ell^i(x')\, n_i(x') > 0 \quad \text{on } \partial\Omega. \tag{16.3b}$$

$$\widetilde{\sigma}(x') = \sigma(x') - \left\langle \ell(x'), \mathbf{n}(x')\right\rangle \leq -\left\langle \ell(x'), \mathbf{n}(x')\right\rangle < 0 \quad \text{on } \partial\Omega. \tag{16.19}$$

Therefore, by applying [153, Theorem 4.1] (cf. [98, Theorem 4.1], [99, Theorem 2.3.5]) with

$$\begin{aligned} &A := \widetilde{\mathcal{L}}, \quad c(x) := \widetilde{c}(x) - \alpha_0, \quad \alpha := 0, \\ &f := f\, e^{d(x)}, \\ &\Gamma := \widetilde{\mathcal{B}}, \quad \gamma(x') := \widetilde{\sigma}(x'), \end{aligned}$$

we obtain that the modified non-homogeneous oblique derivative problem (16.18) has a unique solution $v \in W^{2,p}(\Omega)$ for any function $f\, e^{d(x)} \in L^p(\Omega)$ and any function $\varphi \in B^{1-1/p,p}(\partial\Omega)$.

Namely, the original non-homogeneous oblique derivative problem (16.4) has a unique solution

$$u = v\, e^{-d(x)} \in W^{2,p}(\Omega)$$

for any $f \in L^p(\Omega)$ and any $\varphi \in B^{1-1/p,p}(\partial\Omega)$. Indeed, we have, by assertion (16.8),

$$u = v\, e^{-d(x)} \in W^{2,p}(\Omega) \quad \text{for } v \in W^{2,p}(\Omega).$$

Now the proof of Theorem 16.4 is complete. □

Chapter 17
Oblique Derivative Boundary Conditions

In this chapter, for a given boundary function, we construct an auxiliary function that satisfies an oblique derivative boundary condition. More precisely, for a given boundary function $\varphi(x') \in B^{1-1/p,p}(\partial\Omega)$, we construct a special extension $\phi(x) \in W^{2,p}(\Omega)$ which satisfies the oblique derivative boundary condition (see Fig. 16.1)

$$\frac{\partial \phi}{\partial \ell}(x') = \sum_{i=1}^{n} \ell^i(x') \frac{\partial \phi}{\partial x_i}(x') = \varphi(x') \quad \text{on } \partial\Omega.$$

This result (Lemma 17.1) will allow us to represent, locally near the boundary, the solution of the non-homogeneous oblique derivative problem (16.4) in Chap. 19 (see formula (19.10)). In this way, we are reduced to the study of the *homogeneous* oblique derivative problem:

$$\begin{cases} \mathcal{L}u = \sum_{i,j=1}^{n} a^{ij}(x) \dfrac{\partial^2 u}{\partial x_i \partial x_j} = f & \text{for almost all } x \in \Omega, \\ \mathcal{B}u = \frac{\partial u}{\partial \ell} + \sigma(x')u = 0 & \text{in the sense of traces on } \partial\Omega. \end{cases}$$

17.1 Construction of Auxiliary Functions

Let $\widetilde{\Gamma}$ be a bounded portion of the hyperplane

$$\begin{aligned} &\{x_n = 0\} \\ &:= \left\{x = \left(x', x_n\right) \in \mathbf{R}^n : x' = (x_1, x_2, \ldots, x_{n-1}) \in \mathbf{R}^{n-1},\ x_n = 0\right\}. \end{aligned}$$

K. Taira, *Real Analysis Methods for Markov Processes*,
https://doi.org/10.1007/978-981-97-3659-1_17

Let $\widetilde{\varphi}(x')$ be a function defined on $\widetilde{\Gamma}$ which belongs to $B^{1-1/p,p}(\widetilde{\Gamma})$. The space $B^{1-1/p,p}(\widetilde{\Gamma})$, $1 < p < \infty$, is a Banach space equipped with the norm (cf. the norm (7.11) with $m := 0$ and $\theta := 1 - 1/p$)

$$\|\widetilde{\varphi}\|^*_{B^{1-1/p,p}(\widetilde{\Gamma})} = \left(\int_{\widetilde{\Gamma}} |\widetilde{\varphi}(x')|^p \, dx' \right)^{1/p} + d^{1/2} \left(\int_{\widetilde{\Gamma}} \int_{\widetilde{\Gamma}} \frac{|\widetilde{\varphi}(x') - \widetilde{\varphi}(y')|^p}{|x' - y'|^{p+n-2}} \, dx' \, dy' \right)^{1/p}, \tag{17.1}$$

where $d = \operatorname{diam} \widetilde{\Gamma}/2$.

First, following the proof of [59, Theorem 6.26] we take a bell-shaped function $\zeta(x')$ on $\mathbf{R}^{n-1}$ which satisfies the following four conditions:

$$\zeta(x') \in C_0^2(\mathbf{R}^{n-1}). \tag{17.2a}$$

$$\zeta(x') \geq 0 \quad \text{on } \mathbf{R}^{n-1}. \tag{17.2b}$$

$$\operatorname{supp} \zeta \subset \left\{ x' \in \mathbf{R}^{n-1} : |x'| \leq 1 \right\}. \tag{17.2c}$$

$$\int_{\mathbf{R}^{n-1}} \zeta(x') \, dx' = 1. \tag{17.2d}$$

We take an arbitrary point $x_0 = (x_0', 0)$ of the hyperplane $\{x_n = 0\}$ and $R > 0$, and let (see Fig. 17.1)

$$B_R^+ := B_R(x_0) \cap \{x_n > 0\} = \left\{ x = (x', x_n) \in \mathbf{R}^n : |x - x_0| < R, \ x_n > 0 \right\},$$

and

$$\Gamma_R := B_R(x_0) \cap \{x_n = 0\} = \left\{ x = (x', 0) \in \mathbf{R}^n : |x' - x_0'| < R \right\}.$$

Without loss of generality, we may take Γ_R instead of $\widetilde{\Gamma}$ in the above definition of the norm $\|\widetilde{\varphi}\|^*_{B^{1-1/p,p}(\widetilde{\Gamma})}$, and take $d = R$. For a given function $\widetilde{\varphi} \in B^{1-1/p,p}(\Gamma_R)$,

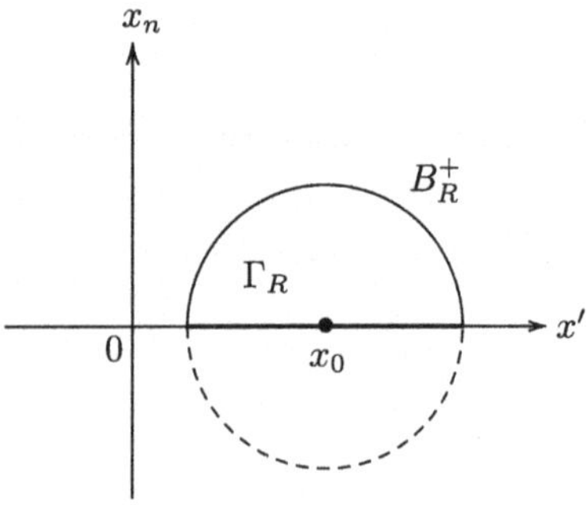

Fig. 17.1 The semi-ball B_R^+ in $\mathbf{R}_+^n$ and the ball $\Gamma_R = B_R \cap \{x_n = 0\}$ in $\mathbf{R}^{n-1}$

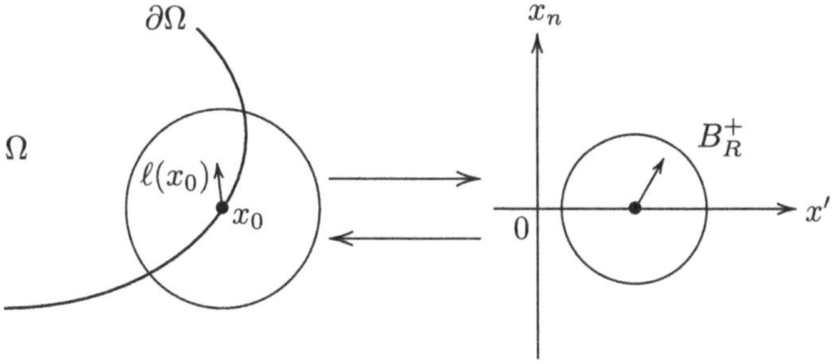

Fig. 17.2 The semi-ball B_R^+ in $\mathbf{R}_+^n$ and the oblique vector field $\ell(x_0)$ at x_0

we may assume that $\widetilde{\varphi}(x')$ can be extended to the whole hyperplane $\{x_n = 0\}$ as a function with compact support, preserving its $B^{1-1/p,p}$-norm.

Assuming that the boundary $\partial\Omega$ is locally flattened out near the point x_0 such that $\Omega \subset \{x_n > 0\}$ (see Fig. 17.2), we remark that the *regular* oblique derivative boundary condition (16.3b) implies that

$$\ell^n(x_0) > 0. \tag{17.3}$$

Consider now the function

$$\phi(x) := \frac{x_n}{\ell^n(x_0)} \int_{\mathbf{R}^{n-1}} \widetilde{\varphi}(x' - x_n y')\zeta(y')\,dy' \quad \text{for } x = \left(x', x_n\right) \in \mathbf{R}_+^n. \tag{17.4}$$

The next lemma is an essential step in our further considerations:

Lemma 17.1 *Let $1 < p < \infty$. The function $\phi(x)$, defined by formula (17.4), belongs to the space $W^{2,p}(B_R^+)$ and satisfies the conditions*

$$\phi(x', 0) = 0 \quad on\ \Gamma_R, \tag{17.5a}$$

$$\frac{\partial \phi}{\partial x_n}(x', 0) = \frac{\widetilde{\varphi}(x')}{\ell^n(x_0)} \quad on\ \Gamma_R. \tag{17.5b}$$

If we introduce a norm

$$\|\phi\|^*_{W^{2,p}(B_R^+)} := \|\phi\|_{L^p(B_R^+)} + R\left\|\nabla^2\phi\right\|_{L^p(B_R^+)}, \tag{17.6}$$

then we have the estimate

$$\|\phi\|^*_{W^{2,p}(B_R^+)} \le C\,R^{1/2}\|\widetilde{\varphi}\|^*_{B^{1-1/p,p}(\Gamma_R)}, \tag{17.7}$$

with a positive constant $C = C(n, p, \lambda, \ell, \zeta)$.

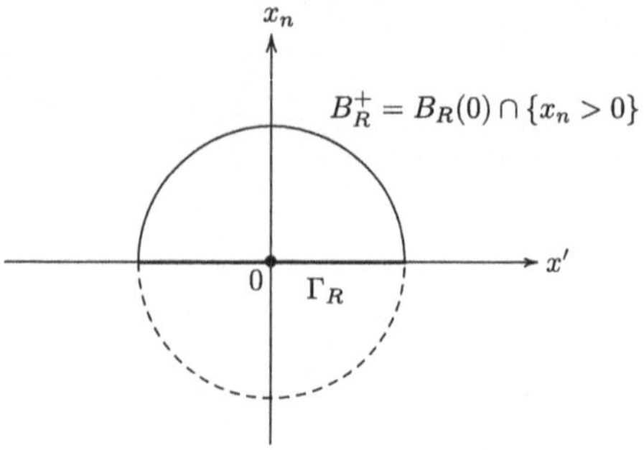

Fig. 17.3 The semi-ball B_R^+ in $\mathbf{R}_+^n$ and the ball $\Gamma_R = B_R \cap \{x_n = 0\}$ in $\mathbf{R}^{n-1}$

Proof The proof of Lemma 17.1 is divided into three steps.

Step 1: First, we prove that, for some positive constant $C_1 = C_1(n, p, \ell, \zeta)$,

$$\|\phi\|_{L^p(B_R^+)} \leq C_1 R^{1+1/p} \|\widetilde{\varphi}\|^*_{B^{1-1/p,p}(\Gamma_R)}. \tag{17.8}$$

Without loss of generality, we may assume (see Fig. 17.3) that

$$\begin{aligned} & x_0 = 0, \\ & B_R^+ = B_R(0) \cap \{x_n > 0\} = \left\{x = \left(x', x_n\right) \in \mathbf{R}^n : |x| < R, x_n > 0\right\}, \end{aligned}$$

and

$$\Gamma_R = B_R(0) \cap \{x_n = 0\} = \left\{x = \left(x', 0\right) \in \mathbf{R}^n : |x'| < R\right\}.$$

By using Minkowski's inequality for integrals (Theorem 3.29) and Fubini's theorem (Theorem 3.20), we obtain that

$$\begin{aligned} & \int_{B_R^+} |\phi(x)|^p \, dx \\ & = \frac{1}{|\ell^n(x_0)|^p} \int_{B_R^+} \left| x_n \int_{\mathbf{R}^{n-1}} \widetilde{\varphi}(x' - x_n y')\zeta(y') \, dy' \right|^p \, dx \\ & \leq C(n, p, \ell, \zeta) \int_{\mathbf{R}^{n-1}} |\zeta(y')|^p \left(\int_{B_R^+} x_n^p |\widetilde{\varphi}(x' - x_n y')|^p \, dx \right) dy' \\ & = C(n, p, \ell, \zeta) \int_{\mathbf{R}^{n-1}} |\zeta(y')|^p I_{B_R^+}(y') \, dy', \end{aligned} \tag{17.9}$$

where

$$I_{B_R^+}(y') := \int_{B_R^+} x_n^p |\widetilde{\varphi}(x' - x_n y')|^p \, dx.$$

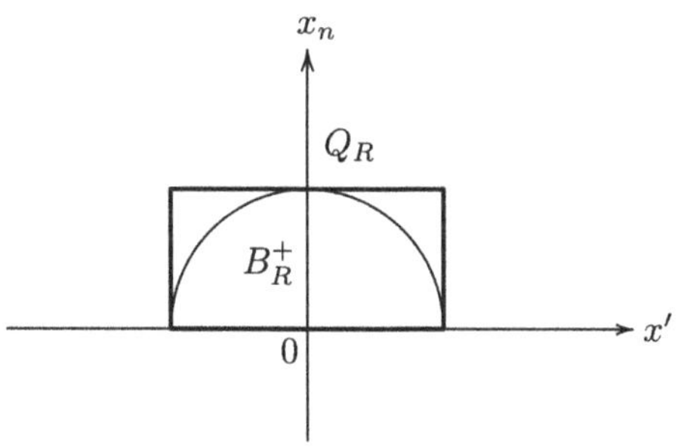

Fig. 17.4 The semi-ball B_R^+ and the cube Q_R in $\mathbf{R}_+^n$

However, by letting (see Fig. 17.4)

$$\begin{aligned} Q_R := \big\{ x = (x_1, x_2, \ldots, x_{n-1}, x_n) \in \mathbf{R}^n : |x_1| \leq R, |x_2| \leq R, \ldots, \\ |x_{n-1}| \leq R,\ 0 \leq x_n \leq R \big\}, \end{aligned}$$

we obtain that

$$\begin{aligned} I_{B_R^+}(y') \leq I_{Q_R}(y') &= \int_{Q_R} x_n^p |\widetilde{\varphi}(x' - x_n y')|^p \, dx' \, dx_n \\ &\leq \int_0^R x_n^p \left(\int_{Q'_R(x_n)} |\widetilde{\varphi}(z')|^p \, dz' \right) dx_n, \end{aligned}$$

where

$$\begin{aligned} z' &= (z_1, z_2, \ldots, z_{n-1}) \\ &= (x_1 - x_n y_1, x_2 - x_n y_2, \ldots, x_{n-1} - x_n y_{n-1}), \end{aligned}$$

and

$$\begin{aligned} Q'_R(x_n) = \big\{ z' = (z_1, z_2, \ldots, z_{n-1}) \in \mathbf{R}^{n-1} : |z_1 + x_n y_1| \leq R, \\ |z_2 + x_n y_2| \leq R, \ldots, |z_n + x_n y_{n-1}| \leq R \big\}. \end{aligned}$$

Since we have, for some positive constant c,

$$\int_{Q'_R(x_n)} |\widetilde{\varphi}(z')|^p \, dz' \leq c \left(\|\widetilde{\varphi}\|^*_{B^{1-1/p,p}(\Gamma_R)} \right)^p \quad \text{for } 0 \leq x_n \leq R,$$

it follows that

$$
\begin{aligned}
I_{B_R^+}(y') &= \int_{B_R^+} x_n^p |\widetilde{\varphi}(x' - x_n y')|^p \, dx \qquad (17.10)\\
&\leq c \left(\|\widetilde{\varphi}\|^*_{B^{1-1/p,p}(\Gamma_R)} \right)^p \int_0^R x_n^p \, dx_n \\
&= \frac{cR^{p+1}}{p+1} \left(\|\widetilde{\varphi}\|^*_{B^{1-1/p,p}(\Gamma_R)} \right)^p \quad \text{for } y' \in \mathbf{R}^{n-1}.
\end{aligned}
$$

Therefore, by combining estimates (17.9) and (17.10) we obtain that

$$
\begin{aligned}
&\int_{B_R^+} |\phi(x)|^p \, dx \\
&\leq C(n, p, \ell, \zeta) \frac{cR^{p+1}}{p+1} \int_{\mathbf{R}^{n-1}} |\zeta(y')|^p \, dy' \left(\|\widetilde{\varphi}\|^*_{B^{1-1/p,p}(\Gamma_R)} \right)^p.
\end{aligned}
$$

This proves the desired estimate (17.8), with

$$
C_1 := \left(\frac{c\, C(n, p, \ell, \zeta)}{p+1} \right)^{1/p} \|\zeta\|_{L^p(\mathbf{R}^{n-1})}.
$$

Step 2: Secondly, we prove that

$$
R \left\| \nabla^2 \phi \right\|_{L^p(B_R^+)} \leq C_2 R^{1/2} \|\widetilde{\varphi}\|^*_{B^{1-1/p,p}(\Gamma_R)}, \qquad (17.11)
$$

for some a positive constant $C_2 = C_2(n, p, \ell, \zeta)$.

To do this, we calculate now the first and second derivatives of the function $\phi(x)$ defined by formula (17.4).

We use the shorthand

$$
D_i = \frac{\partial}{\partial x_i}, \quad D_{ij} = \frac{\partial^2}{\partial x_i \partial x_j}
$$

for derivatives on $\mathbf{R}^n$.

By letting $z' = x' - x_n y'$ in formula (17.4), it follows that

$$
\phi(x', x_n) = \frac{x_n^{2-n}}{\ell^n(x_0)} \int_{\mathbf{R}^{n-1}} \widetilde{\varphi}(z') \zeta \left(\frac{x' - z'}{x_n} \right) dz',
$$

so that

$$
\begin{aligned}
D_i\phi(x', x_n) &= \frac{x_n^{1-n}}{\ell^n(x_0)} \int_{\mathbf{R}^{n-1}} \widetilde{\varphi}(z') \frac{\partial \zeta}{\partial x_i} \left(\frac{x' - z'}{x_n} \right) dz' \quad \text{for } 1 \le i \le n-1; \\
D_n\phi(x', x_n) &= \frac{(2-n)x_n^{1-n}}{\ell^n(x_0)} \int_{\mathbf{R}^{n-1}} \widetilde{\varphi}(z') \zeta \left(\frac{x' - z'}{x_n} \right) dz' \\
&\quad - \frac{x_n^{-n}}{\ell^n(x_0)} \int_{\mathbf{R}^{n-1}} \widetilde{\varphi}(z') \nabla' \zeta \left(\frac{x' - z'}{x_n} \right) \cdot (x' - z')\, dz,
\end{aligned}
$$

where

$$
\nabla' = \left(\frac{\partial}{\partial y_1}, \frac{\partial}{\partial y_2}, \dots, \frac{\partial}{\partial y_{n-1}} \right)
$$

and

$$
x' \cdot y' = \sum_{j=1}^{n-1} x_j y_j.
$$

Hence we have, for the gradient $\nabla \phi = (\partial\phi/\partial x_1, \partial\phi/\partial x_2, \dots, \partial\phi/\partial x_n)$,

$$
\begin{aligned}
D_i\phi(x', x_n) &= \frac{1}{\ell^n(x_0)} \int_{\mathbf{R}^{n-1}} \widetilde{\varphi}(x' - x_n y') D_i \zeta(y')\, dy' \qquad (17.12a)\\
&\quad \text{for } 1 \le i \le n-1; \\
D_n\phi(x', x_n) &= \frac{2-n}{\ell^n(x_0)} \int_{\mathbf{R}^{n-1}} \widetilde{\varphi}(x' - x_n y') \zeta(y')\, dy' \qquad (17.12b)\\
&\quad - \frac{1}{\ell^n(x_0)} \int_{\mathbf{R}^{n-1}} \widetilde{\varphi}(x' - x_n y') \nabla' \zeta(y') \cdot y'\, dy'.
\end{aligned}
$$

However, by conditions (17.2) it follows from an application of the divergence theorem (Theorem 5.3) that

$$
\begin{aligned}
\int_{\mathbf{R}^{n-1}} D_i \zeta(y')\, dy' &= \int_{\mathbf{R}^{n-1}} D_{ij} \zeta(y')\, dy' \qquad (17.13a)\\
&= \int_{\mathbf{R}^{n-1}} \nabla'(D_i\zeta)(y') \cdot y'\, dy' = 0 \quad \text{for } 1 \le i, j \le n-1; \\
\int_{\mathbf{R}^{n-1}} \zeta(y')\, dy' = 1, &\quad \int_{\mathbf{R}^{n-1}} \nabla' \zeta(y') \cdot y'\, dy' = 1 - n, \qquad (17.13b)
\end{aligned}
$$

and

$$
\int_{\mathbf{R}^{n-1}} \nabla' \left(\nabla' \zeta(y') \cdot y' \right) \cdot y'\, dy' = (1-n)^2. \qquad (17.13c)
$$

Therefore, the desired formulas (17.5) follow from formulas (17.4), (17.12b) and (17.13) if we take $x_n = 0$.

Furthermore, since $\zeta(y') \in C_0^2(\mathbf{R}^{n-1})$ we can differentiate formulas (17.12) once again to obtain that

$$
\begin{aligned}
D_{ij}\phi(x', x_n) &= \frac{1}{\ell^n(x_0)} \frac{1}{x_n} \int_{\mathbf{R}^{n-1}} \widetilde{\varphi}(x' - x_n y') D_{ij}\zeta(y')\, dy' \\
&\quad \text{for } 1 \leq i, j \leq n-1; \\
D_{in}\phi(x', x_n) &= \frac{1-n}{\ell^n(x_0)} \frac{1}{x_n} \int_{\mathbf{R}^{n-1}} \widetilde{\varphi}(x' - x_n y') D_i\zeta(y')\, dy' \\
&\quad - \frac{1}{\ell^n(x_0)} \frac{1}{x_n} \int_{\mathbf{R}^{n-1}} \widetilde{\varphi}(x' - x_n y') \nabla'(D_i\zeta)(y') \cdot y'\, dy' \\
&\quad \text{for } 1 \leq i \leq n-1; \\
D_{nn}\phi(x', x_n) &= \frac{(2-n)(1-n)}{\ell^n(x_0)} \frac{1}{x_n} \int_{\mathbf{R}^{n-1}} \widetilde{\varphi}(x' - x_n y') \zeta(y')\, dy' \\
&\quad + \frac{2n-3}{\ell^n(x_0)} \frac{1}{x_n} \int_{\mathbf{R}^{n-1}} \widetilde{\varphi}(x' - x_n y') \nabla'\zeta(y') \cdot y'\, dy' \\
&\quad + \frac{1}{\ell^n(x_0)} \frac{1}{x_n} \int_{\mathbf{R}^{n-1}} \widetilde{\varphi}(x' - x_n y') \nabla' \left(\nabla'\zeta(y') \cdot y'\right) \cdot y'\, dy'.
\end{aligned}
$$

Hence, by combining these three formulas and formulas (17.13) we obtain that

$$
\begin{aligned}
& D_{ij}\phi(x', x_n) \qquad (17.14a) \\
&= \frac{1}{\ell^n(x_0)} \frac{1}{x_n} \int_{\mathbf{R}^{n-1}} \left\{\widetilde{\varphi}(x' - x_n y') - \widetilde{\varphi}(x')\right\} D_{ij}\zeta(y')\, dy' \\
&\quad \text{for } 1 \leq i, j \leq n-1; \\
& D_{in}\phi(x', x_n) \qquad (17.14b) \\
&= \frac{1-n}{\ell^n(x_0)} \frac{1}{x_n} \int_{\mathbf{R}^{n-1}} \left\{\widetilde{\varphi}(x' - x_n y') - \widetilde{\varphi}(x')\right\} D_i\zeta(y')\, dy' \\
&\quad - \frac{1}{\ell^n(x_0)} \frac{1}{x_n} \int_{\mathbf{R}^{n-1}} \left\{\widetilde{\varphi}(x' - x_n y') - \widetilde{\varphi}(x')\right\} \nabla'(D_i\zeta)(y') \cdot y'\, dy' \\
&\quad \text{for } 1 \leq i \leq n-1;
\end{aligned}
$$

and

$$
\begin{aligned}
& D_{nn}\phi(x', x_n) \qquad (17.14c) \\
&= \frac{(2-n)(1-n)}{\ell^n(x_0)} \frac{1}{x_n} \int_{\mathbf{R}^{n-1}} \left\{\widetilde{\varphi}(x' - x_n y') - \widetilde{\varphi}(x')\right\} \zeta(y')\, dy' \\
&\quad + \frac{2n-3}{\ell^n(x_0)} \frac{1}{x_n} \int_{\mathbf{R}^{n-1}} \left\{\widetilde{\varphi}(x' - x_n y') - \widetilde{\varphi}(x')\right\} \nabla'\zeta(y') \cdot y'\, dy' \\
&\quad + \frac{1}{\ell^n(x_0)} \frac{1}{x_n} \int_{\mathbf{R}^{n-1}} \left\{\widetilde{\varphi}(x' - x_n y') - \widetilde{\varphi}(x')\right\} \nabla' \left(\nabla'\zeta(y') \cdot y'\right) \cdot y'\, dy'.
\end{aligned}
$$

Here it should be noticed that the integrals in formulas (17.14) are all of the type

$$\psi(x) = \psi(x', x_n) = \frac{1}{x_n} \int_{\mathbf{R}^{n-1}} \left\{ \widetilde{\varphi}(x' - x_n y') - \widetilde{\varphi}(x') \right\} \mu(y')\, dy', \qquad (17.15)$$

where $\mu(y')$ is (modulo a constant multiplier) one of the functions $\zeta(y')$, $D_i\zeta(y')$, $y' \cdot \nabla'\zeta(y')$, $D_{ij}\zeta(y')$ and $\nabla' \left(\nabla'\zeta \cdot y'\right) \cdot y'$.

However, we can prove that, for some positive constant C_3,

$$\int_{B_R^+} |\psi(x)|^p \, dx \leq C_3 \int_{\Gamma_R} \int_{\Gamma_R} \frac{|\widetilde{\varphi}(x') - \widetilde{\varphi}(y')|^p}{|x' - y'|^{p+n-2}} \, d\sigma_{x'} \, d\sigma_{y'}. \qquad (17.16)$$

The proof of estimate (17.16) will be given in the next Sect. 17.2, due to its length.

In view of definition (17.1), we obtain from formulas (17.14) and (17.15) that, for some positive constants C_4 and C_5,

$$R \left\| \nabla^2 \phi \right\|_{L^p(B_R^+)} \leq C_4 R \|\psi\|_{L^p(B_R^+)} \leq C_5 R^{1/2} \|\widetilde{\varphi}\|^*_{B^{1-1/p,p}(\Gamma_R)}.$$

This proves the desired estimate (17.11).

Step 3: Finally, our main estimate (17.7) follows by combining estimates (17.8) and (17.11).

The proof of Lemma 17.1 is now complete, apart from the proof of estimate (17.16).

17.2 Proof of Estimate (17.16)

The proof is divided into five steps.

Step (I): First, we can rewrite formula (17.15) in the form

$$\begin{aligned}
\psi(x', x_n) &= \frac{1}{x_n} \int_{\mathbf{R}^{n-1}} \left\{ \widetilde{\varphi}(x' - x_n y') - \widetilde{\varphi}(x') \right\} \mu(y')\, dy' \\
&= \frac{1}{x_n^n} \int_{\mathbf{R}^{n-1}} \left\{ \widetilde{\varphi}(x' - z') - \widetilde{\varphi}(x') \right\} \mu\left(\frac{z'}{x_n}\right) dz' \\
&= \frac{1}{x_n^n} \int_{|z'| \leq x_n} \left\{ \widetilde{\varphi}(x' - z') - \widetilde{\varphi}(x') \right\} \mu\left(\frac{z'}{x_n}\right) dz',
\end{aligned}$$

since we have the assertion

$$\operatorname{supp} \mu \subset \left\{ x' \in \mathbf{R}^{n-1} : |x'| \leq 1 \right\}.$$

Therefore, we obtain that

$$\begin{aligned} \left|\psi(x', x_n)\right| &\leq \|\mu\|_{L^\infty(\mathbf{R}^{n-1})} \frac{1}{x_n^n} \\ &= \frac{1}{x_n^n} \int_{|z'| \leq x_n} \left\{\widetilde{\varphi}(x' - z') - \widetilde{\varphi}(x')\right\} dz'. \end{aligned} \tag{17.17}$$

Step (II): Now, if we let

$$w(y') := \|\widetilde{\varphi}(\cdot - y) - \widetilde{\varphi}(\cdot)\|_{L^p(\mathbf{R}^{n-1})} \quad \text{for } y' \in \mathbf{R}^{n-1},$$

then, by applying Minkowski's inequality for integrals (Theorem 3.29) to inequality (17.17) we obtain that

$$\|\psi(\cdot, x_n)\|_{L^p(\mathbf{R}^{n-1})} \leq \|\mu\|_{L^\infty(\mathbf{R}^{n-1})} \frac{1}{x_n^n} \int_{|z'| \leq x_n} w(z')\, dz'. \tag{17.18}$$

Therefore, we obtain from inequality (17.18) that

$$\|\psi(\cdot, \cdot)\|^p_{L^p(\mathbf{R}^n_+)} \leq C_1^p \left\{\int_0^\infty \frac{1}{\rho^{pn}} \left(\int_{|y'| \leq \rho} w(y')\, dy\right)^p d\rho\right\}, \tag{17.19}$$

where

$$C_1 = \|\mu\|_{L^\infty(\mathbf{R}^{n-1})}.$$

Step (III): We estimate the last integral of inequality (17.19). To do this, we rewrite the integral in the form

$$\begin{aligned} &\int_0^\infty \frac{1}{\rho^{pn}} \left(\int_{|y'| \leq \rho} w(y')\, dy'\right)^p d\rho \\ &= \int_0^\infty \frac{1}{\rho^{pn}} \left\{\int_0^\rho \left(\int_{\Sigma_{n-2}} w\,(r\sigma)\, d\sigma\right) r^{n-2}\, dr\right\}^p d\rho \\ &= \int_0^\infty \left(\rho^{-n+1} \int_0^\rho r^{n-2} \frac{h(r)}{\rho^\alpha}\, dr\right)^p d\rho, \end{aligned}$$

where Σ_{n-2} is the unit sphere in $\mathbf{R}^{n-1}$ and

$$h(r) := \int_{\Sigma_{n-2}} w\,(r\sigma)\, d\sigma.$$

Since we have, for $0 < r \leq \rho$,

$$\frac{1}{\rho} \leq \frac{1}{r},$$

it follows that

$$\begin{aligned}
&\int_0^\infty \frac{1}{\rho^{pn}} \left(\int_{|y'|\le\rho} w(y')\,dy' \right)^p d\rho \\
&= \int_0^\infty \left(\rho^{-n+1} \int_0^\rho r^{n-2}\, \frac{h(r)}{\rho}\, dr \right)^p d\rho \\
&\le \int_0^\infty \left(\rho^{-n+1+1/p} \int_0^\rho r^{n-3}\, h(r)\, dr \right)^p \frac{d\rho}{\rho}.
\end{aligned} \tag{17.20}$$

By applying Hardy's inequality (Theorem 3.31) with

$$\begin{aligned}
\gamma &:= -n+1+\frac{1}{p}, \\
f(r) &:= r^{n-3}\, h(r),
\end{aligned}$$

we can estimate the last integral of inequality (17.20) as follows:

$$\begin{aligned}
&\int_0^\infty \left(\rho^{-n+1+1/p} \int_0^\rho r^{n-3}\, h(r)\, dr \right)^p \frac{d\rho}{\rho} \\
&\le \frac{1}{|-n+1+1/p|^p} \left(\int_0^\infty \frac{h(r)^p}{r^p}\, dr \right).
\end{aligned} \tag{17.21}$$

Step (IV): Moreover, we estimate the last integral of inequality (17.21). Since we have, by Hölder's inequality (Theorem 3.24),

$$\begin{aligned}
h(r) &= \int_{\Sigma_{n-2}} w\,(r\sigma)\; d\sigma \\
&\le \left(\int_{\Sigma_{n-2}} w\,(r\sigma)^p\; d\sigma \right)^{1/p} \left(\int_{\Sigma_{n-2}} 1^q\, d\sigma \right)^{1/q} \\
&= \omega_{n-1}^{1/q} \left(\int_{\Sigma_{n-2}} w\,(r\sigma)^p\; d\sigma \right)^{1/p},
\end{aligned}$$

it follows that

$$\begin{aligned}
\int_0^\infty \frac{h(r)^p}{r^p}\, dr &\le \omega_{n-1}^{p/q} \int_0^\infty \frac{1}{r^p} \left(\int_{\Sigma_{n-2}} w\,(r\sigma)^p\; d\sigma \right) dr \\
&= \omega_{n-1}^{p/q} \int_0^\infty \int_{\Sigma_{n-2}} \left(\frac{w(r\sigma)^p}{r^{n+p-2}} \right) r^{n-2}\, dr\, d\sigma \\
&= \omega_{n-1}^{p/q} \int_{\mathbf{R}^{n-1}} \frac{w(y')^p}{|y'|^{n+p-2}}\, dy',
\end{aligned} \tag{17.22}$$

where ω_{n-1} is the surface area of the unit sphere Σ_{n-2}:

$$\omega_{n-1} := |\Sigma_{n-2}| = \frac{2\pi^{(n-1)/2}}{\Gamma((n-1)/2)}.$$

Step (V): By combining inequalities (17.19) through (17.22), we obtain that

$$\begin{aligned}
\|\psi\|^p_{L^p(\mathbf{R}^n_+)} &\le C_1^p \int_0^\infty \frac{1}{\rho^{pn}} \left(\int_{|y'|\le\rho} w(y')\, dy' \right)^p d\rho \\
&\le C_1^p \int_0^\infty \left(\rho^{-n+1+1/p} \int_0^\rho r^{n-3}\, h(r)\, dr \right)^p \frac{d\rho}{\rho} \\
&\le \frac{C_1^p}{|-n+1+1/p|^p} \int_0^\infty \frac{h(r)^p}{r^p}\, dr \\
&\le \frac{C_1^p \omega_{n-1}^{p/q}}{|-n+1+1/p|^p} \int_{\mathbf{R}^{n-1}} \frac{w(y')^p}{|y'|^{n+p-2}}\, dy'.
\end{aligned}$$

However, we have, by Fubini's theorem (Theorem 3.20),

$$\begin{aligned}
\int_{\mathbf{R}^{n-1}} \frac{w(y')^p}{|y'|^{n+p-2}}\, dy' &= \int_{\mathbf{R}^{n-1}} \frac{\|\widetilde{\varphi}(\cdot - y') - \widetilde{\varphi}(\cdot)\|^p_{L^p(\mathbf{R}^{n-1})}}{|y'|^{n+p-2}}\, dy' \\
&= \int_{\mathbf{R}^{n-1}} \int_{\mathbf{R}^{n-1}} \frac{|\widetilde{\varphi}(x'-y') - \widetilde{\varphi}(x')|^p}{|y'|^{n+p-2}}\, dx'\, dy'.
\end{aligned}$$

Summing up, we have proved that

$$\|\psi\|_{L^p(\mathbf{R}^n_+)} \le C_2 \left(\int_{\mathbf{R}^{n-1}} \int_{\mathbf{R}^{n-1}} \frac{\left|\widetilde{\varphi}(x'-y') - \widetilde{\varphi}(x')\right|^p}{|y'|^{n+p-2}}\, dx'\, dy' \right)^{1/p}, \tag{17.23}$$

where

$$C_2 := \frac{C_1\, \omega_{n-1}^{1/q}}{n-1-1/p}.$$

The desired estimate (17.16)

$$\int_{B_R^+} |\psi(x)|^p\, dx \le C_3 \int_{\Gamma_R} \int_{\Gamma_R} \frac{|\widetilde{\varphi}(x') - \widetilde{\varphi}(y')|^p}{|x'-y'|^{p+n-2}}\, d\sigma_{x'}\, d\sigma_{y'}$$

follows from inequality (17.23).

The proof of estimate (17.16) (and hence that of Lemma 17.1) is now complete. □

Chapter 18
Boundary Representation Formula for Solutions

In this chapter we prove boundary representation formulas for solutions of problem (16.4), by using the half space Green function for the uniformly elliptic differential operator

$$\mathcal{L} = \sum_{i,j=1}^{n} a^{ij}(x) \frac{\partial^2}{\partial x_i \partial x_j}.$$

The first step toward the boundary representation formulas is to derive a boundary representation formula (18.4) for the solution of the non-homogeneous oblique derivative problem (16.4) with the constant coefficients differential operator $\mathcal{L}_0$ and the constant coefficients boundary operator $\mathcal{B}_0$ (Lemma 18.1). The second step is to derive integral representation formulas (18.8) for the second derivatives of solutions of the oblique derivative problem for the variable coefficients differential operator $\mathcal{L}$ and the constant coefficients boundary operator $\mathcal{B}_0$ (Theorem 18.3). The third step for the general couple $(\mathcal{L}, \mathcal{B})$ will be carried out in Chap. 19 (Lemma 19.1).

Now we take an arbitrary point $x_0 = (x_0', 0)$ of the hyperplane $\{x_n = 0\}$ and $r > 0$, and we let (see Fig. 18.1 below)

$$\begin{aligned} B_r &:= B_r(x_0) = \left\{x \in \mathbf{R}^n : |x - x_0| < r\right\}, \\ B_r^+ &:= B_r(x_0) \cap \{x_n > 0\} \\ &= \left\{x = (x_1, x_2, \dots, x_n) \in \mathbf{R}^n : |x - x_0| < r,\ x_n > 0\right\}, \end{aligned}$$

and

$$\begin{aligned} C_r &:= B_r(x_0) \cap \{x_n = 0\} \\ &= \left\{x = (x_1, x_2, \dots, x_{n-1}, 0) \in \mathbf{R}^n : |x' - x_0'| < r\right\}. \end{aligned}$$

K. Taira, *Real Analysis Methods for Markov Processes*,
https://doi.org/10.1007/978-981-97-3659-1_18

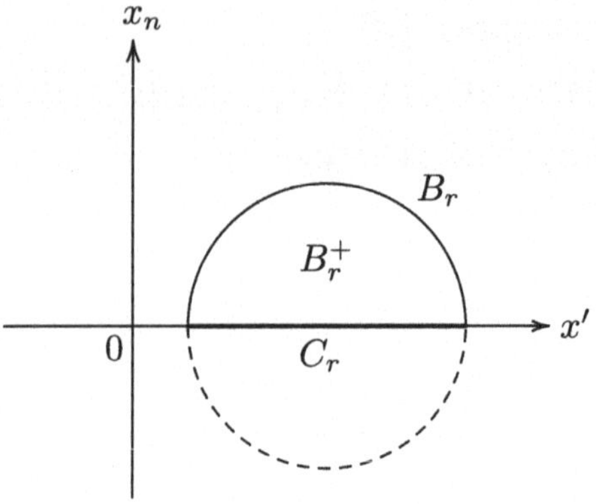

Fig. 18.1 The semi-ball B_r^+ in $\mathbf{R}^n_+$ and the ball $C_r = B_r \cap \{x_n = 0\}$ in $\mathbf{R}^{n-1}$

We consider the second order, uniformly elliptic differential operator

$$\mathcal{L}_0 := \sum_{i,j=1}^{n} a^{ij}(x_0) \frac{\partial^2}{\partial x_i \partial x_j}$$

with constant coefficients, and the first order boundary operator

$$\mathcal{B}_0 u := \frac{\partial u}{\partial \ell(x_0')} + \sigma(x_0')\, u = \sum_{i=1}^{n} \ell^i(x_0') \frac{\partial u}{\partial x_i} + \sigma(x_0')\, u$$

prescribed by a directional derivative with respect to the constant vector field

$$\ell(x_0') = \left(\ell^1(x_0'), \ell^2(x_0'), \ldots, \ell^n(x_0')\right).$$

Here we assume that (see conditions (16.3c) and (16.3b))

$$\sigma(x_0') < 0, \quad \ell^n(x_0') > 0. \tag{18.1}$$

In the following we shall denote the matrix $(a^{ij}(x_0))$ by $\boldsymbol{a}(x_0)$ and its inverse matrix $(A^{ij}(x_0))$ by $\boldsymbol{A}(x_0)$, respectively. Then the fundamental solution $\Gamma(x_0, \xi)$ of $\mathcal{L}_0$ is given by the formula

$$\Gamma(x_0, \xi) = \frac{1}{(2-n)\omega_n \sqrt{\det \boldsymbol{a}(x_0)}} \left(\sum_{i,j=1}^{n} A^{ij}(x_0) \xi_i \xi_j \right)^{(2-n)/2},$$

where

$$\omega_n := |\Sigma_{n-1}| = \frac{2\pi^{n/2}}{\Gamma(n/2)}$$

is the surface area of the unit sphere Σ_{n-1} in $\mathbf{R}^n$. By arguing just as in Gilbarg–Trudinger [59, Sect. 6.7], we can verify that the half space Green function $G(x_0, x, y)$ for the constant coefficients elliptic differential operator $\mathcal{L}_0$ is given by the formula (cf. [59, Sect. 6.7, formula (6.62)])

$$\begin{aligned} &G(x_0, x, y) \\ &:= \Gamma(x_0, x - y) - \Gamma(x_0, T(x; x_0) - y) + \theta(x_0, T(x; x_0) - y), \end{aligned} \tag{18.2}$$

where

$$T(x; x_0) := x - \frac{2x_n}{a^{nn}(x_0)} \boldsymbol{a}^n(x_0) = \begin{pmatrix} x_1 - 2x_n \frac{a^{1n}(x_0)}{a^{nn}(x_0)} \\ x_2 - 2x_n \frac{a^{2n}(x_0)}{a^{nn}(x_0)} \\ \vdots \\ -x_n \end{pmatrix}, \quad x = \begin{pmatrix} x_1 \\ x_2 \\ \vdots \\ x_n \end{pmatrix},$$

and

$$\begin{aligned} &\theta(x_0, \xi) \\ &= \frac{2}{\omega_n \sqrt{\det \boldsymbol{a}(x_0)}} \frac{\ell^n(x_0')}{a^{nn}(x_0)} \\ &\times \int_0^\infty \frac{e^{\sigma(x_0')s} (\xi_n + s\, T_n(\ell(x_0')))}{\left(\sum_{i,j=1}^n A^{ij}(x_0)(\xi_i + s\, T_i(\ell(x_0')))(\xi_j + s\, T_j(\ell(x_0')))\right)^{n/2}} \, ds, \end{aligned} \tag{18.3}$$

with

$$\begin{aligned} \xi + s\, T\left(\ell(x_0')\right) &= \begin{pmatrix} \xi_1 + s\, T_1\left(\ell(x_0')\right) \\ \xi_2 + s\, T_2\left(\ell(x_0')\right) \\ \vdots \\ \xi_n + s\, T_n\left(\ell(x_0')\right) \end{pmatrix} \\ &= \begin{pmatrix} \xi_1 + s\, \ell^1(x_0') - 2s\, \ell^n(x_0') \frac{a^{1n}(\ell(x_0'))}{a^{nn}(\ell(x_0'))} \\ \xi_2 + s\, \ell^2(x_0') - 2s\, \ell^n(x_0') \frac{a^{2n}(\ell(x_0'))}{a^{nn}(\ell(x_0'))} \\ \vdots \\ \xi_n - s\, \ell^n(x_0') \end{pmatrix}. \end{aligned}$$

More precisely, we can prove the following lemma (see [40, Lemma 3.1], [99, Chap. 2, Lemma 2.2.5]):

Lemma 18.1 *Assume that condition* (18.1) *is satisfied. If a function* $u \in C_0^\infty(B_{2r})$, *with* $\operatorname{supp} u \subset B_r$, *is a solution of the homogeneous oblique derivative problem (see Fig. 18.2 below)*

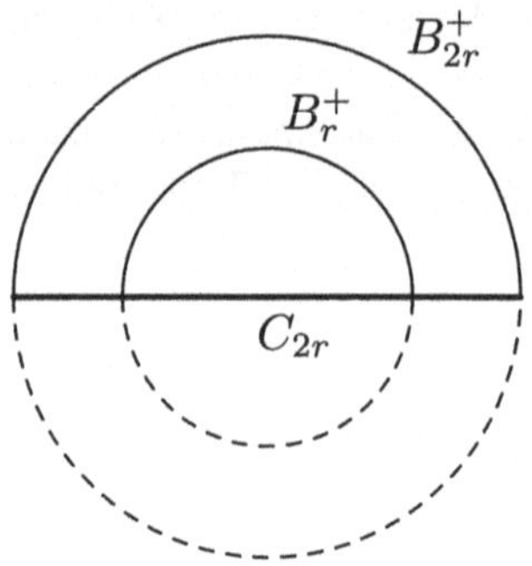

Fig. 18.2 The semi-ball B_{2r}^+ in $\mathbf{R}_+^n$ and the ball $C_{2r} = B_{2r} \cap \{x_n = 0\}$ in $\mathbf{R}^{n-1}$

$$\begin{cases} \mathcal{L}_0 u = f & \text{in } B_{2r}^+, \\ \mathcal{B}_0 u = 0 & \text{on } C_{2r}, \end{cases}$$

then we have, for all $x \in B_r^+$,

$$\begin{aligned} u(x) & \qquad\qquad (18.4) \\ &= \int_{B_{2r}^+} G(x_0, x, y) f(y)\, dy \\ &= \int_{B_{2r}^+} \left(\Gamma(x_0, x-y) - \Gamma(x_0, T(x; x_0) - y) + \theta(x_0, T(x; x_0) - y)\right) f(y)\, dy. \end{aligned}$$

Let B_r be an open ball of radius r, and we assume that the functions $a^{ij} \in \mathrm{VMO} \cap L^\infty(\Omega)$ satisfy the following two conditions:

(i) $a^{ij}(x) = a^{ji}(x)$ for almost all $x \in B_r$.
(ii) There exists a positive constant λ such that

$$\frac{1}{\lambda} |\xi|^2 \leq \sum_{i,j=1}^n a^{ij}(x)\xi_i\xi_j \leq \lambda |\xi|^2 \quad \text{for almost all } x \in B_r \text{ and all } \xi \in \mathbf{R}^n.$$

If $\widetilde{B}_r$ is the subset of B_r where conditions (i) and (ii) hold true, then we let

$$\Gamma(x, \xi) := \frac{1}{(2-n)\omega_n} \frac{1}{\sqrt{\det \boldsymbol{a}(x)}} \left(\sum_{i,j=1}^n A_{ij}(x)\xi_i\xi_j\right)^{(2-n)/2}, \quad \text{for all } x \in \widetilde{B}_r \text{ and all } \xi \in \mathbf{R}^n \setminus \{0\},$$

and

$$\theta(x, \xi)$$

$$= \frac{2}{\omega_n\sqrt{\det \boldsymbol{a}(x)}}\frac{\ell^n(x')}{a^{nn}(x)}$$
$$\times \int_0^\infty \frac{e^{\sigma(x')s}(\xi_n + sT_n\left(\ell(x')\right))}{\left(\sum_{i,j=1}^n A^{ij}(x)\left(\xi_i + sT_i\left(\ell(x')\right)\right)\left(\xi_j + sT_j\left(\ell(x')\right)\right)\right)^{n/2}}\, ds.$$

Here:

$$(A_{ij}(x)) = \text{ the inverse matrix of } \mathbf{a}(x) = (a^{ij}(x)),$$
$$\omega_n := |\Sigma_{n-1}| = \frac{2\pi^{n/2}}{\Gamma(n/2)} \quad \text{(the surface area of the unit sphere } \Sigma_{n-1} \text{ in } \mathbf{R}^n\text{)},$$

and

$$\xi + sT\left(\ell(x')\right) = \begin{pmatrix} \xi_1 + sT_1\left(\ell(x')\right) \\ \xi_2 + sT_2\left(\ell(x')\right) \\ \vdots \\ \xi_n + sT_n\left(\ell(x')\right) \end{pmatrix}$$
$$= \begin{pmatrix} \xi_1 + s\ell^1(x') - 2s\ell^n(x')\frac{a^{1n}(\ell(x'))}{a^{nn}(\ell(x'))} \\ \xi_2 + s\ell^2(x_0') - 2s\ell^n(x')\frac{a^{2n}(\ell(x'))}{a^{nn}(\ell(x'))} \\ \vdots \\ \xi_n - s\ell^n(x') \end{pmatrix}.$$

The next lemma states an important property of the function $\theta(x, \xi)$ (see [40, Remark 3.1]):

Lemma 18.2 *The function $\theta\left(x, T(x; x_0) - y\right)$ satisfies the estimate*

$$\left|D_\xi^\alpha \theta\left(x, T(x; x_0) - y\right)\right| \le \frac{C_\alpha}{|T(x; x_0) - y|^{n-2+|\alpha|}}, \tag{18.5}$$

with a positive constant $C_\alpha = C(n, |\alpha|, \ell, \boldsymbol{a})$.

Proof The proof is divided into three steps.

Step 1: First, we let

$$\psi(x; \tau, \eta)$$
$$:= \frac{2}{\omega_n\sqrt{\det \boldsymbol{a}(x)}}\frac{\ell^n(x')}{a^{nn}(x)}$$
$$\times \int_0^\infty \frac{e^{\sigma(x')\tau t}\left(\eta_n + tT_n\left(\ell(x')\right)\right)}{\left(\sum_{i,j=1}^n A^{ij}(x)\left(\eta_i + tT_i\left(\ell(x')\right)\right)\left(\eta_j + tT_j\left(\ell(x')\right)\right)\right)^{n/2}}\, dt,$$

where

$$\eta_i := \frac{\xi_i}{|T(x; x_0) - y|} \quad \text{for } 1 \leq i \leq n.$$

Then it is easy to see that

$$\begin{aligned}
&\int_0^\infty \frac{e^{\sigma(x')s}\left(\xi_n + sT_n\left(\ell(x')\right)\right)}{\left(\sum_{i,j=1}^n A^{ij}(x)\left(\xi_i + sT_i\left(\ell(x')\right)\right)\left(\xi_j + sT_j(\ell(x'))\right)\right)^{n/2}}\, ds \\
&= \frac{1}{|T(x; x_0) - y|^{n-2}} \\
&\quad \times \int_0^\infty \frac{e^{\sigma(x')|T(x;x_0)-y|t}\left(\eta_n + tT_n\left(\ell(x')\right)\right)}{\left(\sum_{i,j=1}^n A^{ij}(x)\left(\eta_i + tT_i\left(\ell(x')\right)\right)\left(\eta_j + tT_j\left(\ell(x')\right)\right)\right)^{n/2}}\, dt.
\end{aligned}$$

This proves that the function $\theta(x, \xi)$ is expressed in the form

$$\theta(x, \xi) = \frac{1}{|T\,(x; x_0) - y|^{n-2}} \psi\,(x; |T(x; x_0) - y|, \eta), \tag{18.6}$$

with

$$\eta = \frac{\xi}{|T(x; x_0) - y|}. \tag{18.7}$$

Step 2: Secondly, by condition (18.1) it follows that the angle between the vectors η and $T(\ell(x'))$ is less than π for all $\eta \in \mathbf{R}^n_- = \{x_n < 0\}$ (see Fig. 18.3 below). Hence we have, for some constant $0 < \delta_0 < 1$,

$$\left\langle \eta, T(\ell(x'))\right\rangle \geq -|\eta|\, |T(\ell(x'))|\delta_0,$$

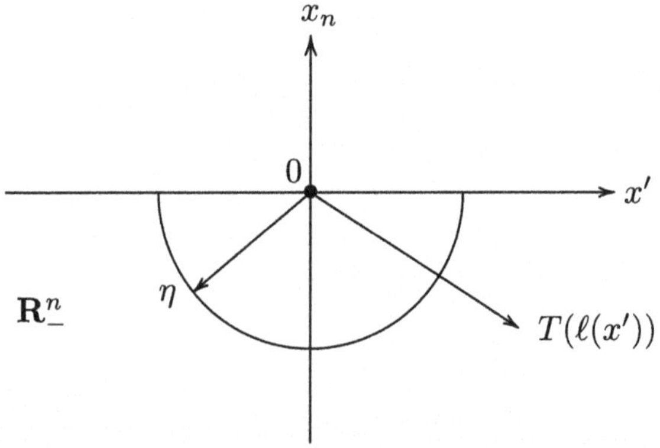

Fig. 18.3 The angle between the vectors $T(\ell(x'))$ and $\eta \in \mathbf{R}^n_- = \{x_n < 0\}$

and so

$$\begin{aligned}|\eta + tT(\ell(x'))|^2 &= |\eta|^2 + 2t\left\langle \eta, T(\ell(x'))\right\rangle + t^2|T\left(\ell(x')\right)|^2 \\ &\geq |\eta|^2 - 2t|\eta|\, |T\left(\ell(x')\right)|\delta_0 + t^2|T\left(\ell(x')\right)|^2 \\ &= |T\left(\ell(x')\right)|^2\left(t - \frac{\delta_0}{|T\left(\ell(x')\right)|}|\eta|\right)^2 + \left(1-\delta_0^2\right)|\eta|^2 \\ &\geq \left(1-\delta_0^2\right)|\eta|^2.\end{aligned}$$

Moreover, we have, for all $t \geq 4|\eta|\delta_0/|T(\ell(x'))|$,

$$\begin{aligned}&|\eta + tT(\ell(x'))|^2 \\ &= |\eta|^2 + 2t\left\langle \eta, T(\ell(x'))\right\rangle + t^2|T(\ell(x'))|^2 \\ &\geq |\eta|^2 - 2t|\eta||T(\ell(x'))|\delta_0 + t^2|T(\ell(x'))|^2 \\ &= |\eta|^2 + \frac{t}{2}\left(t|T(\ell(x'))|^2 - 4|\eta|\delta_0|T(\ell(x'))|\right) + \frac{t^2}{2}|T(\ell(x'))|^2 \\ &\geq |\eta|^2 + \frac{t^2}{2}\left|T(\ell(x'))\right|^2.\end{aligned}$$

By the positivity of $\boldsymbol{A}$, we obtain that, for some positive constant c_0,

$$\begin{aligned}&\sum_{i,j=1}^{n} A^{ij}(x)(\eta_i + tT_i(\ell(x')))(\eta_j + tT_j(\ell(x'))) \\ &\geq c_0|\eta + tT(\ell(x'))|^2 \\ &\geq \begin{cases} c_0\left(1-\delta_0^2\right)|\eta|^2 & \text{for all } t \geq 0, \\ c_0\left(|\eta|^2 + \frac{t^2}{2}|T(\ell(x'))|^2\right) & \text{for all } t \geq \frac{4|\eta|\delta_0}{|T(\ell(x'))|}. \end{cases}\end{aligned}$$

Therefore, we find that $\psi(x;\tau,\eta)$ is a smooth function of (τ,η) for $\tau > 0$ and $\eta \in \mathbf{R}^n_-$, since the denominator of the integrand is bounded away from zero and since $n \geq 3$.

Step 3: Finally, it is easy to see that, for some positive constant $C'_\alpha = C'(n, |\alpha|, \ell, \boldsymbol{a})$,

$$\left| D_\eta^\alpha \psi\left(x; |T(x;x_0) - y|, \frac{T(x;x_0) - y}{|T(x;x_0) - y|}\right)\right| \leq C'_\alpha.$$

This proves the desired estimate (18.5), since we have, by formulas (18.6) and (18.7),

$$D_\xi^\alpha \theta(x,\xi) = \frac{1}{|T(x;x_0) - y|^{|\alpha|+n-2}} D_\eta^\alpha \psi\left(x; |T(x;x_0) - y|, \eta\right).$$

The proof of Lemma 18.2 is complete. □

In the following we shall use the notation

$$\Gamma_i(x,\xi) = \frac{\partial}{\partial \xi_i}\Gamma(x,\xi),$$
$$\Gamma_{ij}(x,\xi) = \frac{\partial^2}{\partial \xi_i \partial \xi_j}\Gamma(x,\xi);$$

and

$$\theta_i(x,\xi) = \frac{\partial}{\partial \xi_i}\theta(x,\xi),$$
$$\theta_{ij}(x,\xi) = \frac{\partial^2}{\partial \xi_i \partial \xi_j}\theta(x,\xi).$$

The next theorem gives integral representation formulas for the second derivatives of solutions of the homogeneous oblique derivative problem for the variable coefficients, uniformly elliptic differential operator

$$\mathcal{L} = \sum_{i,j=1}^{n} a^{ij}(x)\frac{\partial^2}{\partial x_i \partial x_j}$$

and the constant coefficients boundary operator

$$\mathcal{B}_0 = \frac{\partial}{\partial \ell(x_0')} + \sigma(x_0') = \sum_{i=1}^{n} \ell^i(x_0')\frac{\partial}{\partial x_i} + \sigma(x_0').$$

Theorem 18.3 *Let $\widetilde{B}_r$ be the subset of B_r where conditions (i) and (ii) hold true, and let $f \in L^p\left(B_r^+\right)$ for $1 < p < \infty$. If a function $u \in W^{2,p}\left(B_r^+\right)$ is a solution of the oblique derivative problem*

$$\begin{cases} \mathcal{L}u = f & \text{in } B_r^+, \\ \mathcal{B}_0 u = 0 & \text{on } C_r, \end{cases}$$

then we have, for all $x \in \widetilde{B}_r^+ := \widetilde{B}_r \cap \{x_n > 0\}$,

$$\begin{aligned} &\frac{\partial^2 u}{\partial x_i \partial x_j}(x) \qquad (18.8)\\ &= \text{v.p.} \int_{B_r^+} \Gamma_{ij}(x_0, x-y) \\ &\times \left\{ \sum_{h,k=1}^{n} \left\{a^{hk}(x_0) - a^{hk}(y)\right\} \frac{\partial^2 u}{\partial x_h \partial x_k}(y) + f(y) \right\} dy \\ &+ c_{ij}(x_0) f(x) - I_{ij}(x; x_0) + J_{ij}(x; x_0) \quad \text{for } 1 \le i, j \le n. \end{aligned}$$

Here:

$$c_{ij}(x_0) := \int_{|\xi|=1} \Gamma_i(x_0, t) t_j \, d\sigma_t; \tag{18.9}$$

and the terms $I_{ij}(x; x_0)$ are defined respectively as follows:

$$\begin{aligned} &I_{ij}(x; x_0) \quad (1 \leq i, j \leq n-1) \\ &:= \int_{B_r^+} \Gamma_{ij}\,(x_0, T(x; x_0) - y) \\ &\quad \times \left\{ \sum_{h,k=1}^{n} \left\{a^{hk}(x_0) - a^{hk}(y)\right\} \frac{\partial^2 u}{\partial x_h \partial x_k}(y) + f(y) \right\} dy. \end{aligned} \tag{18.10a}$$

$$\begin{aligned} &I_{in}(x; x_0) = I_{ni}(x; x_0) \quad (1 \leq i \leq n-1) \\ &:= \int_{B_r^+} \sum_{\ell=1}^{n} \Gamma_{i\ell}(x_0, T(x; x_0) - y) B_\ell(x_0) \\ &\quad \times \left\{ \sum_{h,k=1}^{n} \left\{a^{hk}(x_0) - a^{hk}(y)\right\} \frac{\partial^2 u}{\partial x_h \partial x_k}(y) + f(y) \right\} dy. \end{aligned} \tag{18.10b}$$

$$\begin{aligned} I_{nn}(x; x_0) := &\int_{B_r^+} \sum_{\ell,m=1}^{n} \Gamma_{\ell m}\,(x_0, T(x; x_0) - y)\, B_\ell(x_0) B_m(x_0) \\ &\times \left\{ \sum_{h,k=1}^{n} \left\{a^{hk}(x_0) - a^{hk}(y)\right\} \frac{\partial^2 u}{\partial x_h \partial x_k}(y) + f(y) \right\} dy. \end{aligned} \tag{18.10c}$$

The terms $J_{ij}(x; x_0)$ are defined respectively as follows:

$$\begin{aligned} &J_{ij}(x; x_0) \quad (1 \leq i, j \leq n-1) \\ &:= \int_{B_r^+} \theta_{ij}\,(x_0, T(x; x_0) - y) \\ &\quad \times \left\{ \sum_{h,k=1}^{n} \left\{a^{hk}(x_0) - a^{hk}(y)\right\} \frac{\partial^2 u}{\partial x_h \partial x_k}(y) + f(y) \right\} dy. \end{aligned} \tag{18.11a}$$

$$\begin{aligned} &J_{in}(x; x_0) = J_{ni}(x; x_0) \quad (1 \leq i \leq n-1) \\ &:= \int_{B_r^+} \sum_{j=1}^{n} \theta_{ij}(x_0, T(x; x_0) - y) B_j(x_0) \\ &\quad \times \left\{ \sum_{h,k=1}^{n} \left\{a^{hk}(x_0) - a^{hk}(y)\right\} \frac{\partial^2 u}{\partial x_h \partial x_k}(y) + f(y) \right\} dy. \end{aligned} \tag{18.11b}$$

$$J_{nn}(x; x_0) := \int_{B_r^+} \sum_{i,j=1}^{n} \theta_{ij}(x_0, T(x; x_0) - y) B_i(x_0) B_j(x_0) \tag{18.11c}$$
$$\times \left\{ \sum_{h,k=1}^{n} \left\{a^{hk}(x_0) - a^{hk}(y)\right\} \frac{\partial^2 u}{\partial x_h \partial x_k}(y) + f(y) \right\} dy.$$

Moreover, the map $T\,(x; x_0)$ is defined by the formula

$$T\,(x; x_0) := \begin{pmatrix} x_1 - 2x_n \frac{a^{1n}(x_0)}{a^{nn}(x_0)} \\ x_2 - 2x_n \frac{a^{2n}(x_0)}{a^{nn}(x_0)} \\ \vdots \\ -x_n \end{pmatrix}, \quad x = \begin{pmatrix} x_1 \\ x_2 \\ \vdots \\ x_n \end{pmatrix},$$

and the vector $B(x_0)$ is defined by the formula

$$B(x_0) = \begin{pmatrix} B_1(x_0) \\ B_2(x_0) \\ \vdots \\ B_n(x_0) \end{pmatrix} := \frac{\partial T}{\partial x_n}(x; x_0) = \begin{pmatrix} -2\frac{a^{1n}(x_0)}{a^{nn}(x_0)} \\ -2\frac{a^{2n}(x_0)}{a^{nn}(x_0)} \\ \vdots \\ -1 \end{pmatrix}.$$

Proof By a density argument, it suffices to prove formula (18.8) for all $u \in C^\infty\left(B_r^+\right)$. Indeed, the general case can be proved by using Theorem 14.5 (and Remark 14.12). More precisely, we obtain the following three assertions:

(I) If we let

$$\widetilde{K}_{ij} f(x) := \int_{\mathbf{R}_+^n} \Gamma_{ij}\,(x, T(x; x_0) - y)\, f(y)\, dy,$$

then there exists a positive constant $C_1 = C_1(n, p, \lambda, M)$ such that (Theorem 14.5)

$$\left\| \widetilde{K}_{ij} f \right\|_{L^p(\mathbf{R}_+^n)} \leq C_1 \, \|f\|_{L^p(\mathbf{R}_+^n)} \quad \text{for all } f \in L^p(\mathbf{R}_+^n).$$

(II) Let $a \in L^\infty(\mathbf{R}^n)$. If we let

$$\widetilde{C}\left[a, K_{ij}\right] f(x) := \int_{\mathbf{R}_+^n} \Gamma_{ij}\,(x, T(x; x_0) - y)\, |a(x) - a(y)| f(y)\, dy,$$

then there exists a positive constant $C_2 = C_2(n, p, \lambda, M)$ such that (Theorem 14.5)

$$\left\| \widetilde{C}[a, K_{ij}] f \right\|_{L^p(\mathbf{R}_+^n)} \leq C_2 \, \|a\|_* \, \|f\|_{L^p(\mathbf{R}_+^n)} \quad \text{for all } f \in L^p(\mathbf{R}_+^n).$$

(III) Let $a \in L^\infty(\mathbf{R}^n)$. If we let

$$\widetilde{C}[a, \Theta_{ij}]f(x) := \int_{\mathbf{R}^n_+} \theta_{ij}(x, T(x; x_0) - y)[a(x) - a(y)]f(y)\, dy,$$

then there exists a positive constant $C_3 = C_3(n, p, \lambda, M)$ such that

$$\|\widetilde{C}[a, \Theta_{ij}]f\|_{L^p(\mathbf{R}^n_+)} \leq C_3\|a\|_*\|f\|_{L^p(\mathbf{R}^n_+)} \quad \text{for all } f \in L^p(\mathbf{R}^n_+).$$

Indeed, it suffices to note that we have, by Lemma 18.2 with $|\alpha| := 2$ and Lemma 19.2 in the next Chap. 19,

$$\left|\theta_{ij}(x_0, T(x; x_0) - y)\right| \leq \frac{C}{|T(x; x_0) - y|^n} \leq \frac{C'}{|\tilde{x} - y|^n},$$

with some positive constants C and C'.

Let x_0 be an arbitrary point of $\widetilde{B}_r$. Since we have the formula

$$\mathcal{L}u(x) = \sum_{i,j=1}^n a^{ij}(x) \frac{\partial^2 u}{\partial x_i \partial x_j}(x) = f(x),$$

it follows that

$$\begin{aligned} \mathcal{L}_0 u(x) &= \sum_{i,j=1}^n a^{ij}(x_0) \frac{\partial^2 u}{\partial x_i \partial x_j}(x) \\ &= \sum_{i,j=1}^n \left\{a^{ij}(x_0) - a^{ij}(x)\right\} \frac{\partial^2 u}{\partial x_i \partial x_j}(x) + f(x) \quad \text{in } B_r^+, \end{aligned}$$

and further that

$$\mathcal{B}_0 u(x') = 0 \quad \text{on } C_r.$$

However, it should be noticed that the half space Green function $G(x_0, x, y)$ for the operator $\mathcal{L}_0$ is given by formula (18.2)

$$\begin{aligned} G(x_0, x, y) := \Gamma(x_0, x - y) - \Gamma(x_0, T(x; x_0) - y) \\ + \theta(x_0, T(x; x_0) - y), \end{aligned} \tag{18.2}$$

Hence we have the formula

$$
\begin{aligned}
& u(x) \qquad (18.12)\\
&= \int_{B_r^+} G(x_0, x, y) \left(\sum_{i,j=1}^{n} \left\{ a^{ij}(x_0) - a^{ij}(y) \right\} \frac{\partial^2 u}{\partial x_i \partial x_j}(y) + f(y) \right) dy \\
&= \int_{B_r^+} \Gamma(x_0, x - y) \left(\sum_{i,j=1}^{n} \left\{ a^{ij}(x_0) - a^{ij}(y) \right\} \frac{\partial^2 u}{\partial x_i \partial x_j}(y) + f(y) \right) dy \\
&\quad - \int_{B_r^+} \Gamma(x_0, T(x; x_0) - y) \\
&\quad - \int_{B_r^+} \times \left(\sum_{i,j=1}^{n} \left\{ a^{ij}(x_0) - a^{ij}(y) \right\} \frac{\partial^2 u}{\partial x_i \partial x_j}(y) + f(y) \right) dy \\
&\quad + \int_{B_r^+} \theta(x_0, T(x; x_0) - y) \\
&\quad + \int_{B_r^+} \times \left(\sum_{i,j=1}^{n} \left\{ a^{ij}(x_0) - a^{ij}(y) \right\} \frac{\partial^2 u}{\partial x_i \partial x_j}(y) + f(y) \right) dy \\
&:= H(x; x_0) - I(x; x_0) + J(x; x_0).
\end{aligned}
$$

We can differentiate the first term

$$
\begin{aligned}
& H(x; x_0) \\
&= \int_{B_r^+} \Gamma(x_0, x - y) \left(\sum_{i,j=1}^{n} \left\{ a^{ij}(x_0) - a^{ij}(y) \right\} \frac{\partial^2 u}{\partial x_i \partial x_j}(y) + f(y) \right) dy
\end{aligned}
$$

in formula (18.12) twice to obtain that

$$
\begin{aligned}
& \frac{\partial^2 H}{\partial x_i \partial x_j}(x; x_0) \qquad (18.13)\\
&= \text{v.p.} \int_{B_r^+} \Gamma_{ij}(x_0, x - y) \\
&\quad \times \left(\sum_{h,k=1}^{n} \left\{ a^{hk}(x_0) - a^{hk}(y) \right\} \frac{\partial^2 u}{\partial x_h \partial x_k}(y) + f(y) \right) dy \\
&\quad + f(x) \left(\int_{|t|=1} \Gamma_i(x_0, t) t_j \, d\sigma_t \right) \quad \text{for } 1 \le i, j \le n.
\end{aligned}
$$

As for the two terms

$$I(x; x_0) = \int_{B_r^+} \left(\sum_{i,j=1}^{n} \left\{a^{ij}(x_0) - a^{ij}(y)\right\} \frac{\partial^2 u}{\partial x_i \partial x_j}(y) + f(y) \right) dy,$$

$$J(x; x_0) = \int_{B_r^+} \left(\sum_{i,j=1}^{n} \left\{a^{ij}(x_0) - a^{ij}(y)\right\} \frac{\partial^2 u}{\partial x_i \partial x_j}(y) + f(y) \right) dy$$

in formula (18.12), we can differentiate them under the integral sign to obtain that

$$\begin{aligned} &\frac{\partial^2 I}{\partial x_i \partial x_j}(x; x_0) \qquad (18.14a) \\ &= \int_{B_r^+} \Gamma_{ij}(x_0, T(x; x_0) - y) \\ &\quad \times \left\{ \sum_{h,k=1}^{n} \left\{a^{hk}(x_0) - a^{hk}(y)\right\} \frac{\partial^2 u}{\partial x_h \partial x_k}(y) + f(y) \right\} dy \\ &= I_{ij}(x; x_0) \quad \text{for } 1 \leq i, j \leq n-1; \end{aligned}$$

$$\begin{aligned} &\frac{\partial^2 I}{\partial x_i \partial x_n}(x; x_0) \qquad (18.14b) \\ &= \int_{B_r^+} \sum_{\ell=1}^{n} \Gamma_{i\ell}(x_0, T(x; x_0) - y) \frac{\partial T_\ell}{\partial x_n}(x; x_0) \\ &\quad \times \left\{ \sum_{h,k=1}^{n} \left\{a^{hk}(x_0) - a^{hk}(y)\right\} \frac{\partial^2 u}{\partial x_h \partial x_k}(y) + f(y) \right\} dy \\ &= I_{in}(x; x_0) \quad \text{for } 1 \leq i \leq n-1; \end{aligned}$$

and

$$\begin{aligned} &\frac{\partial^2 I}{\partial x_n \partial x_n}(x; x_0) \qquad (18.14c) \\ &= \int_{B_r^+} \sum_{\ell,m=1}^{n} \Gamma_{\ell m}(x_0, T(x; x_0) - y) \frac{\partial T_\ell}{\partial x_n}(x; x_0) \frac{\partial T_m}{\partial x_n}(x; x_0) \\ &\quad \times \left\{ \sum_{h,k=1}^{n} \left\{a^{hk}(x_0) - a^{hk}(y)\right\} \frac{\partial^2 u}{\partial x_h \partial x_k}(y) + f(y) \right\} dy \\ &= I_{nn}(x; x_0). \end{aligned}$$

and

$$\frac{\partial^2 J}{\partial x_i \partial x_j}(x; x_0) \qquad (18.15a)$$

$$
\begin{aligned}
&= \int_{B_r^+} \theta_{ij}(x_0, T(x; x_0) - y) \\
&\quad \times \left\{ \sum_{h,k=1}^{n} \left\{ a^{hk}(x_0) - a^{hk}(y) \right\} \frac{\partial^2 u}{\partial x_h \partial x_k}(y) + f(y) \right\} dy \\
&= J_{ij}(x; x_0) \quad \text{for } 1 \leq i, j \leq n-1;
\end{aligned}
$$

$$
\begin{aligned}
&\frac{\partial^2 J}{\partial x_i \partial x_n}(x; x_0) \\
&= \int_{B_r^+} \sum_{\ell=1}^{n} \theta_{i\ell}(x_0, T(x; x_0) - y) \frac{\partial T_\ell}{\partial x_n}(x; x_0) \\
&\quad \times \left\{ \sum_{h,k=1}^{n} \left\{ a^{hk}(x_0) - a^{hk}(y) \right\} \frac{\partial^2 u}{\partial x_h \partial x_k}(y) + f(y) \right\} dy \\
&= J_{in}(x; x_0) \quad \text{for } 1 \leq i \leq n-1;
\end{aligned}
\tag{18.15b}
$$

and

$$
\begin{aligned}
&\frac{\partial^2 J}{\partial x_n \partial x_n}(x; x_0) \\
&= \int_{B_r^+} \sum_{\ell,m=1}^{n} \theta_{\ell m}(x_0, T(x; x_0) - y) \frac{\partial T_\ell}{\partial x_n}(x; x_0) \frac{\partial T_m}{\partial x_n}(x; x_0) \\
&\quad \times \left\{ \sum_{h,k=1}^{n} \left\{ a^{hk}(x_0) - a^{hk}(y) \right\} \frac{\partial^2 u}{\partial x_h \partial x_k}(y) + f(y) \right\} dy \\
&= J_{nn}(x; x_0).
\end{aligned}
\tag{18.15c}
$$

Therefore, the desired formulas (18.8)

$$
\begin{aligned}
&\frac{\partial^2 u}{\partial x_i \partial x_j}(x) \\
&= \text{v.p.} \int_{B_r^+} \Gamma_{ij}(x_0, x - y) \\
&\quad \times \left\{ \sum_{h,k=1}^{n} \left\{ a^{hk}(x_0) - a^{hk}(y) \right\} \frac{\partial^2 u}{\partial x_h \partial x_k}(y) + f(y) \right\} dy \\
&\quad + c_{ij}(x_0) f(x) - I_{ij}(x; x_0) + J_{ij}(x; x_0) \quad \text{for } 1 \leq i, j \leq n
\end{aligned}
$$

follow from formulas (18.13), (18.14) and (18.15).

The proof of Theorem 18.3 is complete. □

Chapter 19
Boundary Regularity of Solutions

The purpose of this chapter is to prove boundary Sobolev regularity of the solutions of problem (16.4) toward the proof of Theorem 16.1 (Lemma 19.1). A combination of this regularity result with the interior regularity (Theorem 12.1) will prove the main result in the next Chap. 20.

Without loss of generality, we may assume (see Fig. 19.1) that $x_0 = 0$, and we let

$$B_r := B_r(0) = \{x \in \mathbf{R}^n : |x| < r\},$$
$$B_r^+ := B_r \cap \{x_n > 0\} = \{x = (x_1, x_2, \ldots, x_n) \in \mathbf{R}^n : |x| < r,\ x_n > 0\},$$

and

$$C_r := B_r \cap \{x_n = 0\} = \{x = (x_1, x_2, \ldots, x_{n-1}, 0) \in \mathbf{R}^n : |x'| < r\}.$$

As in the previous chapter, we assume that the boundary $\partial\Omega$ is locally flattened near a point $x_0 \in \partial\Omega$ such that

$$\Omega \subset \mathbf{R}^n_+ = \{x = (x', x_n) \in \mathbf{R}^n : x_n > 0\}$$

(see Fig. 17.2). The following result implies the boundary regularizing property of the couple $(\mathcal{L}, \mathcal{B})$ in the framework of Sobolev spaces:

Lemma 19.1 *Let $1 < q < p < \infty$, and assume that conditions (16.1) and (16.3) are satisfied. If $r > 0$ and $u \in W^{2,q}(B_r^+)$ is a solution of the non-homogeneous oblique derivative problem*

$$\begin{cases} \mathcal{L}u = f & \text{in } B_r^+, \\ \mathcal{B}u = \varphi & \text{on } C_r, \end{cases} \tag{19.1}$$

K. Taira, *Real Analysis Methods for Markov Processes*,
https://doi.org/10.1007/978-981-97-3659-1_19

Fig. 19.1 The semi-ball B_r^+ in $\mathbf{R}_+^n$ and the ball $C_r = B_r \cap \{x_n = 0\}$ in $\mathbf{R}^{n-1}$

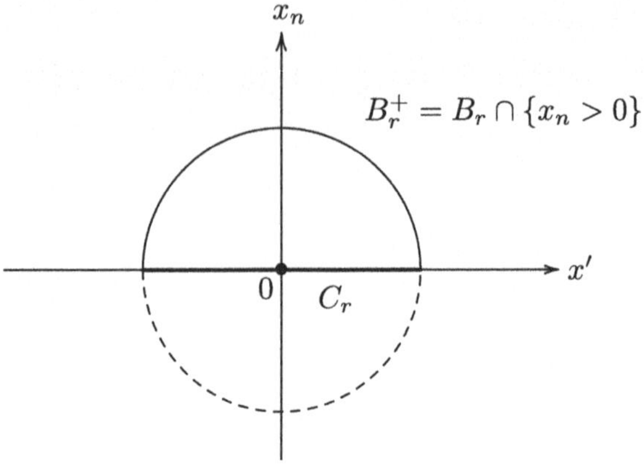

Fig. 19.2 The semi-ball B_R^+ in $\mathbf{R}_+^n$ and the ball $C_R = B_R \cap \{x_n = 0\}$ in $\mathbf{R}^{n-1}$

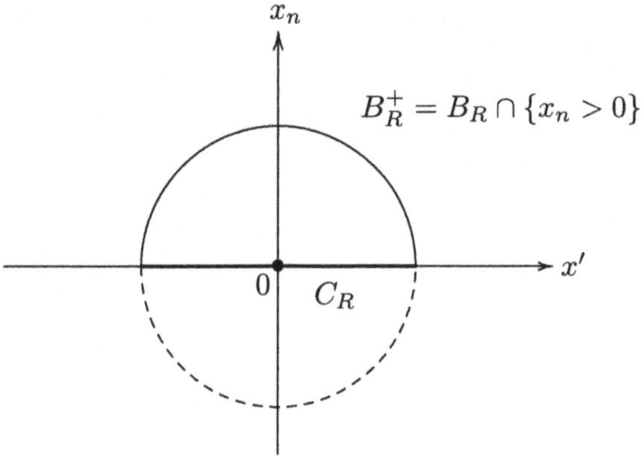

with $f \in L^p\left(B_r^+\right)$ *and* $\varphi \in B^{1-1/p,p}(C_r)$*, then there exists a constant* $0 < R < r$ *sufficiently small so that* $u \in W^{2,p}\left(B_R^+\right)$ *(see Fig. 19.2). Moreover, we have the boundary* a priori *estimate*

$$\left\| \frac{\partial^2 u}{\partial x_i \partial x_j} \right\|_{L^p(B_R^+)} \leq C\left(\|u\|_{L^p(B_R^+)} + \|f\|_{L^p(B_R^+)} + \|\varphi\|_{B^{1-1/p,p}(C_R)} \right), \tag{19.2}$$

with a positive constant $C = C(n, p, \lambda, \eta, \ell, \sigma, \partial\Omega)$.

Proof We make use of the explicit representation formula (18.8) of the second derivatives $\nabla^2 u$. The proof is divided into four steps.

Step 1: Without loss of generality, we may assume that the ball B_r is centered at the origin. Let $x_0 = (x_0', x_{0n})$ be an arbitray point of B_r^+, with $x_0' \in \mathbf{R}^{n-1}$. Then we have the formulas

$$\begin{aligned}\mathcal{L}_0 u(x) &= \sum_{i,j=1}^{n} a^{ij}(x_0)\frac{\partial^2 u}{\partial x_i \partial x_j}(x) \\ &= \sum_{i,j=1}^{n} \left\{a^{ij}(x_0) - a^{ij}(x)\right\} \frac{\partial^2 u}{\partial x_i \partial x_j}(x) + f(x) \quad \text{in } B_r^+,\end{aligned} \tag{19.3}$$

and

$$\begin{aligned}\mathcal{B}_0 u(x') &= \frac{\partial u}{\partial \ell(x_0')}(x') + \sigma(x_0')u(x') \\ &= \sum_{i=1}^{n} \ell^i(x_0')\frac{\partial u}{\partial x_i}(x') + \sigma(x_0')u(x') \\ &= \sum_{i=1}^{n} \left\{\ell^i(x_0') - \ell^i(x')\right\} \frac{\partial u}{\partial x_i}(x') + \left\{\sigma(x_0') - \sigma(x')\right\} u(x') \\ &\quad + \varphi(x') \quad \text{on } C_r.\end{aligned} \tag{19.4}$$

Now we let

$$\begin{aligned}\widetilde{\varphi}(x') := \widetilde{\varphi}(x', u) = &\sum_{i=1}^{n} \left\{\ell^i(x_0') - \ell^i(x')\right\} \frac{\partial u}{\partial x_i}(x') \\ &+ \left\{\sigma(x_0') - \sigma(x')\right\} u(x') + \varphi(x'),\end{aligned} \tag{19.5}$$

and define a function $\phi(x) = \phi(x, u)$ by formula (17.4), that is,

$$\begin{aligned}&\phi(x) = \phi(x, u) \\ &:= \frac{x_n}{\ell^n(x_0)} \int_{\mathbf{R}^{n-1}} \widetilde{\varphi}(x' - x_n y')\zeta(y')\, dy' \\ &= \frac{x_n}{\ell^n(x_0)} \sum_{i=1}^{n} \int_{\mathbf{R}^{n-1}} \Big(\left\{\ell^i(x_0') - \ell^i(x' - x_n y')\right\} \frac{\partial u}{\partial x_i}(x')(x' - x_n y') \\ &\quad + \left\{\sigma(x_0') - \sigma(x' - x_n y')\right\} u(x' - x_n y') + \varphi(x') \Big) \zeta(y')\, dy',\end{aligned} \tag{19.6}$$

where $\zeta(x')$ is a bell-shaped function on $\mathbf{R}^{n-1}$ which satisfies the four conditions (17.2).

Then, by formulas (17.5) it follows that

$$\begin{aligned}\mathcal{B}_0 \phi(x') &= \frac{\partial \phi}{\partial \ell(x_0')}(x') + \sigma(x_0')\phi(x') \\ &= \sum_{i=1}^{n} \ell^i(x_0')\frac{\partial \phi}{\partial x_i}(x') + \sigma(x_0')\phi(x')\end{aligned} \tag{19.7}$$

$$= \widetilde{\varphi}(x') \quad \text{on } C_r.$$

Hence, by combining formulas (19.3), (19.4) and (19.7) we find that the function $u(x) - \phi(x)$ satisfies the conditions

$$\begin{aligned} \mathcal{L}_0(u-\phi)(x) &= \sum_{i,j=1}^{n} a^{ij}(x_0) \frac{\partial^2 (u-\phi)}{\partial x_i \partial x_j}(x) \\ &= \sum_{i,j=1}^{n} \left\{ a^{ij}(x_0) - a^{ij}(x) \right\} \frac{\partial^2 u}{\partial x_i \partial x_j}(x) + f(x) \\ &\quad - \sum_{i,j=1}^{n} a^{ij}(x_0) \frac{\partial^2 \phi}{\partial x_i \partial x_j}(x) \quad \text{in } B_r^+, \end{aligned} \tag{19.8}$$

and

$$\begin{aligned} \mathcal{B}_0(u-\phi)(x') &= \frac{\partial (u-\phi)}{\partial \ell(x_0')}(x') + \sigma(x_0')(u(x') - \phi(x')) \\ &= 0 \quad \text{on } C_r. \end{aligned} \tag{19.9}$$

Therefore, by applying Lemma 18.1 to the function $u(x) - \phi(x)$ we obtain from formulas (19.8) and (19.9) that the solution of the non-homogeneous oblique derivative problem (19.1) can be expressed as follows:

$$\begin{aligned} &u(x) \\ &= \phi(x,u) + \int_{B_r^+} G(x_0, x, y) \left\{ \left\{ \sum_{i,j=1}^{n} a^{ij}(x_0) - a^{ij}(y) \right\} \frac{\partial^2 u}{\partial x_i \partial x_j}(y) \right. \\ &\quad \left. + f(y) - \sum_{i,j=1}^{n} a^{ij}(x_0) \frac{\partial^2 \phi}{\partial x_i \partial x_j}(y,u) \right\} dy, \end{aligned} \tag{19.10}$$

where $G(x_0, x, y)$ is the half space Green function for the elliptic operator $\mathcal{L}_0$ given by the formula (18.2)

$$G(x_0, x, y) = \Gamma(x_0, x - y) - \Gamma(x_0, T(x; x_0) - y) + \theta(x_0, T(x; x_0) - y).$$

Here it should be emphasized that the function $\phi(x, u)$ defined by formula (19.6) depends *affinely* on u.

On the other hand, by applying Theorem 18.3 to the function $u(x) - \phi(x)$ we obtain from formulas (19.8) and (19.9) that, for almost all $x \in B_r^+$,

$$\frac{\partial^2 u}{\partial x_i \partial x_j}(x) \tag{19.11}$$

$$= \frac{\partial^2 \phi}{\partial x_i \partial x_j}(x, u) + \text{v.p.} \int_{B_r^+} \Gamma_{ij}(x_0, x - y)$$

$$\times \left\{ \sum_{h,k=1}^{n} \{a^{hk}(x_0) - a^{hk}(y)\} \frac{\partial^2 u}{\partial x_h \partial x_k}(y) + f(y) - \mathcal{L}_0 \phi(y, u) \right\} dy$$

$$+ c_{ij}(x_0)\,(f(x) - \mathcal{L}\phi(x, u)) - \widetilde{I}_{ij}(x; x_0) + \widetilde{J}_{ij}(x; x_0),$$

where the function $c_{ij}(x_0)$ is defined by the formula (18.9)

$$c_{ij}(x_0) = \int_{|\xi|=1} \Gamma_i(x_0, t) t_j \, d\sigma_t,$$

and the terms $\widetilde{I}_{ij}(x; x_0)$ are defined respectively as follows (see formulas (18.10)):

$$\widetilde{I}_{ij}(x; x_0) \quad (1 \leq i, j \leq n-1) \tag{19.12a}$$

$$:= \int_{B_r^+} \Gamma_{ij}(x_0, T(x; x_0) - y)$$

$$\times \left\{ \sum_{h,k=1}^{n} \{a^{hk}(x_0) - a^{hk}(y)\} \frac{\partial^2 u}{\partial x_h \partial x_k}(y) + f(y) - \mathcal{L}_0 \phi(y, u) \right\} dy.$$

$$\widetilde{I}_{in}(x; x_0) = \widetilde{I}_{ni}(x; x_0) \quad (1 \leq i \leq n-1) \tag{19.12b}$$

$$:= \int_{B_r^+} \sum_{\ell=1}^{n} \Gamma_{i\ell}(x_0, T(x; x_0) - y) B_\ell(x_0)$$

$$\times \left\{ \sum_{h,k=1}^{n} \{a^{hk}(x_0) - a^{hk}(y)\} \frac{\partial^2 u}{\partial x_h \partial x_k}(y) + f(y) - \mathcal{L}_0 \phi(y, u) \right\} dy.$$

$$\widetilde{I}_{nn}(x; x_0) \tag{19.12c}$$

$$:= \int_{B_r^+} \sum_{\ell,m=1}^{n} \Gamma_{\ell m}(x_0, T(x; x_0) - y) B_\ell(x_0) B_m(x_0)$$

$$\times \left\{ \sum_{h,k=1}^{n} \{a^{hk}(x_0) - a^{hk}(y)\} \frac{\partial^2 u}{\partial x_h \partial x_k}(y) + f(y) - \mathcal{L}_0 \phi(y, u) \right\} dy.$$

The terms $\widetilde{J}_{ij}(x; x_0)$ are defined respectively as follows (see formulas (18.10)):

$$\widetilde{J}_{ij}(x; x_0) \quad (1 \leq i, j \leq n-1) \tag{19.13a}$$

$$:= \int_{B_r^+} \theta_{ij}(x_0, T(x; x_0) - y)$$

$$\times \left\{ \sum_{h,k=1}^{n} \left\{ a^{hk}(x_0) - a^{hk}(y) \right\} \frac{\partial^2 u}{\partial x_h \partial x_k}(y) + f(y) - \mathcal{L}_0 \phi(y, u) \right\} dy.$$

$$\begin{aligned} \widetilde{J}_{in}(x; x_0) &= \widetilde{J}_{ni}(x; x_0) \quad (1 \leq i \leq n-1) \\ &:= \int_{B_r^+} \sum_{j=1}^{n} \theta_{ij}(x_0, T(x; x_0) - y) B_j(x_0) \\ &\times \left\{ \sum_{h,k=1}^{n} \left\{ a^{hk}(x_0) - a^{hk}(y) \right\} \frac{\partial^2 u}{\partial x_h \partial x_k}(y) + f(y) - \mathcal{L}_0 \phi(y, u) \right\} dy. \end{aligned} \tag{19.13b}$$

$$\begin{aligned} &\widetilde{J}_{nn}(x; x_0) \\ &:= \int_{B_r^+} \sum_{i,j=1}^{n} \theta_{ij}(x_0, T(x; x_0) - y) B_i(x_0) B_j(x_0) \\ &\times \left\{ \sum_{h,k=1}^{n} \left\{ a^{hk}(x_0) - a^{hk}(y) \right\} \frac{\partial^2 u}{\partial x_h \partial x_k}(y) + f(y) - \mathcal{L}_0 \phi(y, u) \right\} dy. \end{aligned} \tag{19.13c}$$

Here recall that the vector $B(x_0)$ is defined by the formula

$$B(x_0) = \begin{pmatrix} B_1(x_0) \\ B_2(x_0) \\ \vdots \\ B_n(x_0) \end{pmatrix} := \frac{\partial T}{\partial x_n}(x; x_0) = \begin{pmatrix} -2\frac{a^{1n}(x_0)}{a^{nn}(x_0)} \\ -2\frac{a^{2n}(x_0)}{a^{nn}(x_0)} \\ \vdots \\ -1 \end{pmatrix}.$$

Step 2: Now we assume that $q < p$, and let $q \leq s \leq p$. If w is a function in $W^{2,s}\left(B_r^+\right)$, then we define a mapping $\mathcal{S}$ by the formula (cf. formula (19.10))

$$\begin{aligned} \mathcal{S}w =&\phi(x, w) \\ &+ \int_{B_r^+} G(x_0, x, y) \left\{ \sum_{i,j=1}^{n} \left\{ a^{ij}(x_0) - a^{ij}(y) \right\} \frac{\partial^2 w}{\partial x_i \partial x_j}(y) \right. \\ &\left. + f(y) - \sum_{i,j=1}^{n} a^{ij}(x_0) \frac{\partial^2 \phi}{\partial x_i \partial x_j}(y, w) \right\} dy. \end{aligned} \tag{19.14}$$

Here the function $\phi(x, w)$ is defined by formula (19.6), that is,

$$\phi(x, w) := \frac{x_n}{\ell^n(x_0)} \int_{\mathbf{R}^{n-1}} \widetilde{\varphi}(x' - x_n y', w) \zeta(y') \, dy', \tag{19.15}$$

where

$$\widetilde{\varphi}(x', w) := \sum_{i=1}^{n} \left\{\ell^i(x'_0) - \ell^i(x')\right\} \frac{\partial w}{\partial x_i}(x') + \left\{\sigma(x'_0) - \sigma(x')\right\} w(x'). \tag{19.16}$$

We show that $\mathcal{S}$ is a *contraction mapping* of $W^{2,s}\left(B_R^+\right)$ into itself, for each $s \in [q, p]$, if $0 < R < r$ is sufficiently small.

In the following we shall use the shorthand

$$D_i = \frac{\partial}{\partial x_i},$$
$$D_{ij} = \frac{\partial^2}{\partial x_i \partial x_j}$$

for derivatives on $\mathbf{R}^n$.

Step 2-1: Let w_1, w_2 be two arbitrary functions in the space $W^{2,s}\left(B_r^+\right)$. By letting

$$w := w_1 - w_2,$$

we obtain from formula (19.14) that

$$\begin{aligned} &\mathcal{S}w_1 - \mathcal{S}w_2 \\ &= \phi(x, w) + \int_{B_r^+} G(x_0, x, y) \\ &\quad \times \left\{ [\sum_{i,j=1}^{n} \left\{a^{ij}(x_0) - a^{ij}(y)\right\} D_{ij} w(y) - \sum_{i,j=1}^{n} a^{ij}(x_0) D_{ij}\phi(y, w) \right\} dy. \end{aligned} \tag{19.17}$$

Then we have, for some positive constant $C(n, \lambda)$,

$$\begin{aligned} &\|\mathcal{S}w_1 - \mathcal{S}w_2\|_{L^s(B_r^+)} \\ &\leq \|\phi(\cdot, w)\|_{L^s(B_r^+)} \\ &\quad + \left\| \int_{B_r^+} G(x_0, \cdot, y) \sum_{i,j=1}^{n} \left\{a^{ij}(x_0) - a^{ij}(y)\right\} D_{ij} w(y)\, dy \right\|_{L^s(B_r^+)} \\ &\quad + C(n, \lambda) \left\| \int_{B_r^+} G(x_0, \cdot, y) \sum_{i,j=1}^{n} D_{ij}\phi(y, w)\, dy \right\|_{L^s(B_r^+)}. \end{aligned}$$

However, by Lemma 18.2 it follows that

$$G(x_0, x, y) = O(|x - y|^{2-n}) \quad \text{as } |x - y| \to 0.$$

Hence we find that the two integrals

$$\int_{B_r^+} G(x_0, x, y) \sum_{i,j=1}^{n} \left\{a^{ij}(x_0) - a^{ij}(y)\right\} D_{ij}w(y)\, dy,$$

$$\int_{B_r^+} G(x_0, x, y) \sum_{i,j=1}^{n} D_{ij}\phi(y, w)\, dy$$

are Riesz potentials.

Now, we need potential estimates in the classical potential theory (see Sect. 3.10). Since $a^{ij} \in L^\infty(\Omega)$, it follows from an application of Theorem 3.49 with

$$\Omega := B_r^+, \quad \mu := \frac{2}{n}, \quad p = q := s, \quad \delta := 0$$

that we have, for some positive constants $C_1(n, s, \lambda)$ and $C_2(n, s, \lambda)$,

$$\left\| \int_{B_r^+} G(x_0, \cdot, y) \sum_{i,j=1}^{n} \left\{a^{ij}(x_0) - a^{ij}(y)\right\} D_{ij}w(y)\, dy \right\|_{L^s(B_r^+)}$$
$$\leq C_1(n, s, \lambda) r^2 \|\nabla^2 w\|_{L^s(B_r^+)},$$

and

$$\left\| \int_{B_r^+} G(x_0, \cdot, y) \sum_{i,j=1}^{n} D_{ij}\phi(\cdot, w)\, dy \right\|_{L^s(B_r^+)}$$
$$\leq C_2(n, s, \lambda) r^2 \|\nabla^2 \phi(\cdot, w)\|_{L^s(B_r^+)}.$$

Therefore, we have proved that

$$\|\mathcal{S}w_1 - \mathcal{S}w_2\|_{L^s(B_r^+)} \tag{19.18}$$
$$\leq \|\phi(\cdot, w)\|_{L^s(B_r^+)} + C(n, s, \lambda) r^2 \left(\left\|\nabla^2\phi(\cdot, w)\right\|_{L^s(B_r^+)} + \left\|\nabla^2 w\right\|_{L^s(B_r^+)} \right)$$
$$\text{for all } w_1,\ w_2 \in W^{2,s}\left(B_r^+\right).$$

Step 2-2: Furthermore, we have, for almost all $x \in B_r^+$,

$$D_{ij}(\mathcal{S}w_1 - \mathcal{S}w_2)(x)$$
$$= D_{ij}\phi(x, w)$$
$$+ \text{v.p.} \int_{B_r^+} \Gamma_{ij}(x_0, x - y)$$

$$\times \sum_{k,\ell=1}^{n} \left\{\left\{a^{k\ell}(x_0) - a^{k\ell}(y)\right\} D_{k\ell} w(y) - \mathcal{L}_0 \phi(y, w)\right\} dy$$
$$- c_{ij}(x_0) \mathcal{L}\phi(x, w) + \widetilde{I}_{ij}(x; x_0, w) + \widetilde{J}_{ij}(x; x_0, w),$$

where the terms $\widetilde{I}_{ij}(x; x_0, w)$ and $\widetilde{J}_{ij}(x; x_0, w)$ are defined respectively as in formulas (19.12) and (19.13) with u replaced by w and with $f(y) \equiv 0$.

Since the $\Gamma_{ij}(x_0, \xi)$ are Calderón–Zygmund kernels in the variable ξ, it follows from an application of Corollary 14.9 and Remark 14.10 that

$$\left\| \text{v.p.} \int_{B_r^+} \Gamma_{ij}(x_0, \cdot - y) \sum_{k,\ell=1}^{n} \left\{a^{k\ell}(x_0) - a^{k\ell}(y)\right\} D_{k\ell} w(y)\, dy \right\|_{L^s(B_r^+)} \tag{19.19a}$$
$$\leq C_1(n, s, \lambda, \eta, M, \partial\Omega) \eta(r) \left\| \nabla^2 w \right\|_{L^s(B_r^+)},$$

and that

$$\left\| \text{v.p.} \int_{B_r^+} \Gamma_{ij}(x_0, \cdot - y) \mathcal{L}_0 \phi(y, w)\, dy \right\|_{L^s(B_r^+)} \tag{19.19b}$$
$$\leq C_2(n, s, \lambda, \eta, M, \partial\Omega) \left\| \mathcal{L}_0 \phi(\cdot, w) \right\|_{L^s(B_r^+)}$$
$$\leq C_3(n, s, \lambda, \eta, M, \partial\Omega) \left\| \nabla^2 \phi(\cdot, w) \right\|_{L^s(B_r^+)}.$$

Here we recall that

$$M := \max_{1 \leq i,j \leq n} \max_{|\alpha| \leq 2n} \left\| \frac{\partial^\alpha \Gamma_{ij}}{\partial \xi^\alpha}(\cdot, \cdot) \right\|_{L^\infty(\Omega \times \Sigma_n)},$$
$$\eta(r) := \left(\sum_{k,\ell=1}^{n} \eta^{k\ell}(r)^2 \right)^{1/2},$$

where $\eta^{k\ell}(r)$ is the VMO modulus of $a^{k\ell}$.

Step 2-3: Before the proof of Theorem 16.1, we need the following geometric properties of the map $T(x; x_0)$ analogous to Lemma 14.13:

Lemma 19.2 *For the map $T(x; x_0)$, there exist positive constants c_1, c_2 such that, for all $y \in \mathbf{R}_+^n$ and all $x \in \mathbf{R}_+^n$ for which $T(x; x_0)$ is defined, we have the inequalities*

$$c_1 |\tilde{x} - y| \leq |T(x; x_0) - y| \leq c_2 |\tilde{x} - y|, \tag{19.20}$$

where

$$\tilde{x} = \left(x', -x_n\right) \quad \textit{for}\; x = \left(x', x_n\right) \in \mathbf{R}_+^n.$$

Proof The proof of Lemma 19.2 is divided into two steps.

(1) First, we have, for all $y = \left(y', y_n\right) \in \mathbf{R}^n_+$,

$$\begin{aligned} &T(x; x_0) - y \\ &= (T_1(x; x_0) - y_1, T_2(x; x_0) - y_2, \ldots, T_{n-1}(x; x_0) - y_{n-1}, -x_n - y_n) \\ &:= (T(x; x_0)' - y', -x_n - y_n), \end{aligned}$$

and so

$$|T(x; x_0) - y| \geq x_n + y_n \geq x_n.$$

Hence it follows that, for all $y \in \mathbf{R}^n_+$ and all $x \in \mathbf{R}^n_+$ for which $T(x)$ is defined,

$$\begin{aligned} \frac{|T(x; x_0) - \tilde{x}|}{|T(x; x_0) - y|} \leq \frac{|T(x; x_0) - \tilde{x}|}{x_n} & \qquad (19.21) \\ = \frac{1}{x_n} \left|(x', 0) + x_n A(x_0) - (x', -x_n)\right| & \\ = \left|(0', 1) + A(x_0)\right|, & \end{aligned}$$

where the vector $A(x_0)$ is defined by the formula

$$A(x_0) = \left(-2\frac{a^{1n}(x_0)}{a^{nn}(x_0)}, -2\frac{a^{2n}(x_0)}{a^{nn}(x_0)}, \ldots, -1\right).$$

However, we can find a positive constant $C_1(n, \mu)$ such that, for all $x_0 \in \mathbf{R}^n_+$,

$$\left|(0', 1) + A(x_0)\right| \leq C_1(n, \mu).$$

Hence we have, for all $y \in \mathbf{R}^n_+$ and all $x \in \mathbf{R}^n_+$ for which $T(x)$ is defined,

$$\frac{|T(x; x_0) - \tilde{x}|}{|T(x; x_0) - y|} \leq C_1(n, \mu). \qquad (19.22)$$

Therefore, we obtain from inequalities (19.21) and (19.22) that, for all $y \in \mathbf{R}^n_+$ and for all $x \in \mathbf{R}^n_+$ for which $T(x; x_0)$ is defined,

$$\begin{aligned} |\tilde{x} - y| &\leq |T(x; x_0) - \tilde{x}| + |T(x; x_0) - y| \\ &\leq |T(x; x_0) - y| \left(1 + \frac{|T(x; x_0) - \tilde{x}|}{|T(x; x_0) - y|}\right) \\ &\leq (1 + C_1(n, \mu)) \, |T(x; x_0) - y| \, . \end{aligned}$$

This proves the desired inequality (19.20) with

$$c_1 := \frac{1}{1 + C_1(n, \mu)}.$$

(2) On the other hand, it follows that, for all $y = \left(y', y_n\right) \in \mathbf{R}^n_+$ and for all $x = \left(x', x_n\right) \in \mathbf{R}^n_+$,

$$|\tilde{x} - y| = \sqrt{|x' - y'|^2 + (x_n + y_n)^2},$$
$$|T(x; x_0) - y| = \sqrt{|T(x; x_0)' - y'|^2 + (x_n + y_n)^2},$$

where

$$\begin{aligned} &T(x; x_0)' - y' \\ &= x' - y' - \left(2x_n \frac{a^{1n}(x_0)}{a^{nn}(x_0)}, 2x_n \frac{a^{2n}(x_0)}{a^{nn}(x_0)}, \dots, 2x_n \frac{a^{n-1n}(x_0)}{a^{nn}(x_0)}\right). \end{aligned}$$

We remark that, for some constant $C_2(n, \mu) > 0$,

$$\begin{aligned} \left|T(x; x_0)' - y'\right| &\leq \left|x' - y'\right| + 2C_2(n, \mu)x_n \\ &\leq \left|x' - y'\right| + 2C_2(n, \mu)\,(x_n + y_n)\,. \end{aligned}$$

Hence we have, for all $y \in \mathbf{R}^n_+$ and for all $x \in \mathbf{R}^n_+$ for which $T(x; x_0)$ is defined,

$$\begin{aligned} |T(x; x_0) - y| &\leq \left|T(x; x_0)' - y'\right| + (x_n + y_n) \\ &\leq \left|x' - y'\right| + (1 + 2C_2(n, \mu))\,(x_n + y_n) \\ &\leq (2C_2(n, \mu) + 1)\left(\left|x' - y\right| + (x_n + y_n)\right) \\ &\leq \sqrt{2}\,(2C_2(n, \mu) + 1)\sqrt{|x' - y|^2 + (x_n + y_n)^2} \\ &= \sqrt{2}\,(2C_2(n, \mu) + 1)\,|\tilde{x} - y|. \end{aligned}$$

This proves the desired inequality (19.20) with

$$c_2 := \sqrt{2}\,(2C_2(n, \mu) + 1).$$

The proof of Lemma 19.2 is complete. □

Step 2-4: Therefore, by applying Theorems 14.5 and 14.5 (and Remark 14.12) to our situation we obtain that

$$\begin{aligned} &\left\|\widetilde{I}_{ij}(\cdot; x_0, w)\right\|_{L^s(B_r^+)}, \quad \left\|\widetilde{J}_{ij}(\cdot; x_0, w)\right\|_{L^s(B_r^+)} \qquad (19.23) \\ &\leq C_1(n, s, \lambda, \eta, M, \partial\Omega)\left\|\sum_{h,k=1}^{n} \widetilde{C}(a^{hk}, D_{hk}w) + \widetilde{K}(\mathcal{L}\phi(\cdot, w))\right\|_{L^s(B_r^+)} \\ &\leq C_2(n, s, \lambda, \eta, M, \partial\Omega)\left(\eta(r)\|\nabla^2 w\|_{L^s(B_r^+)} + \|\mathcal{L}\phi(\cdot, w)\|_{L^s(B_r^+)}\right) \\ &\leq C_3(n, s, \lambda, \eta, M, \partial\Omega)\left(\eta(r)\|\nabla^2 w\|_{L^s(B_r^+)} + \|\nabla^2\phi(\cdot, w)\|_{L^s(B_r^+)}\right). \end{aligned}$$

Indeed, it suffices to note that we have, by Lemma 18.2 with $|\alpha| := 2$ and Lemma 19.2,

$$\left|\theta_{ij}(x_0, T(x; x_0) - y)\right| \leq \frac{C}{|T(x; x_0) - y|^n} \leq \frac{C'}{|\tilde{x} - y|^n},$$

with some positive constants C and C'.

Finally, it follows that

$$\left\| c_{ij}(x_0)\mathcal{L}\phi(\cdot, w)\right\|_{L^s(B_r^+)} \leq C_3 \left\|\nabla^2\phi(\cdot, w)\right\|_{L^s(B_r^+)}. \tag{19.24}$$

By combining inequalities (19.19), (19.20), (19.23) and (19.24), we have proved that

$$\begin{aligned}
&\left\|\nabla^2(\mathcal{S}w_1 - \mathcal{S}w_2)\right\|_{L^s(B_r^+)} \qquad (19.25)\\
&\leq \left\|\nabla^2\phi(\cdot, w)\right\|_{L^s(B_r^+)} + C_1\left(\eta(r)\left\|\nabla^2 w\right\|_{L^s(B_r^+)} + \left\|\nabla^2\phi(\cdot, w)\right\|_{L^s(B_r^+)}\right)\\
&\quad + C_2\left(\eta(r)\left\|\nabla^2 w\right\|_{L^s(B_r^+)} + \left\|\nabla^2\phi(\cdot, w)\right\|_{L^s(B_r^+)}\right)\\
&\quad + C_3\left\|\nabla^2\phi(\cdot, w)\right\|_{L^s(B_r^+)}.
\end{aligned}$$

Step 2-5: Therefore, we obtain from inequalities (19.18) and (19.25) that

$$\begin{aligned}
&\|\mathcal{S}w_1 - \mathcal{S}w_2\|^*_{W^{2,s}(B_r^+)} \qquad (19.26)\\
&= \|\mathcal{S}w_1 - \mathcal{S}w_2\|_{L^s(B_r^+)} + r\left\|\nabla^2(\mathcal{S}w_1 - \mathcal{S}w_2)\right\|_{L^s(B_r^+)}\\
&\leq C_4\Big(r\eta(r)\left\|\nabla^2(w_1 - w_2)\right\|_{L^s(B_r^+)} + r^2\left\|\nabla^2(w_1 - w_2)\right\|_{L^s(B_r^+)}\\
&\quad + r\left\|\nabla^2\phi(\cdot, w)\right\|_{L^s(B_r^+)} + \|\phi(\cdot, w)\|_{L^s(B_r^+)}\Big)\\
&= C_4\Big((\eta(r) + r)\, r\left\|\nabla^2(w_1 - w_2)\right\|_{L^s(B_r^+)}\\
&\quad + r\left\|\nabla^2\phi(\cdot, w)\right\|_{L^s(B_r^+)} + \|\phi(\cdot, w)\|_{L^s(B_r^+)}\Big)\\
&\leq C_4\Big((\eta(r) + r)\,\|w_1 - w_2\|^*_{W^{2,s}(B_r^+)}\\
&\quad + r\left\|\nabla^2\phi(\cdot, w)\right\|_{L^s(B_r^+)} + \|\phi(\cdot, w)\|_{L^s(B_r^+)}\Big) \quad \text{for all } w_1, w_2 \in W^{2,s}\left(B_r^+\right),
\end{aligned}$$

with a positive constant $C_4 = C_4(n, s, \lambda, \eta, M, \partial\Omega)$.

Step 2-6: Now we estimate the last two terms of inequality (19.26) in terms of the norm

$$\|w\|^*_{W^{2,s}(B_r^+)} = \|w_1 - w_2\|^*_{W^{2,s}(B_r^+)}.$$

To do this, by applying Lemma 17.1 with $p := s$ and $R := r$ we obtain that

$$\begin{aligned}\|\phi(\cdot, w)\|^*_{W^{2,s}(B_r^+)} &:= \|\phi(\cdot, w)\|_{L^s(B_r^+)} + r \left\|\nabla^2\phi(\cdot, w)\right\|_{L^s(B_r^+)} \\ &\leq C r^{1/2} \|\widetilde{\varphi}(\cdot, w)\|^*_{B^{1-1/s,s}(C_r)}.\end{aligned} \tag{19.27}$$

On the other hand, we find from formula (19.16) that

$$\begin{aligned}&r^{1/2} \|\widetilde{\varphi}(\cdot, w)\|^*_{B^{1-1/s,s}(C_r)} \\ &\leq r^{1/2} C_5 \Bigg(\sum_{i=1}^n \left\| \left(\ell^i(x_0') - \ell^i(\cdot)\right) D_i w \right\|^*_{B^{1-1/s,s}(C_r)} \\ &\quad + \left\| \left(\sigma(x_0') - \sigma(\cdot)\right) w \right\|^*_{B^{1-1/s,s}(C_r)} \Bigg).\end{aligned} \tag{19.28}$$

Here we recall that the norm $\| \cdot \|^*_{B^{1-1/s,s}(C_r)}$ is given by the formula (see the norm (17.1) and the norm (7.11) with $m := 0$ and $p := s$)

$$\begin{aligned}\|\widetilde{\psi}\|^*_{B^{1-1/s,s}(C_r)} &= \left(\int_{C_r} |\widetilde{\psi}(x')|^s \, dx' \right)^{1/s} \\ &\quad + r^{1/2} \left(\int_{C_r} \int_{C_r} \frac{|\widetilde{\psi}(x') - \widetilde{\psi}(y')|^s}{|x' - y'|^{s+n-2}} \, dx' \, dy' \right)^{1/s}.\end{aligned}$$

In order to estimate the first and second terms on inequality (19.28), we make use of Rademacher's theorem. Since the functions $\ell^i(x')$ are

Lipschitz continuous, it follows from an application of Rademacher's theorem (Theorem 5.2) and the trace theorem (Theorem 7.6) that

the first term on inequality (19.28) is estimated as follows:

$$\begin{aligned}&r^{1/2} \sum_{i=1}^n \left\| (\ell^i(x_0') - \ell^i(\cdot)) D_i w \right\|^*_{B^{1-1/s,s}(C_r)} \\ &\leq C_6 \Bigg(r^{3/2} \|D_i w\|_{L^s(C_r)} \\ &\quad + r^{1+1/s} \left(\int_{C_r} \int_{C_r} \frac{|D_i w(x') - D_i w(y')|^s}{|x' - y'|^{s+n-2}} \, dx' \, dy' \right)^{1/s} \Bigg) \\ &\leq C_6 \left(r^{3/2} + r^{1+1/s} \right) \|\nabla w\|_{W^{1,s}(B_r^+)}.\end{aligned} \tag{19.29}$$

Moreover, by applying interpolation inequality (13.17) with $\varepsilon := r^{1/2}$ (Theorem 13.4) we obtain that

$$\|\nabla w\|_{L^s(B_r^+)} \leq r^{1/2} \left\|\nabla^2 w\right\|_{L^s(B_r^+)} + \frac{C}{r^{1/2}} \|w\|_{L^s(B_r^+)}.$$

Hence we have the inequality

$$\begin{aligned}
& r^{3/2}\,\|\nabla w\|_{W^{1,s}(B_r^+)} \qquad (19.30)\\
&\le r^{3/2} C_7 \left(\left\|\nabla^2 w\right\|_{L^s(B_r^+)} + \|\nabla w\|_{L^s(B_r^+)}\right)\\
&\le C_7 \left(r^{3/2}\left\|\nabla^2 w\right\|_{L^s(B_r^+)} + r^2\left\|\nabla^2 w\right\|_{L^s(B_r^+)} + r\,\|w\|_{L^s(B_r^+)}\right)\\
&\le C_7 r^{1/2}\,\|w\|^*_{W^{2,s}(B_r^+)} \quad \text{for all } w \in W^{2,s}\left(B_r^+\right).
\end{aligned}$$

Similarly, we have, by interpolation inequality (13.17) with $\varepsilon := r^{1/s}$,

$$\begin{aligned}
& r^{1+1/s}\,\|\nabla w\|_{W^{1,s}(B_r^+)} \qquad (19.31)\\
&\le r^{1+1/s} C_8 \left(\left\|\nabla^2 w\right\|_{L^s(B_r^+)} + \|\nabla w\|_{L^s(B_r^+)}\right)\\
&\le C_8 \left(r^{1+1/s}\left\|\nabla^2 w\right\|_{L^s(B_r^+)} + r^{1+2/s}\left\|\nabla^2 w\right\|_{L^s(B_r^+)} + r\,\|w\|_{L^s(B_r^+)}\right)\\
&\le C_8 r^{1/s}\,\|w\|^*_{W^{2,s}(B_r^+)} \quad \text{for all } w \in W^{2,s}(B_r^+).
\end{aligned}$$

By combining inequalities (19.29) through (19.31), we obtain that

$$\begin{aligned}
& r^{1/2}\sum_{i=1}^{n}\left\|\left(\ell^i(x_0') - \ell^i(\cdot)\right) D_i w\right\|^*_{B^{1-1/s,s}(C_r)} \qquad (19.32)\\
&\le C'\left(r^{1/2} + r^{1/s}\right)\|w\|^*_{W^{2,s}(B_r^+)}\\
&= C'\left(r^{1/2} + r^{1/s}\right)\|w_1 - w_2\|^*_{W^{2,s}(B_r^+)} \quad \text{for all } w_1,\ w_2 \in W^{2,s}\left(B_r^+\right).
\end{aligned}$$

On the other hand, since the function $\sigma(x')$ is Lipschitz continuous, it follows from an application of Rademacher's theorem (Theorem 5.2) and the trace theorem (Theorem 7.6) that the second term on inequality (19.28) is estimated as follows:

$$\begin{aligned}
& r^{1/2}\left\|\left(\sigma(x_0') - \sigma(\cdot)\right) w\right\|^*_{B^{1-1/s,s}(C_r)} \qquad (19.33)\\
&\le C''\left(r^{2/s+1}\left\|\nabla^2 w\right\|_{L^s(B_r^+)} + r\,\|w\|_{L^s(B_r^+)}\right)\\
&\le C'' r^{1/s}\left(r\left\|\nabla^2 w\right\|_{L^s(B_r^+)} + r^{1-1/s}\,\|w\|_{L^s(B_r^+)}\right)\\
&\le C'' r^{1/s}\,\|w\|^*_{W^{2,s}(B_r^+)}\\
&= C'' r^{1/s}\,\|w_1 - w_2\|^*_{W^{2,s}(B_r^+)} \quad \text{for all } w_1,\ w_2 \in W^{2,s}\left(B_r^+\right).
\end{aligned}$$

Therefore, by combining inequalities (19.27), (19.28), (19.32) and (19.33) we can estimate the last two terms of inequality (19.26) as follows:

$$\|\phi(\cdot, w)\|^*_{W^{2,s}(B_r^+)} \qquad (19.34)$$

$$\begin{aligned}
&:= \|\phi(\cdot, w)\|_{L^s(B_r^+)} + r \left\|\nabla^2\phi(\cdot, w)\right\|_{L^s(B_r^+)} \\
&\leq C r^{1/2} \|\widetilde{\varphi}(\cdot, w)\|^*_{B^{1-1/s,s}(C_r)} \\
&\leq C C_5 \left(C' r^{1/s} + C'' \left(r^{1/2} + r^{1/s}\right)\right) \|w\|^*_{W^{2,s}(B_r^+)} \\
&= C C_5 \left(C' r^{1/s} + C'' \left(r^{1/2} + r^{1/s}\right)\right) \|w_1 - w_2\|^*_{W^{2,s}(B_r^+)} \\
&\quad \text{for all } w_1,\ w_2 \in W^{2,s}\left(B_r^+\right).
\end{aligned}$$

Step 2-7: By combining estimates (19.26) and (19.34), we obtain that

$$\|\mathcal{S}w_1 - \mathcal{S}w_2\|^*_{W^{2,s}(B_r^+)} \leq C(r) \|w_1 - w_2\|^*_{W^{2,s}(B_r^+)} \quad \text{for all } w_1,\ w_2 \in W^{2,s}\left(B_r^+\right),$$

with a positive constant

$$\begin{aligned}
C(r) &:= C_4 (\eta(r) + r) + C C_5 \left(C' r^{1/s} + C'' \left(r^{1/2} + r^{1/s}\right)\right) \\
&= o(1) \qquad \text{as } r \downarrow 0.
\end{aligned}$$

Hence we can take $r := R$ so small that

$$C(R) < 1.$$

This proves that $\mathcal{S}$ is a contraction mapping of $W^{2,s}\left(B_R^+\right)$, equipped with the norm $\|\cdot\|^*_{W^{2,s}(B_R^+)}$, into itself,

for each $s \in [q, p]$.

Step 3: Now we assume that a function $u \in W^{2,q}\left(B_R^+\right)$ with $q < p$ is a solution of problem (19.1). Then we obtain from formulas (19.10) and (19.14) that $u \in W^{2,q}\left(B_R^+\right)$ is a fixed point of $\mathcal{S}$, that is,

$$u = \mathcal{S}u,$$

and

$$W^{2,p}\left(B_R^+\right) \subset W^{2,q}\left(B_R^+\right).$$

Hence, by the uniqueness of fixed points of $\mathcal{S}$ it follows that

$$u \in W^{2,p}\left(B_R^+\right).$$

Step 4: To prove estimate (19.2), it suffices to take the L^p-norm of both sides of formula (19.10), just as in the proof of estimate (19.26). More precisely, taking $w_1 := u$ and $w_2 := 0$ we obtain that

$$\|\nabla^2 u\|_{L^p(B_R^+)} = \|\nabla^2(\mathcal{S}u)\|_{L^p(B_R^+)} \tag{19.35}$$

$$\leq \|\nabla^2(\mathcal{S}u - \mathcal{S}0)\|_{L^p(B_R^+)} + \|\nabla^2(\mathcal{S}0)\|_{L^p(B_R^+)}$$
$$\text{for all } u \in W^{2,p}\left(B_R^+\right).$$

Step 4-1: In order to estimate the first term on inequality (19.35), we remark that the term

$$\begin{aligned}
&\mathcal{S}u - \mathcal{S}0 \\
&= \phi(x, u) + \int_{B_r^+} G(x_0, x, y) \\
&\times \left\{ \sum_{i,j=1}^{n} \{a^{ij}(x_0) - a^{ij}(y)\} D_{ij}u(y) - \sum_{i,j=1}^{n} a^{ij}(x_0) D_{ij}\phi(y, u) \right\} dy
\end{aligned}$$

is estimated exactly as in estimate (19.25) with $w = w_1 := u$ and $w_2 := 0$ as follows:

$$\begin{aligned}
\|\nabla^2(\mathcal{S}u - \mathcal{S}0)\|_{L^p(B_R^+)} \leq C_1 \Big(& R\eta(R)\|\nabla^2 u\|_{L^p(B_R^+)} + R^2\|\nabla^2 u\|_{L^p(B_R^+)} \\
& + R\|\nabla^2\phi(\cdot, u)\|_{L^p(B_R^+)} + \|\phi(\cdot, u)\|_{L^p(B_R^+)} \Big) \\
& \text{for all } u \in W^{2,p}\left(B_R^+\right),
\end{aligned}$$

with some positive constant $C_1 = C_1(n, s, \lambda, \eta, M, \partial\Omega)$.

Therefore, by applying inequality (17.6) we obtain that

$$\begin{aligned}
\|\nabla^2(\mathcal{S}u - \mathcal{S}0)\|_{L^p(B_R^+)} \leq\ & C_1 \left(R\eta(R) + R^2\right) \|\nabla^2 u\|_{L^p(B_R^+)} \\
& + C_2 R^{1/2} \|\widetilde{\varphi}(\cdot, u)\|^*_{B^{1-1/p,p}(C_R)}.
\end{aligned} \tag{19.36}$$

However, by formula (19.16) with $w := u$ it follows that

$$\widetilde{\varphi}(x', u) = \mathcal{B}_0 u(x') - \mathcal{B}u(x') = \mathcal{B}_0 u(x') - \varphi(x'),$$

so that, by the trace theorem (Theorem 7.6),

$$\begin{aligned}
&\|\widetilde{\varphi}(\cdot, u)\|^*_{B^{1-1/p,p}(C_R)} \\
&\leq \|\mathcal{B}_0 u\|^*_{B^{1-1/p,p}(C_R)} + \|\varphi\|^*_{B^{1-1/p,p}(C_R)} \\
&\leq C_3 \left(\|u\|_{W^{2,p}(B_R^+)} + \|\varphi\|_{B^{1-1/p,p}(C_R)} \right) \\
&\leq C_3 \left(\|u\|_{L^p(B_R^+)} + \|\nabla u\|_{L^p(B_R^+)} + \|\nabla^2 u\|_{L^p(B_R^+)} \right) \\
&\quad + C_3 \|\varphi\|_{B^{1-1/p,p}(C_R)}.
\end{aligned}$$

Hence, we have, by interpolation inequality (13.17) with $\varepsilon := 1/2$,

$$
\begin{aligned}
&\|\widetilde{\varphi}(\cdot, u)\|^{*}_{B^{1-1/p,p}(C_R)} \\
&\le C_3\left(\|u\|_{L^p(B_R^+)} + \|\nabla u\|_{L^p(B_R^+)} + \|\nabla^2 u\|_{L^p(B_R^+)}\right) \\
&\quad + C_3\|\varphi\|_{B^{1-1/p,p}(C_R)} \\
&\le C_4\left(\|u\|_{L^p(B_R^+)} + \|\nabla^2 u\|_{L^p(B_R^+)} + \|\varphi\|_{B^{1-1/p,p}(C_R)}\right).
\end{aligned}
\tag{19.37}
$$

By combining inequalities (19.36) and (19.37), we have proved that

$$
\begin{aligned}
&\|\nabla^2(\mathcal{S}u - \mathcal{S}0)\|_{L^p(B_R^+)} \\
&\le C_1\left(R\eta(R) + R^2\right)\|\nabla^2 u\|_{L^p(B_R^+)} \\
&\quad + C_2C_4R^{1/2}\left(\|u\|_{L^p(B_R^+)} + \|\nabla^2 u\|_{L^p(B_R^+)} + \|\varphi\|_{B^{1-1/p,p}(C_R)}\right) \\
&\le \left(C_1\left(R\eta(R) + R^2\right) + C_2C_4R^{1/2}\right)\|\nabla^2 u\|_{L^p(B_R^+)} \\
&\quad + C_5\left(\|u\|_{L^p(B_R^+)} + \|\varphi\|_{B^{1-1/p,p}(C_R)}\right) \\
&\text{for all } u \in W^{2,p}\left(B_R^+\right).
\end{aligned}
\tag{19.38}
$$

Step 4-2: In order to estimate the second term on inequality (19.35), we remark that the term

$$
\mathcal{S}0 = \int_{B_r^+} G(x_0, x, y) f(y)\, dy
$$

is estimated as follows:

$$
\|\nabla^2(\mathcal{S}0)\|_{L^p(B_R^+)} \le C_2\|f\|_{L^p(B_R^+)}, \tag{19.39}
$$

with some positive constant $C_2 = C_2(n, s, \lambda, \eta, M, \partial\Omega)$.

Step 4-3: Summing up, we obtain from inequalities (19.35), (19.38) and (19.39) that

$$
\begin{aligned}
\|\nabla^2 u\|_{L^p(B_R^+)} &= \|\nabla^2(\mathcal{S}u)\|_{L^p(B_R^+)} \\
&\le \|\nabla^2(\mathcal{S}u - \mathcal{S}0)\|_{L^p(B_R^+)} + \|\nabla^2(\mathcal{S}0)\|_{L^p(B_R^+)} \\
&\le C(R)\|\nabla^2 u\|_{L^p(B_R^+)} \\
&\quad + C_5\left(\|u\|_{L^p(B_R^+)} + \|\varphi\|_{B^{1-1/p,p}(C_R)} + \|f\|_{L^p(B_R^+)}\right) \\
&\quad \text{for all } u \in W^{2,p}\left(B_R^+\right),
\end{aligned}
$$

with a positive constant

$$
C(R) := C_1\left(R\eta(R) + R^2\right) + C_2C_4R^{1/2} = o(1) \quad \text{as } R \downarrow 0.
$$

Therefore, if we take R so small that

$$C(R) < \frac{1}{2},$$

then we have the desired boundary *a priori* estimate (19.2)

$$\left\|\nabla^2 u\right\|_{L^p(B_R^+)} \leq C\left(\|u\|_{L^p(B_R^+)} + \|f\|_{L^p(B_R^+)} + \|\varphi\|_{B^{1-1/p,p}(C_R)}\right)$$
$$\text{for } u \in W^{2,p}\left(B_R^+\right).$$

Now the proof of Lemma 19.1 is complete. □

Chapter 20
Proof of Theorems 16.1 and 16.2

This chapter is devoted to the study of the following *non-homogeneous* oblique derivative problem (16.4)

$$\begin{cases} \mathcal{L}u(x) := \sum_{i,j=1}^{n} a^{ij}(x) \frac{\partial^2 u}{\partial x_i \partial x_j}(x) = f(x) \text{ for almost all } x \in \Omega, \\ \mathcal{B}u(x') := \frac{\partial u}{\partial \ell}(x') + \sigma(x')u(x') = \varphi(x') \text{ in the sense of traces on } \partial\Omega. \end{cases}$$

Concerning the operator $\mathcal{L}$, we assume that the coefficients $a^{ij}(x)$ are the components of a symmetric contravariant tensor of type $\binom{2}{0}$ in $\mathbf{R}^n$ and satisfy the following three conditions:

(1) $a^{ij} \in \mathrm{VMO} \cap L^\infty(\Omega)$.
(2) $a^{ij}(x) = a^{ji}(x)$ for almost all $x \in \Omega$.
(3) There exists a positive constant λ such that (the uniform ellipticity condition)

$$\frac{1}{\lambda}|\xi|^2 \leq \sum_{i,j=1}^{n} a^{ij}(x)\xi_i\xi_j \leq \lambda|\xi|^2 \tag{16.1}$$

for almost all $x \in \Omega$ and for all $\xi \in \mathbf{R}^n$.

Concerning the boundary operator $\mathcal{B}$, we assume that the following three conditions are satisfied (cf. condition (18.1)):

$$\ell^i(x') \text{and } \sigma(x') \text{ are Lipschitz continuous functions on } \partial\Omega. \tag{16.3a}$$

K. Taira, *Real Analysis Methods for Markov Processes*,
https://doi.org/10.1007/978-981-97-3659-1_20

$$\langle \ell(x'), \mathbf{n}(x') \rangle := \sum_{i=1}^{n} \ell^i(x') n_i(x') > 0 \quad \text{on } \partial\Omega. \tag{16.3b}$$

$$\sigma(x') < 0 \quad \text{on } \partial\Omega. \tag{16.3c}$$

We prove the regularity, existence and uniqueness theorems (Theorems 16.1 and 16.2) for problem (16.4) due to Di Fazio–Palagachev [40]. By Lemma 17.1, for any given function $\varphi \in B^{1-1/p,p}(\partial\Omega)$ we can construct a function $v \in W^{2,p}(\Omega)$ such that $\mathcal{B}v = \varphi$ on $\partial\Omega$ and that, for some positive constant C,

$$\|v\|_{W^{2,p}(\Omega)} \leq C \|\varphi\|_{B^{1-1/p,p}(\partial\Omega)}. \tag{20.1}$$

Hence, by letting

$$w = u - v,$$

we are reduced to the study of the following *homogeneous* oblique derivative problem:

$$\begin{cases} \mathcal{L}w = \mathcal{L}u - \mathcal{L}v = f - \mathcal{L}v & \text{in } \Omega, \\ \mathcal{B}w = \mathcal{B}u - \mathcal{B}v = 0 & \text{on } \partial\Omega. \end{cases} \tag{20.2}$$

Our proof is based on some interior and boundary *a priori* estimates for the solutions of the homogeneous oblique derivative problem (20.2) (Theorem 12.1 and Lemma 19.1). Both the interior and boundary *a priori* estimates are consequences of explicit representation formulas (19.10) and (19.11) for the solutions of problem (19.2) and also of the L^p-boundedness of Calderón–Zygmund singular integral operators and boundary commutators appearing in those representation formulas (Theorems 14.2 and 14.5). It should be emphasized that the vanishing mean oscillation (VMO) assumption on the coefficients a^{ij} is of the greatest relevance in the study of singular commutators.

20.1 Proof of Theorem 16.1

The purpose of this section is to prove Theorem 16.1. The proof of Theorem 16.1 is divided into two steps (see [33–35, 99]).

Step 1: First, in view of Lemma 19.1, we obtain that if a function $w \in W^{2,q}(\Omega)$, $1 < q < p < \infty$, is a solution of problem (20.2) with $f - \mathcal{L}v \in L^p(\Omega)$, then it follows that

$$w = u - v \in W^{2,p}(\Omega).$$

This proves that if a function $u \in W^{2,q}(\Omega)$, $1 < q < p < \infty$, is a solution of problem (16.4) with $f \in L^p(\Omega)$ and $\varphi \in B^{1-1/p,p}(\partial\Omega)$, then it follows that $u \in W^{2,p}(\Omega)$.

Step 2: Secondly, we prove the global *a priori* estimate (16.5). However, by estimate (20.1) we may assume that $u \in W^{2,p}(\Omega)$ is a solution of the homogeneous oblique derivative problem

$$\begin{cases} \mathcal{L}u = f & \text{in } \Omega, \\ \mathcal{B}u = 0 & \text{on } \partial\Omega. \end{cases} \tag{20.3}$$

Therefore, it suffices to prove the following global *a priori* estimate:

$$\|u\|_{W^{2,p}(\Omega)} \leq c_1 \left(\|f\|_{L^p(\Omega)} + \|u\|_{L^p(\Omega)}\right) \quad \text{for all } u \in \mathfrak{B}. \tag{20.4}$$

Here

$$\mathfrak{B} := \left\{u \in W^{2,p}(\Omega) : \mathcal{B}u = 0 \text{ on } \partial\Omega\right\},$$

and $c_1 > 0$ is a constant depending on the coefficients a^{ij} only through the ellipticity constant λ, the bound on the norms $\|a^{ij}\|_{L^p(\Omega)}$ and the VMO moduli of the a^{ij}. The proof of estimate (20.4) is carried out in a standard way from the interior *a priori* estimate (Theorem 12.1) and the boundary *a priori* estimate (Lemma 19.1) by a covering argument, by flattening the boundary $\partial\Omega$ and by using interpolation inequalities (Theorem 13.4), just as in the proof of Theorem 12.2.

Step 2-1: Now we choose a finite covering $\{U_j\}_{j=1}^N$ of the boundary $\partial\Omega$ by open subsets of $\mathbf{R}^n$ and $C^{1,1}$-diffeomorphisms G_j of $U_j \cap \Omega$ onto B_r^+ in each of which the boundary *a priori* estimate (19.2) holds true (see Fig. 20.1 below). Furthermore, we choose an open subset U_0 of Ω, bounded away from $\partial\Omega$, such that (see Fig. 20.2 below)

$$\Omega \subset U_0 \cup \left(\bigcup_{j=1}^N U_j\right).$$

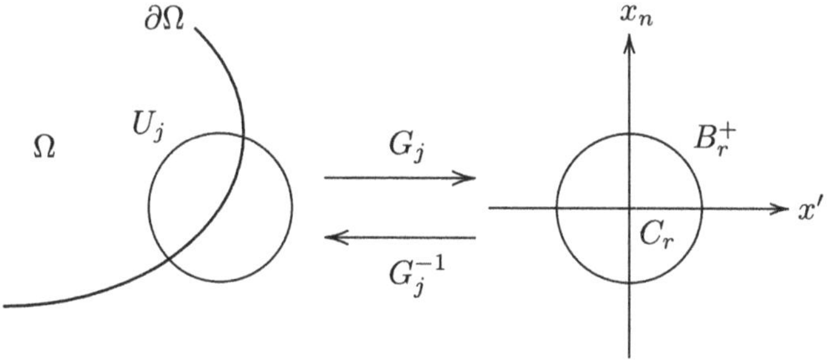

Fig. 20.1 The covering $\{U_j\}$ of $\partial\Omega$ and the $C^{1,1}$-diffeomorphism G_j of $U_j \cap \Omega$ onto B_r^+

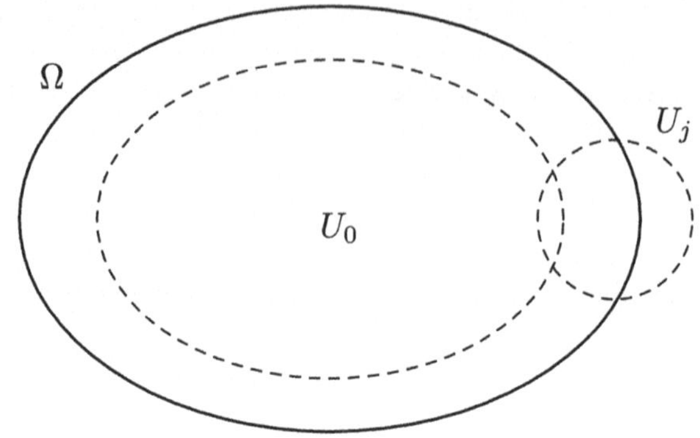

Fig. 20.2 The open covering $\{U_j\}$ of $\partial\Omega$ and the open set U_0 bounded away from $\partial\Omega$

Step 2-2: We take a partition of unity $\{\alpha_k\}_{k=0}^N$ subordinate to the open covering $\{U_k\}_{k=0}^N$ of Ω. Then, by applying the interior *a priori* estimate (12.2) to the function $\alpha_0 u$ we obtain that

$$\begin{aligned}
\|u\|_{W^{2,p}(\Omega)} &\leq \sum_{k=0}^{N} \|\alpha_k u\|_{W^{2,p}(\Omega)} \\
&\leq \|\alpha_0 u\|_{W^{2,p}(\Omega)} \\
&\quad + \sum_{k=1}^{N} \left(\|\nabla^2(\alpha_k u)\|_{L^p(\Omega)} + \|\nabla(\alpha_k u)\|_{L^p(\Omega)} + \|\alpha_k u\|_{L^p(\Omega)}\right) \\
&\leq C_0 \left(\|u\|_{L^p(\Omega)} + \|f\|_{L^p(\Omega)}\right) \\
&\quad + \sum_{k=1}^{N} \left(\|\nabla^2(\alpha_k u)\|_{L^p(U_k\cap\Omega)} + \|\nabla(\alpha_k u)\|_{L^p(U_k\cap\Omega)} + \|u\|_{L^p(\Omega)}\right).
\end{aligned} \tag{20.5}$$

(I) In order to estimate the terms $\nabla(\alpha_k u)$, $1 \leq k \leq N$, we recall the interpolation inequality (Theorem 13.4)

$$\|\nabla v\|_{L^p(\Omega)} \leq \varepsilon \, \|\nabla^2 v\|_{L^p(\Omega)} + \frac{C}{\varepsilon} \, \|v\|_{L^p(\Omega)} \quad \text{for all } v \in W^{2,p}(\Omega). \tag{13.17}$$

Since we have the formula

$$\nabla(\alpha_k u) = \alpha_k \, \nabla u + u(\nabla \alpha_k),$$

by applying inequality (13.17) to the function $u \in W^{2,p}(\Omega)$, we obtain that, for some positive constants C_1 and C_2,

$$\|\nabla(\alpha_k u)\|_{L^p(U_k\cap\Omega)} \tag{20.6}$$

$$\leq \|\alpha_k(\nabla u)\|_{L^p(U_k\cap\Omega)} + \|u(\nabla\alpha_k)\|_{L^p(U_k\cap\Omega)}$$
$$\leq \|\nabla u\|_{L^p(\Omega)} + C_1\|u\|_{L^p(\Omega)}$$
$$\leq \varepsilon\, \|\nabla^2 u\|_{L^p(\Omega)} + \frac{C_2}{\varepsilon}\|u\|_{L^p(\Omega)} + C_1\|u\|_{L^p(\Omega)}.$$

(II) Moreover, by applying the boundary *a priori* estimate (19.2) to the functions $\alpha_k u \in W^{2,p}(U_k \cap \Omega)$, $1 \leq k \leq N$, we obtain that

$$\|\nabla^2(\alpha_k u)\|_{L^p(U_k\cap\Omega)} \tag{20.7}$$
$$\leq C\left(\|\alpha_k u\|_{L^p(U_k\cap\Omega)} + \|\mathcal{L}(\alpha_k u)\|_{L^p(U_k\cap\Omega)} + \|\mathcal{B}(\alpha_k u)\|_{B^{1-1/p,p}(U_k\cap\partial\Omega)}\right).$$

However, since u is a solution of the homogeneous oblique derivative problem

$$\begin{cases} \mathcal{L}u = f & \text{in } \Omega, \\ \mathcal{B}u = 0 & \text{on } \partial\Omega, \end{cases} \tag{20.3}$$

it follows that

$$\mathcal{L}(\alpha_k u) = \alpha_k(\mathcal{L}u) + u\,\mathcal{L}(\alpha_k) + 2\sum_{i,j=1}^{n} a^{ij}(x)\frac{\partial\alpha_k}{\partial x_i}\frac{\partial u}{\partial x_j} \tag{20.8a}$$
$$= \alpha_k f + u\,\mathcal{L}(\alpha_k) + 2\sum_{i,j=1}^{n} a^{ij}(x)\frac{\partial\alpha_k}{\partial x_i}\frac{\partial u}{\partial x_j} \quad \text{in } \Omega,$$

and that

$$\mathcal{B}(\alpha_k u) = \alpha_k(\mathcal{B}u) + \frac{\partial\alpha_k}{\partial\ell}u = \frac{\partial\alpha_k}{\partial\ell}u \quad \text{on } \partial\Omega. \tag{20.8b}$$

Hence, we have, by inequality (20.7) and formulas (20.8),

$$\|\nabla^2(\alpha_k u)\|_{L^p(U_k\cap\Omega)} \tag{20.9}$$
$$\leq C\left(\|\alpha_k u\|_{L^p(U_k\cap\Omega)} + \|\mathcal{L}(\alpha_k u)\|_{L^p(U_k\cap\Omega)} + \|\mathcal{B}(\alpha_k u)\|_{B^{1-1/p,p}(U_k\cap\partial\Omega)}\right)$$
$$\leq C\Bigg(\|\alpha_k f\|_{L^p(U_k\cap\Omega)} + \|u\,\mathcal{L}(\alpha_k)\|_{L^p(U_k\cap\Omega)}$$
$$+2\sum_{i,j=1}^{n}\left\|a^{ij}\frac{\partial\alpha_k}{\partial x_i}\frac{\partial u}{\partial x_j}\right\|_{L^p(U_k\cap\Omega)}\Bigg) + C_3\left\|u\frac{\partial\alpha_k}{\partial\ell}\right\|_{B^{1-1/p,p}(U_k\cap\partial\Omega)}$$
$$\leq C_4\left(\|f\|_{L^p(\Omega)} + \|u\|_{L^p(\Omega)} + \|\nabla u\|_{L^p(\Omega)}\right) + C_5\,\|u\|_{B^{1-1/p,p}(\partial\Omega)}.$$

On the other hand, by applying the trace theorem (Theorem 7.6) we obtain that

$$\|u\|_{B^{1-1/p,p}(\partial\Omega)} \leq C_6 \|u\|_{W^{1,p}(\Omega)} \qquad (20.10)$$
$$\leq C_7 \left(\|u\|_{L^p(\Omega)} + \|\nabla u\|_{L^p(\Omega)}\right).$$

Therefore, it follows from inequalities (20.9), (20.10) and (13.17) that

$$\begin{aligned} &\|\nabla^2(\alpha_k u)\|_{L^p(U_k \cap \Omega)} \qquad (20.11) \\ &\leq C_4 \left(\|f\|_{L^p(\Omega)} + \|u\|_{L^p(\Omega)} + \|\nabla u\|_{L^p(\Omega)}\right) \\ &\quad + C_5 C_7 \left(\|u\|_{L^p(\Omega)} + \|\nabla u\|_{L^p(\Omega)}\right) \\ &\leq \varepsilon \, \|\nabla^2 u\|_{L^p(\Omega)} + \frac{C_8}{\varepsilon} \|u\|_{L^p(\Omega)} + C_9 \left(\|f\|_{L^p(\Omega)} + \|u\|_{L^p(\Omega)}\right). \end{aligned}$$

(III) By combining inequalities (20.5), (20.6) and (20.11), we obtain that

$$\begin{aligned} \|u\|_{W^{2,p}(\Omega)} &\leq 2N\varepsilon \, \|\nabla^2 u\|_{L^p(\Omega)} + (C_0 + NC_9)\left(\|f\|_{L^p(\Omega)} + \|u\|_{L^p(\Omega)}\right) \\ &\quad + \frac{N(C_2 + C_8)}{\varepsilon} \|u\|_{L^p(\Omega)} + (NC_1 + 1) \, \|u\|_{L^p(\Omega)}. \end{aligned}$$

This proves the desired global *a priori* estimate (20.4), if we take

$$\varepsilon = \frac{1}{4N}.$$

The proof of Theorem 16.1 is now complete. □

20.2 Proof of Theorem 16.2

This section is devoted to the proof of Theorem 16.2. More precisely, we prove the following existence and uniqueness theorem for the homogeneous oblique derivative problem (20.3) due to Di Fazio–Palagachev [40, Theorem 1.2]:

Theorem 20.1 *Let $n < p < \infty$ and assume that conditions (16.1) and (16.3) are satisfied. For any given function $f \in L^p(\Omega)$, there exists a unique solution $u \in W^{2,p}(\Omega)$ of the homogeneous oblique derivative problem* (20.3). *Moreover, we have the* a priori *estimate*

$$\|u\|_{W^{2,p}(\Omega)} \leq C_3 \|f\|_{L^p(\Omega)}, \qquad (20.12)$$

with a positive constant $C_3 = C_3(n, p, \lambda, \eta, \ell, \sigma, \partial\Omega)$.

Proof The proof of Theorem 20.1 is divided into three steps.

Step 1: First, the *uniqueness result* of problem (20.3) follows from an application of the Bakel'man and Aleksandrov maximum principle (see [92, Corollary 2.4]):

Theorem 20.2 (Bakel'man–Aleksandrov) *Assume that a function $u \in W^{2,p}(\Omega)$, $n < p < \infty$, satisfies the conditions*

$$\begin{cases} \mathcal{L}u \leq 0 & \textit{almost everywhere in } \Omega, \\ \mathcal{B}u \leq 0 & \textit{on } \partial\Omega. \end{cases}$$

Then it follows that $u(x) \geq 0$ *in* Ω.

Proof First, it should be noticed that we have, by Sobolev's imbedding theorem (Theorem 7.6),

$$W^{2,p}(\Omega) \subset C^1(\overline{\Omega}),$$

since $2 - n/p > 1$ for $n < p < \infty$.

The proof of Theorem 20.2 is divided into the following two cases.

Case (a): If $u(x)$ is a constant function in Ω, then it follows that

$$\sigma(x')u = \mathcal{B}u \leq 0 \quad \text{on } \partial\Omega.$$

This implies that the constant u is non-negative, since $\sigma(x') < 0$ on $\partial\Omega$.

Case (b): Now we consider the case where $u(x)$ is not a constant function in Ω. The proof is based on a reduction to absurdity. Assume, to the contrary, that

$$\min_{\overline{\Omega}} u < 0.$$

Then, by applying the strong maximum principle (Theorem 8.9) to the function $-u(x)$ we obtain that, for some boundary point $x_0' \in \partial\Omega$,

$$0 < -u(x_0') = \max_{\overline{\Omega}}(-u).$$

Moreover, it follows from an application of Hopf's boundary point lemma (Lemma 8.8) that

$$-\frac{\partial u}{\partial \mathbf{n}}(x_0') = \frac{\partial(-u)}{\partial \mathbf{n}}(x_0') < 0,$$

so that

$$\frac{\partial u}{\partial \mathbf{n}}(x_0') > 0.$$

However, by conditions (16.3b) and (16.3c) this implies that

$$0 \geq \mathcal{B}u(x_0') = \frac{\partial u}{\partial \ell}(x_0') + \sigma(x_0')u(x_0') > 0.$$

This contradiction proves that $u(x) \geq 0$ in Ω.

The proof of Theorem 20.2 is complete. □

By applying Theorem 20.2 to the functions $\pm u(x)$, we obtain from condition (16.3c) that

$$\begin{cases} \mathcal{L}u = 0 & \text{almost everywhere in } \Omega, \\ \mathcal{B}u = 0 & \text{on } \partial\Omega \end{cases} \Longrightarrow u = 0 \quad \text{in } \Omega.$$

Namely, the mapping

$$\mathcal{A} = (\mathcal{L}, \mathcal{B}) : W^{2,p}(\Omega) \longrightarrow L^p(\Omega) \oplus B^{1-1/p,p}(\partial\Omega)$$

is *injective* for $n < p < \infty$. This proves the uniqueness of solutions of the homogeneous oblique derivative problem (20.3).

Step 2: In order to prove the *existence result* of problem (20.3), we make use of the method of continuity (Theorem 2.16). We shall apply Theorem 2.16 with

$$\begin{aligned} \mathfrak{B} &:= \left\{ u \in W^{2,p}(\Omega) : \mathcal{B}u = 0 \right\}, \\ \mathfrak{V} &:= L^p(\Omega), \\ \mathcal{L}_1 &:= \mathcal{L} = \sum_{i,j=1}^{n} a^{ij}(x) \frac{\partial^2}{\partial x_i \partial x_j}, \\ \mathcal{L}_0 &:= \Delta. \end{aligned}$$

Here it should be noticed that the space $\mathfrak{B}$ is a closed subspace of $W^{2,p}(\Omega)$, since the trace theorem (Theorem 7.6) asserts that the boundary operator

$$\mathcal{B} = \frac{\partial}{\partial \ell} + \sigma(x') : W^{2,p}(\Omega) \longrightarrow B^{1-1/p,p}(\partial\Omega)$$

is continuous.

Step 2-1: The essential step in our proof is how to show inequality (15.5) for a family of elliptic differential operators

$$\begin{aligned} \mathcal{L}_t &:= t\mathcal{L} + (1-t)\Delta \\ &= \sum_{i,j=1}^{n} \left(t\, a^{ij}(x) + (1-t)\delta^{ij} \right) \frac{\partial^2}{\partial x_i \partial x_j} \quad \text{for } 0 \leq t \leq 1. \end{aligned}$$

More precisely, we consider, instead of the original oblique derivative problem (20.3)

$$\begin{cases} \mathcal{L}u = f & \text{in } \Omega, \\ \mathcal{B}u = 0 & \text{on } \partial\Omega, \end{cases}$$

a family of oblique derivative problems

$$\begin{cases} \mathcal{L}_t u = (t\mathcal{L} + (1-t)\Delta)\, u = f & \text{in } \Omega, \\ \mathcal{B}u = 0 & \text{on } \partial\Omega. \end{cases}$$

By the uniqueness result of problem (20.3), we can get rid of the term $\|u\|_{L^p(\Omega)}$ on the right-hand side of estimate (20.4). Namely, we shall prove the *a priori* estimate (corresponding to inequality (15.5))

$$\|u\|_{W^{2,p}(\Omega)} \leq c_2 \|\mathcal{L}_t u\|_{L^p(\Omega)} \quad \text{for all } u \in \mathfrak{B}. \tag{20.13}$$

Here $c_2 > 0$ is a constant depending on the coefficients a^{ij} only through the ellipticity constant λ, the bound on the norms $\|a^{ij}\|_{L^p(\Omega)}$ and the VMO moduli of the a^{ij}.

Our proof of the *a priori* estimate (20.13) is based on a reduction to absurdity. First, it should be noticed that the coefficients

$$a^{ij}_{(t)}(x) := t\, a^{ij}(x) + (1-t)\delta^{ij} \quad \text{for } 0 \leq t \leq 1,$$

satisfy the following three conditions:

(i) $a^{ij}_{(t)} \in \mathrm{VMO} \cap L^\infty(\mathbf{R}^n)$. Indeed, we have, for all $0 \leq t \leq 1$,

$$\|a^{ij}_{(t)}\|_{L^\infty(\mathbf{R}^n)} \leq \|a^{ij}\|_{L^\infty(\mathbf{R}^n)} + 1,$$

and

$$\|a^{ij}_{(t)}\|_* \leq t\|a^{ij}\|_* \leq \|a^{ij}\|_*.$$

(ii) $a^{ij}_{(t)}(x) = a^{ji}_{(t)}(x)$ for almost all $x \in \Omega$.

(iii) We have, for almost all $x \in \Omega$ and all $\xi \in \mathbf{R}^n$,

$$\frac{1}{\lambda+1}|\xi|^2 \leq \sum_{i,j=1}^n a^{ij}_{(t)}(x)\xi_i\xi_j \leq (\lambda+1)\,|\xi|^2,$$

where λ is the same constant as in condition (16.1).

Now we assume, to the contrary, that the *a priori* estimate (20.13) does not hold true. Then we can find a sequence of uniformly elliptic differential operators

$$\mathcal{L}^{(m)} = \sum_{i,j=1}^n a^{ij}_{(m)}(x)\frac{\partial^2}{\partial x_i \partial x_j},$$

and a sequence of functions $\{u^{(m)}\}$ such that the coefficients $a^{ij}_{(m)}(x)$ and the functions $\{u^{(m)}\}$ satisfy the following four conditions:

(i) $a^{ij}_{(m)} \in \mathrm{VMO} \cap L^\infty(\mathbf{R}^n)$ and

$$\|a^{ij}_{(m)}\|_{L^\infty(\mathbf{R}^n)} \leq \|a^{ij}\|_{L^\infty(\mathbf{R}^n)} + 1, \tag{20.14a}$$

$$\eta^{ij}_{(m)}(r) \leq \eta^{ij}(r). \tag{20.14b}$$

Here we recall that

$$\eta^{ij}_{(m)}(r) = \sup_{\rho \le r} \frac{1}{|B|} \int_B \left| a^{ij}_{(m)}(y) - \left(a^{ij}_{(m)}\right)_B \right| dy,$$

$$\eta^{ij}(r) = \sup_{\rho \le r} \frac{1}{|B|} \int_B \left| a^{ij}(y) - \left(a^{ij}\right)_B \right| dy,$$

where the supremum is taken over all balls B with radius $\rho \le r$.

(ii) $a^{ij}_{(m)}(x) = a^{ji}_{(m)}(x)$ for almost all $x \in \Omega$.

(iii) We have, for almost all $x \in \Omega$ and all $\xi \in \mathbf{R}^n$,

$$\frac{1}{\lambda + 1} |\xi|^2 \le \sum_{i,j=1}^{n} a^{ij}_{(m)}(x) \xi_i \xi_j \le (\lambda + 1) |\xi|^2. \tag{20.15}$$

(iv) $u^{(m)} \in W^{2,p}(\Omega)$ and

$$Bu^{(m)} = 0 \quad \text{on } \partial\Omega, \tag{20.16a}$$

$$\|u^{(m)}\|_{W^{2,p}(\Omega)} = 1, \tag{20.16b}$$

$$\|\mathcal{L}^{(m)} u^{(m)}\|_{L^p(\Omega)} \longrightarrow 0. \tag{20.16c}$$

Step 2-1a: First, it follows that the sequence $\{(a^{ij}_{(m)})_B\}$ is bounded for every $1 \le i, j \le n$. Indeed, we have, by inequality (15.9a),

$$\left|(a^{ij}_{(m)})_B\right| \le \|a^{ij}_{(m)}\|_{L^\infty(\mathbf{R}^n)} \le \|a^{ij}\|_{L^\infty(\mathbf{R}^n)} + 1.$$

Moreover, we have the following lemma:

Lemma 20.3 *For any ball B in $\mathbf{R}^n$, every sequence*

$$\left\{ a^{ij}_{(m)}(x) - \left(a^{ij}_{(m)}\right)_B \right\} \quad \text{for } 1 \le i, j \le n,$$

is compact *in the space $L^1(B)$.*

Proof First, we take a function $\varphi(x)$ on $\mathbf{R}^n$ which satisfies the following four conditions:

$$\varphi \in C_0^\infty(\mathbf{R}^n).$$

$$\varphi(x) \ge 0 \quad \text{on } \mathbf{R}^n.$$

$$\operatorname{supp} \varphi \subset B(0, 1) := \left\{x \in \mathbf{R}^n : |x| \le 1\right\}.$$

$$\int_{\mathbf{R}^n} \varphi(x)\, dx = 1.$$

For each $\varepsilon > 0$, we let

$$\varphi_\varepsilon(x) := \frac{1}{\varepsilon^n}\varphi\left(\frac{x}{\varepsilon}\right).$$

Then it is easy to verify that

$$\varphi_\varepsilon \in C_0^\infty(\mathbf{R}^n).$$
$$\varphi_\varepsilon(x) \geq 0 \quad \text{on } \mathbf{R}^n.$$
$$\operatorname{supp}\varphi_\varepsilon \subset B(0,\varepsilon) := \left\{x \in \mathbf{R}^n : |x| \leq \varepsilon\right\},$$
$$\int_{\mathbf{R}^n} \varphi_\varepsilon(x)\,dx = 1.$$

The functions $\{\varphi_\varepsilon\}$ are called *Friedrichs' mollifiers*.

(1) For simplicity, we write

$$a_{(m)}(x) := a^{ij}_{(m)}(x),$$

and let

$$f_{(m)}(x) := a_{(m)}(x) - \left(a_{(m)}\right)_B, \tag{20.17a}$$
$$f^\varepsilon_{(m)}(x) := a_{(m)} * \varphi_\varepsilon(x) - \left(a_{(m)} * \varphi_\varepsilon\right)_B, \tag{20.17b}$$

where

$$a_{(m)} * \varphi_\varepsilon(x) = \int_{\mathbf{R}^n} a_{(m)}(x-y)\varphi_\varepsilon(y)\,dy$$

is a mollification of $a_{(m)}$.

Then we can prove the following two claims 20.4 and 20.5 (see Claims 15.6 and 15.7):

Claim 20.4 The sequence $\{f^\varepsilon_{(m)}\}$ is *uniformly bounded* and *equicontinuous* in B, for each $\varepsilon > 0$.

Proof (a) The *uniform boundedness* of $\{f^\varepsilon_{(m)}\}$: First, we have, for all $x \in B$,

$$|a_{(m)} * \varphi_\varepsilon(x)| \leq \int_{\mathbf{R}^n} |a_{(m)}(x-y)|\,\varphi_\varepsilon(y)\,dy \leq \|a_{(m)}\|_{L^\infty(\mathbf{R}^n)}$$
$$\leq \|a\|_{L^\infty(\mathbf{R}^n)} + 1,$$

and

$$|\left(a_{(m)} * \varphi_\varepsilon\right)_B| \leq \frac{1}{|B|}\int_{\mathbf{R}^n} |a_{(m)}(x-y)|\,\varphi_\varepsilon(y)\,dy \leq \|a_{(m)}\|_{L^\infty(\mathbf{R}^n)}$$
$$\leq \|a\|_{L^\infty(\mathbf{R}^n)} + 1.$$

Hence it follows that

$$
\begin{aligned}
|f^{\varepsilon}_{(m)}(x)| &\le |a_{(m)} * \varphi_{\varepsilon}(x)| + |\left(a_{(m)} * \varphi_{\varepsilon}\right)_B| \\
&\le 2\left(\|a\|_{L^\infty(\mathbf{R}^n)} + 1\right) \quad \text{for } x \in B.
\end{aligned}
$$

(b) The *equicontinuity* of $\{f^{\varepsilon}_{(m)}\}$: It suffices to show that $\{\nabla f^{\varepsilon}_{(m)}\}$ is uniformly bounded in B, for each $\varepsilon > 0$.

We have, for all $x \in B$,

$$
\begin{aligned}
&\left|\nabla f^{\varepsilon}_{(m)}(x)\right| \qquad (20.18)\\
&\le \frac{1}{\varepsilon^{n+1}} \int_{B(x,\varepsilon)} \left|a_{(m)}(y) - \left(a_{(m)}\right)_{B(x,\varepsilon)}\right| \left|\nabla\varphi\left(\frac{x-y}{\varepsilon}\right)\right| dy \\
&\le \frac{1}{\varepsilon^{n+1}} \left(\sup_{\mathbf{R}^n} |\nabla\varphi|\right) \int_{B(x,\varepsilon)} \left|a_{(m)}(y) - \left(a_{(m)}\right)_{B(x,\varepsilon)}\right| dy \\
&\le \frac{1}{\varepsilon^{n+1}} \left(\sup_{\mathbf{R}^n} |\nabla\varphi|\right) |B(x,\varepsilon)|\, \eta_{(m)}(\varepsilon) = \frac{1}{\varepsilon^{n+1}} \left(\sup_{\mathbf{R}^n} |\nabla\varphi|\right) \left(\varepsilon^n \omega_n\right) \eta_{(m)}(\varepsilon) \\
&\le \frac{\omega_n}{\varepsilon} \left(\sup_{\mathbf{R}^n} |\nabla\varphi|\right) \eta(\varepsilon).
\end{aligned}
$$

Here we recall that

$$
\begin{aligned}
\omega_n &= \frac{2\pi^{n/2}}{\Gamma(n/2)}, \\
\eta_{(m)}(r) = \eta^{ij}_{(m)}(r) &:= \sup_{\rho \le r} \frac{1}{|B|} \int_B \left|a^{ij}_{(m)}(y) - \left(a^{ij}_{(m)}\right)_B\right| dy, \\
\eta(r) &:= \left(\sum_{i,j=1}^{n} \eta^{ij}_{(m)}(r)^2\right)^{1/2}.
\end{aligned}
$$

Thus, we obtain from inequality (20.18) that $\{\nabla f^{\varepsilon}_{(m)}\}$ is uniformly bounded in B, for each $\varepsilon > 0$.

Claim 20.5 There exists a positive constant C, independent of ε, such that

$$
\int_B \left|f^{\varepsilon}_{(m)}(x) - f_{(m)}(x)\right| dx \le C|B|\eta(\varepsilon). \qquad (20.19)
$$

Proof First, by Fubini's theorem (Theorem 3.20) it follows that

$$
\begin{aligned}
\left(a_{(m)} * \varphi_{\varepsilon}\right)_B &= \frac{1}{|B|} \int_B a_{(m)} * \varphi_{\varepsilon}(x)\, dx \qquad (20.20)\\
&= \frac{1}{|B|} \int_B \left(\int_{B(0,\varepsilon)} a_{(m)}(x-y)\varphi_{\varepsilon}(y)\, dy\right) dx
\end{aligned}
$$

$$
\begin{aligned}
&= \int_{B(0,\varepsilon)} \left(\frac{1}{|B|} \int_B a_{(m)}(x-y)\,dx \right) \varphi_\varepsilon(y)\,dy \\
&= \int_{B(0,\varepsilon)} \left(\frac{1}{|B-y|} \int_{B-y} a_{(m)}(z)\,dz \right) \varphi_\varepsilon(y)\,dy \\
&= \int_{B(0,\varepsilon)} \left(a_{(m)}\right)_{B-y} \varphi_\varepsilon(y)\,dy,
\end{aligned}
$$

where $B - y$ is the translation of the ball B by y-units

$$
B - y = \{x - y : x \in B\}.
$$

Hence we have, by formulas (20.17) and (20.20),

$$
\begin{aligned}
&\int_B \left| f^{\varepsilon}_{(m)}(x) - f_{(m)}(x) \right| dx \qquad (20.21)\\
&= \int_B \left| a_{(m)} * \varphi_\varepsilon(x) - \left(a_{(m)} * \varphi_\varepsilon\right)_B - \left(a_{(m)}(x) - \left(a_{(m)}\right)_B\right) \right| dx \\
&= \int_B \left| \int_{B(0,\varepsilon)} \left(a_{(m)}(x-y) - a_{(m)}(x)\right) \right. \\
&\qquad \left. - \left(\left(a_{(m)}\right)_{B-y} - \left(a_{(m)}\right)_B\right) \varphi_\varepsilon(y)\,dy \right| dx \\
&\leq \int_B \int_{B(0,\varepsilon)} \left| \left(a_{(m)}(x-y) - a_{(m)}(x)\right) \right. \\
&\qquad \left. - \left(\left(a_{(m)}\right)_{B-y} - \left(a_{(m)}\right)_B\right) \right| \varphi_\varepsilon(y)\,dy\,dx \\
&\leq \int_{B(0,\varepsilon)} \varphi_\varepsilon(y) \left(\int_B \left| \left(a_{(m)}(x-y) - a_{(m)}(x)\right) \right. \right. \\
&\qquad \left. \left. - \left(\left(a_{(m)}\right)_{B-y} - \left(a_{(m)}\right)_B\right) \right| dx \right) dy.
\end{aligned}
$$

However, we have the formula

$$
\begin{aligned}
\left(a_{(m)}\right)_{B-y} &= \frac{1}{|B-y|} \int_{B-y} a_{(m)}(w)\,dw = \frac{1}{|B|} \int_B a_{(m)}(z-y)\,dz \\
&= a_{(m)}(\cdot - y)_B.
\end{aligned}
$$

Hence it follows from an application of inequality (4.9) with $f := a_{(m)}$ that

$$
\begin{aligned}
&\frac{1}{|B|} \int_B \left| \left(a_{(m)}(x-y) - a_{(m)}(x)\right) - \left(\left(a_{(m)}\right)_{B-y} - \left(a_{(m)}\right)_B\right) \right| dx \qquad (20.22)\\
&= \frac{1}{|B|} \int_B \left| \left(a_{(m)}(x-y) - a_{(m)}(x)\right) - \left(a_{(m)}(\cdot - y) - a_{(m)}\right)_B \right| dx
\end{aligned}
$$

$$
\begin{aligned}
&\leq \|a_{(m)}(\cdot - y) - a_{(m)}(\cdot)\|_* \\
&\leq C\eta_{(m)}(\varepsilon) \quad \text{for } |y| < \varepsilon.
\end{aligned}
$$

Therefore, by combining inequalities (20.21) and (20.22) we obtain that

$$
\begin{aligned}
\int_B \left| f^{\varepsilon}_{(m)}(x) - f_{(m)}(x) \right| dx &\leq C \left(\int_{\{|y| \leq \varepsilon\}} \varphi_\varepsilon(y) \eta_{(m)}(\varepsilon)\, dy \right) |B| \\
&= C\, \eta_{(m)}(\varepsilon)\, |B| \\
&\leq C|B|\eta(\varepsilon) \quad \text{for all } \varepsilon > 0.
\end{aligned}
$$

This proves the desired inequality (20.19).

(2) By combining Claims 20.4 and 20.5 and the Ascoli–Arzelá theorem, we find that the sequence $\{f_{(m)}\}$ is *totally bounded* in $L^1(B)$, so that it is compact in $L^1(B)$.

The proof of Lemma 20.3 is complete. □

Step 2-1b: By using Lemma 20.3 and the Bolzano–Weierstrass theorem, we can choose a subsequence of the sequence

$$
a^{ij}_{(m)}(x) = \left(a^{ij}_{(m)}(x) - (a^{ij}_{(m)})_B \right) + (a^{ij}_{(m)})_B
$$

which converges almost everywhere in B. Therefore, by considering an exhaustive sequence of balls of $\mathbf{R}^n$ we can choose a subsequence of $a^{ij}_{(m)}$, denoted again by $a^{ij}_{(m)}$, which converges to a function α^{ij} almost everywhere in $\mathbf{R}^n$, as $m \to \infty$. Then it is easy to verify the following three assertions (i), (ii) and (iii):

(i) $\alpha^{ij} \in \text{VMO} \cap L^\infty(\mathbf{R}^n)$. Indeed, since we have the inequality

$$
\|a^{ij}_{(m)}\|_{L^\infty(\mathbf{R}^n)} \leq \|a^{ij}\|_{L^\infty(\mathbf{R}^n)} + 1,
$$

it follows that

$$
\|\alpha^{ij}\|_{L^\infty(\mathbf{R}^n)} \leq \|a^{ij}\|_{L^\infty(\mathbf{R}^n)} + 1.
$$

Moreover, since we have, for all balls B with radius $\rho \leq r$,

$$
\frac{1}{|B|} \int_B |a^{ij}_{(m)}(x) - (a^{ij}_{(m)})_B|\, dx \leq \eta^{ij}_{(m)}(r) \leq \eta^{ij}(r),
$$

and since we have, by the Lebesgue dominated convergence theorem (Theorem 3.13),

$$
(a^{ij}_{(m)})_B = \frac{1}{|B|} \int_B a^{ij}_{(m)}(y)\, dy \longrightarrow \frac{1}{|B|} \int_B \alpha^{ij}(y)\, dy = (\alpha^{ij})_B \quad \text{as } m \to \infty,
$$

it follows that, for all balls B with radius $\rho \leq r$,

$$\frac{1}{|B|}\int_B |\alpha^{ij}(x) - (\alpha^{ij})_B|\,dx = \lim_{m\to\infty} \frac{1}{|B|}\int_B |a^{ij}_{(m)}(x) - (a^{ij}_{(m)})_B|\,dx$$
$$\leq \eta^{ij}(r).$$

This proves that the VMO modulus of α^{ij} is dominated by $\eta^{ij}(r)$:

$$\sup_{\rho\leq r} \frac{1}{|B|}\int_B |\alpha^{ij}(x) - (\alpha^{ij})_B|\,dx \leq \eta^{ij}(r).$$

(ii) $\alpha^{ij}(x) = \alpha^{ji}(x)$ for almost all $x \in \Omega$.
Indeed, we have, for almost all $x \in \Omega$,

$$\alpha^{ij}(x) = \lim_{m\to\infty} a^{ij}_{(m)}(x) = \lim_{m\to\infty} a^{ji}_{(m)}(x) = \alpha^{ji}(x).$$

(iii) We have, for almost all $x \in \Omega$ and all $\xi \in \mathbf{R}^n$,

$$\frac{1}{\lambda+1}|\xi|^2 \leq \sum_{i,j=1}^{n} \alpha^{ij}(x)\xi_i\xi_j \leq (\lambda+1)|\xi|^2.$$

These inequalities can be obtained by passing to the limit in inequalities (20.15).

We introduce a new second order, uniformly elliptic differential operator $\mathcal{A}$ by the formula

$$\mathcal{A} := \sum_{i,j=1}^{n} \alpha^{ij}(x)\frac{\partial^2}{\partial x_i \partial x_j}.$$

Step 2-1c: By using the Eberlein–Shmulyan theorem (Theorem 2.32) and the Rellich–Kondrachov theorem (Theorem 7.8), we can obtain the following two assertions:

(A) Let X be a reflexive Banach space, and let $\{x_n\}$ be any sequence which is norm bounded. Then there exists a subsequence $\{x_{n'}\}$ which converges *weakly* to an element of X.
(B) The injection $W^{2,p}(\Omega) \to W^{1,p}(\Omega)$ is *compact*.

Therefore, we can find a subsequence of $\{u^{(m)}\}$, denoted again by $\{u^{(m)}\}$, which converges weakly to a function $v \in W^{2,p}(\Omega)$ in the space $W^{2,p}(\Omega)$ and also converges strongly to v in the space $L^p(\Omega)$:

$$u^{(m)} \longrightarrow v \quad \text{weakly in } W^{2,p}(\Omega) \text{ as } m \to \infty,$$
$$u^{(m)} \longrightarrow v \quad \text{in } L^p(\Omega) \text{ as } m \to \infty.$$

We shall prove that $v = 0$, that is,

$$u^{(m)} \longrightarrow 0 \quad \text{weakly in } W^{2,p}(\Omega) \text{ as } m \to \infty, \tag{20.23a}$$

$$u^{(m)} \longrightarrow 0 \quad \text{in } L^p(\Omega) \text{ as } m \to \infty. \tag{20.23b}$$

(I) First, we prove that

$$Bv = 0 \quad \text{on } \partial\Omega. \tag{20.24}$$

To do this, it should be noticed that the set

$$\mathfrak{B} := \left\{u \in W^{2,p}(\Omega) : \mathcal{B}u = 0 \text{ on } \partial\Omega\right\}$$

is balanced, convex and *strongly closed* in $W^{2,p}(\Omega)$. Indeed, it suffices to note that if $\{u_j\}$ is any sequence in $\mathfrak{B}$ which converges strongly to a function u in $W^{2,p}(\Omega)$, then we have, by the continuity of $\mathcal{B}\colon W^{2,p}(\Omega) \to B^{1-1/p,p}(\partial\Omega)$ (Theorem 7.6),

$$\mathcal{B}u = \lim_{j\to\infty} \mathcal{B}u_j = 0 \quad \text{in } B^{1-1/p,p}(\partial\Omega).$$

Hence, it follows that the set $\mathfrak{B}$ is *weakly closed* in $W^{2,p}(\Omega)$, by applying Mazur's theorem (Theorem 2.23) with

$$X := W^{2,p}(\Omega), \quad M := \mathfrak{B}.$$

Therefore, we obtain assertion (20.24), that is, $v \in \mathfrak{B}$. Indeed, it suffices to note by assertion (20.16a) that $u^{(m)} \in \mathfrak{B}$ and further that $\{u^{(m)}\}$ converges weakly to $v \in W^{2,p}(\Omega)$.

(II) Next, let φ be an arbitrary function in $L^q(\Omega)$ with $q = p/(p-1)$. Then we have, by Hölder's inequality (Theorem 3.24) and condition (20.16b),

$$\begin{aligned}
&\left|\int_\Omega \left(\mathcal{L}^{(m)}u^{(m)} - \mathcal{A}v\right)\varphi\,dx\right| \qquad (20.25)\\
&= \left|\sum_{i,j=1}^{n}\int_\Omega \left(a^{ij}_{(m)}(x)\frac{\partial^2 u^{(m)}}{\partial x_i \partial x_j} - \alpha^{ij}(x)\frac{\partial^2 v}{\partial x_i \partial x_j}\right)\varphi\,dx\right|\\
&\leq \sum_{i,j=1}^{n}\left|\int_\Omega \left(a^{ij}_{(m)}(x)\frac{\partial^2 u^{(m)}}{\partial x_i \partial x_j} - \alpha^{ij}(x)\frac{\partial^2 v}{\partial x_i \partial x_j}\right)\varphi\,dx\right|\\
&\leq \sum_{i,j=1}^{n}\left|\int_\Omega \left(a^{ij}_{(m)}(x) - \alpha^{ij}(x)\right)\frac{\partial^2 u^{(m)}}{\partial x_i \partial x_j}\varphi\,dx\right|\\
&\quad + \sum_{i,j=1}^{n}\left|\int_\Omega \alpha^{ij}(x)\left(\frac{\partial^2 u^{(m)}}{\partial x_i \partial x_j} - \frac{\partial^2 v}{\partial x_i \partial x_j}\right)\varphi\,dx\right|
\end{aligned}$$

$$
\begin{aligned}
&\leq \sum_{i,j=1}^{n} \left\| \nabla^2 u^{(m)} \right\|_{L^p(\Omega)} \left\| \left(a^{ij}_{(m)} - \alpha^{ij} \right) \varphi \right\|_{L^q(\Omega)} \\
&\quad + \sum_{i,j=1}^{n} \left| \int_\Omega \alpha^{ij}(x)\varphi(x) \cdot \left(\frac{\partial^2 u^{(m)}}{\partial x_i \partial x_j} - \frac{\partial^2 v}{\partial x_i \partial x_j} \right) dx \right| \\
&\leq \sum_{i,j=1}^{n} \left\| \left(a^{ij}_{(m)} - \alpha^{ij} \right) \varphi \right\|_{L^q(\Omega)} \\
&\quad + \sum_{i,j=1}^{n} \left| \int_\Omega \alpha^{ij}(x)\varphi(x) \cdot \left(\frac{\partial^2 u^{(m)}}{\partial x_i \partial x_j} - \frac{\partial^2 v}{\partial x_i \partial x_j} \right) dx \right|.
\end{aligned}
$$

However, by condition (20.14a) it follows from an application of the Lebesgue dominated convergence theorem (Theorem 3.13) that the first term on the last inequality (20.25) tends to zero as $m \to \infty$:

$$
\sum_{i,j=1}^{n} \left\| \left(a^{ij}_{(m)} - \alpha^{ij} \right) \varphi \right\|_{L^q(\Omega)} \longrightarrow 0.
$$

Moreover, we recall that $\{u^{(m)}\}$ converges weakly to a function $v \in W^{2,p}(\Omega)$. Since $\alpha^{ij}(x)\varphi \in L^q(\Omega)$, we find that the second term on the last inequality (20.25) tends to zero as $m \to \infty$:

$$
\sum_{i,j=1}^{n} \left| \int_\Omega \alpha^{ij}(x)\varphi(x) \cdot \left(\frac{\partial^2 u^{(m)}}{\partial x_i \partial x_j} - \frac{\partial^2 v}{\partial x_i \partial x_j} \right) dx \right| \longrightarrow 0.
$$

Hence we have, by inequality (20.25),

$$
\int_\Omega \mathcal{L}^{(m)} u^{(m)} \cdot \varphi \, dx \longrightarrow \int_\Omega \mathcal{A} v \cdot \varphi \, dx \quad \text{as } m \to \infty.
$$

On the other hand, by Hölder's inequality (Theorem 3.24) and condition (20.16c) it follows that

$$
\left| \int_\Omega \mathcal{L}^{(m)} u^{(m)} \cdot \varphi \, dx \right| \leq \| \mathcal{L}^{(m)} u^{(m)} \|_{L^p(\Omega)} \cdot \| \varphi \|_{L^q(\Omega)} \longrightarrow 0 \quad \text{as } m \to \infty.
$$

This proves that

$$
\int_\Omega \mathcal{A} v \cdot \varphi \, dx = \lim_{n \to \infty} \int_\Omega \mathcal{L}^{(m)} u^{(m)} \cdot \varphi \, dx = 0 \quad \text{for all } \varphi \in L^q(\Omega).
$$

Summing up, we have proved that

$$\begin{aligned}
&v \in W^{2,p}(\Omega),\\
&\mathcal{A}v = 0 \quad \text{in } \Omega,\\
&\mathcal{B}v = 0 \quad \text{on } \partial\Omega.
\end{aligned}$$

By applying the uniqueness result of problem (20.3) to the operator $\mathcal{A}$, we obtain that

$$v = 0 \quad \text{in } \Omega.$$

This proves the desired assertions (20.23).

Step 2-1d: By combining assertions (20.16c) and (20.23b), we have proved that

$$\begin{aligned}
&\mathcal{L}^{(m)}u^{(m)} \longrightarrow 0 \quad \text{in } L^p(\Omega),\\
&u^{(m)} \longrightarrow 0 \quad \text{in } L^p(\Omega).
\end{aligned}$$

Therefore, by applying estimate (20.4) to the operators $\{L^{(m)}\}$ we obtain that

$$\|u^{(m)}\|_{W^{2,p}(\Omega)} \leq c_1 \left(\|\mathcal{L}^{(m)}u^{(m)}\|_{L^p(\Omega)} + \|u^{(m)}\|_{L^p(\Omega)}\right),$$

so that

$$u^{(m)} \longrightarrow 0 \quad \text{in } W^{2,p}(\Omega).$$

However, this assertion contradicts condition (20.16b):

$$\|u^{(m)}\|_{W^{2,p}(\Omega)} = 1.$$

This contradiction proves the desired *a priori* estimate (20.13).

Step 2-2: By virtue of estimate (20.13), we can apply Theorem 15.1 to obtain that the oblique derivative problem is uniquely solvable for the operator $\mathcal{L}_0$ if and only if it is uniquely solvable for the operator $\mathcal{L}_1$. However, it is known (see [172, Theorem 3.29]) that the oblique derivative problem is uniquely solvable for the Laplace operator $\mathcal{L}_0 = \Delta$: More precisely, for any $f \in L^p(\Omega)$, there exists a unique solution $u \in W^{2,p}(\Omega)$ of the homogeneous oblique derivative problem

$$\begin{cases}
\Delta u = f & \text{in } \Omega,\\
\mathcal{B}u = 0 & \text{on } \partial\Omega.
\end{cases}$$

In other words, the operator $\mathcal{L}_0 := \Delta$ maps $\mathfrak{B} := \{u \in W^{2,p}(\Omega) : \mathcal{B}u = 0\}$ onto $L^p(\Omega)$.

Therefore, it follows from an application of Theorem 15.1 that the operator $\mathcal{L}_1 := \mathcal{L}$ maps $\mathfrak{B}$ onto $L^p(\Omega)$. Namely, for any $f \in L^p(\Omega)$, there exists a unique solution $u \in W^{2,p}(\Omega)$ of problem (20.3).

Step 3: Finally, the *a priori* estimate (20.12) follows from the *a priori* estimate (20.13) with $t := 1$.

Now the proof of Theorem 20.1 (and hence that of Theorem 16.2) is complete. □

Part VI
Construction of Feller Semigroups with Discontinuous Coefficients

Chapter 21
Markov Processes and Feller Semigroups

This chapter is devoted to the functional analytic approach to the study of Markov processes. In Sect. 21.1, we summarize the basic definitions and results about Markov processes, and formulate Markov processes in terms of transition functions. From the viewpoint of functional analysis, the transition function is something more convenient than the Markov process itself (Theorem 21.1). Indeed, we can associate with each transition function in a natural way a family of bounded linear operators acting on the space of continuous functions on the state space, and the so-called Markov property implies that this family forms a semigroup. The semigroup approach to Markov processes can be traced back to the pioneering work of W. Feller [47, 48] in the early 1950s. Transition functions and their associated semigroups are studied in Sect. 21.2 (Theorem 21.11). These semigroups are called Feller semigroups. In Sect. 21.3, by using the Hille–Yosida theory of semigroups we characterize Feller semigroups in terms of their infinitesimal generators. In particular, we prove a version of the Hille–Yosida theorem adapted to the present context (Theorem 21.12), which forms a functional analytic background for the proof of Theorem 1.1 (Theorem 21.1). Moreover, we give two useful criteria in order that a linear operator is the infinitesimal generator of a Feller semigroup in terms of the *positive maximum principle* (Theorem 21.14 and Corollary 21.15).

The construction of Feller semigroups will be carried out in the final Chap. 25.

For more thorough treatments of this subject, the reader might refer to Blumenthal–Getoor [12], Dynkin [42, 43], Ethier–Kurtz [45], Garroni–Menaldi [58], Lamperti [87], Revuz–Yor [119], Taira [154, 157, 162].

21.1 Markov Processes and Transition Functions

In this section, we summarize the basic definitions and results about Markov processes, and formulate Markov processes in terms of transition functions.

K. Taira, *Real Analysis Methods for Markov Processes*,
https://doi.org/10.1007/978-981-97-3659-1_21

21.1.1 Definition of a Markov Process

Let K be a locally compact, separable metric space and $\mathcal{B}$ the σ-algebra of all Borel sets in K, that is, the smallest σ-algebra containing all open sets in K. Let $(\Omega, \mathcal{F}, P)$ be a probability space. A function X defined on Ω taking values in K is called a *random variable* if it satisfies the condition

$$X^{-1}(E) = \{X \in E\} \in \mathcal{F} \quad \text{for all } E \in \mathcal{B}.$$

We express this by saying that X is $\mathcal{F}/\mathcal{B}$-measurable. A family $\{x_t\}_{t \geq 0}$ of random variables is called a *stochastic process*, and it may be thought of as the motion in time of a physical particle. The space K is called the *state space* and Ω the *sample space*. For a fixed $\omega \in \Omega$, the function $x_t(\omega)$, $t \geq 0$, defines in the state space K a *trajectory* or *path* of the process corresponding to the sample point ω.

Now we introduce a class of *Markov processes* which we will deal with in this book.

Definition 21.1 Assume that we are given the following:

(1) A locally compact, separable metric space K and the σ-algebra $\mathcal{B}$ of all Borel sets in K. A point ∂ is adjoined to K as the *point at infinity* if K is not compact, and as an isolated point if K is compact. We let

$$\begin{aligned} K_\partial &= K \cup \{\partial\}, \\ \mathcal{B}_\partial &= \text{the } \sigma-\text{algebra in } K_\partial \text{ generated by } \mathcal{B}. \end{aligned}$$

(2) The space Ω of all mappings $\omega \colon [0, \infty] \to K_\partial$ such that $\omega(\infty) = \partial$ and that if $\omega(t) = \partial$ then $\omega(s) = \partial$ for all $s \geq t$. Let ω_∂ be the constant map $\omega_\partial(t) = \partial$ for all $t \in [0, \infty]$.
(3) For each $t \in [0, \infty]$, the coordinate map x_t defined by $x_t(\omega) = \omega(t)$, $\omega \in \Omega$.
(4) For each $t \in [0, \infty]$, a pathwise shift mapping $\theta_t : \Omega \to \Omega$ defined by the formula $\theta_t\omega(s) = \omega(t+s)$ for $\omega \in \Omega$. We remark that $\theta_\infty\omega = \omega_\partial$ and that $x_t \circ \theta_s = x_{t+s}$ for all $t, s \in [0, \infty]$.
(5) A σ-algebra $\mathcal{F}$ in Ω and an increasing family $\{\mathcal{F}_t\}_{0 \leq t \leq \infty}$ of sub-σ-algebras of $\mathcal{F}$.
(6) For each $x \in K_\partial$, a probability measure P_x on $(\Omega, \mathcal{F})$.

We say that these elements define a (temporally homogeneous) *Markov process* $\mathcal{X} = (x_t, \mathcal{F}, \mathcal{F}_t, P_x)$ if the following four conditions are satisfied:

(i) For each $0 \leq t < \infty$, the function x_t is $\mathcal{F}_t/\mathcal{B}_\partial$- measurable, that is,

$$\{x_t \in E\} \in \mathcal{F}_t \quad \text{for all } E \in \mathcal{B}_\partial.$$

(ii) For each $0 \leq t < \infty$ and $E \in \mathcal{B}$, the function

$$p_t(x, E) = P_x\{x_t \in E\}$$

is a Borel measurable function of $x \in K$.

(iii) $P_x\{\omega \in \Omega : x_0(\omega) = x\} = 1$ for each $x \in K_\partial$.
(iv) For all $t, h \in [0, \infty]$, $x \in K_\partial$ and $E \in \mathcal{B}_\partial$, we have the formula

$$P_x\{x_{t+h} \in E \mid \mathcal{F}_t\} = p_h(x_t, E) \quad \text{a. e.},$$

or equivalently

$$P_x(A \cap \{x_{t+h} \in E\}) = \int_A p_h(x_t(\omega), E)\, dP_x(\omega) \quad \text{for all } A \in \mathcal{F}_t.$$

Here is an intuitive way of thinking about the above definition of a Markov process. The sub-σ-algebra $\mathcal{F}_t$ may be interpreted as the collection of events which are observed during the time interval $[0, t]$. The value $P_x(A)$, $A \in \mathcal{F}$, may be interpreted as the probability of the event A under the condition that a particle starts at position x; hence the value $p_t(x, E)$ expresses the transition probability that a particle starting at position x will be found in the set E at time t. The function $p_t(x, \cdot)$ is called the *transition function* of the process $\mathcal{X}$. The transition function $p_t(x, \cdot)$ specifies the probability structure of the process. The intuitive meaning of the crucial condition (iv) is that the future behavior of a particle, knowing its history up to time t, is the same as the behavior of a particle starting at $x_t(\omega)$, that is, a particle starts afresh.

A Markovian particle moves in the space K until it "dies" or "disappears" at the time when it reaches the point ∂; hence the point ∂ is called the *terminal point* or *cemetery*. With this interpretation in mind, we let

$$\zeta(\omega) = \inf\{t \in [0, \infty] : x_t(\omega) = \partial\}.$$

The random variable ζ is called the *lifetime* of the process $\mathcal{X}$. The process $\mathcal{X}$ is said to be *conservative* if it satisfies the condition

$$P_x\{\zeta = \infty\} = 1 \quad \text{for all } x \in K.$$

21.1.2 Markov Transition Functions

Our first job is to give the precise definition of a transition function adapted to the Hille–Yosida theory of semigroups:

Definition 21.2 Let (K, ρ) be a locally compact, separable metric space and $\mathcal{B}$ the σ-algebra of all Borel sets in K. A function $p_t(x, E)$, defined for all $t \geq 0$, $x \in K$ and $E \in \mathcal{B}$, is called a (temporally homogeneous) *Markov transition function* on K if it satisfies the following four conditions:

(a) $p_t(x, \cdot)$ is a non-negative measure on $\mathcal{B}$ and $p_t(x, K) \leq 1$ for each $t \geq 0$ and each $x \in K$.
(b) $p_t(\cdot, E)$ is a Borel measurable function for each $t \geq 0$ and each $E \in \mathcal{B}$.
(c) $p_0(x, \{x\}) = 1$ for each $x \in K$.
(d) (The *Chapman–Kolmogorov equation*) For any $t, s \geq 0$, any $x \in K$ and any $E \in \mathcal{B}$, we have the formula

$$\boxed{p_{t+s}(x, E) = \int_K p_t(x, dy)\, p_s(y, E).} \tag{21.1}$$

It is just condition (d) which reflects the Markov property that a particle starts afresh (see [32, 79]). Here is an intuitive way of thinking about the above definition of a Markov transition function. The value $p_t(x, E)$ expresses the transition probability that a physical particle starting at position x will be found in the set E at time t. Equation (21.1) expresses the idea that a transition from the position x to the set E in time $t + s$ is composed of a transition from x to some position y in time t, followed by a transition from y to the set E in the remaining time s; the latter transition has probability $p_s(y, E)$ which depends only on y. Thus a physical particle "starts afresh"; this property is called the *Markov property*.

A Markov transition function $p_t(x, \cdot)$ is said to be *normal* if it satisfies the condition

$$p_{+0}(x, K) = \lim_{t \downarrow 0} p_t(x, K) = 1 \quad \text{for all } x \in K.$$

The next theorem, due to Dynkin [42, Chap. 4, Sect. 2], justfies the definition of a transition function, and hence it will be fundamental for our further study of Markov processes:

Theorem 21.1 (Dynkin) *For every Markov process, the function p_t, defined by the formula*

$$p_t(x, E) = P_x\{x_t \in E\} \quad \textit{for } x \in K, E \in \mathcal{B} \textit{ and } t \geq 0,$$

is a Markov transition function. Conversely, every normal Markov transition function corresponds to some Markov process.

21.1.3 Feller Transition Functions

Let (K, ρ) be a locally compact, separable metric space. Let $C(K)$ be the space of real-valued, bounded continuous functions on K; $C(K)$ is a Banach space with the supremum norm

$$\|f\|_\infty = \sup_{x \in K} |f(x)|.$$

We say that a function $f \in C(K)$ converges to 0 as $x \to \partial$ if, for each $\varepsilon > 0$, there exists a compact subset E of K such that

$$|f(x)| < \varepsilon \quad \text{for all } x \in K \setminus E,$$

and write

$$\lim_{x \to \partial} f(x) = 0.$$

Let $C_0(K)$ be the subspace of $C(K)$ which consists of all functions satisfying $\lim_{x \to \partial} f(x) = 0$; $C_0(K)$ is a closed subspace of $C(K)$. We remark that $C_0(K)$ may be identified with $C(K)$ if K is compact.

Now we introduce some condition on the measures $p_t(x, \cdot)$ related to continuity in $x \in K$ for every fixed $t \geq 0$.

Definition 21.3 A transition function p_t is called a *Feller function* if the function

$$\boxed{T_t f(x) = \int_K p_t(x, dy)\, f(y)}$$

is a continuous function of $x \in K$ whenever f is bounded and continuous on K, that is, $C(K)$ is an invariant space for the operators T_t. We say that p_t is a C_0*-function* if $C_0(K)$ is an invariant subspace of $C(K)$ for the operators T_t.

Remark 21.2 The Feller property is equivalent to saying that the measures $p_t(x, \cdot)$ depend continuously on $x \in K$ in the usual *weak topology*, for every fixed $t \geq 0$.

21.1.4 Path Functions of Markov Processes

A Markov process $\mathcal{X} = (x_t, \mathcal{F}, \mathcal{F}_t, P_x)$ is said to be *right-continuous* provided that, for each $x \in K$,

$$P_x\{\omega \in \Omega : \text{the mapping } t \mapsto x_t(\omega) \text{ is a right-continuous function from } [0, \infty) \text{ into } K_\partial\} = 1.$$

Furthermore, we say that $\mathcal{X}$ is *continuous* provided that, for each $x \in K$,

$$P_x\{\omega \in \Omega : \text{the mapping } t \mapsto x_t(\omega) \text{ is a continuous function from } [0, \zeta) \text{ into } K_\partial\} = 1.$$

Here ζ is the lifetime of the process $\mathcal{X}$.

Now we give some useful criteria for path-continuity in terms of transition functions (see Dynkin [42, Chap. 6], [43, Chap. 3, Sect. 2]):

Theorem 21.3 (Dynkin) *Let p_t be a normal transition function on K.*

(i) Assume that the following two conditions are satisfied:

(L) For each $s > 0$ and each compact $E \subset K$, we have the assertion

$$\lim_{x \to \partial} \sup_{0 \le t \le s} p_t(x, E) = 0.$$

(M) For each $\varepsilon > 0$ and each compact $E \subset K$, we have the assertion

$$\lim_{t \downarrow 0} \sup_{x \in E} p_t(x, K \setminus U_\varepsilon(x)) = 0,$$

where $U_\varepsilon(x) = \{y \in K : \rho(y, x) < \varepsilon\}$ is an ε-neighborhood of x.

Then there exists a Markov process $\mathcal{X}$ with transition function p_t whose paths are right-continuous on $[0, \infty)$ and have left-hand limits on $[0, \zeta)$ almost surely.

(ii) Assume that condition (L) and the following condition (N) (replacing condition (M)) are satisfied:

(N) For each $\varepsilon > 0$ and each compact $E \subset K$, we have the assertion

$$\lim_{t \downarrow 0} \frac{1}{t} \sup_{x \in E} p_t(x, K \setminus U_\varepsilon(x)) = 0.$$

Then there exists a Markov process $\mathcal{X}$ with transition function p_t whose paths are almost surely continuous on $[0, \zeta)$.

Remark 21.4 (1) Condition (L) is trivially satisfied, if the state space K is compact.

(2) It is known (see Dynkin [42, Lemma 6.2]) that if the paths of a Markov process are right-continuous, then the transition function p_t satisfies the condition

$$\lim_{t \downarrow 0} p_t(x, U_\varepsilon(x)) = 1 \quad \text{for every } x \in K.$$

21.1.5 Strong Markov Processes

A Markov process is called a *strong Markov process* if the "starting afresh" property holds not only for every fixed moment but also for suitable random times.

We formulate precisely this "strong" Markov property. Let

$$\mathcal{X} = (x_t, \mathcal{F}, \mathcal{F}_t, P_x)$$

be a Markov process. A mapping $\tau\colon \Omega \to [0, \infty]$ is called a *stopping time* or *Markov time* with respect to $\{\mathcal{F}_t\}$ if it satisfies the condition

$$\{\tau \le t\} \in \mathcal{F}_t \quad \text{for all } t \in [0, \infty).$$

Intuitively, this means that the events $\{\tau \le t\}$ depend on the process only up to time t, but not on the "future" after time t. It should be noticed that any non-negative constant mapping is a stopping time.

If τ is a stopping time with respect to $\{\mathcal{F}_t\}$, we let

$$\mathcal{F}_\tau = \{A \in \mathcal{F} : A \cap \{\tau \le t\} \in \mathcal{F}_t \quad \text{for all } t \in [0, \infty)\}.$$

Intuitively, we may think of $\mathcal{F}_\tau$ as the "past" up to the random time τ. It is easy to verify that $\mathcal{F}_\tau$ is a σ-algebra. If $\tau \equiv t_0$ for some constant $t_0 \ge 0$, then $\mathcal{F}_\tau$ reduces to $\mathcal{F}_{t_0}$.

For each $t \in [0, \infty]$, we define a mapping

$$\Phi_t\colon [0, t] \times \Omega \longrightarrow K_\partial$$

by the formula

$$\Phi_t(s, \omega) = x_s(\omega).$$

A Markov process $\mathcal{X} = (x_t, \mathcal{F}, \mathcal{F}_t, P_x)$ is said to be *progressively measurable* with respect to $\{\mathcal{F}_t\}$ if the mapping Φ_t is $\mathcal{B}_{[0,t]} \times \mathcal{F}_t/\mathcal{B}_\partial$-measurable for each $t \in [0, \infty]$, that is, if we have the condition

$$\Phi_t^{-1}(E) = \{\Phi_t \in E\} \in \mathcal{B}_{[0,t]} \times \mathcal{F}_t \quad \text{for all } E \in \mathcal{B}_\partial.$$

Here $\mathcal{B}_{[0,t]}$ is the σ-algebra of all Borel sets in the interval $[0, t]$ and $\mathcal{B}_\partial$ is the σ-algebra in K_∂ generated by $\mathcal{B}$. It should be noticed that if $\mathcal{X}$ is progressively measurable and if τ is a stopping time, then the mapping $x_\tau : \omega \mapsto x_{\tau(\omega)}(\omega)$ is $\mathcal{F}_\tau/\mathcal{B}_\partial$- measurable.

Definition 21.4 We say that a progressively measurable Markov process $\mathcal{X} = (x_t, \mathcal{F}, \mathcal{F}_t, P_x)$ has the *strong Markov property* with respect to $\{\mathcal{F}_t\}$ if the following condition is satisfied:

For all $h \ge 0$, $x \in K_\partial$, $E \in \mathcal{B}_\partial$ and all stopping times τ, we have the formula

$$P_x\{x_{\tau+h} \in E \mid \mathcal{F}_\tau\} = p_h(x_\tau, E),$$

or equivalently (cf. formula (21.1)),

$$\boxed{P_x(A \cap \{x_{\tau+h} \in E\}) = \int_A p_h(x_{\tau(\omega)}(\omega), E)\, dP_x(\omega) \quad \text{for every } A \in \mathcal{F}_\tau.}$$

This expresses the idea of "starting afresh" at random times.

The next result gives a useful criterion for the strong Markov property (see [42, Theorem 5.10]).

Theorem 21.5 (Dynkin) *Every right-continuous Markov process is a strong Markov process if its transition function has C_0-property.*

We state a simple criterion for the strong Markov property in terms of transition functions. To do this, we introduce the following.

Definition 21.5 A transition function p_t on K is said to be *uniformly stochastically continuous* on K if it satisfies the following condition:

For each $\varepsilon > 0$ and each compact $E \subset K$, we have the assertion

$$\boxed{\lim_{t \downarrow 0} \sup_{x \in E} [1 - p_t(x, U_\varepsilon(x))] = 0,} \tag{21.2}$$

where $U_\varepsilon(x) = \{y \in K : \rho(y, x) < \varepsilon\}$ is an ε-neighborhood of x.

For example, the transition functions of Poisson process, Brownian motion and Cauchy process satisfy the condition (21.2). It should be emphasized that every uniformly stochastically continuous transition function is normal and satisfies condition (M) in Theorem 21.3.

By combining part (i) of Theorems 21.3 and 21.5, we obtain the following theorem (see [42, Theorem 6.3]).

Theorem 21.6 (Dynkin) *If a uniformly stochastically continuous, C_0 transition function p_t satisfies condition (L), then it is the transition function of some strong Markov process whose paths are right-continuous. and have no discontinuities other than jumps.*

We remark that Theorem 21.6 can be visualized as in Fig. 21.1.

A continuous strong Markov process is called a *diffusion process*.

The next result states a sufficient condition for the existence of a diffusion process with a prescribed transition function.

Theorem 21.7 (Dynkin) *If a uniformly stochastically continuous, C_0 transition function p_t satisfies conditions (L) and (N), then it is the transition function of some diffusion process.*

This is an immediate consequence of part (ii) of Theorems 21.3 and 21.6.

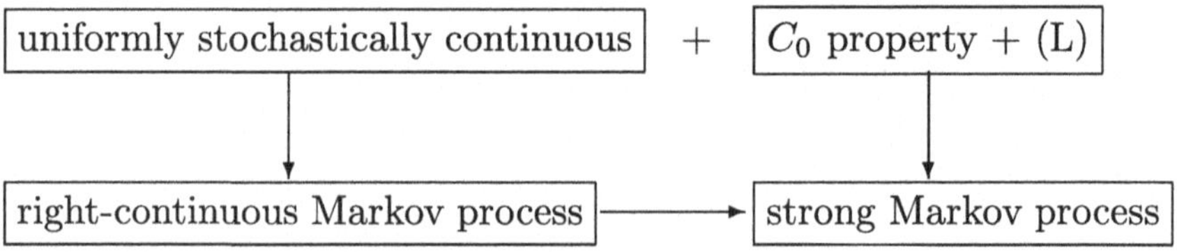

Fig. 21.1 A functional analytic approach to strong Markov processes

21.2 Transition Functions and Feller Semigroups

In this section we study the semigroups associated with Feller transition functions.

Let (K, ρ) be a locally compact, separable metric space and let $C(K)$ be the Banach space of real-valued, bounded continuous functions on K.

Then we have the following (see [154, Theorems 9.31 and 9.32]).

Theorem 21.8 *If p_t is a Feller transition function on a locally compact, separable metric space K, then the associated operators $\{T_t\}_{t\geq 0}$, defined by the formula*

$$T_t f(x) = \int_K p_t(x, dy) f(y) \quad \text{for } f \in C(K), \tag{21.3}$$

form a non-negative and contraction semigroup on $C(K)$:

(i) $T_{t+s} = T_t \cdot T_s$ *for* $t, s \geq 0$ *(the semigroup property);* $T_0 = I$.
(ii) $f \in C(K)$, $0 \leq f(x) \leq 1$ *on* $K \Longrightarrow 0 \leq T_t f(x) \leq 1$ *on* K.

Conversely, if $\{T_t\}_{t\geq 0}$ is a non-negative and contraction semigroup on the space $C_0(K)$, then there exists a unique C_0-transition function p_t on K such that formula (21.3) *holds true.*

It should be emphasized that the C_0-property deals with continuity of a transition function $p_t(x, E)$ in x, and does not, by itself, have no concern with continuity in t.

Now we give a necessary and sufficient condition on $p_t(x, E)$ in order that its associated semigroup $\{T_t\}_{t\geq 0}$ be strongly continuous in t on the space $C_0(K)$ (see [154, Theorem 9.33]):

$$\boxed{\lim_{s\downarrow 0} \|T_{t+s} f - T_t f\|_\infty = 0 \quad \text{for every } f \in C_0(K).} \tag{21.4}$$

Theorem 21.9 *Let p_t be a C_0-transition function on a locally compact, separable metric space K. Then the associated semigroup $\{T_t\}_{t\geq 0}$, defined by formula* (21.3), *is strongly continuous in t on $C_0(K)$ if and only if p_t is uniformly stochastically continuous on K and satisfies condition (L) of Theorem 21.3.*

Remark 21.10 Since the semigroup $\{T_t\}_{t\geq 0}$ is a contraction semigroup, we find that the strong continuity (21.4) of $\{T_t\}$ in t for $t \geq 0$ is equivalent to the strong continuity at $t = 0$:

$$\lim_{t\downarrow 0} \|T_t f - f\|_\infty = 0 \quad \text{for every } f \in C_0(K). \tag{21.4$'$}$$

Definition 21.6 A family $\{T_t\}_{t\geq 0}$ of bounded linear operators acting on $C_0(K)$ is called a *Feller semigroup* on K if it satisfies the following three conditions:

(i) $T_{t+s} = T_t \cdot T_s$, $t, s \geq 0$ (the semigroup property); $T_0 = I$.

(ii) $\{T_t\}$ is strongly continuous in t for $t \geq 0$:

$$\lim_{s \downarrow 0} \|T_{t+s}f - T_t f\|_\infty = 0, \quad f \in C_0(K).$$

(iii) $\{T_t\}$ is non-negative and contractive on $C_0(K)$:

$$f \in C_0(K), \ 0 \leq f(x) \leq 1 \quad \text{on } K \implies 0 \leq T_t f(x) \leq 1 \quad \text{on } K.$$

By combining Theorems 21.8 and 21.9, we obtain the following (see [157, Theorem 3.31]).

Theorem 21.11 *If p_t is a uniformly stochastically continuous, C_0-transition function on a locally compact, separable metric space K which satisfies condition (L) of Theorem 21.3, then its associated operators $\{T_t\}_{t\geq 0}$, defined by formula* (21.3), *form a Feller semigroup on K. Conversely, if $\{T_t\}_{t\geq 0}$ is a Feller semigroup on K, then there exists a uniformly stochastically continuous, C_0-transition p_t on K, satisfying condition (L), such that formula* (21.3) *holds true.*

21.3 Feller Semigroups and Their Infinitesimal Generators

Let K be a locally compact, separable metric space. If $\{T_t\}_{t\geq 0}$ is a Feller semigroup on K, we define its infinitesimal generator $\mathfrak{A}$ by the formula

$$\boxed{\mathfrak{A}u = \lim_{t \downarrow 0} \frac{T_t u - u}{t},} \tag{21.5}$$

provided that the limit (21.5) exists in $C_0(K)$. More precisely, the generator $\mathfrak{A}$ is a linear operator from $C_0(K)$ into itself defined as follows.

(1) The domain $D(\mathfrak{A})$ of $\mathfrak{A}$ is the set

$$D(\mathfrak{A}) = \{u \in C_0(K) : \text{the limit (21.5) exists}\}.$$

(2) $\mathfrak{A}u = \lim_{t\downarrow 0} \dfrac{T_t u - u}{t}$ for every $u \in D(\mathfrak{A})$.

The next theorem is a version of the Hille–Yosida theorem adapted to the present context (see [157, Theorem 3.34]).

Theorem 21.12 (Hille–Yosida) *Let K be a locally compact, separable metric space. Then we have the following two assertions (i) and (ii):*

(i) Let $\{T_t\}_{t\geq 0}$ be a Feller semigroup on K and let $\mathfrak{A}$ be its infinitesimal generator. Then we have the following four assertions:

(a) *The domain* $D(\mathfrak{A})$ *is dense in the space* $C_0(K)$.

(b) *For each* $\alpha > 0$, *the equation* $(\alpha I - \mathfrak{A})u = f$ *has a unique solution* u *in* $D(\mathfrak{A})$ *for any* $f \in C_0(K)$. *Hence, for each* $\alpha > 0$, *the* Green operator

$$G_\alpha := (\alpha I - \mathfrak{A})^{-1} : C_0(K) \longrightarrow C_0(K)$$

can be defined by the formula

$$u = G_\alpha f = (\alpha I - \mathfrak{A})^{-1} f \quad \text{for } f \in C_0(K).$$

(c) *For each* $\alpha > 0$, *the Green operator* $G_\alpha = (\alpha I - \mathfrak{A})^{-1}$ *is non-negative on* $C_0(K)$:

$$f \in C_0(K),\ f(x) \geq 0 \quad \text{on } K \implies G_\alpha f(x) \geq 0 \quad \text{on } K.$$

(d) *For each* $\alpha > 0$, *the Green operator* $G_\alpha = (\alpha I - \mathfrak{A})^{-1}$ *is bounded on* $C_0(K)$ *with norm*

$$\|G_\alpha\| \leq \frac{1}{\alpha}.$$

(ii) *Conversely, if* $\mathfrak{A}$ *is a linear operator from* $C_0(K)$ *into itself satisfying condition (a) and if there is a constant* $\alpha_0 \geq 0$ *such that, for all* $\alpha > \alpha_0$, *conditions (b) through (d) are satisfied, then* $\mathfrak{A}$ *is the infinitesimal generator of some Feller semigroup* $\{T_t\}_{t\geq 0}$ *on* K.

Now, let K be a *compact* metric space. We remark that the space $C_0(K)$ may be identified with $C(K)$. Then we have the following corollary (see [124, p. 534, Corollary]; [157, Corollary 3.35]).

Corollary 21.13 (Sato–Ueno) *Let* K *be a* compact *metric space and let* $\mathfrak{A}$ *be the infinitesimal generator of a Feller semigroup on* K. *Assume that the constant function* 1 *belongs to the domain* $D(\mathfrak{A})$ *of* $\mathfrak{A}$ *and that we have, for some constant* c,

$$\mathfrak{A}1 \leq -c \quad \text{on } K.$$

Then the operator $\mathfrak{A}' = \mathfrak{A} + cI$ *is the infinitesimal generator of some Feller semigroup* $\{T_t\}_{t\geq 0}$ *on* K.

Although Theorem 21.12 tells us precisely when a linear operator $\mathfrak{A}$ is the infinitesimal generator of some Feller semigroup, it is usually difficult to verify conditions (b) through (d). So we give useful criteria in terms of the *positive maximum principle* which have evolved from the ideas of Bony–Courrège–Priouret[15], Dynkin [43] and Sato–Ueno [124] (see [71, Theorem 39.1], [179, Lemma 1 and Theorem 2], [157, Theorem 3.36 and Corollary 3.37]).

Theorem 21.14 (Hille–Yosida–Ray) *Let* K *be a* compact *metric space. Then we have the following two assertions:*

(i) Let B be a linear operator from $C(K) = C_0(K)$ into itself, and assume that the following two conditions are satisfied:

(α) *The domain $D(B)$ of B is dense in $C(K)$.*

(β) *There exists an open and dense subset K_0 of K such that if $u \in D(B)$ takes a* positive *maximum at a point x_0 of K_0, then we have the inequality*

$$Bu(x_0) \leq 0.$$

Then the operator B is closable *in $C(K)$.*

(ii) Let B be as in part (i), and further assume that the following two conditions are satisfied:

(β') *If $u \in D(B)$ takes a* positive *maximum at a point x' of K, then we have the inequality*

$$Bu(x') \leq 0.$$

(γ) *For some $\alpha_0 \geq 0$, the range $R(\alpha_0 I - B)$ of $\alpha_0 I - B$ is dense in $C(K)$.*

Then the minimal closed extension $\overline{B}$ of B is the infinitesimal generator of some Feller semigroup on K.

Corollary 21.15 (Sato–Ueno) *Let M be a bounded linear operator on $C(K)$ into itself. If $\mathfrak{A}$ generates a Feller semigroup $\{T_t\}_{t \geq 0}$ on K and if either M or $\mathfrak{A}' = \mathfrak{A} + M$ satisfies condition (β') of Theorem 21.14, then the operator $\mathfrak{A}'$ is the infinitesimal generator of some Feller semigroup on K.*

Chapter 22
Feller Semigroups with Dirichlet Condition

This chapter is devoted to the proof of Theorem 1.5. Namely, we prove that the closed realization $\mathfrak{A}_D$ is the infinitesimal generator of a Feller semigroup on $C_0(\overline{\Omega})$ associated with absorption phenomenon at the boundary.

To do so, we consider the Dirichlet problem for the diffusion operator with vanishing mean oscillation (VMO) coefficients in the framework of L^p Sobolev spaces, and prove an existence and uniqueness theorem for the Dirichlet problem (Theorem 22.2). The uniqueness result in Theorem 22.2 follows from a variant of the Bakel'man–Aleksandrov maximum principle in the framework of Sobolev spaces due to Bony [13] (Theorem 22.7). In this way, we construct a Feller semigroup associated with absorption phenomenon at the boundary (Theorem 1.5). Finally, we prove Remark 1.6, that is, the domain $D(\mathfrak{A}_D)$ is independent of $N < p < \infty$.

An intuitive meaning of Theorem 1.5 can be visualized as in Fig. 22.1.

22.1 Formulation of the Dirichlet Problem

Let Ω be a bounded domain in Euclidean space $\mathbf{R}^N$, $N \geq 3$, with boundary $\partial\Omega$ of class $C^{1,1}$. If $1 < p < \infty$ and if k is a non-negative integer, then the Sobolev space $W^{k,p}(\Omega)$ is defined to be the space of those functions $u \in L^p(\Omega)$ such that $D^\alpha u \in L^p(\Omega)$ for $|\alpha| \leq k$, and the norm $\|u\|_{W^{k,p}(\Omega)}$ is defined by the formula

$$\|u\|_{W^{k,p}(\Omega)} = \left(\sum_{|\alpha| \leq k} \int_\Omega |D^\alpha u(x)|^p \, dx \right)^{1/p}.$$

K. Taira, *Real Analysis Methods for Markov Processes*,
https://doi.org/10.1007/978-981-97-3659-1_22

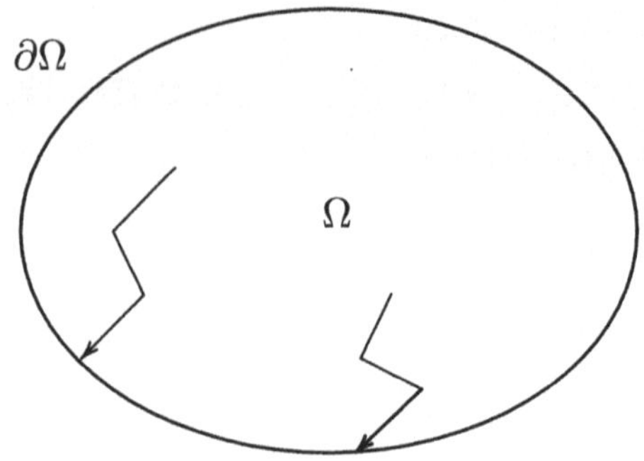

Fig. 22.1 A Markovian particle moves by continuous paths in the state space until it dies at the time when it reaches the boundary

We define the boundary space of traces

$$\begin{aligned} B^{k-1/p,p}(\partial\Omega) := &\text{ the space of distributions } \varphi \in \mathcal{D}'(\partial\Omega) \text{ such that} \\ &\text{there exists a function } u \in W^{k,p}(\Omega) \text{ with } \gamma_0 u = \varphi. \end{aligned}$$

In the space $B^{k-1/p,p}(\partial\Omega)$, we introduce a norm

$$|\varphi|_{B^{k-1/p,p}(\partial\Omega)} = \inf\left\{\|u\|_{W^{k,p}(\Omega)} : u \in W^{k,p}(\Omega),\ \gamma_0 u = \varphi \text{ on } \partial\Omega\right\},$$

where the infimum is taken over all functions $u \in W^{k,p}(\Omega)$ that equal φ on the boundary $\partial\Omega$. The space $B^{k-1/p,p}(\partial\Omega)$ is a Banach space with respect to this norm. More precisely, the space $B^{k-1/p,p}(\partial\Omega)$ is a Besov space (see the trace theorem (Theorem 7.6)).

Let A be a second order, uniformly elliptic differential operator with real *discontinuous* coefficients of the form

$$Au(x) := \sum_{i,j=1}^{N} a^{ij}(x)\frac{\partial^2 u}{\partial x_i \partial x_j}(x) + \sum_{i=1}^{N} b^i(x)\frac{\partial u}{\partial x_i}(x) + c(x)u(x).$$

Here the coefficients $a^{ij}(x)$, $b^i(x)$ and $c(x)$ satisfy the following three conditions:

(1) The $a^{ij}(x)$ are the components of a symmetric contravariant tensor of type $\binom{2}{0}$ in $\mathbf{R}^N$ and $a^{ij} \in \mathrm{VMO} \cap L^\infty(\Omega)$,
and there exists a positive constant λ such that (the uniform ellipticity condition)

$$\frac{1}{\lambda}|\xi|^2 \leq \sum_{i,j=1}^{N} a^{ij}(x)\xi_i\xi_j \leq \lambda|\xi|^2 \quad \text{for almost all } x \in \Omega \text{ and all } \xi \in \mathbf{R}^N.$$

(2) $b^i \in L^\infty(\Omega)$ for $1 \leq i \leq N$.
(3) $c \in L^\infty(\Omega)$ and $c(x) \leq 0$ for almost all $x \in \Omega$.

In this section we consider the following non-homogeneous Dirichlet boundary value problem: Given functions $f(x)$ and $\varphi(x')$ defined in Ω and on $\partial\Omega$, respectively, find a function $u(x)$ in Ω such that

$$\boxed{\begin{cases} Au = f & \text{in } \Omega, \\ \gamma_0 u = \varphi & \text{on } \partial\Omega. \end{cases}} \tag{22.1}$$

The first main result of this chapter is stated as follows (cf. [176, Theorem 2.2]).

Theorem 22.1 (the regularity theorem) *Let $1 < p < \infty$. Assume that conditions (16.1) and (16.3) are satisfied. If a function $u \in W^{2,q}(\Omega)$, $1 < q < p < \infty$, is a solution of the non-homogeneous Dirichlet problem* (22.1) *with $f \in L^p(\Omega)$ and $\varphi \in B^{2-1/p,p}(\partial\Omega)$, then it follows that $u \in W^{2,p}(\Omega)$. Moreover, we have the global* a priori *estimate*

$$\|u\|_{W^{2,p}(\Omega)} \leq C_1 \left(\|u\|_{L^p(\Omega)} + \|f\|_{L^p(\Omega)} + \|\varphi\|_{B^{2-1/p,p}(\partial\Omega)} \right), \tag{22.2}$$

with a positive constant $C_1 = C_1(n, p, \lambda, \eta, \ell, \sigma, \partial\Omega)$.

The second main result of this chapter is a generalization of Bony [13, Théorème 3] to the VMO case (cf. [176, Theorems 2.3 and 2.4]).

Theorem 22.2 (the existence and uniqueness theorem) *Let $1 < p < \infty$. Assume that*

$$c(x) \leq 0 \quad \textit{for almost all } x \in \Omega.$$

Then the non-homogeneous Dirichlet problem (22.1) *has a unique solution $u \in W^{2,p}(\Omega)$ for any $f \in L^p(\Omega)$ and any $\varphi \in B^{2-1/p,p}(\partial\Omega)$.*

Remark 22.3 Theorem 22.2 plays an essential role in the study of the existence of positive solutions of semilinear Dirichlet eigenvalue problems for diffusive logistic equations with discontinuous coefficients which model population dynamics in environments with spatial heterogeneity (see [151]).

If we associate with problem (22.1) a linear operator

$$\mathcal{A}_D = (A, \gamma_0) : W^{2,p}(\Omega) \longrightarrow L^p(\Omega) \oplus B^{2-1/p,p}(\partial\Omega),$$

then we obtain from the trace theorem (Theorem 7.6) and Theorem 22.2 that the operator $\mathcal{A}_D$ is continuous and *bijective* for all $1 < p < \infty$.

22.2 Proof of Theorem 22.1

The proof of the regularity theorem for the Dirichlet problem (Theorem 22.1) is divided into three steps.

Step 1: For any $\varphi \in B^{2-1/p,p}(\partial\Omega)$, we can find a function $v \in W^{2,p}(\Omega)$ such that $\gamma_0 v = \varphi$ on $\partial\Omega$, and further that the mapping

$$B^{2-1/p,p}(\partial\Omega) \ni \varphi \longmapsto v \in W^{2,p}(\Omega) \tag{22.3}$$

is continuous (see Stein [136, Theorem]). On the other hand, it should be noticed (see [2, Theorem 5.37]) that the closure $W_0^{1,p}(\Omega)$ of $C_0^\infty(\Omega)$ in $W^{1,p}(\Omega)$ may be characterized as follows:

$$W_0^{1,p}(\Omega) = \left\{u \in W^{1,p}(\Omega) : \gamma_0 u = 0 \text{ on } \partial\Omega\right\} \quad \text{for all } 1 < p < \infty.$$

Therefore, we have only to prove Theorem 22.1 in the case where $\varphi := 0$ and $1 < q < p < \infty$:

$$u \in W^{2,q}(\Omega) \cap W_0^{1,q}(\Omega),\ Au = f \in L^p(\Omega) \implies u \in W^{2,p}(\Omega). \tag{22.4}$$

Step 2: In order to prove assertion (22.4), we need the Sobolev imbedding theorems (Theorem 7.6):

$$W^{2,p}(\Omega) \subset \begin{cases} L^{Nq/(N-2q)}(\Omega) & \text{if } 1 < q < N/2, \\ L^r(\Omega) \quad \text{for all } N/2 \leq r < \infty & \text{if } q = N/2, \\ L^\infty(\Omega) & \text{if } N/2 < q < \infty. \end{cases} \tag{22.5}$$

$$W^{1,q}(\Omega) \subset \begin{cases} L^{Nq/(N-q)}(\Omega) & \text{if } 1 < q < N, \\ L^r(\Omega) \quad \text{for all } N \leq r < \infty & \text{if } q = N, \\ L^\infty(\Omega) & \text{if } N < q < \infty. \end{cases} \tag{22.6}$$

We consider the following three cases (A), (B) and (C).
(A) The case where $1 < q < N/2$: By assertions (22.5) and (22.6), it follows that

$$\begin{cases} u \in W^{2,q}(\Omega) \subset L^{Nq/(N-2q)}(\Omega), \\ \frac{\partial u}{\partial x_i} \in W^{1,q}(\Omega) \subset L^{Nq/(N-q)}(\Omega) \quad \text{for all } 1 \leq i \leq N. \end{cases}$$

Hence we have the assertion

$$\begin{aligned} A_0 u &:= \sum_{i,j=1}^N a^{ij}(x) \frac{\partial^2 u}{\partial x_i \partial x_j} \\ &= f - \sum_{i=1}^N b^i(x) \frac{\partial u}{\partial x_i} - c(x) u \in L^{Nq/(N-q)}(\Omega). \end{aligned}$$

By applying the global regularity theorem (Theorem 12.2), we obtain that

$$u \in L^{p_1}(\Omega),$$

where

$$p_1 = \min\left\{p, \frac{Nq}{N-q}\right\}.$$

It is easy to see that

$$p_1 > q,$$

and further that

$$p_1 - q > \frac{1}{N-1}.$$

Therefore, by making use of a standard "bootstrap argument" we can conclude that

$$u \in L^p(\Omega).$$

(B) The case where $N/2 \leq q < N$: By assertions (22.5) and (22.6), it follows that

$$\begin{cases} u \in W^{2,q}(\Omega) \subset L^p(\Omega), \\ \frac{\partial u}{\partial x_i} \in W^{1,q}(\Omega) \subset L^{Nq/(N-q)}(\Omega) \quad \text{for all } 1 \leq i \leq N. \end{cases}$$

Hence we have the assertion

$$A_0 u = f - \sum_{i=1}^{N} b^i(x)\frac{\partial u}{\partial x_i} - c(x)u \in L^{Nq/(N-q)}(\Omega).$$

Therefore, by making use of a standard "bootstrap argument" we can conclude that

$$u \in L^p(\Omega).$$

(C) The case where $N \leq q < \infty$: By assertions (22.5) and (22.6), it follows that

$$\begin{cases} u \in W^{2,q}(\Omega) \subset L^\infty(\Omega), \\ \frac{\partial u}{\partial x_i} \in W^{1,q}(\Omega) \subset L^p(\Omega) \quad \text{for all } 1 \leq i \leq N. \end{cases}$$

Hence we have the assertion

$$A_0 u = f - \sum_{i=1}^{N} b^i(x)\frac{\partial u}{\partial x_i} - c(x)u \in L^p(\Omega).$$

By applying the global regularity theorem (Theorem 12.2), we obtain that

$$u \in L^p(\Omega).$$

Summing up, we have proved the desired assertion (22.4).

Step 3: Finally, it remains to prove the global *a priori* estimate (22.2).

First, by using the global *a priori* estimate (12.3) we obtain that

$$\|u\|_{W^{2,p}(\Omega)} \leq C_1 \left(\|u\|_{L^p(\Omega)} + \|A_0 u\|_{L^p(\Omega)} + \|\varphi\|_{B^{2-1/p,p}(\partial\Omega)}\right), \tag{22.7}$$

with a positive constant C_1. Indeed, it suffices to note that the mapping (22.3) is continuous.

On the other hand, we have the inequality

$$\begin{aligned} &\|A_0 u\|_{L^p(\Omega)} \\ &\leq \left\| f - \sum_{i=1}^{N} b^i(x) \frac{\partial u}{\partial x_i} - c(x) u \right\|_{L^p(\Omega)} \\ &\leq \|f\|_{L^p(\Omega)} + \sum_{i=1}^{N} \|b^i\|_{L^\infty(\Omega)} \left\| \frac{\partial u}{\partial x_i} \right\|_{L^p(\Omega)} + \|c\|_{L^\infty(\Omega)} \|u\|_{L^p(\Omega)}. \end{aligned} \tag{22.8}$$

Hence, by combining two inequalities (22.7) and (22.8) we obtain that

$$\begin{aligned} &\|u\|_{W^{2,p}(\Omega)} \\ &\leq C_2 \left(\|u\|_{L^p(\Omega)} + \|f\|_{L^p(\Omega)} + \|\varphi\|_{B^{2-1/p,p}(\partial\Omega)} + \|\nabla u\|_{L^p(\Omega)}\right), \end{aligned} \tag{22.9}$$

with a positive constant C_2.

However, recall the *interpolation inequality* (13.17)

$$\|\nabla u\|_{L^p(\Omega)} \leq \varepsilon \, \|\nabla^2 u\|_{L^p(\Omega)} + \frac{C_3}{\varepsilon} \, \|u\|_{L^p(\Omega)} \quad \text{for all } \varepsilon > 0. \tag{22.10}$$

Therefore, the desired estimate (22.2) follows from inequalities (22.9) and (22.10) if we take

$$\varepsilon = \frac{1}{2C_2}.$$

Now the proof of Theorem 22.1 is complete. □

Let $\mathcal{A}_D$ be a continuous linear operator defined by the formula

$$\mathcal{A}_D = (A, \gamma_0) : W^{2,p}(\Omega) \longrightarrow L^p(\Omega) \oplus B^{2-1/p,p}(\partial\Omega) \quad \text{for } 1 < p < \infty.$$

Then we have the following:

Corollary 22.4 *The null space*

$$\begin{aligned} N(\mathcal{A}_D) &= \left\{u \in W^{2,p}(\Omega) : Au = 0 \ in \ \Omega, \ \gamma_0 u = 0\right\} \\ &= \left\{u \in W^{2,p}(\Omega) \cap W_0^{1,p}(\Omega) : Au = 0 \ in \ \Omega\right\} \end{aligned}$$

of $\mathcal{A}_D$ *is* independent *of* p *for* $1 < p < \infty$.

Proof In the proof, we let

$$N(\mathcal{A}_D(p)) := \{u \in W^{2,p}(\Omega) : Au = 0 \text{ in } \Omega,\ \gamma_0 u = 0\}.$$

If $1 < p_1 < p_2 < \infty$, then it follows that

$$N(\mathcal{A}_D(p_2)) \subset N(\mathcal{A}_D(p_1)).$$

Conversely, if $u \in N(\mathcal{A}_D(p_1))$, then we have, by assertion (22.4),

$$\begin{cases} u \in W^{2,p_1}(\Omega) \cap W_0^{1,p_1}(\Omega), \\ Au = 0 \in L^{p_2}(\Omega) \end{cases} \implies u \in W^{2,p_2}(\Omega).$$

This proves that

$$N(\mathcal{A}_D(p_1)) \subset N(\mathcal{A}_D(p_2)).$$

The proof of Corollary 22.4 is complete. □

22.3 Proof of Theorem 22.2

The proof of the unique solvability theorem for the non-homogeneous Dirichlet problem (Theorem 22.2) is divided into four steps.

Step 1: Our proof is based on the following existence and uniqueness theorem for the *homogeneous* Dirichlet problem (see Theorem 15.1).

Theorem 22.5 *Let $N \geq 3$ and $1 < p < \infty$. Then, for any $f \in L^p(\Omega)$ there exists a unique solution $u \in W^{2,p}(\Omega) \cap W_0^{1,p}(\Omega)$ of the homogeneous Dirichlet problem*

$$\begin{cases} A_0 u = f & \text{in } \Omega, \\ \gamma_0 u = 0 & \text{on } \partial\Omega. \end{cases} \tag{22.11}$$

Moreover, we have the a priori *estimate*

$$\|u\|_{W^{2,p}(\Omega)} \leq C \|f\|_{L^p(\Omega)}, \tag{22.12}$$

with a positive constant $C = C(N, p, \lambda, \eta, M, \partial\Omega)$.

Now, for any $\varphi \in B^{2-1/p,p}(\partial\Omega)$ we can find a function $v \in W^{2,p}(\Omega)$ such that $\gamma_0 v = \varphi$ on $\partial\Omega$ (see Theorem 7.6). Hence we have the following existence and uniqueness theorem for the non-homogeneous Dirichlet problem.

Corollary 22.6 *Let $N \geq 3$ and $1 < p < \infty$. For any $f \in L^p(\Omega)$ and any $\varphi \in B^{2-1/p,p}(\partial\Omega)$, there exists a unique solution $u \in W^{2,p}(\Omega)$ of the Dirichlet problem*

$$\begin{cases} A_0 u = f & \text{in } \Omega, \\ \gamma_0 u = \varphi & \text{on } \partial\Omega. \end{cases} \tag{22.13}$$

We recall (see Sect. 2.7) that a linear operator T from a Banach space X into a Banach space Y is called a

Fredholm operator if it satisfies the following five conditions:

(i) The domain $D(T)$ of T is dense in X.
(ii) T is a closed operator.
(iii) The null space $N(T) = \{x \in D(T) : Tx = 0\}$ of T has finite dimension, that is, $\dim N(T) < \infty$.
(iv) The range $R(T) = \{Tx : x \in D(T)\}$ of T is closed in Y.
(v) The range $R(T)$ of T has finite codimension, that is, $\operatorname{codim} R(T) = \dim Y/R(T) < \infty$.

Then the *index* of T is defined by the formula

$$\boxed{\operatorname{ind} T := \dim N(T) - \operatorname{codim} R(T).}$$

If we associate with problem (22.13) a continuous linear operator

$$\mathcal{A}_0 = (A_0, \gamma_0) : W^{2,p}(\Omega) \longrightarrow L^p(\Omega) \oplus B^{2-1/p,p}(\partial\Omega),$$

then Corollary 22.6 asserts that the mapping $\mathcal{A}_0$ is an algebraic and topological *isomorphism* for all $1 < p < \infty$. In particular, we have the assertion

$$\operatorname{ind} \mathcal{A}_0 = 0. \tag{22.14}$$

Step 2: If we let

$$Bu := \sum_{i=1}^{N} b^i(x) \frac{\partial u}{\partial x_i} + c(x)u,$$

then it is clear that the operator

$$B \colon W^{2,p}(\Omega) \longrightarrow W^{1,p}(\Omega)$$

is continuous for all $1 < p < \infty$. Moreover, it follows from an application of the Rellich–Kondrachov theorem (Theorem 7.8) that the injection

$$W^{1,p}(\Omega) \longrightarrow L^p(\Omega)$$

is compact for all $N < p < \infty$. Hence we find that the mapping

$$B \colon W^{2,p}(\Omega) \longrightarrow L^p(\Omega)$$

is *compact* for all $N < p < \infty$. However, it follows from an application of Theorem 2.53 that the index is *stable* under compact perturbations.

Therefore, we obtain that the mapping

$$\begin{aligned} \mathcal{A}_D &= (A, \gamma_0) = (A_0 + B, \gamma_0) = (A_0, \gamma_0) + (B, 0) \\ &= \mathcal{A}_0 + (B, 0) \colon W^{2,p}(\Omega) \longrightarrow L^p(\Omega) \oplus B^{2-1/p,p}(\partial\Omega) \end{aligned}$$

is a Fredholm operator with index *zero* for all $N < p < \infty$, since we have, by assertion (22.14),

$$\operatorname{ind} \mathcal{A}_D = \operatorname{ind} \mathcal{A}_0 = 0 \quad \text{for all } N < p < \infty. \tag{22.15}$$

Step 3: In order to prove that

$$\dim N(\mathcal{A}_D) = 0 \quad \text{for all } N < p < \infty, \tag{22.16}$$

we need the following weak maximum principle (Theorem 8.1), due to Bony [13].

Theorem 22.7 (the weak maximum principle) *Let $N < p < \infty$, and assume that*

$$c(x) \leq 0 \quad \textit{for almost all } x \in \Omega.$$

If a function $u \in W^{2,p}(\Omega)$ satisfies the condition

$$Au(x) \geq 0 \quad \textit{for almost all } x \in \Omega,$$

then we have the inequality

$$\max_{\overline{\Omega}} u \leq \max_{\partial\Omega} u^+,$$

where

$$u^+(x) = \max\{u(x), 0\} \quad \textit{for } x \in \overline{\Omega}.$$

Remark 22.8 More precisely, Bony proved this maximum principle under the weaker condition that $a^{ij} \in L^\infty(\Omega)$ (see the proof of Theorem 8.1).

By applying Theorem 22.7 to the two functions $\pm u(x)$, we obtain that

$$\begin{cases} u \in W^{2,p}(\Omega) & \text{for } N < p < \infty, \\ Au = 0 & \text{in } \Omega, \\ \gamma_0 u = 0 & \text{on } \partial\Omega \end{cases} \Longrightarrow u = 0 \quad \text{in } \Omega.$$

This proves the desired assertion (22.16).

In view of Corollary 22.4, we have proved that

$$\dim N(\mathcal{A}_D) = 0 \quad \text{for all } 1 < p < \infty,$$

that is, the mapping $\mathcal{A}_D$ is *injective* for all $1 < p < \infty$.

Therefore, it is also *surjective* for $1 < p < \infty$, since we have, by assertion (22.15),

$$\operatorname{ind} \mathcal{A}_D = \dim N(\mathcal{A}_D) - \operatorname{codim} R(\mathcal{A}_D) = 0.$$

Step 4: Summing up, we have proved that that is, the mapping

$$\mathcal{A}_D = (A, \gamma_0) : W^{2,p}(\Omega) \longrightarrow L^p(\Omega) \oplus B^{2-1/p,p}(\partial\Omega)$$

is an algebraic and topological *isomorphism* for all $1 < p < \infty$. Indeed, the continuity of the inverse of $\mathcal{A}_D$ follows immediately from an application of Banach's open mapping theorem (Theorem 2.38).

Now the proof of Theorem 22.2 is complete. □

22.4 Proof of Theorem 1.5

This section is devoted to the proof of Theorem 1.5. To do so, we shall apply a version of the Hille–Yosida theorem (Theorem 21.12).

22.4.1 The Space $C_0(\overline{\Omega})$

First, we consider a one-point compactification $K_\partial = K \cup \{\partial\}$ of the locally compact space $K = \Omega$. We say that two points x and y of $\overline{\Omega}$ are equivalent modulo $\partial\Omega$ if $x = y$ or $x, y \in \partial\Omega$. We denote by $\overline{\Omega}/\partial\Omega$ the totality of equivalence classes modulo $\partial\Omega$. On the set $\overline{\Omega}/\partial\Omega$ we define the quotient topology induced by the projection

$$q : \overline{\Omega} \longrightarrow \overline{\Omega}/\partial\Omega.$$

Then it is easy to see that the topological space $\overline{\Omega}/\partial\Omega$ is a *one-point compactification* K_∂ of the space Ω and that the *point at infinity* ∂ corresponds to the boundary $\partial\Omega$ (see Fig. 22.2):

$$K_\partial = \overline{\Omega}/\partial\Omega,$$
$$\partial = \partial\Omega.$$

Furthermore, we have the following two assertions:

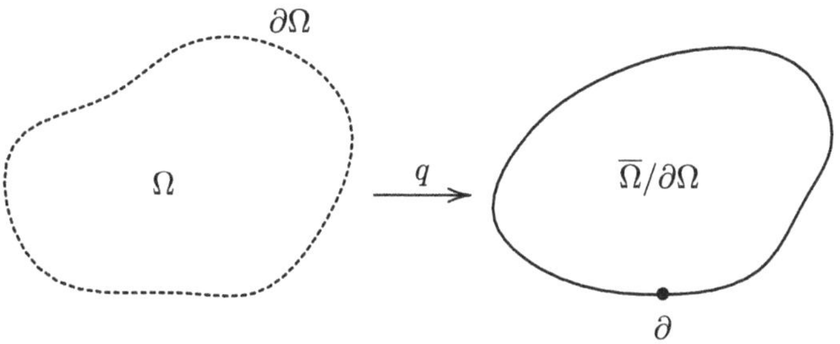

Fig. 22.2 The compactification $\overline{\Omega}/\partial\Omega$ of Ω

(i) If $\tilde{u}$ is a continuous function defined on $\overline{\Omega}/\partial\Omega$, then the function $\tilde{u} \circ q$ is continuous on $\overline{\Omega}$ and constant on $\partial\Omega$.
(ii) Conversely, if u is a continuous function defined on $\overline{\Omega}$ and constant on $\partial\Omega$, then it defines a continuous function $\tilde{u}$ on $\overline{\Omega}/\partial\Omega$.

In other words, we have the isomorphism

$$C(K_\partial) \cong \left\{u \in C(\overline{\Omega}) : \gamma_0 u \text{ is constant on } \partial\Omega\right\}. \tag{22.17}$$

Now we introduce a closed subspace of $C(K_\partial)$ as follows:

$$C_0(K) = \{u \in C(K_\partial) : u(\partial) = 0\}.$$

Then we have, by assertion (22.17),

$$\boxed{C_0(K) \cong C_0(\overline{\Omega}) = \left\{u \in C(\overline{\Omega}) : \gamma_0 u = 0 \text{ on } \partial\Omega\right\}.}$$

22.4.2 End of Proof of Theorem 1.5

Now it remains to verify all conditions (a) through (d) in part (ii) of Theorem 21.12 with

$$\begin{aligned} K &:= \Omega, \\ C_0(K) &:= C_0(\overline{\Omega}), \\ \mathfrak{A} &:= \mathfrak{A}_D. \end{aligned}$$

where

$$D(\mathfrak{A}_D) = \left\{u \in C_0(\overline{\Omega}) \cap W^{2,p}(\Omega) : Au \in C_0(\overline{\Omega})\right\}.$$

The proof is divided into four steps.

Step 1: (b) For each $\alpha > 0$, the equation $(\alpha I - \mathfrak{A}_D)\, u = f$ has a unique solution $u \in D(\mathfrak{A}_D)$ for any $f \in C_0(\overline{\Omega})$.

Since we have the inequality

$$c(x) - \alpha \leq -\alpha \quad \text{for almost all } x \in \Omega,$$

by applying Theorem 22.2 to the operator $A - \alpha$ we obtain that the Dirichlet problem

$$\begin{cases} (\alpha - A)\, u = f & \text{almost everywhere in } \Omega, \\ \gamma_0 u = 0 & \text{on } \partial\Omega \end{cases}$$

has a unique solution $u \in W^{2,p}(\Omega)$ for any $f \in L^p(\Omega)$ with $N < p < \infty$. In particular, for any $f \in C_0(\overline{\Omega})$ there exists a unique function $u \in W^{2,p}(\Omega) \cap C_0(\overline{\Omega})$ such that

$$(\alpha - A)\, u = f \quad \text{in } \Omega.$$

Hence we have the assertion

$$Au = \alpha u - f \in C_0(\overline{\Omega}).$$

By formula (1.5), this proves that

$$\begin{cases} u \in D(\mathfrak{A}_D), \\ (\alpha I - \mathfrak{A}_D)\, u = f. \end{cases}$$

Step 2: (c) For each $\alpha > 0$, the *Green operator* $G_\alpha^0 = (\alpha I - \mathfrak{A}_D)^{-1}$ is non-negative on the space $C_0(\overline{\Omega})$:

$$f \in C_0(\overline{\Omega}),\ f(x) \geq 0 \ \text{ in } \Omega \implies u(x) = G_\alpha^0 f(x) \geq 0 \ \text{ in } \Omega.$$

Indeed, if we let

$$v(x) := -u(x) = -G_\alpha^0 f(x),$$

then it follows that

$$\begin{cases} (A - \alpha)\, v = f \geq 0 & \text{in } \Omega, \\ \gamma_0 v = 0 & \text{on } \partial\Omega. \end{cases}$$

Therefore, by applying Theorem 22.7 to our situation we obtain that

$$v(x) \leq 0 \quad \text{in } \Omega,$$

so that

$$u(x) = -v(x) \geq 0 \quad \text{in } \Omega.$$

Step 3: (d) For each $\alpha > 0$, the Green operator $G_\alpha^0 = (\alpha I - \mathfrak{A}_D)^{-1}$ is bounded on the space $C_0(\overline{\Omega})$ with norm $1/\alpha$: $\|G_\alpha^0\| \leq 1/\alpha$.

Let $f(x)$ be an arbitrary function in $C_0(\overline{\Omega})$. If we let

$$u_\pm(x) = \pm\alpha G_\alpha^0 f(x) - \|f\|_{C(\overline{\Omega})} \in W^{2,p}(\Omega),$$

it suffices to show that

$$u_\pm(x) \leq 0 \quad \text{in } \Omega. \tag{22.18}$$

Indeed, it follows that

$$\begin{aligned}(A - \alpha)\, u_\pm(x) &= \mp\alpha f(x) + (\alpha - c(x))\, \|f\|_{C(\overline{\Omega})} \\ &= \alpha\left(\|f\|_{C(\overline{\Omega})} \mp f(x)\right) + (-c(x))\, \|f\|_{C(\overline{\Omega})} \\ &\geq 0 \quad \text{in } \Omega.\end{aligned}$$

Thus, by applying Theorem 22.7 to the operator $A - \alpha$ we obtain that the function $u_\pm(x)$ may take its positive maximum only on $\partial\Omega$. This proves assertion (22.18), since we have the inequality

$$\gamma_0(u_\pm) = -\|f\|_{C(\overline{\Omega})} < 0 \quad \text{on } \partial\Omega.$$

Step 4: (a) The domain $D(\mathfrak{A}_D)$ is dense in $C_0(\overline{\Omega})$. More precisely, we prove that, for each $u \in C_0(\overline{\Omega})$, we have the assertion

$$\lim_{\alpha\to+\infty} \|\alpha G_\alpha^0 u - u\|_{C(\overline{\Omega})} = 0. \tag{22.19}$$

To do this, we introduce an extension $\widetilde{G_\alpha^0}$ of the Green operator G_α^0 to the space $L^\infty(\Omega)$ as in Fig. 22.3.

More precisely. by Theorem 22.2 we find that the Dirichlet problem

$$\begin{cases}(\alpha - A)\, u = f & \text{almost everywhere in } \Omega, \\ \gamma_0 u = 0 & \text{on } \partial\Omega\end{cases}$$

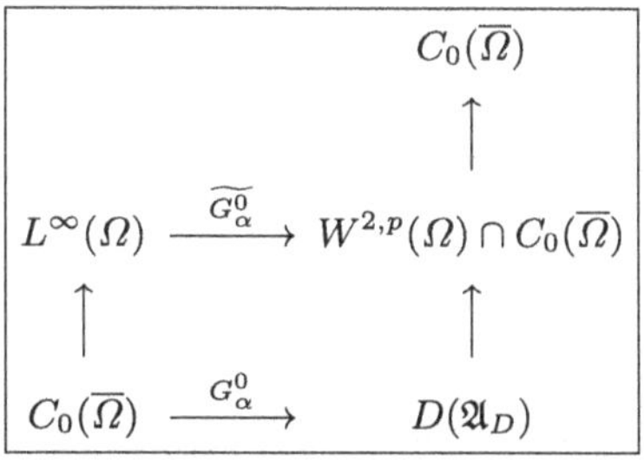

Fig. 22.3 The operators G_α^0 and $\widetilde{G_\alpha^0}$

has a unique solution $u \in W^{2,p}(\Omega) \cap C_0(\overline{\Omega})$ for any $f \in L^\infty(\Omega)$, where $N < p < \infty$. If we let

$$u = \widetilde{G^0_\alpha} f,$$

then it is easy to verify that the operator $\widetilde{G^0_\alpha}$ is an extension of G^0_α to $L^\infty(\Omega)$. Moreover, just as in Steps 2 and 3, we can prove the following two assertions:

(A) The operator $\widetilde{G^0_\alpha}: L^\infty(\Omega) \to C_0(\overline{\Omega})$ is non-negative.
(B) The operator $\widetilde{G^0_\alpha}: L^\infty(\Omega) \to C_0(\overline{\Omega})$ is bounded with norm $1/\alpha$: $\|\widetilde{G^0_\alpha}\| \leq 1/\alpha$.

Since the space $C^2_0(\overline{\Omega}) := C^2(\overline{\Omega}) \cap C_0(\overline{\Omega})$ is dense in $C_0(\overline{\Omega})$, it suffices to prove assertion (22.19) for any $u \in C^2_0(\overline{\Omega})$.

First, since $a^{ij}, b^i, c \in L^\infty(\Omega)$, it follows that

$$Au = \sum_{i,j=1}^N a^{ij}(x) \frac{\partial^2 u}{\partial x_i \partial x_j} + \sum_{i=1}^N b^i(x) \frac{\partial u}{\partial x_i} + c(x)u \quad \text{for } u \in C^2_0(\overline{\Omega}). \tag{22.20}$$

Therefore, we obtain from formula (22.20) that

$$Au \in L^\infty(\Omega) \quad \text{for } u \in C^2_0(\overline{\Omega}).$$

Now, if we let

$$w = \alpha G^0_\alpha u - \widetilde{G^0_\alpha}(Au),$$

then we have the assertions

$$\begin{cases} w \in W^{2,p}(\Omega) \cap C_0(\overline{\Omega}), \\ (A - \alpha)\, w = -\alpha u + Au = (A - \alpha)\, u \quad \text{in } \Omega, \end{cases}$$

and so

$$\begin{cases} w - u \in W^{2,p}(\Omega) \cap C_0(\overline{\Omega}), \\ (A - \alpha)\,(w - u) = 0 \quad \text{in } \Omega. \end{cases}$$

By Theorem 22.2, this implies that $w - u = 0$ in Ω, that is,

$$u = w = \alpha G^0_\alpha u - \widetilde{G^0_\alpha}(Au).$$

Therefore, assertion (22.19) for any $u \in C^2_0(\overline{\Omega})$ follows from an application of assertion (B), since we have, for all $\alpha > 0$,

$$\|u - \alpha G^0_\alpha u\|_{C(\overline{\Omega})} = \|\widetilde{G^0_\alpha}(Au)\|_{C(\overline{\Omega})} \leq \frac{1}{\alpha} \|Au\|_{L^\infty(\Omega)}.$$

Now the proof of Theorem 1.5 is complete. □

22.5 Proof of Remark 1.6

Finally, we prove that the domain

$$D(\mathfrak{A}_D) = \left\{u \in C_0(\overline{\Omega}) \cap W^{2,p}(\Omega) : Au \in C_0(\overline{\Omega})\right\}$$

is *independent* of $N < p < \infty$.

We let

$$\mathcal{D}_p := \left\{u \in W^{2,p}(\Omega) \cap C_0(\overline{\Omega}) : Au \in C_0(\overline{\Omega})\right\}.$$

In order to prove Remark 1.6, it suffices to show that

$$\mathcal{D}_{p_1} = \mathcal{D}_{p_2} \quad \text{for } N < p_1 < p_2 < \infty.$$

First, it follows that

$$\mathcal{D}_{p_2} \subset \mathcal{D}_{p_1},$$

since we have $L^{p_2}(\Omega) \subset L^{p_1}(\Omega)$ for $p_2 > p_1$.

Conversely, let v be an arbitrary element of $\mathcal{D}_{p_1}$:

$$v \in W^{2,p_1}(\Omega) \cap C_0(\overline{\Omega}), \quad Av \in C_0(\overline{\Omega}).$$

Then, since we have v, $Av \in C_0(\overline{\Omega}) \subset L^{p_2}(\Omega)$, it follows from an application of Theorem 22.2 with $p := p_2$ that there exists a unique function $u \in W^{2,p_2}(\Omega)$ such that

$$\begin{cases} Au = Av & \text{in } \Omega, \\ \gamma_0 u = 0 & \text{on } \partial\Omega. \end{cases}$$

Hence we have the assertions

$$\begin{cases} u - v \in W^{2,p_1}(\Omega), & \\ A(u - v) = 0 & \text{in } \Omega, \\ \gamma_0(u - v) = 0 & \text{on } \partial\Omega. \end{cases}$$

Therefore, by applying again Theorem 22.2 with $p := p_1$ we obtain that $u - v = 0$, so that $v = u \in W^{2,p_2}(\Omega)$. This implies that $v \in \mathcal{D}_{p_2}$.

The proof of Remark 1.6 is complete. □

22.6 Notes and Comments

(1) Under the *uniform continuity* condition on the leading terms $a^{ij} \in C(\overline{\Omega})$ for $1 \leq i, j \leq N$ where $N \geq 2$, Gilbarg and Trudinger proved Theorems 22.1 and 22.2 (see [59, Theorems 9.13 and 9.15]).
(2) Theorem 1.5 is generalized to Feller semigroups with *jump phenomena* by [163, Corollary 1.1].

Chapter 23
Feller Semigroups with an Oblique Derivative Condition

This chapter is devoted to the proof of Theorem 1.3. Namely, we prove that the closed realization $\mathfrak{A}_N$ is the infinitesimal generator of a Feller semigroup on $C(\overline{\Omega})$ associated with absorption, reflection and drift phenomena at the boundary (see Fig. 1.7).

In Sect. 23.1 we formulate a regularity theorem for the oblique derivative problem (Theorem 23.1) and an existence and uniqueness theorem for the non-homogeneous oblique derivative problem (Theorem 23.2) in the framework of L^p Sobolev spaces. The regularity theorem for the oblique derivative problem is proved in Sect. 23.2 (Theorem 23.1). We prove Theorem 23.1, by making use of a standard "bootstrap argument" based on Theorem 16.1 and the Sobolev imbedding theorems (Theorem 7.6). In Sect. 23.3 we prove the existence and uniqueness theorem for the oblique derivative problem with vanishing mean oscillation (VMO) coefficients (Theorem 23.4). The proof of Theorem 23.4 is based on the existence and uniqueness theorem for the non-homogeneous oblique derivative problem proved in Chap. 16 (Theorem 16.2 and the proof of Theorem 16.4). The uniqueness result in Theorem 23.2 follows from a variant of the Bakel'man–Aleksandrov maximum principle in the framework of L^p Sobolev spaces due to Lieberman [92] (Theorem 23.5). Section 23.4 is devoted to the proof of Theorem 1.3 where the proof of Theorem 16.4 plays an important role (Theorem 23.6). Finally, we prove Remark 1.4, that is, the domain $D(\mathfrak{A}_N)$ does not depend on p, for $N < p < \infty$.

23.1 Formulation of the Oblique Derivative Problem

Let A be a second order, uniformly elliptic differential operator with real *discontinuous* coefficients of the form

K. Taira, *Real Analysis Methods for Markov Processes*,
https://doi.org/10.1007/978-981-97-3659-1_23

$$Au(x) := \sum_{i,j=1}^{N} a^{ij}(x)\frac{\partial^2 u}{\partial x_i \partial x_j}(x) + \sum_{i=1}^{N} b^i(x)\frac{\partial u}{\partial x_i}(x) + c(x)u(x). \tag{1.1}$$

Here the functions $a^{ij}(x)$, $b^i(x)$ and $c(x)$ satisfy the following three conditions:

(1) $a^{ij} \in \mathrm{VMO} \cap L^\infty(\Omega)$, $a^{ij}(x) = a^{ji}(x)$ for almost all $x \in \Omega$ and there exist a constant $\lambda > 0$ such that (the uniform ellipticity condition)

$$\frac{1}{\lambda}|\xi|^2 \leq \sum_{i,j=1}^{N} a^{ij}(x)\xi_i\xi_j \leq \lambda|\xi|^2 \quad \text{for almost all } x \in \Omega \text{ and all } \xi \in \mathbf{R}^N.$$

(2) $b^i \in L^\infty(\Omega)$ for $1 \leq i \leq N$.
(3) $c \in L^\infty(\Omega)$ and $c(x) \leq 0$ for almost all $x \in \Omega$.

In this chapter, we consider an oblique derivative boundary operator of the form

$$L_0 u(x') := \mu(x')\frac{\partial u}{\partial \mathbf{n}}(x') + \beta(x') \cdot u(x') + \gamma(x')u(x'). \tag{23.1}$$

Here the functions $\mu(x')$, $\beta(x')$ and $\gamma(x')$ satisfy the following three conditions (see Fig. 23.1):

(4) $\mu(x')$ is a Lipschitz continuous function on $\partial\Omega$ and $\mu(x') \geq 0$ on $\partial\Omega$.
(5) $\beta(x')$ is a Lipschitz continuous vector field on $\partial\Omega$.
(6) $\gamma(x')$ is a Lipschitz continuous function on $\partial\Omega$ and $\gamma(x') \leq 0$ on $\partial\Omega$.
(7) $\mathbf{n} = (n_1, n_2, \ldots, n_N)$ is the unit inward normal to the boundary $\partial\Omega$.

The purpose of this section is to prove an existence and uniqueness theorem for the following non-homogeneous oblique derivative problem in the framework of L^p Sobolev spaces:

$$\boxed{\begin{cases} Au = f & \text{in } \Omega, \\ L_0 u = \varphi & \text{on } \partial\Omega. \end{cases}} \tag{23.2}$$

The first main result of this chapter is the following regularity theorem for the oblique derivative problem (23.2) (cf. [99, Theorem 2.2.1]).

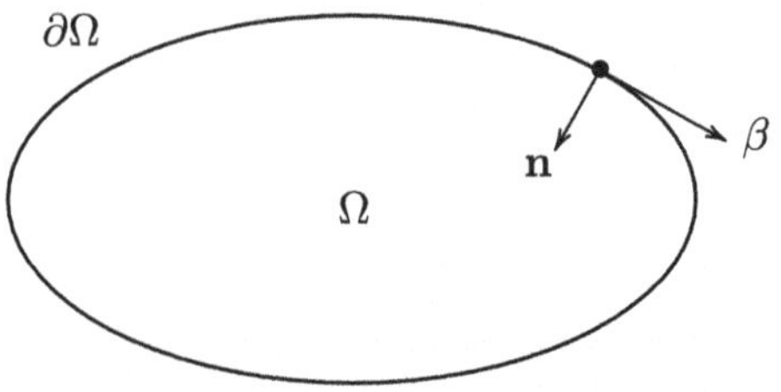

Fig. 23.1 The vector field β and the unit inward normal $\mathbf{n}$

Theorem 23.1 (the regularity theorem) *Let $N \geq 3$ and $1 < p < \infty$. Assume that the functions $\mu(x')$ and $\gamma(x')$ satisfy the following conditions:*

$$\mu(x') > 0 \quad on\ \partial\Omega. \tag{H.1}$$

$$\gamma(x') \leq 0 \quad on\ \partial\Omega. \tag{H.2$'$}$$

If a function $u \in W^{2,q}(\Omega)$, $1 < q < p < \infty$, is a solution of the oblique derivative problem (23.2) with $f \in L^p(\Omega)$ and $\varphi \in B^{1-1/p,p}(\partial\Omega)$, then it follows that $u \in W^{2,p}(\Omega)$. Moreover, we have the global a priori *estimate*

$$\|u\|_{W^{2,p}(\Omega)} \leq C_1 \left(\|u\|_{L^p(\Omega)} + \|f\|_{L^p(\Omega)} + \|\varphi\|_{B^{1-1/p,p}(\partial\Omega)}\right), \tag{23.3}$$

with a positive constant $C_1 = C_1(n, p, \lambda, \eta, \mu, \beta, \gamma, \partial\Omega)$.

It should be emphasized that condition (H.1) is equivalent to condition (16.3b) for $\mathcal{B}$ and that condition $(H.2')$ is equivalent to condition (16.7) for $\mathcal{B}$ in Chap. 18. Indeed, by virtue of Fig. 23.1 it follows that

$$\begin{aligned}\langle \ell(x'), \mathbf{n}(x')\rangle &= \langle \mu(x')\mathbf{n} + \beta(x'), \mathbf{n}\rangle = \mu(x')\langle \mathbf{n}, \mathbf{n}\rangle \\ &= \mu(x') \quad \text{for } x' \in \partial\Omega,\end{aligned}$$

since $\mathbf{n}$ is the unit inward normal to the boundary $\partial\Omega$.

The second main result of this chapter is the following existence and uniqueness theorem for the modified oblique derivative problem (23.2) (cf. [98, Theorem 4.1], [99, Theorem 2.2.2], [160, Theorem 5.1]).

Theorem 23.2 (the existence and uniqueness theorem) *Let $N \geq 3$ and $N < p < \infty$ and $\alpha > 0$. Assume that conditions (H.1) and (H.2$'$) are satisfied. Then the non-homogeneous oblique derivative problem*

$$\boxed{\begin{cases}(A - \alpha)\, u = f & in\ \Omega, \\ L_0 u = \varphi & on\ \partial\Omega\end{cases}} \tag{23.4}$$

has a unique solution $u \in W^{2,p}(\Omega)$ for any $f \in L^p(\Omega)$ and any $\varphi \in B^{1-1/p,p}(\partial\Omega)$. Moreover, we have the global a priori *estimate*

$$\|u\|_{W^{2,p}(\Omega)} \leq C_2 \left(\|f\|_{L^p(\Omega)} + \|\varphi\|_{B^{1-1/p,p}(\partial\Omega)}\right), \tag{23.5}$$

with a positive constant $C_2(\alpha) = C_2(n, p, \lambda, \eta, \mu, \beta, \gamma, \alpha, \partial\Omega)$.

If we associate with problem (23.4) a continuous linear operator

$$\mathcal{A}_N(\alpha) := (A - \alpha, L_0) : W^{2,p}(\Omega) \longrightarrow L^p(\Omega) \oplus B^{1-1/p,p}(\partial\Omega),$$

then we obtain from the trace theorem (Theorem 7.6) and Theorem 23.2 that the operator $\mathcal{A}_N(\alpha)$ is continuous and *bijective* for all $N < p < \infty$.

23.2 Proof of Theorem 23.1

The proof of the regularity theorem for the oblique derivative problem (Theorem 23.1) is divided into three steps.

Step 1: For any $\varphi \in B^{1-1/p,p}(\partial\Omega)$, we can find a function $v \in W^{2,p}(\Omega)$ such that $L_0 v = \varphi$ on $\partial\Omega$, and further that the mapping

$$B^{1-1/p,p}(\partial\Omega) \ni \varphi \longmapsto v \in W^{2,p}(\Omega) \tag{23.6}$$

is continuous (see Lemma 17.1).

Therefore, we have only to prove Theorem 23.1 in the case where $\varphi := 0$ and $1 < q < p < \infty$:

$$\begin{cases} u \in W^{2,q}(\Omega), \\ Au = f \in L^p(\Omega), \\ L_0 u = 0 \text{ on } \partial\Omega \end{cases} \tag{23.7}$$
$$\Longrightarrow u \in W^{2,p}(\Omega).$$

Step 2: In order to prove assertion (23.7), we need the Sobolev imbedding theorems (Theorem 7.6):

$$W^{2,p}(\Omega) \subset \begin{cases} L^{Nq/(N-2q)}(\Omega) & \text{if } 1 < q < N/2, \\ L^r(\Omega) \quad \text{for all } N/2 \leq r < \infty & \text{if } q = N/2, \\ L^\infty(\Omega) & \text{if } N/2 < q < \infty. \end{cases} \tag{22.5}$$

$$W^{1,q}(\Omega) \subset \begin{cases} L^{Nq/(N-q)}(\Omega) & \text{if } 1 < q < N, \\ L^r(\Omega) \quad \text{for all } N \leq r < \infty & \text{if } q = N, \\ L^\infty(\Omega) & \text{if } N < q < \infty. \end{cases} \tag{22.6}$$

We consider the following three cases (A), (B) and (C).

(A) The case where $1 < q < N/2$: By assertions (22.5) and (22.6), it follows that

$$\begin{cases} u \in W^{2,q}(\Omega) \subset L^{Nq/(N-2q)}(\Omega), \\ \frac{\partial u}{\partial x_i} \in W^{1,q}(\Omega) \subset L^{Nq/(N-q)}(\Omega) \quad \text{for all } 1 \leq i \leq N. \end{cases}$$

Hence we have the assertion

$$
\begin{aligned}
A_0 u &:= \sum_{i,j=1}^{N} a^{ij}(x) \frac{\partial^2 u}{\partial x_i \partial x_j} \\
&= f - \sum_{i=1}^{N} b^i(x) \frac{\partial u}{\partial x_i} - c(x)u \in L^{Nq/(N-q)}(\Omega).
\end{aligned}
$$

By applying the global regularity theorem (Theorem 16.1), we obtain that

$$
u \in L^{p_1}(\Omega),
$$

where

$$
p_1 = \min\left\{p, \frac{Nq}{N-q}\right\}.
$$

It is easy to see that

$$
p_1 > q,
$$

and further that

$$
p_1 - q > \frac{1}{N-1}.
$$

Therefore, by making use of a standard "bootstrap argument" we can conclude that

$$
u \in L^p(\Omega).
$$

(B) The case where $N/2 \leq q < N$: By assertions (22.5) and (22.6), it follows that

$$
\begin{cases}
u \in W^{2,q}(\Omega) \subset L^p(\Omega), \\
\frac{\partial u}{\partial x_i} \in W^{1,q}(\Omega) \subset L^{Nq/(N-q)}(\Omega) \quad \text{for all } 1 \leq i \leq N.
\end{cases}
$$

Hence we have the assertion

$$
A_0 u = f - \sum_{i=1}^{N} b^i(x) \frac{\partial u}{\partial x_i} - c(x)u \in L^{Nq/(N-q)}(\Omega).
$$

Therefore, by making use of a standard "bootstrap argument" we can conclude that

$$
u \in L^p(\Omega).
$$

(C) The case where $N \leq q < \infty$: By assertions (22.5) and (22.6), it follows that

$$
\begin{cases}
u \in W^{2,q}(\Omega) \subset L^\infty(\Omega), \\
\frac{\partial u}{\partial x_i} \in W^{1,q}(\Omega) \subset L^p(\Omega) \quad \text{for all } 1 \leq i \leq N.
\end{cases}
$$

Hence we have the assertion

$$A_0 u = f - \sum_{i=1}^{N} b^i(x) \frac{\partial u}{\partial x_i} - c(x)u \in L^p(\Omega).$$

By applying the global regularity theorem (see Theorem 16.1 and the proof of Theorem 16.4), we obtain that

$$u \in L^p(\Omega).$$

Summing up, we have proved the desired assertion (23.7).

Step 3: Finally, it remains to prove the global *a priori* estimate (23.3).

First, by using the global *a priori* estimate (16.5) we obtain that

$$\|u\|_{W^{2,p}(\Omega)} \leq C_1 \left(\|u\|_{L^p(\Omega)} + \|A_0 u\|_{L^p(\Omega)} + \|\varphi\|_{B^{1-1/p,p}(\partial\Omega)} \right), \tag{23.8}$$

with a positive constant C_1. Indeed, it suffices to note that the mapping (23.6) is continuous.

On the other hand, we have the inequality

$$\begin{aligned} &\|A_0 u\|_{L^p(\Omega)} \\ &\leq \left\| f - \sum_{i=1}^{N} b^i(x) \frac{\partial u}{\partial x_i} - c(x)u \right\|_{L^p(\Omega)} \\ &\leq \|f\|_{L^p(\Omega)} + \sum_{i=1}^{N} \|b^i\|_{L^\infty(\Omega)} \left\| \frac{\partial u}{\partial x_i} \right\|_{L^p(\Omega)} + \|c\|_{L^\infty(\Omega)} \|u\|_{L^p(\Omega)} . \end{aligned} \tag{23.9}$$

Hence, by combining two inequalities (23.8) and (23.9) we obtain that

$$\begin{aligned} &\|u\|_{W^{2,p}(\Omega)} \\ &\leq C_2 \left(\|u\|_{L^p(\Omega)} + \|f\|_{L^p(\Omega)} + \|\varphi\|_{B^{1-1/p,p}(\partial\Omega)} + \|\nabla u\|_{L^p(\Omega)} \right), \end{aligned} \tag{23.10}$$

with a positive constant C_2.

However, recall the *interpolation inequality* (13.17)

$$\|\nabla u\|_{L^p(\Omega)} \leq \varepsilon \left\|\nabla^2 u\right\|_{L^p(\Omega)} + \frac{C_3}{\varepsilon} \|u\|_{L^p(\Omega)} \quad \text{for all } \varepsilon > 0. \tag{22.10}$$

Therefore, the desired estimate (23.3) follows from inequalities (23.10) and (22.10) if we take

$$\varepsilon = \frac{1}{2C_2}.$$

Now the proof of Theorem 23.1 is complete. □

Let $\mathcal{A}_O$ be a continuous linear operator defined by the formula

$$\mathcal{A}_N := (A, L_0) : W^{2,p}(\Omega) \longrightarrow L^p(\Omega) \oplus B^{1-1/p,p}(\partial\Omega) \quad \text{for } 1 < p < \infty.$$

Then we have the following.

Corollary 23.3 *The null space*

$$N(\mathcal{A}_N) = \left\{u \in W^{2,p}(\Omega) : Au = 0 \text{ in } \Omega,\ L_0 u = 0 \text{ on } \partial\Omega\right\}$$

of $\mathcal{A}_N$ *is* independent *of* p *for* $1 < p < \infty$.

Proof In this proof, we let

$$\mathcal{A}_N(p) := (A, L_0) : W^{2,p}(\Omega) \longrightarrow L^p(\Omega) \oplus B^{1-1/p,p}(\partial\Omega).$$

If $1 < p_1 < p_2 < \infty$, then it follows that

$$N(\mathcal{A}_N(p_2)) \subset N(\mathcal{A}_N(p_1)).$$

Conversely, if $u \in N(\mathcal{A}_N(p_1))$, then we have, by assertion (23.7),

$$\begin{cases} u \in W^{2,p_1}(\Omega), \\ Au = 0 \in L^{p_2}(\Omega), \\ L_0 u = 0 \text{ on } \partial\Omega \end{cases}$$
$$\Longrightarrow u \in W^{2,p_2}(\Omega).$$

This proves that

$$N(\mathcal{A}_N(p_1)) \subset N(\mathcal{A}_N(p_2)).$$

The proof of Corollary 23.3 is complete. □

23.3 Proof of Theorem 23.2

The proof of Theorem 23.2 is divided into four steps.

Step 1: Our proof is based on the following existence and uniqueness theorem for the *non-homogeneous* oblique derivative problem (see Theorem 16.2 and the proof of Theorem 16.4).

Theorem 23.4 *Let* $N \geq 3$ *and* $1 < p < \infty$ *and*

$$A_0 u = \sum_{i,j=1}^{N} a^{ij}(x) \frac{\partial^2 u}{\partial x_i \partial x_j}.$$

Assume that the conditions (H.1) and (H.2′) are satisfied. Then there exists a constant $\alpha_0 > 0$ *(see formula (16.14) below) such that the non-homogeneous oblique derivative problem*

$$\begin{cases} (A_0 - \alpha_0)\, u = f & \text{in } \Omega, \\ L_0 u = \varphi & \text{on } \partial\Omega \end{cases} \tag{23.11}$$

has a unique solution $u \in W^{2,p}(\Omega)$ *for any function* $f \in L^p(\Omega)$ *and any function* $\varphi \in B^{1-1/p,p}(\partial\Omega)$.

Moreover, we have the global a priori *estimate*

$$\|u\|_{W^{2,p}(\Omega)} \leq C_2 \left(\|f\|_{L^p(\Omega)} + \|\varphi\|_{B^{1-1/p,p}(\partial\Omega)} \right), \tag{23.12}$$

with a positive constant $C_2 = C_2(n, p, \lambda, \eta, \mu, \beta, \gamma, \partial\Omega)$.

If we associate with problem (23.11) a continuous linear operator

$$\mathcal{A}_O(\alpha_0) := (A_0 - \alpha_0, L_0) : W^{2,p}(\Omega) \longrightarrow L^p(\Omega) \oplus B^{1-1/p,p}(\partial\Omega),$$

then we obtain from the trace theorem (Theorem 7.6) and Theorem 23.4 that the mapping $\mathcal{A}_0(\alpha_0)$ is an algebraic and topological *isomorphism*, for all $1 < p < \infty$. In particular, we have the assertion

$$\begin{aligned} \operatorname{ind} \mathcal{A}_O(\alpha_0) &= \dim N\left(\mathcal{A}_O(\alpha_0)\right) - \operatorname{codim} R\left(\mathcal{A}_O(\alpha_0)\right) \\ &= 0 \quad \text{for all } 1 < p < \infty. \end{aligned} \tag{23.13}$$

Step 2: If we let

$$B(\alpha)u := \sum_{i=1}^{N} b^i(x) \frac{\partial u}{\partial x_i} + (c(x) + \alpha)\, u \quad \text{for } \alpha > 0,$$

then it follows that the operator

$$B(\alpha) \colon W^{2,p}(\Omega) \longrightarrow W^{1,p}(\Omega)$$

is continuous for all $1 < p < \infty$. Moreover, it follows from an application of the Rellich–Kondrachov theorem (Theorem 7.8) that the injection

$$W^{1,p}(\Omega) \longrightarrow L^p(\Omega)$$

is compact for all $N < p < \infty$. Hence we find that the mapping

$$B(\alpha) \colon W^{2,p}(\Omega) \longrightarrow L^p(\Omega)$$

is *compact* for all $N < p < \infty$ and $\alpha > 0$. It should be noticed that

$$\begin{aligned}\mathcal{A}_N(\alpha) &:= (A - \alpha, L_0) \\ &= (A_0 - \alpha_0, L_0) + (B(\alpha_0 - \alpha), 0) \\ &= \mathcal{A}_O(\alpha_0) + (B\,(\alpha_0 - \alpha)\,, 0).\end{aligned}$$

However, it follows from an application of Theorem 2.53 that the index is *stable* under compact perturbations.

Therefore, we obtain that the mapping

$$\mathcal{A}_N(\alpha) = \mathcal{A}_O(\alpha_0) + (B\,(\alpha_0 - \alpha)\,, 0) : W^{2,p}(\Omega) \longrightarrow L^p(\Omega) \oplus B^{1-1/p,p}(\partial\Omega)$$

is a Fredholm operator with index *zero* for all $N < p < \infty$ and $\alpha > 0$, since we have, by assertion (23.13),

$$\operatorname{ind} \mathcal{A}_N(\alpha) = \operatorname{ind} \mathcal{A}_O(\alpha_0) = 0. \tag{23.14}$$

Step 3: In order to prove that

$$\dim N\,(\mathcal{A}_N(\alpha)) = 0 \quad \text{for all } \alpha > 0 \text{ and } N < p < \infty, \tag{23.15}$$

we need the following Bakel'man and Aleksandrov maximum principle (see [92, Corollary 2.4], [173]).

Theorem 23.5 (Bakel'man–Aleksandrov) *Let $N < p < \infty$ and $\alpha > 0$. Assume that conditions (H.1) and (H.2') are satisfied. If a function $u \in W^{2,p}(\Omega)$ satisfies the conditions*

$$\begin{cases} (A - \alpha)\,u \leq 0 & \text{almost everywhere in } \Omega, \\ L_0 u \leq 0 & \text{on } \partial\Omega, \end{cases}$$

then it follows that either $u(x)$ is a non-negative constant function or $u(x) > 0$ on $\overline{\Omega}$.

Proof First, it should be noticed that we have, by Sobolev's imbedding theorem (see [2, Theorem 4.12, Part II]; [155, Theorem 4.17]),

$$W^{2,p}(\Omega) \subset C^1(\overline{\Omega}),$$

since $2 - n/p > 1$ for $N < p < \infty$.

The proof of Theorem 23.5 is divided into the following two cases.

Case (a): If u is a constant in Ω, then we have the inequality

$$(A - \alpha)\,u = (c(x) - \alpha)\,u \leq 0 \quad \text{in } \Omega.$$

This implies that the constant u is non-negative, since $c(x) \leq 0$ in Ω and $\alpha > 0$.

Case (b): Now we consider the case when u is not a constant. The proof is based on a reduction to absurdity. Assume, to the contrary, that

$$\min_{\overline{\Omega}} u \leq 0.$$

If we let

$$v(x) = -u(x),$$

then we have the assertions

$$\begin{cases} v \in W^{2,p}(\Omega) & \text{for } N < p < \infty, \\ (A - \alpha)\, v = -\,(A - \alpha)\, u \geq 0 & \text{in } \Omega, \\ L_0 v = -L_0 u \geq 0 & \text{on } \partial\Omega. \end{cases}$$

Hence, by applying the strong maximum principle (Theorem 8.9) to the function $v(x)$ we obtain that, for some boundary point $x_0' \in \partial\Omega$,

$$v(x_0') = \max_{\overline{\Omega}}(-u) \geq 0.$$

Moreover, it follows from an application of Hopf's boundary point lemma (Lemma 8.8) that

$$\frac{\partial v}{\partial \mathbf{n}}(x_0') < 0.$$

By conditions (H.1) and (H.2′), this implies that

$$\begin{aligned} 0 \leq L_0 v(x_0') &= \mu(x_0') \frac{\partial u}{\partial \mathbf{n}}(x_0') + \gamma(x_0') u(x_0') \\ &\leq \mu(x_0') \frac{\partial u}{\partial \mathbf{n}}(x_0') < 0. \end{aligned}$$

This contradiction proves that $u(x) > 0$ on $\overline{\Omega}$.

The proof of Theorem 23.5 is complete. □

By applying Theorem 23.5 to the two functions $\pm u(x)$, we find that

$$\begin{cases} u \in W^{2,p}(\Omega) & \text{for } N < p < \infty, \\ (A - \alpha)\, u = 0 & \text{in } \Omega, \\ L_0 u = 0 & \text{on } \partial\Omega \end{cases}$$
$$\Longrightarrow u = 0 \quad \text{in } \Omega.$$

This proves the desired assertion (23.15).

In view of Corollary 23.3 with $A := A - \alpha$, we have proved that

$$\dim N\,(\mathcal{A}_N(\alpha)) = 0 \quad \text{for all } N < p < \infty,$$

that is, the mapping $\mathcal{A}_N(\alpha)$ is *injective* for all $N < p < \infty$ and $\alpha > 0$.

Therefore, it is also *surjective* for all $N < p < \infty$ and $\alpha > 0$, since we have, by assertion (23.14),

$$\operatorname{ind} \mathcal{A}_N(\alpha) = \dim N\,(\mathcal{A}_N(\alpha)) - \operatorname{codim} R\,(\mathcal{A}_N(\alpha)) = 0.$$

Step 4: Summing up, we have proved that the mapping

$$\mathcal{A}_N(\alpha) = (A - \alpha, L_0) : W^{2,p}(\Omega) \longrightarrow L^p(\Omega) \oplus B^{1-1/p,p}(\partial\Omega)$$

is an algebraic and topological *isomorphism* for all $N < p < \infty$ and $\alpha > 0$. Indeed, the continuity of the inverse of $\mathcal{A}_N(\alpha)$ follows immediately from an application of Banach's open mapping theorem (Theorem 2.38).

The proof of Theorem 23.2 is complete. □

23.4 Proof of Theorem 1.3

This section is devoted to the proof of Theorem 1.3. To do so, we have only to verify all conditions (a) through (d) in part (ii) of Theorem 21.12 with

$$\begin{aligned} K &:= \overline{\Omega}, \\ C_0(K) &:= C(\overline{\Omega}), \\ \mathfrak{A} &:= \mathfrak{A}_N, \end{aligned}$$

where

$$D(\mathfrak{A}_N) = \left\{u \in W^{2,p}(\Omega) : Au \in C(\overline{\Omega}),\ L_0 u = 0 \text{ on } \partial\Omega\right\}.$$

The proof is divided into four steps.

Step 1: First, by arguing just as in the proof of Theorem 16.4 we can obtain the following theorem.

Theorem 23.6 *Let $N \geq 3$ and $N < p < \infty$ and $\alpha > 0$. Assume that the boundary condition* (23.1)

$$L_0 u = \mu(x')\,\frac{\partial u}{\partial \mathbf{n}} + \beta(x') \cdot u + \gamma(x') u$$

satisfies the following two conditions:

$$\mu(x') > 0 \quad on\ \partial\Omega. \tag{H.1}$$

$$\gamma(x') \leq 0 \quad on\ \partial\Omega. \tag{H.2$'$}$$

Then, for each $\alpha > 0$ the non-homogeneous oblique derivative problem

$$\begin{cases} (A - \alpha)\, u = f & in\ \Omega, \\ L_0 u = \varphi & on\ \partial\Omega \end{cases}$$

has a unique solution $u \in W^{2,p}(\Omega)$ for any function $f \in L^p(\Omega)$ and any function $\varphi \in B^{1-1/p,p}(\partial\Omega)$.

Now we are in a position to prove that, for each $\alpha > 0$ the equation $(\alpha I - A)u = f$ has a unique solution $u \in D(\mathfrak{A}_N)$ for any $f \in C(\overline{\Omega})$.

By applying Theorem 23.6, we obtain that, for each $\alpha > 0$ the oblique derivative problem

$$\begin{cases} (\alpha - A)\, u = f & \text{almost everywhere in } \Omega, \\ L_0 u = 0 & \text{on } \partial\Omega \end{cases}$$

has a unique solution $u \in W^{2,p}(\Omega)$ for any $f \in L^p(\Omega)$ with $N < p < \infty$. In particular, for any $f \in C(\overline{\Omega})$ there exists a function $u \in W^{2,p}(\Omega)$ such that

$$(\alpha - A)\, u = f \quad \text{in } \Omega.$$

Hence we have the assertion

$$Au = \alpha u - f \in C(\overline{\Omega}).$$

By formula (1.4), this proves that

$$\begin{cases} u \in D(\mathfrak{A}_N), \\ (\alpha I - \mathfrak{A}_N)u = f. \end{cases}$$

Step 2: Secondly, we prove that, for each $\alpha > 0$ the *Green operator* $G_\alpha^N = (\alpha I - \mathfrak{A}_N)^{-1}$ is non-negative on the space $C(\overline{\Omega})$:

$$f \in C(\overline{\Omega}),\ f(x) \geq 0 \quad \text{in } \Omega \implies u(x) = G_\alpha^N f(x) \geq 0 \quad \text{in } \Omega.$$

More precisely, we prove the following assertion:

$$\begin{aligned} &f \in C(\overline{\Omega}),\ f(x) \geq 0,\ f(x) \not\equiv 0 \quad \text{in } \Omega \\ &\implies u(x) = G_\alpha^N f(x) > 0 \quad \text{on } \overline{\Omega}. \end{aligned} \tag{23.16}$$

Since we have the assertions

$$\begin{cases} u \in W^{2,p}(\Omega) & \text{for } N < p < \infty, \\ (A - \alpha)\, u = -f \leq 0 & \text{almost everywhere in } \Omega, \\ L_0 u = 0 & \text{on } \partial\Omega, \end{cases}$$

by arguing just as in the proof of Theorem 23.5 we obtain that either $u(x)$ is a non-negative constant function or $u(x) > 0$ on $\overline{\Omega}$. Indeed, it suffices to note the following fact in Case (a): If u is a constant in Ω, then we have the inequality

$$(A - \alpha)\, u = (c(x) - \alpha)\, u = -f \leq 0 \quad \text{on } \partial\Omega.$$

This implies that the constant u is non-negative, since $c(x) \leq 0$ on $\partial\Omega$ and $\alpha > 0$.

However, if $u(x) \equiv 0$ in Ω, then it follows that

$$f(x) = (\alpha - A)\, u(x) \equiv 0 \quad \text{in } \Omega.$$

This contradiction proves that either $u(x)$ is a positive constant function or $u(x) > 0$ on $\overline{\Omega}$. Namely, we have the desired assertion

$$f \in C(\overline{\Omega}),\ f(x) \geq 0,\ f(x) \not\equiv 0 \quad \text{in } \Omega \implies G_\alpha^N f(x) > 0 \quad \text{on } \overline{\Omega}.$$

Step 3: Thirdly, we prove that, for each $\alpha > 0$ the Green operator $G_\alpha^N = (\alpha I - \mathfrak{A}_N)^{-1}$ is bounded on the space $C(\overline{\Omega})$ with norm $1/\alpha$: $\|G_\alpha^N\| \leq 1/\alpha$ for all $\alpha > 0$.

By assertion (23.16), it suffices to show that

$$\alpha G_\alpha^N 1(x) \leq 1 \quad \text{on } \overline{\Omega}, \text{ for all } \alpha > 0.$$

If we let

$$v(x) := \alpha G_\alpha^N 1(x) - 1,$$

then we have the assertions

$$\begin{cases} u \in W^{2,p}(\Omega) & \text{for } N < p < \infty, \\ (A - \alpha)\, v = 0 & \text{in } \Omega, \\ L_0 v = 0 & \text{on } \partial\Omega. \end{cases}$$

By applying Theorem 23.5 to the function $-v(x)$ as above, we arrive at a contradiction that

$$\max_{\overline{\Omega}} v > 0 \implies v(x) \equiv 0 \quad \text{in } \Omega.$$

This proves that

$$\max_{\overline{\Omega}} v \leq 0,$$

that is,

$$\alpha G_\alpha^N 1(x) \le 1 \quad \text{on } \overline{\Omega}.$$

Step 4: The *closedness* of $\mathfrak{A}_N$ is an immediate consequence of the boundedness of $G_\alpha^N = (\alpha I - \mathfrak{A}_N)^{-1}$. Indeed, it suffices to note the formula

$$\mathfrak{A}_N = \alpha I - \left(G_\alpha^N\right)^{-1}.$$

Step 5: Finally, we prove that the domain $D(\mathfrak{A}_N)$ is *dense* in $C(\overline{\Omega})$. More precisely, we prove that, for each $u \in C(\overline{\Omega})$, we have the assertion

$$\lim_{\alpha \to +\infty} \|\alpha G_\alpha^N u - u\|_{C(\overline{\Omega})} = 0. \tag{23.17}$$

Step 5-1: It suffices to prove assertion (23.17) for any $v \in C^2(\overline{\Omega})$ such that $L_0 v = 0$ on $\partial\Omega$. In fact, we have the following:

Lemma 23.7 *Let $u \in C(\overline{\Omega})$. For any given $\varepsilon > 0$, we can find a function $v \in C^2(\overline{\Omega})$ such that*

$$\begin{cases} \|u - v\|_{C(\overline{\Omega})} < \varepsilon, \\ L_0 v = 0 \quad \text{on } \partial\Omega. \end{cases} \tag{23.18}$$

Proof First, it follows from an application of the Weierstrass approximation theorem that there exists a polynomial $g(x)$ such that

$$\|u - g\|_{C(\overline{\Omega})} < \frac{\varepsilon}{2}.$$

Secondly, we can construct a function $h \in C^2(\overline{\Omega})$ such that (see Lemma 17.1)

$$\begin{aligned} & h = 0 \quad \text{on } \partial\Omega, \\ & \frac{\partial h}{\partial \mathbf{n}} = \frac{1}{\mu(x')} L_0 g \quad \text{on } \partial\Omega, \\ & \|h\|_{C(\overline{\Omega})} < \frac{\varepsilon}{2}. \end{aligned}$$

This implies that

$$L_0 h = \mu(x') \frac{\partial h}{\partial \mathbf{n}} + \beta(x') \cdot h + \gamma(x') h = \mu(x') \frac{\partial h}{\partial \mathbf{n}} = L_0 g \quad \text{on } \partial\Omega.$$

Therefore, it is easy to verify that the function $v(x) = g(x) - h(x)$ satisfies the desired conditions (23.18).

The proof of Lemma 23.7 is complete. □

Step 5-2:To prove assertion (23.17) for any $v \in C^2(\overline{\Omega})$ such that $L_0 v = 0$, we introduce an extension $\widetilde{G_\alpha^N}$ of the Green operator G_α^N to the space $L^\infty(\Omega)$ as in Fig. 23.2.

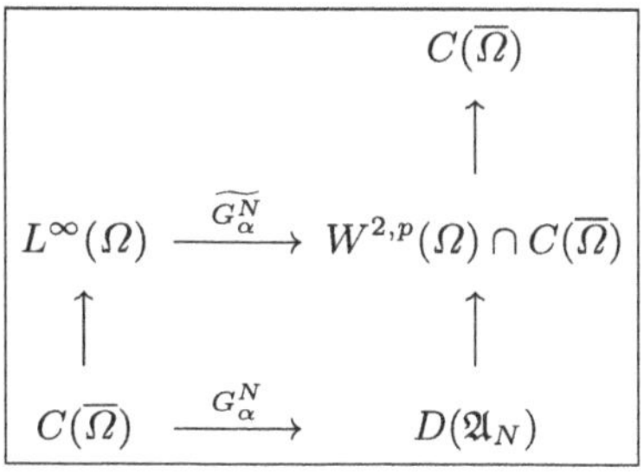

Fig. 23.2 The operators G_α^N and $\widetilde{G_\alpha^N}$

More precisely, by Theorem 23.6 we find that, for each $\alpha > 0$ the oblique derivative problem

$$\begin{cases} (\alpha - A)\, u = f & \text{almost everywhere in } \Omega, \\ L_0 u = 0 & \text{on } \partial\Omega \end{cases}$$

has a unique solution $u \in W^{2,p}(\Omega)$ for any $f \in L^\infty(\Omega)$, where $N < p < \infty$. If we let

$$u := \widetilde{G_\alpha^N} f \quad \text{for } \alpha > 0,$$

then it is easy to verify that the operator $\widetilde{G_\alpha^N}$ is an extension of G_α^N to $L^\infty(\Omega)$. Moreover, just as in Steps 2 and 3 we can prove the following two assertions:

(A) The operator $\widetilde{G_\alpha^N} : L^\infty(\Omega) \to C(\overline{\Omega})$ is non-negative.
(B) The operator $\widetilde{G_\alpha^N} : L^\infty(\Omega) \to C(\overline{\Omega})$ is bounded with norm $1/\alpha$: $\|\widetilde{G_\alpha^N}\| \leq 1/\alpha$ for all $\alpha > 0$.

First, since $a^{ij}, b^i, c \in L^\infty(\Omega)$ and $v \in C^2(\overline{\Omega})$, it follows that

$$Av = \sum_{i,j=1}^{N} a^{ij}(x) \frac{\partial^2 v}{\partial x_i \partial x_j} + \sum_{i=1}^{N} b^i(x) \frac{\partial v}{\partial x_i} + c(x) v \in L^\infty(\Omega).$$

Thus, if we let

$$w := \alpha G_\alpha^N v + \widetilde{G_\alpha^N}(Av) \quad \text{for } \alpha > 0,$$

then we have the assertions

$$\begin{cases} w \in W^{2,p}(\Omega) \cap C^1(\overline{\Omega}) & \text{for } N < p < \infty, \\ (A - \alpha)\, w = (A - \alpha)\, v & \text{almost everywhere in } \Omega, \\ L_0 w = 0 & \text{on } \partial\Omega, \end{cases}$$

and so

$$\begin{cases} w - v \in W^{2,p}(\Omega) \cap C(\overline{\Omega}) & \text{for } N < p < \infty, \\ (A - \alpha)(w - v) = 0 & \text{almost everywhere in } \Omega, \\ L_0(w - v) = 0 & \text{on } \partial\Omega. \end{cases}$$

By applying Theorem 23.6 to the function $w(x) - v(x)$, we obtain that $w - v = 0$ in Ω. This implies that

$$v = w = \alpha G_\alpha^N v + \widetilde{G_\alpha^N}(Av) \quad \text{for } \alpha > 0.$$

Therefore, the desired assertion (23.17) for any $v \in C^2(\overline{\Omega})$ such that $L_0 v = 0$ follows from an application of assertion (B), since we have the assertion

$$\begin{aligned} \left\| v - \alpha G_\alpha^N v \right\|_{C(\overline{\Omega})} &= \left\| \widetilde{G_\alpha^N}(Av) \right\|_{C(\overline{\Omega})} \\ &\leq \frac{1}{\alpha} \|Av\|_{L^\infty(\Omega)} \longrightarrow 0 \quad \text{as } \alpha \to \infty. \end{aligned}$$

Now the proof of Theorem 1.3 is complete. □

23.5 Proof of Remark 1.4

Finally, we prove that the domain

$$D(\mathfrak{A}_N) = \left\{ u \in W^{2,p}(\Omega) : Au \in C(\overline{\Omega}),\ L_0 u = 0 \text{ on } \partial\Omega \right\}$$

is *independent* of $N < p < \infty$.

We let

$$\mathcal{D}_p := \left\{ u \in W^{2,p}(\Omega) : Au \in C(\overline{\Omega}),\ L_0 u = 0 \text{ on } \partial\Omega \right\}.$$

In order to prove Remark 1.4, it suffices to show that

$$\mathcal{D}_{p_1} = \mathcal{D}_{p_2} \quad \text{for } N < p_1 < p_2 < \infty.$$

First, it follows that

$$\mathcal{D}_{p_2} \subset \mathcal{D}_{p_1},$$

since we have $L^{p_2}(\Omega) \subset L^{p_1}(\Omega)$ for $p_2 > p_1$.

Conversely, let v be an arbitrary element of $\mathcal{D}_{p_1}$:

$$v \in W^{2,p_1}(\Omega), \quad Av \in C(\overline{\Omega}),\ L_0 u = 0 \text{ on } \partial\Omega.$$

Then, since we have v, $Av \in C(\overline{\Omega}) \subset L^{p_2}(\Omega)$, it follows from an application of Theorem 23.2 with $p := p_2$ that there exists a unique function $u \in W^{2,p_2}(\Omega)$ such that

$$\begin{cases} (A - \alpha)\, u = (A - \alpha)\, v & \text{in } \Omega, \\ L_0 u = 0 & \text{on } \partial\Omega. \end{cases}$$

Hence we have the assertions

$$\begin{cases} u - v \in W^{2,p_1}(\Omega), \\ (A - \alpha)\,(u - v) = 0 & \text{in } \Omega, \\ L_0(u - v) = 0 & \text{on } \partial\Omega. \end{cases}$$

Therefore, by applying again Theorem 8.1 with $p := p_1$ we obtain that $u - v = 0$, so that $v = u \in W^{2,p_2}(\Omega)$. This implies that $v \in \mathcal{D}_{p_2}$.

The proof of Remark 1.4 is complete. □

23.6 Notes and Comments

(1) Under the *Lipschitz continuity* condition on the leading terms $a^{ij} \in C^{0,1}(\overline{\Omega})$ for $1 \leq i, j \leq N$ where $N \geq 2$, Troianiello proved Theorems 23.1 and 23.2 for $2 \leq p < \infty$ (see [172, Lemmas 3.18 and 3.19]).

(2) Theorem 1.3 is generalized to Feller semigroups with *jump phenomena* by [163, Corollaries 1.2 and 1.3].

Chapter 24
Feller Semigroups and Boundary Value Problems

The purpose of this chapter is to prove a general existence theorem for Feller semigroups in terms of boundary value problems (Theorem 24.9), following the main idea of Taira [148, Sect. 9.6] (cf. Bony–Courrège–Priouret [15] and Sato–Ueno [124]). Intuitively, Theorem 24.9 asserts that we can "piece together" a Markov process on the boundary $\partial\Omega$ with A-diffusion in the interior Ω to construct a Markov process on the closure $\overline{\Omega} = \Omega \cup \partial\Omega$ (see Remark 24.5).

Let Ω be a bounded domain in Euclidean space $\mathbf{R}^N$, $N \geq 3$, with boundary $\partial\Omega$ of class $C^{1,1}$. Let A be a second order, uniformly elliptic differential operator with real *discontinuous* coefficients of the form

$$Au(x) := \sum_{i,j=1}^{N} a^{ij}(x) \frac{\partial^2 u}{\partial x_i \partial x_j}(x) + \sum_{i=1}^{N} b^i(x) \frac{\partial u}{\partial x_i}(x) + c(x)u(x). \tag{1.1}$$

We assume that the coefficients $a^{ij}(x)$, $b^i(x)$ and $c(x)$ of the differential operator A satisfy the following three conditions:

(1) $a^{ij} \in \mathrm{VMO} \cap L^\infty(\Omega)$, $a^{ij}(x) = a^{ji}(x)$ for almost all $x \in \Omega$ and there exist a constant $\lambda > 0$ such that (the uniform ellipticity condition)

$$\frac{1}{\lambda}|\xi|^2 \leq \sum_{i,j=1}^{N} a^{ij}(x)\xi_i\xi_j \leq \lambda|\xi|^2 \quad \text{for almost all } x \in \Omega \text{ and all } \xi \in \mathbf{R}^N.$$

(2) $b^i \in L^\infty(\Omega)$ for $1 \leq i \leq N$.
(3) $c \in L^\infty(\Omega)$ and $c(x) \leq 0$ for almost all $x \in \Omega$.

K. Taira, *Real Analysis Methods for Markov Processes*,
https://doi.org/10.1007/978-981-97-3659-1_24

Let L be a first order, *Ventcel' (Wentzell) boundary condition* of the form

$$\boxed{Lu(x') := \mu(x')\frac{\partial u}{\partial \mathbf{n}}(x') + \beta(x')\cdot u(x') + \gamma(x')u(x') - \delta(x')\left(Au(x')\right).} \tag{1.2}$$

We assume that the coefficients $\mu(x'), \beta(x'), \gamma(x')$ and $\delta(x')$ of the boundary operator L satisfy the following four conditions (see Fig. 23.1):

(4) $\mu(x')$ is a Lipschitz continuous function on $\partial\Omega$ and $\mu(x') \geq 0$ on $\partial\Omega$.
(5) $\beta(x')$ is a Lipschitz continuous vector field on $\partial\Omega$.
(6) $\gamma(x')$ is a Lipschitz continuous function on $\partial\Omega$ and $\gamma(x') \leq 0$ on $\partial\Omega$.
(7) $\delta(x')$ is a Lipschitz continuous function on $\partial\Omega$ and $\delta(x') \geq 0$ on $\partial\Omega$.
(8) $\mathbf{n} = (n_1, n_2, \ldots, n_N)$ is the unit inward normal to the boundary $\partial\Omega$.

Now we are interested in the following functional analytic problem of construction of strong Markov processes with boundary conditions in probability theory:

Problem Given a differential operator A and a Ventcel' boundary condition L, can we construct a Feller semigroup $\{T_t\}_{t\geq 0}$ on $\overline{\Omega}$ whose infinitesimal generator $\mathfrak{A}$ is characterized by the data (A, L) ?

24.1 Green Operators and Harmonic Operators

Let $N < p < \infty$ and $\alpha > 0$. Since we have the inequality

$$c(x) - \alpha \leq -\alpha \quad \text{for almost all } x \in \Omega$$

and since we have the inclusions

$$\begin{aligned} C(\overline{\Omega}) &\subset L^p(\Omega), \\ C^2(\partial\Omega) &\subset B^{2-1/p,p}(\partial\Omega), \end{aligned}$$

it follows from an application of Theorem 22.2 to the operator $A - \alpha$ that the Dirichlet problem

$$\begin{cases} (\alpha - A)\, u = f & \text{almost everywhere in } \Omega, \\ \gamma_0 u = \varphi & \text{on } \partial\Omega \end{cases} \tag{24.1}$$

has a unique solution $u \in W^{2,p}(\Omega)$ for any $f \in C(\overline{\Omega})$ and any $\varphi \in C^2(\partial\Omega)$. Therefore, we can introduce two linear operators

$$G_\alpha^0 \colon C(\overline{\Omega}) \longrightarrow C(\overline{\Omega}),$$

and

$$H_\alpha : C^2(\partial\Omega) \longrightarrow C(\overline{\Omega})$$

as follows:

(a) For any $f \in C(\overline{\Omega})$, the function $G_\alpha^0 f \in W^{2,p}(\Omega) \cap W_0^{1,p}(\Omega)$ is the unique solution of the problem

$$\begin{cases} (\alpha - A)\, G_\alpha^0 f = f & \text{in } \Omega, \\ \gamma_0 \left(G_\alpha^0 f \right) = 0 & \text{on } \partial\Omega. \end{cases} \tag{24.2}$$

(b) For any $\varphi \in C^2(\partial\Omega)$, the function $H_\alpha \varphi \in W^{2,p}(\Omega)$ is the unique solution of the problem

$$\begin{cases} (\alpha - A)\, H_\alpha \varphi = 0 & \text{in } \Omega, \\ \gamma_0 \left(H_\alpha \varphi \right) = \varphi & \text{on } \partial\Omega. \end{cases} \tag{24.3}$$

The operators $G_\alpha^0 : C(\overline{\Omega}) \to C(\overline{\Omega})$ and $H_\alpha : C^2(\partial\Omega) \to C(\overline{\Omega})$ can be visualized as in Figs. 24.1 and 24.2.

Here it should be noticed that we have, by Sobolev's imbedding theorem (see [2, Theorem 4.12, Part II]; [155, Theorem 4.17]),

$$W^{2,p}(\Omega) \subset C^1(\overline{\Omega}),$$

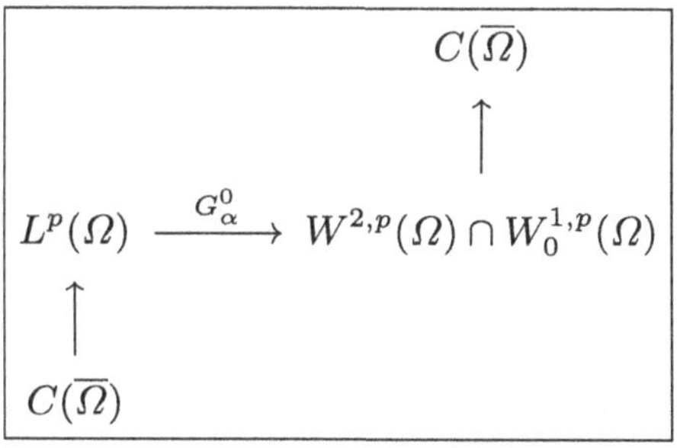

Fig. 24.1 The definition of the Green operator G_α^0

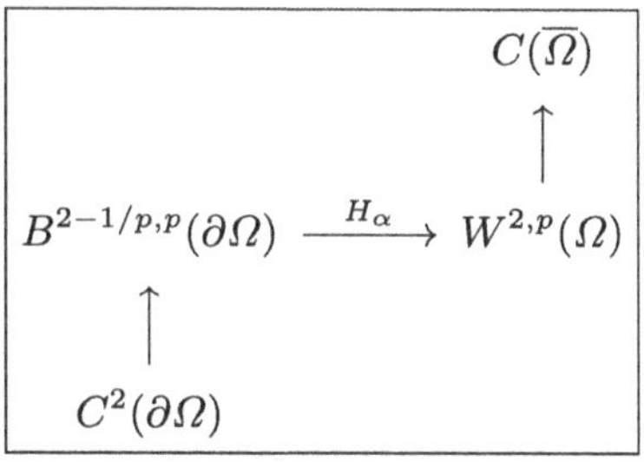

Fig. 24.2 The definition of the harmonic operator H_α

and, by an imbedding theorem for Besov spaces (see [2, Theorem 7.34]),

$$C^2(\partial\Omega) \subset B^{2-1/p,p}(\partial\Omega) \subset C^1(\partial\Omega),$$

since $2 - N/p > 1$ and $(1 - 1/p)p = p - 1 > N - 1$ for $N < p < \infty$.

The operator G_α^0 is called the *Green operator* and the operator H_α is called the *harmonic operator*, respectively.

Then we have the following fundamental results for the operators G_α^0 and H_α:

Theorem 24.1 *Let $N < p < \infty$ and $\alpha > 0$. Then we have the following two assertions (i) and (ii):*

(i) (a) The Green operators G_α^0 are non-negative and bounded with norm

$$\|G_\alpha^0\| = \|G_\alpha^0 1\|_{C(\overline{\Omega})} \leq \frac{1}{\alpha}. \tag{24.4}$$

(b) For any $f \in C(\overline{\Omega})$, we have the assertion

$$\gamma_0 \left(G_\alpha^0 f\right) = 0 \quad on\ \partial\Omega.$$

(c) For all α, $\beta > 0$, the resolvent equation *holds true:*

$$G_\alpha^0 f - G_\beta^0 f + (\alpha - \beta)\, G_\alpha^0 (G_\beta^0 f) = 0 \quad for\ f \in C(\overline{\Omega}). \tag{24.5}$$

(d) For any $f \in C(\overline{\Omega})$, we have the assertion

$$\lim_{\alpha \to +\infty} \alpha G_\alpha^0 f(x_0) = f(x_0) \quad for\ x_0 \in \Omega. \tag{24.6}$$

Furthermore, if $f|_{\partial\Omega} = 0$, that is, if $f \in C_0(\overline{\Omega})$, then this convergence is uniform in $x \in \overline{\Omega}$. In other words, we have the assertion

$$\lim_{\alpha \to +\infty} \alpha G_\alpha^0 f = f \quad in\ C_0(\overline{\Omega}). \tag{24.7}$$

(ii) (e) The harmonic operators H_α, $\alpha > 0$, can be uniquely extended to non-negative, bounded linear operators on $C(\partial\Omega)$ into $C(\overline{\Omega})$, denoted again by H_α, with norm $\|H_\alpha\| = 1$.

(f) For any $\varphi \in C(\partial\Omega)$, we have the assertion

$$\gamma_0 (H_\alpha \varphi) = \varphi \quad on\ \partial\Omega.$$

(g) For all α, $\beta > 0$, we have the equation

$$H_\alpha \varphi - H_\beta \varphi + (\alpha - \beta)\, G_\alpha^0 (H_\beta \varphi) = 0 \quad for\ \varphi \in C(\partial\Omega). \tag{24.8}$$

Proof (i) *Assertion (a)*: First, we show that the operators G_α^0 are non-negative for all $\alpha > 0$:

$$f \in C(\overline{\Omega}), \ f(x) \geq 0 \ \text{ in } \Omega \implies G_\alpha^0 f(x) \geq 0 \ \text{ in } \Omega.$$

If we let

$$v(x) := -G_\alpha^0 f(x),$$

then it follows that

$$\begin{cases} (A - \alpha)\, v = f \geq 0 & \text{in } \Omega, \\ \gamma_0 v = 0 & \text{on } \partial\Omega. \end{cases}$$

Therefore, by applying the weak maximum principle (Theorem 22.7) with $A := A - \alpha$ we obtain that

$$v(x) \leq 0 \ \text{ in } \Omega,$$

so that

$$G_\alpha^0 f(x) = -v(x) \geq 0 \ \text{ in } \Omega.$$

Secondly, we show that the operators G_α^0 are bounded with norm $1/\alpha$, for all $\alpha > 0$. To do this, it suffices to show that

$$G_\alpha^0 1(x) \leq \frac{1}{\alpha} \ \text{ on } \overline{\Omega}. \tag{24.4'}$$

since G_α^0 are non-negative on $C(\overline{\Omega})$.

If we let

$$u(x) := \alpha G_\alpha^0 1(x) - 1 \in W^{2,p}(\Omega),$$

then it follows that

$$(A - \alpha)\, u(x) = -\alpha + (\alpha - c(x)) = -c(x) \geq 0 \ \text{ in } \Omega,$$

and that

$$\gamma_0 u = -1 \ \text{ on } \partial\Omega.$$

Thus, by applying Theorem 22.7 with $A := A - \alpha$ we obtain that

$$\alpha G_\alpha^0 1(x) - 1 = u(x) \leq 0 \ \text{ on } \overline{\Omega}.$$

This proves the desired assertion (24.4′).

Assertion (b): It suffices to note that the function $G_\alpha^0 f \in W^{2,p}(\Omega) \cap W_0^{1,p}(\Omega)$ is the unique solution of the Dirichlet problem (24.2).

Assertion (c): This is an immediate consequence of the uniqueness theorem for problem (24.1) (Theorem 23.2). Indeed, it follows that the function

$$u := G_\alpha^0 f - G_\beta^0 f + (\alpha - \beta)\, G_\alpha^0 \left(G_\beta^0 f\right) \in W^{2,p}(\Omega)$$

satisfies the equation

$$\begin{aligned}(\alpha - A)\, u &= f - (\alpha - A)\, G_\beta^0 f + (\alpha - \beta)\, G_\beta^0 f \\ &= f - (\beta - A + \alpha - \beta)\, G_\beta^0 f + (\alpha - \beta)\, G_\beta^0 f \\ &= f - f - (\alpha - \beta)\, G_\beta^0 f + (\alpha - \beta)\, G_\beta^0 f \\ &= 0 \quad \text{in } \Omega,\end{aligned}$$

and the boundary condition

$$\gamma_0 u = 0 \quad \text{on } \partial\Omega.$$

By applying Theorem 23.2 to the operator $A - \alpha$, we obtain that

$$u = 0 \quad \text{in } \Omega.$$

This proves the resolvent Eq. (24.5) for $f \in C(\overline{\Omega})$.

Assertion (d): First, let $f(x)$ be an arbitrary function in $C(\overline{\Omega})$ satisfying $f = 0$ on $\partial\Omega$. Then it follows from the uniqueness theorem for problem (24.1) that we have, for all α, β,

$$f - \alpha G_\alpha^0 f = G_\alpha^0 \left((\beta - A)\, f\right) - \beta G_\alpha^0 f.$$

Thus we have, by estimate (24.4),

$$\| f - \alpha G_\alpha^0 f \|_{C(\overline{\Omega})} \leq \frac{1}{\alpha} \| (\beta - A)\, f \|_{C(\overline{\Omega})} + \frac{\beta}{\alpha} \| f \|_{C(\overline{\Omega})},$$

so that

$$\lim_{\alpha \to +\infty} \| f - \alpha G_\alpha^0 f \|_{C(\overline{\Omega})} = 0.$$

To prove assertion (24.6), let $f(x)$ be an arbitrary function in $C(\overline{\Omega})$ and let x_0 be an arbitrary point of Ω. Take a function $\psi \in C(\overline{\Omega})$ such that

$$\begin{cases} 0 \leq \psi(x) \leq 1 & \text{on } \overline{\Omega}, \\ \psi(x) = 0 & \text{in a neighborhood of } x_0, \\ \psi(x) = 1 & \text{near the boundary } \partial\Omega. \end{cases}$$

Then it follows from the non-negativity of G_α^0 and estimate (24.4) that

$$0 \leq \alpha G_{\alpha}^{0}\psi(x_0) + \alpha G_{\alpha}^{0}(1-\psi)(x_0) = \alpha G_{\alpha}^{0}1(x_0) \leq 1. \tag{24.9}$$

However, by applying assertion (24.7) to the function $1 - \psi(x)$ we obtain that

$$\lim_{\alpha \to +\infty} \alpha G_{\alpha}^{0}(1-\psi)(x_0) = 1 - \psi(x_0) = 1.$$

In view of inequalities (24.9), this implies that

$$\lim_{\alpha \to +\infty} \alpha G_{\alpha}^{0}\psi(x_0) = 0.$$

Thus, since we have the inequality

$$-\|f\|_{C(\overline{\Omega})}\psi \leq f\psi \leq \|f\|_{C(\overline{\Omega})}\psi \quad \text{on } \overline{\Omega},$$

it follows that

$$|\alpha G_{\alpha}^{0}(f\psi)(x_0)| \leq \|f\|_{C(\overline{\Omega})}\, \alpha G_{\alpha}^{0}\psi(x_0) \to 0 \quad \text{as } \alpha \to +\infty.$$

Therefore, by applying assertion (24.7) to the function $(1 - \psi(x))\, f(x)$ we obtain that

$$\begin{aligned} f(x_0) &= ((1-\psi) f)(x_0) \\ &= \lim_{\alpha \to +\infty} \alpha G_{\alpha}^{0}((1-\psi) f)(x_0) \\ &= \lim_{\alpha \to +\infty} \alpha G_{\alpha}^{0} f(x_0). \end{aligned}$$

This proves the desired assertion (24.6).

(ii) *Assertion (e)*: First, let $\varphi(x')$ be an arbitrary function in $C^2(\partial\Omega)$ such that $\varphi \geq 0$ on $\partial\Omega$. Then we have the assertions

$$\begin{cases} (A-\alpha)(-H_{\alpha}\varphi) = 0 & \text{in } \Omega, \\ -H_{\alpha}\varphi = -\varphi \leq 0 & \text{on } \partial\Omega. \end{cases}$$

Therefore, by applying Theorem 22.7 with $A := A - \alpha$ to the function $u := -H_{\alpha}\varphi$ we obtain that

$$H_{\alpha}\varphi \geq 0 \quad \text{on } \overline{\Omega}.$$

This proves the non-negativity of H_{α}.

In order to prove the boundedness of H_{α}

$$\|H_{\alpha}\| = 1,$$

it suffices to show that

$$H_{\alpha}1(x) \leq 1 \quad \text{in } \Omega,$$

since H_α is non-negative.

To do this, we remark that the function $H_\alpha 1 - 1$ satisfies the conditions

$$\begin{cases} (A - \alpha)\,(H_\alpha 1 - 1) = -c(x) + \alpha \geq 0 & \text{in } \Omega, \\ H_\alpha 1 - 1 = 0 & \text{on } \partial\Omega. \end{cases}$$

Therefore, by applying Theorem 22.7 with $A := A - \alpha$ and $u := H_\alpha 1 - 1$ we obtain that

$$H_\alpha 1(x) - 1 \leq 0 \quad \text{in } \Omega.$$

Since the space $C^2(\partial\Omega)$ is dense in $C(\partial\Omega)$, it follows that the operator $H_\alpha \colon C^2(\partial\Omega) \to C(\overline{\Omega})$ can be uniquely extended to a non-negative, bounded linear operator, denoted again by $H_\alpha \colon C(\partial\Omega) \to C(\overline{\Omega})$.

Assertion (f): This assertion follows from formula (24.3), since the space $C^2(\partial\Omega)$ is dense in $C(\partial\Omega)$ and since the operator $H_\alpha \colon C(\partial\Omega) \to C(\overline{\Omega})$ is bounded.

Assertion (g): We find from the uniqueness theorem for problem (24.3) (Theorem 23.2) that the desired Eq. (24.8) holds true for all $\varphi \in C^2(\partial\Omega)$. Hence it holds true for all $\varphi \in C(\partial\Omega)$, since the space $C^2(\partial\Omega)$ is dense in $C(\partial\Omega)$ and since the operators G_α^0 and H_α are bounded.

The proof of Theorem 24.1 is now complete.

Summing up, we have the following mapping properties of the Green operators and the harmonic operators as in Figs. 24.3 and 24.4.

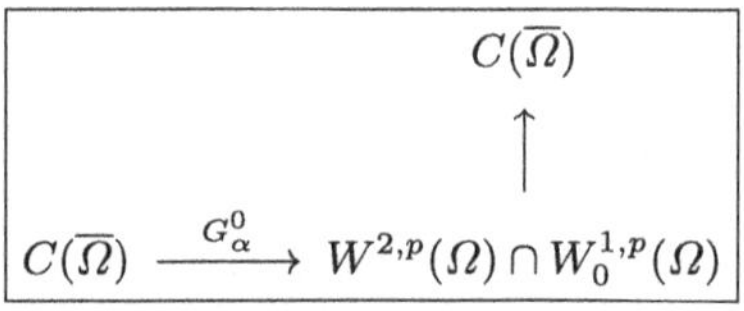

Fig. 24.3 The mapping properties of the Green operator G_α^0

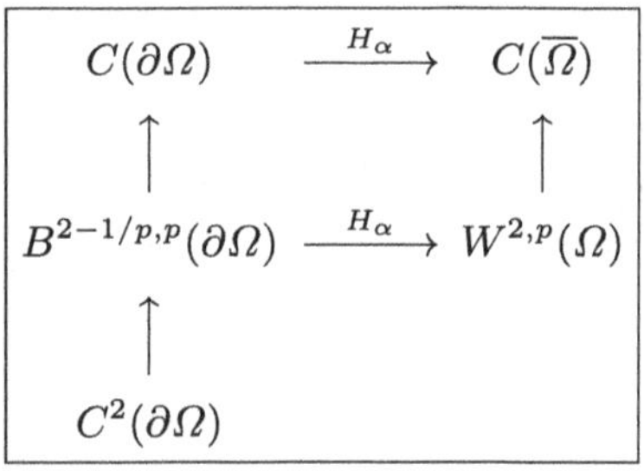

Fig. 24.4 The mapping properties of the harmonic operator H_α

24.2 General Boundary Value Problems

Now we consider the following general boundary value problem in the framework of the spaces of *continuous functions*:

$$\begin{cases} (\alpha - A)\, u = f & \text{in } \Omega, \\ Lu = 0 & \text{on } \partial\Omega. \end{cases} \tag{24.10}$$

To do this, we introduce *three* linear operators associated with problem (24.10).

(I) First, we introduce a linear operator

$$\overline{\mathcal{A}} \colon C(\overline{\Omega}) \longrightarrow C(\overline{\Omega})$$

as follows (see Fig. 24.5):

(a) The domain $D(\overline{\mathcal{A}})$ of $\overline{\mathcal{A}}$ is the space

$$D\left(\overline{\mathcal{A}}\right) = \left\{u \in W^{2,p}(\Omega) : Au \in C(\overline{\Omega})\right\} \quad (N < p < \infty).$$

(b) $\overline{\mathcal{A}}u = Au$ for every $u \in D(\overline{\mathcal{A}})$.

Then we have the following:

Lemma 24.2 *The operator $\overline{\mathcal{A}}$ is a densely defined, closed linear operator in the space $C(\overline{\Omega})$.*

Proof First, by the definition of $\overline{\mathcal{A}}$ and $\mathfrak{A}_N$ it follows that

$$\mathfrak{A}_N \subset \overline{\mathcal{A}}.$$

This proves the density of the domain $D(\overline{\mathcal{A}})$ in $C(\overline{\Omega})$, since the domain $D(\mathfrak{A}_N)$ is dense in $C(\overline{\Omega})$ (see assertion (23.7) and Fig. 24.6).

Now, let (u, v) be an arbitrary element of the product space $C(\overline{\Omega}) \oplus C(\overline{\Omega})$ such that there exists a sequence $\{u_n\} \subset D(\overline{\mathcal{A}})$ which satisfies the conditions

$$u_n \longrightarrow u \quad \text{in } C(\overline{\Omega}),$$
$$Au_n \longrightarrow v \quad \text{in } C(\overline{\Omega}).$$

$$\begin{array}{ccc} W^{2,p}(\Omega) & \xrightarrow{\ A\ } & L^p(\Omega) \\ \uparrow & & \uparrow \\ D\left(\overline{\mathcal{A}}\right) & \xrightarrow[\ \overline{\mathcal{A}}\ ]{} & C(\overline{\Omega}) \end{array}$$

Fig. 24.5 The definition of the operator $\overline{\mathcal{A}}$

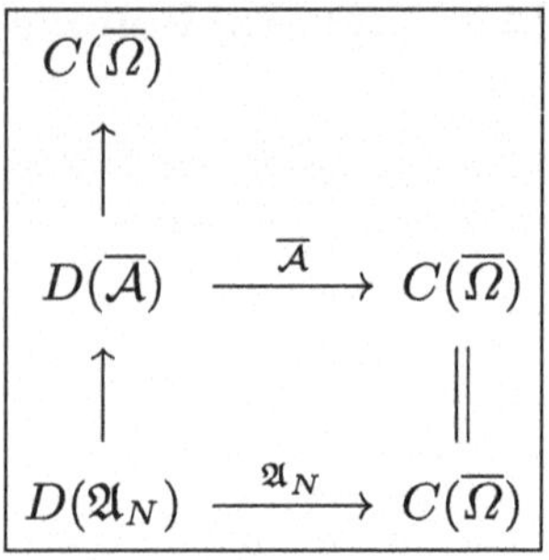

Fig. 24.6 The definition of the operators $\overline{\mathcal{A}}$ and $\mathfrak{A}_N$

Then we have, by the boundedness of G_α^0,

$$G_\alpha^0(Au_n) = \alpha G_\alpha^0 u_n - u_n \longrightarrow \alpha G_\alpha^0 u - u \quad \text{in } C(\overline{\Omega}),$$

and also

$$G_\alpha^0(Au_n) \longrightarrow G_\alpha^0 v \quad \text{in } C(\overline{\Omega}).$$

This proves that

$$u = \alpha G_\alpha^0 u - G_\alpha^0 v \in W^{2,p}(\Omega). \tag{24.11}$$

Thus, by applying the operator $\alpha - A$ to the both hand sides of formula (24.11) we obtain that

$$(\alpha - A)\, u = \alpha\,(\alpha - A)\, G_\alpha^0 u - (\alpha - A)\, G_\alpha^0 v = \alpha u - v,$$

so that

$$Au = v \in C(\overline{\Omega}).$$

Summing up, we have proved that

$$u \in D(\overline{\mathcal{A}}),$$
$$\overline{\mathcal{A}} u = v.$$

This proves the closedness of $\overline{\mathcal{A}}$.

The proof of Lemma 24.2 is complete. □

Remark 24.1 The domain $D(\overline{\mathcal{A}})$ does not depend on p, for $N < p < \infty$ (see Sect. 24.4).

The extended operators $G_\alpha^0 \colon C(\overline{\Omega}) \to C(\overline{\Omega})$ and $H_\alpha \colon C(\partial\Omega) \to C(\overline{\Omega})$ $(\alpha > 0)$ still satisfy formulas (24.2) and (24.3) respectively in the following sense:

Lemma 24.3 *(i) For any* $f \in C(\overline{\Omega})$*, we have the assertions*

$$\begin{cases} G_\alpha^0 f \in D(\overline{\mathcal{A}}), \\ \left(\alpha I - \overline{\mathcal{A}}\right) G_\alpha^0 f = f. \end{cases} \tag{24.12}$$

(ii) For any $\varphi \in C(\partial\Omega)$*, we have the assertions*

$$\begin{cases} H_\alpha \varphi \in D(\overline{\mathcal{A}}), \\ \left(\alpha I - \overline{\mathcal{A}}\right) H_\alpha \varphi = 0. \end{cases} \tag{24.13}$$

Proof *Assertion (i)*: If $f \in C(\overline{\Omega})$, then it follows from the definition of G_α^0 that

$$\begin{aligned} &G_\alpha^0 f \in W^{2,p}(\Omega), \\ &A(G_\alpha^0 f) = \alpha G_\alpha^0 f - f \in C(\overline{\Omega}). \end{aligned}$$

This proves the desired assertions (24.12).

Assertion (ii): If $\varphi \in C(\partial\Omega)$, we can find a sequence $\{\varphi_j\}$ in the space $C^2(\partial\Omega)$ such that

$$\varphi_j \longrightarrow \varphi \quad \text{in } C(\partial\Omega).$$

Hence, we have, by the boundedness of H_α,

$$H_\alpha \varphi_j \longrightarrow H_\alpha \varphi \quad \text{in } C(\overline{\Omega}).$$

However, it follows that

$$\begin{aligned} &H_\alpha \varphi_j \in W^{2,p}(\Omega), \\ &A(H_\alpha \varphi_j) = \alpha H_\alpha \varphi_j \in C(\overline{\Omega}), \end{aligned}$$

so that

$$H_\alpha \varphi_j \in D(\overline{\mathcal{A}}).$$

Therefore, we have proved that

$$\begin{aligned} &H_\alpha \varphi_j \in D(\overline{\mathcal{A}}), \\ &H_\alpha \varphi_j \longrightarrow H_\alpha \varphi \quad \text{in } C(\overline{\Omega}), \\ &A(H_\alpha \varphi_j) \longrightarrow \alpha H_\alpha \varphi \quad \text{in } C(\overline{\Omega}). \end{aligned}$$

This proves the desired assertions (24.13), since the operator $\overline{\mathcal{A}}$ is closed.

The proof of Lemma 24.3 is complete. □

Corollary 24.4 *Every function u in* $D(\overline{\mathcal{A}})$ *can be written in the form*

$$u = G_\alpha^0\left(\left(\alpha I - \overline{\mathcal{A}}\right) u\right) + H_\alpha(u|_{\partial\Omega}) \quad \textit{for each } \alpha > 0. \tag{24.14}$$

Proof We let

$$w := u - G_\alpha^0\left(\left(\alpha I - \overline{\mathcal{A}}\right) u\right) - H_\alpha\left(u|_{\partial\Omega}\right).$$

Then it follows from Lemma 24.3 that the function w is in $D(\overline{\mathcal{A}})$ and satisfies the conditions

$$\begin{cases} \left(\alpha I - \overline{\mathcal{A}}\right) w = 0 & \text{in } \Omega, \\ \gamma_0 w = 0 & \text{on } \partial\Omega. \end{cases}$$

Therefore, we can apply Theorem 23.2 to the operator $A - \alpha$ to obtain that

$$w = 0.$$

This proves the desired formula (24.14).

The proof of Corollary 24.4 is complete. □

(II) Secondly, we introduce a linear operator

$$\overline{LG_\alpha^0} : C(\overline{\Omega}) \longrightarrow C(\partial\Omega)$$

as follows:

(a) The domain $D\left(\overline{LG_\alpha^0}\right)$ of $\overline{LG_\alpha^0}$ is the space $C(\overline{\Omega})$.

(b) $\overline{LG_\alpha^0} f = L(G_\alpha^0 f) = \mu(x')\dfrac{\partial}{\partial \mathbf{n}}(G_\alpha^0 f) + \delta(x')(f|_{\partial\Omega})$ for every $f \in D\left(\overline{LG_\alpha^0}\right)$.

The operator $\overline{LG_\alpha^0} : C(\overline{\Omega}) \to C(\partial\Omega)$ can be visualized as in Fig. 24.7.

Here it should be emphasized that we have, by Sobolev's imbedding theorem (see [2, Theorem 4.12, Part II]; [155, Theorem 4.17]),

$$G_\alpha^0 f \in W^{2,p}(\Omega) \subset C^1(\overline{\Omega}),$$

since $2 - N/p > 1$ for $N < p < \infty$.

Then we have the following:

Lemma 24.5 *The operators* $\overline{LG_\alpha^0} : C(\overline{\Omega}) \to C(\partial\Omega)$ *are non-negative and bounded for all* $\alpha > 0$.

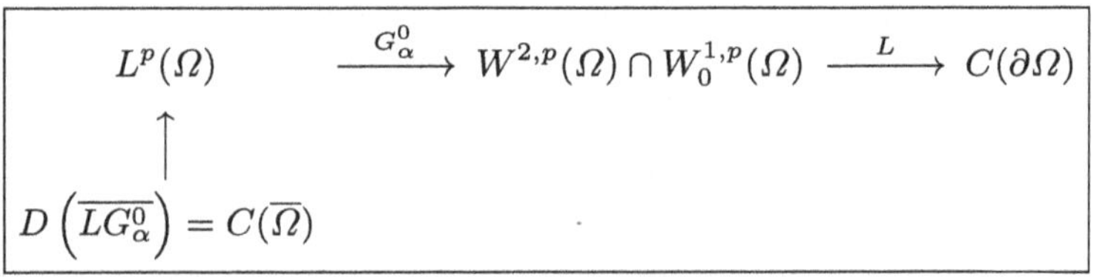

Fig. 24.7 The definition of the operator $\overline{LG_\alpha^0}$

Proof Let f be an arbitrary function in $D(\overline{LG^0_\alpha}) = C(\overline{\Omega})$ such that $f(x) \geq 0$ on $\overline{\Omega}$. Then we have the assertions

$$\begin{cases} G^0_\alpha f \in C^1(\overline{\Omega}), \\ G^0_\alpha f \geq 0 & \text{on } \overline{\Omega}, \\ G^0_\alpha f = 0 & \text{on } \partial\Omega, \end{cases}$$

and so

$$\begin{aligned} \overline{LG^0_\alpha} f(x') &= L(G^0_\alpha f)(x') \\ &= \mu(x') \frac{\partial}{\partial \mathbf{n}} (G^0_\alpha f)(x') + \delta(x') f(x') \\ &\geq 0 \quad \text{on } \partial\Omega. \end{aligned}$$

This proves that the operator $\overline{LG^0_\alpha}$ is non-negative.

By the non-negativity of $\overline{LG^0_\alpha}$, we have, for all $f \in D(\overline{LG^0_\alpha})$,

$$-\overline{LG^0_\alpha} \|f\|_{C(\overline{\Omega})} \leq \overline{LG^0_\alpha} f \leq \overline{LG^0_\alpha} \|f\|_{C(\overline{\Omega})} \quad \text{on } \partial\Omega.$$

This implies the boundedness of $\overline{LG^0_\alpha}$ with norm

$$\|\overline{LG^0_\alpha}\| = \|L(G^0_\alpha 1)\|_{C(\partial\Omega)}.$$

The proof of Lemma 24.5 is complete. □

Remark 24.2 Similarly, we can prove that the operators

$$\widetilde{LG^0_\alpha} : L^\infty(\Omega) \longrightarrow C(\partial\Omega)$$

are non-negative and bounded for all $\alpha > 0$, with norm

$$\left\| \widetilde{LG^0_\alpha} \right\| = \left\| L(G^0_\alpha 1) \right\|_{C(\partial\Omega)}.$$

The operators $\overline{LG^0_\alpha}$ and $\widetilde{LG^0_\alpha}$ can be visualized as in Fig. 24.8.

The next lemma states a fundamental relationship between the operators $\overline{LG^0_\alpha}$ and $\overline{LG^0_\beta}$ for $\alpha, \beta > 0$:

Lemma 24.6 *For any $\alpha, \beta > 0$, we have the equation*

$$\overline{LG^0_\alpha} f - \overline{LG^0_\beta} f + (\alpha - \beta)\, \overline{LG^0_\alpha} (G^0_\beta f) = 0 \quad \text{for } f \in C(\overline{\Omega}). \tag{24.15}$$

Proof We have, by the *resolvent equation*

$$
\begin{array}{ccccc}
D\left(\widetilde{LG_\alpha^0}\right) = L^\infty(\Omega) & \xrightarrow{G_\alpha^0} & W^{2,p}(\Omega) \cap W_0^{1,p}(\Omega) & \xrightarrow{L} & C(\partial\Omega) \\
\uparrow & & & & \| \\
D\left(\overline{LG_\alpha^0}\right) = C(\overline{\Omega}) & \xrightarrow{\overline{LG_\alpha^0}} & & & C(\partial\Omega)
\end{array}
$$

Fig. 24.8 The mapping properties of the operators $\overline{LG_\alpha^0}$ and $\widetilde{LG_\alpha^0}$ in Lemma 24.5 and Remark 24.2

$$G_\alpha^0 f - G_\beta^0 f + (\alpha - \beta)\, G_\alpha^0(G_\beta^0 f) = 0. \tag{24.5}$$

Therefore, the desired formula (24.15) follows by applying the operator L to the both hand sides of Eq. (24.5).

The proof of Lemma 24.6 is complete. □

Remark 24.3 The Eq. (24.15) remains valid for $f \in L^\infty(\Omega)$:

$$LG_\alpha^0 f - LG_\beta^0 f + (\alpha - \beta)\, \overline{LG_\alpha^0}\,(G_\beta^0 f) = 0, \quad f \in L^\infty(\Omega). \tag{24.15$'$}$$

Indeed, it suffices to note that the function

$$u := G_\alpha^0 f - G_\beta^0 f + (\alpha - \beta)\, G_\alpha^0 \left(G_\beta^0 f\right) \in W^{2,p}(\Omega)$$

is a unque solution of the Dirichlet problem

$$\begin{cases} (\alpha - A)\, u = 0 & \text{almost everywhere in } \Omega, \\ \gamma_0 u = 0 & \text{on } \partial\Omega. \end{cases}$$

(III) Finally, we introduce a linear operator

$$LH_\alpha \colon C(\partial\Omega) \longrightarrow C(\partial\Omega)$$

as follows:

(a) The domain $D\,(LH_\alpha)$ of LH_α is the space $B^{2-1/p,p}(\partial\Omega)$.

(b) $LH_\alpha\psi = L(H_\alpha\psi) = \mu(x')\dfrac{\partial}{\partial \mathbf{n}}(H_\alpha\psi) + \beta(x') \cdot \psi + \gamma(x')\psi - \alpha\delta(x')\psi$ for every $\psi \in D\,(LH_\alpha)$.

Then we have the following lemma (see Fig. 24.9):

Lemma 24.7 *The operator LH_α, $\alpha > 0$, has its minimal closed extension $\overline{LH_\alpha}$ in the space $C(\partial\Omega)$.*

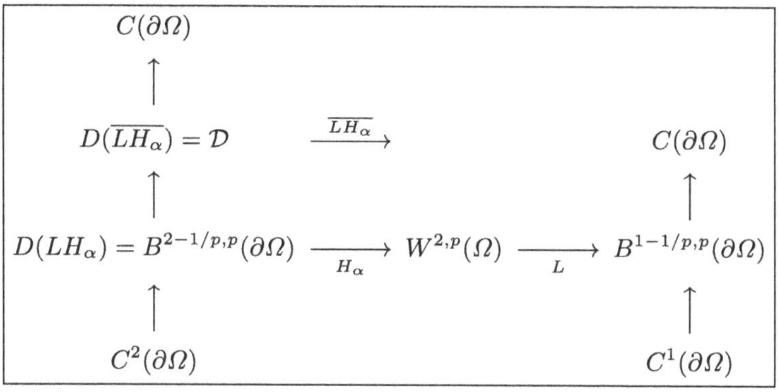

Fig. 24.9 The mapping properties of the operators LH_α and $\overline{LH_\alpha}$ in Lemmas 24.7 and 24.8

Proof We apply part (i) of Theorem 21.14 (the Hille–Yosida–Ray theorem) with $K := \partial\Omega$ to the operator $B := LH_\alpha$. To do this, it suffices to show that the operator LH_α satisfies condition (β') with $K := \partial\Omega$ (or condition (β) with $K := K_0 = \partial\Omega$) of the same theorem.

Assume that a function φ in the domain $D(LH_\alpha) = B^{2-1/p,p}(\partial\Omega)$ takes its positive maximum at some point x'_0 of $\partial\Omega$. Since the function $H_\alpha\varphi$ is in $W^{2,p}(\Omega)$ and satisfies the conditions

$$\begin{cases} (A-\alpha)\, H_\alpha\varphi = 0 & \text{in } \Omega, \\ \gamma_0\,(H_\alpha\varphi) = \varphi & \text{on } \partial\Omega, \end{cases}$$

by applying the weak maximum principle (Theorem 22.7) with $A := A - \alpha$ to the function $H_\alpha\psi$, we find that the function $H_\alpha\varphi$ takes its positive maximum at a boundary point $x'_0 \in \partial\Omega$. Thus we can apply Hopf's boundary point lemma (Lemma 8.8) to obtain that

$$\frac{\partial}{\partial \mathbf{n}}(H_\alpha\varphi)(x'_0) < 0. \tag{24.16}$$

However, it should be noticed that the coefficients of the boundary condition L satisfy the three conditions:

- $\mu(x') > 0$ on $\partial\Omega$,
- $\gamma(x') < 0$ on $\partial\Omega$,
- $\delta(x') \geq 0$ on $\partial\Omega$.

Hence we have the inequality

$$\begin{aligned} LH_\alpha\varphi(x_0') &= \mu(x_0')\frac{\partial}{\partial \mathbf{n}}(H_\alpha\varphi)(x_0') + \beta(x_0')\cdot\varphi(x_0') \\ &\quad + \gamma(x_0')\varphi(x_0') - \alpha\delta(x_0')\varphi(x_0') \\ &= \mu(x_0')\frac{\partial}{\partial \mathbf{n}}(H_\alpha\varphi)(x_0') + \gamma(x_0')\varphi(x_0') - \alpha\delta(x_0')\varphi(x_0') \\ &< 0. \end{aligned}$$

This verifies condition (β') of Theorem 21.14.

Therefore, Lemma 24.7 follows from an application of Theorem 21.14 (the Hille–Yosida–Ray theorem).

Remark 24.4 The operator $\overline{LH_\alpha}$ enjoys the *positive maximum principle* (see [162, p. 622, assertion (12.45)]):

If a function φ in the domain $D\left(\overline{LH_\alpha}\right)$ takes its *positive* maximum at some point x_0' of $\partial\Omega$, then we have the inequality $\left(\overline{LH_\alpha}\varphi\right)(x_0') \leq 0.$ (24.17)

The next lemma states a fundamental relationship between the operators $\overline{LH_\alpha}$ and $\overline{LH_\beta}$ for $\alpha,\ \beta > 0$ (see Fig. 24.9):

Lemma 24.8 *The domain $D(\overline{LH_\alpha})$ of $\overline{LH_\alpha}$ does not depend on $\alpha > 0$; so we denote by $\mathcal{D}$ the common domain. Then we have, for all $\alpha,\ \beta > 0$,*

$$\overline{LH_\alpha}\varphi - \overline{LH_\beta}\varphi + (\alpha-\beta)\,\overline{LG_\alpha^0}\,(H_\beta\varphi) = 0, \quad \varphi \in \mathcal{D}. \tag{24.18}$$

Proof Let φ be an arbitrary function in $D(\overline{LH_\beta})$, and choose a sequence $\{\varphi_j\}$ in $D(LH_\beta) = B^{2-1/p,p}(\partial\Omega)$ such that

$$\begin{cases} \varphi_j \longrightarrow \varphi & \text{in } C(\partial\Omega), \\ LH_\beta\varphi_j \longrightarrow \overline{LH_\beta}\varphi & \text{in } C(\partial\Omega). \end{cases}$$

Then it follows from the boundedness of H_β and $\overline{LG_\alpha^0}$ that

$$\overline{LG_\alpha^0}(H_\beta\varphi_j) \longrightarrow \overline{LG_\alpha^0}(H_\beta\varphi) \quad \text{in } C(\partial\Omega).$$

Therefore, by using formula (24.8) with $\varphi := \varphi_j$ we obtain that

$$\begin{aligned} LH_\alpha\varphi_j &= LH_\beta\varphi_j - (\alpha-\beta)\,\overline{LG_\alpha^0}(H_\beta\varphi_j) \\ &\longrightarrow \overline{LH_\beta}\varphi - (\alpha-\beta)\,\overline{LG_\alpha^0}(H_\beta\varphi) \quad \text{in } C(\partial\Omega). \end{aligned}$$

Since the operator $\overline{LH_\alpha}$ is closed, it follows that

$$\begin{cases} \varphi \in D(\overline{LH_\alpha}), \\ \overline{LH_\alpha}\varphi = \overline{LH_\beta}\varphi - (\alpha - \beta)\, \overline{LG_\alpha^0}(H_\beta\varphi). \end{cases}$$

This proves the desired Eq. (24.18).

Conversely, we have, by interchanging α and β,

$$D(\overline{LH_\alpha}) \subset D(\overline{LH_\beta}),$$

and so

$$D(\overline{LH_\alpha}) = D(\overline{LH_\beta}).$$

The proof of Lemma 24.8 is complete. □

24.3 General Existence Theorem for Feller Semigroups

Now we can give a general existence theorem for Feller semigroups on $\partial\Omega$ in terms of boundary value problem (24.10). The next theorem asserts that the operator $\overline{LH_\alpha}$ is the infinitesimal generator of some Feller semigroup on $\partial\Omega$ if and only if problem (24.10) is solvable for sufficiently *many* functions φ in the space $C(\partial\Omega)$:

Theorem 24.9 *Let $N < p < \infty$ and $\alpha > 0$. Then we have the following two assertions:*

(i) If the operator $\overline{LH_\alpha}$, $\alpha > 0$, is the infinitesimal generator of a Feller semigroup on $\partial\Omega$, then, for each constant $\lambda > 0$ the boundary value problem

$$\begin{cases} (\alpha - A)\, u = 0 & \text{in } \Omega, \\ (\lambda - L)\, u = \varphi & \text{on } \partial\Omega \end{cases} \tag{24.19}$$

has a solution $u \in W^{2,p}(\Omega)$ for any φ in some dense *subset of $C(\partial\Omega)$.*

(ii) Conversely, if, for some constant $\lambda \geq 0$, problem (24.19) *has a solution $u \in W^{2,p}(\Omega)$, $N < p < \infty$, for any φ in some* dense *subset of $C(\partial\Omega)$, then the operator $\overline{LH_\alpha}$ is the infinitesimal generator of some Feller semigroup on $\partial\Omega$.*

Proof *Assertion (i)*: If the operator $\overline{LH_\alpha}$ generates a Feller semigroup on $\partial\Omega$, by applying part (i) of Theorem 21.14 (the Hille–Yosida–Ray theorem) with $K := \partial\Omega$ to the operator $\overline{LH_\alpha}$ we obtain that

$$R\left(\lambda I - \overline{LH_\alpha}\right) = C(\partial\Omega) \quad \text{for each } \lambda > 0.$$

This implies that the range $R(\lambda I - LH_\alpha)$ is a dense subset of $C(\partial\Omega)$ for each $\lambda > 0$. However, if $\varphi \in C(\partial\Omega)$ is in the range $R(\lambda I - LH_\alpha)$, and if $\varphi = (\lambda I - LH_\alpha)\,\psi$

with $\psi \in B^{2-1/p,p}(\partial\Omega)$, then the function $u = H_\alpha \psi \in W^{2,p}(\Omega)$ is a solution of problem (24.19). This proves the desired assertion (i).

Assertion (ii): We apply part (ii) of Theorem 21.14 (the Hille–Yosida–Ray theorem) with $K := \partial\Omega$ to the operator LH_α. To do this, it suffices to show that the operator LH_α satisfies condition (γ) of the same theorem, since it satisfies condition (β'), as is shown in the proof of Lemma 24.7.

By the uniqueness theorem for problem (24.1) (Theorem 23.2), it follows that every function $u \in W^{2,p}(\Omega)$ which satisfies the equation

$$(\alpha - A)\, u = 0 \quad \text{in } \Omega$$

can be written in the form

$$u = H_\alpha\, (u|_{\partial\Omega}), \quad u|_{\partial\Omega} \in B^{2-1/p,p}(\partial\Omega) = D\,(LH_\alpha)\,.$$

Thus we find that if there exists a solution $u \in W^{2,p}(\Omega)$ of problem (24.19) for a function $\varphi \in C(\partial\Omega)$, then we have the formula

$$(\lambda I - LH_\alpha)\,(u|_{\partial\Omega}) = \varphi,$$

and so

$$\varphi \in R\,(\lambda I - LH_\alpha).$$

Therefore, if there exists a constant $\lambda \geq 0$ such that problem (24.19) has a solution u in $W^{2,p}(\Omega)$ for any φ in some dense subset of $C(\partial\Omega)$, then the range $R\,(\lambda I - LH_\alpha)$ is dense in $C(\partial\Omega)$. This verifies condition (γ) of Theorem 21.14 with $\alpha_0 := \lambda$. Hence the desired assertion (ii) follows from an application of the Hille–Yosida–Ray theorem.

The proof of Theorem 24.9 is complete. □

Remark 24.5 Intuitively, Theorem 24.9 asserts that we can "piece together" a Markov process on the boundary $\partial\Omega$ with A-diffusion in the interior Ω to construct a Markov process on the closure $\overline{\Omega} = \Omega \cup \partial\Omega$. It seems that our method of construction of Feller semigroups is, in spirit, not far away from the probabilistic method of construction of diffusion processes by means of Poisson point processes of *Brownian excursions* used by Watanabe [178]. The situation may be represented schematically by Fig. 24.10.

We conclude this section by giving a precise meaning to the boundary conditions Lu for functions u in the domain $D(\overline{\mathcal{A}})$.

We let

$$\begin{aligned} D(L) &= \{u \in D(\overline{\mathcal{A}}) : u|_{\partial\Omega} \in \mathcal{D}\} \\ &= \{u \in W^{2,p}(\Omega) : Au \in C(\overline{\Omega}),\ u|_{\partial\Omega} \in \mathcal{D}\}, \end{aligned}$$

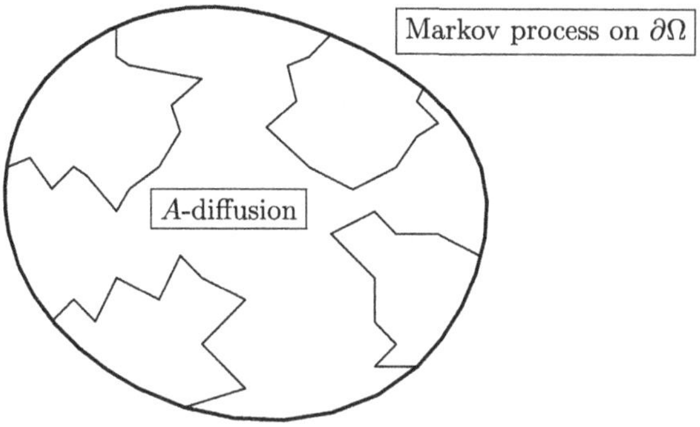

Fig. 24.10 A Markov process on $\partial\Omega$ can be "pieced together" with A-diffusion in Ω

where $\mathcal{D}$ is the common domain of the operators $\overline{LH_\alpha}$, $\alpha > 0$ (see Lemma 24.8). It should be noticed that the domain $D(L)$ contains $W^{2,p}(\Omega)$ for $N < p < \infty$, since $B^{2-1/p,p}(\partial\Omega) = D\left(LH_\alpha\right) \subset \mathcal{D}$. Moreover, Corollary 24.4 asserts that every function u in $D(L) \subset D(\overline{\mathcal{A}})$ can be written in the form

$$u = G_\alpha^0\left((\alpha I - \overline{\mathcal{A}})u\right) + H_\alpha\left(u|_{\partial\Omega}\right) \quad \text{for each } \alpha > 0. \tag{24.14}$$

Then we define the boundary condition Lu by the formula

$$Lu = \overline{LG_\alpha^0}\left((\alpha I - \overline{\mathcal{A}})u\right) + \overline{LH_\alpha}\left(u|_{\partial\Omega}\right). \tag{24.20}$$

The next lemma justifies the definition (24.20) of Lu for $u \in D(L)$:

Lemma 24.10 *The right-hand side of formula* (24.20) *depends only on u, not on the choice of expression* (24.14).

Proof Assume that

$$\begin{aligned} u &= G_\alpha^0\left((\alpha I - \overline{\mathcal{A}})u\right) + H_\alpha\left(u|_{\partial\Omega}\right) \\ &= G_\beta^0\left((\beta I - \overline{\mathcal{A}})u\right) + H_\beta\left(u|_{\partial\Omega}\right), \end{aligned}$$

where $\alpha > 0$, $\beta > 0$. Then it follows from formula (24.15) with $f := (\alpha I - \overline{\mathcal{A}})u$ and formula (24.18) with $\psi := u|_{\partial\Omega}$ that

$$\begin{aligned} &\overline{LG_\alpha^0}\left(\left(\alpha I - \overline{\mathcal{A}}\right)u\right) + \overline{LH_\alpha}\left(u|_{\partial\Omega}\right) \\ &= \overline{LG_\beta^0}\left(\left(\alpha I - \overline{\mathcal{A}}\right)u\right) - (\alpha - \beta)\,\overline{LG_\alpha^0}G_\beta^0\left(\left(\alpha I - \overline{\mathcal{A}}\right)u\right) \\ &\quad + \overline{LH_\beta}\left(u|_{\partial\Omega}\right) - (\alpha - \beta)\,\overline{LG_\alpha^0}H_\beta\left(u|_{\partial\Omega}\right) \\ &= \overline{LG_\beta^0}\left((\beta I - A)u\right) + \overline{LH_\beta}\left(u|_{\partial\Omega}\right) \\ &\quad + (\alpha - \beta)\left\{\overline{LG_\beta^0}u - \overline{LG_\alpha^0}G_\beta^0\left(\alpha I - \overline{\mathcal{A}}\right)u - \overline{LG_\alpha^0}H_\beta(u|_{\partial\Omega})\right\}. \end{aligned} \tag{24.21}$$

However, the last term of formula (24.21) vanishes. Indeed, it follows from formula (24.14) with $\alpha := \beta$ and formula (24.15) with $f := u$ that

$$\begin{aligned}
&\overline{LG_{\beta}^{0}}u - \overline{LG_{\alpha}^{0}}\left(G_{\beta}^{0}\left(\alpha I - \overline{\mathcal{A}}\right)u\right) - \overline{LG_{\alpha}^{0}}H_{\beta}\left(u|_{\partial\Omega}\right) \\
&= \overline{LG_{\beta}^{0}}u - \overline{LG_{\alpha}^{0}}\left(G_{\beta}^{0}(\beta I - \overline{\mathcal{A}})u + H_{\beta}\left(u|_{\partial\Omega}\right) + (\alpha - \beta)\, G_{\beta}^{0}u\right) \\
&= \overline{LG_{\beta}^{0}}u - \overline{LG_{\alpha}^{0}}u - (\alpha - \beta)\overline{LG_{\alpha}^{0}}G_{\beta}^{0}u \\
&= 0.
\end{aligned}$$

Therefore, we obtain from formula (24.21) that

$$\overline{LG_{\alpha}^{0}}\left((\alpha I - \overline{\mathcal{A}})u\right) + \overline{LH_{\alpha}}\left(u|_{\partial\Omega}\right) = \overline{LG_{\beta}^{0}}\left((\beta I - \overline{\mathcal{A}})u\right) + \overline{LH_{\beta}}\left(u|_{\partial\Omega}\right).$$

This proves Lemma 24.10.

24.4 Proof of Remark 24.1

Finally, we prove that the domain

$$D(\overline{\mathcal{A}}) = \left\{u \in W^{2,p}(\Omega) : Au \in C(\overline{\Omega})\right\}$$

is *independent* of $N < p < \infty$.

We let

$$\mathcal{D}_p := \left\{u \in W^{2,p}(\Omega) : Au \in C(\overline{\Omega})\right\}.$$

In order to prove Remark 24.1, it suffices to show that

$$\mathcal{D}_{p_1} = \mathcal{D}_{p_2} \quad \text{for } N < p_1 < p_2 < \infty.$$

First, it follows that

$$\mathcal{D}_{p_2} \subset \mathcal{D}_{p_1},$$

since we have $L^{p_2}(\Omega) \subset L^{p_1}(\Omega)$ for $p_2 > p_1$.

Conversely, let v be an arbitrary element of $\mathcal{D}_{p_1}$:

$$v \in W^{2,p_1}(\Omega), \quad Av \in C(\overline{\Omega}), \ \ Lu = 0 \text{ on } \partial\Omega.$$

Then, since we have v, $Av \in C(\overline{\Omega}) \subset L^{p_2}(\Omega)$, it follows from an application of Theorem 23.2 with $p := p_2$ that

$$G_{\alpha}^{0}\left((\alpha - A)\, v\right) \in W^{2,p_2}(\Omega).$$

Moreover, we can find a sequence $\{\varphi_j\}$ in $C^2(\partial\Omega)$ such that

$$\varphi_j \longrightarrow v|_{\partial\Omega} \quad \text{in } C(\partial\Omega).$$

Then we have the assertions

$$\begin{aligned} &H_\alpha \varphi_j \in W^{2,p_2}(\Omega), \\ &H_\alpha \varphi_j \longrightarrow H_\alpha(v|_{\partial\Omega}) \quad \text{in } C(\overline{\Omega}), \\ &A(H_\alpha \varphi_j) = \alpha H_\alpha \varphi_j \longrightarrow \alpha H_\alpha(v|_{\partial\Omega}) \quad \text{in } C(\overline{\Omega}). \end{aligned}$$

However, since $\overline{\mathcal{A}}\colon \mathcal{D}_{p_2} \to C(\overline{\Omega})$ is closed, it follows that

$$H_\alpha\left(v|_{\partial\Omega}\right) \in \mathcal{D}_{p_2} \subset W^{2,p_2}(\Omega).$$

Therefore, by applying Corollary 24.4 with $u := v$ we obtain that

$$v = G_\alpha^0\left((\alpha - A)\, v\right) + H_\alpha\left(v|_{\partial\Omega}\right) \in W^{2,p_2}(\Omega).$$

This implies that $v \in \mathcal{D}_{p_2}$.

The proof of Remark 24.1 is complete. □

24.5 Notes and Comments

The general existence theorem for Feller semigroups in terms of boundary value problems (Theorem 24.9) is generalized to Feller semigroups with *jump phenomena* by [163, Theorem 9.2].

Chapter 25
Feller Semigroups with a First Order Ventcel' Boundary Condition

Chapter 25 is devoted to the proof of Theorem 1.1. Namely, we prove that the closed realization $\mathfrak{A}_L$ is the infinitesimal generator of a Feller semigroup on $C(\overline{\Omega})$ corresponding to such a diffusion phenomenon that a Markovian particle moves by continuous paths in the state space, with absorption, reflection, drift and sticking phenomena at the boundary. Moreover, we prove Remark 1.2, that is, the domain $D(\mathfrak{A}_L)$ does not depend on p, for $N < p < \infty$.

An intuitive meaning of Theorem 1.1 can be visualized as in Fig. 25.1.

The crucial point in the proof is that we consider the term $\delta(x')\,(Au|_{\partial\Omega})$ of sticking phenomenon in the Ventcel' (Wentzell) boundary condition (1.2)

$$Lu(x') = L_0 u(x') - \delta(x')\left(Au(x')\right) \quad \text{on } \partial\Omega$$

as a term of "perturbation" of the oblique derivative boundary condition (1.3) (or the boundary condition (23.1))

$$L_0 u(x') := \mu(x')\frac{\partial u}{\partial \mathbf{n}}(x') + \beta(x')\cdot u(x') + \gamma(x')u(x') \quad \text{on } \partial\Omega.$$

More precisely, we make use of a generation theorem for Feller semigroups with oblique derivative boundary condition L_0 to verify all the conditions of a version of the Hille–Yosida theorem (Theorem 21.14) adapted to the present context for the operator $\mathfrak{A}_L$ defined by formula (1.3).

Let Ω be a bounded domain in Euclidean space $\mathbf{R}^N$, $N \geq 3$, with boundary $\partial\Omega$ of class $C^{1,1}$. Let A be a second order, uniformly elliptic differential operator with real discontinuous coefficients of the form

$$Au(x) := \sum_{i,j=1}^{N} a^{ij}(x)\frac{\partial^2 u}{\partial x_i \partial x_j}(x) + \sum_{i=1}^{N} b^i(x)\frac{\partial u}{\partial x_i}(x) + c(x)u(x). \tag{1.1}$$

K. Taira, *Real Analysis Methods for Markov Processes*,
https://doi.org/10.1007/978-981-97-3659-1_25

Fig. 25.1 A Markovian particle moves by continuous paths in the state space with absorption, reflection, drift and sticking phenomena at the boundary

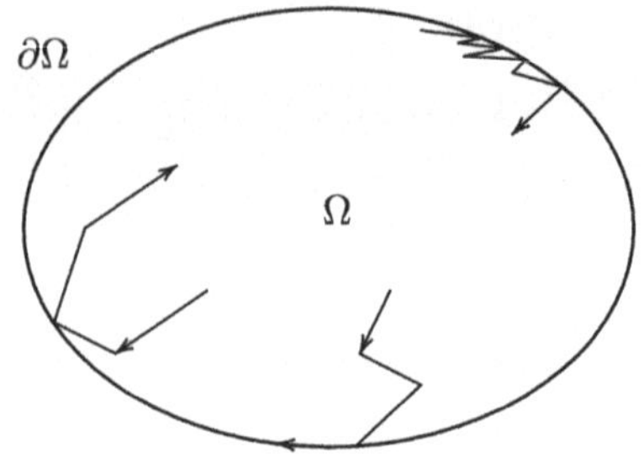

Here the coefficients $a^{ij}(x)$, $b^i(x)$ and $c(x)$ satisfy the following three conditions:

(1) The $a^{ij}(x)$ are the components of a symmetric contravariant tensor of type $\binom{2}{0}$ in $\mathbf{R}^n$ and $a^{ij} \in \mathrm{VMO} \cap L^\infty(\Omega)$, and there exists a constant $\lambda > 0$ such that (the uniform ellipticity condition)

$$\frac{1}{\lambda}|\xi|^2 \leq \sum_{i,j=1}^{N} a^{ij}(x)\xi_i\xi_j \leq \lambda|\xi|^2 \quad \text{for almost all } x \in \Omega \text{ and all } \xi \in \mathbf{R}^N.$$

(2) $b^i \in L^\infty(\Omega)$ for $1 \leq i \leq N$.
(3) $c \in L^\infty(\Omega)$ and $c(x) \leq 0$ for almost all $x \in \Omega$.

Let L be a first order, Ventcel' (Wentzell) boundary condition of the form

$$\begin{aligned} Lu(x') &:= \mu(x')\frac{\partial u}{\partial \mathbf{n}}(x') + \beta(x') \cdot u(x') + \gamma(x')u(x') - \delta(x')\left(Au(x')\right) \\ &= L_0 u(x') - \delta(x')\left(Au(x')\right) \quad \text{on} \partial\Omega. \end{aligned} \tag{1.2}$$

Here the functions $\mu(x')$, $\beta(x')$, $\gamma(x')$ and $\delta(x')$ satisfy the following four conditions (see Fig. 23.1):

(4) $\mu(x')$ is a Lipschitz continuous function on $\partial\Omega$ and $\mu(x') \geq 0$ on $\partial\Omega$.
(5) $\beta(x')$ is a Lipschitz continuous vector field on $\partial\Omega$.
(6) $\gamma(x')$ is a Lipschitz continuous function on $\partial\Omega$ and $\gamma(x') \leq 0$ on $\partial\Omega$.
(7) $\delta(x')$ is a Lipschitz continuous function on $\partial\Omega$ and $\delta(x') \geq 0$ on $\partial\Omega$.
(8) $\mathbf{n} = (n_1, n_2, \ldots, n_N)$ is the unit inward normal to the boundary $\partial\Omega$.

25.1 Proof of Theorem 1.1

This section is devoted to the proof of Theorem 1.1. We shall apply part (ii) of Theorem 21.14 (the Hille–Yosida–Ray theorem) to the operator $\mathfrak{A}_L$ defined by formula (1.3). The proof is divided into eight steps.

Step 1: First, we prove that:

> The operator $\overline{L_0 H_\alpha}$ is the generator of some Feller semigroup on $\partial\Omega$, for any sufficiently large $\alpha > 0$.

To do this, we apply Theorem 24.9 with $L := L_0$.

By applying Theorem 23.2, we obtain that the oblique derivative problem

$$\begin{cases} (A - \alpha)\, u = 0 & \text{in } \Omega, \\ L_0 u = \varphi & \text{on } \partial\Omega \end{cases}$$

has a unique function $u \in W^{2,p}(\Omega)$ for any function $\varphi \in B^{1-1/p,p}(\partial\Omega)$, if $N < p < \infty$. Here it should be noticed that we have, by an imbedding theorem for Besov spaces (see [2, Theorem 7.34]),

$$C^1(\partial\Omega) \subset B^{1-1/p,p}(\partial\Omega) \subset C(\partial\Omega),$$

since $(1 - 1/p)p = p - 1 > N - 1$ for $N < p < \infty$.

Therefore, it follows that, for any function $\varphi \in B^{1-1/p,p}(\partial\Omega)$ there exists a unique function $\psi \in B^{2-1/p,p}(\partial\Omega)$ such that $L_0(H_\alpha \varphi) = \psi$. This implies that the range $R(L_0 H_\alpha)$ is a *dense* subset of $C(\partial\Omega)$. Hence, by applying part (ii) of Theorem 24.9 with $\lambda := 0$ we obtain that the operator $\overline{L_0 H_\alpha}$ generates a Feller semigroup on $\partial\Omega$, for any $\alpha > 0$.

Step 2: Next, we prove that:

> The operator $\overline{L H_\alpha}$ generates a Feller semigroup on $\partial\Omega$, for any $\alpha > 0$.

To do this, we apply Corollary 21.15 (Sato–Ueno) with $K := \partial\Omega$ to the operator $\overline{L H_\alpha}$ for $\alpha > 0$.

By formula (24.13), it follows that the operator $\overline{L H_\alpha}$ can be written as

$$\overline{L H_\alpha} = \overline{L_0 H_\alpha} - \alpha\, \delta(x') := \overline{L_0 H_\alpha} + M,$$

where

$$M := -\alpha\, \delta(x')$$

is a bounded linear operator on $C(\partial\Omega)$ into itself. However, we find that the operator M satisfies condition (β') of Theorem 21.14 (the Hille–Yosida–Ray theorem):

(β') If $\psi \in C(\partial\Omega)$ takes a positive maximum at a point x_0' of $\partial\Omega$, then we have the assertion

$$M\psi(x_0') = -\alpha\delta(x_0')\psi(x_0') \leq 0.$$

Therefore, it follows from an application of Corollary 21.15 (Sato–Ueno) with $\mathfrak{A} := \overline{L_0 H_\alpha}$ and $M := -\alpha\delta(x')$ that the operator $\overline{LH_\alpha}$ also generates a Feller semigroup on $\partial\Omega$.

Step 3: Now we prove that:

The equation (25.1)

$$\overline{LH_\alpha}\,\psi = \varphi$$

has a unique solution ψin $D(\overline{LH_\alpha})$for any $\varphi \in C(\partial\Omega)$; hence the inverse $\overline{LH_\alpha}^{-1}$of $\overline{LH_\alpha}$ can be defined on the whole space $C(\partial\Omega)$. Furthermore, the operator $-\overline{LH_\alpha}^{-1}$ is non-negative and bounded on $C(\partial\Omega)$.

Since the function $H_\alpha 1$ takes its positive maximum 1 only on the boundary $\partial\Omega$, we can apply Hopf's boundary point lemma (Lemma 8.8) to obtain that

$$\frac{\partial}{\partial \mathbf{n}}(H_\alpha 1) < 0 \quad \text{on } \partial\Omega. \tag{25.2}$$

However, it should be noticed that the coefficients of the boundary condition L satisfy the three conditions:

- $\mu(x') > 0 \quad \text{on } \partial\Omega,$
- $\gamma(x') < 0 \quad \text{on } \partial\Omega,$
- $\delta(x') \geq 0 \quad \text{on } \partial\Omega.$

Hence it follows from inequality (25.2) that

$$\begin{aligned} LH_\alpha 1(x') &= \mu(x')\frac{\partial}{\partial \mathbf{n}}(H_\alpha 1)(x') + \gamma(x') - \alpha\delta(x') \\ &< 0 \quad \text{on } \partial\Omega, \end{aligned}$$

so that

$$\ell_\alpha = -\sup_{x' \in \partial\Omega} LH_\alpha 1(x') > 0.$$

Furthermore, by using Corollary 21.13 (Sato–Ueno) with

$$K := \partial\Omega, \quad \mathfrak{A} := \overline{LH_\alpha}, \quad c := \ell_\alpha,$$

we obtain that the operator $\overline{LH_\alpha} + \ell_\alpha I$ is the infinitesimal generator of some Feller semigroup on $\partial\Omega$. Therefore, since $\ell_\alpha > 0$, it follows from an application of part (i) of Theorem 21.12 (the Hille–Yosida theorem) with $\mathfrak{A} := \overline{LH_\alpha} + \ell_\alpha I$ that the equation

$$-\overline{LH_\alpha}\,\psi = \left(\ell_\alpha I - (\overline{LH_\alpha} + \ell_\alpha I)\right)\psi = \varphi$$

has a unique solution $\psi \in D(\overline{LH_\alpha})$ for any $\varphi \in C(\partial\Omega)$, and further that the operator $-\overline{LH_\alpha}^{-1} = \left(\ell_\alpha I - (\overline{LH_\alpha} + \ell_\alpha I)\right)^{-1}$ is non-negative and bounded on the space $C(\partial\Omega)$ with norm

$$\begin{aligned}\left\|-\overline{LH_\alpha}^{-1}\right\| &= \left\|\left(\ell_\alpha I - (\overline{LH_\alpha} + \ell_\alpha I)\right)^{-1}\right\| \\ &\le \frac{1}{\ell_\alpha}.\end{aligned}$$

Step 4: By assertion (25.1), we can define the *Green operator* G_α^L, $\alpha > 0$, by the formula

$$G_\alpha^L f = G_\alpha^0 f - H_\alpha\left(\overline{LH_\alpha}^{-1}\left(\overline{LG_\alpha^0 f}\right)\right) \quad \text{for } f \in C(\overline{\Omega}). \tag{25.3}$$

We prove that

$$G_\alpha^L = (\alpha I - \mathfrak{A}_L)^{-1} \quad \text{for } \alpha > 0, \tag{25.4}$$

where $\mathfrak{A}_L$ is a linear operator from $C(\overline{\Omega})$ into itself defined as follows (see formula (1.3)):

(a) The domain $D(\mathfrak{A}_L)$ is the set

$$D(\mathfrak{A}_L) = \left\{u \in W^{2,p}(\Omega) : u \in D(\overline{A}),\ u|_{\partial\Omega} \in \mathcal{D},\ Lu = 0 \text{ on } \partial\Omega\right\}. \tag{25.5}$$

(b) $\mathfrak{A}_L u = Au$ for all $u \in D(\mathfrak{A}_L)$.

Here $\mathcal{D}$ is the common domain of the operators $\overline{LH_\alpha}$, $\alpha > 0$ (see Lemma 24.8).

In view of Lemmas 24.3 and 24.8, it follows that we have, for any $f \in C(\overline{\Omega})$,

$$\begin{cases} G_\alpha^L f = G_\alpha^0 f - H_\alpha\left(\overline{LH_\alpha}^{-1}\left(\overline{LG_\alpha^0 f}\right)\right) \in D(\overline{A}), \\ G_\alpha^L f|_{\partial\Omega} = -\overline{LH_\alpha}^{-1}\left(\overline{LG_\alpha^0 f}\right) \in D\left(\overline{LH_\alpha}\right) = \mathcal{D}, \\ L(G_\alpha^L f) = \overline{LG_\alpha^0} f - \overline{LH_\alpha}\left(\overline{LH_\alpha}^{-1}\left(\overline{LG_\alpha^0 f}\right)\right) = 0, \end{cases}$$

and also

$$\left(\alpha I - \overline{A}\right)\left(G_\alpha^L f\right) = f.$$

This proves that

$$\begin{cases} G_\alpha^L f \in D(\mathfrak{A}_L), \\ (\alpha I - \mathfrak{A}_L)\left(G_\alpha^L f\right) = f, \end{cases}$$

that is,

$$(\alpha I - \mathfrak{A}_L)\, G_\alpha^L = I \quad \text{on } C(\overline{\Omega}).$$

Therefore, in order to prove formula (25.4) it suffices to show the injectivity of the operator $\alpha I - \mathfrak{A}_L$ for $\alpha > 0$.

Assume that

$$\begin{cases} u \in D(\mathfrak{A}_L), \\ (\alpha I - \mathfrak{A}_L)\, u = 0. \end{cases}$$

Then, by Corollary 24.4 it follows that the function u can be written as

$$\begin{cases} u = H_\alpha\,(u|_{\partial\Omega}), \\ u|_{\partial\Omega} \in \mathcal{D} = D\left(\overline{LH_\alpha}\right). \end{cases}$$

Thus we have the assertion

$$\overline{LH_\alpha}\,(u|_{\partial\Omega}) = Lu = 0.$$

In view of assertion (25.1), this implies that

$$u|_{\partial\Omega} = 0,$$

so that

$$u = H_\alpha\,(u|_{\partial\Omega}) = 0 \quad \text{in } \Omega.$$

Step 5: The *non-negativity* of G_α^L for $\alpha > 0$ follows immediately from formula (25.3), since the operators G_α^0, H_α, $-\overline{LH_\alpha}^{-1}$ and $\overline{LG_\alpha^0}$ are all non-negative.

Step 6: We prove that the operator G_α^L is *bounded* on the space $C(\overline{\Omega})$ with norm

$$\left\|G_\alpha^L\right\| \leq \frac{1}{\alpha} \quad \text{for } \alpha > 0. \tag{25.6}$$

To do this, it suffices to show that

$$G_\alpha^L 1 \leq \frac{1}{\alpha} \quad \text{on } \overline{\Omega}. \tag{25.7}$$

since G_α is non-negative on $C(\overline{\Omega})$.

First, it follows from the uniqueness property of solutions of problem (25.1) (Theorem 23.2) that

$$\alpha\left(G_\alpha^0 1\right)(x) + (H_\alpha 1)(x) = 1 + \left(G_\alpha^0 c\right)(x) \quad \text{on } \overline{\Omega}. \tag{25.8}$$

Indeed, it suffices to note that both sides of formula (25.8) have the same boundary value 1 and satisfy the same equation: $(\alpha - A)\, u = \alpha$ in Ω.

Applying the operator L to both sides of formula (25.8), we obtain that

$$\begin{aligned}
-L(H_\alpha 1) &= -L1 - L\left(G_\alpha^0 c\right) + \alpha L\left(G_\alpha^0 1\right) \\
&= -(\gamma(x') - \delta(x')c(x')) - \left(\mu(x')\frac{\partial}{\partial \mathbf{n}}\left(G_\alpha^0 c\right) + \delta(x')c(x')\right) \\
&\quad + \alpha L\left(G_\alpha^0 1\right) \\
&= -\gamma(x') - \mu(x')\frac{\partial}{\partial \mathbf{n}}\left(G_\alpha^0 c\right) + \alpha L\left(G_\alpha^0 1\right) \\
&\geq \alpha L\left(G_\alpha^0 1\right) \quad \text{on } \partial\Omega,
\end{aligned}$$

since $G_\alpha^0 c\big|_{\partial\Omega} = 0$ and $G_\alpha^0 c \leq 0$ on $\overline{\Omega}$. Hence we have, by the non-negativity of $-\overline{LH_\alpha}^{-1}$,

$$-\overline{LH_\alpha}^{-1}\left(L(G_\alpha^0 1)\right) \leq \frac{1}{\alpha} \quad \text{on } \partial\Omega. \tag{25.9}$$

By using formula (25.3) with $f := 1$, inequality (25.9) and formula (25.8), we obtain that

$$\begin{aligned}
G_\alpha^L 1 &= G_\alpha^0 1 + H_\alpha\left(-\overline{LH_\alpha}^{-1}\left(L(G_\alpha^0 1)\right)\right) \\
&\leq G_\alpha^0 1 + \frac{1}{\alpha} H_\alpha 1 \\
&= \frac{1}{\alpha} + \frac{1}{\alpha}(G_\alpha^0 c)(x) \\
&\leq \frac{1}{\alpha} \quad \text{on } \overline{\Omega},
\end{aligned}$$

since both the operators H_α and G_α^0 are non-negative.

Step 7: Finally, we prove that

$$\text{The domain } D\left(\mathfrak{A}_L\right) \text{ is } \textit{dense} \text{ in the space } C(\overline{\Omega}). \tag{25.10}$$

Step 7-1: Before the proof, we need some lemmas on the behavior of G_α^0, H_α and $-\overline{LH_\alpha}^{-1}$ as $\alpha \to +\infty$:

Lemma 25.1 *For all $f \in C(\overline{\Omega})$, we have the assertion*

$$\lim_{\alpha \to +\infty}\left[\alpha G_\alpha^0 f + H_\alpha\left(f|_{\partial\Omega}\right)\right] = f \quad \text{in } C(\overline{\Omega}). \tag{25.11}$$

Proof Choose a constant $\beta > 0$ and let

$$g = f - H_\beta(f|_{\partial\Omega}).$$

Then, by using formula (25.8) with $\varphi := f|_{\partial\Omega}$ we obtain that

$$\alpha G_\alpha^0 g - g = \left[\alpha G_\alpha^0 f + H_\alpha(f|_{\partial\Omega}) - f\right] - \beta G_\alpha^0 H_\beta(f|_{\partial\Omega}). \tag{25.12}$$

However, we have, by estimate (25.4),

$$\lim_{\alpha\to+\infty} G_\alpha^0 H_\beta\,(f|_{\partial\Omega}) = 0 \quad \text{in } C(\overline{\Omega}),$$

and, by assertion (25.7),

$$\lim_{\alpha\to+\infty} \alpha G_\alpha^0 g = g \quad \text{in } C(\overline{\Omega}),$$

since $g|_{\partial\Omega} = 0$. Therefore, the desired assertion (25.11) follows by letting $\alpha \to +\infty$ in formula (25.12).

The proof of Lemma 25.1 is complete.

Lemma 25.2 *The function*

$$\frac{\partial}{\partial \mathbf{n}}\,(H_\alpha 1)\,(x') \quad \textit{for } x' \in \partial\Omega,$$

diverges to $-\infty$ *uniformly and monotonically as* $\alpha \to +\infty$.

Proof First, formula (25.8) with $\varphi := 1$ gives that

$$H_\alpha 1 = H_\beta 1 - (\alpha - \beta)\,G_\alpha^0 H_\beta 1.$$

Thus, in view of the non-negativity of G_α^0 and H_α it follows that

$$\alpha \geq \beta \implies H_\alpha 1 \leq H_\beta 1 \quad \text{on } \overline{\Omega}.$$

Since $H_\alpha 1|_{\partial\Omega} = H_\beta 1|_{\partial\Omega} = 1$, this implies that the functions

$$\frac{\partial}{\partial \mathbf{n}}(H_\alpha 1)(x') \quad \text{for } x' \in \partial\Omega,$$

are monotonically non-increasing in α. Furthermore, by using formula (25.6) with $f := H_\beta 1$ we find that the function

$$H_\alpha 1(x) = H_\beta 1(x) - \left(1 - \frac{\beta}{\alpha}\right)\alpha G_\alpha^0 H_\beta 1(x)$$

converges to zero monotonically as $\alpha \to +\infty$, for each interior point x of Ω.

Now, for any given constant $K > 0$ we can construct a function $u \in W^{2,p}(\Omega)$ such that

$$u = 1 \quad \text{on } \partial\Omega, \tag{25.13a}$$

$$\frac{\partial u}{\partial \mathbf{n}} \leq -K \quad \text{on } \partial\Omega. \tag{25.13b}$$

Indeed, by applying Theorem 23.2 to our situation we obtain that, for any integer $m > 0$ the function

$$u = (H_1 1)^m$$

belongs to $W^{2,p}(\Omega)$ for $N < p < \infty$ and satisfies condition (25.13a), since we have the formula

$$u = (H_1 1)^m = 1 \quad \text{on } \partial\Omega,$$

and also the assertions

- $u = (H_1 1)^m \in L^p(\Omega)$,
- $\dfrac{\partial u}{\partial x_i} = m(H_1 1)^{m-1}(H_1 1)_{x_i} \in L^p(\Omega) \quad \text{for } 1 \leq i \leq N$,
- $\dfrac{\partial^2 u}{\partial x_i \partial x_j} = m(H_\alpha 1)^{m-1}(H_1 1)_{x_i x_j}$
  $+ m(m-1)(H_\alpha 1)^{m-2}(H_1 1)_{x_i}(H_1 1)_{x_j} \in L^p(\Omega) \quad \text{for } 1 \leq i, j \leq N.$

Moreover, we obtain that

$$\frac{\partial u}{\partial \mathbf{n}} = m \frac{\partial}{\partial \mathbf{n}} (H_1 1) \leq m \sup_{x' \in \partial\Omega} \frac{\partial}{\partial \mathbf{n}} (H_1 1)(x').$$

In view of inequality (25.16) with $\varphi := 1$, this implies that the function $u = (H_1 1)^m$ satisfies condition (25.13b) for m sufficiently large.

Finally, it is easy to verify that

$$Au = A\left((H_1 1)^m\right) \in L^\infty(\Omega). \tag{25.13c}$$

Indeed, since we have the assertions

$$(A - 1)\, H_1 1 = 0 \quad \text{in } \Omega,$$
$$H_1 1 \in C^1(\overline{\Omega}),$$

it suffices to note that

$$
\begin{aligned}
Au &= m\,(H_1 1)^{m-1} \sum_{i,j=1}^{N} a^{ij}(x)(H_1 1)_{x_i x_j} \\
&\quad + m(m-1)\,(H_1 1)^{m-2} \sum_{i,j=1}^{N} a^{ij}(x)\,(H_1 1)_{x_i}\,(H_1 1)_{x_j} \\
&\quad + m\,(H_1 1)^{m-1} \sum_{i=1}^{N} b^{i}(x)\,(H_1 1)_{x_i} + c(x)\,(H_1 1)^{m} \\
&= m\,(H_1 1)^{m-1}\,(-(H_1 1) - c(x)\,(H_1 1)) \\
&\quad + m(m-1)\,(H_1 1)^{m-2} \sum_{i,j=1}^{N} a^{ij}(x)\,(H_1 1)_{x_i}\,(H_1 1)_{x_j} \\
&= -m\,(H_1 1)^{m} - mc(x)\,(H_1 1)^{m} + c(x)\,(H_1 1)^{m} \\
&\quad + m(m-1)\,(H_1 1)^{m-2} \sum_{i,j=1}^{N} a^{ij}(x)\,(H_1 1)_{x_i}\,(H_1 1)_{x_j} \in L^{\infty}(\Omega).
\end{aligned}
$$

We take a function $u \in W^{2,p}(\Omega)$ which satisfies conditions (25.13a), (25.13b) and (25.13c), and choose a neighborhood U of $\partial\Omega$, relative to the closure $\overline{\Omega}$, with smooth boundary ∂U such that (see Fig. 25.2)

$$
u \geq \frac{1}{2} \quad \text{on } U. \tag{25.14}
$$

We recall that the function $H_\alpha 1$ converges to zero in Ω monotonically as $\alpha \to +\infty$. Since we have the formula

$$
u|_{\partial\Omega} = H_\alpha 1|_{\partial\Omega} = 1,
$$

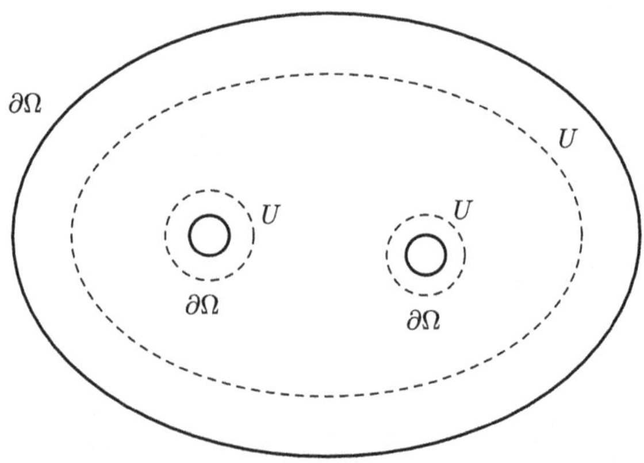

Fig. 25.2 A neighborhood U of $\partial\Omega$, relative to $\overline{\Omega}$, with smooth boundary ∂U

by using *Dini's theorem* we can find a constant $\alpha > 1$ (depending on u and hence on K) such that

$$H_\alpha 1 \leq u \quad \text{on } \partial U \setminus \partial \Omega, \tag{25.15a}$$
$$\alpha > 2\|Au\|_{L^\infty(\Omega)}. \tag{25.15b}$$

It follows from inequalities (25.14) and (25.15b) that

$$\begin{aligned}(A-\alpha)(H_\alpha 1 - u) &= \alpha u - Au \\ &\geq \frac{\alpha}{2} - \|Au\|_{L^\infty(\Omega)} \\ &> 0 \quad \text{in } U.\end{aligned}$$

Thus, by applying the weak maximum principle (Theorem 22.7) with $A := A - \alpha$ to the function $H_\alpha 1 - u$ we obtain that the function $H_\alpha 1 - u$ may take its positive maximum only on the boundary ∂U. However, conditions (25.13a) and (25.15a) imply that

$$H_\alpha 1 - u \leq 0 \quad \text{on } \partial U = (\partial U \setminus \partial \Omega) \cup \partial \Omega.$$

Therefore, we have the inequality

$$H_\alpha 1 \leq u \quad \text{on } \overline{U} = U \cup \partial U,$$

and so

$$\frac{\partial}{\partial \mathbf{n}}(H_\alpha 1) \leq \frac{\partial u}{\partial \mathbf{n}} \leq -K \quad \text{on } \partial \Omega,$$

since $u|_{\partial\Omega} = H_\alpha 1|_{\partial\Omega} = 1$ on $\partial\Omega$.

The proof of Lemma 25.2 is complete.

Corollary 25.3 $\lim_{\alpha\to+\infty} \| - \overline{LH_\alpha}^{-1} \| = 0.$

Proof Since $\mu(x') > 0$ on $\partial\Omega$, it follows from an application of Lemma 25.2 that the function

$$LH_\alpha 1(x') = \mu(x')\frac{\partial}{\partial \mathbf{n}}(H_\alpha 1)(x') + \gamma(x') - \alpha\delta(x') \quad \text{for } x' \in \partial\Omega,$$

diverges to $-\infty$ monotonically as $\alpha \to +\infty$. By Dini's theorem, this convergence is uniform in $x' \in \partial\Omega$. Hence we obtain that the function

$$\frac{1}{LH_\alpha 1(x')}$$

converges to zero uniformly in $x' \in \partial\Omega$ as $\alpha \to +\infty$. This implies that

$$\begin{aligned}\left\| -\overline{LH_\alpha}^{-1} \right\| &= \left\| -\overline{LH_\alpha}^{-1} 1 \right\|_{C(\partial\Omega)} \\ &\leq \left\| \frac{1}{LH_\alpha 1} \right\|_{C(\partial\Omega)} \longrightarrow 0 \quad \text{as } \alpha \to +\infty,\end{aligned}$$

since we have the assertion

$$1 = \frac{-LH_\alpha 1(x')}{|LH_\alpha 1(x')|} \leq \left\| \frac{1}{LH_\alpha 1} \right\|_{C(\partial\Omega)} \left(-LH_\alpha 1(x') \right) \quad \text{for } x' \in \partial\Omega.$$

The proof of Corollary 25.3 is complete.

Step 7-2: Proof of Assertion (25.10)
In view of formula (25.4) and inequality (25.6), it suffices to prove that

$$\lim_{\alpha \to +\infty} \|\alpha\, G_\alpha^L f - f\|_{C(\overline{\Omega})} = 0 \quad \text{for } f \in C^2(\overline{\Omega}), \tag{25.16}$$

since the space $C^2(\overline{\Omega})$ is dense in $C(\overline{\Omega})$.
First, we remark that

$$\begin{aligned}\left\|\alpha G_\alpha^L f - f\right\|_{C(\overline{\Omega})} &= \left\|\alpha G_\alpha^0 f - \alpha H_\alpha \left(\overline{LH_\alpha}^{-1}\left(\overline{LG_\alpha^0} f\right)\right) - f\right\|_{C(\overline{\Omega})} \\ &\leq \left\|\alpha G_\alpha^0 f + H_\alpha\left(f|_{\partial\Omega}\right) - f\right\|_{C(\overline{\Omega})} \\ &\quad + \left\|-\alpha H_\alpha\left(\overline{LH_\alpha}^{-1}\left(\overline{LG_\alpha^0} f\right)\right) - H_\alpha(f|_{\partial\Omega})\right\|_{C(\overline{\Omega})} \\ &\leq \left\|\alpha G_\alpha^0 f + H_\alpha(f|_{\partial\Omega}) - f\right\|_{C(\overline{\Omega})} \\ &\quad + \left\|-\alpha\overline{LH_\alpha}^{-1}\left(\overline{LG_\alpha^0} f\right) - f|_{\partial\Omega}\right\|_{C(\partial\Omega)}.\end{aligned}$$

Thus, in view of assertion (25.11) it suffices to show that

$$\lim_{\alpha \to +\infty} \left[-\alpha\overline{LH_\alpha}^{-1}\left(\overline{LG_\alpha^0} f\right) - f|_{\partial\Omega}\right] = 0 \quad \text{in } C(\partial\Omega). \tag{25.17}$$

Take a constant β such that $0 < \beta < \alpha$, and write

$$f = G_\beta^0 g + H_\beta \varphi,$$

where (cf. formula (25.14)):

$$\begin{cases} g = (\beta - A)\, f \in L^\infty(\Omega), \\ \varphi = f|_{\partial\Omega} \in C^2(\partial\Omega). \end{cases}$$

Then we have the assertion

$$L\left(G^0_\beta g\right) = L\left(G^0_\beta (\beta - A) f\right) = Lf \in C(\partial\Omega),$$

and, by Lemma 24.6 and Remark 24.3,

$$\begin{aligned} L\left(G^0_\alpha g\right) &= L\left(G^0_\beta g\right) - (\alpha - \beta)\,\overline{LG^0_\alpha}(G^0_\beta g) \\ &= Lf - (\alpha - \beta)\,\overline{LG^0_\alpha}(Lf) \in C(\partial\Omega). \end{aligned}$$

Moreover, by using the resolvent Eq. (25.5) with $f := g \in L^\infty(\Omega)$ and the Eq. (25.8) we obtain that

$$\begin{aligned} G^0_\alpha f &= G^0_\alpha\left(G^0_\beta g\right) + G^0_\alpha H_\beta \varphi \\ &= \frac{1}{\alpha - \beta}\left(G^0_\beta g - G^0_\alpha g + H_\beta\varphi - H_\alpha\varphi\right). \end{aligned}$$

Therefore, it follows that

$$\begin{aligned} &\left\| -\alpha \overline{LH_\alpha}^{-1}\left(\overline{LG^0_\alpha f}\right) - f|_{\partial\Omega} \right\|_{C(\partial\Omega)} \qquad (25.18) \\ &= \left\| \frac{\alpha}{\alpha - \beta}\left(-\overline{LH_\alpha}^{-1}\right)\left(L(G^0_\beta g) - L(G^0_\alpha g) + L(H_\beta\varphi)\right) + \frac{\alpha}{\alpha - \beta}\varphi - \varphi \right\|_{C(\partial\Omega)} \\ &= \left\| \frac{\alpha}{\alpha - \beta}\left(-\overline{LH_\alpha}^{-1}\right)\left((Lf + L(H_\beta\varphi)) - L(G^0_\alpha g)\right) + \frac{\beta}{\alpha - \beta}\varphi \right\|_{C(\partial\Omega)} \\ &\le \frac{\alpha}{\alpha - \beta}\left\| -\overline{LH_\alpha}^{-1} \right\| \cdot \left\| Lf + L(H_\beta\varphi) \right\|_{C(\partial\Omega)} \\ &\quad + \frac{\alpha}{\alpha - \beta}\left\| -\overline{LH_\alpha}^{-1} \right\| \cdot \left\| LG^0_\alpha \right\| \cdot \|g\|_{L^\infty(\Omega)} \\ &\quad + \frac{\beta}{\alpha - \beta}\|\varphi\|_{C(\partial\Omega)} \quad \text{for all } \alpha > \beta. \end{aligned}$$

By Corollary 25.3, it follows that the first term on the last inequality (25.18) converges to zero as $\alpha \to +\infty$:

$$\frac{\alpha}{\alpha - \beta}\left\| -\overline{LH_\alpha}^{-1} \right\| \cdot \left\| Lf + L(H_\beta\varphi) \right\|_{C(\partial\Omega)} \longrightarrow 0 \quad \text{as } \alpha \to +\infty.$$

For the second term, by using the resolvent Eq. (25.5) with $f := 1$ and the non-negativity of G^0_β and LG^0_α (Remark 24.2) we find that

$$\begin{aligned}\|LG_\alpha^0\| &= \|L(G_\alpha^0 1)\|_{C(\partial\Omega)} \\ &= \|L(G_\beta^0 1) - (\alpha - \beta)\,\overline{LG_\alpha^0}\,(G_\beta^0 1)\,\|_{C(\partial\Omega)} \\ &\leq \|L\left(G_\beta^0 1\right)\|_{C(\partial\Omega)} \quad \text{for all } \alpha > \beta.\end{aligned}$$

Hence the second term on the last inequality (25.18) also converges to zero as $\alpha \to +\infty$:

$$\begin{aligned}&\frac{\alpha}{\alpha - \beta}\left\|-\overline{LH_\alpha}^{-1}\right\| \cdot \left\|LG_\alpha^0\right\| \cdot \|g\|_{L^\infty(\Omega)} \\ &\leq \frac{\alpha}{\alpha - \beta}\left\|-\overline{LH_\alpha}^{-1}\right\| \cdot \|L\left(G_\beta^0 1\right)\|_{C(\partial\Omega)} \cdot \|g\|_{L^\infty(\Omega)} \quad \text{as } \alpha \to +\infty.\end{aligned}$$

It is clear that the third term on the last inequality (25.18) converges to zero as $\alpha \to +\infty$.

In this way, we can prove assertion (25.17) and hence the desired assertion (25.16).

Step 8: Summing up, we have proved that the operator $\mathfrak{A}_L$, defined by formula (25.5), satisfies conditions (a) through (d) in Theorem 21.14. Hence it follows from an application of the Hille–Yosida–Ray theorem that the operator $\mathfrak{A}_L$ is the infinitesimal generator of some Feller semigroup on $\overline{\Omega}$.

The proof of Theorem 1.1 is now complete. □

25.2 Notes and Comments

Theorem 1.1 is generalized to Feller semigroups with *jump phenomena* by [163, Theorem 1.4].

Chapter 26
Concluding Remarks

This book is devoted to a careful and accessible exposition of real and functional analytic methods for the problem of constructing strong Markov processes on a bounded domain Ω in Euclidean space $\mathbf{R}^N$ with boundary $\partial\Omega$. For more thorough treatments of this subject, the reader might refer to recent works [157, 162, 163].

26.1 Functional Analytic Approach to Markov Processes

The functional analytic approach consists of two problems (Bony–Courrège–Priouret [15, p. 369]):

(I) The *direct problem*: For any given Feller semigroup $\{T_t\}_{t\geq 0}$ defined on the closure $\overline{\Omega}$, chraracterize concretely its infinitesimal generator $\mathfrak{A}$ in terms of elliptic boundary value problems.

Analytically, a Markovian particle on the closure $\overline{\Omega}$ is governed by an integro-differential operator W, called the Waldenfels operator, in the interior Ω, and it obeys a boundary condition L, called the Ventcel' boundary condition, on the boundary $\partial\Omega$. The Waldenfels operator W takes the form of the sum of a differential operator A and an integro-differential operator S. The operator A is called a diffusion operator which describes analytically a strong Markov process with continuous paths in the interior Ω of the domain, while the operator S is called a Lévy operator which is supposed to correspond to a jump phenomenon in the interior Ω of the domain.

Probabilistically, a Markovian particle moves both by jumps and by continuous paths in the state space and it obeys the Ventcel' boundary condition L, which consists of six terms corresponding to diffusion along the boundary, an absorption phenomenon, a reflection phenomenon, a sticking (or viscosity) phenomenon and a jump phenomenon on the boundary and an inward jump phenomenon from the boundary.

K. Taira, *Real Analysis Methods for Markov Processes*,
https://doi.org/10.1007/978-981-97-3659-1_26

(II) The *inverse problem*: Conversely, for any given analytic data (W, L), construct a Feller semigroup $\{T_t\}_{t\geq 0}$ whose infinitesimal generator $\mathfrak{A}$ is characterized by (W, L).

In this book we prove existence theorems for Feller semigroups with Dirichlet boundary condition, oblique derivative boundary condition and first order Ventcel' boundary condition for second order, uniformly elliptic differential operators with vanishing mean oscillation (VMO) coefficients. Our approach is distinguished by the extensive use of the ideas and techniques characteristic of the recent developments in the Calderón–Zygmund theory of singular integral operators with non-smooth kernels. It should be emphasized that singular integral operators with non-smooth kernels provide a powerful tool to deal with smoothness of solutions of partial differential equations, with minimal assumptions of regularity on the coefficients. See also Stroock Sk1975, Komatsu [81, 82] and Lenhart [88].

26.2 Feller Semigroups and Pseudo-Differential and Singular Integral Operators

For general results on generation theorems for Feller semigroups through pseudo-differential operators and singular integral operators, we give two overviews as in Tables 26.1 and 26.2 below.

Here the boundary condition L_1 of *Robin type* is given by the formula

$$L_1 u := \mu(x') \frac{\partial u}{\partial \mathbf{n}} + \gamma(x') u \quad \text{on } \partial\Omega,$$

where the coefficients $\mu(x')$ and $\gamma(x')$ satisfy the following three conditions:

(i) $\mu(x')$ is a smooth function on $\partial\Omega$ and $\mu(x') \geq 0$ on $\partial\Omega$.
(ii) $\gamma(x')$ is a smooth function on $\partial\Omega$ and $\gamma(x') \leq 0$ on $\partial\Omega$.
(iii) $\mu(x') + |\gamma(x')| > 0$ on $\partial\Omega$.

Table 26.1 Generation theorems for Feller semigroups in the smooth and VMO coefficient cases

Diffusion operator A	Lévy operator S	Proved by
Smooth coefficient case	$S \equiv 0$	[147, 148, 162]
Smooth coefficient case	General case	[149, 154]
Smooth coefficient case	Hölder continuous case	[150, 157]
VMO coefficient case	General case	[152, 156]
VMO coefficient case	$S \equiv 0$	[153, 161, 163]

Table 26.2 Generation theorems for Feller semigroups in terms of pseudo-differential and singular integral operators

Ventcel' (Wentzell) condition L	Using the theory of	Proved by
Second order case	Hypoelliptic pseudo-differential operators of second order	[147, 148, 162]
General case	Boutet de Monvel calculus with transmission property	[149, 154]
Robin condition L_1	Hypoelliptic pseudo-differential operators of first order	[150, 157]
Dirichlet case	Singular integrals of Calderón and Zygmund	[152, 156]
First order case	Singular integrals of Calderón and Zygmund	[153, 161, 163]

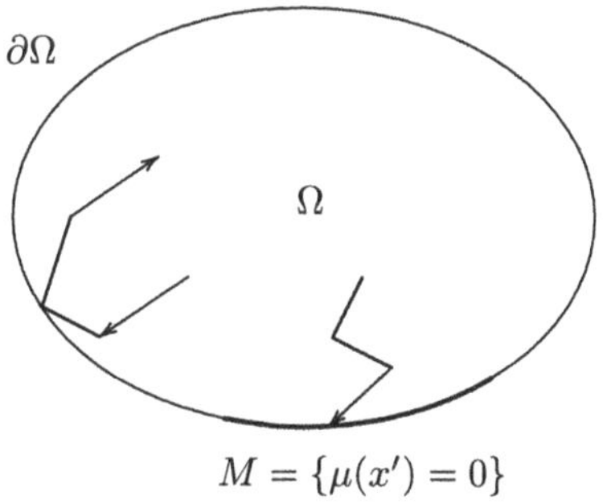

Fig. 26.1 A Markovian particle dies at the time when it reaches the set M

We remark that L_1 becomes a *degenerate* boundary condition from an analytical point of view. This is due to the fact that the so-called Shapiro–Lopatinskii complementing condition is violated at the points $x' \in \partial\Omega$ where $\mu(x') = 0$ (see [4, 67, 93, 96, 135, 181]). The intuitive meaning of hypothesis (iii) is that either the reflection phenomenon or the absorption phenomenon occurs at each point of the boundary $\partial\Omega$. More precisely, condition (iii) implies that absorption phenomenon occurs at each point of the set

$$M = \{x' \in \partial\Omega : \mu(x') = 0\},$$

while reflection phenomenon occurs at each point of the set

$$\partial\Omega \setminus M = \{x' \in \partial\Omega : \mu(x') > 0\}.$$

In other words, a Markovian particle moves by continuous paths in the space $\overline{\Omega} \setminus M$ until it "dies" at the time when it reaches the set M (see Fig. 26.1).

Finally, it should be emphasized that Maz'ya et al. [100, Theorem 1.1] proved the existence and uniqueness theorem for the Dirichlet problem for higher order, strongly elliptic systems in divergence form with complex-valued, bounded measurable coefficients in Lipschitz domains in the framework of Sobolev and Besov spaces under sharp smoothness assumptions.

26.3 Stochastic Analysis Methods for Ventcel' Boundary Value Problem

We state a brief history of the *stochastic analysis* methods for Ventcel' boundary value problems. More precisely, we remark that Ventcel' boundary value problems are studied by Anderson [6, 7], Cattiaux [31] and Takanobu–Watanabe [164] from the viewpoint of stochastic analysis (see also Ikeda–Watanabe [68, Chap. IV, Sect. 7]).

(1) Anderson [6, 7] studies the non-degenerate case under low regularity in the framework of the submartingale problem and shows the existence and uniqueness of solutions to the considered submartingale problem.
(2) Takanobu and Watanabe [164] study certain cases of both degenerate interior and boundary operators under minimal assumptions of regularity based on the theory of stochastic differential equations, and they show the existence and uniqueness of solutions. Such existence and uniqueness results on the diffusion processes corresponding to the boundary value problems imply the existence and uniqueness of the associated Feller semigroups on the space of continuous functions.
(3) Cattiaux [31] studies the hypoellipticity for diffusions with Ventcel' boundary conditions. By making use of a variant of the Malliavin calculus under Hömander's type conditions ([66]), he proves that some laws and conditional laws of such diffusions have a smooth density with respect to the Lebesgue measure. See also Komatsu–Takeuchi [83].

Appendix A
Change-of-Variables Formulas in the n-Dimensional Lebesgue Integral

In this appendix we set forth the basic concepts of Lebesgue integral theory on the Euclidean space $\mathbf{R}^n$. In particular, following Schwartz [130, Theorem] and Folland [51, Theorem 2.47] we prove two change-of-variables formulas under linear transformations and diffeomorphisms (Theorems A.7 and A.14). For more thorough treatments of this subject, the reader might refer to Federer [46], Folland [51], Rudin [120] and Stein–Shakarchi [141].

We denote by $(\mathbf{R}^n, \mathcal{L}^n, m)$ and $(\mathbf{R}^n, \mathcal{B}(\mathbf{R}^n), \mu)$ the Lebesgue and Borel measure spaces, respectively (see Table A.1).

A.1 Half-open Intervals and Figures

A set of the form

$$I = (a_1, b_1] \times \ldots \times (a_n, b_n] \quad \text{for } -\infty \leq a_j < b_j \leq \infty$$

is called a *half-open interval* in Euclidean space $\mathbf{R}^n$, where

$$(a, b] = \{x : a < x < b\} = (a, \infty) \quad \text{if } b = \infty.$$

The collection of such intervals in $\mathbf{R}^n$ is denoted by $\mathcal{I}_n$.

A set of finite disjoint unions of such intervals

$$F = I_1 \cup \cdots \cup I_N$$

is called a *figure* in $\mathbf{R}^n$. The collection of figures in $\mathbf{R}^n$ is denoted by $\mathcal{F}_n$ (see Table A.1 and Fig. A.1 below).

K. Taira, *Real Analysis Methods for Markov Processes*,
https://doi.org/10.1007/978-981-97-3659-1

Table A.1 A bird's-eye view of Lebesgue and Borel measure spaces on the Euclidean space $\mathbf{R}^n$

Euclidean space	Families of subsets	Measures
$\mathbf{R}^n$	$\mathcal{P}(\mathbf{R}^n)$ (all subsets)	μ^* (outer measure)
$\mathbf{R}^n$	$\mathcal{L}^n$ (Lebesgue measurable sets)	$m = \overline{\mu} = \mu^*\lvert_{\mathcal{L}^n}$ (Lebesgue measure)
$\mathbf{R}^n$	$\mathcal{B}(\mathbf{R}^n) = \sigma(\mathcal{F}_n)$ (Borel sets)	$\mu = \mu^*\lvert_{\mathcal{B}(\mathbf{R}^n)}$ (Borel measure)
$\mathbf{R}^n$	$\mathcal{F}_n$ (figures)	μ_0 (premeasure)

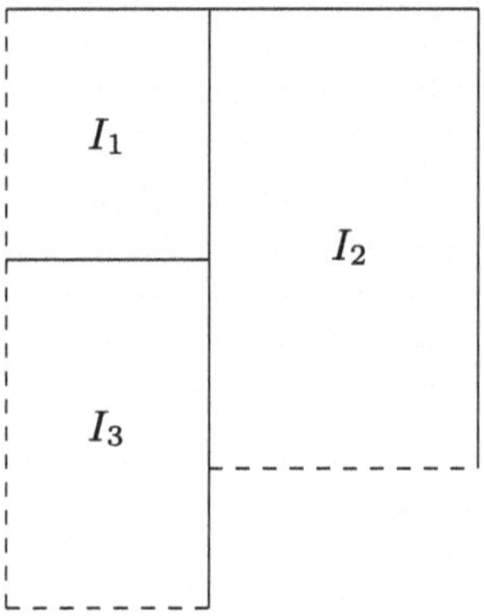

Fig. A.1 An example of a figure $F = I_1 + I_2 + I_3$ in the plane $\mathbf{R}^2$

A.2 Translation Invariance Formula for the Lebesgue Integral

First, we prove that Lebesgue integral is *translation invariant*. More precisely, we start with the following theorem (see [51, Theorem 2.42]).

Theorem A.1 *If $a \in \mathbf{R}^n$, we define a translation map τ_a as follows:*

$$\begin{aligned} \tau_a : \mathbf{R}^n &\longrightarrow \mathbf{R}^n \\ x &\longmapsto x + a. \end{aligned}$$

Then we have the following assertions:

(i) If $E \in \mathcal{L}^n$, then $\tau_a(E) \in \mathcal{L}^n$ and we have the formula

$$m\left(\tau_a(E)\right) = m(E). \tag{A.1}$$

(ii) If $f(y)$ is a Lebesgue measurable function on $\mathbf{R}^n$, then so is the function $f(\tau_a(x))$. Moreover, if either

(a) $f(y) \geq 0$ *on* $\mathbf{R}^n$ *or*
(b) $f(y)$ *is a Lebesgue integrable function on* $\mathbf{R}^n$*, then we have the formula*

$$\int_{\mathbf{R}^n} f(\tau_a(x))\, dm(x) = \int_{\mathbf{R}^n} f(y)\, dm(y). \tag{A.2}$$

Roughly speaking, we have the change-of-variables formula for Lebesgue measures

$$dm(y) = dm(\tau_a(x)) = dm(x).$$

Proof The proof of Theorem A.1 is divided into four steps.

Step 1: Since the translation maps τ_a and τ_{-a} are both continuous, they preserve the class $\mathcal{F}_n$ of figures and the class of $\mathcal{B}(\mathbf{R}^n)$ of Borel sets:

$$\tau_{\pm a} : \mathcal{F}_n \longrightarrow \mathcal{F}_n,$$
$$\tau_{\pm a} : \mathcal{B}\left(\mathbf{R}^n\right) \longrightarrow \mathcal{B}\left(\mathbf{R}^n\right)$$

and further, the formulas

$$\mu_0(\tau_{\pm a}(F)) = \mu_0(F)$$

hold true for figures $F \in \mathcal{F}_n$. Hence we obtain that

$$\begin{aligned}
\mu(\tau_{\pm a}(E)) &= \inf\left\{\sum_{j=1}^{\infty} \mu_0\left(F_j\right) : \tau_{\pm a}(E) \subset \bigcup_{j=1}^{\infty} F_j,\ F_j \in \mathcal{F}_n\right\} \qquad (A.3)\\
&= \inf\left\{\sum_{j=1}^{\infty} \mu_0\left(F_j\right) : E \subset \bigcup_{j=1}^{\infty} \tau_{\mp a}\left(F_j\right),\ F_j \in \mathcal{F}_n\right\}\\
&= \inf\left\{\sum_{j=1}^{\infty} \mu_0\left(\tau_{\mp a}\left(F_j\right)\right) : E \subset \bigcup_{j=1}^{\infty} \tau_{\mp a}\left(F_j\right),\ F_j \in \mathcal{F}_n\right\}\\
&= \mu(E) \quad \text{for all } E \in \mathcal{B}\left(\mathbf{R}^n\right).
\end{aligned}$$

This formula implies that the collection of *Borel null sets* is invariant under τ_a and τ_{-a}:

$$E \in \mathcal{B}\left(\mathbf{R}^n\right),\ \mu(E) = 0 \Longrightarrow \mu(\tau_{\pm a}(E)) = 0. \tag{A.4}$$

Step 2: For the proof of assertion (A.4) for Lebesgue measurable sets, we need the following lemma.

Lemma A.2 *For every set* $A \subset \mathbf{R}^n$*, we have the formula*

$$\mu^*(A) = \inf\{\mu(G) : A \subset G, \; G \textit{ is open}\}. \tag{A.5}$$

Proof For any $\varepsilon > 0$, there is a sequence $\{I_n\}_{n=1}^{\infty}$ of half-open intervals such that

$$A \subset \bigcup_{n=1}^{\infty} I_n, \tag{A.6a}$$

$$\sum_{n=1}^{\infty} \mu(I_n) \leq \mu^*(A) + \frac{\varepsilon}{2}. \tag{A.6b}$$

For each half-open interval I_n, there is a sequence $\{G_n\}_{n=1}^{\infty}$ of open sets such that

$$I_n \subset G_n, \tag{A.7a}$$

$$\mu(G_n) \leq \mu(I_n) + \frac{\varepsilon}{2^{n+1}}. \tag{A.7b}$$

Hence, if we let

$$G := \bigcup_{n=1}^{\infty} G_n,$$

we obtain from assertions (A.6) and (A.7) that

$$A \subset \bigcup_{n=1}^{\infty} I_n \subset G = \bigcup_{n=1}^{\infty} G_n$$

and further that

$$\begin{aligned} \mu(G) &\leq \sum_{n=1}^{\infty} \mu(G_n) \leq \sum_{n=1}^{\infty} \left(\mu(I_n) + \frac{\varepsilon}{2^{n+1}}\right) = \sum_{n=1}^{\infty} \mu(I_n) + \frac{\varepsilon}{2} \\ &\leq \mu^*(A) + \varepsilon. \end{aligned}$$

This implies that

$$\inf\{\mu(G) : A \subset G, \; G \text{ is open}\} \leq \mu^*(A) + \varepsilon. \tag{A.8}$$

However, since we have the inequality

$$\mu^*(A) \leq \mu^*(G) = \mu(G),$$

and so

$$\mu^*(A) \leq \inf\{\mu(G) : A \subset G, \; G \text{ is open}\}. \tag{A.9}$$

Therefore, the desired formula (A.5) follows by combining inequalities (A.8) and (A.9), since $\varepsilon > 0$ is arbitrary.

The proof of Lemma A.2 is complete. □

The next corollary asserts that all Borel sets are of reasonably simple from modulo sets of measure zero.

Corollary A.3 *(i) For every set $A \subset \mathbf{R}^n$, there exists a Borel set B such that*

$$\begin{cases} B \supset A, \\ \mu(B) = \mu^*(A). \end{cases} \tag{A.10}$$

(ii) For every set $E \in \mathcal{L}^n$, there exists a Borel set B such that

$$\begin{cases} B \supset E, \\ \mu(B) = m(E). \end{cases} \tag{A.11}$$

Proof (i) By virtue of formula (A.5), for each integer $k \in \mathbf{N}$ we can find an open set $G_k \in \mathcal{B}(\mathbf{R}^n)$ such that

$$\begin{cases} G_k \supset A, \\ \mu(G_k) = \mu^*(A) + \frac{1}{k}. \end{cases} \tag{A.12}$$

If we let

$$B := \bigcap_{k=1}^{\infty} G_k,$$

then we have, by assertion (A.12),

$$\begin{aligned} &B \in \mathcal{B}\left(\mathbf{R}^n\right), \\ &B \supset A, \\ &\mu(B) \geq \mu^*(A), \end{aligned}$$

and so

$$\mu^*(A) \leq \mu(B) \leq \mu(G_k) = \mu^*(A) + \frac{1}{k} \quad \text{for all } k \in \mathbf{N}.$$

This proves that

$$\mu(B) = \mu^*(A).$$

(ii) Part (ii) is a special case of part (i), since we have, for $A := E \in \mathcal{L}^n$,

$$m(E) = \mu^*(E) = \mu(B).$$

The proof of Corollaey A.3 is complete. □

The next corollary asserts that the outer measure μ^* is tranlation invariant.

Corollary A.4 *For all $B \subset \mathbf{R}^n$, we have the formulas*

$$\mu^* \left(\tau_a(B)\right) = \mu^*(B), \tag{A.13a}$$
$$\mu^* \left(\tau_{-a}(B)\right) = \mu^*(B). \tag{A.13b}$$

Proof Since the collection of open sets in $\mathbf{R}^n$ is invariant under $\tau_{\pm a}$ and since open sets are Borel sets, we obtain from formulas (A.5) and (A.3) that

$$\begin{aligned} &\mu^* \left(\tau_{\pm a}(B)\right) = \inf \{\mu(G) : \tau_{\pm a}(B) \subset G,\ G \text{ is open}\} \\ &= \inf \{\mu(G) : B \subset \tau_{\mp a}(G),\ \tau_{\mp a}(G) \text{ is open}\} \\ &= \inf \{\mu \left(\tau_{\mp a}(G)\right) : B \subset \tau_{\mp a}(G),\ \tau_{\mp a}(G) \text{ is open}\} \\ &= \mu^*(B) \quad \text{for all } B \subset \mathbf{R}^n. \end{aligned}$$

The proof of Corollaey A.4 is complete. □

Now we are in a position to prove assertion (A.4) for Lebesgue measurable sets.

Theorem A.5 *Let $\tau_{\pm a} \colon \mathbf{R}^n \to \mathbf{R}^n$ be translation maps of $\mathbf{R}^n$. If $E \in \mathcal{L}^n$, then $\tau_{\pm a}(E) \in \mathcal{L}^n$ and we have the formula*

$$m \left(\tau_{\pm a}(E)\right) = m(E). \tag{A.14}$$

Proof (1) If $E \in \mathcal{L}^n$, we have the formula

$$\mu^*(A) = \mu^* \left(A \cap E\right) + \mu^* \left(A \setminus E\right) \quad \text{for all } A \subset \mathbf{R}^n.$$

By taking

$$A := \tau_{\mp a}(B) \quad \text{with } B \subset \mathbf{R}^n,$$

we have the formulas

$$\begin{aligned} A \cap E &= \tau_{\mp a} \left(B \cap \tau_{\pm a}(E)\right), \\ A \setminus E &= \tau_{\mp a} \left(B \setminus \tau_{\pm a}(E)\right), \end{aligned}$$

and so

$$\begin{aligned} &\mu^* \left(\tau_{\mp a}(B)\right) \\ &= \mu^* \left(\tau_{\mp a} \left(B \cap \tau_{\mp a}(E)\right)\right) + \mu^* \left(\tau_{\mp a} \left(B \setminus \tau_{\pm a}(E)\right)\right) \quad \text{for all } B \subset \mathbf{R}^n. \end{aligned}$$

Therefore, by using formulas (A.13) we obtain that

$$\mu^*(B) = \mu^* \left(B \cap \tau_{\pm a}(E)\right) + \mu^* \left(B \setminus \tau_{\pm a}(E)\right) \quad \text{for all } B \subset \mathbf{R}^n.$$

This proves that

$$\tau_{\pm a}(E) \in \mathcal{L}^n \quad \text{for all } E \in \mathcal{L}^n.$$

(2) Secondly, the desired formula (A.14) follows from formulas (A.13):

$$\begin{aligned} m\left(\tau_{\pm a}(E)\right) &= \mu^*\left(\tau_{\pm a}(E)\right) = \mu^*(E) \\ &= m(E) \quad \text{for all } E \in \mathcal{L}^n. \end{aligned}$$

The proof of Theorem A.5 is complete. □

The next claim proves assertion (A.6) for Lebesgue measurable sets:

Claim A.6 *The class of* Lebesgue null sets *is invariant under* $\tau_{\pm a}$*:*

$$E \in \mathcal{L}^n, \ m(E) = 0 \implies m\left(\tau_{\pm a}(E)\right) = 0. \tag{A.15}$$

Proof By using Lemma A.2, for any $\varepsilon > 0$ we can find an open set G such that

$$\begin{aligned} &E \subset G, \\ &\mu(G) = \mu^*(G) < \varepsilon. \end{aligned}$$

Hence, we have, by formulas (A.10) with $B := G$,

$$\begin{aligned} &\tau_{\pm a}(E) \subset \tau_{\pm a}(G), \\ &\mu\left(\tau_{\pm a}(G)\right) = \mu(G) < \varepsilon, \end{aligned}$$

and so

$$m\left(\tau_{\pm a}(E)\right) \leq \mu\left(\tau_{\pm a}(G)\right) < \varepsilon.$$

This proves the desired assertion (A.15) for $\tau_{\pm a}(E)$, since $\varepsilon > 0$ is arbitrary.
The proof of Claim A.6 is complete. □

Step 3: Let $B \in \mathcal{B}(\mathbf{R}^n)$. If $f(y)$ is a real-valued Lebesgue mesurable function, then it follows from an application of Theorem 3.2 that

$$\begin{aligned} &f^{-1}(B) = E \cup N, \\ &E \in \mathcal{B}\left(\mathbf{R}^n\right), \\ &N \in \mathcal{L}^n \text{ with } m(N) = 0, \end{aligned}$$

since the σ-algebra $\mathcal{L}^n$ is the completion of the σ-algebra $\mathcal{B}(\mathbf{R}^n)$. Then it follows that

$$\begin{aligned} (f \circ \tau_a)^{-1}(B) &= \tau_a^{-1}\left(f^{-1}(B)\right) = \tau_{-a}\left(f^{-1}(B)\right) \\ &= \tau_{-a}(E) \cup \tau_{-a}(N), \end{aligned}$$

where we have, by Theorem A.5,

$$\tau_{-a}(E) \in \mathcal{B}\left(\mathbf{R}^n\right),$$
$$m\left(\tau_{-a}(N)\right) = m(N) = 0.$$

This proves that

$$(f \circ \tau_a)^{-1}(B) = \tau_{-a}(E) \cup \tau_{-a}(N) \in \mathcal{L}^n \quad \text{for all } B \in \mathcal{B}\left(\mathbf{R}^n\right).$$

Namely, the mapping

$$f \circ \tau_a \colon \left(\mathbf{R}^n, \mathcal{L}^n, m\right) \longrightarrow \left(\mathbf{R}^n, \mathcal{B}\left(\mathbf{R}^n\right), \mu\right)$$

is measurable.

Step 4: It remains to prove the formula

$$\int_{\mathbf{R}^n} f\left(\tau_a(x)\right) dm(x) = \int_{\mathbf{R}^n} f(y)\, dm(y) \tag{A.2}$$

in the framework of *Lebesgue* measurable functions and sets.

First, we consider a characteristic function

$$f(y) = \chi_E(y) = \begin{cases} 1 & \text{for } y \in E, \\ 0 & \text{for } y \in \mathbf{R}^n \setminus E \end{cases}$$

for $E \in \mathcal{L}^n$, then we have the formula

$$\int_{\mathbf{R}^n} f(y)\, dm(y) = m(E)$$

and

$$\int_{\mathbf{R}^n} f\left(\tau_a(x)\right) dm(x) = m\left(\tau_{-a}(E)\right).$$

Therefore, the desired formula (A.2) for $f = \chi_E$ reduces to formula (A.14).

Secondly, we consider a simple function of the form (see Fig. A.2 below)

$$f(y) = \sum_{j=1}^{N} c_j\, \chi_{E_j}(y),$$

where

$$E_j \in \mathcal{L}^n \quad \text{for } 1 \leq j \leq N,$$

and

$$c_j > 0 \quad \text{for } 1 \leq j \leq N.$$

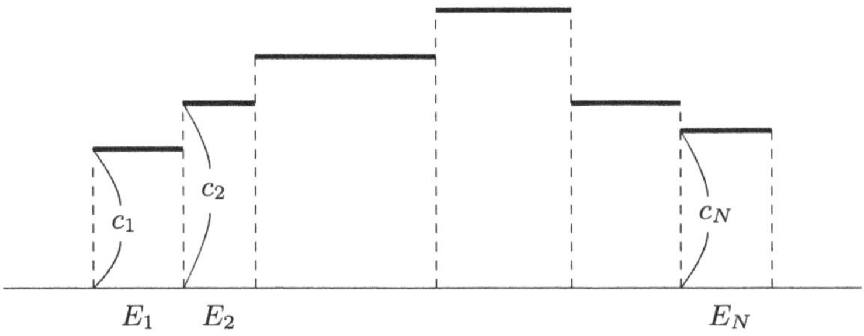

Fig. A.2 A simple function $f(y) = \sum_{j=1}^{N} c_j \, \chi_{E_j}(y)$

Then we have the formulas

$$\int_{\mathbf{R}^n} f(y)\, dm(y) = \sum_{j=1}^{N} c_j \, m(E_j)$$

and

$$\int_{\mathbf{R}^n} f\left(\tau_a(x)\right) dm(x) = \sum_{j=1}^{N} m\left(\tau_{-a}(E_j)\right) = \sum_{j=1}^{N} c_j \, m(E_j).$$

Namely, the desired formula (A.2) holds true for non-negative simple functions:

$$\int_{\mathbf{R}^n} f\left(\tau_a(x)\right) dm(x) = \sum_{j=1}^{N} c_j \, m(E_j) = \int_{\mathbf{R}^n} f(y)\, dm(y).$$

Finally, if either $f(y) \geq 0$ on $\mathbf{R}^n$ or $f(y)$ is Lebesgue integrable on $\mathbf{R}^n$, then the desired formula (A.2) follows by approximating $f(y)$ from below by simple functions (see the proof [51, Theorem 2.10]).

Now the proof of Theorem A.1 is complete. □

A.3 Change-of-Variables Formula Under Linear Transformations

In this section, we consider the Lebesgue integral under linear transformations. More precisely, we prove the following theorem (see [51, Theorem 2.44]).

Theorem A.7 (the change-of-variables formula) *Let*

$$\begin{aligned} T : \mathbf{R}^n &\longrightarrow \mathbf{R}^n \\ x &\longmapsto T(x) \end{aligned}$$

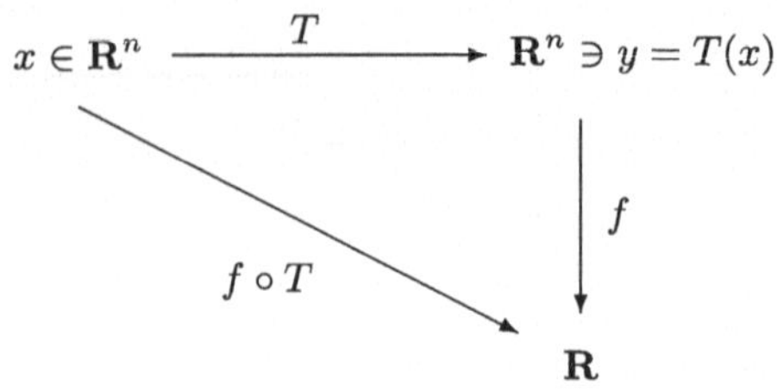

Fig. A.3 The functions $f(y)$ and $f(T(x))$

be an invertible *linear transformation of* $\mathbf{R}^n$. *Then we have the following assertions (see Fig. A.3 above):*

(1) If $f(y)$ *is a Lebesgue measurable function on* $\mathbf{R}^n$, *then so is the function* $f(T(x))$. *Moreover, if either*

(a) $f(y) \geq 0$ *on* $\mathbf{R}^n$ *or*
(b) $f(y)$ *is a Lebesgue integrable function on* $\mathbf{R}^n$, *then we have the formula*

$$\int_{\mathbf{R}^n} f(y)\, dm(y) = |\det T| \int_{\mathbf{R}^n} f(T(x))\, dm(x). \tag{A.16}$$

(2) If E *is a Lebesgue measurable set of* $\mathbf{R}^n$, *then so is the set* $T(E)$, *and we have the formula*

$$m(T(E)) = |\det T|\, m(E). \tag{A.17}$$

Roughly speaking, we have the change-of-variables formula for Lebesgue measures

$$dm(y) = dm(T(x)) = |\det T|\, dm(x).$$

The detailed proof of Theorem A.7 will be given in Sects. A.3.2 and A.3.3, due to its length.

A.3.1 Left Elementary Transformations in Linear Algebra

In order to prove Theorem A.7, we recall the following *left elementary transformations* T_1, T_2 and T_3 in linear algebra:

(1) T_1: Multiply a row by a non-zero constant c.
(2) T_2: Add a row by a multiplied another row.
(3) T_3: Interchange two rows.

Example A.8 Here are concrete examples of T_1, T_2 and T_3 for $n = 4$:

$$T_1 = \begin{pmatrix} 1 & 0 & 0 & 0 \\ 0 & 1 & 0 & 0 \\ 0 & 0 & c & 0 \\ 0 & 0 & 0 & 1 \end{pmatrix}, \quad T_2 = \begin{pmatrix} 1 & 0 & 0 & 0 \\ 0 & 1 & 0 & 0 \\ 0 & a & 1 & 0 \\ 0 & 0 & 0 & 1 \end{pmatrix}, \quad T_3 = \begin{pmatrix} 1 & 0 & 0 & 0 \\ 0 & 0 & 0 & 1 \\ 0 & 0 & 1 & 0 \\ 0 & 1 & 0 & 0 \end{pmatrix}.$$

Then the well-known *Gaussian elimination* reads as follows.

Theorem A.9 (Gauss) *An $n \times n$ matrix*

$$A = \begin{pmatrix} a_{11} & a_{12} & \dots & a_{1n} \\ a_{21} & a_{22} & \dots & a_{2n} \\ \vdots & \vdots & \ddots & \vdots \\ a_{n1} & a_{n2} & \dots & a_{nn} \end{pmatrix}$$

is invertible if and only if an $n \times 2n$ matrix

$$(A \ I) = \begin{pmatrix} a_{11} & a_{12} & \dots & a_{1n} & 1 & 0 & \dots & 0 \\ a_{21} & a_{22} & \dots & a_{2n} & 0 & 1 & \dots & 0 \\ \vdots & \vdots & \ddots & \vdots & \vdots & \vdots & \ddots & 0 \\ a_{n1} & a_{n2} & \dots & a_{nn} & 0 & 0 & \dots & 1 \end{pmatrix}$$

can be transformed into an $n \times 2n$ matrix

$$(I \ B) = \begin{pmatrix} 1 & 0 & \dots & 0 & b_{11} & b_{12} & \dots & b_{1n} \\ 0 & 1 & \dots & 0 & b_{21} & b_{22} & \dots & b_{2n} \\ \vdots & \vdots & \ddots & \vdots & \vdots & \vdots & \ddots & \vdots \\ 0 & 0 & \dots & 1 & b_{n1} & b_{n2} & \dots & b_{nn} \end{pmatrix}$$

after left elementary transformations. Here the $n \times n$ matrix

$$B = \begin{pmatrix} b_{11} & b_{12} & \dots & b_{1n} \\ b_{21} & b_{22} & \dots & b_{2n} \\ \vdots & \vdots & \ddots & \vdots \\ b_{n1} & b_{n2} & \dots & b_{nn} \end{pmatrix} = A^{-1}$$

is the inverse matrix of A.

If the change-of-variables formula (A.16) holds true for transformations T and S, then it remains valid for the composition $T \circ S$. Indeed, it suffices to note that

$$\begin{aligned}\int_{\mathbf{R}^n} f(y)\, dm(y) &= |\det T| \int_{\mathbf{R}^n} f\,(T(x))\, dm(x) \\ &= |\det T| \left(|\det S| \int_{\mathbf{R}^n} f\,(T\,(S(z)))\; dm(z) \right) \\ &= |\det(T \circ S)| \left(\int_{\mathbf{R}^n} f\,(((T \circ S))\,(z)) \right) dm(z).\end{aligned}$$

Therefore, by virtue of Gauss' theorem (Theorem A.9), we have only to prove the desired formula (A.16) when T is of the types T_1, T_2 and T_3.

A.3.2 *Proof of Theorem A.7*

First, we prove Theorem A.7 in the framework of *Borel* measurable functions and sets.

(I) We remark that if E is a Borel set, so is the set $T(E)$, since T is continuous. Similarly, if $f(x)$ is a Borel measurable function, so is the function $f\,(T(x))$, since T is continuous.

For the left elementary transformation T_3 (see Example A.8), by using Fubini's theorem for Borel sets we can interchange the order of the integration in the variables x_j and x_k in the change-of-variables formula (A.16), where

$$\det T_3 = -1.$$

For the left elementary transformation T_2 (see Example A.8), by using Theorem A.1 we have the change-of-variables formula (A.16). More precisely, for T_2 we integrate first with respect to x_j and use the one-dimensional formula

$$\int_{\mathbf{R}} \varphi(t)\, dt = \int_{\mathbf{R}} \varphi(t + a)\, dt = |\det T_2| \int_{\mathbf{R}} \varphi(t + a)\, dt,$$

where

$$\det T_2 = 1.$$

Similarly, for the left elementary transformation T_1 (see Example A.8) we integrate first with respect to x_j and use the one-dimensional formula

$$\int_{\mathbf{R}} \varphi(t)\, dt = |c| \int_{\mathbf{R}} \varphi(cs)\, ds = |\det T_1| \int_{\mathbf{R}} \varphi(cs)\, ds,$$

where

$$\det T_1 = c.$$

In this way, we have proved the change-of-variables formula (A.16) under linear transformations in the framework of *Borel* measurable functions and sets.

(II) If E is a Borel set, by taking

$$f(y) := \chi_{T(E)}(y) = \begin{cases} 1 & \text{if } y \in T(E), \\ 0 & \text{if } y \in \mathbf{R}^n \setminus T(E) \end{cases}$$

in the change-of-variables formula (A.16), we obtain that

$$\begin{aligned} \mu(T(E)) &= \int_{\mathbf{R}^n} f(y)\, d\mu(y) = |\det T| \int_{\mathbf{R}^n} f(T(x))\, d\mu(x) \\ &= |\det T|\, \mu(E) \quad \text{for all Borel set } E. \end{aligned} \tag{A.18}$$

This formula implies that the class of *Borel null sets* is invariant under T and T^{-1}:

$$E \in \mathcal{B}(\mathbf{R}^n),\ \mu(E) = 0 \Longrightarrow \mu(T(E)) = 0 \text{ and } \mu\left(T^{-1}(E)\right) = 0. \tag{A.19}$$

Moreover, we need the following claim for Lebesgue measurable sets:

Claim A.10 *For every set $B \subset \mathbf{R}^n$, we have the formulas*

$$\mu^*\left(T^{-1}(B)\right) = |\det T|^{-1}\, \mu^*(B). \tag{A.20a}$$

$$\mu^*(T(B)) = |\det T|\, \mu^*(B). \tag{A.20b}$$

Proof Since the collection of open sets in $\mathbf{R}^n$ is invariant under T and T^{-1} and since open sets are Borel sets, we obtain from formulas (A.5) and (A.18) that

$$\begin{aligned} &\mu^*\left(T^{\mp 1}(B)\right) = \inf\left\{\mu(G) : T^{\mp}(B) \subset G,\ G \text{ is open}\right\} \\ &= \inf\left\{\mu(G) : B \subset T^{\pm}(G),\ T^{\pm}(G) \text{ is open}\right\} \\ &= |\det T|^{\mp} \inf\left\{\mu\left(T^{\pm}(G)\right) : B \subset T^{\pm}(G),\ T^{\pm}(G) \text{ is open}\right\} \\ &= |\det T|^{\mp}\, \mu^*(B). \end{aligned}$$

The proof of Claim A.10 is complete. □

Now we are in a position to prove assertion (A.18) for Lebesgue measurable sets.

Theorem A.11 *Let $T : \mathbf{R}^n \to \mathbf{R}^n$ be an* invertible *linear transformation of $\mathbf{R}^n$. If $E \in \mathcal{L}^n$, then $T(E) \in \mathcal{L}^n$ and we have the formula*

$$m(T(E)) = |\det T|\, m(E). \tag{A.21}$$

Proof (1) If $E \in \mathcal{L}^n$, we have the formula

$$\mu^*(A) = \mu^*(A \cap E) + \mu^*(A \setminus E) \quad \text{for all } A \subset \mathbf{R}^n.$$

By taking

$$A := T^{-1}(B) \quad \text{with } B \subset \mathbf{R}^n,$$

we have the formulas

$$\begin{aligned} A \cap E &= T^{-1}\left(B \cap T(E)\right), \\ A \setminus E &= T^{-1}\left(B \setminus T(E)\right). \end{aligned}$$

and so

$$\begin{aligned} &\mu^*\left(T^{-1}(B)\right) \\ &= \mu^*\left(T^{-1}\left(B \cap T(E)\right)\right) + \mu^*\left(T^{-1}\left(B \setminus T(E)\right)\right) \quad \text{for all } B \subset \mathbf{R}^n. \end{aligned}$$

Therefore, by using formula (A.20b) we obtain that

$$\mu^*(B) = \mu^*\left(B \cap T(E)\right) + \mu^*\left(B \setminus T(E)\right) \quad \text{for all } B \subset \mathbf{R}^n.$$

This proves that

$$T(E) \in \mathcal{L}^n \quad \text{for } E \in \mathcal{L}^n.$$

(2) Secondly, the desired formula (A.21) follows from formula (A.20b):

$$m\left(T(E)\right) = \mu^*\left(T(E)\right) = |\det T|\, \mu^*(E) = |\det T|\, m(E) \quad \text{for } E \in \mathcal{L}^n.$$

The proof of Theorem A.11 is complete. □

The next claim asserts that the collection of Lebesgue null sets is invariant under T and T^{-1}:

Claim A.12 *The class of* Lebesgue null sets *is invariant under T and T^{-1}:*

$$E \in \mathcal{L}^n, \ m(E) = 0 \Longrightarrow m\left(T(E)\right) = 0 \text{ and } m\left(T^{-1}(E)\right) = 0. \tag{A.22}$$

Proof We only prove the assertion (A.22) for the transformation T:

$$m\left(T(E)\right) = 0.$$

By using Lemma A.2, for any $\varepsilon > 0$ we can find an open set G such that

$$\begin{aligned} &E \subset G, \\ &\mu(G) = \mu^*(G) < \varepsilon. \end{aligned}$$

Hence, we have, by formula (A.20b) with $B := G$,

$$\begin{aligned}&T(E) \subset T(G),\\&\mu(T(G)) = |\det T|\,\mu(G) < \varepsilon\,|\det T|,\end{aligned}$$

and so

$$m(T(E)) \leq m(T(G)) = \mu(T(G)) < \varepsilon\,|\det T|.$$

This proves the desired assertion (A.22) for $T(E)$, since $\varepsilon > 0$ is arbitrary.
The proof of Claim A.12 is complete. □

A.3.3 *End of Proof of Theorem A.7*

It remains to prove the formula

$$\int_{\mathbf{R}^n} f(y)\,dm(y) = |\det T| \int_{\mathbf{R}^n} f(T(x))\,dm(x) \tag{A.16}$$

in the framework of *Lebesgue* measurable functions and sets.
First, we consider a characteristic function

$$f(y) = \chi_E(y) = \begin{cases} 1 & \text{for } y \in E,\\ 0 & \text{for } y \in \mathbf{R}^n \setminus E\end{cases}$$

for $E \in \mathcal{L}^n$, then we have the formula

$$\int_{\mathbf{R}^n} f(y)\,dm(y) = m(E)$$

and

$$\int_{\mathbf{R}^n} f(T(x))\,dm(x) = m\left(T^{-1}(E)\right) = |\det T|^{-1}\,m(E),$$

so that

$$|\det T| \int_{\mathbf{R}^n} f(T(x))\,dm(x) = m(E) = \int_{\mathbf{R}^n} f(y)\,dm(y).$$

Therefore, the desired formula (A.16) for $f = \chi_E$ reduces to formula (A.21) for T^{-1}.
Secondly, we consider a simple function of the form (see Fig. A.2)

$$f(y) = \sum_{j=1}^{N} c_j\,\chi_{E_j}(y),$$

where

$$E_j \in \mathcal{L}^n \quad \text{for } 1 \leq j \leq N,$$

and

$$c_j > 0 \quad \text{for } 1 \leq j \leq N.$$

Then we have the formulas

$$\int_{\mathbf{R}^n} f(x)\, dm(x) = \sum_{j=1}^{N} c_j\, m(E_j)$$

and

$$\int_{\mathbf{R}^n} f(T(x))\, dm(x) = \sum_{j=1}^{N} m\left(T^{-1}(E_j)\right) = |\det T|^{-1} \sum_{j=1}^{N} c_j\, m(E_j),$$

so that

$$|\det T| \int_{\mathbf{R}^n} f(T(x))\, dm(x) = \sum_{j=1}^{N} c_j\, m(E_j) = \int_{\mathbf{R}^n} f(y)\, dm(y).$$

Namely, the desired formula (A.16) holds true for non-negative simple functions $f = \sum_{j=1}^{N} c_j\, \chi_{E_j}$.

Finally, if either $f(y) \geq 0$ on $\mathbf{R}^n$ or $f(y)$ is Lebesgue integrable on $\mathbf{R}^n$, then the desired formula (A.16) follows by approximating $f(y)$ from below by simple functions (see the proof [51, Theorem 2.10]).

Now the proof of Theorem A.7 is complete. □

A.4 Change-of-Variables Formula Under Diffeomorphisms

Let Ω be an open set of Euclidean space $\mathbf{R}^n$. In this section we consider a C^1 map

$$g = \begin{pmatrix} g_1 \\ g_2 \\ \vdots \\ g_n \end{pmatrix} : \Omega \longrightarrow \mathbf{R}^n,$$

where the components g_j of g are functions of class C^1 in Ω, and we denote by $Jg(x)$ a linear transformation defined by the *Jacobian matrix* of $g(x)$ at each point $x \in \Omega$:

$$Jg(x) := (j_{ik}(x)) = \begin{pmatrix} \frac{\partial g_1}{\partial x_1}(x) \dots \frac{\partial g_1}{\partial x_n}(x) \\ \frac{\partial g_2}{\partial x_1}(x) \dots \frac{\partial g_2}{\partial x_n}(x) \\ \vdots \qquad\qquad \vdots \\ \frac{\partial g_n}{\partial x_1}(x) \dots \frac{\partial g_n}{\partial x_n}(x) \end{pmatrix}.$$

The map g is called a C^1-*diffeomorphism* if the map

$$g(x) \colon \Omega \longrightarrow \mathbf{R}^n$$

is *injective* and the Jacobian matrix

$$Jg(x) \colon \mathbf{R}^n \longrightarrow \mathbf{R}^n$$

is *invertible* for all $x \in \Omega$.

Then, by using the inverse function theorem we have the following elementary claim:

Claim A.13 (the multiplicative principle) *Let*

$$\det (Jg(x))$$

be the Jacobian determinant *of the transformation g at a point $x \in \Omega$. Then we have the formula*

$$\det \left(Jg^{-1}(y)\right) \cdot \det \left(Jg\left(g^{-1}(y)\right)\right) = 1 \quad \text{at every point } y \in g(\Omega).$$

The next theorem generalizes Theorem A.7 to differentiable maps (see [130, Theorem], [120, Theorem 7.26], [51, Theorem 2.47], [110, Theorem 1]).

Theorem A.14 (the change-of-variables formula) *Let Ω be an open set in Euclidean space $\mathbf{R}^n$ and let*

$$g \colon \Omega \longrightarrow \mathbf{R}^n$$

be a C^1-diffeomorphism. *Then we have the following assertions (see Fig. A.4 below):*

(1) If $f(y)$ is a Lebesgue measurable function on $g(\Omega)$, then the function $f(g(x))$ is a Lebesgue measurable function on Ω. Moreover, if either

(a) $f(y) \geq 0$ on $g(\Omega)$ or

(b) $f(y)$ is a Lebesgue integrable function on $g(\Omega)$, then we have the formula

$$\int_{g(\Omega)} f(y)\, dm(y) = \int_{\Omega} f(g(x))\, |\det (Jg(x))|\, dm(x). \tag{A.23}$$

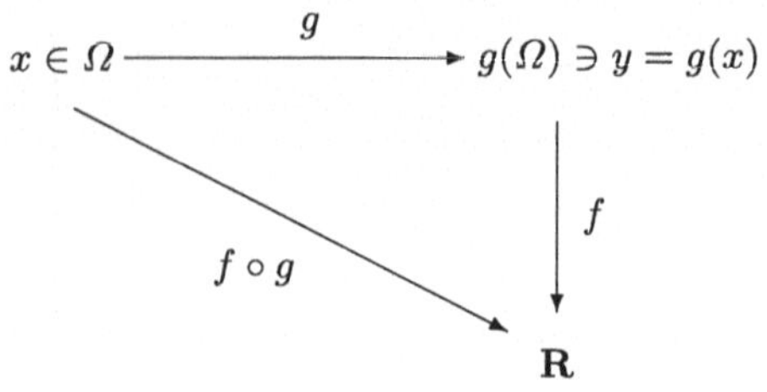

Fig. A.4 The functions $f(y)$ and $f(g(x))$

(2) *If E is a Lebesgue measurable set of Ω, then $g(E)$ is a Lebesgue measurable set of $\mathbf{R}^n$, and we have the formula*

$$m\,(g(E)) = \int_E |\det\,(Jg(x))|\,dm(x). \tag{A.24}$$

Roughly speaking, we have the change-of-variables formula for Lebesgue measures

$$dm(y) = dm\,(g(x)) = |\det\,(Jg(x))|\,dm(x).$$

A.4.1 *Proof of Theorem A.14*

In this subsection we only consider Borel measurable functions and sets. Since g and g^{-1} are both continuous, we have no measurability problems in this case.

The general case follows just as in the proof of Theorem A.7. Moreover, formula (A.24) follows by taking

$$f(y) := \chi_{g(E)}(y) = \begin{cases} 1 & \text{for } y \in g(E), \\ 0 & \text{for } y \in \Omega \setminus g(E). \end{cases}$$

We start with some notational conventions (see Schwartz [130, Theorem]). Let

$$x = (x_1, \ldots, x_n)$$

be a point of Euclidean space $\mathbf{R}^n$ with the maximum norm

$$\|x\|_\infty := \max_{1 \leq i \leq n} |x_i|.$$

Let

$$T : \mathbf{R}^n \longrightarrow \mathbf{R}^n$$

be a linear transformation represented by the matrix (t_{ij})

$$T = \begin{pmatrix} t_{11} & t_{12} & \dots & t_{1n} \\ t_{21} & t_{22} & \dots & t_{2n} \\ \vdots & \vdots & \ddots & \vdots \\ t_{n1} & t_{n2} & \dots & t_{nn} \end{pmatrix},$$

with the norm

$$\|T\|_\infty := \max_{1 \le i \le n} \sum_{j=1}^{n} |t_{ij}|.$$

Then we have the inequality

$$\|T(x)\|_\infty \le \|T\|_\infty \|x\|_\infty \quad \text{for all } x \in \mathbf{R}^n.$$

A cube Q with center a and side length $2h$ can be expressed as follows:

$$Q = Q(a, h) = \left\{x \in \mathbf{R}^n : \|x - a\|_\infty \le h\right\}.$$

In the proof we write the volume of a set S in Euclidean space $\mathbf{R}^n$ as $\mu(S)$, so that

$$\mu(Q) = (2h)^n.$$

The proof of Theorem A.14 is divided into three steps.

Step (I): Now we have, by the mean value theorem,

$$g_i(x) - g_i(a) = \sum_{k=1}^{n} j_{ik} \left(a + \theta_i(x)(x - a)\right) (x_k - a_k),$$

where

$$0 < \theta_i(x) < 1 \quad \text{for } 1 \le i \le n.$$

Hence it follows that

$$\|g(x) - g(a)\|_\infty \le h \max_{y \in Q} \|Jg(y)\|_\infty \quad \text{for all } x \in Q.$$

Since $g(Q)$ is entirely contained in the cube defined by the formula

$$\|z - g(a)\|_\infty \le h \max_{y \in Q} \|Jg(y)\|_\infty,$$

we find that

$$\mu\left(g(Q)\right) \leq \left\{\max_{y\in Q} \|Jg(y)\|_{\infty}\right\}^{n} \mu(Q).$$

If T is an invertible linear transformation, by applying this formula with

$$g := T^{-1}g$$

and then by using formula (A.18) with $T := T^{-1}$ and $E := g(Q)$ we obtain that

$$\begin{aligned}|\det T|^{-1}\,\mu(g(Q)) &= \left|\det\left(T^{-1}\right)\right|\mu(g(Q)) = \mu\left(T^{-1}(g(Q))\right)\\ &\leq \left\{\max_{y\in Q} \|T^{-1}\left(Jg(y)\right)\|_{\infty}\right\}^{n} \mu(Q) \quad \text{for any cube } Q.\end{aligned}$$

This proves that

$$\mu(g(Q)) \leq |\det T| \left\{\max_{y\in Q} \|T^{-1}\left(Jg(y)\right)\|_{\infty}\right\}^{n} \mu(Q) \quad \text{for any cube } Q. \tag{A.25}$$

Since $Jg(x)$ is a matrix-valued, continuous function in Ω, for any $\varepsilon > 0$ we can choose a constant $\delta = \delta(\varepsilon) > 0$ such that

$$\left\{\left\|(Jg(z))^{-1}\left(Jg(y)\right)\right\|_{\infty}\right\}^{n} \leq 1 + \varepsilon \quad \text{if } y, z \in Q \text{ and } \|y - z\|_{\infty} \leq \delta. \tag{A.26}$$

Now we subdivide the cube Q into a finite set $Q_1, \ldots, Q_N$ of non-overlapping cubes with centers $x_1, \ldots, x_N$, respectively:

$$Q = \bigcup_{j=1}^{N} Q_j,$$

where δ is greater than the length of a side of any of $Q_1, \ldots, Q_N$.

By applying inequality (A.25) with

$$\begin{aligned}Q &:= Q_j,\\ T &:= Jg(x_j),\end{aligned}$$

we obtain from assertions (A.25) and (A.26) that

$$
\begin{aligned}
\mu\left(g(Q)\right) &\leq \sum_{j=1}^{N} \mu\left(g\left(Q_j\right)\right) \qquad\qquad (A.27) \\
&\leq \sum_{j=1}^{N} \left|\det\left(Jg(x_j)\right)\right| \left\{\max_{y\in Q_j} \left\|\left(Jg(x_j)\right)^{-1}\left(Jg(y)\right)\right\|_{\infty}\right\}^{n} \mu(Q_j) \\
&\leq (1+\varepsilon) \sum_{j=1}^{N} \left|\det\left(Jg(x_j)\right)\right| \mu(Q_j).
\end{aligned}
$$

We remark that the last term

$$\sum_{j=1}^{N} \left|\det\left(Jg(x_j)\right)\right| \mu(Q_j)$$

is the Lebesgue integral of the simple function

$$\sum_{j=1}^{N} \left|\det\left(Jg(x_j)\right)\right| \chi_{Q_j}(x),$$

which tends uniformly on Q to the function

$$|\det\left(Jg(x)\right)| \quad \text{as } \delta \downarrow 0,$$

since $Jg(x)$ is a matrix-valued, continuous function in Ω.

Therefore, by letting $\varepsilon \downarrow 0$ and $\delta \downarrow 0$ in inequality (A.27) we find that

$$\mu\left(g(Q)\right) \leq \int_{Q} |\det\left(Jg(x)\right)| \, d\mu(x) \quad \text{for any cube } Q \subset \Omega. \qquad (A.28)$$

Step (II): We prove that inequality (A.28) remains valid for any Borel set $E \subset \Omega$. The proof is divided into four substeps.

Substep (II)-1: If U is an open set in $\mathbf{R}^n$, it is known (see [51, Lemma 2.43]) that U can be expressed as a countable union of cubes with disjoint interiors:

$$U = \bigcup_{j=1}^{\infty} Q_j.$$

We remark that the boundaries of cubes Q_j have Lebesgue measure zero, and further that $g(U)$ is a Borel set, since g is a diffeomorphism.

By using inequality (A.28) with

$$Q := Q_j \quad \text{for } j = 1, 2, \ldots,$$

we have the inequality

$$\begin{aligned}
\mu\left(g(U)\right) &= \mu\left(\bigcup_{j=1}^{\infty} g\left(Q_j\right)\right) \leq \sum_{j=1}^{\infty} \mu\left(g(Q_j)\right) \\
&\leq \sum_{j=1}^{\infty} \int_{Q_j} |\det\left(Jg(x)\right)| \, d\mu(x) \\
&= \sum_{j=1}^{\infty} \int_{\Omega} \chi_{Q_j}(x) \, |\det\left(Jg(x)\right)| \, d\mu(x) \\
&= \int_{\Omega} \sum_{j=1}^{\infty} \chi_{Q_j}(x) \, |\det\left(Jg(x)\right)| \, d\mu(x) \\
&= \int_{\Omega} \chi_U(x) \, |\det\left(Jg(x)\right)| \, d\mu(x) \\
&= \int_{U} |\det\left(Jg(x)\right)| \, d\mu(x),
\end{aligned}$$

so that

$$\mu\left(g(U)\right) \leq \int_{U} |\det\left(Jg(x)\right)| \, d\mu(x) \quad \text{for any open set } U \subset \Omega. \tag{A.29}$$

Substep (II)-2: Let E is a Borel set in Ω with

$$\mu(E) < \infty.$$

By using Lemma A.2, we can find a sequence of open sets $U_j \subset \Omega$ such that

$$\begin{aligned}
& U_j \supset E, \\
& \mu(E) \leq \mu\left(U_j\right) \leq \mu(E) + \frac{1}{j} \quad \text{for } j = 1, 2, \ldots, \\
& U_j \searrow U \quad \text{as } j \to \infty,
\end{aligned}$$

Hence, if we let

$$U = \bigcap_{j=1}^{\infty} U_j,$$

then it follows that U is a Borel set in Ω such that

$$\begin{aligned}
& U \supset E, \\
& \mu(U) = \lim_{j \to \infty} \mu\left(U_j\right) = \mu(E).
\end{aligned}$$

By using inequality (A.29) with

$$U := U_j \quad \text{for } j = 1, 2, \ldots,$$

we have the assertions

$$g\left(U_j\right) \searrow g(U) \quad \text{as } j \to \infty,$$
$$\mu(U \setminus E) = 0,$$

and so the inequality

$$\begin{aligned}
&\mu\left(g(E)\right) \leq \mu\left(g(U)\right) = \lim_{j\to\infty} \mu\left(g\left(U_j\right)\right) \\
&\leq \lim_{j\to\infty} \int_{U_j} \left|\det\left(Jg(x)\right)\right| d\mu(x) \\
&= \lim_{j\to\infty} \int_{\Omega} \chi_{U_j}(x) \left|\det\left(Jg(x)\right)\right| d\mu(x) \\
&= \int_{\Omega} \chi_{U}(x) \left|\det\left(Jg(x)\right)\right| d\mu(x) \\
&= \int_{U} \left|\det\left(Jg(x)\right)\right| d\mu(x) \\
&= \int_{E} \left|\det\left(Jg(x)\right)\right| d\mu(x) + \int_{U\setminus E} \left|\det\left(Jg(x)\right)\right| d\mu(x) \\
&= \int_{E} \left|\det\left(Jg(x)\right)\right| d\mu(x).
\end{aligned}$$

This proves that

$$\mu\left(g(E)\right) \leq \int_{E} \left|\det\left(Jg(x)\right)\right| d\mu(x) \tag{A.30}$$
$$\text{for any Borel set } E \subset \Omega \text{ with } \mu(E) < \infty.$$

Substep (II)-3: Let E is a Borel set in Ω. If we let

$$B_\ell := \text{the open ball of radius } \ell \text{ about the origin},$$

then we have the assertions

$$E = \bigcup_{\ell=1}^{\infty} E \cap B_\ell,$$
$$\mu\left(E \cap B_\ell\right) < \infty \quad \text{for } \ell = 1, 2, \ldots.$$

Hence, by applying inequality (A.30) with

$$E_j := E \cap B_j,$$

we find that inequality (A.30) holds true for any Borel set $E \subset \Omega$.

Substep (II)-4: If $f(x)$ is a simple function of the form (cf. Fig. A.2)

$$f(y) = \sum_{j=1}^{N} a_j \chi_{A_j}(y), \tag{A.31}$$

where

$$A_j \in \mathcal{B}\left(\mathbf{R}^n\right) \quad \text{for } 1 \le j \le N,$$

and

$$a_j > 0 \quad \text{for } 1 \le j \le N,$$

then we have, by inequality (A.30) with $E_j := g^{-1}(A_j)$,

$$\begin{aligned}
\int_{g(\Omega)} f(y)\, d\mu(y) &= \sum_{j=1}^{N} a_j\, \mu(A_j) = \sum_{j=1}^{N} a_j\, \mu\left(g\left(g^{-1}\left(A_j\right)\right)\right) \\
&\le \sum_{j=1}^{N} a_j \int_{g^{-1}(A_j)} |\det\left(Jg(x)\right)|\, d\mu(x) \\
&= \sum_{j=1}^{N} a_j \int_{\Omega} \chi_{g^{-1}(A_j)}(x)\, |\det\left(Jg(x)\right)|\, d\mu(x) \\
&= \int_{\Omega} f(g(x))\, |\det\left(Jg(x)\right)|\, d\mu(x),
\end{aligned}$$

since we have, by formula (A.31) for $y = g(x)$,

$$f\left(g(x)\right) = \sum_{j=1}^{N} a_j\, \chi_{A_j}\left(g(x)\right).$$

Namely, the desired formula (A.30) remains valid for non-negative simple functions $f = \sum_{j=1}^{N} a_j\, \chi_{A_j}$:

$$\int_{g(\Omega)} f(y)\, dm(y) \le \int_{\Omega} f(g(x))\, |\det\left(Jg(x)\right)|\, d\mu(x). \tag{A.32}$$

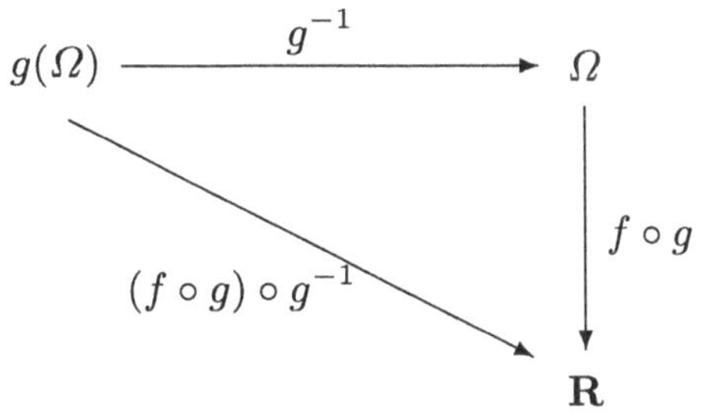

Fig. A.5 The functions $f \circ g$ and $f = (f \circ g) \circ g^{-1}$

Moreover, if either $f(y) \geq 0$ on $\mathbf{R}^n$ or $f(y)$ is Borel integrable on $\mathbf{R}^n$, then the desired inequality (A.31) follows by approximating $f(y)$ from below by simple functions (see the proof [51, Theorem 2.10]).

Step (III): Now, by applying inequality (A.32) with (see Fig. A.5)

$$\begin{aligned} \Omega &:= g(\Omega), \\ g &:= g^{-1}, \\ f &:= f \circ g, \end{aligned}$$

we obtain from the multiplicative principle (Claim A.13) that

$$\begin{aligned} &\int_{\Omega} f(g(x)) \, |\det (Jg(x))| \, d\mu(x) \qquad \text{(A.33)} \\ &\leq \int_{g(\Omega)} (f \circ g) \left(g^{-1}y\right) \left|\det \left(Jg \left(g^{-1}y\right)\right)\right| \left|\det \left(Jg^{-1}(y)\right)\right| \, d\mu(y) \\ &= \int_{g(\Omega)} f(y) \, dm(y). \end{aligned}$$

Therefore, the desired change-of-variables formula (A.23) follows by combining inequalities (A.32) and (A.33).

The proof of Theorem A.14 is complete. □

A.5 Integration in Polar Coordinates

We denote the unit sphere in $\mathbf{R}^n$ by Σ_{n-1}:

$$\Sigma_{n-1} := \left\{x \in \mathbf{R}^n : |x| = 1\right\}.$$

If $x \in \mathbf{R}^n \setminus \{0\}$, the *polar coordinates* of x are given as follows:

$$r = |x| \in (0, \infty),$$
$$x' = \frac{x}{|x|} \in \Sigma_{n-1}.$$

The map

$$\begin{aligned} \Phi\colon \mathbf{R}^n \setminus \{0\} &\longrightarrow (0, \infty) \times \Sigma_{n-1} \\ x &\longmapsto \left(r, x'\right) \end{aligned}$$

is a continuous bijection of $\mathbf{R}^n \setminus \{0\}$ onto $(0, \infty) \times \Sigma_{n-1}$.

We denote by m_* the Borel measure on $(0, \infty) \times \Sigma_{n-1}$ induced by the map Φ from the Lebesgue measure m on $\mathbf{R}^n$:

$$m_*(E) = m\left(\Phi^{-1}(E)\right) \quad \text{for } E \subset (0, \infty) \times \Sigma_{n-1}.$$

The situation can be visualized as follows:

$$\boxed{\begin{array}{ccc} (\mathbf{R}^n, m) & & (\mathbf{R}^n, m_*) \\ \uparrow & & \uparrow \\ \Phi^{-1}(E) \subset \mathbf{R}^n \setminus \{0\} & \xleftarrow[\Phi^{-1}]{} & ((0, \infty) \times \Sigma_{n-1}, m_*) \supset E \end{array}}$$

Moreover, we define the measure ρ on the interval $(0, \infty)$ by the formula

$$\rho = r^{n-1}\, dr \quad \text{on } (0, \infty).$$

Then we have the following formula in polar coordinates (see [51, Theorem 2.49]).

Theorem A.15 *There exists a unique Borel measure σ on the unit sphere Σ_{n-1} such that*

$$d\mu_*\left(r, x'\right) = r^{n-1}\, dr\, d\sigma(x') \quad \text{on } (0, \infty) \times \Sigma_{n-1}.$$

Let $f(x)$ be a Borel measurable function on $\mathbf{R}^n$. If $f(x)$ is either non-negative or integrable on $\mathbf{R}^n$, then we have the formula

$$\int_{\mathbf{R}^n} f(x)\, d\mu(x) = \int_0^\infty r^{n-1}\, dr \int_{\Sigma_{n-1}} f(r\, x')\, d\sigma(x'). \tag{A.34}$$

Remark A.16 Formula (A.34) can be extended to Lebesgue measurable functions by considering the completion of the Borel measure σ.

Corollary A.17 *Let $f(x)$ be a Lebesgue measurable function on $\mathbf{R}^n$ such that*

$$f(x) = g\left(|x|\right)$$

for some Lebesgue measurable function on the interval $(0, \infty)$. *If* $f(x)$ *is either non-negative or integrable on* $\mathbf{R}^n$, *then we have the formula*

$$\int_{\mathbf{R}^n} f(x)\, d\mu(x) = |\Sigma_{n-1}| \int_0^\infty r^{n-1}\, g(r)\, dr, \tag{A.35}$$

where

$$|\Sigma_{n-1}| = \frac{2\pi^{n/2}}{\Gamma(n/2)}$$

is the surface area of the unit sphere Σ_{n-1}.

Appendix B
A Short Course to the Potential Theoretic Approach

In this appendix, following faithfully Gilbarg–Trudinger [59], we present a brief introduction to the classical *potential theoretic approach* to the Dirichlet problem for the usual Laplacian

$$\Delta = \sum_{j=1}^{n} \frac{\partial^2}{\partial x_j^2} = \frac{\partial^2}{\partial x_1^2} + \ldots + \frac{\partial^2}{\partial x_n^2}.$$

Ladyzhenskaya–Ural'tseva [84] and Miranda [106] are the classics for the theory of elliptic partial differential equations.

The approach here can be traced back to the pioneering work of Juliusz Schauder, [127] and [128], on the Dirichlet problem for second order elliptic partial differential operators. This appendix is included for the sake of completeness and most of the material will be quite familiar to the reader and may be omitted.

The appendix may be summarized as in Table B.1 below.

B.1 Hölder Continuity and Hölder Spaces

Let Ω be an open set in Euclidean space $\mathbf{R}^n$. First, we let

$$C(\Omega) = \text{ the space of continuous functions on } \Omega.$$

If k is a positive integer, we let

$$C^k(\Omega) = \text{ the space of functions of class } C^k \text{ on } \Omega.$$

K. Taira, *Real Analysis Methods for Markov Processes*,
https://doi.org/10.1007/978-981-97-3659-1

Table B.1 A bird's eye view of the classical potential theoretic approach and the singular integral approach to the Dirichlet problem

Potential theoretic approach	Singular integral approach
Theorem B.9	Lemma 13.2
Theorem B.11	Theorem 13.3 (Theorem 12.1)
Theorem B.18	Lemma 14.13 (Lemma B.19)
Theorem B.20	Theorem 12.2
Theorem B.21	Theorem 15.1

Furthermore, we let

$$C(\overline{\Omega}) = \text{the space of functions in } C(\Omega) \text{ having continuous extensions to the closure } \overline{\Omega} \text{ of } \Omega.$$

If k is a positive integer, we let

$$C^k(\overline{\Omega}) = \text{the space of functions in } C^k(\Omega) \text{ all of whose derivatives of order } \leq k \text{ have continuous extensions to } \overline{\Omega}.$$

Let $0 < \theta < 1$. A function u defined on Ω is said to be *uniformly Hölder continuous* with exponent θ in Ω if the quantity

$$[u]_{\theta;\Omega} = \sup_{\substack{x,y\in\Omega \\ x\neq y}} \frac{|u(x) - u(y)|}{|x - y|^\theta} \tag{B.1}$$

is finite. We say that u is *locally Hölder continuous* with exponent θ in Ω if it is uniformly Hölder continuous with exponent θ on compact subsets of Ω.

If $0 < \theta < 1$, we define the Hölder space $C^\theta(\Omega)$ as follows:

$$C^\theta(\Omega) = \text{the space of functions in } C(\Omega) \text{ which are locally Hölder continuous with exponent } \theta \text{ on } \Omega.$$

If k is a positive integer and $0 < \theta < 1$, we define the Hölder space $C^{k+\theta}(\Omega)$ as follows:

$$C^{k+\theta}(\Omega) = \text{the space of functions in } C^k(\Omega) \text{ all of whose } k\text{-th order derivatives are locally Hölder continuous with exponent} \theta \text{ on } \Omega.$$

Furthermore, we let

$$C^{\theta}(\overline{\Omega}) = \text{the space of functions in } C(\overline{\Omega}) \text{ which are Hölder continuous with exponent } \theta \text{ on } \overline{\Omega},$$

and

$$C^{k+\theta}(\overline{\Omega}) = \text{the space of functions in } C^k(\overline{\Omega}) \text{ all of whose } k\text{-th order derivatives are Hölder continuous with exponent } \theta \text{ on } \overline{\Omega}.$$

Let k be a non-negative integer and $0 < \theta < 1$. We introduce various seminorms on the spaces $C^k(\Omega)$ and $C^{k+\theta}(\Omega)$ as follows:

$$[u]_{k,0;\Omega} = \left|D^k u\right|_{0;\Omega} = \sup_{x\in\Omega} \sup_{|\alpha|=k} |D^\alpha u(x)|\,, \tag{B.2a}$$

$$[u]_{k,\theta;\Omega} = \left[D^k u\right]_{\theta;\Omega} = \sup_{|\alpha|=k} \left[D^\alpha u\right]_{\theta;\Omega}\,. \tag{B.2b}$$

We can define the associated norms on the spaces $C^k(\overline{\Omega})$ and $C^{k+\theta}(\overline{\Omega})$ as follows:

$$\|u\|_{C^k(\Omega)} = |u|_{k;\Omega} = \sum_{j=0}^{k} \left|D^j u\right|_{0;\Omega}\,, \tag{B.3a}$$

$$\|u\|_{C^{k+\theta}(\Omega)} = |u|_{k,\theta;\Omega} = |u|_{k;\Omega} + \left[D^k u\right]_{\theta;\Omega}\,. \tag{B.3b}$$

Moreover, if Ω is a *bounded* domain in $\mathbf{R}^n$ with the *diameter*

$$d := \text{diam } \Omega = \sup_{x,y\in\Omega} |x-y|\,,$$

then we can introduce two non-dimensional norms

$$\|u\|_{C^k(\Omega)}\,, \quad \|u\|_{C^{k+\theta}(\Omega)}$$

equivalent respectively to the norms

$$\|u\|_{C^k(\Omega)}\,, \quad \|u\|_{C^{k+\theta}(\Omega)}$$

as follows:

$$\|u\|'_{C^k(\Omega)} = \sum_{j=0}^{k} d^j \left|D^j u\right|_{0;\Omega}, \tag{B.4a}$$

$$\|u\|'_{C^{k+\theta}(\Omega)} = \sum_{j=0}^{k} d^j \left|D^j u\right|_{0;\Omega} + d^{k+\theta}\left[D^k u\right]_{\theta;\Omega}. \tag{B.4b}$$

Then we have the following claims:

Claim B.1 *If Ω is bounded, then the space $C^k(\overline{\Omega})$ is a Banach space with the norms $\|\cdot\|_{C^k(\Omega)}$ and $\|\cdot\|'_{C^k(\Omega)}$.*

Claim B.2 *If Ω is bounded, then the Hölder space $C^{k+\theta}(\overline{\Omega})$ is a Banach space with the norms $\|\cdot\|_{C^{k+\theta}(\Omega)}$ and $\|\cdot\|'_{C^{k+\theta}(\Omega)}$.*

Claim B.3 *Let Ω be a* bounded *domain in $\mathbf{R}^n$ with smooth boundary $\partial\Omega$. Let k, j be non-negative integers and $0 < \theta, \tau < 1$. Assume that*

$$j + \tau < k + \theta.$$

Then the injection

$$C^{k+\theta}(\overline{\Omega}) \subset C^{j+\tau}(\overline{\Omega})$$

is compact *(or* completely continuous*) ([59, Chap. 6, Lemma 6.36]).*

B.2 Interior Estimates for Harmonic Functions

First, by differentiating the Poisson integral we can obtain the following interior derivative estimates for harmonic functions ([59, Chap. 2, Theorem 2.10]).

Theorem B.4 *Let Ω be an open set in $\mathbf{R}^n$ and let Ω' be open subset of Ω that has compact closure in Ω:*

$$\Omega' \Subset \Omega.$$

If $u(x)$ is harmonic *in Ω, then we have, for any multi-index α,*

$$\sup_{x\in\Omega'} |D^\alpha u(x)| \leq \left(\frac{n|\alpha|}{d}\right)^{|\alpha|} \sup_{x\in\Omega} |u(x)|, \tag{B.5}$$

where

$$d = \operatorname{dist}\left(\Omega', \partial\Omega\right).$$

Proof We only prove the case where $|\alpha| = 1$. We assume that

$$B := B(y, R) \subset \Omega' \Subset \Omega,$$

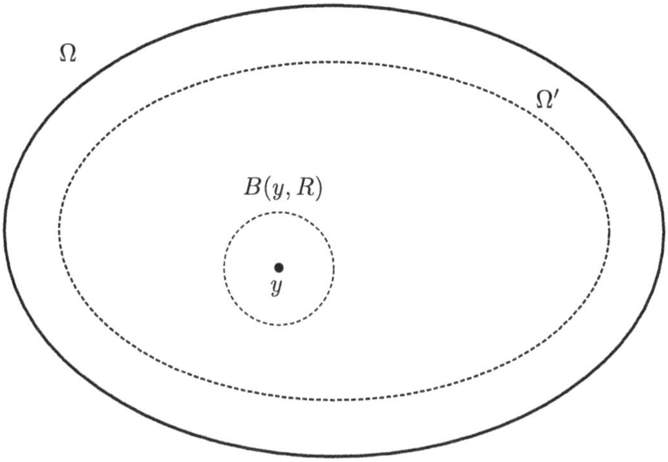

Fig. B.1 The domains Ω and Ω'

where $B = B(y, R)$ is the open ball of radius R about y (see Fig. B.1).

First, since $u(x)$ is harmonic in Ω, it follows that

$$\Delta (D_i u) = D_i (\Delta u) = 0 \quad \text{in } \Omega \text{ for each } 1 \leq i \leq n.$$

Hence, by applying the mean value theorem for harmonic functions to each function $D_i u$ (see [50, Corollary 2.9]) and then the divergence theorem (Theorem 5.3) we obtain that

$$\begin{aligned} D_i u(y) &= \frac{1}{R^n \omega_n} \int_B D_i u(x)\, dx \\ &= \frac{n}{\omega_n R^n} \int_{\partial B} u(z)\, \nu_i \, d\sigma(z) \quad \text{for each } 1 \leq i \leq n, \end{aligned}$$

where $d\sigma$ is the surface element of ∂B,

$$\omega_n = \frac{2\pi^{n/2}}{\Gamma(n/2)}$$

is the surface area of the unit ball in $\mathbf{R}^n$ and

$$\boldsymbol{\nu} = (\nu_1, \nu_2, \ldots, \nu_n)$$

is the unit outward normal to ∂B.

Then we have the inequality

$$\begin{aligned}
|D_i u(y)| &\le \frac{n}{R^n \omega_n} \int_{\partial B} |u(z)\, \nu_i|\, d\sigma(z) \\
&\le \frac{n}{R^n \omega_n} \sup_{z \in \partial B} |u(z)| \cdot \int_{\partial B} d\sigma(z) \\
&= \frac{n}{R} \sup_{z \in \partial B} |u(z)| \\
&\le \frac{n}{R} \sup_{x \in B} |u(x)| \\
&\le \frac{n}{R} \sup_{x \in \Omega} |u(x)| .
\end{aligned}$$

By letting $R \uparrow d_y = \operatorname{dist}\,(y, \partial\Omega)$, we obtain that

$$|D_i u(y)| \le \frac{n}{d_y} \sup_{x \in \Omega} |u(x)| \quad \text{for all } y \in \Omega.$$

We remark that

$$\frac{1}{d_z} \le \frac{1}{d} \quad \text{for all } z \in \Omega'.$$

Hence we have the inequality

$$|D_i u(z)| \le \frac{n}{d} \sup_{x \in \Omega} |u(x)| \quad \text{for all } z \in \Omega'.$$

This proves the desired interior estimate (B.5) for $|\alpha| = 1$.

The proof of Theorem B.4 is complete. □

B.3 Hölder Regularity for the Newtonian Potential

We consider the fundamental solution $\Gamma(x - y)$ for the Laplacian in the case $n \ge 3$:

$$\Gamma(x - y) = \Gamma(|x - y|) = \frac{1}{(2 - n)\omega_n} |x - y|^{2-n} . \tag{B.6}$$

Then we have the following formulas for the fundamental solution $\Gamma(x-y)$:

$$D_i\Gamma(x-y) = \frac{\partial\Gamma}{\partial x_i}(x,y) = \frac{1}{\omega_n}\,(x_i - y_i)\,|x-y|^{-n}, \tag{B.7a}$$

$$\begin{aligned} D_iD_j\Gamma(x-y) &= \frac{\partial^2\Gamma}{\partial x_i \partial x_j}(x,y) \\ &= \frac{1}{\omega_n}\left\{|x-y|^2\delta_{ij} - n\,(x_i-y_i)\left(x_j-y_j\right)\right\}|x-y|^{-n-2}. \end{aligned} \tag{B.7b}$$

We remark that the fundamental solution $\Gamma(x-y)$ is *harmonic* for $x \neq y$:

$$\Delta_x\Gamma(x-y) = \sum_{i=1}^{n}\frac{\partial^2\Gamma}{\partial x_i^2}(x-y) = 0 \quad \text{for } x \neq y.$$

By formulas (B.7), we have the following estimates for the fundamental solution $\Gamma(x,y)$:

$$|D_i\Gamma(x-y)| \leq \frac{1}{\omega_n}\,|x-y|^{1-n}, \tag{B.8a}$$

$$\left|D_iD_j\Gamma(x-y)\right| \leq \frac{n}{\omega_n}\,|x-y|^{-n}. \tag{B.8b}$$

Claim B.5 *Let Ω be a* smooth *domain with boundary $\partial\Omega$. If $u \in C^2(\overline{\Omega})$, then we have the* Green representation formula

$$\begin{aligned} &u(y) \\ &= \int_{\partial\Omega}\left(u(x)\frac{\partial\Gamma}{\partial\boldsymbol{\nu}}(x,y) - \Gamma(x,y)\frac{\partial u}{\partial\boldsymbol{\nu}}(x)\right)d\sigma(x) + \int_{\Omega}\Gamma(x,y)\Delta u(x)\,dx \end{aligned} \tag{B.9}$$

for $y \in \Omega$.

Proof We shall apply Green's second identity (5.4b) for the fundamental solution $\Gamma(x,y)$.

Let y be an arbitrary point of Ω. We replace the domain Ω by the punctured domain $\Omega \setminus \overline{B_\rho}$, where $B_\rho = B(y,\rho)$ is a sufficiently small open ball of radius ρ about y (see Fig. B.2 below). By applying Green's second identity (5.4b) to our situation, we obtain that

$$\begin{aligned} &\int_{\Omega\setminus B_\rho}\Gamma(x,y)\Delta u(x)\,dx \\ &= \int_{\partial\Omega}\left(\Gamma(x,y)\frac{\partial u}{\partial\boldsymbol{\nu}}(x) - u(x)\frac{\partial\Gamma}{\partial\boldsymbol{\nu}}(x,y)\right)d\sigma(x) \\ &\quad + \int_{\partial B_\rho}\left(\Gamma(x,y)\frac{\partial u}{\partial\boldsymbol{\nu}}(x) - u(x)\frac{\partial\Gamma}{\partial\boldsymbol{\nu}}(x,y)\right)d\sigma(x). \end{aligned} \tag{B.10}$$

Fig. B.2 The punctured domain $\Omega \setminus \overline{B_\rho}$

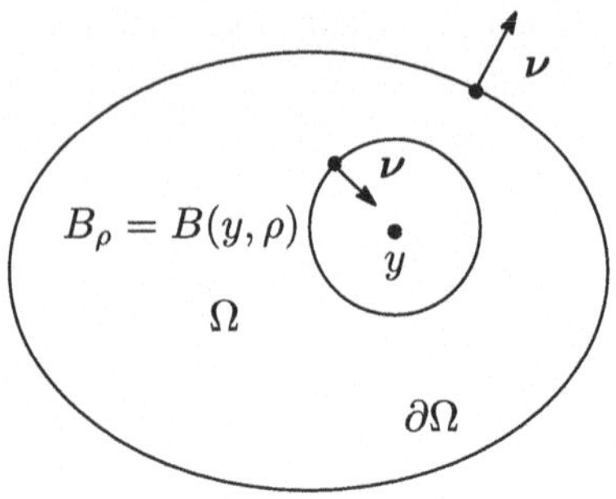

However, we have, as $\rho \downarrow 0$,

$$\bullet \int_{\partial B_\rho} \Gamma(x, y) \frac{\partial u}{\partial \boldsymbol{\nu}}(x)\, d\sigma(x)$$
$$= \Gamma(\rho) \int_{\partial B_\rho} \frac{\partial u}{\partial \boldsymbol{\nu}}(x)\, d\sigma(x)$$
$$\leq \frac{\rho^{2-n}}{(2-n)\omega_n}\, \omega_n \rho^{n-1} \left(\sup_{B_\rho} |Du| \right)$$
$$= \frac{\rho}{2-n} \left(\sup_{B_\rho} |Du| \right) \longrightarrow 0.$$

On the other hand, since we have the formula

$$\frac{\partial}{\partial \boldsymbol{\nu}} = \frac{1}{\rho} \sum_{i=1}^{n} (y_i - x_i) \frac{\partial}{\partial x_i} \quad \text{on the sphere } \partial B_\rho = \partial B(y, \rho),$$

it follows that

$$\frac{\partial \Gamma}{\partial \boldsymbol{\nu}}(x, y)$$
$$= \frac{1}{(2-n)\omega_n \rho} \sum_{i=1}^{n} (y_i - x_i) \frac{\partial}{\partial x_i} \left(\sum_{i=1}^{n} (x_i - y_i)^2 \right)^{1-n/2}$$
$$= -\frac{1}{\omega_n \rho} \left(\sum_{i=1}^{n} (x_i - y_i)^2 \right) \left(\sum_{i=1}^{n} (x_i - y_i)^2 \right)^{-n/2} = -\frac{1}{\omega_n \rho} \rho^2 \rho^{-n}$$
$$= -\frac{1}{\omega_n \rho^{n-1}} \quad \text{on } \partial B_\rho.$$

Hence we have, as $\rho \downarrow 0$,

- $$\int_{\partial B_\rho} u(x)\frac{\partial \Gamma}{\partial \boldsymbol{\nu}}(x,y)d\sigma(x) = -\frac{1}{\omega_n \rho^{n-1}}\int_{\partial B_\rho} u(x)\,d\sigma(x) \longrightarrow \quad -u(y).$$

Indeed, by the continuity of u it suffices to note that

$$\begin{aligned}
&\left|\frac{1}{\omega_n \rho^{n-1}}\int_{\partial B_\rho} u(x)\,d\sigma(x) - u(y)\right| \\
&= \frac{1}{\omega_n \rho^{n-1}}\left|\int_{\partial B_\rho} (u(x)-u(y))\,d\sigma(x)\right| \\
&\leq \frac{1}{\omega_n \rho^{n-1}}\int_{\partial B_\rho} |u(x)-u(y)|\,d\sigma(x) \\
&\leq \frac{1}{\omega_n \rho^{n-1}}\left(\sup_{|x-y|=\rho} |u(x)-u(y)|\right)\int_{\partial B_\rho} d\sigma(x) \\
&= \sup_{|x-y|=\rho} |u(x)-u(y)| \longrightarrow 0 \quad \text{as } \rho \downarrow 0.
\end{aligned}$$

Therefore, by letting $\rho \downarrow 0$ in formula (B.10) we obtain the desired Green representation formula

$$\begin{aligned}
&u(y) \\
&= \int_{\partial\Omega}\left(u(x)\frac{\partial \Gamma}{\partial \boldsymbol{\nu}}(x,y) - \Gamma(x,y)\frac{\partial u}{\partial \boldsymbol{\nu}}(x)\right) d\sigma(x) + \int_\Omega \Gamma(x,y)\Delta u(x)\,dx.
\end{aligned}$$

The proof of Claim B.5 is complete. □

Now we study some differentiability properties of the *Newtonian potential* of a function $f(x)$

$$w(x) := (\Gamma * f)(x) = \int_\Omega \Gamma(x-y)\,f(y)\,dy$$

in an open subset Ω of Euclidean space $\mathbf{R}^n$.

First, we obtain the following ([59, Chap. 4, Lemma 4.1]).

Lemma B.6 *Let $f(x)$ be a bounded and integrable function in Ω. Then it follows that*

$$w(x) = \int_\Omega \Gamma(x-y)\,f(y)\,dy \in C^1(\Omega),$$

and we have, for any $x \in \Omega$,

$$D_i w(x) = \int_\Omega D_i\Gamma(x-y)\,f(y)\,dy \quad \textit{for } 1 \leq i \leq n. \tag{B.11}$$

Secondly, we obtain the following ([59, Chap. 4, Lemma 4.2]).

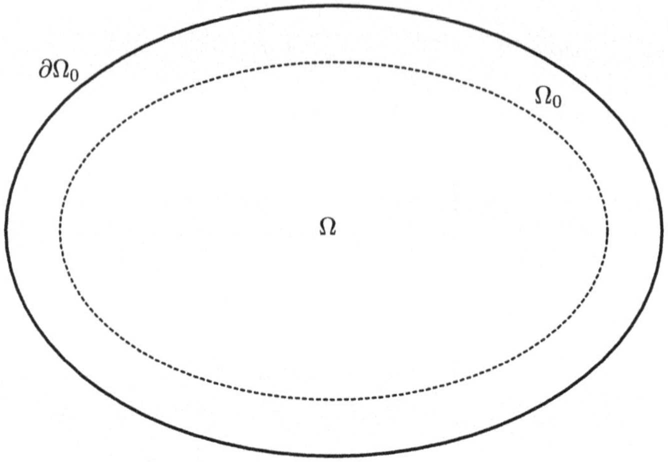

Fig. B.3 The domains Ω and Ω_0

Lemma B.7 *Let $f(x)$ be a bounded and locally Hölder continuous function with exponent $0 < \theta \leq 1$ in Ω. Then it follows that*

$$\begin{cases} w(x) = \int_\Omega \Gamma(x-y)\, f(y)\, dy \in C^2(\Omega), \\ \Delta w = f \quad \text{in } \Omega. \end{cases}$$

Moreover, take an arbitrary smooth domain Ω_0 that contains Ω, and let $\nu = (\nu_1, \nu_2, \ldots, \nu_n)$ be the unit outward normal to $\partial\Omega_0$ and let $d\sigma$ be the surface element on $\partial\Omega_0$ (see Fig. B.3). Then we have, for any $x \in \Omega$,

$$\begin{aligned} D_i D_j w(x) = &\int_{\Omega_0} D_i D_j \Gamma(x-y) \left(f^0(y) - f(y)\right) dy \\ &- f(x) \int_{\partial\Omega_0} D_i \Gamma(x-y)\, \nu_j(y)\, d\sigma(y) \quad \text{for } 1 \leq i, j \leq n, \end{aligned} \tag{B.12}$$

where f^0 is the zero-extension *of f outside Ω:*

$$f^0(x) = \begin{cases} f(x) & \text{in } \Omega, \\ 0 & \text{in } \Omega_0 \setminus \Omega. \end{cases}$$

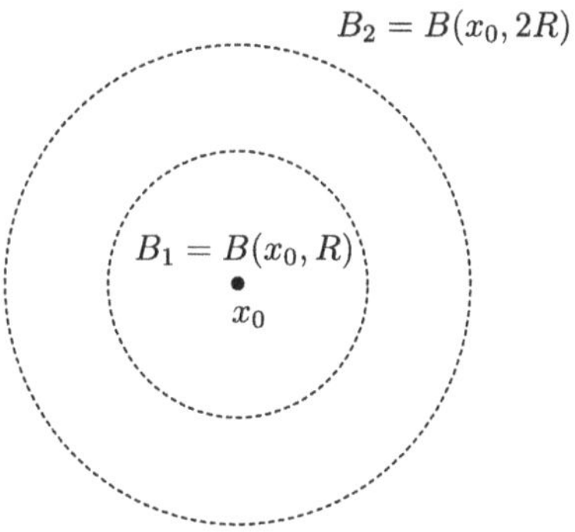

Fig. B.4 The concentric balls B_1 and B_2

B.4 Hölder Estimates for the Second Derivatives

We start with the following basic Hölder estimates fo the Newtonian potential ([59, Chap. 4, Lemma 4.4]).

Lemma B.8 *Let $B_1 = B(x_0, R)$ and $B_2 = B(x_0, 2R)$ be concentric balls in $\mathbf{R}^n$ (see Fig. B.4). For a function $f \in C^\theta(\overline{B_2})$ with $0 < \theta < 1$, let $w(x)$ be the* Newtonian potential *of f in B_2:*

$$w(x) = \int_{B_2} \Gamma(x - y) f(y) \, dy.$$

Then it follows that

$$w \in C^{2+\theta}(\overline{B_1}),$$

and we have the interior estimate

$$\left|D^2 w\right|'_{0,\theta;B_1} \leq C_1 \, |f|'_{0,\theta;B_2}\,, \tag{B.13}$$

that is,

$$\left|D^2 w\right|_{0,B_1} + (2R)^\theta \left[D^2 w\right]_{\theta:B_1} \leq C_1 \left(|f|_{0,B_2} + (4R)^\theta \, [f]_{\theta;B_2}\right),$$

with a constant $C_1 = C_1(n, \theta) > 0$.

We can prove the following *interior Hölder estimate* for solutions of Poisson's equation ([59, Chap. 4, Theorem 4.6]).

Theorem B.9 *Let Ω be a domain in $\mathbf{R}^n$ and let $f \in C^\theta(\Omega)$ with $0 < \theta < 1$. If a function $u \in C^2(\Omega)$ satisfies Poisson's equation*

$$\Delta u = f \quad in\ \Omega,$$

then it follows that

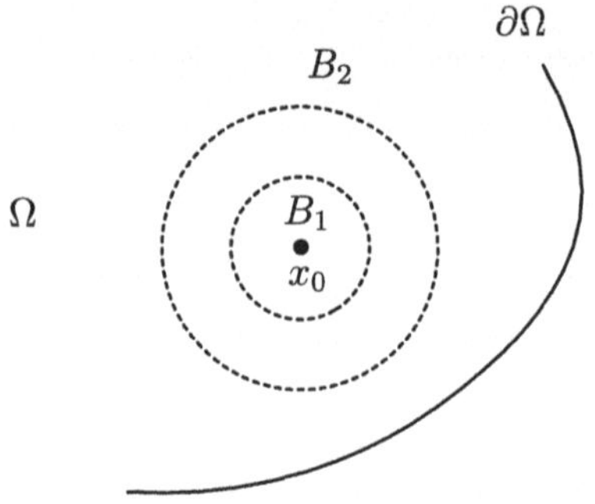

Fig. B.5 The concentric balls B_1 and B_2 in Ω

$$u \in C^{2+\theta}(\Omega).$$

Moreover, for any two concentric balls (see Fig. B.5)

$$B_1 = B(x_0, R), \quad B_2 = B(x_0, 2R) \Subset \Omega,$$

we have the estimate

$$|u|'_{2,\theta;B_1} \leq C_2 \left(|u|_{0;B_2} + (4R)^2 \, |f|'_{0,\theta;B_2} \right), \tag{B.14}$$

that is,

$$\begin{aligned} &|u|_{0;B_1} + (2R)\,|Du|_{0;B_1} + (2R)^2 \left|D^2 u\right|_{0;B_1} + (2R)^{2+\theta} \left[D^2 u\right]_{\theta;B_1} \\ &\leq C_2 \left(|u|_{0;B_2} + (4R)^2 \, |f|_{0,B_2} + (4R)^{2+\theta} \, [f]_{\theta;B_2} \right), \end{aligned}$$

with a constant $C_2 = C_2(n, \theta) > 0$.

Let Ω be an open set in $\mathbf{R}^n$. For $x, y \in \Omega$, we let

$$\begin{aligned} d_x &= \text{dist}\,(x, \partial\Omega), \\ d_{x,y} &= \min\left(d_x, d_y\right). \end{aligned}$$

If k is a non-negative integer and $0 < \theta < 1$, then we introduce various interior seminorms and norms on the Hölder spaces $C^k(\Omega)$ and $C^{k+\theta}(\Omega)$ as follows:

$$[u]^*_{k,0;\Omega} = [u]^*_{k;\Omega} = \sup_{x\in\Omega} \sup_{|\alpha|=k} d_x^k \, |D^\alpha u(x)| \,, \tag{B.15a}$$

$$|u|^*_{k;\Omega} = \sum_{j=0}^{k} [u]^*_{j;\Omega} \,, \tag{B.15b}$$

$$[u]^*_{k,\theta;\Omega} = \sup_{x,y\in\Omega} \sup_{|\alpha|=k} d_{x,y}^{k+\theta} \frac{|D^\alpha u(x) - D^\alpha u(y)|}{|x-y|^\theta}, \tag{B.15c}$$

$$|u|^*_{k,\theta;\Omega} = |u|^*_{k;\Omega} + [u]^*_{k,\theta;\Omega} \,. \tag{B.15d}$$

Claim B.10 *(i) If Ω is bounded with $d = \operatorname{diam}\,\Omega$, then we have the inequality*

$$|u|^*_{k,\theta;\Omega} \le \max\left(1, d^{\theta+k}\right) |u|_{k,\theta;\Omega}. \tag{B.14$'$}$$

(ii) If $\Omega' \Subset \Omega$ with $\sigma = \operatorname{dist}\left(\Omega', \partial\Omega\right)$, then we have the inequality

$$\min\left(1, \sigma^{\theta+k}\right) |u|_{k,\theta;\Omega'} \le |u|^*_{k,\theta;\Omega}. \tag{B.14$''$}$$

Moreover, we introduce a seminorm on the Hölder space $C^\theta(\Omega)$ as follows:

$$|f|^{(k)}_{0,\theta;\Omega} = \sup_{x\in\Omega} d_x^k \, |f(x)| + \sup_{x,y\in\Omega} d_{x,y}^{k+\theta} \frac{|f(x) - f(y)|}{|x-y|^\theta}. \tag{B.16}$$

Then, by using Theorem B.9 we can obtain a *Schauder interior estimate* for a general domain Ω ([59, Chap. 4, Theorem 4.8]).

Theorem B.11 (the Schauder interior estimate) *Assume that a function $u \in C^2(\Omega)$ satisfies the equation*

$$\Delta u = f \quad in\ \Omega$$

for a function $f \in C^\theta(\Omega)$. Then we have the interior estimate

$$|u|^*_{2,\theta;\Omega} \le C_3 \left(|u|_{0;\Omega} + |f|^{(2)}_{0,\theta;\Omega}\right) \tag{B.17}$$

with a constant $C_3 = C_3(n,\theta) > 0$.

Proof We have only to consider the case where

$$|u|_{0;\Omega} < \infty, \quad |f|^{(2)}_{0,\theta;\Omega} < \infty.$$

The proof is divided into two steps. In the following the letter C denotes a generic positive constant independent of u and f.

Step 1: For each point x of Ω, we let (see Fig. B.6 below)

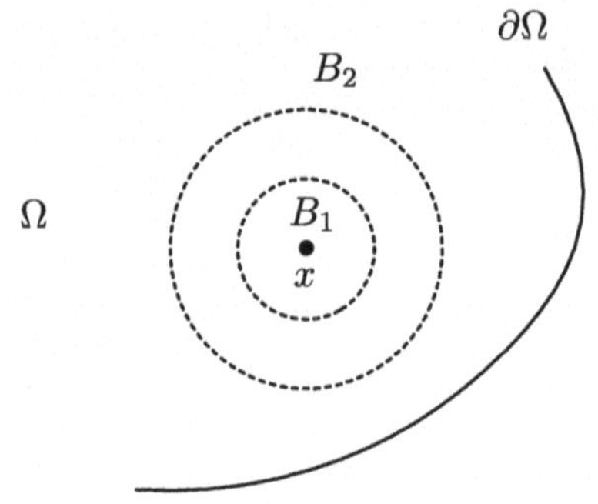

Fig. B.6 The concentric balls B_1 and B_2 in Ω

$$
\begin{aligned}
R &:= \frac{1}{3}\, d_x = \frac{1}{3}\, \operatorname{dist}(x, \partial\Omega),\\
B_1 &:= B(x, R),\\
B_2 &:= B(x, 2R).
\end{aligned}
$$

Then we have, by estimate (B.14),

$$
\begin{aligned}
d_x\, |Du(x)| + d_x^2 \left|D^2u(x)\right| &\leq 3R\, |Du|_{0; B_1} + (3R)^2 \left|D^2 u\right|_{0, B_1} \\
&\leq C\left(|u|_{0; B_2} + R^2\, |f|'_{0,\theta; B_2}\right) \\
&\leq C\left(|u|_{0; \Omega} + |f|^{(2)}_{0,\theta; B_2}\right).
\end{aligned} \tag{B.18}
$$

However, we have the estimate

$$
R^2\, |f|'_{0,\theta; B_2} = R^2\left(|f|_{0; B_2} + (4R)^{\theta}\, [f]_{\theta; B_2}\right) \leq C\, |f|^{(2)}_{0,\theta; \Omega}\,. \tag{B.19}
$$

Indeed, since we have the inequality

$$
R = \frac{1}{3} d_x \leq d_z \quad \text{for all } z \in B_2,
$$

it follows that

$$
R^2 \sup_{z \in B_2} |f(z)| \leq \sup_{z \in B_2} d_z^2\, |f(z)| \leq \sup_{x \in \Omega} d_x^2\, |f(x)|\,.
$$

Similarly, since we have the inequality

$$
R = \frac{1}{3} d_x \leq d_{z,w} = \min\left(d_z, d_w\right) \quad \text{for all } z, w \in B_2,
$$

it follows that

$$
\begin{aligned}
R^{2+\theta} \frac{|f(z) - f(w)|}{|z - w|^{\theta}} &\leq \sup_{z, w \in B_2} d_{z,w}^{2+\theta} \frac{|f(z) - f(w)|}{|z - w|^{\theta}} \\
&\leq \sup_{x, y \in \Omega} d_{x,y}^{2+\theta} \frac{|f(z) - f(w)|}{|z - w|^{\theta}}.
\end{aligned}
$$

Hence we have the inequality

$$
R^2 \, |f|'_{0,\theta; B_2} \leq C \, |f|^{(2)}_{0,\theta;\Omega} \, .
$$

Therefore, by combining inequalities (B.18) and (B.19) we obtain that

$$
\begin{aligned}
|u|^*_{2;\Omega} &= |u|_{0;\Omega} + \sup_{x \in \Omega} d_x \, |Du(x)| + \sup_{x \in \Omega} d_x^2 \left|D^2 u(x)\right| \\
&\leq C \left(|u|_{0;\Omega} + |f|^{(2)}_{0,\theta;\Omega} \right).
\end{aligned} \tag{B.20}
$$

Step 2: We assume that $d_x \leq d_y$ for $x, y \in \Omega$, so that

$$
d_x = d_{x,y} = 3R \quad \text{for } x, y \in \Omega.
$$

Then it follows that

$$
\frac{|D^2 u(x) - D^2 u(y)|}{|x - y|^{\theta}} \leq \begin{cases} \left[D^2 u\right]_{\theta; B_1} & \text{if } y \in B_1, \\ \frac{1}{R^{\theta}} \left(\left|D^2 u(x)\right| + \left|D^2 u(y)\right| \right) & \text{if } y \in \Omega \setminus B_1. \end{cases}
$$

Hence we have, for $x, y \in \Omega$,

$$
\begin{aligned}
d_{x,y}^{2+\theta} \frac{|D^2 u(x) - D^2 u(y)|}{|x - y|^{\theta}} &\leq (3R)^{2+\theta} \left[D^2 u\right]_{\theta; B_1} \\
&+ 3^{\theta} \, (3R)^2 \left(\left|D^2 u(x)\right| + \left|D^2 u(y)\right| \right).
\end{aligned} \tag{B.21}
$$

However, by using inequalities (B.14) and (B.19) we can estimate the first term on the right-hand side of inequality (B.21) as follows:

$$
\begin{aligned}
&(3R)^{2+\theta} \left[D^2 u\right]_{\theta; B_1} \\
&\leq C \left(|u|_{0; B_2} + R^2 \, |f|'_{0,\theta; B_2} \right) \leq C \left(|u|_{0,\Omega} + |f|^{(2)}_{0,\theta;\Omega} \right).
\end{aligned}
$$

On the other hand, by using inequality (B.20) we can estimate the second term on the right-hand side of inequality (B.21) as follows:

$$
\begin{aligned}
3^{\theta} \, (3R)^2 \left(\left|D^2 u(x)\right| + \left|D^2 u(y)\right| \right) &\leq 6 \sup_{x \in \Omega} d_x^2 \left|D^2 u(x)\right| \leq 6 \, |u|^*_{2;\Omega} \\
&\leq C \left(|u|_{0;\Omega} + |f|^{(2)}_{0,\theta;\Omega} \right).
\end{aligned}
$$

Summing up, we obtain that

$$\sup_{x,y\in\Omega} d_{x,y}^{2+\theta} \frac{|D^2u(x) - D^2u(y)|}{|x-y|^\theta} \leq C\left(|u|_{0;\Omega} + |f|_{0,\theta;\Omega}^{(2)}\right). \tag{B.22}$$

The desired interior estimate (B.17) follows by combining inequalities (B.20) and (B.22).

The proof of Theorem B.11 is complete. □

The next corollary provides *equicontinuity* of solutions of Poisson's equation and their derivatives up to the second order on compact subsets ([59, Chap. 6, Corollary 6.3]).

Corollary B.12 *Let Ω be an open set in $\mathbf{R}^n$ and let Ω' be open subset of Ω that has compact closure in Ω:*

$$\Omega' \Subset \Omega.$$

Assume that a function $u \in C^2(\Omega)$ satisfies the equation

$$\Delta u = f \quad \text{in } \Omega$$

for a function $f \in C^\theta(\Omega)$. Then we have, for any $0 < d \leq \operatorname{dist}\left(\Omega', \partial\Omega\right)$,

$$d\,|Du|_{0;\Omega'} + d^2\left|D^2u\right|_{0;\Omega'} + d^{2+\theta}\left[D^2u\right]_{\theta;\Omega'} \leq C_4\left(|u|_{0;\Omega} + |f|_{0,\theta;\Omega}^{(2)}\right), \tag{B.23}$$

with a constant $C_4 = C_4(n,\theta) > 0$.

Proof For all $x, y \in \Omega'$, we have the assertions

$$\begin{aligned} &d_x \geq d, \quad d_y \geq d, \\ &d_{x,y} \geq d. \end{aligned}$$

Hence, by using estimate (B.17) we obtain that

$$\begin{aligned} &d\,|Du|_{0;\Omega'} + d^2\left|D^2u\right|_{0;\Omega'} + d^{2+\theta}\left[D^2u\right]_{\theta;\Omega'} \\ &\leq \sup_{x\in\Omega'} d_x\,|Du(x)| + \sup_{x\in\Omega'} d_x^2\left|D^2u(x)\right| + \sup_{x,y\in\Omega'} d_{x,y}^{2+\theta}\frac{|D^2u(x) - D^2u(y)|}{|x-y|^\theta} \\ &\leq |u|_{2,\theta;\Omega'}^* \leq |u|_{2,\theta;\Omega}^* \\ &\leq C\left(|u|_{0;\Omega} + |f|_{0,\theta;\Omega}^{(2)}\right). \end{aligned}$$

This proves the desired estimate (B.23).

The proof of Corollary B.12 is complete. □

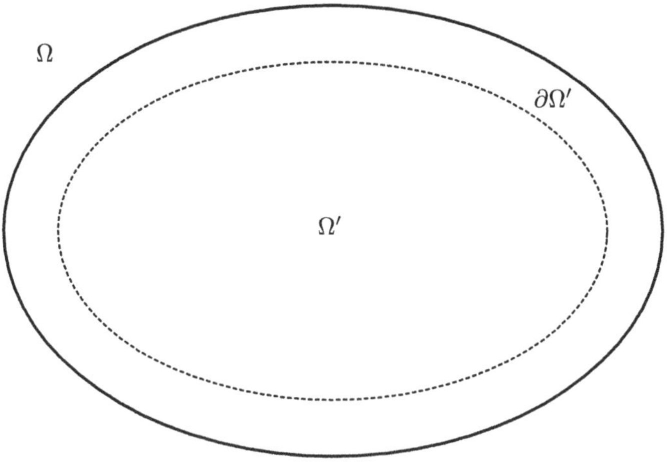

Fig. B.7 The domains Ω and Ω'

Fig. B.8 The domains B_1^+, B_2^+ and T in $\overline{\mathbf{R}_+^n}$

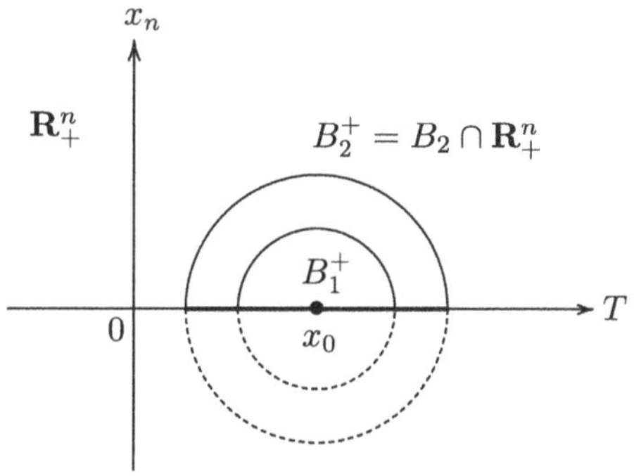

Rephrased, this corollary provides a bound on the seminorms $|Du|_{0;\Omega'}$, $|D^2u|_{0;\Omega'}$ and $[D^2u]_{\theta;\Omega'}$ in any subset Ω' of Ω for which dist $(\Omega', \partial\Omega) \geq d$ (see Fig. B.7).

B.5 Hölder Estimates at the Boundary

The purpose of this section is to prove regularity results for solutions of Poisson's equation at a hyperplane portion of the boundary (see Theorems B.14 and B.18).

First, we introduce some notation (see Fig. B.8):

$$\mathbf{R}_+^n = \{(x', x_n) \in \mathbf{R}^n : x_n > 0\},$$
$$T = \{(x', x_n) \in \mathbf{R}^n : x_n = 0\},$$
$$B_1 = B(x_0, R) = \{x \in \mathbf{R}^n : |x - x_0| < R\} \quad \text{for } x_0 \in \overline{\mathbf{R}_+^n},$$
$$B_2 = B(x_0, 2R) = \{x \in \mathbf{R}^n : |x - x_0| < 2R\} \quad \text{for } x_0 \in \overline{\mathbf{R}_+^n},$$

$$B_1^+ = B_1 \cap \mathbf{R}_+^n = B(x_0, R) \cap \mathbf{R}_+^n,$$
$$B_2^+ = B_2 \cap \mathbf{R}_+^n = B(x_0, 2R) \cap \mathbf{R}_+^n.$$

The next lemma is a version of Lemma B.8 applicable to the intersection of a domain Ω and the half space $\mathbf{R}_+^n$ ([59, Chap. 4, Lemma 4.10]).

Lemma B.13 *If $f \in C^\theta(\overline{B_2^+})$, we let w be the Newtonian potential of f in B_2^+:*

$$w(x) = \int_{B_2^+} \Gamma(x-y)\, f(y)\, dy.$$

Then it follows that

$$w \in C^{2+\theta}(\overline{B_1^+}),$$

and we have the boundary a priori *estimate*

$$\left|D^2 w\right|'_{0,\theta; B_1^+} \leq C_5 \, |f|'_{0,\theta; B_2^+}, \tag{B.24}$$

that is,

$$\left|D^2 w\right|_{0; B_1^+} + R^\theta \left[D^2 w\right]_{\theta; B_1^+} \leq C_5 \left(|f|_{0, B_2^+} + R^\theta\, [f]_{\theta; B_2^+}\right),$$

with a constant $C_5 = C_5(n, \theta) > 0$.

Proof First, by applying Lemma B.7 with

$$\Omega := B_1^+, \quad \Omega_0 := B_2^+,$$

we obtain that

$$\begin{aligned} D_i D_j w(x) &= \int_{B_2^+} D_i D_j \Gamma(x-y)\,(f(y) - f(x))\, dy \\ &\quad - f(x) \int_{\partial B_2^+} D_i \Gamma(x-y)\, \nu_j(y)\, d\sigma(y) \quad \text{in } B_1^+. \end{aligned}$$

However, it follows from an application of the divergence theorem (Theorem 5.3) that

$$\begin{aligned} &\int_{\partial B_2^+} D_i \Gamma(x-y)\, \nu_j(y)\, d\sigma(y) \\ &= \int_{B_2^+} D_j D_i \Gamma(x-y) = \int_{B_2^+} D_i D_j \Gamma(x-y) \\ &= \int_{\partial B_2^+} D_j \Gamma(x-y)\, \nu_i(y)\, d\sigma(y). \end{aligned}$$

Fig. B.9 The domain B_1^+ and the boundary $\partial B_2^+ \setminus T$

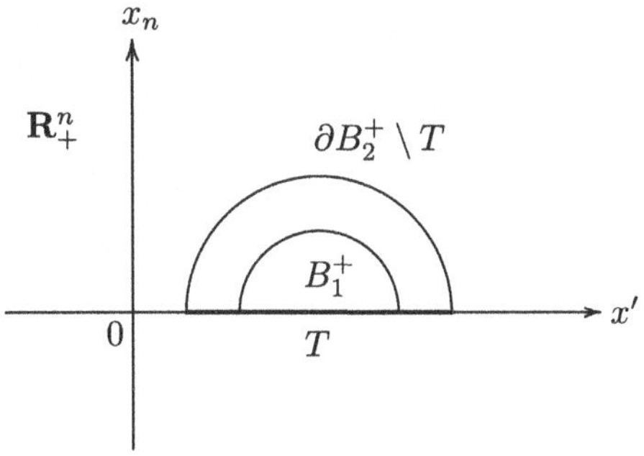

Since we have the formula

$$\nu = (0, \ldots, 0, -1) \quad \text{on } T,$$

we have, for $i \neq n$ or $j \neq n$,

$$\begin{aligned} D_i D_j w(x) &= \int_{B_2^+} D_i D_j \Gamma(x-y)\,(f(y) - f(x))\,dy \\ &\quad - f(x) \int_{\partial B_2^+ \setminus T} D_i \Gamma(x-y)\,\nu_j(y)\,d\sigma(y) \quad \text{in } B_1^+. \end{aligned}$$

We remark that (see Fig. B.9)

$$R \leq |x-y| \leq 3R \quad \text{for } x \in B_1^+ \text{ and } y \in \partial B_2^+ \setminus T.$$

Hence we can estimate the derivatives $D_i D_j w$ for $i \neq n$ or $j \neq n$ as follows:

$$\left| D_i D_j w \right|'_{0,\theta; B_1^+} \leq C \, |f|'_{0,\theta; B_2^+}.$$

Moreover, it follows from an application of Lemma B.7 that the function

$$w(x) = \int_{B_2^+} \Gamma(x-y)\, f(y)\, dy$$

satisfies the equation

$$D_n D_n w(x) = f(x) - \sum_{k=1}^{n-1} D_k D_k w(x) \quad \text{in } B_1^+.$$

Therefore, we can estimate the derivatives $D_n D_n w$ as follows:

$$\begin{aligned} |D_n D_n w|'_{0,\theta; B_1^+} &\leq |f|'_{0,\theta; B_2^+} + \sum_{k=1}^{n-1} |D_k D_k w|'_{0,\theta; B_1^+} \\ &\leq C \, |f|'_{0,\theta; B_2^+}. \end{aligned}$$

Here and in the following the letter C denotes a generic positive constant independent of w and f.

The proof of Lemma B.13 is complete. □

The next theorem is a generalization of Theorem B.9 to the half space $\mathbf{R}^n_+$ ([59, Chap. 4, Theorem 4.11]).

Theorem B.14 *Assume that a function $u \in C^2(B_2^+) \cap C(\overline{B_2^+})$ is a solution of the Dirichlet problem*

$$\begin{cases} \Delta u = f & \text{in } \Omega, \\ u = 0 & \text{on } T \end{cases}$$

for a function $f \in C^\theta\left(\overline{B_2^+}\right)$. Then it follows that

$$u \in C^{2+\theta}(\overline{B_2^+}),$$

and we have the a priori *estimate*

$$|u|'_{2,\theta;B_1^+} \leq C_6 \left(|u|_{0;B_2^+} + R^2 \, |f|'_{0,\theta;B_2^+} \right) \tag{B.25}$$

with a constant $C_6 = C_6(n, \theta) > 0$.

Proof The proof of Theorem B.14 is divided into five steps.

Step (I): In the proof we make use of a method based on *reflection*. Namely, for each point $x = \left(x', x_n\right) \in B_2^+$ we define the point (see Fig. B.10)

$$x^* = \left(x', -x_n\right) \in B_2^- := B(x_0, 2R) \cap \mathbf{R}^n_-.$$

Fig. B.10 The mapping $x \longmapsto x^*$

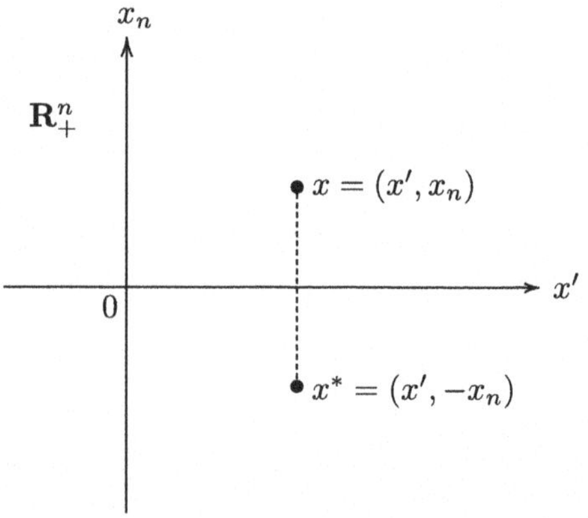

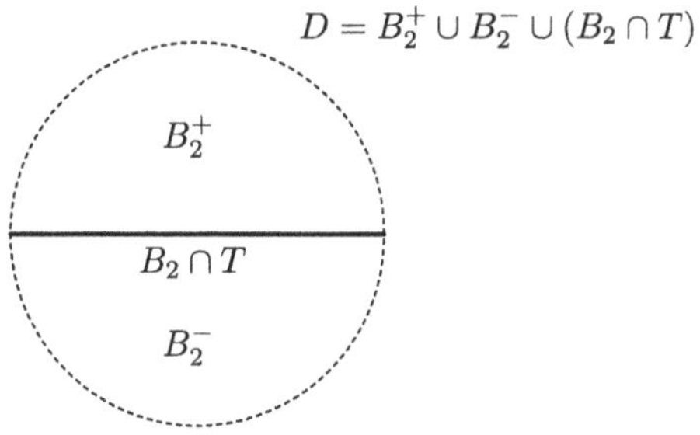

Fig. B.11 The domains $B_2^\pm$, T and D in $\mathbf{R}^n$

Then we have the following:

Claim B.15 *For a function $f \in C^\theta\left(\overline{B_2^+}\right)$, we consider a function $f^*(x)$ defined on the domain (see Fig. B.11)*

$$D = B_2^+ \cup B_2^- \cup (B_2 \cap T)$$

as follows:

$$f^*(x) = f^*(x', x_n) = \begin{cases} f(x', x_n) & \text{for } x_n > 0, \\ f(x', -x_n) & \text{for } x_n < 0. \end{cases}$$

Then it follows that

$$f^* \in C^\theta\left(\overline{D}\right),$$

and we have the estimate

$$\left|f^*\right|'_{0,\theta;D} \leq 4\,|f|'_{0,\theta;B_2^+}\,.$$

Proof Indeed, it suffices to note that

$$\begin{aligned}
\left|f^*\right|'_{0,\theta;D} &= \left|f^*\right|'_{0;D} + (\text{diam } D)^\theta \left[f^*\right]_{0,\theta;D} \\
&= \left|f^*\right|_{0;D} + (\text{diam } D)^\theta \left[f^*\right]_{\theta;D} \\
&\leq |f|_{0;B_2^+} + \left(2 \text{ diam } B_2^+\right)^\theta 2\left[f^*\right]_{\theta;B_2^+} \\
&\leq 4\left(|f|_{0;B_2^+} + \left(\text{diam } B_2^+\right)^\theta \left[f^*\right]_{\theta;B_2^+}\right) \\
&= 4\,|f|'_{0,\theta;B_2^+}\,.
\end{aligned}$$

The proof of Claim B.15 is complete. □

Stpe (II): We let

$$w(x) := \int_{B_2^+} \left(\Gamma(x-y) - \Gamma(x^*-y)\right) f(y)\,dy.$$

Since we have the formula

$$\left|x^* - y\right| = \left|(x' - y', x_n + y_n)\right| = \left|x - y^*\right|,$$

it follows that

$$w(x) = \int_{B_2^+} \left(\Gamma(x-y) - \Gamma(x-y^*)\right) f(y)\, dy.$$

Moreover, we have the following:

Claim B.16 *The function $w(x)$ is a solution of the Dirichlet problem*

$$\begin{cases} \Delta w = f & \text{in } B_2^+, \\ w = 0 & \text{in } T. \end{cases}$$

Proof (1) First, we have the assertion

$$\begin{aligned} &\left(\Gamma(x-y) - \Gamma(x-y^*)\right) |f(y)| \\ &= \frac{1}{(n-2)\omega_n} \left(\frac{1}{|x-y|^n} - \frac{1}{|x^*-y|^n} \right) |f(x)| \longrightarrow 0 \quad \text{as } x_n \downarrow 0. \end{aligned}$$

Moreover, we have the inequality

$$\begin{aligned} &\int_{B_2^+} \left(\Gamma(x-y) - \Gamma(x-y^*)\right) |f(y)|\, dy \\ &\leq \sup_{z \in B_2^+} |f(z)| \left[\int_{B_2^+} |\Gamma(x-y)|\, dy + \int_{B_2^+} \left|\Gamma(x^*-y)\right| dy \right]. \end{aligned}$$

However, we have, for $x, y \in B_2^+$,

$$|x-y| \leq 4R, \quad \left|x^* - y\right| \leq 8R.$$

Hence we have the inequality

$$\begin{aligned} &\int_{B_2^+} \left(\Gamma(x-y) - \Gamma(x-y^*)\right) |f(y)| dy \\ &\leq \frac{1}{(n-2)\omega_n} \int_{B_2^+} \left(\frac{1}{|x-y|^n} + \frac{1}{|x^*-y|^n} \right) |f(y)|\, dy \\ &\leq \frac{1}{(n-2)\omega_n} \sup_{z \in B_2^+} |f(z)| \left[\int_0^{4R} r^{n-2} r^{n-1}\, dr + \int_0^{8R} r^{n-2} r^{n-1}\, dr \right] \omega_n \\ &= \frac{40R^2}{n-2} |f|_{0;B_2^+}. \end{aligned}$$

Therefore, by applying Beppo Levi's theorem (Theorem 3.14) we obtain that

$$\begin{aligned}&\limsup_{x_n \downarrow 0} \left| \int_{B_2^+} \left(\Gamma(x-y) - \Gamma(x-y^*)\right) f(y)\, dy \right| \\ &\le \lim_{x_n \downarrow 0} \int_{B_2^+} \left(\Gamma(x^*-y) - \Gamma(x-y)\right) |f(x)|\, dy = 0.\end{aligned}$$

This proves that

$$w(x) = \int_{B_2^+} \left(\Gamma(x-y) - \Gamma(x-y^*)\right) f(y)\, dy = 0 \quad \text{on } T = \{x_n = 0\}.$$

(2) Secondly, we show that

$$\Delta_x \Gamma(x^* - y) = 0 \quad \text{for } x, y \in B_2^+.$$

Indeed, we have the formulas

$$\frac{\partial^2}{\partial x_n^2} \left(|x^* - y|^{2-n}\right) = (n-2) \left[n \left(x_n^2 + y_n^2\right) - |x^* - y|^2\right] |x^* - y|^{-n-2},$$

and

$$\begin{aligned}&\frac{\partial^2}{\partial x_j^2} \left(|x^* - y|^{2-n}\right) \\ &= (n-2) \left[n \left(x_j - y_j\right)^2 - |x^* - y|^2\right] |x^* - y|^{-n-2}, \quad 1 \le j \le n-1.\end{aligned}$$

This proves that

$$\begin{aligned}\Delta_x \Gamma(x^* - y) &= (n-2) \left[n\, |x^* - y|^2 - n\, |x^* - y|^2\right] |x^* - y|^{-n-2} \\ &= 0 \quad \text{for } x, y \in B_2^+,\end{aligned}$$

since we have the formula

$$|x^* - y|^2 = \sum_{j=1}^{n-1} \left(x_j - y_j\right)^2 + \left(x_n^2 + y_n^2\right).$$

(3) Summing up, we find from Lemma B.7 that

$$
\begin{aligned}
&\Delta w(x) \\
&= \Delta_x \left(\int_{B_2^+} \Gamma(x-y)\, f(y)\, dy \right) - \Delta_x \left(\int_{B_2^+} \Gamma(x-y^*)\, f(y)\, dy \right) \\
&= \Delta_x \left(\int_{B_2^+} \Gamma(x-y)\, f(y)\, dy \right) \\
&= f(x) \quad \text{in } B_2^+.
\end{aligned}
$$

The proof of Claim B.16 is complete. □

Step (III): Since we have the assertion

$$
y \in B_2^+ \Longleftrightarrow y^* \in B_2^-, \quad f(y) = f^*(y^*),
$$

we have the formula

$$
\begin{aligned}
w(x) &= \int_{B_2^+} \left(\Gamma(x-y) - \Gamma(x^*-y) \right) f(y)\, dy \\
&= 2 \int_{B_2^+} \Gamma(x-y)\, f(y)\, dy \\
&\quad - \left(\int_{B_2^+} \Gamma(x-y)\, f^*(y)\, dy + \int_{B_2^-} \Gamma(x-y)\, f(y)\, dy \right) \\
&= 2\, (\Gamma * f)\,(x) - \int_D \Gamma(x-y)\, f^*(y)\, dy.
\end{aligned}
$$

However, it follows from Claim B.15 that

$$
f^* \in C^\theta(\overline{D}).
$$

If we let

$$
w^*(x) := \int_D \Gamma(x-y)\, f^*(y)\, dy,
$$

then it follows from an application of Lemma B.8 that (see Fig. B.12 below)

$$
\left| D^2 w^* \right|'_{0,\theta; B_1^+} \leq C\, |f|'_{0,\theta; D} \leq 2C\, |f|'_{0,\theta; B_2^+}.
$$

Therefore, by combining this inequality with Lemma B.13 we obtain that

$$
\begin{aligned}
|w|'_{2,\theta; B_1^+} &\leq 2\, |\Gamma * f|'_{2,\theta; B_1^+} + \left| w^* \right|'_{2,\theta; B_1^+} \leq C \left(|f|_{0, B_2^+} + R^\theta\, [f]_{\theta; B_2^+} \right) \qquad \text{(B.26)} \\
&= C\, |f|'_{0,\theta; B_2^+}.
\end{aligned}
$$

Fig. B.12 The domains B_2^+, B_1^+ and D

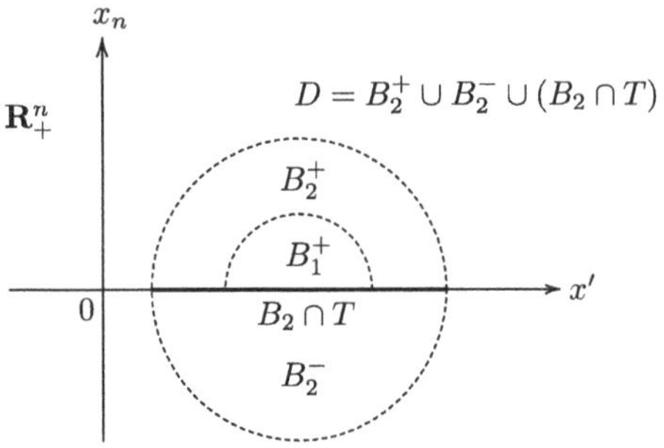

Step (IV): If we let

$$v(x) := u(x) - w(x),$$

it follows that v is a solution of the Dirichlet problem

$$\begin{cases} \Delta v = \Delta u - \Delta w = 0 & \text{in } B_2^+, \\ v = 0 & \text{on } T. \end{cases}$$

Moreover, we have the following:

Claim B.17 *The function*

$$V(x) = V(x', x_n) = \begin{cases} v(x', x_n) & \text{for } x_n \geq 0, \\ -v(x', -x_n) & \text{for } x_n \leq 0. \end{cases}$$

is harmonic *in* D*:*

$$\Delta V = 0 \quad \text{in } D.$$

Moreover, we have the estimate

$$|V|_{0;D} \leq |v|_{0;B_2^+}.$$

Proof First, it follows that

$$\begin{cases} \Delta V = 0 & \text{in } B_2^+, \\ \Delta V = 0 & \text{in } B_2^-, \end{cases}$$

and that

$$V = v = 0 \quad \text{on } B_2 \cap T.$$

Hence we have the assertion

$$V \in C(\overline{D}) \cap C^\infty(B_2^+) \cap C^\infty(B_2^-).$$

Secondly, we show that V is harmonic near T in D. To do so, let

$$B_r := B\left((x',0),r\right) \Subset D$$

be an arbitrary small open ball of radius r about $\left(x',0\right) \in T$. Then we have the formula

$$\begin{aligned}
\int_{\partial B_r} V(y)\,d\sigma(y) &= \int_{\partial B_r^+} V(y',y_n)\,d\sigma(y) - \int_{\partial B_r^-} V(z',-z_n)\,d\sigma(z) \\
&= \int_{\partial B_r^+} V(y',y_n)\,d\sigma(y) - \int_{\partial B_r^+} V(y',y_n)\,d\sigma(y) \\
&= 0.
\end{aligned}$$

Hence there exists a constant $\varepsilon > 0$ such that

$$\int_{\partial B_r} V(y)\,d\sigma(y) = 0 \quad \text{for all } 0 < r < \varepsilon.$$

Therefore, by using Green's first identity (5.4a) to our situation we obtain that

$$\begin{aligned}
0 &= \frac{d}{dr}\left(\frac{1}{r^{n-1}\,\omega_n}\int_{\partial B_r} V(y)\,d\sigma(y)\right) \\
&= \frac{d}{dr}\left(\frac{1}{\omega_n}\int_{S(0,1)} V\left((x',0)+rz\right)\,d\sigma(z)\right) \\
&= \frac{1}{\omega_n}\int_{S(0,1)} z\cdot\nabla V\left((x',0)+rz\right)\,d\sigma(z) \\
&= \frac{1}{\omega_n}\int_{S(0,r)} \frac{y}{r}\cdot\nabla V\left((x',0)+y\right)\,r^{1-n}d\sigma(y) \\
&= \frac{1}{r^{n-1}\,\omega_n}\int_{S(0,r)} \frac{y}{r}\cdot\nabla V\left((x',0)+y\right)\,d\sigma(y) \\
&= \frac{1}{r^{n-1}\,\omega_n}\int_{S(0,r)} \frac{\partial V}{\partial \boldsymbol{\nu}}\left((x',0)+y\right)d\sigma(y) \\
&= \frac{1}{r^{n-1}\,\omega_n}\int_{B_r} \Delta V(x)\,dx,
\end{aligned}$$

where

$$S(0,r) = \left\{z \in \mathbf{R}^n : |z| = r\right\}.$$

Since the integral of ΔV over any ball near T in D vanishes, it follows that

$$\Delta V = 0 \quad \text{near } T \text{ in } D.$$

Summing up, we have proved that V is harmonic in D.
The proof of Claim B.17 is complete. □

Step (V): It remains to prove estimate (B.25).
(1) For the function $v = u - w$, we have the estimate

$$|v|_{0;B_1^+} \leq |u|_{0;B_1^+} + |w|_{0;B_1^+} \leq |u|_{0;B_2^+} + C\,R^2\,|f|_{0;B_2^+}\,.$$

(2) By Theorem B.4, we have the estimate

$$|Dv|_{0;B_1^+} \leq \frac{n}{R}\,|V|_{0;B_2} \leq \frac{n}{R}\,|v|_{0;B_2^+}\,.$$

$$\left|D^2 v\right|_{0;B_1^+} \leq \left(\frac{2n}{R}\right)^2 |V|_{0;B_2} \leq \left(\frac{2n}{R}\right)^2 |v|_{0;B_2^+}.$$

(3) By Theorem B.9, we have the estimate

$$\left[D^2 v\right]_{\theta;B_1^+} \leq C\,R^{-2-\theta}\,|V|_{0;B_2} \leq C\,R^{-2-\theta}\,|v|_{0;B_2^+}\,.$$

(4) By Theorem B.9, we have the estimate

$$|w|_{0;B_2^+} \leq C\,R^2\,|f|_{0;B_2^+}\,.$$

For the function $v = u - w$, we have the estimate

$$|v|_{0;B_2^+} \leq |u|_{0;B_2^+} + |w|_{0;B_2^+} \leq |u|_{0;B_2^+} + C\,R^2\,|f|_{0;B_2^+}\,.$$

Summing up, we obtain the estimate

$$\begin{aligned} |v|'_{2,\theta;B_1^+} &= |v|'_{2;B_1^+} + \left(\text{diam }B_1^+\right)^{2+\theta}\left[D^2 v\right]_{\theta;B_1^+} \\ &\leq C\,|v|_{0;B_2^+} \leq C\left(|u|_{0;B_2^+} + R^2\,|f|_{0;B_2^+}\right) \\ &\leq C\left(|u|_{0;B_2^+} + R^2\,|f|'_{0,\theta;B_2^+}\right). \end{aligned} \tag{B.27}$$

Therefore, the desired estimate (B.25) follows by combining estimates (B.26) and (B.27).

The proof of Theorem B.14 is complete. □

Let Ω be an open set in $\mathbf{R}^n_+$ with open boundary portion T on the set

$$\left\{\left(x', x_n\right) \in \mathbf{R}^n : x_n = 0\right\}$$

(see Fig. B.13 below). For $x,\ y \in \Omega$, we let

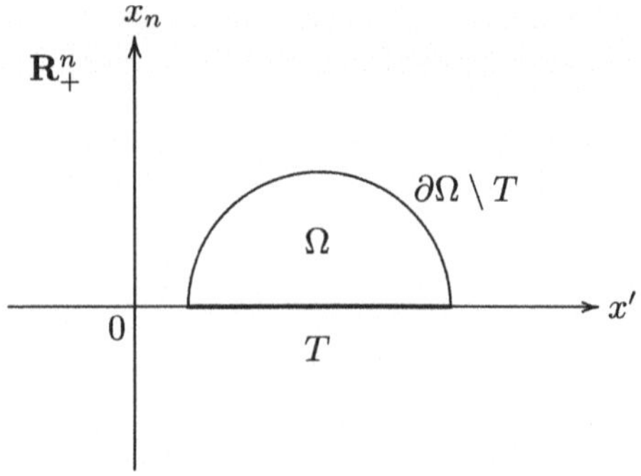

Fig. B.13 The open set Ω with an open boundary portion T

$$\begin{aligned}\overline{d}_x &= \operatorname{dist}(x, \partial\Omega \setminus T),\\ \overline{d}_{x,y} &= \min\left(\overline{d}_x, \overline{d}_y\right).\end{aligned}$$

If k is a non-negative integer and $0 < \theta < 1$, then we introduce various interior seminorms and norms on the Hölder spaces $C^k(\Omega)$ and $C^{k+\theta}(\Omega)$ as follows:

$$[u]^*_{k,0;\Omega\cup T} = [u]^*_{0;\Omega\cup T} = \sup_{x\in\Omega} \sup_{|\alpha|=k} \overline{d}_x^k \, |D^\alpha u(x)| \,, \tag{B.28a}$$

$$|u|^*_{k;\Omega\cup T} = \sum_{j=0}^{k} [u]_{j;\Omega\cup T} \,, \tag{B.28b}$$

$$[u]^*_{k,\theta;\Omega\cup T} = \sup_{x,y\in\Omega} \sup_{|\alpha|=k} \overline{d}_{x,y}^{k+\theta} \frac{|D^\alpha u(x) - D^\alpha u(y)|}{|x-y|^\theta}, \tag{B.28c}$$

$$|u|^*_{k,\theta;\Omega\cup T} = |u|^*_{k;\Omega\cup T} + [u]^*_{k,\theta;\Omega\cup T} \,, \tag{B.28d}$$

and

$$|u|^{(k)}_{0,\theta;\Omega\cup T} = \sup_{x\in\Omega} \overline{d}_x^k \, |u(x)| + \sup_{x,y\in\Omega} \overline{d}_{x,y}^{k+\theta} \frac{|u(x) - u(y)|}{|x-y|^\theta}. \tag{B.28e}$$

The next theorem is a generalization of Theorem B.11 to the half space $\mathbf{R}^n_+$ ([59, Chap. 4, Theorem 4.12]).

Theorem B.18 *Let Ω be an open set in $\mathbf{R}^n_+$ with a boundary portion T on $\{(x', x_n) \in \mathbf{R}^n : x_n = 0\}$. Assume that a function $u \in C^2(\Omega) \cap C(\Omega \cup T)$ is a solution of the Dirichlet problem*

$$\begin{cases} \Delta u = f & \text{in } \Omega,\\ u = 0 & \text{on } T \end{cases}$$

for a function $f \in C^\theta(\Omega \cup T)$. Then we have the boundary estimate

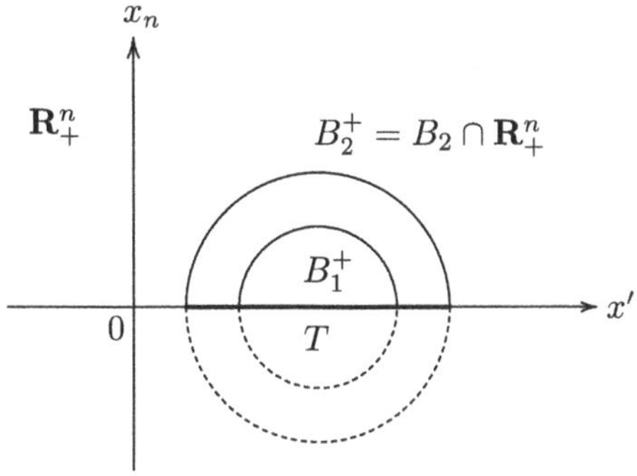

Fig. B.14 The domains B_1 and B_2 in $\mathbf{R}^n_+$

$$|u|^*_{2,\theta;\Omega\cup T} \leq C_7 \left(|u|_{0;\Omega} + |f|^{(2)}_{0,\theta;\Omega\cup T}\right), \tag{B.29}$$

with a constant $C_7 = C_7(n, \theta) > 0$.

Proof The proof of this theorem is similar to that of Theorem B.11.

Step 1: For each point x of Ω, we let (see Fig. B.14)

$$\begin{aligned} R &:= \frac{1}{3}\overline{d}_x, \\ B_1 &:= B(x, R), \\ B_2 &:= B(x, 2R), \end{aligned}$$

and

$$\begin{aligned} B_1^+ &:= B(x, R) \cap \mathbf{R}^n_+, \\ B_2^+ &:= B(x, 2R) \cap \mathbf{R}^n_+. \end{aligned}$$

Then we have, by estimate (B.25),

$$\begin{aligned} \overline{d}_x \, |Du(x)| + \overline{d}_x^2 \left|D^2u(x)\right| &\leq 3R \, |Du|_{0;B_1} + (3R)^2 \left|D^2u\right|_{0,B_1} \\ &\leq C \left(|u|_{0;B_2} + R^2 \, |f|'_{0,\theta;B_2}\right). \end{aligned} \tag{B.30}$$

Moreover, if we assume that $\overline{d}_x \leq \overline{d}_y$ for $x, y \in \Omega$, then it follows that

$$\overline{d}_x = \overline{d}_{x,y} = 3R \quad \text{for } x, y \in \Omega.$$

Hence we have the inequalities

$$R^2 \, |f(x)| = \frac{1}{9}\overline{d}_x^2 \, |f(x)| \leq \frac{1}{9} \, |f|^{(2)}_{0,\theta;\Omega\cup T},$$

and

$$R^2 \overline{d}_x^{-\theta} \frac{|f(x) - f(y)|}{|x - y|^\theta} = \frac{1}{9} \overline{d}_{x,y}^{2+\theta} \frac{|f(x) - f(y)|}{|x - y|^\theta} \leq \frac{1}{9} |f|^{(2)}_{0,\theta;\Omega \cup T} .$$

This proves that

$$R^2 |f|'_{0,\theta;B_2} \leq C |f|^{(2)}_{0,\theta;\Omega \cup T} . \tag{B.31}$$

Therefore, by combining estimates (B.30) and (B.31) we obtain that

$$|u|^*_{2;\Omega \cup T} \leq C \left(|u|_{0;\Omega} + |f|^{(2)}_{0,\theta;\Omega \cup T} \right), \tag{B.32}$$

Step 2: Similarly, we assume that $\overline{d}_x \leq \overline{d}_y$ for $x, y \in \Omega$, so that

$$\overline{d}_x = \overline{d}_{x,y} = 3R \quad \text{for } x, y \in \Omega.$$

Then it follows that

$$\frac{|D^2 u(x) - D^2 u(y)|}{|x - y|^\theta} \leq \begin{cases} \left[D^2 u\right]_{\theta; B_1} & \text{if } y \in B_1, \\ \frac{1}{R^\theta} \left(\left|D^2 u(x)\right| + \left|D^2 u(y)\right| \right) & \text{if } y \in \Omega \setminus B_1. \end{cases}$$

Hence we have, for $x, y \in \Omega$,

$$\begin{aligned} &\overline{d}_{x,y}^{2+\theta} \frac{|D^2 u(x) - D^2 u(y)|}{|x - y|^\theta} \\ &\leq (3R)^{2+\theta} \left[D^2 u\right]_{\theta; B_1^+} + 3^\theta (3R)^2 \left(\left|D^2 u(x)\right| + \left|D^2 u(y)\right| \right) \\ &\leq (3R)^{2+\theta} \left[D^2 u\right]_{\theta; B_1^+} + 6 |u|^*_{2;\Omega \cup T} . \end{aligned} \tag{B.33}$$

Moreover, by using estimates (B.25) and (B.32) we obtain that

$$\begin{aligned} \overline{d}_{x,y}^{2+\theta} \frac{|D^2 u(x) - D^2 u(y)|}{|x - y|^\theta} &\leq C \left(|u|_{0;B_2} + R^2 |f|'_{0,\theta;B_2} \right) + 6 |u|^*_{2;\Omega \cup T} \\ &\leq C \left(|u|_{0;\Omega} + |f|^{(2)}_{0,\theta;\Omega \cup T} \right) + 6 |u|^*_{2;\Omega \cup T} , \end{aligned}$$

so that

$$\begin{aligned} &\sup_{x,y \in \Omega \cup T} \overline{d}_{x,y}^{2+\theta} \frac{|D^2 u(x) - D^2 u(y)|}{|x - y|^\theta} \\ &\leq C \left(|u|_{0;\Omega} + |f|^{(2)}_{0,\theta;B_2} \right) + 6 |u|^*_{2;\Omega \cup T} . \end{aligned} \tag{B.34}$$

Therefore, the desired boundary estimate (B.29) follows by combining estimates (B.32), (B.33) and (B.34).

The proof of Theorem B.18 is complete. □

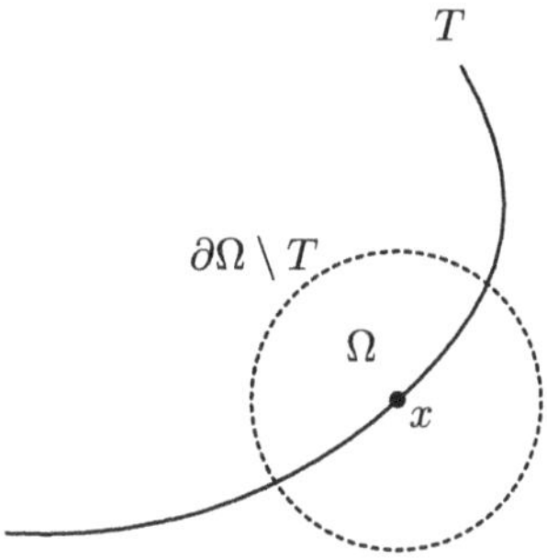

Fig. B.15 The open set Ω with an open boundary portion T

Let Ω be an open set in $\mathbf{R}^n$ with $C^{2+\theta}$ boundary portion T. For $x, y \in \Omega$, we let (see Fig. B.15)

$$\overline{d}_x = \text{dist}\,(x, \partial\Omega \setminus T),$$
$$\overline{d}_{x,y} = \min\left(\overline{d}_x, \overline{d}_y\right).$$

If k is a non-negative integer and $0 < \theta < 1$, then we introduce various boundary seminorms and norms on the Hölder spaces $C^k(\Omega \cup T)$ and $C^{k+\theta}(\Omega \cup T)$ as follows:

$$[u]^*_{k,0;\Omega\cup T} = [u]^*_{0;\Omega\cup T} = \sup_{x\in\Omega} \sup_{|\alpha|=k} \overline{d}_x^k \, |D^\alpha u(x)| \,, \tag{B.35a}$$

$$|u|^*_{k;\Omega\cup T} = \sum_{j=0}^{k} [u]_{j;\Omega\cup T} \,, \tag{B.35b}$$

$$[u]^*_{k,\theta;\Omega\cup T} = \sup_{x,y\in\Omega} \sup_{|\alpha|=k} \overline{d}_{x,y}^{k+\theta} \frac{|D^\alpha u(x) - D^\alpha u(y)|}{|x-y|^\theta}, \tag{B.35c}$$

$$|u|^*_{k,\theta;\Omega\cup T} = |u|^*_{k;\Omega\cup T} + [u]^*_{k,\theta;\Omega\cup T} \,, \tag{B.35d}$$

and

$$|u|^{(k)}_{0,\theta;\Omega\cup T} = \sup_{x\in\Omega} \overline{d}_x^k \, |u(x)| + \sup_{x,y\in\Omega} \overline{d}_{x,y}^{k+\theta} \frac{|u(x) - u(y)|}{|x-y|^\theta}. \tag{B.35e}$$

Then we can prove the following *Schauder local boundary estimate* for solutions of the Dirichlet problem for curved boundaries ([59, Chap. 6, Lemma 6.5]).

Lemma B.19 (the Schauder local boundary estimate) *Let Ω be a $C^{2,\theta}$ domain in $\mathbf{R}^n$ with boundary ∂Ω. Assume that a function $u \in C^{2+\theta}(\overline{\Omega})$ is a solution of the Dirichlet problem*

$$\begin{cases} \Delta u = f & \text{in } \Omega, \\ u = 0 & \text{on } \partial\Omega \end{cases}$$

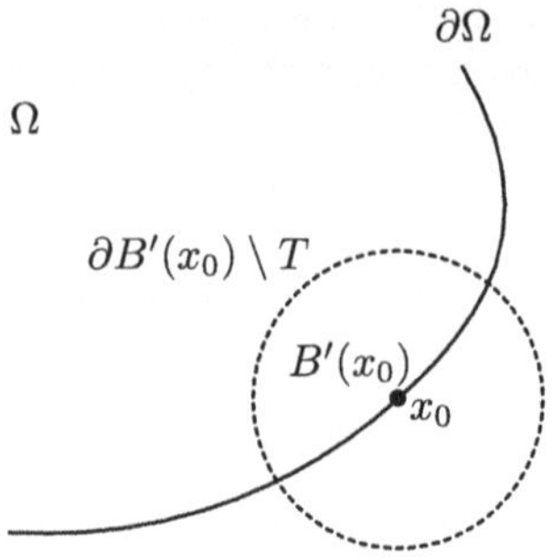

Fig. B.16 The domain $B'(x_0)$ and the boundary $\partial B'(x_0) \setminus T$

for a function $f \in C^\theta(\overline{\Omega})$. *Then, at each boundary point* $x_0 \in \partial\Omega$ *there is a ball* $B = B(x_0, \delta)$ *of radius* $\delta > 0$, *independent of* x_0, *such that we have the* boundary *estimate*

$$|u|_{2,\theta;B\cap\Omega} \leq C_8 \left(|u|_{0;\Omega} + |f|_{0,\theta;\Omega}\right), \tag{B.36}$$

with a constant $C_8 = C_8(n, \theta, \Omega) > 0$.

Proof The proof of Lemma B.19 is divided into three steps.

Step (1): First, we consider the case where

$$x \in B'(x_0) = B(x_0, \rho) \cap \Omega.$$

Since we have the inequality (see Fig. B.16)

$$\overline{d}_x = \text{dist}\left(x, \partial B'(x_0) \setminus T\right) \leq \text{diam}\ \Omega,$$

it follows that

$$\begin{aligned}
&|f|^{(2)}_{2,\theta;B'(x_0)\cup T} \\
&= \sup_{x\in B'(x_0)} \overline{d}_x^{\,2}\, |f(x)| + \sup_{x,y\in B'(x_0)} \overline{d}_{x,y}^{\,2+\theta} \frac{|f(x)-f(y)|}{|x-y|^\theta} \\
&\leq C \sup_{x\in B'(x_0)} |f(x)| + \sup_{x,y\in B'(x_0)} \frac{|f(x)-f(y)|}{|x-y|^\theta} \leq C\, |f|_{0,\theta;B'(x_0)} \\
&\leq C\, |f|_{0,\theta;\Omega}\,.
\end{aligned} \tag{B.37}$$

On the other hand, by applying Theorem B.18 we obtain that

$$\begin{aligned}
|u|^*_{2,\theta;B'(x_0)\cup T} &\leq C\left(|u|_{0;B'(x_0)} + |f|^{(2)}_{2,\theta;B'(x_0)\cup T}\right) \\
&\leq C\left(|u|_{0;\Omega} + |f|^{(2)}_{2,\theta;B'(x_0)\cup T}\right).
\end{aligned}$$

Therefore, we have, by inequality (B.37),

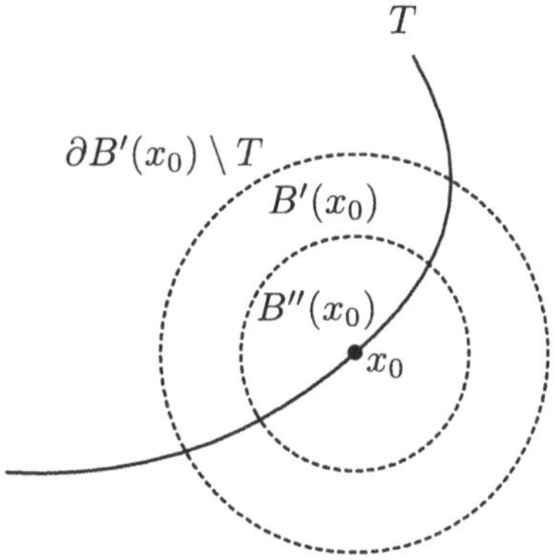

Fig. B.17 The open neighborhoods $B'(x_0)$ and $B''(x_0)$ of x_0 in Ω

$$|u|^*_{2,\theta;B'(x_0)\cup T} \le C_1 \left(|u|_{0;\Omega} + |f|_{0,\theta;\Omega}\right), \tag{B.38}$$

with a positive constant

$$C_1 = C_1 \left(n, \theta, B'(x_0)\right).$$

Step (2): Secondly, we consider the case where

$$x \in B''(x_0) = B\left(x_0, \rho/2\right) \cap \Omega.$$

However, we remark that (see Fig. B.17)

$$\overline{d}_x = \text{dist}\left(x, \partial B'(x_0) \setminus T\right) \ge \frac{\rho}{2} \quad \text{for all } x \in B''(x_0),$$
$$\overline{d}_{x,y} \ge \frac{\rho}{2} \quad \text{for all } x, y \in B''(x_0).$$

Hence we have the inequality

$$\begin{aligned} & |u|^*_{2,\theta;B'(x_0)\cup T} \\ &= \sup_{x\in B'(x_0)} |u(x)| + \sup_{x\in B'(x_0)} \overline{d}_x\,|Du(x)| + \sup_{x\in B'(x_0)} \overline{d}_x^{\,2}\,|D_2 u(x)| \\ &\quad + \sup_{x,y\in B'(x_0)} \overline{d}_{x,y}^{\,2+\theta} \frac{|D^2u(x) - D^2u(y)|}{|x-y|^\theta} \\ &\ge \min\left(1, \frac{\rho}{2}, \left(\frac{\rho}{2}\right)^{2+\theta}\right)\Big(\sup_{x\in B'(x_0)} |u(x)| \\ &\quad + \sup_{x\in B'(x_0)} |Du(x)| + \sup_{x\in B'(x_0)} |D_2 u(x)|\Big) \\ &= \min\left(1, \frac{\rho}{2}, \left(\frac{\rho}{2}\right)^{2+\theta}\right) |u|_{0,\theta;B''(x_0)}. \end{aligned} \tag{B.39}$$

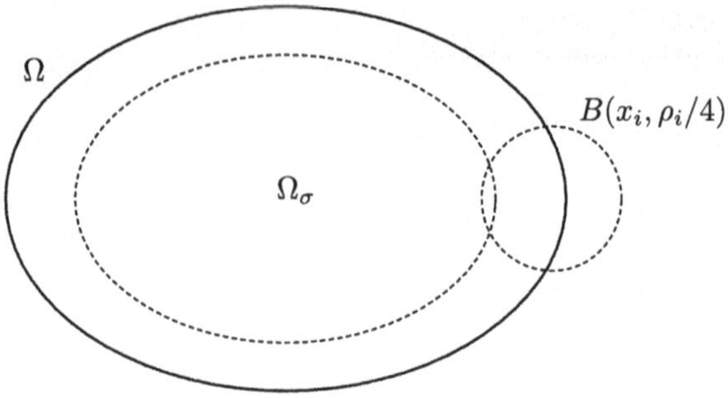

Fig. B.18 The subdomain Ω_σ and the ball $B(x_i, \rho_i/4)$ about x_i

Therefore, by combining inequalities (B.38) and (B.39) we obtain that This proves that

$$|u|_{0,\theta;B''(x_0)} \leq C_2 \left(|u|_{0;\Omega} + |f|_{0,\theta;\Omega}\right), \tag{B.40}$$

with a positive constant

$$C_2 = C_2 \left(n, \theta, B'(x_0), B''(x_0)\right).$$

Step (3): Since the boundary $\partial\Omega$ is compact, we can find a finite number of boundary points $\{x_i\}_{i=1}^N$ and positive numbers $\{\rho_i\}_{i=1}^N$ such that (see Fig. B.18)

$$\partial\Omega \subset \bigcup_{i=1}^{N} B\left(x_i, \rho_i/4\right).$$

We let

$$\delta := \min_{1 \leq i \leq N} \frac{\rho_i}{4},$$

and

$$C := \max_{1 \leq i \leq N} C_2 \left(n, \theta, B'(x_i), B''(x_i)\right).$$

then, for each boundary point $x_0 \in \partial\Omega$ we can find some ball $B(x_i, \rho_i/4)$ such that

$$x_0 \in B\left(x_i, \rho_i/4\right).$$

Hence we have the inequality

$$|x - x_i| \leq |x - x_0| + |x_0 - x_i| \leq \delta + \frac{\rho_i}{4} < \frac{\rho_i}{2} \quad \text{for all } x \in B,$$

and so

$$B \cap \Omega \subset B(x_i, \rho_i/2) \cap \Omega = B''(x_i).$$

By using inequality (B.40), we obtain that

$$|u|_{0,\theta;B\cap\Omega} \leq |u|_{0,\theta;B''(x_i)} \leq C\left(|u|_{0;\Omega} + |f|_{0,\theta;\Omega}\right).$$

This proves the the desired estimate (B.36).

The proof of Lemma B.19 is complete. □

Finally, by using Lemma B.19 we can obtain the following *Schauder global estimate* for solutions with $C^{2+\theta}$ boundary values defined on a $C^{2+\theta}$ domain Ω ([59, Chap. 6, Theorem 6.6]).

Theorem B.20 (the Schauder global estimate) *Let Ω be a $C^{2+\theta}$ domain in $\mathbf{R}^n$ with boundary $\partial\Omega$. For given functions $f \in C^\theta(\Omega)$ and $\varphi \in C^{2+\theta}(\Omega)$, assume that a function $u \in C^{2+\theta}(\overline{\Omega})$ is a solution of the Dirichlet problem for the Laplacian Δ*

$$\begin{cases} \Delta u = f & \text{in } \Omega, \\ u = \varphi & \text{on } \partial\Omega. \end{cases}$$

Then we have the global estimate

$$|u|_{2,\theta;\Omega} \leq C_9\left(|u|_{0;\Omega} + |\varphi|_{2,\theta;\Omega} + |f|_{0,\theta;\Omega}\right), \tag{B.41}$$

with a constant $C_9 = C_9(n, \theta, \Omega) > 0$.

Proof The proof of Theorem B.20 is divided into two steps.

Step I: The homogeneous case where $\varphi = 0$. We show that every solution $u \in C^{2+\theta}(\overline{\Omega})$ of the Dirichlet problem

$$\begin{cases} \Delta u = f & \text{in } \Omega, \\ u = 0 & \text{on } \partial\Omega \end{cases}$$

satisfies the global estimate

$$|u|_{2,\theta;\Omega} \leq C\left(|u|_{0,\theta;\Omega} + |f|_{0,\theta;\Omega}\right), \tag{B.42}$$

with a positive constant $C = C(n, \theta)$.

(1) First, we consider the case where

$$x \in B(x_0, \delta) \cap \Omega.$$

Here δ is the positive constant in Lemma B.19. Then it follows from an application of Lemma B.19 that

$$\begin{aligned} |Du(x)| + \left|D^2u(x)\right| &\leq |u|_{2,\theta;B(x_0,\delta)\cap\Omega} \\ &\leq C_\delta\left(|u|_{0;\Omega} + |f|_{0,\theta;\Omega}\right) \quad \text{for all } x \in B(x_0, \delta) \cap \Omega, \end{aligned} \tag{B.43}$$

with a positive constant $C_\delta = C(n, \theta, \delta)$.

(2) Secondly, we consider the case where

$$x \in \Omega_\sigma := \{x \in \Omega : \operatorname{dist}(x, \partial\Omega) > \sigma\} \quad \text{for } \sigma := \frac{\delta}{2}.$$

Then it follows from an application of estimate (B.23) that

$$\begin{aligned} \sigma \, |Du|_{0;\Omega_\sigma} + \sigma^2 \left|D^2 u\right|_{0;\Omega_\sigma} &\leq C \left(|u|_{0;\Omega} + |f|^{(2)}_{0,\theta;\Omega}\right) \\ &\leq C \left(|u|_{0;\Omega} + |f|_{0,\theta;\Omega}\right). \end{aligned}$$

This proves that

$$|Du(x)| + \left|D^2 u(x)\right| \leq C_\sigma \left(|u|_{0;\Omega} + |f|_{0,\theta;\Omega}\right) \quad \text{for all } x \in \Omega_\sigma, \tag{B.44}$$

with a positive constant $C_\sigma = C(n, \theta, \sigma)$.

Therefore, by combining estimates (B.43) and (B.44) we obtain that

$$|u|_{2;\Omega} \leq C_{\delta,\sigma} \left(|u|_{0;\Omega} + |f|_{0,\theta;\Omega}\right), \tag{B.45}$$

with a positive constant $C_{\delta,\sigma} = C(n, \theta, \delta, \sigma)$.

(3) It remains to estimate the quantity $\left[D^2 u\right]_{\theta;\Omega}$.

(a) First, we consider the case where

$$x, y \in B(x_0, \delta) \cap \Omega.$$

Then it follows from an application of estimate (B.36) that

$$\frac{|D^2 u(x) - D^2 u(y)|}{|x - y|^\theta} \leq |u|_{2,\theta; B(x_0,\delta)\cap\Omega} \leq C_1 \left(|u|_{0;\Omega} + |f|_{0,\theta;\Omega}\right). \tag{B.46}$$

(b) Secondly, we consider the case where

$$x, y \in \Omega_\sigma.$$

It follows from an application of estimate (B.23) that

$$\begin{aligned} \sigma^{2+\theta} \frac{|D^2 u(x) - D^2 u(y)|}{|x - y|^\theta} &\leq C \left(|u|_{0;\Omega} + |f|^{(2)}_{0,\theta;\Omega}\right) \\ &\leq C_2 \left(|u|_{0;\Omega} + |f|_{0,\theta;\Omega}\right). \end{aligned} \tag{B.47}$$

(c) Finally, we consider the case where

$$|x - y| > \sigma \quad \text{either } x \notin \Omega_\sigma \text{ or } y \notin \Omega_\sigma.$$

Then it follows from estimate (B.45) that

$$\frac{|D^2u(x) - D^2u(y)|}{|x-y|^\theta} \leq \sigma^{-\theta}\left(\left|D^2u(x)\right| + \left|D^2u(y)\right|\right) \leq \sigma^{-\theta}\, 2\, |u|_{2;\Omega} \leq C_3\left(|u|_{0;\Omega} + |f|_{0,\theta;\Omega}\right). \tag{B.48}$$

Therefore, we obtain from estimates (B.46), (B.47) and (B.48) that

$$\left[D^2u\right]_{\theta;\Omega} = \sup_{x,y\in\Omega} \frac{|D^2u(x) - D^2u(y)|}{|x-y|^\theta} \leq \left(C_1 + \frac{C_2}{\sigma^{2+\theta}} + \frac{C_3}{\sigma^\theta}\right)\left(|u|_{0;\Omega} + |f|_{0,\theta;\Omega}\right) \quad \text{for } \sigma = \frac{\delta}{2}. \tag{B.49}$$

The desired global estimate (B.42) follows by combining estimates (B.45) and (B.49).

Step II: The non-homogeneous case where $\varphi \in C^{2+\theta}(\overline{\Omega})$. Assume that a function $u \in C^{2+\theta}(\overline{\Omega})$ is a solution of the Dirichlet problem

$$\begin{cases} \Delta u = f & \text{in } \Omega, \\ u = \varphi & \text{on } \partial\Omega. \end{cases}$$

If we let

$$v := u - \varphi \in C^{2+\theta}(\Omega),$$

then the function v is a solution of the homogeneous Dirichlet problem

$$\begin{cases} \Delta v = f - \Delta\varphi & \text{in } \Omega, \\ v = 0 & \text{on } \partial\Omega. \end{cases}$$

Hence, by applying estimate (B.42) to the solution $v = u - \varphi$ we obtain that

$$|u - \varphi|_{2,\theta;\Omega} \leq C\left(|u - \varphi|_{0;\Omega} + |f - \Delta\varphi|_{0,\theta;\Omega}\right). \tag{B.50}$$

However, we have the inequalities

$$|\Delta\varphi|_{0,\theta;\Omega} \leq C\, |\varphi|_{2,\theta;\Omega}\,,$$

and

$$|u - \varphi|_{0,\theta;\Omega} \leq |u|_{0,\theta;\Omega} + |\varphi|_{0,\theta;\Omega} \leq |u|_{0,\theta;\Omega} + |\varphi|_{2,\theta;\Omega}\,.$$

Therefore, we obtain from estimate (B.50) that

$$
\begin{aligned}
|u-\varphi|_{2,\theta;\Omega} &\le C\left(|u-\varphi|_{0;\Omega} + |f|_{0,\theta;\Omega} + |\varphi|_{2,\theta;\Omega}\right) \\
&\le C\left(|u|_{0,\theta;\Omega} + |f|_{0,\theta;\Omega} + |\varphi|_{2,\theta;\Omega}\right).
\end{aligned}
$$

This proves that

$$
\begin{aligned}
|u|_{2,\theta;\Omega} &\le |u-\varphi|_{2,\theta;\Omega} + |\varphi|_{2,\theta;\Omega} \\
&\le C\left(|u|_{0;\Omega} + |f|_{0,\theta;\Omega} + |\varphi|_{2,\theta;\Omega}\right).
\end{aligned}
$$

Now the proof of Theorem B.20 is complete. □

B.6 The General Case

Now let Ω be a bounded $C^{2,\theta}$ domain in Euclidean space $\mathbf{R}^n$ with $0 < \theta < 1$, and let L be a second order *strictly elliptic* differential operator with real coefficients such that

$$
L = \sum_{i,j=1}^{n} a^{ij}(x) \frac{\partial^2}{\partial x_i \partial x_j} + \sum_{i=1}^{n} b^i(x) \frac{\partial}{\partial x_i} + c(x),
$$

where the coefficients a^{ij}, b^i, c satisfy the following conditions:

(1) $a^{ij} \in C^{\theta}(\overline{\Omega})$, $a^{ij}(x) = a^{ji}(x)$ for $1 \le i, j \le n$ and $x \in \overline{\Omega}$ and there exists a constant $\lambda > 0$ such that (the strict ellipticity condition)

$$
\sum_{i,j=1}^{n} a^{ij}(x)\xi_i\xi_j \ge \lambda\, |\xi|^2 \quad \text{for all } (x,\xi) \in T^*\left(\overline{\Omega}\right) = \overline{\Omega} \times \mathbf{R}^n. \tag{B.51}
$$

(2) $b^i \in C^{\theta}(\overline{\Omega})$ for $1 \le i \le n$.
(3) $c \in C^{\theta}(\overline{\Omega})$ and $c(x) \le 0$ in Ω.

Moreover, we assume that there exists a constant $\Lambda > 0$ such that

$$
\left|a^{ij}\right|_{0,\theta;\Omega},\ \left|b^i\right|_{0,\theta;\Omega},\ |c|_{0,\theta;\Omega} \le \Lambda \quad \text{for } 1 \le i, j \le n. \tag{B.52}
$$

We introduce a family of second order *strictly elliptic* differential operators

$$
\begin{aligned}
L_t &:= t\,L + (1-t)\,\Delta \qquad\qquad (\text{B.53}) \\
&= \sum_{i,j=1}^{n} \left(t\,a^{ij}(x) + (1-t)\delta_{ij}\right) \frac{\partial^2}{\partial x_i \partial x_j} + \sum_{i=1}^{n} \left(t\,b^i(x)\right) \frac{\partial}{\partial x_i} + t\,c(x) \\
&\quad \text{for } 0 \le t \le 1,
\end{aligned}
$$

and consider a family of the Dirichlet problem for the operators L_t for $0 \leq t \leq 1$: For any given functions $f \in C^{\theta}(\overline{\Omega})$ and $\varphi \in C^{2+\theta}(\partial\Omega)$, find a solution $u_t \in C^{2+\theta}(\overline{\Omega})$ of the Dirichlet problem

$$\begin{cases} L_t\, u_t = f & \text{in } \Omega, \\ u_t = \varphi & \text{on } \partial\Omega. \end{cases} \tag{$(D)_t$}$$

By conditions (B.51) and (B.52), we remark that the coefficients of the differential operators L_t satisfy the following conditions for all $0 \leq t \leq 1$:

$$\sum_{i,j=1}^{n} \left(t\, a^{ij}(x) + (1-t)\delta_{ij}\right) \xi_i \xi_j \geq \min\{1, \lambda\}\, |\xi|^2 \quad \text{on } T^*\left(\overline{\Omega}\right), \tag{B.54a}$$

$$\left|t\, a^{ij} + (1-t)\delta_{ij}\right|_{0;\theta;\Omega} \leq \max\{1, \Lambda\} \quad \text{for } 1 \leq i, j \leq n, \tag{B.54b}$$

$$\left|t\, b^i\right|_{0;\theta;\Omega} \leq \Lambda \quad \text{for } 1 \leq i \leq n, \tag{B.54c}$$

$$|t\, c|_{0;\theta;\Omega} \leq \Lambda. \tag{B.54d}$$

Now we assume that a function $u_t \in C^{2+\theta}(\overline{\Omega})$ is a solution of the Dirichlet problem $(D)_t$ for $f \in C^{\theta}(\overline{\Omega})$ and $\varphi \in C^{2+\theta}(\partial\Omega)$. Then, by using the *weak maximum principle* in the framework of Hölder spaces (see [59, Chap. 3, Theorem 3.7]) we can find a constant $C_1 > 0$ such that

$$\sup_{\Omega} |u_t| \leq \sup_{\partial\Omega} |\varphi| + \frac{C_1}{\min\{1, \lambda\}} \sup_{\Omega} |f|. \tag{B.55}$$

Here it should be noticed that the structure constant $C_1 = C_1(\lambda, \Lambda, \operatorname{diam}\Omega)$ is independent of t.

On the other hand, we obtain from [59, Chap. 6, Theorem 6.6] that the *global estimate* (B.41) remains valid for the family of elliptic differential operators L_t for all $0 \leq t \leq 1$:

$$|u_t|_{2,\theta;\Omega} \leq C_2 \left(|u_t|_{0;\Omega} + |\varphi|_{2,\theta;\partial\Omega} + |f|_{0,\theta;\Omega}\right), \tag{B.56}$$

where the structure constant $C_2 = C_2(n, \theta, \lambda, \Lambda, \Omega) > 0$ is independent of t.

Therefore, by combining the two estimates (B.55) and (B.56) we obtain the *uniform estimate*

$$|u_t|_{2,\theta;\Omega} \leq C_3 \left(|\varphi|_{2,\theta;\partial\Omega} + |f|_{0,\theta;\Omega}\right) \tag{B.57}$$

for solutions $u_t \in C^{2+\theta}(\overline{\Omega})$ of the Dirichlet problem $(D)_t$ for each $0 \leq t \leq 1$, where the structure constant $C_3 = C_3(n, \theta, \lambda, \Lambda, \Omega) > 0$ is independent of t.

Now, by using the *method of continuity* (Theorem 2.16) we can prove the following theorem for the differential operator L (see [59, Chap. 6, Theorem 6.8]).

Theorem B.21 *Assume that the Dirichlet problem for the Laplacian* Δ

$$\begin{cases} \Delta u = f & \textit{in } \Omega, \\ u = \varphi & \textit{on } \partial\Omega \end{cases} \tag{$(D)_0$}$$

has a solution $u \in C^{2+\theta}(\overline{\Omega})$ *for any* $f \in C^{\theta}(\overline{\Omega})$ *and any* $\varphi \in C^{2+\theta}(\partial\Omega)$. *Then the Dirichlet problem for the operator* L

$$\begin{cases} Lu = f & \textit{in } \Omega, \\ u = \varphi & \textit{on } \partial\Omega \end{cases} \tag{$(D)_1$}$$

also has a solution $u \in C^{2+\theta}(\overline{\Omega})$ *for any* $f \in C^{\theta}(\overline{\Omega})$ *and any* $\varphi \in C^{2+\theta}(\partial\Omega)$.

Indeed, we are reduced to the homogeneous case:

$$\begin{cases} L_t\, u_t = f & \text{in } \Omega, \\ u_t = 0 & \text{on } \partial\Omega. \end{cases}$$

Thus, it suffices to apply Theorem 2.16 with

$$\begin{aligned} \mathcal{B} &:= \left\{ u \in C^{2+\theta}(\overline{\Omega}) : u = 0 \text{ on } \partial\Omega \right\}, \\ \mathcal{V} &:= C^{\theta}(\overline{\Omega}), \\ \mathcal{L}_1 &:= L, \\ \mathcal{L}_0 &:= \Delta, \end{aligned}$$

since the uniform estimate (B.57) with $\varphi := 0$ corresponds to the uniform estimate (2.12).

References

1. Acquistapace, P.: On BMO regularity for linear elliptic systems. Ann. Mat. Pura Appl. **161**, 231–269 (1992)
2. Adams, R.A., Fournier, J.J.F.: Sobolev Spaces. Pure and Applied Mathematics, vol. 140, 2nd edn. Elsevier/Academic Press, Amsterdam Heidelberg New York Oxford (2003)
3. Agmon, S.: Lectures on Elliptic Boundary Value Problems, Revised Edition of the 1965 Original. American Mathematical Society Chelsea Publishing, Providence, Rhode Island (2010)
4. Agmon, S., Douglis, A., Nirenberg, L.: Estimates near the boundary for solutions of elliptic partial differential equations satisfying general boundary conditions I. Comm. Pure Appl. Math. **12**, 623–727 (1959)
5. Amano, K.: Maximum principles for degenerate elliptic-parabolic operators. Indiana Univ. Math. J. **29**, 545–557 (1979)
6. Anderson, R.F.: Diffusions with second order boundary conditions I. Indiana Univ. Math. J. **25**, 367–397 (1976)
7. Anderson, R.F.: Diffusions with second order boundary conditions II. Indiana Univ. Math. J. **25**, 403–441 (1976)
8. Aronszajn, N., Smith, K.T.: Theory of Bessel potentials I. Ann. Inst. Fourier (Grenoble) **11**, 385–475 (1961)
9. Axler, S., Bourdon, P., Ramey, W.: Harmonic Function Theory. 2nd edn. Graduate Texts in Mathematics, No. 137. Springer-Verlag, Berlin Heidelberg New York (2001)
10. Barles, G., Chasseigne, E., Imbert, C.: On the Dirichlet problem for second-order elliptic integro-differential equations. Indiana Univ. Math. J. **57**, 213–246 (2008)
11. Bergh, J., Löfström, J.: Interpolation Spaces, an Introduction, Grundlehren der Mathematischen Wissenschaften, No. 223. Springer-Verlag, Berlin (1976)
12. Blumenthal, R.M., Getoor, R.K.: Markov Processes and Potential Theory. Dover Publications Inc., Mineola, New York (2007)
13. Bony, J.-M.: Principe du maximum dans les espaces de Sobolev, C. R. Acad. Sc. Paris Sér. A-B **265**, A333–A336 (1967)
14. Bony, J.-M.: Problème de Dirichlet et semi-groupe fortement fellérien associés à un opérateur intégro-différentiel, C. R. Acad. Sc. Paris Sér. A-B **265**, A361–A364 (1967)
15. Bony, J.-M., Courrège, P., Priouret, P.: Semi-groupes de Feller sur une variété à bord compacte et problèmes aux limites intégro-différentiels du second ordre donnant lieu au principe du maximum. Ann. Inst. Fourier (Grenoble) **18**, 369–521 (1968)

K. Taira, *Real Analysis Methods for Markov Processes*,
https://doi.org/10.1007/978-981-97-3659-1

16. Bramanti, M.: Commutators of integral operators with positive kernels. Matematiche (Catania) **49**, 149–168 (1994)
17. Bramanti, M., Cerutti, M.C.: $W_p^{1,2}$ solvability for the Cauchy-Dirichlet problem for parabolic equations with VMO coefficients. Comm. Partial Diff. Equ. **18**, 1735–1763 (1993)
18. Brezis, H.: Functional Analysis, Sobolev Spaces and Partial Differential Equations. Universitext. Springer-Verlag, New York (2011)
19. Calderón, A.P.: Uniqueness in the Cauchy problem for partial differential equations. Amer. J. Math. **80**, 16–36 (1958)
20. Calderón, A.P.: Boundary value problems for elliptic equations. In: Outlines Joint Sympos. Partial Differential Equations (Novosibirsk, 1963), Acad. Sci. USSR Siberian Branch, Moscow, pp. 303–304 (1963)
21. Calderón, A.P.: Commutators of singular integral operators. Proc. Nat. Acad. Sci. U.S.A. **53**, 1092–1099 (1965)
22. Calderón, A.P.: Singular integrals. Bull. Amer. Math. Soc. **72**, 427–465 (1966)
23. Calderón, A.P., Zygmund, A.: On the existence of certain singular integrals. Acta Math. **88**, 85–139 (1952)
24. Calderón, A.P., Zygmund, A.: On singular integrals. Amer. J. Math. **78**, 289–309 (1956)
25. Calderón, A.P., Zygmund, A.: Algebras of certain singular operators. Amer. J. Math. **78**, 310–320 (1956)
26. Calderón, A.P., Zygmund, A.: Singular integral operators and differential equations. Amer. J. Math. **79**, 901–921 (1957)
27. Calderón, A.P., Zygmund, A.: Local properties of solutions of elliptic partial differential equations. Studia Math. **20**, 171–225 (1961)
28. Calderón, A.P., Zygmund, A.: On singular integrals with variable kernels. Appl. Anal. **7**, 221–238 (1977/78)
29. Campanato, S.: Equazioni ellittiche del II° ordine e spazi $\mathfrak{L}^{(2,\lambda)}$. Ann. Mat. Pura Appl. **69**, 321–382 (1965)
30. Cancelier, C.: Problèmes aux limites pseudo-différentiels donnant lieu au principe du maximum. Comm. Partial Diff. Equ. **11**, 1677–1726 (1986)
31. Cattiaux, P.: Hypoellipticité et hypoellipticité partielle pour les diffusions avec une condition frontière. Ann. Inst. H. Poincaré Probab. Statist. **22**, 67–112 (1986)
32. Chapman, S.: On the Brownian displacements and thermal diffusion of grains suspended in a non-uniform fluid. Proc. Roy. Soc. London Ser. A. **119**, 34–54 (1928)
33. Chiarenza, F. (1994). L^p regularity for systems of PDEs, with coefficients in VMO. In: Nonlinear Analysis, Function Spaces and Applications, vol. 5, Prometheus, Prague, pp. 1–32 (1994)
34. Chiarenza, F., Frasca, M., Longo, P.: Interior $W^{2,p}$ estimates for nondivergence elliptic equations with discontinuous coefficients. Ricerche Mat. **60**, 149–168 (1991)
35. Chiarenza, F., Frasca, M., Longo, P.: $W^{2,p}$-solvability of the Dirichlet problem for nondivergence elliptic equations with VMO coefficients. Trans. Amer. Math. Soc. **336**, 841–853 (1993)
36. Chicco, M.: Terzo problema al contorno per una classe di equazioni ellittiche del secondo ordine a coefficienti discontinui. Ann. Mat. Pura Appl. **112**, 241–259 (1977)
37. Chicco, M.: Correzione alla Nota "Terzo problema al contorno per una classe di equazioni ellittiche del secondo ordine a coefficienti discontinui.". Ann. Mat. Pura Appl. **112**, 399–401 (1982)
38. Cohn, D.L.: Measure Theory, 2nd edn, Birkhäuser Advanced Texts: Basler Lehrbücher. Birkhäuser/Springer-Verlag, New York (2013)
39. Coifman, R., Rochberg, R., Weiss, G.: Factorization theorems for Hardy spaces in several variables. Ann. of Math. **2**(103), 611–635 (1976)
40. Di Fazio, G., Palagachev, D.K.: Oblique derivative problem for elliptic equations in nondivergence form with VMO coefficients. Comment. Math. Univ. Carolinae **37**, 537–556 (1996)
41. Duoandikoetxea, J.: Fourier Analysis. American Mathematical Society, Providence, Rhode Island (2001)

42. Dynkin, E.B.: Foundations of the theory of Markov processes, (Fizmatgiz, Moscow (in Russian)). English translation: (Pergamon Press, Oxford London New York Paris, 1960); German translation: (Springer-Verlag, Berlin Göttingen Heidelberg, 1961); French translation: (Dunod, Paris, 1963) (1959)
43. Dynkin, E.B.: Markov Processes I, II, Die Grundlehren der mathematischen Wissenschaften, Band 121, 122. Springer-Verlag, Berlin (1965)
44. Dynkin, E.B., Yushkevich, A.A.: Markov Processes: Theorems and Problems. Plenum Press, New York (1969)
45. Ethier, S.N., Kurtz, T.G.: Markov Processes. Characterization and Convergence. John Wiley & Sons Inc., New York Chichester Brisbane Toronto Singapore (1986)
46. Federer, H.: Geometric Measure Theory, reprint of the, 1969th edn. Classics in Mathematics. Springer-Verlag, Berlin Heidelberg New York (1996)
47. Feller, W.: The parabolic differential equations and the associated semigroups of transformations. Ann. Math. **55**, 468–519 (1952)
48. Feller, W.: On second order differential equations. Ann. Math. **61**, 90–105 (1955)
49. Fiorenza, R.: Sui problemi di derivata obliqua per le equazioni ellittiche. Ricerche Mat. **8**, 83–110 (1959)
50. Folland, G.B.: Introduction to Partial Differential Equations, 2nd edn. Princeton University Press, Princeton, New Jersey (1995)
51. Folland, G.B.: Real Analysis, 2nd edn. John Wiley & Sons, New York Chichester Weinheim Brisbane Singapore Toronto (1999)
52. Fortet, R.: Les fonctions aléatoires du type de Markoff associées à certaines équations linéaires aux dérivées partielles du type paraboloque. J. Math. Pures Appl. **22**, 177–243 (1943)
53. Friedman, A.: Remarks on the maximum principle for parabolic equations and its applications. Pacific. J. Math. **8**, 201–211 (1958)
54. Friedman, A.: Partial Differential Equations. Dover Publications Inc., Mineola, New York (1969/2008)
55. Friedman, A.: Foundations of Modern Analysis. Dover Publications Inc., New York (1970/1982)
56. Garcia-Cuerva, J.: Rubio de Francia. Weighted Norm Inequalities and Related Topics. North-Holland, Amsterdam New York Oxford, J.L. (1985)
57. Garnett, J.B.: Bounded Analytic Functions, Revised 1st edn, Graduate Texts in Mathematics, vol. 236. Springer-Verlag, New York (2007)
58. Garroni, M.G., Menaldi, J.L.: Second Order Elliptic Integro-differential Problems, Research Notes in Mathematics, No. 430. Chapman & Hall/CRC, Boca Raton, Florida (2002)
59. Gilbarg, D., Trudinger, N.S.: Elliptic Partial Differential Equations of Second Order, reprint of the, 1998th edn. Classics in Mathematics. Springer-Verlag, New York Berlin Heidelberg Tokyo (2001)
60. Gimbert, F., Lions, P.-L.: Existence and regularity results for solutions of second-order, elliptic, integrodifferential operators. Ricerche Mat. **33**, 315–558 (1984)
61. Gohberg, I.G., Kreĭn, M.G.: The basic propositions on defect numbers, root numbers and indices of linear operators. Amer. Math. Soc. Transl. **2**(13), 185–264 (1960)
62. Große, N., Schneider, C.: Sobolev spaces on Riemannian manifolds with bounded geometry: general coordinates and traces. Math. Nachr. **286**, 1586–1613 (2013)
63. Grubb, G.: Functional Calculus for Pseudodifferential Boundary Value Problems, Progress in Mathematics, vol. 65, 2nd edn. Birkhäuser-Verlag, Boston, Massachusetts (1996)
64. Hille, E., Phillips, R.S.: Functional Analysis and Semi-groups, 1957th edn. American Mathematical Society Colloquium Publications. American Mathematical Society, Providence, Rhode Island (1957)
65. Hopf, E.: Elementare Bemerkungen über die Lösungen partieller Differentialgleichungen zweiter ordnung vom elliptischen Typus, Sitz. Ber. Preuss. Akad. Wissensch. Berlin Math.-Phys. Kl. **19**, 147–152 (1927)
66. Hörmander, L.: Hypoelliptic second order differential equations. Acta Math. **119**, 147–171 (1967)

67. Hörmander, L.: The Analysis of Linear Partial Differential Operators III, Pseudo-differential Operators, reprint of the, 1994th edn. Classics in Mathematics. Springer-Verlag, Berlin Heidelberg New York Tokyo (2007)
68. Ikeda, N., Watanabe, S.: Stochastic Differential Equations and Diffusion Processes, 2nd edn, North-Holland Mathematical Library, vol. 24. North-Holland Publishing, Amsterdam/Kodansha, Tokyo (1989)
69. Ishikawa, Y.: Stochastic calculus of variations for jump processes, 3rd edn, De Gruyter Studies in Mathematics, vol. 54. De Gruyter, Boston, Massachusetts (2023)
70. Ishikawa, Y., Kunita, H.: Malliavin calculus on the Wiener-Poisson space and its application to canonical SDE with jumps. Stochast. Process. Appl. **116**, 1743–1769 (2006)
71. Itô, K.: Stochastic Processes (in Japanese). Iwanami-Shoten, Tokyo (1957)
72. Itô, K., McKean, H.P., Jr.: Diffusion Processes and their Sample Paths, reprint of the, 1974th edn. Classics in Mathematics. Springer-Verlag, Berlin Heidelberg New York (1974)
73. John, F., Nirenberg, L.: On functions of bounded mean oscillation. Comm. Pure Appl. Math. **14**, 415–426 (1961)
74. Jones, P.: Extension theorems for BMO. Indiana Univ. Math. J. **29**, 41–66 (1980)
75. Kakutani, S.: Markoff process and the Dirichlet problem. Proc. Japan Acad. **21**, 227–233 (1945)
76. Kato, T.: Perturbation Theory for Linear Operators, reprint of the, 1980th edn. Classics in Mathematics. Springer-Verlag, Berlin Heidelberg New York (1995)
77. Kinney, J.R.: Continuity properties of Markov processes. Trans. Amer. Math. Soc. **74**, 289–302 (1953)
78. Knight, F.B.: Essentials of Brownian Motion and Diffusion. American Mathematical Society, Providence, Rhode Island (1981)
79. Kolmogorov, A.N.: Über die analytischen Methoden in der Wahrscheinlichkeitsrechnung. Math. Ann. **104**, 415–458 (1931)
80. Kolmogorov, A.N., Fomin, S.V.: Introductory Real Analysis. Silverman (Dover Publications, New York), Translated from the second Russian edition and edited by R. A (1975)
81. Komatsu, T.: Markov processes associated with certain integro-differential operators. Osaka J. Math. **10**, 271–303 (1973)
82. Komatsu, T.: Pseudo-differential operators and Markov processes. J. Math. Soc. Japan **36**, 387–418 (1984)
83. Komatsu, T., Takeuchi, A.: Simplified probabilistic approach to the Hörmander theorem. Osaka J. Math. **38**, 681–691 (2001)
84. Ladyzhenskaya, O.A., Ural'tseva, N.N.: Linear and Quasilinear Elliptic Equations. Translated from the Russian by Scripta Technica, Inc. Translation editor: Leon Ehrenpreis. Academic Press, New York London (1968)
85. Ladyzhenskaya, O.A., Ural'tseva, N.N.: A survey of results on the solubility of boundary-value problems for second-order uniformly elliptic and parabolic quasi-linear equations having unbounded singularities. Russian Math. Surv. **41**, 1–31 (1986)
86. Ladyženskaja, O.A., Solonnikov, V.A., Ural'ceva, N.N.: Linear and Quasilinear Equations of Parabolic Type (Russian). Translated from the Russian by S. Smith. Translations of Mathematical Monographs, vol. 23. American Mathematical Society, Providence, Rhode Island (1968)
87. Lamperti, J.: Stochastic Processes, Applied Mathematical Sciences, vol. 23. Springer-Verlag, New York Heidelberg Berlin (1977)
88. Lenhart, S.: Integro-differential operators associated with diffusion processes with jumps. Appl. Math. Optim. **9**, 177–191 (1982/83)
89. Levi, E.E.: Sulle equazione lineari totalmente ellitiche. Rend. Circ. Mat. Palermo **24**, 275–317 (1907)
90. Lévy, P.: Processus Stochastiques et Mouvement Brownien. Gauthier-Villars, Paris (1948)
91. Lévy, P.: Théorie de l'addition des variables aléatoires, deuxième édition. Gauthier-Villars, Paris (1954)

92. Lieberman, G.M.: Local estimates for subsolutions and supersolutions of oblique derivative problems for general second order elliptic equations. Trans. Amer. Math. Soc. **304**, 343–353 (1987)
93. Lions, J.-L., Magenes, E.: Problèmes aux limites non-homogènes et applications, 1, 2 (Dunod, Paris); English translation: Non-homogeneous boundary value problems and applications, 1, 2. Springer-Verlag, Berlin Heidelberg New York (1968)
94. Lions, P.-L.: A remark on Bony maximum principle. Proc. Amer. Math. Soc. **88**, 503–508 (1983)
95. Lions, P.-L., Sznitman, A.-S.: Stochastic differential equations with reflecting boundary conditions. Comm. Pure Appl. Math. **37**, 511–537 (1984)
96. Lopatinskii, Ya., B.: On a method of reducing boundary value problems for systems of differential equations of elliptic type to regular integral equations (in Russian). Ukrain. Mat. Žurnal **5**, 123–151 (1953)
97. Malý, J., Ziemer, W.P.: Fine Regularity of Solutions of Elliptic Partial Differential Equations. American Mathematical Society, Providence, Rhode Island (1997)
98. Maugeri, A., Palagachev, D.K.: Boundary value problem with an oblique derivative for uniformly elliptic operators with discontinuous coefficients. Forum Math. **10**, 393–405 (1998)
99. Maugeri, A., Palagachev, D.K., Softova, L.G.: Elliptic and Parabolic Equations with Discontinuous Coefficients, Mathematical Research, 109. Wiley-VCH, Berlin (2000)
100. Maz'ya, V., Mitrea, M., Shaposhnikova, T.: The Dirichlet problem in Lipshitz domains for higher order elliptic systems with rough coefficients. J. Anal. Math **110**, 167–239 (2010)
101. McLean, W.: Strongly Elliptic Systems and Boundary Integral Equations. Cambridge University Press, Cambridge (2000)
102. Meyer, Y.: Ondelettes et opérateurs II: Opérateurs de Calderón-Zygmund. Actualités Mathématiques. Hermann, Paris (1990)
103. Meyers, N.: An L^p-estimate for the gradient of solutions of second order elliptic divergence equations. Ann. Scuola Norm. Sup. Pisa **17**, 189–206 (1963)
104. Meyers, N., Serrin, J.: $H = W$. Proc. Nat. Acad. Sci. U.S.A. **51**, 1055–1056 (1964)
105. Miranda, C.: Sulle equazioni ellittiche del secondo ordine di tipo non variazionale a coefficienti discontinui. Ann. Mat. Pura Appl. **63**, 353–386 (1963)
106. Miranda, C.: Partial Differential Equations of Elliptic Type, Second revised edition. Translated from the Italian by Zane C. Motteler. Ergebnisse der Mathematik und ihrer Grenzgebiete, Band 2. Springer-Verlag, New York Berlin (1970)
107. Mizohata, S.: Note sur le traitement par les opérateurs d'intégrale singulière du problème de Cauchy. J. Math. Soc. Japan **11**, 234–240 (1959)
108. Neri, U.: Singular Integrals, Lecture Notes in Mathematics, No. 200. Springer-Verlag, Berlin Heidelberg New York (1971)
109. Neri, U.: Some properties of functions with bounded mean oscillation. Studia Math. **61**, 63–75 (1977)
110. Netuka, I.: The change-of-variables theorem for the Lebesgue integral. Acta Univ. M. Belii Ser. Math. **19**, 37–42 (2011)
111. Nirenberg, L.: A strong maximum principle for parabolic equations. Comm. Pure Appl. Math. **6**, 167–177 (1953)
112. Oleĭnik, O.A.: On properties of solutions of certain boundary problems for equations of elliptic type (in Russian). Mat. Sbornik **30**, 695–702 (1952)
113. Peetre, J.: Rectification à l'article "Une caractérisation des opérateurs différentiels.". Math. Scand. **8**, 116–120 (1960)
114. Peetre, J.: New Thoughts on Besov Spaces, Duke Univ. Math. Ser., No. 1. Duke University, Mathematics Department, Durham, North Carolina (1976)
115. Protter, M.H., Weinberger, H.F.: Maximum Principles in Differential Equations. Corrected second printing. Springer-Verlag, New York (1984)
116. Ray, D.: Stationary Markov processes with continuous paths. Trans. Amer. Math. Soc. **82**, 452–493 (1956)

117. Redheffer, R.M.: The sharp maximum principle for nonlinear inequalities. Indiana Univ. Math. J. **21**, 227–248 (1971)
118. Reed, M., Simon, B.: Methods of Modern Mathematical Physics I: Functional Analysis. Revised and Enlarged Academic Press, New York (1980)
119. Revuz, D., Yor, M.: Continuous Martingales and Brownian Motion, 3rd edn. Springer-Verlag, Berlin New York Heidelberg (1999)
120. Rudin, W.: Real and Complex Analysis, 3rd edn. McGraw-Hill, New York (1987)
121. Runst, T., Youssfi, A.: Boundary value problems for Waldenfels operators. Indiana Univ. Math. J. **54**, 237–255 (2005)
122. Sarason, D.: Functions of vanishing mean oscillation. Trans. Amer. Math. Soc. **207**, 391–405 (1975)
123. Sato, K.: Lévy Processes and Infinitely Divisible Distributions, revised edition of the 1999 English translation, Cambridge Studies in Advanced Mathematics, vol. 68. Cambridge University Press, Cambridge (2013)
124. Sato, K., Ueno, T.: Multi-dimensional diffusion and the Markov process on the boundary. J. Math. Kyoto Univ. **14**, 529–605 (1965)
125. Sawano, Y., Di Fazio, G., Hakim, D.I.: Morrey Spaces, Introduction and Applications to Integral Operators and PDE's, vol. I. Monographs and Research Notes in Mathematics. CRC Press/Chapman & Hall, Boca Raton, Florida (2020)
126. Sawano, Y., Di Fazio, G., Hakim, D.I.: Morrey Spaces, Introduction and Applications to Integral Operators and PDE's, vol. II. Monographs and Research Notes in Mathematics. CRC Press/Chapman & Hall, Boca Raton, Florida (2020)
127. Schauder, J.: Über lineare elliptische Differentialgleichungen zweiter Ordunung. Math. Z. **38**, 257–282 (1934)
128. Schauder, J.: Numerische Abschätzungen in elliptischen linearen Differentialgleichungen. Studia Math. **5**, 34–42 (1935)
129. Schechter, M.: Principles of Functional Analysis, 2nd edn, Graduate Studies in Mathematics, vol. 36. American Mathematical Society, Providence, Rhode Island (2002)
130. Schwartz, L.: The formula for change in variables in a multiple integral. Amer. Math. Monthly **61**, 81–85 (1954)
131. Seeley, R.T.: Singular integrals on compact manifolds. Amer. J. Math. **81**, 658–690 (1959)
132. Seeley, R.T.: Integro-differential operators on vector bundles. Trans. Amer. Math. Soc. **117**, 167–204 (1965)
133. Seeley, R.T.: Singular integrals and boundary value problems. Amer. J. Math. **88**, 781–809 (1966)
134. Seregin, L.V.: Continuity Conditions for Stochastic Processes, Teoriya Veroyat. i ee Primen **6**, 3–30 (Russian). English translation: *Theory Prob. and its Appl.* **6**, 1–26 (1961)
135. Shapiro, Z., Ya.: The first boundary value problem for an elliptic system of differential equations (in Russian). Mat. Sbornik **28**, 55–78 (1951)
136. Stein, E.M.: The characterization of functions arising as potentials II. Bull. Amer. Math. Soc. **68**, 577–582 (1962)
137. Stein, E.M.: Singular Integrals and Differentiability Properties of Functions. Princeton University Press, Princeton (1970)
138. Stein, E.M.: The differentiability of functions in $\mathbf{R}^n$. Ann. Math. **2**(113), 383–385 (1981)
139. Stein, E.M.: Harmonic Analysis: Real-variable Methods, Orthogonality, and Oscillatory Integrals, Princeton Mathematical Series, vol. 43. Monographs in Harmonic Analysis, III. Princeton University Press, Princeton, New Jersey (1993)
140. Stein, E.M.: Singular integrals: the roles of Calderón and Zygmund. Notices Amer. Math. Soc. **45**, 1130–1140 (1998)
141. Stein, E.M., Shakarchi, R.: Real Analysis: Measure Theory, Integration, and Hilbert Spaces, Princeton Lectures in Analysis, vol. III. Princeton University Press, Princeton, New Jersey (2005)
142. Stroock, D.W.: Diffusion processes associated with Lévy generators. Z. Wahrscheinlichkeitstheorie und Verw. Gebiete **32**, 209–244 (1975)

143. Stroock, D.W., Varadhan, S.R.S.: On the Support of Diffusion Processes with Applications to the Strong Maximum Principle. In: Proceedings of the Sixth Berkeley Symposium on Mathematical Statistics and Probability (Univ. California, Berkeley, Calif., 1970/1971), vol. III: *Probability theory*, pp. 333–359. Univ. California Press, Berkeley, California (1972)
144. Stroock, D.W., Varadhan, S.R.S.: On degenerate elliptic-parabolic operators of second order and their associated diffusions. Comm. Pure Appl. Math. **25**, 651–713 (1972)
145. Stroock, D.W., Varadhan, S.R.S.: Multidimensional Diffusion Processes, reprint of the, 1997th edn. Classics in Mathematics. Springer-Verlag, Berlin (2006)
146. Taibleson, M.H.: On the theory of Lipschitz spaces of distributions on Euclidean n-space I. J. Math. Mech. **13**, 407–479 (1964)
147. Taira, K.: Sur l'existence de processus de diffusion. Ann. Inst. Fourier (Grenoble) **29**, 99–126 (1979). http://www.numdam.org/item/AIF_1979__29_4_99_0/
148. Taira, K.: Diffusion Processes and Partial Differential Equations. Academic Press, San Diego New York London Tokyo (1988). http://hdl.handle.net/2241/0002001094 (Tsukuba repository)
149. Taira, K.: On the existence of Feller semigroups with boundary conditions. Mem. Amer. Math. Soc. **99**, 475 (1992), 1–95. https://doi.org/10.1090/memo/0475
150. Taira, K.: Boundary value problems for elliptic integro-differential operators. Math. Z. **222**, 305–327 (1996). https://doi.org/10.1007/BF02621868
151. Taira, K.: Logistic Dirichlet problems with discontinuous coefficients. J. Math. Pures Appl. **82**, 1137–1190 (2003). https://doi.org/10.1016/S0021-7824(03)00058-8
152. Taira, K.: On the existence of Feller semigroups with discontinuous coefficients. Acta Math. Sinica (English Series) **22**, 595–606 (2006). https://doi.org/10.1007/s10114-005-0689-z
153. Taira, K.: On the existence of Feller semigroups with discontinuous coefficients II. Acta Math. Sinica (English Series) **25**, 715–740 (2009). https://doi.org/10.1007/s10114-008-6564-y
154. Taira, K.: Semigroups, Boundary Value Problems and Markov Processes, 2nd edn. Springer Monographs in Mathematics. Springer-Verlag, Berlin Heidelberg New York (2014). https://doi.org/10.1007/978-3-662-43696-7; https://www.springer.com/gp/book/9783662436950
155. Taira, K.: Analytic Semigroups and Semilinear Initial Boundary Value Problems, 2nd edn. London Mathematical Society Lecture Note Series, No. 434. Cambridge University Press, Cambridge (2016). ISBN: 978-1-316-62086-1, https://doi.org/10.1017/CBO9781316729755
156. Taira, K.: Dirichlet problems with discontinuous coefficients and Feller semigroups. Rend. Circ. Mat. Palermo, II. Ser. **69**, 287–323 (2020). https://doi.org/10.1007/s12215-019-00404-5
157. Taira, K.: Boundary Value Problems and Markov Processes: Functional Analysis Methods for Markov Processes. Lecture Notes in Mathematics, vol. 1499, 3rd edn. Springer-Verlag, Berlin Heidelberg New York (2020). ISBN: 978-3-030-48787-4, https://www.springer.com/gp/book/9783030487874
158. Taira, K.: Logistic Neumann problems with discontinuous coefficients. Ann. Univ. Ferrara **66**, 409–485 (2020). https://doi.org/10.1007/s11565-020-00350-6
159. Taira, K.: Ventcel' boundary value problems for elliptic Waldenfels operators. Boll. Unione Mat. Ital. **13**, 213–256 (2020). https://doi.org/10.1007/s40574-019-00214-8
160. Taira, K.: Oblique derivative problems and Feller semigroups with discontinuous coefficients. Ricerche Mat. (2020). https://doi.org/10.1007/s11587-020-00509-5
161. Taira, K.: Singular Integrals and Feller Semigroups: Real Analysis Methods for Markov Processes. Tsukuba Repository, Tsukuba (2021). https://tsukuba.repo.nii.ac.jp/records/56313
162. Taira, K.: Functional Analytic Techniques for Diffusion Processes. Springer Monographs in Mathematics. Springer-Verlag, Singapore (2022). https://link.springer.com/book/9789811910982
163. Taira, K.: Singular integrals and Feller semigroups with jump phenomena. Rend. Circ. Mat. Palermo, II. Ser. (2023). https://doi.org/10.1007/s12215-023-00907-2
164. Takanobu, S., Watanabe, S.: On the existence and uniqueness of diffusion processes with Wentzell's boundary conditions. J. Math. Kyoto Univ. **28**, 71–80 (1988)

165. Talenti, G.: Equazioni lineari ellittiche in due variabili. Matematiche (Catania) **21**, 339–376 (1966)
166. Tanabe, H.: Functional Analysis, I, II. (in Japanese) Jikkyo-Shuppan, Tokyo (1978/1981)
167. Tanabe, H.: Functional Analytic Methods for Partial Differential Equations. Marcel Dekker, New York Basel (1997)
168. Torchinsky, A.: Real-variable Methods in Harmonic Analysis. Dover Publications Inc., Mineola, New York (2004)
169. Transirico, M., Troisi, M., Vitolo, A.: BMO spaces on domains of $\mathbf{R}^n$. Ricerche Mat. **45**, 355–378 (1996)
170. Triebel, H.: Theory of Function Spaces, reprint of the, 1983rd edn. Modern Birkhäuser Classics. Birkhäuser-Verlag, Basel (1983)
171. Triebel, H.: Theory of Function Spaces II, reprint of the, 1992nd edn. Monographs in Mathematics. Birkhäuser-Verlag, Basel (1992)
172. Troianiello, G.M.: Elliptic Differential Equations and Obstacle Problems. The University Series in Mathematics. Plenum Press, New York London (1987)
173. Trudinger, N.S.: Local estimates for subsolutions and supersolutions of general second order elliptic quasilinear equations. Invent. Math. **61**, 67–79 (1980)
174. Višik, M.I.: On general boundary problems for elliptic differential equations. Amer. Math. Soc. Transl. **2**(24), 107–172 (1963)
175. Višik, M.I., Eskin, G.I.: Equations in convolutions in a bounded region. Russ. Math. Surv. **20**, 85–151 (1966)
176. Vitanza, C.: $W^{2,p}$-regularity for a class of elliptic second order equations with discontinuous coefficients. Le Matematiche **47**, 177–186 (1992)
177. von Waldenfels, W.: Positive Halbgruppen auf einem n-dimensionalen Torus. Arch. Math. **15**, 191–203 (1964)
178. Watanabe, S.: Construction of diffusion processes with Wentzell's boundary conditions by means of Poisson point processes of Brownian excursions. In: Probability Theory, pp. 255–271. Banach Center Publications, vol. 5. PWN-Polish Scientific Publishers, Warsaw (1979)
179. Wentzell (Ventcel'), A.D.: On boundary conditions for multidimensional diffusion processes (in Russian). Teoriya Veroyat. i ee Primen. **4**, 172–185. English translation in *Theory Prob. and its Appl.* **4**, 164–177 (1959)
180. Wiener, N.: Differential space. J. Math. Phys. **2**, 131–174 (1923)
181. Wloka, J.: Partial Differential Equations. Cambridge University Press, Cambridge (1980)
182. Yosida, K.: Functional Analysis, reprint of the sixth (1980) edition. Classics in Mathematics. Springer-Verlag, Berlin Heidelberg New York (1995)
183. Ziemer, W.P.: Weakly differentiable functions, Sobolev spaces and functions of bounded variation, Graduate Texts in Mathematics, vol. 120. Springer-Verlag, New York (1989)

Index

K. Taira, *Real Analysis Methods for Markov Processes*,
https://doi.org/10.1007/978-981-97-3659-1

A

B

C

The manufacturer's authorised representative in the EU is Springer Nature Customer Service Centre GmbH, Europaplatz 3, 69115 Heidelberg, Germany. If you have any concerns regarding our products, please contact ProductSafety@springernature.com

Printed and bound by CPI Group (UK) Ltd, Croydon, CR0 4YY

01/10/2026

02225849-0020